Handbook of Research on Green Synthesis and Applications of Nanomaterials

Rajni Garg
Rayat Bahra University, India

Rishav Garg
Galgotias College of Engineering and Technology, India

Nnabuk Okon Eddy
Department of Pure and Industrial Chemistry, University of Nigeria, Nsukka, Nigeria

IGI Global
PUBLISHER of TIMELY KNOWLEDGE

A volume in the Advances in Chemical and Materials Engineering (ACME) Book Series

Published in the United States of America by
IGI Global
Engineering Science Reference (an imprint of IGI Global)
701 E. Chocolate Avenue
Hershey PA, USA 17033
Tel: 717-533-8845
Fax: 717-533-8661
E-mail: cust@igi-global.com
Web site: http://www.igi-global.com

Library of Congress Cataloging-in-Publication Data

Names: Garg, Rajni, 1979- editor. | Garg, Rishav, 1977- editor. | Eddy, Nnabuk Okon, 1971- editor.
Title: Handbook of research on green synthesis and applications of nanomaterials / Rajni Garg, Rishav Garg, and Nnabuk Okon Eddy, editor.
Description: Hershey, PA : Engineering Science Reference, [2022] | Includes bibliographical references and index. | Summary: "With nanotechnology being a multidisciplinary field of research that is providing new avenues of diverse applications, this book raises the awareness about non-toxic, eco-friendly and economical green techniques for synthesis of various nanomaterials"-- Provided by publisher.
Identifiers: LCCN 2021028557 (print) | LCCN 2021028558 (ebook) | ISBN 9781799889366 (hardcover) | ISBN 9781799889380 (ebook)
Subjects: LCSH: Nanotechnology--Industrial applications. | Green chemistry. | Nanostructured materials.
Classification: LCC T174.7 .G74 2022 (print) | LCC T174.7 (ebook) | DDC 620/.5--dc23
LC record available at https://lccn.loc.gov/2021028557
LC ebook record available at https://lccn.loc.gov/2021028558

This book is published in the IGI Global book series Advances in Chemical and Materials Engineering (ACME) (ISSN: 2327-5448; eISSN: 2327-5456)

British Cataloguing in Publication Data
A Cataloguing in Publication record for this book is available from the British Library.

For electronic access to this publication, please contact: eresources@igi-global.com.

Advances in Chemical and Materials Engineering (ACME) Book Series

J. Paulo Davim
University of Aveiro, Portugal

ISSN:2327-5448
EISSN:2327-5456

Mission

The cross disciplinary approach of chemical and materials engineering is rapidly growing as it applies to the study of educational, scientific and industrial research activities by solving complex chemical problems using computational techniques and statistical methods.

The **Advances in Chemical and Materials Engineering (ACME) Book Series** provides research on the recent advances throughout computational and statistical methods of analysis and modeling. This series brings together collaboration between chemists, engineers, statisticians, and computer scientists and offers a wealth of knowledge and useful tools to academics, practitioners, and professionals through high quality publications.

Coverage

- Composites
- Heat Treatments
- Coatings and surface treatments
- Electrochemical and Corrosion
- Thermo-Chemical Treatments
- Nanomaterials
- Sustainable Materials
- Industrial Chemistry
- Ductility and Crack-Resistance
- Fatigue and Creep

IGI Global is currently accepting manuscripts for publication within this series. To submit a proposal for a volume in this series, please contact our Acquisition Editors at Acquisitions@igi-global.com or visit: http://www.igi-global.com/publish/.

The Advances in Chemical and Materials Engineering (ACME) Book Series (ISSN 2327-5448) is published by IGI Global, 701 E. Chocolate Avenue, Hershey, PA 17033-1240, USA, www.igi-global.com. This series is composed of titles available for purchase individually; each title is edited to be contextually exclusive from any other title within the series. For pricing and ordering information please visit http://www.igi-global.com/book-series/advances-chemical-materials-engineering/73687. Postmaster: Send all address changes to above address.

Titles in this Series

For a list of additional titles in this series, please visit: www.igi-global.com/book-series/advances-chemical-materials-engineering/73687

Handbook of Research on Advancements in the Processing, Characterization, and Application of Lightweight Maerials
Kaushik Kumar (Birla Institute of Technology, India) B. Sridhar Babu (CMR Institute of Technology, India) and J. Paulo Davim (University of Aveiro, Portugal)
Engineering Science Reference • © 2022 • 458pp • H/C (ISBN: 9781799878643) • US $285.00

Emerging Applications and Implementations of Metal-Organic Frameworks
Shimaa Mohamed Elsaeed (Egyptian Petroleum Research Institute, Egypt) Elsayed Zaki (Egyptian Petroleum Research Institute, Egypt) and Abdel-Azim Abdel-Azim (Egyptian Petroleum Research Institute, Egypt)
Engineering Science Reference • © 2021 • 254pp • H/C (ISBN: 9781799847601) • US $225.00

Normal Partitions and Hierarchical Fillings of N-Dimensional Spaces
Gennadiy Vladimirovich Zhizhin (Russian Academy of Natural Sciences, Russia)
Engineering Science Reference • © 2021 • 280pp • H/C (ISBN: 9781799867685) • US $195.00

Applications of Nanomaterials in Agriculture, Food Science, and Medicine
Mohd Amin Bhat (Higher Education Department, Government of Jammu and Kashmir, India) Irshad Ahmad Wani (Government Degree College for Boys, Anantnag, India) and Shah Ashraf (National Institute of Technology Srinagar (NIT Srinagar), India)
Engineering Science Reference • © 2021 • 442pp • H/C (ISBN: 9781799855637) • US $215.00

Technology Development for Adsorptive Heat Energy Converters Emerging Research and Opportunities
Kostyantyn M. Sukhyy (Ukrainian State University of Chemical Engineering, Ukraine) Elena A. Belyanovskaya (Ukrainian State University of Chemical Engineering, Ukraine) and Mikhailo P. Sukhyy (Ukrainian State University of Chemical Engineering, Ukraine)
Engineering Science Reference • © 2021 • 305pp • H/C (ISBN: 9781799844327) • US $195.00

Nanotechnologies and Clusters in the Spaces of Higher Dimension Emerging Research and Opportunities
Gennadiy Vladimirovich Zhizhin (Independent Researcher, Russia)
Engineering Science Reference • © 2021 • 286pp • H/C (ISBN: 9781799837848) • US $195.00

Advanced Surface Coating Techniques for Modern Industrial Applications
Supriyo Roy (Haldia Institute of Technology, India) and Goutam Kumar Bose (Haldia Institute of Technology, India)
Engineering Science Reference • © 2021 • 342pp • H/C (ISBN: 9781799848707) • US $215.00

IGI Global
PUBLISHER of TIMELY KNOWLEDGE

701 East Chocolate Avenue, Hershey, PA 17033, USA
Tel: 717-533-8845 x100 • Fax: 717-533-8661
E-Mail: cust@igi-global.com • www.igi-global.com

List of Contributors

Table of Contents

Detailed Table of Contents

The world is going green; hence, environmentally friendly practices that would conserve natural resources for the future generation are encouraged. As a consequence, the world is less concerned about the numerous applications of nanotechnology, especially in the health sector; rather, it is more concerned about the sustainability of functionalized nanomaterials. Thus, the future of nanotechnology depends on its ability to 'go green'. Green nanotechnology attempts to synthesize improved, nontoxic, and biocompatible nanomaterials with sustainable benefits using eco-friendly materials. Although green nanotechnology is considered a sustainable, viable, and biocompatible approach to the production of eco-friendly nanomaterials, there are shortcomings especially in microbial handling and process optimization. In this chapter, the authors aim to appraise not only the use of biocompatible approaches for the synthesis of nanoparticles and/or nanomaterials but also their shortcomings.

This chapter focuses on advancements in the green synthesis approach for nanomaterials and their environmental applications. The eco-friendly, cost-effective, and simple synthesis of inorganic nanoparticles on the environmental applications are discussed. The first section presents an introduction with the basic topics of the green materials synthesis. The second section summarizes the green chemical methods through different routes (precipitation synthesis, electrospinning synthesis, reflux condensation synthesis, hydrothermal, microwave-assisted hydrothermal methods) and the green routes using biological systems from biosynthetic process to synthesize nanoparticles and biosynthesis of passivating nanofilms using spontaneous formation. In the third section, the environmental and energy applications of nanomaterials, including solar cells, photocatalysis, and biohydrogen production are discussed. Therefore, the scientific community can experience the relation of the achieved nanomaterials with their plentiful application possibilities onto the environmental remediation.

Chapter 3
Phyto-Nanofabrication: Plant-Mediated Synthesis of Metal and Metal Oxide Nanoparticles 51
Israt Jahan, Dicle University, Turkey

Green synthesis of metallic nanoparticles through natural entities (i.e., bacteria, actinomycetes, yeast, fungus, microalgae, seaweed, plants, and plant-derived materials) has become an advantageous and ecofriendly approach. However, phytocompounds of plant extract have achieved huge attention since by utilizing them high yield NPs with desirable size and shape, which can be produced through single-step synthesis scheme. Plants retain diverse biochemicals that exhibit strong hyper-accumulating potential, crucial for metallic ion reduction to metallic NPs, like platinum, gold, silver, titanium oxide, iron oxide, copper oxide, zinc oxide, palladium, etc. Here, previously published studies were reviewed for providing the latest scientific evidence on biosynthesis of metal and metal oxide NPs using different plant materials, especially medicinal plants and food and agro-wastes.

Chapter 4
Green Synthesis of Nanocomposites: A Greener Approach for a Cleaner Future................................ 77
Meenal D. Patil, Shivaji University, Kolhapur, India
Suprimkumar D. Dhas, Shivaji University, Kolhapur, India
Annasaheb V. Moholkar, Shivaji University, Kolhapur, India

Nanotechnology has been a dynamic research area over the past few decades because it assures the resolution to the problems that hamper progress. Currently, a new era of 'green synthesis' is an emerging multidisciplinary field in nanotechnology which employs reliable, sustainable, low-cost, non-hazardous, and eco-friendly techniques. Green synthesis is considered a vital tool to reduce the negative impacts accompanying the traditional methods of synthesis for NPs commonly employed in industry and laboratory. This chapter unveils a comprehensive overview of the recent research on available green techniques for the synthesis of various nanocomposites in order to solve future generation challenges. This chapter also focuses on the green synthesis of various nanocomposites, synthesis parameters, potential applications, merits/demerits, and future prospects.

Continuous release of contaminants into the environment and their toxic effects on the living world prompted public and scientific interests in the remediation of pollutants. The release of toxic pollutants not only hinders the soil environments but also disturbs the aquatic environments. The hindrance in soil and aquatic environments causes a serious warning to the productive environment and food security. Thus, excluding and counteracting the pollutants from the environment is necessary to develop a safe and productive environment. Currently, numerous techniques are used to remediate contaminants from the environment; moreover, nanoparticles are one of such most widely used techniques for environmental clean-up. This chapter briefly discusses the need for environmental remediation and nanoparticles for environmental clean-up, especially soil and water. Additionally, the complications and current research trends of using nanoparticles in environmental remediation are briefly discussed for determining future research prospects.

Chapter 11
CaO Nanoparticles: Synthesis and Applications in Water Remediation 247
Nnabuk Okon Eddy, Department of Pure and Industrial Chemistry, University of Nigeria, Nsukka, Nigeria
Rajni Garg, Rayat Bahra University, India

Adsorption is widely acknowledged as one of the best options that are available for the removal of contaminants from water. Contamination of water does not only create water scarcity, but it has the capacity to generate and transfer several environmental problems including threat to public health. This chapter reviewed calcium oxide nanoparticle (CaONP) as a noble metal oxide for the removal of contaminants from water. The review is concentrated in the general overview of water contamination, metal oxide nanoparticles, general application of CaONP, synthetic methods, characterization method, and applications. The chapter observed that little is done on the use of CaONP for the removal of contaminants from water except for dyes, some heavy metal ions, and few organic/inorganic compounds. It is also observed that CaONP can be applied as adsorbent and in photocatalytic degradation of dye. Suggestions are made on the possibility of utilizing local raw materials that are easily accessible, cheap, and environmental sources of raw materials for the synthesis of CaONP.

Chapter 12
Antimicrobial Applications of Nanoparticles 269
Ayesha Kanwal, The Islamia University of Bahawalpur, Pakistan
Zeeshan Ahmad Bhutta, Chungbuk National University, South Korea
Ambreen Ashar, University of Agriculture, Faisalabad, Pakistan
Ashar Mahfooz, University of Agriculture, Faisalabad, Pakistan
Rizwan Ahmed, University of Agriculture, Faisalabad, Pakistan
Muhammad Fakhar-e-Alam Kulyar, Huazhong Agricultural University, China
Kun Li, Nanjing Agricultural University, China

Human mortality due to drug-resistant infections is becoming more prevalent in our society. Antibiotics are impotent due to abuse and/or misuse, leading to new, more expensive, and more effective medicines and treatments. Therefore, it causes many short-term and long-term side effects in the patient. On the other hand, nanoparticles have exhibited antibacterial activity against various pathogens due to their small size and ability to destroy cells by various mechanisms. Unlike antibiotics for the treatment of patients' diseases and infections, nanomaterials provide an exciting way to limit the growth of microorganisms due to infections in humans. This has led to the development of a number of nanoparticles as active

Nanotechnology facilitates exclusive opportunities for investigating and controlling a wide range of biomedical approaches at nano-scale, which could provide a groundbreaking impact on medicine, biology, and other health-related areas. This latest technology offers certain distinctive benefits in different aspects such as nanocarriers of targeted drug delivery system, diagnosis and treatment of cancer, vaccination and immunotherapy, biosensing and imaging for clinical diagnosis. A wide range of nanomaterials with biomedical prominence have already been exploited for drug delivery and cancer therapy comprising, solid lipid nanoparticles, metallic nanoparticles, polymeric nanoparticles, carbon nanotube, nanogel-based compounds, nanocapsules, magnetic nanoparticles, nanofluids, nanowires, liposomes, etc. Therefore, this chapter reviewed the potential use of different nanomaterials for therapeutic applications, especially for drug delivery and cancer diagnosis and therapy by focusing their suitability and biocompatibility for extensive and safer usages.

The rapid growth of nanotechnology towards the development of nanomedicines has improved cancer treatment. Nanomedicine provides the opportunity to implement complex and targeted multifunctional strategies. Today, nanoparticles (NPs) have many uses in a number of scientific fields. In recent years, it has been repeatedly reported that NPs hold a significant place in the regulation of modern medicine by implementing a varying number of clinical approaches like drug carrying substances, genetic material delivery to tumors, as well as in radiography as a contrast media agent. Various nanomaterials based on organic, inorganic, lipid or glycan compounds, and synthetic polymers have been used to develop and improve new cancer treatments. In this chapter, the authors discussed the role of NPs in cancer treatment among various anticancer drug delivery methods.

Since the end of 2019, the SARS-CoV-2 virus, commonly known as COVID-19, has become an infectious disease that has produced a severe global public health problem that requires solutions from different scientific initiatives. Nanomaterials are used to diagnose, treat, and prevent the SARS-CoV-2 virus. Its potential is highlighted through specific applications and developments carried out by researchers

Preface

Materials in the nano dimension have received enormous research and industrial interest, driven by the fact that materials' macro properties can be considered as a component of several microsystems. The term, 'Nano' takes its origin from the word, "Nanos", a Greek word for "dwarf", which refers to an order of magnitude in the range from 1 to 100 nm (1 nm = 10^{-9} m). Nanotechnology is seen as a critical technology with the ability to bring about fundamental improvements in a wide range of technological sectors and has ultimately generated improved and diverse system innovation. Considering the current rate of engagement of nano materials in our society, it is expected to have significant economic, ecological, and societal consequences shortly. As a result, not only are fine-sized objects operated in nanotechnology but it is also distinguished by the fact that unique characteristics and material properties exist in nano-scale. The great expectations for nanotechnology are based on the controlled fabrication of nanostructures with novel types of frameworks and processes by manipulation of the properties of macroscopic bodies to generate new applications in almost all sectors of science and technology. There has been a flurry of cutting-edge nanotechnology research carried out in the scientific and technical communities, and many of the findings have shown enormous potential for industrialization. Carbon nanotubes, nano-polishing liquids, and nano-sensors, among other products, have achieved large-scale production.

The increasing attention towards nanotechnology is beyond its nano-scale but is determined by the importance of nanotechnology in promoting huge changes in human society. Nanotechnology is a concentrated manifestation of the interdisciplinary nature of different fields of science and engineering and is characterized by the integration of multidisciplinary and cross-technology, which must be, backed by basic research and development. It's the basic need of the hour to closely integrate the basic research and application development of nanotechnology for subsequent industrial applications in a timely and effective manner and to integrate the research strength of various disciplines and for finding solutions to major scientific challenges. Hence, the current research practices have shifted application-oriented basic research to applied research and integrated research with suitable technology transfer and transformation.

The development of modern science and technology has almost always achieved innovative breakthroughs in the cross and fringe fields. It is precisely this way that nanotechnology is full of opportunities for original innovation. Once the theories and concepts formed in the process of exploration in this field are widely used in our production and life, it will greatly enrich our cognitive world and bring conceptual changes to human society. With the revolution of human cognition of the objective world, nanotechnology will trigger a new industrial revolution. For example, the computing and storage capabilities of computers manufactured on the nanometer scale will increase exponentially compared with the current computer performance under micron technology. This will be a profound revolution in the information industry and other related industries. Similarly, life technology is also facing changes under the influence of nanotechnology. Therefore, people believe that nanotechnology is the common foundation for the further development of information technology and life technology in the future.

The nanomaterials exhibit a huge application potential including drug delivery, medical imaging, health consumer products, environmental remediation, nanocatalysis, etc. Nanotechnology has the potential to stimulate innovation in energy technology by introducing novel nanomaterials that have the potential to revolutionize chemical catalysis and manufacturing techniques. Nano-based therapeutics and drug carriers are being developed as nanomedicine that can be precisely positioned and delivered while also bypassing the immune system and the blood-brain barrier. In the field of nano-diagnostics, the latest developments include applying flexible electronic sensors to monitor nerve activity; High-speed nanopore DNA sequence technology; Cancer-diagnostics based on micro-nano sensors using liquid biopsy technology, etc. The diagnosis and medical function can be combined with contrast imaging for monitoring the therapeutic effect of nanomedicine. In information and communication technology, significant progress is being made toward developing new computer designs that will ultimately include DNA and quantum computing to replace traditional silicon electronics.

Archaeological explorations have revealed the use of nanomaterials by the Ancient Egyptians in dyes and paints more than 4,000 years ago. The synthetic metallic nanoparticle era has been traced back to the bronze age with the use of various metallic nanoparticles in glass and ceramic structures. Although the modern era started with the synthesis of colloidal gold suspension by Michael Faraday in 1857, Richard Feynman bagged the credit for the formal introduction to the concept of nanotechnology in 1959. However, the term "nanotechnology" was used formally by Norio Taniguchi in 1974 and the invention of the scanning tunneling microscope (STM) by Gerd Binning and Heinrich Rohrer in 1981 progressed the field. Later on, the discovery of fullerenes by Harold Kroto, Robert Curl, and Richard Smalley in 1985, synthesis of carbon nanotubes by Sumio Lijima and his co-workers in 1991, and the discovery of graphene by Andre Geim and Kostya Novoselov in 2004 opened the avenues for innovative nanomaterials.

With the advancement and development of new nanomaterials with the varying origin, morphology, chemical composition including, nanoparticles, nanowires, nanotubes, nanosheets, nanostructured materials, complex extensions flutter nanostructures and even nano sea urchins the need for their classification has ripened. Based upon their source of origin, nanomaterials are classified into natural nanomaterials (existing in living organisms), incidental nanomaterials (produced incidentally in a process), and engineered nanomaterials (manufactured by human beings). Natural nanomaterials include various biomolecules such as carbohydrates (glucose, cellulose, chitin), genetic material (DNA and RNA), proteins, lipids, etc. The incidental nanomaterials are produced during various natural processes (forest fires, dust storms, volcanic eruptions, and cosmic reactions) and anthropogenic activities (burning, engine exhausts, industrial and agricultural operations). On the other hand, engineered nanomaterials are produced by human beings with controlled morphology for specific applications.

Based upon their dimensionality with 3, 2, 1, and 0 dimensions within the nanoscale, nanomaterials are classified as 0D, 1D, 2D, and 3D nanomaterials with electron movement along with dimensionless space, one-axis, two-axis, and all the three axes respectively. 0D nanomaterials include quantum dots, metal and metal oxide nanoparticles with enormous applications that have been extensively explored. ID nanomaterials such as nanotubes, nanorods, nanowires, and nanofibers have been profoundly used in the nanoelectronics and energy sector. Nanosheets, nanoplates, and nano prisms are typical examples of 2D nanomaterials with novel applications in sensors, reactors, and catalysis. Recently, 3D nanomaterials such as nanoflowers, nanoballs, and nano coils have been synthesized with controlled morphological features for extensive research interest. The most widely accepted classification is based upon the chemical composition and categorizes nanomaterials into carbon-based, inorganic-based, organic-based, and composite-based nanomaterials. Fullerenes, graphene, carbon black, carbon nanotubes, and carbon nanofibers are typical examples of carbon-based nanomaterials while the metal and metal oxide nanoparticles constitute the inorganic-based nanomaterials. The organic-based nanomaterials include those excluding the carbon and inorganic-based nanomaterials. The composite-based nanomaterials constitute blends of the first three categories with other frameworks.

The globe is becoming more environmentally friendly; as a result, environmentally friendly behaviours that protect natural resources for future generations are promoted. In particular, the world is less concerned with the myriad uses of nanotechnology, especially in the health sector, than with the sustainability of functionalized nanomaterials. Thus, nanotechnology's future is contingent upon its capacity to 'become green.' Green nanotechnology aims to create enhanced, nontoxic, and biocompatible nanomaterials with long-term advantages utilizing environmentally friendly ingredients. Although green nanotechnology is seen as a sustainable, feasible, and biocompatible method for producing environmentally benign nanomaterials, it has limitations, particularly in terms of microbial treatment and process optimization. The authors of Chapter 1 sought to evaluate not only the utilization of biocompatible techniques for the manufacture of nanoparticles and/or nanomaterials but also their drawbacks.

Green Nanotechnology refers to the process of synthesizing non-toxic and environmentally friendly nanomaterials while minimizing the quantity of nanomaterials preparation and mitigating any environmental harm created during the process. Chapter 2 discusses breakthroughs in the green production of nanomaterials and their applications in the environment to address the environmentally benign, cost-effective, and straightforward synthesis of inorganic nanoparticles for environmental applications. The first part provides an overview of the fundamental concepts of green materials synthesis. The second section summarises green chemical methods via various routes (precipitation synthesis, electrospinning synthesis, reflux condensation synthesis, hydrothermal, and microwave-assisted hydrothermal methods), as well as green biological methods, ranging from biosynthetic processes to biosynthesis of passivating nanofilms via spontaneous formation. The third part discusses nanomaterials' environmental and energy uses, including solar cells, photocatalysis, and biohydrogen generation. As a result, the scientific community can see the relationship between the synthesized nanomaterials and their many application possibilities in environmental cleanup.

In recent years, green production of metallic nanoparticles using natural entities such as bacteria, actinomycetes, yeast, fungus, microalgae, seaweed, plants, and plant-derived materials has shown to be cost-effective and environmentally benign method. However, phytocompounds derived from plant extracts have garnered considerable interest because they enable the production of high yield nanoparticles with desired size and shape using a one-step synthesis strategy. Plants maintain a variety of biochemicals with a high capacity for hyperaccumulation, which is required for the reduction of metallic ions to metallic nanoparticles, such as platinum, gold, silver, titanium oxide, iron oxide, copper oxide, zinc oxide, and palladium. Chapter 3 reviews previously published research to provide the most up-to-date scientific information on the biosynthesis of metal and metal oxide nanoparticles utilizing a variety of plant resources, including medicinal plants, as well as food and agro-waste.

Nanotechnology has been a burgeoning study field over the last several decades due to its promise of resolving issues that impede advancement. At the moment, a new age of 'green synthesis' is developing as a multidisciplinary area of nanotechnology that utilizes dependable, sustainable, low-cost, non-hazardous, and environmentally friendly processes. Green synthesis is seen as a critical tool for mitigating the negative consequences associated with the standard techniques of synthesis for nanoparticles frequently used in various applications. Chapter 4 provides an in-depth summary of current research on environmentally friendly methodologies for the synthesis of diverse nanocomposites to address future generation issues. Additionally, this chapter discusses the synthesis parameters, prospective applications, advantages/disadvantages, and future possibilities.

One of the most promising nanomaterials, graphene is a two-dimensional network of carbon atoms with a single atom thickness and a hexagonal crystalline structure with sp^2 hybridization compressed by covalent connections. Graphene has unusual capabilities due to its structure and geometry, including a large specific surface area, fast mobility of electrons, and excellent thermal and electrical conductivity. However, these properties are constrained by the graphene sheets' structure. As a result, several investigations have been conducted on the synthesis of three-dimensional structures that avoid sheet aggregation and the loss of graphene's characteristics. These three-dimensional structures have a low density, high porosity, and surface area, are mechanically stable, and exhibit excellent mass and electron transport capabilities. Chapter 5 summarises the processes for synthesizing the many three-dimensional structures obtained from graphene, as well as their numerous uses in environmental remediation, sensing, medicinal, and energy-related fields, among others.

At the present, a diverse range of materials is employed as barrier coatings for a variety of purposes. Numerous research has focused on polymers and graphene. Polymers are utilized in a wide variety of sectors owing to their extraordinary qualities, which include resistance to heat degradation, chemical penetration, and excellent mechanical capabilities. On the other hand, graphene, a one-atom-thick two-dimensional nanomaterial, prevents gases and liquid molecules from passing through its plane; as a result, it has been identified as a promising nanomaterial for use in gas and liquid barrier applications, as well as corrosion inhibition coatings, as discussed in Chapter 6.

Nanomaterials are nanostructures that have piqued the scientific community's curiosity owing to their novel functional and physicochemical features. They exhibit excellent stability, sophisticated optical-electronic characteristics, and programmable surface functionalization. These qualities have made them a common material in a variety of sectors, including pharmaceuticals, cosmetics, and electronics. Given their increased use, there is a need for more dependable and advanced quantitative and qualitative characterization techniques. Chapter 7 discusses such characterization approaches for nanomaterials analysis. Numerous associated strategies, both different and combined, are addressed in depth. These tools are arranged in ascending order of complexity, with each stating a distinct feature of nanomaterials such as elemental composition, morphology, crystal structure, magnetic, and strength. Finally, some recent advances in nanostructure characterization approaches are discussed. By better understanding the characteristics of nanomaterials, we can broaden their societal uses.

Petroleum has always been a rapidly developing sector and new technologies are always being created to address the issues that arise. Nanomaterials are being incorporated into these technologies to enhance the efficiency of various operations. Their specific physical and chemical properties enable them to be used in a variety of sectors within the oil and gas industry, including upstream, midstream, and downstream. Chapter 8 focuses on the nanomaterials used in the oil and gas sectors. Additionally, its implementation in a variety of applications has been discussed. Hydrocarbon exploration, well drilling and completion, production operations, increased oil recovery techniques, transportation, and refinery activities are only some of the uses. Additionally, there is a discussion of current issues and potential future applications.

Numerous research investigations have been conducted over the previous decades on the mitigation of various kinds of pollutants using various types of nanoparticles. The treatment mechanisms were determined by the contaminants' properties (e.g., organic vs. inorganic, molecular size, and hydrophobicity, etc.), nanoparticles' properties (e.g., surface charge, area, pore size, and surface functional groups, etc.), and aqueous water's properties (e.g., organic matter types, pH, ionic strength, surfactants, and temperature). As such, Chapter 9 presents a review of the literature on the removal of contaminants using carbon- and metal-based nanoparticles. It also provides a discussion of the controlling mechanisms for the mitigation of contaminants using carbon- and metal-based nanoparticles; and an evaluation of the application and implications of carbon- and metal-based nanoparticles in environmental studies. Additionally, this chapter addresses future research requirements and obstacles in the field of environmental remediation using carbon and metal-based nanoparticles.

Contaminants' continuous release into the environment and damaging impacts on the living world sparked public and scientific interest in pollutant cleanup. Not only does the discharge of hazardous contaminants harm soil settings, but it also disrupts aquatic environments. The impediment in soil and aquatic ecosystems puts the productive environment and food security at risk. Thus, it is vital to eliminate and mitigate contaminants from the environment to create a safe and productive workplace. Numerous strategies are now being employed to remove toxins from the environment; also, nanoparticles are one of the most extensively used ways. Chapter 10 highlights the importance of environmental remediation and the use of nanoparticles to clean up the environment, particularly the soil and water. Additionally, the difficulties and current research trends associated with the use of nanoparticles in environmental remediation have been explored briefly to establish future research directions.

Adsorption is generally recognized as one of the most effective methods for removing pollutants from water. Water contamination does not just contribute to water shortage; it also has the potential to develop and spread a variety of environmental concerns, including threats to public health. Chapter 11 discusses calcium oxide nanoparticles (CaONP) as noble metal oxide that may be used to purify water. The review focuses on a broad overview of water pollution, metal oxide nanoparticles, broad use of CaONP, synthetic techniques, characterization methods, and applications. The article noted that suitable research has been conducted on the use of CaONP to remove pollutants from water such as dyes, heavy metal ions, and other organic/inorganic chemicals. Additionally, it is shown that CaONP may be used as an adsorbent and in photocatalytic dye degradation. Suggestions are provided about the feasibility of synthesizing CaONP using locally sourced raw ingredients that are readily available, inexpensive, and environmentally friendly.

In current civilization, human mortality as a result of drug-resistant illnesses is increasing. Antibiotics become ineffective as a result of overuse and/or misuse, necessitating the development of new, more costly, more effective drugs and therapies. As a result, the patient has several short- and long-term adverse consequences. On the other hand, owing to their tiny size and capacity to damage cells through a variety of methods, nanoparticles have shown antibacterial effectiveness against a variety of diseases. In contrast to antibiotics, which are used to treat patients' ailments and infections, nanomaterials provide an interesting new technique to inhibit the development of bacteria associated with human infections. This has resulted in the creation of antibacterial materials for a variety of applications including nanoparticles acting as active antimicrobial agents. As a result, the Chapter 12 reviews the recent advances in the use of nanomaterials for antibacterial applications and the processes by which they function as an effective alternative antibacterial agent.

There has been tremendous expansion and increased interest in the use of nanoparticles in a variety of biomedical applications in recent years, including medical diagnostics, targeted medication delivery, tissue engineering, regenerative medicine, and biomedical textiles. Nanoparticles functionalized with biological molecules, in particular, exhibit unique features and are very helpful in medical diagnostics. Additionally, nanoparticles have a broad variety of therapeutic uses, including the creation of nano-drug delivery systems, the design of innovative medications, and their role in the design of therapeutic materials. Chapter 13 summarises recent advances in nanoparticle biological applications. Finally, this book chapter discusses the difficulties inherent in the toxicological evaluation of engineered nanoparticles and the critical nature of conducting in-depth studies on the synthesis of future nanomaterials to develop cutting-edge technologies for addressing a wide variety of biomedical issues.

Numerous researchers have been drawn to nanomaterials because of their small size, great stability, affinity, and selectivity. Over the last few decades, extensive research on a variety of metal and metal-oxide nanomaterials has been conducted to determine their significant properties such as size, shape, surface mass proportion, and reactivity; all of these properties are critical for the evolution and use of nanomaterials and nanoscale gadgets in biomedical fields. Additionally, a comprehensive variety of biofunctional applications such as antibacterial, antiviral, antifungal, anticancer, bioimaging, tissue engineering, biosensors, and gene and drug delivery have a large potential. Chapter 14 provides an overview of the nature, properties, and synthesis of nanomaterials, as well as a comprehensive study of biological synthesis. It also discusses in detail how microbes generate nanoparticles, and this section provides updates on the many biological and biomedical uses of these bionanomaterials.

Nanotechnology offers unprecedented chances for researching and manipulating a diverse array of biomedical techniques at nanoscales, which has the potential to have a transformative effect on medicine, biology, and other health-related fields. This cutting-edge technology enables a variety of novel applications, including nanocarriers for targeted drug delivery systems, cancer detection and treatment, vaccination and immunotherapy, and biosensing and imaging for clinical diagnostics. Numerous nanomaterials with significant biomedical applications have already been used for drug delivery and cancer therapy, including solid lipid nanoparticles, metallic nanoparticles, polymeric nanoparticles, carbon nanotubes, nanogel-based compounds, nanocapsules, magnetic nanoparticles, nanofluids, nanowires, and liposomes. Thus, Chapter 15 discusses the prospective uses of various nanomaterials for therapeutic purposes, including drug transport, cancer diagnostics, and treatment, with an emphasis on their appropriateness and biocompatibility for widespread and safer use.

Nanomedicine is another popular application area of nanotechnology. As an emerging interdisciplinary field, nanomedicine is in near-continuous development over the years. nanomedicine has penetrated the biomedical industry by drug discovery, biomedicine, diagnostics, and medical imaging. Common nanomedicine applications include tissue engineering, rapid diagnosis of infectious diseases using nano-sensors, anti-tumor nano-drugs, and controlled drug delivery with nano-drug delivery systems. Nanotechnology's fast advancement toward the creation of nanomedicines has aided in the advancement of cancer therapy. Nanomedicine enables the development of sophisticated and multifunctional techniques that are both precise and targeted. Nanoparticles have a wide variety of applications in a wide variety of scientific domains nowadays. It has been consistently reported in recent years that nanoparticles play a critical role in the regulation of contemporary medicine by facilitating a variety of therapeutic procedures such as medication administration, genetic material transfer to malignancies, and radiography as a contrast media agent. Numerous nanomaterials derived from organic, inorganic, lipid, or glycan substances, as well as synthetic polymers, have been employed to create and enhance novel cancer therapies. Chapter 16 explores the function of nanoparticles in cancer therapy in comparison to other techniques of anticancer medication delivery.

Nanotechnology can be used to reduce potential health and environmental threats in response to sustainable development challenges. Related examples include the use of nanomaterials for the treatment of viral infections and techniques to combat COVID-19. The SARS-CoV-2 virus, popularly referred to as COVID-19, has developed into an infectious illness since the end of the year 2019, creating a serious worldwide public health crisis that demands the collaboration of several scientific projects. To diagnose, treat, and prevent the SARS-CoV-2 infection, nanomaterials have beeb utilized. Its promise is shown by the many applications and advances undertaken by researchers worldwide. Along with presenting the COVID-19 virus's traits and capabilities, Chapter 17 focuses on the function that nanomaterials play and the challenges that must be faced as a result of the pandemic. Future research paths have been discussed to inform readers about the advances that will likely be needed shortly for the diagnosis, treatment, and prevention of SARS-CoV-2 and its variations. Nanomaterials' effect on COVID-19 will shorten the time required for individuals to revert to their pre-pandemic rhythms.

The World Health Organization proclaimed the unexpected coronavirus pneumonia pandemic a worldwide public health threat on January 30, 2020. COVID-19 have been known to induce coagulation problems and acute respiratory distress syndrome in its most severe form. Due to viral mutations and the emergence of novel viral strains, the efficacy of conventional viral infection therapies is quickly eroding. To overcome these constraints and to enhance antiviral therapies, integrated scientific study into nanotechnology therapy studies is predicted for potential application in the prevention and/or treatment of viral infections. Chapter 18 discusses nanotechnology-based diagnostics, followed by a discussion of nanotechnology-based therapies and treatments that have been demonstrated to be successful against viruses closely similar to SARS-CoV-2. Additionally, we hope that the nanotechnology-based methodologies discussed in this chapter will assist researchers in creating novel COVID-19 prevention, diagnostic, and treatment procedures.

Finally, because of the challenges, currently of getting the best results in the sustainable way, this book presents the innovative opportunities of "green nanotechnology" in nanotechnological products, processes, and applications in the synthesis, energy, medical sector as well as the significant contribution to environmental and climate protection. The authors were carefully selected from the various field of nano researches and the content have been properly vetted by experts in the respective fields. We believe that the use of nanomaterials or nanotechnological techniques may decrease environmental effects and save energy and resources, if tailored suitably.

Acknowledgment

We would like to express our profound gratitude for the support and contributions of everyone participating in this effort.

We would like to express our gratitude to the authors, reviewers, Editorial Advisory Board, and copy editors who contributed to this book project. Without their extraordinary help, this book would never have been completed.

First and foremost, we would like to express our heartfelt appreciation to the authors for contributing their expertise and valuable time to this book.

Second, we are grateful for the invaluable contributions of the reviewers in cultivating the quality, coherence, and content presentation of the chapters.

Thirdly, we want to convey our heartfelt appreciation to the Editorial Advisory Board, which contributed invaluable knowledge and guidance in refining the content of the book.

Last, but not least, we appreciate the sincere efforts of copy editors in polishing the book.

Rajni Garg
Rayat Bahra University, India

Rishav Garg
Galgotias College of Engineering and Technology, India

Nnabuk Okon Eddy
Department of Pure and Industrial Chemistry, University of Nigeria, Nsukka, Nigeria

Introduction

Technology is the elaboration, application, and advancement of science for practical purposes. Currently, the term "green technology" has garnered worldwide attention, being more and more explored with innovative applications. Green technology concerns the use of methods and materials to produce clean and environmentally friendly products and energy. Through the application of green technology, it is expected to give birth to innovation and change in human civilization. The pace of development is expected to have an exemplary rise with solutions to the prevailing global issues. The perception of green technology encloses the concept of sustainability, where community needs can be continuously met without destroying or depleting natural resources. In other words, the needs of the present can be met without compromising the ability of future generations to meet their own needs. The concept of recycling, in which the manufacturing process is designed in such a way that it can be recycled or reused. The concept of reducing waste and pollution means changing the pattern of production and consumption in such a way that it produces only a minimal amount of waste and pollution. The concept of innovation, in terms of always trying to develop alternative technologies. The concept of viability, where the technology and production activities should be environmentally friendly. The inclusion of these concepts can lead to environmental as well as social and economic sustainability.

Green technology is mainly comprised of four critical aspects. The most urgent problem for green technology is energy, which includes both the development of alternative fuels and new methods of energy production, such as energy efficiency. "Green building" encompasses all aspects of construction, from material selection to site selection. The techniques and goods utilized in this instance have the least effect on the environment. Green chemistry minimizes the use of harmful hazardous compounds and their derivatives in the discovery, design, and application of chemical processes and products. The term "green nanotechnology" refers to the utilization of green chemistry for manipulation of materials on the nanoscale scale, which is one billionth of a meter. Numerous experts think that with mastery of nanotechnology, many things will be possible in the future. According to many analysts, reforms in the energy industry are necessary not just to integrate renewable energy but also to restructure and optimize the whole power production and distribution system. Green nanotechnology will undoubtedly be critical to the global development of different industrial sectors, and it is anticipated that it will revolutionize everyone's life in the future years.

Green nanotechnology seeks through miniaturization and the use of ultra-small particles obtained by exploring green chemistry to solve many problems or challenges of everyday life. The use of innovative, lightweight, abundant, and affordable materials is a common feature in all products and systems that use green nanotechnology. This discipline aims to improve energy efficiency in all fields and applications, whether at the industrial or domestic level. Achieve better fuels, ways to use clean and renewable energy, and create more modern destination energy models in line with reality as explored in this book. Green nanotechnology can help prevent major energy crises around the world if supported and promoted by its development. The energy sector will benefit a lot from the use of this type of technology which is applied to different energy sources. This book provides an insight on the synthesis and wide applications of nanomaterials in the various sectors with more efficient manufacturing, less pollution, and energy saving. Various increasingly efficient, ecologically sound, and inexpensive techniques for the design and development of nanomaterials have been explored. The possible applications of such novel materials in the field of environmental remediation, biomedical, and pharmaceutical sectors have been reviewed.

Rajni Garg
Rayat Bahra University, India

Rishav Garg
Galgotias College of Engineering and Technology, India

Nnabuk Okon Eddy
Department of Pure and Industrial Chemistry, University of Nigeria, Nsukka, Nigeria

Section 1
Green Synthesis and Characterization of Nanomaterials

Chapter 1
Innovative and Biocompatible Approaches for Nanomaterial Production and Application:
Eco-Friendly Nanoscience

Tobiloba Christiana Elebiyo
Landmark University, Nigeria

Damilare Rotimi
Landmark University, Nigeria

Rotdelmwa Maimako Asaleye
Landmark University, Nigeria

Boluwatife Boluwatife Afolabi
Landmark University, Nigeria

Bukola Taiwo Atunwa
Landmark University, Nigeria

Oluyomi Stephen Adeyemi
Landmark University, Nigeria

ABSTRACT

The world is going green; hence, environmentally friendly practices that would conserve natural resources for the future generation are encouraged. As a consequence, the world is less concerned about the numerous applications of nanotechnology, especially in the health sector; rather, it is more concerned about the sustainability of functionalized nanomaterials. Thus, the future of nanotechnology depends on its ability to 'go green'. Green nanotechnology attempts to synthesize improved, nontoxic, and biocompatible nanomaterials with sustainable benefits using eco-friendly materials. Although green nanotechnology is considered a sustainable, viable, and biocompatible approach to the production of eco-friendly nanomaterials, there are shortcomings especially in microbial handling and process optimization. In this chapter, the authors aim to appraise not only the use of biocompatible approaches for the synthesis of nanoparticles and/or nanomaterials but also their shortcomings.

DOI: 10.4018/978-1-7998-8936-6.ch001

INTRODUCTION

Today's world emphasizes a "green economy" and, as such, the focus must shift from just meeting man's demand for sustenance, shelter, healthcare, transportation, and energy, to ensuring that these commodities and services meet the requirements for sustainable development. Humankind is now on the lookout for green food, green jobs, and green and sustainable technology at the individual, organizational, national, and global levels. As a result, any technology that is to thrive now or shortly must also be environmentally friendly. Product transitions from manufacturing processes to applications and then disposal should have little or no adverse effect on the ecosystem.

Over the years, nanotechnology has shown immense growth and is gradually finding relevance in almost every field of science today (Enescu *et al.*, 2019). By exploring the small size and large surface area of nanoparticles, nanotechnology has proffered green and sustainable solutions to global challenges. Presently, green nanomaterials are synthesized from biomaterials such as plants, algae, fungus, and bacteria, eggshells, honey, and cobwebs using environmentally friendly methods, and they have found sustainable applications in environmental remediation, agriculture, renewable energy, and biomedical research, among others (Rajasekhar & Kanchi, 2018).

Nanotechnology is a multidisciplinary science that encompasses many other disciplines; it is the study and application of elements with nano-sizes ranging from 1 to 100nm. It involves atoms and molecules in the submicron dimension and the resultant utilization in larger systems (Nasrollahzadeh *et al.*, 2019). The word 'nano' is from a Greek term signifying a size of 10-9 which is 1000 times smaller when compared to a micron. The term 'nanoscale' is used when a material has a dimension less than 100nm (Nasrollahzadeh *et al.*, 2019). Thus, nanoparticles can be defined as a compendium of atoms that are bonded together and have a radius of between 1 and 100nm. Sizes beyond the 1-10 nm range, like those in the 1000nm range, are sometimes referred to as nanomaterials (Leon *et al.*, 2020).

Nanoparticles possess an enlarged surface area available for reactions which results in a distinct quantum effect and chemical reactivity (Nasrollahzadeh *et al.*, 2019). Materials at different sizes usually possess different biological, physical, and chemical properties at the nanoscale, and this difference in characteristics is what nanoscience leverages to manufacture unique products that solve life's problems without posing grievous threats to environmental sustainability. The size of nanomaterials is what confers nanoparticles with unique properties that distinguish them from the parent material (Nasrollahzadeh *et al.*, 2019; Chaturvedi & Dave, 2020; Leon *et al.*, 2020).

However, nanomaterials are as old as man and could exist naturally without deliberate fabrication. A good example of naturally occurring nanomaterials is the magnetotactic bacteria, which uses iron oxide nanoparticles to dictate the Earth's magnetic field lines. The use of nanoparticles by man dates back to the fourth century AD, in the form of the Lycurgus cup, which contained metallic nanoparticles that gave the cup either a green or red color depending on the reflected or transmitted light (Leon *et al.*, 2020). Michael Faraday later confirmed this and reported that gold metal had unusual optical properties at a minute size scale. Gustav Mie also reported similar observations in 1908. In the 16th century, a Swiss chemist, Doctor Von Hohenheim, produced and treated his patients with nano-sized gold particles (Nasrollahzadeh *et al.*, 2019; Leon *et al.*, 2020).

Nanoparticles possess catalytic properties and are being employed for energy storage, reacting systems, and drug delivery (Nasrollahzadeh *et al.*, 2019). Their high resistance to tear and wear coupled with their anti-microbial properties makes them a veritable tool in biomedical research for testing and diagnosis through the invention of very specific sensors and quick testing material that could be essential for the health sector (Abdel-Halim *et al.*, 2020). Functionalized nanomaterials composed of iron oxide nanoparticles are used as drug carriers because of their ability to permeate the blood-brain barrier and are being studied in functionalized forms to ascertain biocompatibility and its interaction with other biological components (Sruthi *et al.*, 2018). For instance, silver nanoparticles, for instance, are widely known for their antimicrobial properties, thus informing their use in a wide range of industrial products like medical instruments, cosmetics, materials for food packaging, clothing, and household appliances (Auclair *et al.*, 2019).

Nanoparticles are also candidates for environmental remediation; they can also be used to tailor cleaner industrial processes and green products that pose less risk to the environment. Nanotechnology has been utilized in so many ways to improve the mechanical properties of materials like their strength, malleability, hardness, weight, and even compressibility. They also exhibit improved electrical conductivity that could inform their utilization as new electronic devices (Nasrollahzadeh *et al.*, 2019). Nanotechnology in agriculture could be employed for more sophisticated and advanced genetic modification of animals and crops. It could also be utilized to enhance the farming process to give better yields and quality, as the case may be. It can also be utilized for soil fertilization and in the processing, storage, and packaging of food (Abdel-Halim *et al.*, 2020).

BACKGROUND

There are two concepts to the synthesis of nanomaterials. This category includes the "top-down" and the "bottom-up" approach (See Figure 1).

Figure 1.

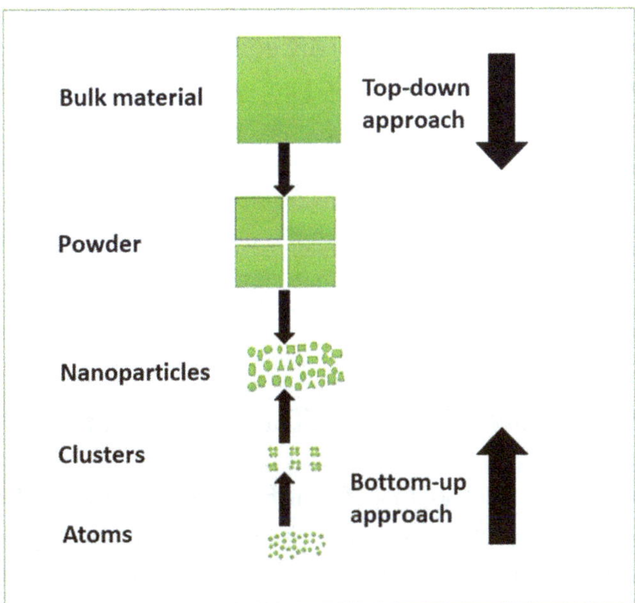

The top-bottom approach involves the manipulation of larger devices to achieve certain technological advantages, while the bottom-up approach involves the sequential assembling of molecules or atoms via sono-decomposition to give a larger one. The bottom-up approach is the type seen in biological systems in processes such as protein synthesis and DNA replication, etc., while industries manufacturing electronic devices utilize the top-down approach (Zhang, Liu, et al., 2016; Nasrollahzadeh *et al.*, 2019). Top-down can be tagged 'miniaturization' and bottom-up is tagged 'self-organization'.

Conventional Methods of Synthesizing Nanoparticles and Their Pitfalls

The synthesis of some noble nanoparticles such as PtNPs, AuNPs, and AgNPs, etc. is one major achievement of nanotechnologists in the past two decades (Zhang, Liu, *et al.*, 2016). The composition, size, and shape of nanoparticles are the key determinants of the physiochemical properties of these nanoparticles (Pareek *et al.*, 2017). Thus, nanotechnologists have developed a reliable and rapid experimental protocol for synthesizing various nanoparticles (Pareek *et al.*, 2017). There are three major conventional methods of synthesizing nanomaterials, namely physical, chemical, and biological methods (Pareek *et al.*, 2017; Teleanu *et al.*, 2019).

The physical method is an example of a top-down approach to synthesizing nanoparticles that involve mechanical grinding (structural decomposition) of grain structures to produce fine particles as a result of severe deformation. This method is relatively simple and inexpensive. However, it has a low yield, and it is difficult to prevent contamination from milling media or the atmosphere. It also has the disadvantage of not being able to produce nanoparticles with an excellent nanocrystalline structure (Teleanu *et al.*, 2019). Some of the strategies involved in the physical method of synthesizing nanoparticles are: pyrolysis, co-precipitation, attrition (Milling), and melt mixing (Pareek *et al.*, 2017).

Chemical methods are bottom-up approaches to synthesizing nanoparticles using reducing agents on metal precursors via organic solvents and stabilizing the end nanoparticles by capping agents. Sol-gel, microemulsion, and hydrothermal methods are being investigated to produce uniform-sized nanoparticles (Du *et al.*, 2014). Although chemical methods usually give higher yields than physical methods, they are capital-intensive and it is also difficult to obtain pure nanoparticles via this method (Subramaniam *et al.*, 2019; Teleanu *et al.*, 2019).

Biological methods are also bottom-up approaches to nanoparticle synthesis. They include the use of strong reducing agents to produce nanoparticles with morphological and structural control from biological sources such as plants, microorganisms, and biomolecules (amino acids and vitamins). These methods are inexpensive, simple, and biocompatible (Subramaniam *et al.*, 2019). However, the synthesis efficiency of these methods is very low (Paosen *et al.*, 2017).

GREEN SYNTHESIS OF FUNCTIONALIZED NANOMATERIALS

Green chemistry, a sustainable approach for manufacturing chemical compounds without posing a threat to the ecosystem, has gained prominence. The advent of green chemistry began over 25 years ago and is seen as the chemistry that employs eco-friendly methods to decrease or outrightly eliminate toxic molecules while creating or using these chemical products (Nasrollahzadeh et al., 2019). Green chemistry is the branch of chemistry that drives the engineering of chemicals and their products by exploring sustainable, harmless, and nontoxic, energy-saving, and cost-effective approaches while avoiding the production of superfluous or non-degradable waste. The sole aim of green chemistry is to invent new chemical processes that do not result in environmental pollution. It is concerned with designing and manufacturing products by adopting procedures that are aimed at plummeting or eradicating the generation and consequent release of dangerous waste. The principles governing green chemistry are shown in Figure 2.

Figure 2.

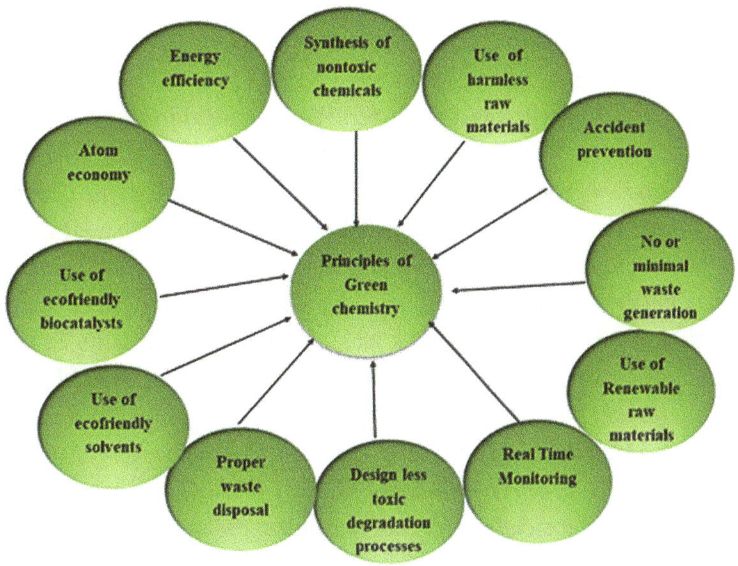

Green Synthesis and Green Nanotechnology

Green synthesis is concerned with the use of resources that are much more environmentally compatible, such as fungi, bacteria, and plants, etc., in the creation of nanoparticles (Mohamad *et al.*, 2013). Unlike the conventional method for synthesizing nanoparticles, green synthesis seeks to circumvent undesirable impacts of nanoparticle fabrication on the ecosystem via two major mechanisms: one, using green materials such as leaves, cobwebs, and honey as reducing and capping or stabilization agents; two, adopting processes that are less energy-consuming or require the excessive manipulation of natural resources in the synthesis process, and third, aspires to ambient synthesis reaction conditions (Wang *et al.*, 2016). The improved environmental impacts observed in green synthesized nanoparticles are largely due to the discontinued usage of traditional chemical reagents such as sodium borohydride in the nanoparticle production process (Wang *et al.*, 2019).

Although nanotechnological advancement has led to innovations that have been channeled to manufacturing devices and products that have novel applications in the manufacturing of communication devices, drug formulation, diagnosis, and targeted drug delivery, etc. However, the 'green economy' is more concerned about the environmental impacts associated with the manufacture and use of functionalized nanomaterials (Bai *et al.*, 2018). To alleviate these concerns and secure the future of nanotechnology, a wide range of green options, one of which is the use of green reducing and capping agents, have been explored for nanotechnology. These green agents are inexpensive compared with the conventional reducing and capping agents; they are biocompatible, non-hazardous, and environmentally friendly. Also, the by-products in the green synthesis of functionalized nanomaterials are harmless to the environment (Khan, 2020). Green nanotechnology makes use of standing 'green' ethics of both chemistry and engineering to produce different nanosized products devoid of toxic ingredients using low temperatures and lesser amounts of energy wherever possible.

Biogenic Materials in the Synthesis of Nanoparticles

Biogenic materials such as plants, animal products, amino acids, polysaccharides, and microorganisms have been explored in the green synthesis of nanomaterials (Table 1). Numerous microorganisms, including bacteria, fungi, yeasts, and viruses, are currently being used in the fabrication of nanoparticles (NPs). Using *Lactobacillus* sp. and *Bacillus* sp., respectively, Jha *et al.* (2009) and Kirthi *et al.* (2011) fabricated TiO_2 NPs. *Fusarium oxysporum* was also used by Bansal *et al.* (2006), to hydrolytically transform hexafluorotitanate complexes into spherical shaped TiO_2NPs. Nadaf and Kanase (2016) and do Nascimento *et al.* (2021) synthesized AuNPs using *Bacillus marisflavi* and *Trichoderma harzianum*, respectively. *Ganoderma enigmaticum* was also used by Krishna *et al.* (2021) to mediate the synthesis of AgNPs. Non-enzymatic fabrication of AgNPs was also carried out using lactic acid bacteria as a reducing agent. The bacteria also serve as stabilizing agents (De Gusseme *et al.*, 2010).

Plants such as *Terminalia chebula,* which contain flavonoids, tannin, glycosides, saponins, and polyphenolic compounds, act as reducing and stabilizing agents, and these have been explored in the synthesis of AuNPs, PdNPs, FeNPs, and AgNPs (Gopinath *et al.*, 2014). Aqueous extract from the leaf and root of *Berberis vulgaris* was used to fabricate silver nanoparticles; the synthesis was clear, inexpensive, and less toxic (Behravan *et al.*, 2019). Extracts from the leaves, seeds, fruits, or peels of *Punica granatum, Trigonella foenum-graecum*, *Alstonia venenata*, *Viburnum opulus*, and *Citrus macroptera* have also been explored in the synthesis of AgNPs ranging from 7–92 nm (David & Moldovan, 2020; Esan *et al.*, 2020; Awad *et al.*, 2021; Khan *et al.*, 2021; Saha *et al.*, 2021; *et al.*, 2021; *et al.*, 2021; *et al.*, 2021; *et al.*, 2021; *et al.*, 2021; *et al.*, 2021; *et al.*, 2021; *et al.*, 2021; *et al.*, 2021; *et al.*, 2021; *et al.*, 2021) AuNPs have also been fabricated using extracts from the parts of *Camellia japonica*, *Prunus amygdalus*, and *Argemone mexicana* (Sharma *et al.*, 2019).

Extracts from potatoes have been used as a reducing and stabilizing agent in the fabrication of Fe_3O_4NPs in the presence of functional groups such as aldehyde groups, hydroxyl groups, and carboxyl (Buazar *et al.*, 2016; Wang *et al.*, 2019). *Cocos nucifera* pulp and *Bauhinia tomentosa* leaves have also been used to synthesize Fe_3O_4NPs with sizes ranging from 90-95 nm (Lakshmnarayanan *et al.*, 2021; Rajendran *et al.*, 2021). Aisida *et al.* (2019) and Qasim *et al.* (2020) also synthesized Fe_3O_4 nanorods from *Moringa oleifera* and *Withania coagulans*, respectively. According to Qasim *et al.* (2020), the synthesis of Fe_3O_4 nanorods using biogenic materials did not only give higher yields compared with the chemical method but also produced Fe_3O_4 nanorods with better antibacterial activity.

Also, products from plants and animals such as gum acacia, eggshells, honey, and spider cobwebs have been explored in the mediation of nanoparticle synthesis (Lateef *et al.*, 2015; Wang *et al.*, 2019). Sinha and Ahmaruzzaman (2015) used extracts from the eggshells of Anas Platyrhynchos to fabricate AuNPs. According to Wang *et al.* (2019), the application of the eggshell as a reducing and stabilizing agent could be due to the presence of amino acids. The researchers reported that the amino and carboxyl groups present in the amino acids facilitate the reduction of $AuCl_4$ and coating of the NP surface

(Maruyama *et al.*, 2015). Lateef *et al.* (2015) used spider for the green synthesis of AgNPs under ambient conditions and observed that the AgNPs obtained from this biomaterial demonstrated good antibacterial activities against multi-drug resistant pressures of bacteria in contrast to the chemically synthesized AgNPs (Lateef *et al.*, 2015).

Table 1. Biogenic materials explored in the synthesis of nanoparticles

	Biogenic Material	Common Name	Size	Shape	Reference
AgNPs					
1	*T. foenum-graecum* seed	Fenugreek	82 nm	Spherical	Awad *et al.* (2021)
2	*A. venenata* Leaf	Devil tree	30 - 40 nm	Spherical	Esan *et al.* (2020)
3	*V. opulus* Fruit	Cranberry bush	7 -26 nm	Spherical	David and Moldovan (2020)
4	*P. granatum* fruit peel	Pomegranate	15 – 30 nm	Spherical	Khan *et al.* (2021)
5	*G. enigmaticum*	White rot fungus	5 -40 nm	Spherical	Krishna *et al.* (2021)
6	*C. macroptera* fruit peel	Wild orange	92 nm	Spherical	Saha *et al.* (2021)
Fe_3O_4NPs					
7	*C. nucifera* pulp	coconut	90 -95 nm	Husked rice-shaped	Rajendran *et al.* (2021)
8	*B. tomentosa* leaves	Yellow Bauhinia	70 nm	–	Lakshmnarayanan *et al.* (2021)
Fe_3O_4 nanorod					
9	*M. oleifera*	Drumstick tree	15.01 ± 6.03 nm	–	Aisida *et al.* (2019)
10	*W. coagulans*	Vegetable rennet	16 ± 2 nm	–	Qasim *et al.* (2020)
AuNPs					
11	*P. amygdalus* seed coat	almond	5 – 15 nm	Oval and spherical	Sadalage *et al.* (2021)
12	*T. harzianum*	fungus	5 – 40 nm	Spherical	do Nascimento *et al.* (2021
13	*A. mexicana*	Mexican poopy	20 – 40 nm	Hexagonal	Datkhile *et al.* (2021do Nascimento *et al.* (2021)
14	*B. marisflavi*	Bacteria	14 nm	Spherical	Nadaf and Kanase (2016
15	*A. Platyrhynchos* eggshells	Duck eggshells	–	Rice-shaped	Sinha and Ahmaruzzaman (2015)

Conventional Nanotechnology vs. Green Nanotechnology

Despite the potential of nanotechnology, there are unknown implications of nanomaterials for the environment and human health. According to Philip (2009), although nanoparticles are not a threat, some aspects of nanotechnology are risky as a result of the increased reactivity and mobility of nanoparticles. The application of nanotechnology is associated with limitations such as increased waste generation, use of non-renewable raw materials, and emissions of greenhouse gases. Consequently, nanotechnology must adopt strategies aimed at reducing the negative impacts of nanotechnology. All processes involved with the fabrication, utilization, and disposal of nanomaterials must be non-toxic and environmentally friendly. The goals of any green technology are to reduce, reuse, recycle, and renew (Maksimović & Omanović-Mikličanin, 2017).

Green chemistry has introduced novel methods for chemical synthesis, processing, and application of products that do not endanger human or environmental health (Verma, 2015). Green nanotechnology implies the application of the principles of green chemistry, which involves improving the efficiency of chemical processes and simultaneously reducing the use of hazardous substances. Green nanotechnology applies the principles of green chemistry to design nanoscale products and develops the production methods of nanomaterials (Maksimović & Omanović-Mikličanin, 2017). Unlike conventional nanotechnology, green nanotechnology offers both environmental and economic benefits, as conventional technologies are capital and energy-intensive and involve the use and generation of toxic materials and waste (Shapira & Youtie, 2012). Contrastingly, green nanotechnology reduces the generation of harmful products and contaminants by directly altering and counteracting nanomaterials that have been generated (Rani & Sridevi, 2017).

SUSTAINABLE APPLICATIONS OF FUNCTIONALIZED GREEN NANOMATERIALS

Functionalized Green Nanomaterials in Environmental Remediation

One of the primary foci of the "green economy" is a cleaner environment, and as a sustainable technology, nanotechnology must strive to create a cleaner environment by eliminating pollutants arising from anthropogenic activities such as overpopulation, urbanization, deforestation, and industrialization. To meet the demand for green products and processes, nanotechnology must be ready to engineer functionalized nanomaterials using green processes that are less toxic, generate minimal waste during manufacturing, and are degradable.

It is no news that repeated manmade activities ultimately result in a drastic reduction in the quality of all the compartments surrounding living organisms (soil, water, and air). Today, the earth is challenged with sporadic climate change, global warming, heightened greenhouse emissions, erosion, contaminated air, and undrinkable water; and a lot of strategies have been put in place for the remediation of the environment using different strategies including absorption, adsorption, photocatalysis, chemical reaction or filtration, bioremediation, phytoremediation, and nano-remediation (Guerra *et al.*, 2018). However, one major disadvantage of all of these clean-up technologies is that they eventually leave scars on Mother Earth. For instance, in bioremediation, micro-organisms tailored to sequester or degrade waste eventually become a source of contamination as microbial growth becomes extremely difficult to control. In the case of phytoremediation, plants used to sequester contaminants such as heavy metals are subjected to burning, which also contributes to their quota of environmental pollution.

Thankfully, proponents of nanotechnology have realized that "green and sustainable development" is the watchword in this new economy. Today, functionalized nanomaterials that could trap and/or eliminate contaminants are synthesized by adopting the basic principles of green chemistry. Furthermore, these green nanomaterials could be recovered and reused again. Nano-sensors that can detect and degrade certain pollutants in the air, water, and soil even at minute concentrations have also been engineered

(Chaudhary *et al.*, 2019). Nanotechnology has also gone the extra mile to design functionalized nanomaterials that minimize the generation of contaminants such as nano-based batteries and even nano-fuels.

Functionalized Nanomaterials in the Removal of Air, Water, and Soil Pollutants

Nanoscience has successfully designed functionalized nanomaterials such as gas nanosensors and nanocatalysts to detect, trap, and mop up pollutants from the soil, water, and air compartments of the environment, even at minute concentrations (Figure 3).

Figure 3.

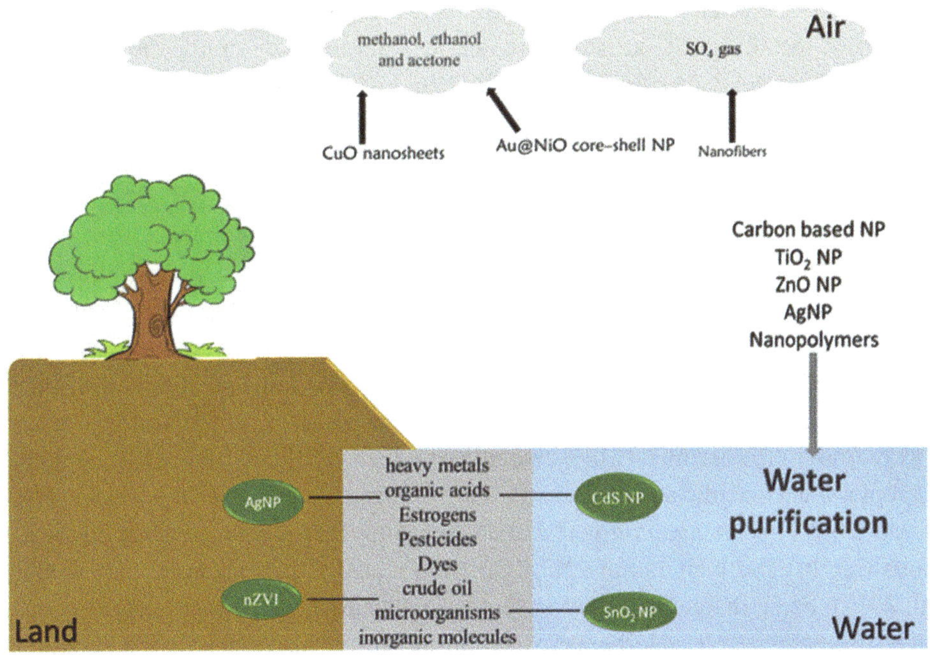

In 2018, Au@NiO core-shell nanoparticles of 70-120nm in size were synthesized and used to design an efficient proximity gas sensor that could detect ethanol at 200 °C without having direct contact with the ethanol itself (Majhi *et al.*, 2018). Nanofibers with metal-organic frameworks have also been used to efficiently remove air pollutants such as SO_2 gas (Zhang, Yuan, *et al.*, 2016). Umar *et al.* (2017), also designed CuO nanosheets to sense the presence of harmful gases such as methanol, ethanol, and acetone.

The photocatalytic activity of nanoparticles and their functionalized nanomaterials are being explored in the degradation of toxic wastes present in soil and water. According to Saikia *et al.* (2019), photocatalysis works by using a nano-catalyst to harness ultraviolet light; this catalyst then uses this energy for the oxidative degradation of organic and inorganic compounds, estrogens, pesticides, dyes, crude oil, microorganisms, and inorganic molecules, etc. Most of these nanocatalysts are now synthesized from biomaterials such as bacteria, fungus, and algae. Consequently, nanotechnology can perform its environmental remediative roles without causing harm to the environment.

Cadmium sulfide nanoparticles (CdSNPs) were synthesized in situ using *Spirulina platensis* and then used for the catalytic degradation of malachite green dye (Mandal *et al.*, 2016). AuNPs and Ag-NPs synthesized from plant sources like *Alpinia nigra* and *Viburnum opulus*, respectively, were used to photocatalytically degrade dyes like direct yellow 12, rhodamine B, and reactive black 5 (Baruah *et al.*, 2018; David & Moldovan, 2020). SnO_2NPs synthesized from the leaf extract of Piper betel (betel pepper) have also been used to photo-catalytically degrade reactive yellow 186 (RY 186 dye) with an efficiency of 92.17% (Singh *et al.*, 2018). Rajendran *et al.* (2021), reported that Fe_2O_3NPs derived from coconut pulp photocatalytically degraded 92% of Rhodamine B dye present in an aqueous solution.

Due to their high absorptive capacity, nanomaterials have also been explored in the removal of heavy metals from soil and water compartments. Heavy metals are non-biodegradable and have the potential to bioaccumulate in soil, plants, animals, and humans; they have a relatively higher density than water and are extremely toxic to life when their threshold dose is exceeded (Das *et al.*, 2019). Paints, dyes, paper, and pulp industries; metallurgical extractions; tanneries; and fertilizer application have all resulted in elevated levels of heavy metals in soil and water (Bakshi & Abhilash, 2020). Arsenic, Cadmium, Chromium, Lead, and Mercury are ranked as five priority metals in public health because they can interact with numerous cellular components, resulting in deregulation of normal physiological processes that could ultimately lead to cancer formation (Tchounwou *et al.*, 2012; Balali-Mood *et al.*, 2012).

The most widely researched nanoparticle for the subtraction of heavy metals from polluted soil and groundwater is the nanoscale zerovalent iron (nZVI) (Hu *et al.*, 2019; Bakshi & Abhilash, 2020). It possesses a large surface area, is highly reactive, and has a strong reducing capability (Bakshi & Abhilash, 2020). Numerous independent researchers have reported that nZVI could remove hexavalent chromium (Cr (VI)) by reducing it to a less trivalent form (Cr (III) in soils contaminated with hexavalent chromium (Chen *et al.*, 2020; Ye *et al.*, 2021; Zhu *et al.*, 2021). According to research, nZVI can successfully reduce the availability of other metals like arsenic (As), lead (Pb), cadmium (Cd), and zinc (Zn) in a multi-metal contaminated acidic and calcareous soil (Cheng *et al.*,2021). Other nanomaterials, such as AgNPs derived from *Ficus benjamina* leaf extract (weeping fig) and proanthocyanidin-AuNPs, have been functionalized to effectively remove Ni^{2+}, Cu^{2+}, Cd^{2+}, and Pb^{2+} from water and aqueous solutions (Al-Qahtani, 2017). Fe_3O_4 magnetic nanorods synthesized using *P. granatum* peel extract as a reducing material can be used as reusable adsorbents for the elimination of heavy metals from contaminated soils or wastewater (Venkateswarlu *et al.*, 2014).

Functionalized Nanomaterials in Water Treatment and Purification

Water is the most precious liquid substance to almost all life forms, especially plants and humans. It is not only required for drinking but also for cooking, processing, and for maintaining proper hygiene. Nevertheless, according to Jose *et al.* (2018), only 3% of the water on the surface of the ground is in the freshwater form, which implies that 97% of the groundwater cannot be consumed by humans and is detrimental to most plants because of its salinity. Statistics have also shown that about 785 million of the world population lack access to clean drinking water facilities (WHO, 2019). The ever-increasing human population has necessitated the need to maximize available water sources through a variety of strategies, one of which is waste recycling (Singh *et al.*, 2020). Nanotechnology is a sustainable technology that is efficient in water purification, and unlike other technologies, nanoparticles are: easy to use; economical; anti-microbial agents; reusable; easy to control; and highly reactive with neighboring atoms. These attributes, coupled with their high reduction potential, make them a perfect candidate for water purification.

One of the widely used functionalized nanomaterials used for water purification is titanium nano-based materials. They have notable photocatalytic activity, they are non-toxic at the stipulated dosage and they possess strong antimicrobial activity against a broad range of water pathogens, some of which include *Mycobacteria, Salmonella, Shigella, Entamoeba histolytica, and Giardia duodenalis* (Khan & Malik, 2019). TiO_2 has also been investigated for photodegradation of contaminants in water such as hexavalent chromium, bromomethane, and rhodamine (Kasinathan *et al.*, 2016). ZnONPs could also be used for water purification as they are said to possess a remarkable anti-microbial activity against *Escherichia coli and Vibrio cholera* (Pirsaheb *et al.*, 2017; Khan, 2020). In the presence of sunlight, ZnO nanoparticles mediate the photolytic degradation of dyes like methylene blue, aniline, and fast green (Kumar *et al.*, 2015). AgNPs have also been reported to possess anti-microbial activities against numerous waterborne pathogens, including *E. coli, Salmonella Typhimurium, Shigella dysenteriae, Enterococcus faecalis, and V. cholerae,* with an efficiency of 100% (Bhatt *et al.*, 2018). Nanopolymers, such as nanocellulose and dendrimers, have also been explored as adsorbents for wastewater treatment. These nano-adsorbents can be easily synthesized, have a high adsorption capacity, and can be regenerated and reused after the desorption of pollutants. This makes nano-adsorbents superior candidates for effluent treatment and water decontamination (Saikia *et al.*, 2019).

With nanotechnology, the consumption of seawater has become feasible as carbon-based nanomaterials such as carbon nanotubes, graphene, and zeolite are currently being used in desalinating seawater (Teow & Mohammad, 2019; Zhang *et al.*, 2019). Carbon nanotubes (CNTs) have the largest weight-to-strength ratio of any known material. They have unique attributes such as long-range porosity, superior thermal and electrical conductivity, and exceptional mechanical strength and stiffness (Teow & Mohammad, 2019). Carbon-based nanomaterials can be fabricated using green approaches such as the fabrication of ultralight graphene–CNT hybrid aerogel monoliths from aqueous gel precursors processed by supercritical CO_2 drying (Sui *et al.*, 2012). Membrane matrices incorporated with CNTs have shown enhanced surface hydrophilicity, permeability, solute rejection, tensile strength, and electrical conductivity; they also possess reduced fouling tendencies (Khalid *et al.*, 2015; Teow & Mohammad, 2019).

Functionalized Nanomaterials in Biomedicine

The basic green chemistry principle has been used to engineer functionalized nanomaterials with immense applications in biomedicine. Because of their unique physicochemical properties, impeccable catalytic activity, optical properties, electronic properties, and magnetic properties, nanoparticles have been employed in diagnostics, catalysis, imaging, and drug delivery (Figure 4) (Zhang, Liu, *et al.*, 2016).

AgNPs from *Alstonia venenata* extract showed moderate larvicidal activity against the malaria vectors, *Culex quinquefasciatus, Aedes aegypti,* and *Anopheles stephensi* (Esan *et al.*, 2020). AgNPs and Fe_2O_3NPs synthesized from *T. foenum-graecum* seed extract and *C. nucifera* pulp showed high anti-bacterial activity against *E. coli, Bacillus cereus,* and *Staphylococcus aureus* (Awad *et al.*, 2021; Rajendran *et al.*, 2021). The potential of nano-antibiotics has the potential to eliminate drug-resistant microbes while simultaneously reducing cellular toxicity that often accompanies the use of associated (Ali *et al.*, 2020). Ebselen, a silver nano-based antimicrobial compound, has shown synergistic antibacterial activity against most clinically resistant strains of *Klebsiella pneumonia, E. coli, Enterobacter cloacae, Acinetobacter baumannii*, and *Pseudomonas aeruginosa* (Zou et al., 2018; Ali *et al.*, 2020). Also, naringin and flavonoid-hesperidin-loaded nanoparticles stabilized with gums from plants are now used as a new antimicrobial agent against multiple drug-resistant bacteria and brain-eating amebae (Anwar *et al.*, 2019).

Figure 4.

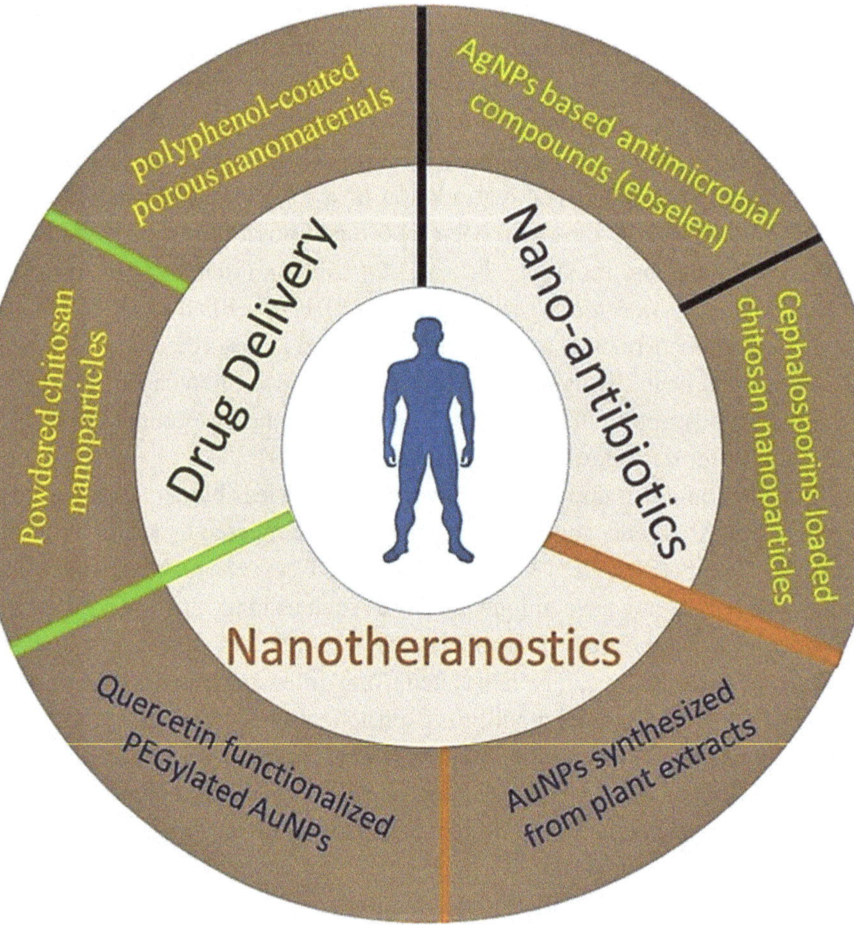

Nanoparticles have been used to tackle the challenges of multiple drug resistance as nanocarriers are synthesized from green materials and used in the targeted delivery of therapeutic agents. Drugs delivered using nanomaterials are more stable than free drugs and have improved bioavailability. According to Su and Kang (2020), via the exploration of green chemistry, biodegradable drug delivery compounds such as polyphenol-coated porous nanomaterials have been successfully fabricated. Chitosan nanoparticles from purified chitin have also shown antimicrobial properties and can be used to facilitate drug delivery. Chitosan nanoparticles have also been functionalized with galactose, trisaccharide, and folate to enhance the recognition and uptake of nanocarriers for targeted delivery of drugs such as rifampicin into target cells (Ali *et al.*, 2020) (Rawal *et al.*, 2017). Also, Cefazolin, a cephalosporin-loaded chitosan nanoparticle, has been used as an antibiotic against drug-resistant strains of *P. aeruginosa* and *K. pneumonia* (Jamil *et al.*, 2016).

According to Weldrick *et al.* (2021), nanoencapsulation of Penicillin G and Oxacillin into shellac nanoparticles and subsequent coating with a serine endo-peptidase protease facilitated the delivery of the antibiotics into Penicillin G-resistant biofilms, stimulating more effective destruction of the bacteria cells. Also, functionalized fucosylated lipid nanocarriers encapsulating the anti-tuberculosis agent levofloxacin were reported to preferentially deliver the drug into the endosomal compartments housing the mycobacteria (Duran *et al.*, 2021). Novel mixed micelles from pluronic F127 and Cremophor EL have also been used to efficiently deliver Norfloxacin, a hydrophobic drug, to clinical isolates of S. aureus 35923, *E. coli* 25922, and *P. aeruginosa* 5399 (Tanase *et al.*, 2021).

The current cancer imaging and diagnostic modalities are still somewhat inefficient. According to Frangioni (2008); Du *et al.* (2014), one of the most difficult cancers to image is pediatric cancer, one of which is retinoblastoma. Because of their size and shape-dependent optical and electrical properties, AuNPs are currently being investigated in cancer diagnosis and imaging (Zhang *et al.*, 2019); and a series of studies have shown that AuNPs, Au nanorods, Au nanoclusters, quantum dots, and Au nanocages can be used for the diagnosis of retinoblastoma (Toda *et al.*, 2019; Wang *et al.*, 2020; Arshad *et al.*,

AgNPs synthesized using *C. quadrangularis, C. maxima, M. oleifera, Syzygium aromaticum,* and *Helminthosporium solani* showed anticancer effects against the human liver, colon, skin, HeLa, lung, HEK293, and breast cancer cell lines (Rajasekharreddy *et al.*, 2019). *P. granatum* and the white rot fungi, *G. enigmaticum* and *T. ljubarskyi,* reportedly showed moderate to high cytotoxic, anti-proliferative, and apoptotic potentials against MDA-MB-231 breast and human lung cancer cells (Khan *et al.*, 2021; Krishna *et al.*, 2021). In addition, coconut pulp-mediated Fe_2O_3NPs demonstrated strong cytotoxic activity against cancer cells in a dose-dependent manner (Rajendran *et al.*,2021). Also, AuNPs, AgNPs, and PtNPs synthesized using tannic acid as reductant and stabilizer exhibited promising cytotoxic properties against breast cancer MDA-MB-231 cell lines (Adeyemi & Otohinoyi, 2018).

Quercetin-functionalized PEGylated AuNPs showed more potent anticancer and anti-inflammatory potential than AuNPs and quercetin alone (Sadalage *et al.*, 2021). According to Zhao *et al.* (2020), cisplatin-functionalized nanoliposomes significantly induced apoptosis in retinoblastoma cells both in-vitro and in-vivo by upregulating the expression and activity of pro-apoptotic factors such as Bax and capase-3 and inhibiting the expression of Bcl-2, an anti-apoptotic factor. These functionalized nanomaterials are effective in target recognition; they have high stability and release loaded drugs in a controlled manner. A nanocarrier encapsulating NLGplatin, a difunctional drug with immunochemotherapeutic properties, effectively delivered the drug to 4T1 murine breast cancer cells and HeLa human cervical cancer cells in-vivo and induced apoptotic responses in the cancer cells by amplifying tumor antigen presentation and stimulating immunogenic responses against the cells (Zhu *et al.*, 2020). A similar study also showed that a mesoporous silica dual nanocarrier encapsulating palbociclib, a cell cycle inhibitor, and hydroxychloroquine, an autophagy inhibitor, effectively co-delivered the drugs into pancreatic ductal adenocarcinoma cell lines, thus, triggering apoptotic responses via the downregulation of Bcl-2 activity (Ji *et al.*, 2020).

Safety Assessment of Green Functionalized Nanomaterials

Nanotechnology is a promising tool in many fields and has been incorporated in health care and biomedical research; food and chemical industries; electronics manufacturing; and environmental remediation. At first, nanomaterials and their associated nanoparticles were synthesized using chemical and physical methods that required harsh and toxic chemicals like hydrazine and dimethylformamide; they were synthesized under extreme temperature and vacuum conditions, and these processes also required equipment that was not pocket-friendly. Today, the face of nanotechnology has been transformed to accommodate the principles of green chemistry, and toxic and expensive synthetic procedures are gradually being replaced by eco-friendly and cost-effective processes (green synthesis).

Table 2. Environmental and biomedical applications of functionalized nanomaterials

Environmental Applications of Functionalized Nanomaterials			
S/n	**Applications**	**Functionalized Nanomaterials**	**Reference**
1	Detection and entrapment of air pollutants such as ethanol, methanol, acetone, and SO_4	Au@NiO -based Proximity gas sensors; Nanofibers; CuO nanosheets sensors	Zhang, Yuan, *et al.* (2016); Umar *et al.* (2017); Majhi *et al.* (2018)
2	Adsorption of heavy metals: Ni^{2+}, Cu^{2+}, Cd^{2+}, Cr(VI), As, Pb, Cd, Zn, and Pb from wastewater and soil	Nanopolymers; AuNPs; nZVI; Fe_3O_4 magnetic nanorod	Venkateswarlu *et al.* (2014); Al-Qahtani (2017); Bakshi and Abhilash (2020); Chen *et al.* (2020); Cheng *et al.* (2021); Zhu *et al.* (2021) Saikia *et al.* (2019)
3	Desalinating of seawater	Carbon nanotubes; graphene; zeolite	Teow and Mohammad (2019)
4	Degradation of dyes such as malachite green dye, direct yellow 12 and 186, methyl orange, methylene blue, Rho B, and reactive black 5	CdSNPs; AuNPs; AgNPs; SnO_2NPs; ZnONPs; Fe_2O_3NPs	Mandal *et al.* (2016); Baruah *et al.* (2018); Singh *et al.* (2018); David and Moldovan (2020); Rajendran *et al.* (2021)
5	Removal of microorganisms: *E. coli, E. histolytica, G. duodenalis, S. Typhimurium, S. dysenteriae, E. faecalis, and V. cholerae*	TiO_2NPs; ZnONPs; AgNPs;	Pirsaheb *et al.* (2017); Bhatt *et al.* (2018); Khan and Malik (2019); Khan (2020)
Biomedical applications of functionalized nanomaterials			
6	Destruction of clinically resistant strains of *K. pneumonia, S. aureus, E. coli, E. cloacae, A. baumannii*, and *P. aeruginosa*	Ag nano-based antimicrobial compounds (ebselen); AgNPs; Fe_2O_3NPs	Zou *et al.* (2018); Anwar *et al.* (2019); Ali *et al.* (2020); Awad *et al.* (2021); Rajendran *et al.* (2021)
7	Larvicidal activity against the mosquito vectors: *C. quinquefasciatus, Ae. aegypti* and *an. Stephensi*	AgNPS	Esan *et al.* (2020)
8	Drug delivery	polyphenol-coated porous nanomaterials; functionalized chitosan nanocarriers; functionalized shellac NPs; fucosylated lipid nanocarriers; polymeric mixed micelles	Jamil *et al.* (2016); Rawal *et al.* (2017); Ali *et al.* (2020); Su and Kang (2020); Duran *et al.* (2021); Tanase *et al.* (2021); Weldrick *et al.* (2021)amil *et al.* (2016); Rawal *et al.* (2017); Ali *et al.* (2020); Su and Kang (2020); Duran *et al.* (2021); Tanase *et al.* (2021); Weldrick *et al.* (2021)
9	Anti-cancer activity against human cancer cell lines	AgNPs; Fe_2O_3NPs; AuNPs; PtNPs; Quercetin functionalized PEGylated AuNPs; cisplatin functionalized nanoliposomes; NLGplatin encapsulating nanocarrier; dual mesoporous silica nanocarrier encapsulating Palbociclib and hydroxychloroquine	Rajasekharreddy *et al.* (2019); (Rajendran *et al.*, 2021) Khan *et al.* (2021); Krishna *et al.* (2021); Adeyemi and Otohinoyi (2018); Ji *et al.* (2020); Zhao *et al.* (2020); Zhu *et al.* (2020); Sadalage *et al.* (2021)
	Cancer diagnostics and imaging	AuNPs; Au nanorods; Au nanoclusters; Au nanocages; quantum dots	Zhang *et al.* (2019) Toda *et al.* (2019); Wang *et al.* (2020); Arshad *et al.* (2021)

Green synthesis of nanoparticles is the safe synthesis of nanomaterials from biogenic sources such as plants, microorganisms, and other biogenic materials including eggshells, honey, and cobwebs. By adopting a biological approach for synthesizing nanomaterials rather than physical and chemical methods, nanoscience has successfully synthesized several nanoparticles, including gold, silver, silver oxide, titanium oxide, cadmium disulfide, palladium and platinum, zinc oxide, and iron oxides from bacteria, fungi, viruses, leaf and root extracts, bulbs, the latex of plants, honey, and ascorbic acid (Balasooriya *et al.*, 2017).

The broad toxicity investigation of nanomaterials indicates that nanotoxicity results in the heightened generation of reactive oxygen species. Oxidative stress ultimately leads to cancer initiation, excessive apoptosis, cytotoxicity, inflammation, DNA damage, and deregulated cell signaling (Bai *et al.*, 2018). Although the use of biocompatible, nontoxic agents makes the synthesis of green nanoparticles ecologically friendly and economical, there is a need to investigate the possible toxicity of green functionalized nanomaterials. The toxicity of nanomaterials is usually analyzed by studying the chemical and physical properties of these materials. For instance, comparing the toxicological profiles of nanomaterials based on their different sizes indicates that small-sized nanomaterials could possess lesser toxicity in comparison to their larger counterparts (Bai *et al.*, 2018).

The use of biogenic materials such as extracts from plants, bacteria, fungi is non-toxic, cost-effective, biodegradable, and eco-friendly. For instance, green ZnONPs are environmentally friendly and much safer than engineered ZnONPs (Kalpana & Devi Rajeswari, 2018).

Saif *et al.* (2016), carried out an acute toxicity study on engineered CuONPs and plant-mediated CuONPs. Different CuONPs were subjected to the same experimental conditions. It was observed that the plant-mediated CuONPs have an EC_{50} of 0.69±0.226 mg/L; this was greater than the EC_{50} of engineered CuONPs (0.102±0.019 mg/L). Their result also indicated that engineered CuONPs accumulation in Daphnia was significantly higher than the accumulation of plant-mediated CuONPs. Plant-synthesized CuONPs were found to be more stable when compared with engineered nanoparticles. The acute nanotoxicity test also proved that plant-mediated CuONPs are less toxic than engineered CuONPs (Saif *et al.*, 2016).

The same nanoparticles generated from different biogenic sources could also have different toxicological profiles. Ronavari *et al.* (2017) synthesized AgNPs using coffee and green tea extracts, and they observed that the different AgNPs behaved differently when introduced to biological systems. Based on their findings, although the green tea mediated-AgNPs showed excellent antimicrobial activities, its potential applications were limited because it was highly toxic to mouse and human cells. In contrast, the coffee mediated-AgNPs showed substantial inhibitory action on *E. coli* as well as *Cryptococcus neoformans*, and were non-toxic to mouse and human cells, signifying an outstanding capacity to discriminate between potential mammalian and pathogen cells (Ronavari *et al.*, 2017).

Challenges to the Adoption of Green Nanotechnology

Green nanotechnology is not without challenges, especially in the areas of testing, regulatory policies, and the processes needed for commercial-scale synthesis (Rani & Sridevi, 2017). Generally, the use of green technology compared with conventional technology is more expensive. Also, many processes and principles involved in green technology have not been fully elucidated as green nanotechnology is a relatively new technology. Consequently, the training and development costs in this industry are higher compared with the already-established technologies (Iravani *et al.*, 2017).

Among the obstacles to developing and commercializing green nanotechnology are the following:

1. The initial discovery phases of green nanotechnology showed no clear design guidelines for researchers.
2. There is a need for new commercial production techniques in the synthesizing of different green nanomaterials. This necessitates the coordination of both engineering and basic research on one hand and the scientific and industrial communities on the other.
3. There is limited expertise in the development of green nanotechnology.
4. The analysis protocols and toxicology of the products require more development and constant updating to reflect advances in science.
5. There is persistency in uncertain regulation, and green products encounter more and higher regulatory barriers than conventional or existing materials.
6. There are uncertainties in end-market demand, particularly since there are limited commercial-grade materials that can be compared to conventional products in performance.

Many other obstacles limit the adoption and circulation of green technologies. Some of them include institutional barriers involving inadequate appropriate regulatory framework; financial, technological, legal, political, and cultural barriers; absence of alternative raw materials or chemical inputs; lack of data and information; the uncertainty involved in the performance impacts of the products; high implementing costs; limited skilled personnel; and limited human resources. The solution to overcome these barriers may be a complex approach. Encouraging the use of green technology must first identify and remove these obstacles to prevent the large-scale dissemination of clean technology (Ghanshyam, 2015).

FUTURE PERSPECTIVE

Although recent advancements in nanoscience have brought about the biogenic synthesis of many nanomaterials, which have been explored in environmental remediation and targeted drug delivery (Table 1), it is challenging to ascertain the safety of these green nanomaterials. This is largely because each biogenic material mediates the synthesis of nanomaterials with a distinct size, shape, and activity. This implies that the same nanoparticles generated from different biogenic sources could possess different toxicological profiles. Consequently, regulatory policies must be put in place to enforce adequate toxicological profiling of nanomaterials synthesized from different biogenic materials before such materials are commercialized or allowed into society.

CONCLUSION

Green nanotechnology has become a blessing to current industrialization in the world. So, the adoption of these technological methods will reduce environmental pollution and the recycling of natural resources. However, the toxicological profile of all the nanomaterials synthesized from different biogenic sources must be fully elucidated as the same nanomaterial from different sources will exhibit different properties. Apart from the poor acceptability of green nanotechnology arising from the unknown negative health and environmental impacts, the commercialization of green nanotechnology is challenged by issues of poor implementation of regulatory policies and other political, technological, and financial barriers. is challenging, and to overcome these barriers, it will require much effort from communities such as governmental and research.

REFERENCES

Abdel-Halim, K. Y., Osman, S. R., & Abdou, G. Y. (2020). In vivo evaluation of oxidative stress and biochemical alteration as biomarkers in glass clover snail, Monacha cartusiana exposed to zinc oxide nanoparticles. *Environ Pollut, 257*, 113120. doi:10.1016/j.envpol.2019.113120 PMID:31753629

Adeyemi, O. S., & Otohinoyi, D. A. (2018). Inorganic nanoparticles restrict viability of metastatic breast cancer cells in vitro. *Comparative Clinical Pathology, 28*(4), 949–954. doi:10.100700580-018-2841-z

Aisida, S. O., Madubuonu, N., Alnasir, M. H., Ahmad, I., Botha, S., Maaza, M., & Ezema, F. I. (2019). Biogenic synthesis of iron oxide nanorods using Moringa oleifera leaf extract for antibacterial applications. *Applied Nanoscience, 10*(1), 305–315. doi:10.100713204-019-01099-x

Al-Qahtani, K. M. (2017). Cadmium removal from aqueous solution by green synthesis zero valent silver nanoparticles with Benjamina leaves extract. *Egyptian Journal of Aquatic Research, 43*(4), 269–274. doi:10.1016/j.ejar.2017.10.003

Ali, R., Batool, T., Manzoor, B., Waseem, H., Mehmood, S., Kabeer, A., Ali, Z., Habib, S., Rashid, U., & Iqbal, M. J. (2020). Nanobiotechnology-based drug delivery strategy as a potential weapon against multiple drug-resistant pathogens. In Antibiotics and Antimicrobial Resistance Genes in the Environment (pp. 350-368). doi:10.1016/B978-0-12-818882-8.00022-X

Anwar, A., Masri, A., Rao, K., Rajendran, K., Khan, N. A., Shah, M. R., & Siddiqui, R. (2019). Antimicrobial activities of green synthesized gums-stabilized nanoparticles loaded with flavonoids. *Scientific Reports, 9*(1), 3122. doi:10.103841598-019-39528-0 PMID:30816269

Arshad, R., Barani, M., Rahdar, A., Sargazi, S., Cucchiarini, M., Pandey, S., & Kang, M. (2021). Multi-Functionalized Nanomaterials and Nanoparticles for Diagnosis and Treatment of Retinoblastoma. *Biosensors (Basel), 11*(4), 97. Advance online publication. doi:10.3390/bios11040097 PMID:33810621

Auclair, J., Turcotte, P., Gagnon, C., Peyrot, C., Wilkinson, K. J., & Gagne, F. (2019). The influence of surface coatings on the toxicity of silver nanoparticle in rainbow trout. *Comparative Biochemistry and Physiology. Toxicology & Pharmacology: CBP, 226*, 108623. doi:10.1016/j.cbpc.2019.108623 PMID:31505268

Awad, M. A. A., Hendi, A., Ortashi, K. M., Alzahrani, B., Soliman, D., Alanazi, A., Alenazi, W., Taha, R. M., Ramadan, R., El-Tohamy, M., AlMasoud, N., & Alomar, T. S. (2021). Biogenic synthesis of silver nanoparticles using Trigonella foenum-graecum seed extract: Characterization, photocatalytic and antibacterial activities. *Sensors and Actuators. A, Physical, 323*. Advance online publication. doi:10.1016/j.sna.2021.112670

Bai, R. G., Sabouni, R., & Husseini, G. (2018). Green Nanotechnology—A Road Map to Safer Nanomaterials. In Applications of Nanomaterials (pp. 133-159). doi:10.1016/B978-0-08-101971-9.00006-5

Bakshi, M., & Abhilash, P. C. (2020). Nanotechnology for soil remediation: Revitalizing the tarnished resource. In *Nano-Materials as Photocatalysts for Degradation of Environmental Pollutants* (pp. 345–370). Elsevier Inc. doi:10.1016/B978-0-12-818598-8.00017-1

Balali-Mood, M., Naseri, K., Tahergorabi, Z., Khazdair, M. R., & Sadeghi, M. (2021). Toxic Mechanisms of Five Heavy Metals: Mercury, Lead, Chromium, Cadmium, and Arsenic. *Frontiers in Pharmacology*, *12*, 643972. doi:10.3389/fphar.2021.643972 PMID:33927623

Balasooriya, E. R., Jayasinghe, C. D., Jayawardena, U. A., Ruwanthika, R. W. D., Mendis de Silva, R., & Udagama, P. V. (2017). Honey Mediated Green Synthesis of Nanoparticles: New Era of Safe Nanotechnology. *Journal of Nanomaterials*, *2017*, 1–10. doi:10.1155/2017/5919836

Bansal, V., Ahmad, A., & Sastry, M. (2006). Fungus-mediated biotransformation of amorphous silica in rice husk to nanocrystalline silica. *Journal of the American Chemical Society*, *128*(43), 14059–14066. doi:10.1021/ja062113+ PMID:17061888

Baruah, D., Goswami, M., Yadav, R. N. S., Yadav, A., & Das, A. M. (2018). Biogenic synthesis of gold nanoparticles and their application in photocatalytic degradation of toxic dyes. *Journal of Photochemistry and Photobiology. B, Biology*, *186*, 51–58. doi:10.1016/j.jphotobiol.2018.07.002 PMID:30015060

Behravan, M., Hossein Panahi, A., Naghizadeh, A., Ziaee, M., Mahdavi, R., & Mirzapour, A. (2019). Facile green synthesis of silver nanoparticles using Berberis vulgaris leaf and root aqueous extract and its antibacterial activity. *International Journal of Biological Macromolecules*, *124*, 148–154. doi:10.1016/j.ijbiomac.2018.11.101 PMID:30447360

Bhatt, D., Gupta, E., Kaushik, S., Srivastava, V. K., Saxena, J., & Jyoti, A. (2018). Bio-fabrication of silver nanoparticles by Pseudomonas aeruginosa: Optimisation and antibacterial activity against selected waterborne human pathogens. *IET Nanobiotechnology / IET*, *12*(7), 981–986. doi:10.1049/iet-nbt.2018.0051 PMID:30247141

Buazar, F., Baghlani-Nejazd, M. H., Badri, M., Kashisaz, M., Khaledi-Nasab, A., & Kroushawi, F. (2016). Facile one-pot phytosynthesis of magnetic nanoparticles using potato extract and their catalytic activity. *Stärke*, *68*(7-8), 796–804. doi:10.1002tar.201500347

Chaturvedi, S., & Dave, P. N. (2020). Application of nanotechnology in foods and beverages. In Nanoengineering in the Beverage Industry (pp. 137-162). doi:10.1016/B978-0-12-816677-2.00005-3

Chaudhary, S., Sharma, P., Chauhan, P., Kumar, R., & Umar, A. (2019). Functionalized nanomaterials: A new avenue for mitigating environmental problems. *International Journal of Environmental Science and Technology*, *16*(9), 5331–5358. doi:10.100713762-019-02253-2

Chen, X., Li, X., Xu, D., Yang, W., & Bai, S. (2020). Application of nanoscale zero-valent iron in hexavalent chromium-contaminated soil: A review. *Nanotechnology Reviews*, *9*(1), 736–750. doi:10.1515/ntrev-2020-0059

Cheng, P., Zhang, S., Wang, Q., Feng, X., Zhang, S., Sun, Y., & Wang, F. (2021). Contribution of Nano-Zero-Valent Iron and Arbuscular Mycorrhizal Fungi to Phytoremediation of Heavy Metal-Contaminated Soil. *Nanomaterials (Basel, Switzerland)*, *11*(5), 1264. Advance online publication. doi:10.3390/nano11051264 PMID:34065026

Das, A., Kamle, M., Bharti, A., & Kumar, P. (2019). Nanotechnology and it's applications in environmental remediation: An overview. *Vegetos, 32*(3), 1–11. doi:10.100742535-019-00040-5

David, L., & Moldovan, B. (2020). Green Synthesis of Biogenic Silver Nanoparticles for Efficient Catalytic Removal of Harmful Organic Dyes. *Nanomaterials (Basel, Switzerland), 10*(2), 202. Advance online publication. doi:10.3390/nano10020202 PMID:31991548

De Gusseme, B., Sintubin, L., Baert, L., Thibo, E., Hennebel, T., Vermeulen, G., Uyttendaele, M., Verstraete, W., & Boon, N. (2010). Biogenic silver for disinfection of water contaminated with viruses. *Applied and Environmental Microbiology, 76*(4), 1082–1087. doi:10.1128/AEM.02433-09 PMID:20038697

do Nascimento, J. M., Cruz, N. D., de Oliveira, G. R., Sá, W. S., de Oliveira, J. D., Ribeiro, P. R. S., & Leite, S. G. F. (2021). Evaluation of the kinetics of gold biosorption processes and consequent biogenic synthesis of AuNPs mediated by the fungus Trichoderma harzianum. *Environmental Technology & Innovation, 21*, 101238. Advance online publication. doi:10.1016/j.eti.2020.101238

Du, J.-J., Chen, C., Gan, Y.-L., Zhang, R.-H., Yang, C.-Y., & Zhou, X.-W. (2014). Facile one-pot hydrothermal synthesis of Pt nanoparticles and their electrocatalytic performance. *International Journal of Hydrogen Energy, 39*(31), 17634–17637. doi:10.1016/j.ijhydene.2014.08.044

Duran, V., Grabski, E., Hozsa, C., Becker, J., Yasar, H., Monteiro, J. T., Costa, B., Koller, N., Lueder, Y., Wiegmann, B., Brandes, G., Kaever, V., Lehr, C. M., Lepenies, B., Tampe, R., Forster, R., Bosnjak, B., Furch, M., Graalmann, T., & Kalinke, U. (2021). Fucosylated lipid nanocarriers loaded with antibiotics efficiently inhibit mycobacterial propagation in human myeloid cells. *Journal of Controlled Release, 334*, 201–212. doi:10.1016/j.jconrel.2021.04.012 PMID:33865899

Enescu, D., Cerqueira, M. A., Fucinos, P., & Pastrana, L. M. (2019). Recent advances and challenges on applications of nanotechnology in food packaging. A literature review. *Food and Chemical Toxicology, 134*, 110814. doi:10.1016/j.fct.2019.110814 PMID:31520669

Esan, V., Mahboob, S., Al-Ghanim, K. A., Elanchezhiyan, C., Al-Misned, F., Ahmed, Z., & Govindarajan, M. (2020). Novel Biogenic Synthesis of Silver Nanoparticles Using Alstonia venenata Leaf Extract: An Enhanced Mosquito Larvicidal Agent with Negligible Impact on Important Eco-biological Fish and Insects. *Journal of Cluster Science, 32*(2), 489–497. doi:10.100710876-020-01808-5

Frangioni, J. V. (2008). New technologies for human cancer imaging. *Journal of Clinical Oncology, 26*(24), 4012–4021. doi:10.1200/JCO.2007.14.3065 PMID:18711192

Ghanshyam, D. S. (2015). Advantages of green technology. *Social Issues and Environmental Problems, 3*(9), 2394–3629.

Gopinath, K., Gowri, S., Karthika, V., & Arumugam, A. (2014). Green synthesis of gold nanoparticles from fruit extract of Terminalia arjuna, for the enhanced seed germination activity of Gloriosa superba. *Journal of Nanostructure in Chemistry, 4*(3), 115. Advance online publication. doi:10.100740097-014-0115-0

Guerra, F. D., Attia, M. F., Whitehead, D. C., & Alexis, F. (2018). Nanotechnology for Environmental Remediation: Materials and Applications. *Molecules (Basel, Switzerland), 23*(7), 1760. Advance online publication. doi:10.3390/molecules23071760 PMID:30021974

Hu, Y. B., Zhang, M., & Li, X. Y. (2019). Improved longevity of nanoscale zero-valent iron with a magnesium hydroxide coating shell for the removal of Cr(VI) in sand columns. *Environ Int, 133*(Pt B), 105249. doi:10.1016/j.envint.2019.105249

Jamil, B., Habib, H., Abbasi, S., Nasir, H., Rahman, A., Rehman, A., Bokhari, H., & Imran, M. (2016). Cefazolin loaded chitosan nanoparticles to cure multi drug resistant Gram-negative pathogens. *Carbohydrate Polymers, 136*, 682–691. doi:10.1016/j.carbpol.2015.09.078 PMID:26572401

Jha, A. K., Prasad, K., & Kulkarni, A. R. (2009). Synthesis of TiO2 nanoparticles using microorganisms. *Colloids and Surfaces. B, Biointerfaces, 71*(2), 226–229. doi:10.1016/j.colsurfb.2009.02.007 PMID:19285838

Ji, Y., Liu, X., Li, J., Xie, X., Huang, M., Jiang, J., Liao, Y. P., Donahue, T., & Meng, H. (2020). Use of ratiometrically designed nanocarrier targeting CDK4/6 and autophagy pathways for effective pancreatic cancer treatment. *Nature Communications, 11*(1), 4249. doi:10.103841467-020-17996-7 PMID:32843618

Jose, A. J., Jacob, A. M., Manjush, K. C., & Kappen, J. (2018). Chitosan in water purification technology. In S. Ahmad & C. M. Hussain (Eds.), *Green and sustainable advanced materials: Applications.* doi:10.1002/9781119528463.ch5

Kalpana, V. N., & Devi Rajeswari, V. (2018). A Review on Green Synthesis, Biomedical Applications, and Toxicity Studies of ZnO NPs. *Bioinorganic Chemistry and Applications, 3569758*, 1–12. Advance online publication. doi:10.1155/2018/3569758 PMID:30154832

Kasinathan, K., Kennedy, J., Elayaperumal, M., Henini, M., & Malik, M. (2016). Photodegradation of organic pollutants RhB dye using UV simulated sunlight on ceria based TiO2 nanomaterials for antibacterial applications. *Scientific Reports, 6*(1), 38064. doi:10.1038rep38064 PMID:27901105

Khalid, A., Al-Juhani, A. A., Al-Hamouz, O. C., Laoui, T., Khan, Z., & Atieh, M. A. (2015). Preparation and properties of nanocomposite polysulfone/multi-walled carbon nanotubes membranes for desalination. *Desalination, 367*, 134–144. doi:10.1016/j.desal.2015.04.001

Khan, A. A., Alanazi, A. M., Alsaif, N., Wani, T. A., & Bhat, M. A. (2021). Pomegranate peel induced biogenic synthesis of silver nanoparticles and their multifaceted potential against intracellular pathogen and cancer. *Saudi Journal of Biological Sciences, 28*(8), 4191–4200. doi:10.1016/j.sjbs.2021.06.022 PMID:34354399

Khan, S. H. (2020). Green Nanotechnology for the Environment and Sustainable Development. In M. Naushad & E. Lichtfouse (Eds.), *Green Materials for Wastewater Treatment* (pp. 13–46). Springer Nature. doi:10.1007/978-3-030-17724-9_2

Khan, S. T., & Malik, A. (2019). Engineered nanomaterials for water decontamination and purification: From lab to products. *Journal of Hazardous Materials, 363*, 295–308. doi:10.1016/j.jhazmat.2018.09.091 PMID:30312926

Kirthi, A., Rahuman, A. A., Rajakumar, G., Marimuthu, S., Santhoshkumar, T., Jayaseelan, C., Elango, G., Zahir, A. A., Kamaraj, C., & Bagavan, A. (2011). Biosynthesis of titanium dioxide nanoparticles using bacterium Bacillus subtilis. *Materials Letters, 65*(17-18), 2745–2747. doi:10.1016/j.matlet.2011.05.077

Krishna, G., Srileka, V., Singara Charya, M. A., Abu Serea, E. S., & Shalan, A. E. (2021). Biogenic synthesis and cytotoxic effects of silver nanoparticles mediated by white rot fungi. *Heliyon*, *7*(3), e06470. doi:10.1016/j.heliyon.2021.e06470 PMID:33786393

Kumar, S., Thakur, A., Rangra, V. S., & Sharma, S. (2015). Synthesis and Use of Low-Band-Gap ZnO Nanoparticles for Water Treatment. *Arabian Journal for Science and Engineering*, *41*(7), 2393–2398. doi:10.100713369-015-1852-1

Lakshmnarayanan, S., Shereen, M. F., Niraimathi, K. L., Brindha, P., & Arumugam, A. (2021). One-pot green synthesis of iron oxide nanoparticles from Bauhinia tomentosa: Characterization and application towards synthesis of 1, 3 diolein. *Scientific Reports*, *11*(1), 8643. doi:10.103841598-021-87960-y PMID:33883589

Lateef, A., Ojo, S. A., Azeez, M. A., Asafa, T. B., Yekeen, T. A., Akinboro, A., Oladipo, I. C., Gueguim-Kana, E. B., & Beukes, L. S. (2015). Cobweb as novel biomaterial for the green and eco-friendly synthesis of silver nanoparticles. *Applied Nanoscience*, *6*(6), 863–874. doi:10.100713204-015-0492-9

Leon, L., Chung, E. J., & Rinaldi, C. (2020). A brief history of nanotechnology and introduction to nanoparticles for biomedical applications. In Nanoparticles for Biomedical Applications (pp. 1-4). doi:10.1016/B978-0-12-816662-8.00001-1

Majhi, S., Naik, G., Lee, H., Song, H., Lee, C., Lee, I., & Yu, Y. (2018). Au@NiO Core-shell nanoparticles as a P-type gas sensor: Novel synthesis, characterization, and their gas sensing properties with sensing mechanism. *Sensors and Actuators. B, Chemical*, *268*, 223–231. doi:10.1016/j.snb.2018.04.119

Maksimović, M., & Omanović-Mikličanin, E. (2017). Towards green nanotechnology: Maximizing benefits and minimizing harm. *IFMBE Proceedings*, *62*, 164–170. doi:10.1007/978-981-10-4166-2_26

Mandal, R. P., Sekh, S., Sarkar, N. S., Chattopadhyay, D., & De, S. (2016). Algae mediated synthesis of cadmium sulphide nanoparticles and their application in bioremediation. *Materials Research Express*, *3*(5), 055007. doi:10.1088/2053-1591/3/5/055007

Maruyama, T., Fujimoto, Y., & Maekawa, T. (2015). Synthesis of gold nanoparticles using various amino acids. *Journal of Colloid and Interface Science*, *447*, 254–257. doi:10.1016/j.jcis.2014.12.046 PMID:25591824

Mohamad, N. A. N., Arham, N. A., Jai, J., & Hadi, A. (2013). *Plant Extract as Reducing Agent in Synthesis of Metallic Nanoparticles: A Review*. International Conference 4th Nanoscience and Nanotechnology (NANO-SciTech 2013), Shah Alam, Malaysia.

Nadaf, N. Y., & Kanase, S. S. (2016). Biosynthesis of gold nanoparticles by Bacillus marisflavi and its potential in catalytic dye degradation. *Arabian Journal of Chemistry*, *12*(8), 4806–4814. doi:10.1016/j.arabjc.2016.09.020

Nasrollahzadeh, M., Sajadi, S. M., Sajjadi, M., & Issaabadi, Z. (2019). An Introduction to Nanotechnology. In An introduction to green nanotechnology (Vol. 28., pp. 1-27). Elsevier. doi:10.1016/B978-0-12-813586-0.00001-8

Paosen, S., Saising, J., Septama, A., & Voravuthikunchai, S. (2017). Green synthesis of silver nanoparticles using plants from Myrtaceae family and characterization of their antibacterial activity. *Materials Letters*, *209*, 201–206. Advance online publication. doi:10.1016/j.matlet.2017.07.102

Pareek, V., Bhargava, A., Gupta, R., Jain, N., & Panwar, J. (2017). Synthesis and Applications of Noble Metal Nanoparticles: A Review. *Advanced Science, Engineering and Medicine*, *9*(7), 527–544. doi:10.1166/asem.2017.2027

Philip, D. (2009). Biosynthesis of Au, Ag and Au-Ag nanoparticles using edible mushroom extract. *Spectrochimica Acta. Part A: Molecular and Biomolecular Spectroscopy*, *73*(2), 374–381. doi:10.1016/j.saa.2009.02.037 PMID:19324587

Pirsaheb, M., Shahmoradi, B., Beikmohammadi, M., Azizi, E., Hossini, H., & Md Ashraf, G. (2017). Photocatalytic degradation of Aniline from aqueous solutions under sunlight illumination using immobilized Cr:ZnO nanoparticles. *Scientific Reports*, *7*(1), 1473. doi:10.103841598-017-01461-5 PMID:28469186

Qasim, S., Zafar, A., Saif, M. S., Ali, Z., Nazar, M., Waqas, M., Haq, A. U., Tariq, T., Hassan, S. G., Iqbal, F., Shu, X. G., & Hasan, M. (2020). Green synthesis of iron oxide nanorods using Withania coagulans extract improved photocatalytic degradation and antimicrobial activity. *Journal of Photochemistry and Photobiology. B, Biology*, *204*, 111784. doi:10.1016/j.jphotobiol.2020.111784 PMID:31954266

Rajasekhar, C., & Kanchi, S. (2018). Green Nanomaterials for Clean Environment. In L. Martínez, O. Kharissova, & B. Kharisov (Eds.), *Handbook of Ecomaterials*. Springer. doi:10.1007/978-3-319-48281-1_73-1

Rajasekharreddy, P., Huang, C., Busi, S., Rajkumari, J., Tai, M. H., & Liu, G. (2019). Green Synthesized Nanomaterials as Theranostic Platforms for Cancer Treatment: Principles, Challenges and the Road Ahead. *Current Medicinal Chemistry*, *26*(8), 1311–1327. doi:10.2174/0929867324666170309124327 PMID:28294042

Rajendran, A., Alsawalha, M., & Alomayri, T. (2021). Biogenic synthesis of husked rice-shaped iron oxide nanoparticles using coconut pulp (Cocos nucifera L.) extract for photocatalytic degradation of Rhodamine B dye and their in vitro antibacterial and anticancer activity. *Journal of Saudi Chemical Society*, *25*(9), 101307. Advance online publication. doi:10.1016/j.jscs.2021.101307

Rani, K., & Sridevi, V. (2017). An Overview on Role of Nanotechnology in Green and Clean Technology. *Austin Environ Sci*, *2*(3), 1026.

Rawal, T., Parmar, R., Tyagi, R. K., & Butani, S. (2017). Rifampicin loaded chitosan nanoparticle dry powder presents an improved therapeutic approach for alveolar tuberculosis. *Colloids and Surfaces. B, Biointerfaces*, *154*, 321–330. doi:10.1016/j.colsurfb.2017.03.044 PMID:28363192

Ronavari, A., Kovacs, D., Igaz, N., Vagvolgyi, C., Boros, I. M., Konya, Z., Pfeiffer, I., & Kiricsi, M. (2017). Biological activity of green-synthesized silver nanoparticles depends on the applied natural extracts: A comprehensive study. *International Journal of Nanomedicine*, *12*, 871–883. doi:10.2147/IJN.S122842 PMID:28184158

Sadalage, P. S., Patil, R. V., Havaldar, D. V., Gavade, S. S., Santos, A. C., & Pawar, K. D. (2021). Optimally biosynthesized, PEGylated gold nanoparticles functionalized with quercetin and camptothecin enhance potential anti-inflammatory, anti-cancer and anti-angiogenic activities. *Journal of Nanobiotechnology*, *19*(1), 84. doi:10.118612951-021-00836-1 PMID:33766058

Saha, P., Mahiuddin, M., Islam, A., & Ochiai, B. (2021). Biogenic Synthesis and Catalytic Efficacy of Silver Nanoparticles Based on Peel Extracts of Citrus macroptera Fruit. *ACS Omega*, *6*(28), 18260–18268. doi:10.1021/acsomega.1c02149 PMID:34308057

Saif, S., Tahir, A., Asim, T., & Chen, Y. (2016). Plant Mediated Green Synthesis of CuO Nanoparticles: Comparison of Toxicity of Engineered and Plant Mediated CuO Nanoparticles towards Daphnia magna. *Nanomaterials (Basel, Switzerland)*, *6*(11), 205. Advance online publication. doi:10.3390/nano6110205 PMID:28335333

Saikia, J., Gogoi, A., & Baruah, S. (2019). Nanotechnology for Water Remediation. In Environmental Nanotechnology (pp. 195-211). doi:10.1007/978-3-319-98708-8_7

Shapira, P., & Youtie, J. (2012). *The Economic Contributions of Nanotechnology to green and sustainable growth. International Symposium on Assessing the Economic Impact of Nanotechnology, Washington, DC.*

Sharma, T. S. K., Selvakumar, K., Hwa, K. Y., Sami, P., & Kumaresan, M. (2019). Biogenic fabrication of gold nanoparticles using Camellia japonica L. leaf extract and its biological evaluation. *Journal of Materials Research and Technology*, *8*(1), 1412–1418. doi:10.1016/j.jmrt.2018.10.006

Singh, Bhat, Dervash, Qadri, Mehmood, Dar, Hameed, & Rashid. (2020). Wonders of Nanotechnology for Remediation of Polluted Aquatic Environs. In Fresh Water Pollution Dynamics and Remediation (pp. 319-339). Springer Nature Singapore Pte Ltd. doi:10.1007/978-981-13-8277-2_17

Singh, J., Kaur, N., Kaur, P., Kaur, S., Kaur, J., Kukkar, P., Kumar, V., Kukkar, D., & Rawat, M. (2018). Piper betle leaves mediated synthesis of biogenic SnO2 nanoparticles for photocatalytic degradation of reactive yellow 186 dye under direct sunlight. *Environmental Nanotechnology, Monitoring & Management*, *10*, 331–338. doi:10.1016/j.enmm.2018.07.001

Sinha, T., & Ahmaruzzaman, M. (2015). A novel green and template free approach for the synthesis of gold nanorice and its utilization as a catalyst for the degradation of hazardous dye. *Spectrochimica Acta. Part A: Molecular and Biomolecular Spectroscopy*, *142*, 266–270. doi:10.1016/j.saa.2015.02.020 PMID:25706595

Sruthi, S., Maurizi, L., Nury, T., Sallem, F., Boudon, J., Riedinger, J. M., Millot, N., Bouyer, F., & Lizard, G. (2018). Cellular interactions of functionalized superparamagnetic iron oxide nanoparticles on oligodendrocytes without detrimental side effects: Cell death induction, oxidative stress and inflammation. *Colloids and Surfaces. B, Biointerfaces*, *170*, 454–462. doi:10.1016/j.colsurfb.2018.06.041 PMID:29958160

Su, S., & Kang, P. M. (2020). Systemic Review of Biodegradable Nanomaterials in Nanomedicine. *Nanomaterials (Basel, Switzerland)*, *10*(4), 656. Advance online publication. doi:10.3390/nano10040656 PMID:32244653

Subramaniam, M. N., Goh, P. S., Lau, W. J., & Ismail, A. F. (2019). The Roles of Nanomaterials in Conventional and Emerging Technologies for Heavy Metal Removal: A State-of-the-Art Review. *Nanomaterials (Basel, Switzerland)*, *9*(4), 625. Advance online publication. doi:10.3390/nano9040625 PMID:30999639

Sui, Z., Meng, Q., Zhang, X., Ma, R., & Cao, B. (2012). Green synthesis of carbon nanotube–graphene hybrid aerogels and their use as versatile agents for water purification. *Journal of Materials Chemistry*, *22*(18), 8767. Advance online publication. doi:10.1039/c2jm00055e

Tanase, M. A., Raducan, A., Oancea, P., Ditu, L. M., Stan, M., Petcu, C., Scomoroscenco, C., Ninciuleanu, C. M., Nistor, C. L., & Cinteza, L. O. (2021). Mixed Pluronic-Cremophor Polymeric Micelles as Nanocarriers for Poorly Soluble Antibiotics-The Influence on the Antibacterial Activity. *Pharmaceutics*, *13*(4), 435. Advance online publication. doi:10.3390/pharmaceutics13040435 PMID:33804932

Tchounwou, P. B., Yedjou, C. G., Patlolla, A. K., & Sutton, D. J. (2012). Heavy metal toxicity and the environment. *Exp*, *101*(Suppl 101), 133–164. doi:10.1007/978-3-7643-8340-4_6 PMID:22945569

Teleanu, D. M., Chircov, C., Grumezescu, A. M., & Teleanu, R. I. (2019). Neurotoxicity of Nanomaterials: An Up-to-Date Overview. *Nanomaterials (Basel, Switzerland)*, *9*(1), 96. Advance online publication. doi:10.3390/nano9010096 PMID:30642104

Teow, Y. H., & Mohammad, A. W. (2019). New generation nanomaterials for water desalination: A revie. *Desalination*, *451*, 2–17. doi:10.1016/j.desal.2017.11.041

Toda, M., Yukawa, H., Yamada, J., Ueno, M., Kinoshita, S., Baba, Y., & Hamuro, J. (2019). In Vivo Fluorescence Visualization of Anterior Chamber Injected Human Corneal Endothelial Cells Labeled With Quantum Dots. *Investigative Ophthalmology & Visual Science*, *60*(12), 4008–4020. doi:10.1167/iovs.19-27788 PMID:31560364

Umar, A., Alshahrani, A., Algarni, H., & Kumar, R. (2017). CuO nanosheets as potential scaffolds for sensing applications. *Sensors and Actuators. B, Chemical*, *250*, 24–31. doi:10.1016/j.snb.2017.04.062

Venkateswarlu, S., Kumar, B. N., Prathima, B., SubbaRao, Y., & Jyothi, N. V. V. (2014). A novel green synthesis of Fe3O4 magnetic nanorods using Punica Granatum rind extract and its application for removal of Pb(II) from aqueous environment. *Arabian Journal of Chemistry*, *12*(4), 588–596. doi:10.1016/j.arabjc.2014.09.006

Verma, G. (2015). Various areas of green chemistry and safer environment an overview. *International Journal of Research in Applied. Natural and Social Sciences.*, *3*(3), 87–94.

Wang, M., Yang, Q., Li, M., Zou, H., Wang, Z., Ran, H., Zheng, Y., Jian, J., Zhou, Y., Luo, Y., Ran, Y., Jiang, S., & Zhou, X. (2020). Multifunctional Nanoparticles for Multimodal Imaging-Guided Low-Intensity Focused Ultrasound/Immunosynergistic Retinoblastoma Therapy. *ACS Applied Materials & Interfaces*, *12*(5), 5642–5657. doi:10.1021/acsami.9b22072 PMID:31940169

Wang, P., Lombi, E., Zhao, F. J., & Kopittke, P. M. (2016). Nanotechnology: A New Opportunity in Plant Sciences. *Trends in Plant Science*, *21*(8), 699–712. doi:10.1016/j.tplants.2016.04.005 PMID:27130471

Wang, Y., O'Connor, D., Shen, Z., Lo, I. M. C., Tsang, D. C. W., Pehkonen, S., Pu, S., & Hou, D. (2019). Green synthesis of nanoparticles for the remediation of contaminated waters and soils: Constituents, synthesizing methods, and influencing factors. *Journal of Cleaner Production*, *226*, 540–549. doi:10.1016/j.jclepro.2019.04.128

Weldrick, P. J., Hardman, M. J., & Paunov, V. N. (2021). Smart active antibiotic nanocarriers with protease surface functionality can overcome biofilms of resistant bacteria. *Materials Chemistry Frontiers*, *5*(2), 961–972. doi:10.1039/D0QM00874E

WHO. (2019). *Drinking-water*. https://www.who.int/news-room/fact-sheets/detail/drinking-water

Ye, J., Wang, Y., Xu, Q., Wu, H., Tong, J., & Shi, J. (2021). Removal of hexavalent chromium from wastewater by Cu/Fe bimetallic nanoparticles. *Scientific Reports*, *11*(1), 10848. doi:10.103841598-021-90414-0 PMID:34035405

Zhang, X. F., Liu, Z. G., Shen, W., & Gurunathan, S. (2016). Silver Nanoparticles: Synthesis, Characterization, Properties, Applications, and Therapeutic Approaches. *International Journal of Molecular Sciences*, *17*(9), 1534. Advance online publication. doi:10.3390/ijms17091534 PMID:27649147

Zhang, Y., Li, M., Gao, X., Chen, Y., & Liu, T. (2019). Nanotechnology in cancer diagnosis: Progress, challenges and opportunities. *Journal of Hematology & Oncology*, *12*(1), 137. doi:10.118613045-019-0833-3 PMID:31847897

Zhang, Y., Yuan, S., Feng, X., Li, H., Zhou, J., & Wang, B. (2016). Preparation of Nanofibrous Metal-Organic Framework Filters for Efficient Air Pollution Control. *Journal of the American Chemical Society*, *138*(18), 5785–5788. doi:10.1021/jacs.6b02553 PMID:27090776

Zhao, L., Wang, F., & Fan, W. (2020). Cisplatin Nano-Liposomes Promoting Apoptosis of Retinoblastoma Cells Both In Vivo and In Vitro. *Nanoscience and Nanotechnology Letters*, *12*(4), 536–542. doi:10.1166/nnl.2020.3127

Zhu, F., Liu, T., Zhang, Z., & Liang, W. (2021). Remediation of hexavalent chromium in column by green synthesized nanoscale zero-valent iron/nickel: Factors, migration model and numerical simulation. *Ecotoxicology and Environmental Safety*, *207*, 111572. doi:10.1016/j.ecoenv.2020.111572 PMID:33254420

Zhu, Y., Xing, L., Zheng, X., Yang, C. X., He, Y. J., Zhou, T. J., Jin, Q. R., & Jiang, H. L. (2020). Amplification of tumor antigen presentation by NLGplatin to improve chemoimmunotherapy. *International Journal of Pharmaceutics*, *573*, 118736. doi:10.1016/j.ijpharm.2019.118736 PMID:31756442

Zou, L., Wang, J., Gao, Y., Ren, X., Rottenberg, M. E., Lu, J., & Holmgren, A. (2018). Synergistic antibacterial activity of silver with antibiotics correlating with the upregulation of the ROS production. *Scientific Reports*, *8*(1), 11131. doi:10.103841598-018-29313-w PMID:30042429

ADDITIONAL READING

Adeyemi, O. S., Molefe, N. I., Awakan, O. J., Nwonuma, C. O., Alejolowo, O. O., Olaolu, T., . . . Kato, K. (2018). Metal nanoparticles restrict the growth of protozoan parasites. *Artif Cells Nanomed Biotechnol, 46*(sup3), S86-S94. doi:10.1080/21691401.2018.1489267

Bapat, M. S., Singh, H., Shukla, S. K., Singh, P. P., Vo, D. N., Yadav, A., Goyal, A., Sharma, A., & Kumar, D. (2021). Evaluating green silver nanoparticles as prospective biopesticides: An environmental standpoint. *Chemosphere, 286*(Pt 2), 131761. doi:10.1016/j.chemosphere.2021.131761 PMID:34375828

Ettadili, F. E., Aghris, S., Laghrib, F., Farahi, A., Saqrane, S., Bakasse, M., Lahrich, S., & El Mhammedi, M. A. (2022). Recent advances in the nanoparticles synthesis using plant extract: Applications and future recommendations. *Journal of Molecular Structure, 1248*, 131538. Advance online publication. doi:10.1016/j.molstruc.2021.131538

KEY TERMS AND DEFINITIONS

Bio-Factory: Any living organism that produces significant quantities of therapeutic compounds.

Functionalization: The process of optimizing an existing feature of a material by changing its surface chemistry.

Nanocapsulation: The act of improving nanoparticle cellular uptake and availability by packaging the nanoparticles within a secondary material.

Nanoremediation: The use of nanoparticles in the removal of contaminants from the soil, air, and water compartments.

Nanotheranostics: The technology that explores nanoparticles in the simultaneous diagnosis and treatment of diseases.

Chapter 2
Green Synthesis of Semiconductors and Environmental Applications:
From Single to High Added Value Products

Fernanda da Costa Romeiro
https://orcid.org/0000-0002-9632-4290
São Paulo State University (UNESP), Brazil

Aline Varella Rodrigues
São Paulo State University (UNESP), Brazil

ABSTRACT

This chapter focuses on advancements in the green synthesis approach for nanomaterials and their environmental applications. The eco-friendly, cost-effective, and simple synthesis of inorganic nanoparticles on the environmental applications are discussed. The first section presents an introduction with the basic topics of the green materials synthesis. The second section summarizes the green chemical methods through different routes (precipitation synthesis, electrospinning synthesis, reflux condensation synthesis, hydrothermal, microwave-assisted hydrothermal methods) and the green routes using biological systems from biosynthetic process to synthesize nanoparticles and biosynthesis of passivating nanofilms using spontaneous formation. In the third section, the environmental and energy applications of nanomaterials, including solar cells, photocatalysis, and biohydrogen production are discussed. Therefore, the scientific community can experience the relation of the achieved nanomaterials with their plentiful application possibilities onto the environmental remediation.

INTRODUCTION

Nanomaterial is commonly defined as a material that presents, at least, one-dimension smaller in size than 100 nanometers (nm). Nowadays, nanomaterials have been used in large applications, such as cos-

DOI: 10.4018/978-1-7998-8936-6.ch002

metics, medicine, space applications, and others (Zhang et al., 2020). This is considered a relevant topic, which has attracted great interest from researchers worldwide because of its scientific and industrial significance. Because of their tiny size, high surface area, and quantum effects, these materials present innovative properties that are not normally observed in bulk materials. Although humanity has been exposed to nanomaterials since all its history, Richard Zsigmondy was the first researcher to propose the term "nanometer" in 1925. The important idea of nanotechnology was discussed for the first time in the famous lecture "There is plenty of room at the bottom" presented by Richard Feynman (1959), Nobel Prize Laureate in 1965. Feynman introduced the concept of manipulating individual atoms and molecules using precise tools. The science of nanomaterials and nanotechnology was further advanced with the contribution of several scientists over the recent years (Hulla, et al., 2015).

As discussed above, due to the higher surface area and larger surface energy, the nanomaterials present interesting properties in physical, chemical, optical, magnetic, and biological scientific areas (Gour & Jain, 2019; Romeiro et al., 2013). The potential benefits of nanomaterials and nanotechnology are many and diverse, however, the potential health and environmental risks are still in a discussion because of extensive human exposure to nanoparticles. For instance, some toxic intermediates or products, even as wastes can be produced during the chemical processes. During the nanomaterial's synthesis, many chemicals as solvents, precursors, and template materials are employed. Thus, to reduce or mitigate these undesirable by-products, new concepts of semiconductors materials and nanocomposites syntheses have been introduced into chemical science. The eco-friendly, non-toxic, and cost-effective biological methods of synthesis, known as green synthesis, have been developed, aiming to apply the principles of green chemistry. Several inorganic nanomaterials with a well-defined structure, chemical composition, and morphology have been obtained by using diverse microorganisms, plants extract, and enzymes. This chapter highlights the current advances of the eco-friendly, cost-effective, and simple chemical methods and bio-synthesis of inorganic nanoparticles. This chapter brings the most recent and trending environmental applications of nanoparticles. This special study focuses on the strong correlation between diverse nanoparticles (metallic nanoparticles, nanocomposites, oxide nanoparticles) with their potential application that encompasses this field.

BACKGROUND

Why Is Green Nanotechnology Important?

The notion of green chemistry was introduced to the scientific community and industry to reduce the formation of hazardous materials in chemical processes. For instance, in several clinical and biomedical applications, the possibility of chemical contamination by toxic compounds restrains the use of nanomaterials (Hua et al., 2018). The green synthesis processes can improve the environmental performance of nanoparticles for industries and can reduce the consumption of resources. Some synthesis approaches are well known for their environmentally friendly character, such as the precipitation method, electrospinning synthesis, hydrothermal and microwave-assisted hydrothermal methods (chemical methods), and nanoparticle synthesis using microorganisms or plants by sustainable synthesis methodology (biological methods).

Aspects of Green Techniques for Nanoparticles Synthesis

The green methods use mild reaction conditions and non-toxic precursors in the obtainment of nanomaterials. Among many different nanoparticles, the green synthesis of titanium dioxide (TiO_2), (Goutam et al., 2018) zinc oxide (ZnO) (Marinho et al., 2012) copper oxide (CuO) (Velsankar et al., 2020), or aluminum oxide (Al_2O_3) (Ansari et al., 2015) are being extensively studied. There are some interesting green chemical approaches to obtain these nanomaterials. For their production, it needed to be highly considered the chosen solvent, the reducing agent used, natural materials as templates and precursors, the use of microorganisms, and a simpler synthesis process (Gonçalves et al., 2021). Considering this, investigations of non-toxic solvents and natural template materials have been stated. The green electrospinning represents a convenient and simple process to produce ultrathin nanofibers. The solution employed in the electrospinning process can be non-toxic and pollution-free. The non-toxic aqueous solution can be used as a chemical component or solvent in the hydrothermal and microwave-hydrothermal methods, which classifies these chemical approaches as environmentally friendly (Romeiro et al., 2017; G. Yang & Park, 2019). Silver nanoparticles (AgNPs) were prepared by biosynthesis method using *L. camara* LE by Ajitha et al. (2015) for the evaluation of an efficient bactericidal activity. In this case, synthesis using plant extracts presented advantages due to the reduced risk of contamination by decreasing the reaction time and keeping the cell structure intact. The strategies for the synthesis of nanomaterials using non-toxic and biological materials by eco-friendly approaches will be discussed in the subsequent topics.

TYPES OF GREEN TECHNIQUES

Precipitation Synthesis

Low cost, simple, easy controlling, and wide variability are some features of the precipitation process that leads to the morphology control, e.g. a uniform size and shape of materials, usually controlled by heating temperatures. Associated with lower synthesis time and aging temperature, precursor, solvent, precipitant, and dispersing agent, the precipitation guides the particles' growth toward the smaller surface energy, once the particles agglomerations are prevented, thus producing high specific surface areas, and consequently improving many materials properties. Therefore, the precipitation technique is widely used as a chemical route for nanomaterials synthesis (A. V. Rodrigues & Orlandi, 2019; A. V. Rodrigues & Sabino, 2019).

With good repeatability and excellent operation, precipitation is a convenient method because it does not use sophisticated equipments, and it is capable to synthesize high-quality ultra-fine materials, with a wide variability of nanostructures, such as nanospheres, nanowires, nano lamellae, nanoparticles, nanorods, nanodots, and nanosheets. Especially for nanoparticles, nanorods, and nanosheets, different morphologies materials can be formed (Kabcum et al., 2016; Shan et al., 2017).

Accordingly, the precipitation comprehends the processes of the chemical synthesis in an aqueous solution and under thermal treatment, which define the chemical composition, crystal structure, and morphology of the nanomaterials. Precursor materials are mixed under atmospheric conditions to low temperatures at long periods of time. Concentrated bases are introduced into the solution, precipitating the materials in the solvent. The capping agents or electrostatic stabilization, e.g. ionic liquids, enable the stabilization of the NPs, control their crystal growth, and the particles growth from hours to days of synthesis when the precursor is under elevated temperature to obtain the desired material. The differences between the nanostructured metallic oxides obtained, using different procedures of syntheses. i.e. the process speed, heating uniformity, size distribution, and shape control, are affected by the method applied and reaction temperature (Mirzaei & Neri, 2016; Terraschke et al., 2018). Amorphous and microporous materials are achieved when the precipitation occurs at lower pH and temperature without aging, in which non-uniform heating, wide size distribution, and poor shape control are obtained (Li et al., 2012).

Electrospinning Synthesis

Ultra-fine and long fibers from dozen of nanometers to micrometers, with diameter, composition, surface-to-volume ratio, and porosity controlled using electrospinning are developed to achieve one-dimensional (1D) nanomaterials, whose properties are influenced by this technique (A. V. Rodrigues & Sabino, 2019a; Wang et al., 2009; J. Zhao et al., 2011). Due to their unique properties, the 1D nanomaterials obtained from the electrospinning technology, i.e. polymer, polymer/inorganic, composites, and inorganic materials with nanorods, nanowires, and nanofibers structures, are attractive in optoelectronic devices (A. V. Rodrigues & Sabino, 2019a, 2019b).

It is simpler and more cost-effective than other techniques, which present many disadvantages as a long annealing time, high annealing temperature, impurities formed during the chemical process that is difficult to be removed, and gases that should be used in the synthesis. Through it, previous solutions are prepared by dissolving a polymer and other requested materials in a solvent, achieving the desired viscosity. They are injected into a plastic syringe with a stainless needle, connected with a peristaltic pump, and linked to a high voltage power supply, achieving a plate of aluminum foil located at a distance from the needle tip. From this technique, the nanofibers are generated as dense mats (A. V. Rodrigues & Sabino, 2019b; J. Zhao et al., 2011), with smaller electrospun fiber diameter (A. V. Rodrigues & Sabino, 2019b; Yoon et al., 2018). The collection process of the nanofibers is depicted in Figure 1.

Usually, inorganic crystals grow up by classic routes, in which the crystal is obtained by nucleation. Here, the crystal growth rates from crystallographic orientation and are correlated to the surface energy. In other routes, instead, nanocrystallites aggregate into large and disordered particles. During the crystallization, shells are formed with high crystallinity and density, in which the crystallization extension occurs from the surface to the core (Greer et al., 2011).

Also named reversed crystal growth mechanism, the nanofiber's pores and uniform lengths are regulated through that surface-to-core extension, accompanied by thermal treatments: burning organic component. The Ostwald ripening is the phenomenon that explains the process of crystallization, that occurs from the surface to the core and pores of nanostructures, and this process decreases at a slower heating rate. The surface-to-core extension or crystal growth mechanism can be found in zeolites, perovskites, metals, and metal oxides. Besides, electrospinning is convenient for ceramic nanofibers precursors, producing oxide and non-oxide nanofibers. Thus, electrospinning produces nanofibers of polymers, composite materials, and ceramics (Greer et al., 2011; A. V. Rodrigues & Sabino, 2019b; J.-G. Zhao et al., 2008) with a high aspect ratio, once an external electric field is enforced on a polymeric liquid or a sol-gel precursor solution (Zhang et al., 2007).

Figure 1. The collection of nanofibers, which are obtained from the reversed nanofibers growth or surface-to-core-extension process

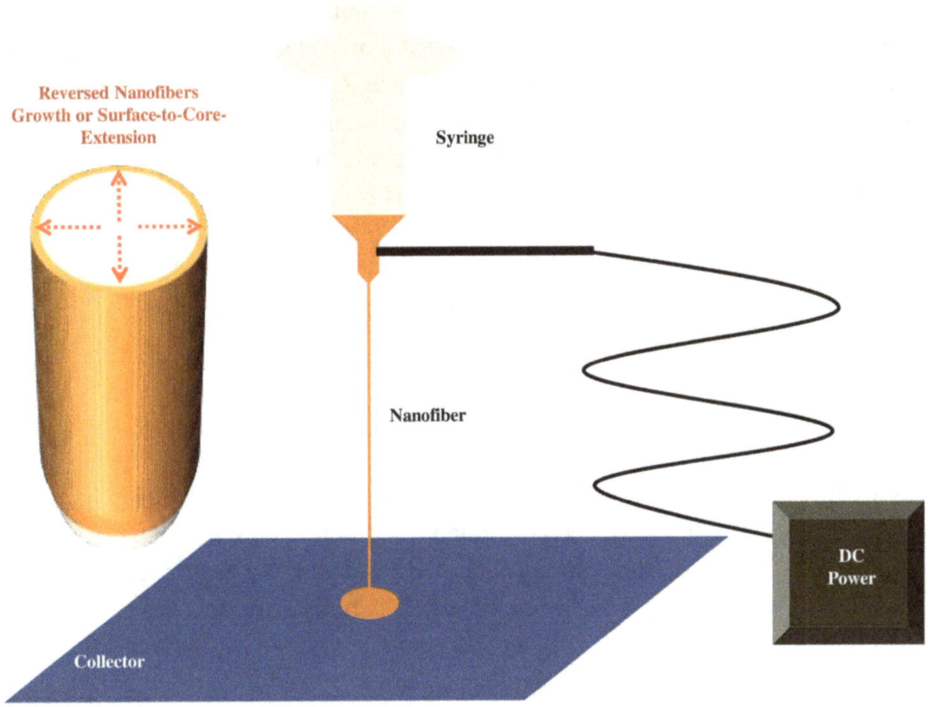

Reflux Condensation Synthesis

Reflux condensation is considered a soft-chemistry route that works at low temperature, and precisely controls the morphology and surface properties from some reaction parameters, thus a great variety of 1D nanostructures can be developed, which includes nanocrystalline oxide materials. Due to its economic, energy efficiency, and environmentally-friendly features, this technique has advantages over others that involve complicated procedures (El-Sayed et al., 2011; Girija et al., 2013; A. V. Rodrigues & Sabino, 2019a, 2019b). The reflux condensation involves the reflux and condensation of volatile species, without losing analyte due to the volatilization, in which fewer reactants are consumed and the contamination risk is reduced. (da Rosa et al., 2016).

A bigger tube of glass is closed with a smaller glass and filled with cold water. Thereby, the solution is recirculated within the bigger tube, with an increase of the heating temperature above the boiling point of the solution, whose condensation occurs by reflux condensation (da Rosa et al., 2016). Thus, the chemical reaction occurs in this aqueous medium. The energy necessary for this process comes from the heating of the chemical solution over a long period of time. Time of reflux as reaction time, the concentration of precursor, type of solvent, and cooling rate can influence the desired phases and morphologies of the nanostructures, i.e. nanoparticles, nanowires, nanorods, nanourchins, core-shell nanostructures, and hierarchical nanostructures. The order of precursor added also can affect the reaction: a secondary phase has a great potential to be formed whether the product is quaternary. Therefore, the secondary phases, and their reactivities must be identified, in order to obtain the desired single phase. In order to achieve a pure and desired phase, the time of reflux must be optimized: neither the reaction can stop earlier than initial nucleation and grain growth, nor the reaction can take a long time to avoid the formation of larger grain sizes and then bulk crystals. Also, regarding the cooling rate, it must be slow in order to form particles with fewer defects (Aditha et al., 2016). During the cooling, the reflux extraction also enables the re-absorption of the analytes (George et al., 2019). The temperature strongly influences the reaction kinetics, and consequently the nucleation, aggregation, coalescence, and growth rate. The temperature must promote enough pressure in the solution that induces enough kinetic movements of cations in order to deposit the analyte (Kamble et al., 2021).

The heating mantle is an apparatus connected to the reflux condensation system. There is a collector in the middle of the system, and the round-bottom flask contains the solvent mixture with an agent. This mixture is refluxed using the heating mantle and a closed-loop system where the water is circulated. This sort of system is an alternative for drying volatile solvents and making O_2-free organic solvents to use in laboratories in general (Simas et al., 2009).

Hydrothermal Synthesis

Nanoparticles with different physical and chemical properties have been obtained in the last years. Several processing methodologies have been developed to generate homogeneous structures with high crystalline quality and promising applications. The hydrothermal synthesis method is considered an interesting and inexpensive cost approach that provides stability to the synthesis by using a closed system, temperature and pressure controlled, and also the size and nucleation control of nanomaterials (Gonçalves et al., 2021). The nanoparticles obtained by the green hydrothermal syntheses are obtained from non-toxic solvents, such as water, decreasing the production of not desirable or harmful by-products by this synthesis procedure (Singh et al., 2018).

Due to the fact that these reactions are performed in a sealed autoclave, the solubility of reactants is enhanced, favoring chemical reactions at low temperatures (Razavi et al., 2015). In this method, the reactants are dissolved and enter the medium as ions or molecules, and then these ions/molecules are transported to a low-temperature region in the solution, where the crystal seed can grow to form a supersaturated solution (G. Yang & Park, 2019). In order to predict the crystal growth mechanism of a specific morphology, it is necessary an extensive study on the synthesis parameters. Figure 2 shows a representation of the parameters that can affect the synthesis of nanomaterials using the hydrothermal method. Wang et al., (2017) described that zinc oxide, with diverse morphologies, was prepared using the hydrothermal method, which was regulated by modifying the concentration of citric acid. On the other hand, some modifications can be observed on the morphology of nanomaterials, by varying the reaction parameters. The metal precursors, the concentration of acid/alkali solutions, temperature, and time of the synthesis can affect the growth of the nanoparticles. For instance, temperatures of 130, 150, and 170° C, and different times of 3, 7, and 12 hours have been studied, and it was observed that as the temperature increased, the TiO_2 nanorods have decreased in size (Navab et al., 2018). The crystalline phase and morphology of zinc oxide can be defined by changing the hydrothermal autoclave treatments at 100–120° C for 6 h (or 170° C for 0 h) to treatment at 170° C for 3–6 h, in which the products changed from cubic ZnO_2 nanocrystals to hexagonal ZnO nanorods (Guo et al., 2011).

During the hydrothermal synthesis using water as the solvent, the high temperature and pressure can affect the characteristics of water. As described by Yang et. al (2019), the hydrolysis and ion reactions rates will raise using water as the solvent. In addition, the water viscosity decreases, facilitating molecular and ionic mobility, thus promoting a faster nanoparticle crystal growth. The chance of collision among

Figure 2. Parameters of synthesis for the hydrothermal method

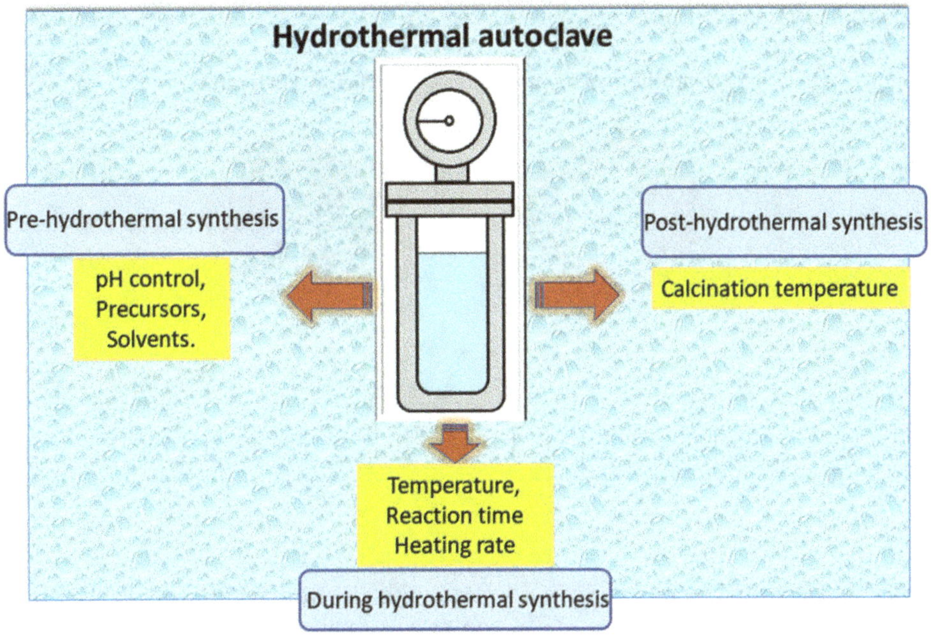

molecules during hydrothermal synthesis increases because the vapor pressure of water increases and accelerates the reactions inside the autoclave.

Microwave-Assisted Hydrothermal Synthesis

In the last years, researchers have focused their attention on the use of microwaves to design several inorganic nanomaterials, such as oxide materials, nanocomposites, metal chalcogenides, or quantum dots (de Medeiros et al., 2019; Lemos et al., 2019; Romeiro et al., 2020). The microwave-assisted hydrothermal method is considered a promising green synthesis chemistry approach, once it is possible the use of green solvents, it is faster and more economical than the hydrothermal method and presents high energy efficiency (Mulvihill et al., 2011). Komarneni was the pioneer researcher that compared the differences between the conventional hydrothermal method and the microwave hydrothermal synthesis, which expanded the use of microwaves to obtain nanoparticles (Komarneni et al., 1992). In the microwave-assisted hydrothermal method, the microwaves can penetrate the reaction medium, heating the solution at the same time, and the microwaves can be absorbed by the reaction medium and then, converted into heat, avoiding thermal gradient (Kumar et al., 2020). Therefore, this process provides homogeneous nucleation and growth of nanoparticles with uniform size distribution. The conventional hydrothermal method requires an external heating process, conducting the exterior heating source to the chemical solution, which is inside the autoclave. In this case, the products are formed by convection or conduction mechanisms (Figure 3a) (G. Yang & Park, 2019). When compared to the conventional hydrothermal method, the microwave-assisted hydrothermal method presents a faster heating speed and uniform heating system, once it is no longer conduction heating, but it is heated by microwaves.

Figure 3. (a) Temperature profile by the treatment in conventional and microwave heating, and (b) representation of electromagnetic espectrum, emphasizing the microwave region of electromagnetic spectrum
Source: Adapted from Kumar et al., 2020

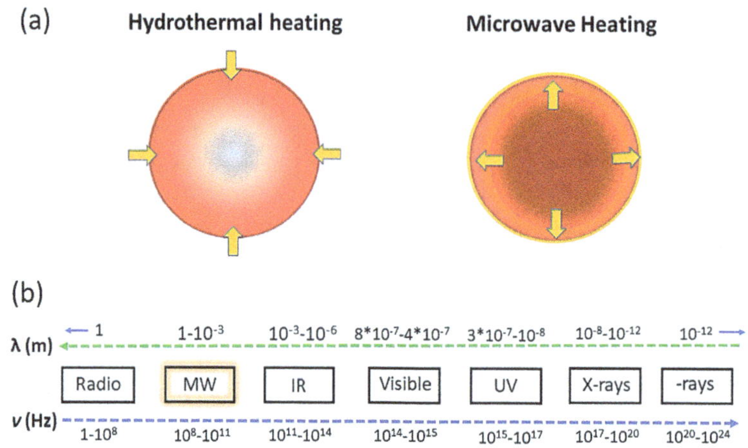

A microwave is electromagnetic energy associated with electromagnetic fields. The frequency and wavelength of the microwave radiation are from 300 MHz and 300 GHz and 1 mm to 1 m, respectively (Figure 3b). The ideal frequency of the microwaves for heating is concentrated between 900 MHz and 2.45 GHz, in which the medium is heated by the dipole polarization and ion conduction mechanisms.

During a reaction, the microwaves provide energy higher than necessary for breaking a chemical bond, showing that they do not affect the structure of the molecule. Water is non-toxic, available at low cost, non-flammable, and environmentally benign molecule, providing promising use for green synthesis of materials. The direct coupling of microwave irradiation with molecules (solvents, such as water; reagents) produces internal heating (Dallinger & Kappe, 2007). These molecules present in the reaction medium are in a disorderly state, before microwave heating. When the microwaves are irradiated in a very high frequency, the microwave energy field changes constantly, and the reactants or solvents dipoles tend to align with this oscillating electric field. Therefore, higher mobility of dissolved ions and molecules results in higher collision and friction of molecules, and then the microwave energy is converted into heat energy (Kumar et al., 2020). An important consequence of this fast and effective heating is the enhancement of the thermal and kinetic effect when polar molecules are under a microwave field (Dallinger & Kappe, 2007; Kappe, 2004). The frequency of 2.45 GHz provides the reaction dipoles of molecules adequate time to orient them with the electric field in all commercial devices.

Nanoparticles synthesized using the microwave-assisted hydrothermal method normally present controlled morphology, purity, and a narrow particles grain size distribution. Recently, small ZrO_2 nanoparticles supported on carbon black Printex L6 were obtained by fast (<1h) microwave-assisted hydrothermal synthesis route (Kronka et al., 2021). The characteristics, such as particles dispersion, size, and preferential crystallography of this material were adjusted by modifying the synthesis parameters (precursor concentration, time of synthesis, temperature, and pressure) (Kronka et al., 2021). Another study showed that the synthesis conditions (solvents, pH, or the use of surfactant) affect the structural phase and morphology of $BiVO_4$ nanomaterials obtained through the microwave-assisted method, in which the monoclinic and tetragonal $BiVO_4$ phases were observed in acidic media (Rodrigues et al., 2020). All this information can provide useful insights into the synthesis of nanomaterials using the microwave-assisted hydrothermal method. The solvent, pH, pressure, and temperature conditions play an essential role in the properties of the nanomaterials and they need to be considered during future synthesis planning.

Biological Synthesis

The biosynthetic process is a safe and environment-friendly approach to synthesize nanoparticles for various applications, including biomedical, biosensors, and photocatalytic purposes. Some of the green synthesis methodologies of nanoparticles are processes in which plant extracts, microorganisms, and enzymes act an important role in the nanomaterial's synthesis (El-Rafie et al., 2013; Fatima et al., 2020; Liu et al., 2020; Sivaraj et al., 2020). These methodologies depend on several parameters, such as temperature, pH, and solvent. In addition, the morphology and size of nanoparticles can be controlled by the investigation of the experimental conditions. For instance, under acidic conditions, the formation of gold nanoparticles can be observed, however with a low stabilization rate. On the other hand, under alkaline conditions, a faster process for stable gold nanoparticles formation is observed (Shou et al., 2011). The increasing temperature synthesis of silver nanoparticles can lead to a rapid reduction rate of the Ag^+ and homogeneous nucleation, with small size, of silver nuclei (Khalil et al., 2014).

In the specific use of plants, various parts such as fruit, leaf, stem, and root have been largely employed during the green synthesis of nanoparticles. During the synthesis process, the selected fragment of the plant can be washed and boiled in water. After some processes such as grinding, filtering, and the addition of nanoparticle precursors, the formation of nanomaterials can start. The synthesis using plant extract is considered an eco-friendly, cheap, and non-toxic process. In general, the plant extract can be simply added to a solution containing the metal salt at room temperature, during nanoparticle synthesis (Mittal et al., 2013). This process involves three main steps: activation, growth, and process termination step. In the first step, the reduction of metal ions can occur, leading to the formation of self-organized structures, while the second step involves the growth of the new structures and the reduction of the metal ions. In the last synthesis process step, the nanoparticles achieve a higher stability and desirable morphology.

Considering the efficiency of this sustainable approach, different materials can be synthesized, such as oxide materials (Ngom et al., 2021), silver nanoparticles (Jebril et al., 2020), and sulfide material (Bai & Zhang, 2009). For the synthesis of silver nanoparticles with morphology control, bio-molecules found in the plant extracts act as capping and reducing agents (Jadoun et al., 2021). Similarly, the reduction of Ag^+ ions to nanoparticles, by the use of *Desmodium triforuum* extract, was related to the H^+ ions, NAD^+ and ascorbic acid found in the extract (Ahmad et al., 2011). Table 1 summarizes some reports regarding nanoparticles synthesis mediated by biological methods.

Biosynthesis of Nanofilms

A natural exposure of unalloyed commercially (cp)-Ti and Ti alloys to a physiological medium facilitates the improvement of several properties. Therefore, these materials are considered biomaterials (A. V. Rodrigues et al., 2015).

Approved by the norm ASTM F 2066, Ti-Mo could be implanted in humans. Some researchers demonstrated the possible application of Ti–Mo alloys in biomedical devices and dental implants. Mo affects the properties of alloys, e.g. increasing their corrosion resistance. The literature reported that 4–20 wt% Mo in Ti-Mo enables the spontaneous passivation in different solutions, causing the prevention of pitting

Table 1. Biosynthesis of different nanoparticles and their morphologies and applications

Biosynthesis Precursor	Nanomaterial	Morphology	Application	Reference
Ureolytic bacteria	ZnO	Hairy shaped	Photocatalysis	(Dhandapani et al., 2020)
Leaf extract of *Jatropha curcas L.*	TiO_2	Spherical shaped	Photocatalysis	(Goutam et al., 2018)
Allium sativum extract	CuO	Spherical, circular, and rectangular shaped	Antimicrobial, antioxidant, and antilarvicidal activities	(Velsankar et al., 2020)
Lantana camara leaf extract	Ag	Spherical shaped	Antibacterial activity	(Ajitha et al., 2015)
Red algae *Portieria hornemannii*	Ag	Spherical shape	Antibacterial activity	(Fatima et al., 2020)
Commercial yeast extract	AgCl	Spherical shaped	Anti-mycobacterial Activity	(Sivaraj et al., 2020)
Polysaccharides from marine macro-algae	Ag	Spherical shaped	Antimicrobial activity	(El-Rafie et al., 2013)
Fruit extract of *Amomum longiligulare*	ZnO	-	Photocatalysis	(Liu et al., 2020)
Extract from *Moringa Oleifera*	ZnO	-	Optical properties	(Ngom et al., 2021)
Moringa oleifera seed	Ag	Spherical shaped	Antimicrobial and photocatalysis	(Mehwish et al., 2021)
Olive leaf extract	Ag	Quasi spherical shaped	Antibacterial activity	(Khalil et al., 2014)
Extract from alga, *Padina boryana*	Pd	Spherical shaped	Against cancer cells	(Sonbol et al., 2021)
Melia azedarach leaf extract	Ag	Spherical shaped	Antifungal activities	(Jebril et al., 2020)

(Bolat et al., 2013; A. V. Rodrigues et al., 2015). When introduced onto the physiological environment, a nanofilm is formed on the surface of alloys, spontaneously, consisting of amorphous TiO_2 with the role of protecting the implant against corrosion due to their physicochemical, electrochemical, stability, and biocompatibility properties (González & Mirza-Rosca, 1999; Mareci et al., 2009).

Cytotoxic effects to the human body, as well as undesirable reactions with incompatibility of elastic modulus between the biomaterial and bone, are due to alloy elements present in small amounts, which cause stress shield and premature failure in the implant (Eisenbarth et al., 2004).

Fluoride ions do not avoid the formation of the nanofilm, probably due to the small amount of HF in the solution. Meanwhile, they dissolve the nanofilm because TiF_3, TiF_4, and $TiOF_2$ are formed (Oufella & Benchettara, 2016), as demonstrated in (1), (2), and (3) reactions:

$$TiO_2 + 2\ HF \rightarrow TiOF_2 + H_2O \quad (1)$$

$$TiO_2 + 4\ HF \rightarrow TiF_4 + 2\ H_2O \quad (2)$$

$$Ti_2O_3 + 6\,HF \rightarrow 2\,TiF_3 + 3\,H_2O \quad (3)$$

Electrochemical techniques enable the observation of the spontaneous formation, as well as the growth of nanoscaled films on the biomaterial surface, besides the natural oxidation or dissolution of this protector film in a medium that contains chloride and fluoride ions. Accordingly, electrochemical stability and corrosion resistance of the (cp)-Ti and Ti-Mo alloys decrease in aqueous environments containing those ions, mainly in the fluoride medium. The highest corrosion protection is from the alloy with higher Mo content concentration (A. V. Rodrigues et al., 2015) (A. V. Rodrigues et al., 2015; A. V. Rodrigues & Guastaldi, 2018).

In the open circuit potential technique, the potential value raises with the formation and thickening of the TiO_2 nanofilm on the (cp)-Ti and Ti alloy surfaces, stopping at a maximum value, demonstrating protection against corrosion. In the cyclic voltammetry technique, the current density also raises with the formation and growth of the anodic nanofilm, stabilizing when Ti materials are passivated, i.e. the oxide nanofilm is thickened enough that the protection against the corrosion is achieved. The current density returns to increase due to the reaction of the water oxidation and the O_2 release, as the Pourbaix diagram shows when the water oxidation reactions reach above 1.2 V values. The electrochemical impedance spectroscopy is used after the materials are immersed in the aqueous environment containing the corrosive ions, simulating the physiological medium. The X-ray electron spectroscopy is used after that, which identifies another protective layer besides TiO_2: MoO_2 passivating nanofilm (A. V. Rodrigues & Guastaldi, 2018).

Mo has a high solubility in Ti and a larger atomic radius (Gabriel et al., 2015). Inert and insoluble in water and alkaline media, MoO_2 is the most stable nanofilm of Mo, and it enables the production of solid solutions with TiO_2. The MoO_2 nanofilm is the first barrier attacked by the aggressive ions, and TiO_2 is the second one. Even the aggressive ions attack both nanofilms, there is still the alloy surface protection (Braithwaite, 1994; A. V. Rodrigues et al., 2015).

ENVIRONMENTAL AND ENERGY APPLICATIONS OF NANOMATERIALS

Solar Cells

Increasing population, technology, and economic development are the main factors that have contributed to the growing worldwide demand for more energy in order to create a better life. Nonetheless, climate change due to global warming is one of the environmental damages caused by fossil fuels as one of the energy sources (Sampaio & González, 2017).

Solar energy consumption has been increased in the world due to the benefits of the photovoltaic systems, such as converting sunlight from the solar panels to electricity without heat, its design with little maintenance, and working with outputs from microwatts to megawatts. Therefore, from 2004 to 2014, the capacity of solar energy was achieved from 34 GW to 177 GW. This is the reason for new investments in renewable energy, especially in developed and developing countries (Parida et al., 2011; Razmjoo et al., 2017). Even countries with low solar irradiation can promote an increase of 32% of their energy consumption from renewable sources (Milewicz et al., 2021).

Module, charge controller, inverter, and battery assemble the photovoltaic system that enables the achievement of clean, reliable, scalable, and affordable electricity (Gul et al., 2016) by converting the

Figure 4. The luminescent solar concentrator system consists of a transparent matrix and luminophores (blue spheres) added to the waveguide material: a fraction of absorbed sunlight (solid yellow arrow) irradiates the luminophore, travels into the material (dot blue arrows) by total reflection, is waveguided to the photovoltaic cells, excites electrons from the semiconductor host, and is converted to electricity, and another fraction of light is lost to out (solid blue arrows) the scape cone (dot yellow lines)
Source: Adapted from (Erickson et al., 2014)

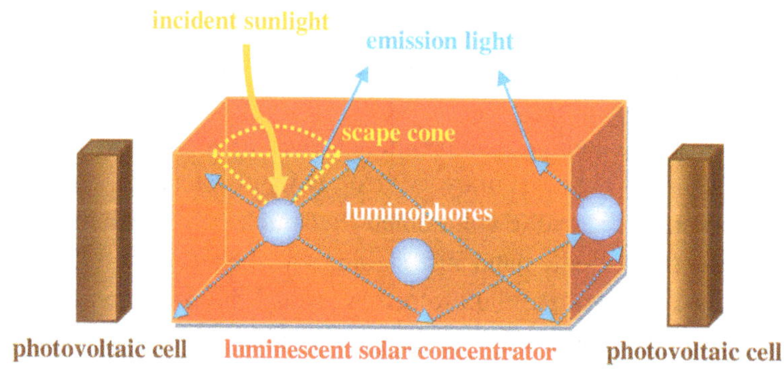

direct current to alternating current electricity or storing the energy in order to use it later. Even with expensive costs to be produced in comparison to the traditional sources, the solar photovoltaic market is very competitive, especially in Europe, China, the United States, and BRICs' countries, which represents 32% of the world energy demand (Ferreira et al., 2018).

There are energy losses detailed by Shockley and Queisser: approximately 20% were low-energy photons as incomplete absorption of light trapping or angle restrictions, 33% were high-energy photons as excess energy in thermalization of hot carriers, and 15% were extractions from recombination of charge carrier (Day et al., 2019; Goodwin et al., 2018). This spectral mismatch is diagnosed when that losses occur by a lack of photon energy ($E < E_g$) and excessive photon energy ($E > E_g$), i.e. an incompatibility between the material bandgap used in the module and the photon energy of the incident solar spectrum, which causes a degradation of the photovoltaic system, and consequently, its efficiency. Usually, it is used a Si cell connected with the photovoltaic system. This traditional module has its bandgap of 1.1 eV, and the conversion efficiency through it could be achieved as 33.7% (Goodwin et al., 2018; Lee et al., 2015).

Recent researches have been developing new technologies connected with photovoltaics in order to mitigate or eliminate that mismatch effects, by working on different irradiance, temperature, dirt, or shadows of objects, thus converting sunlight to electricity (Orduz et al., 2013). A luminescent solar concentrator system is depicted in Figure 4.

Photocatalytic Applications: From Degradation to Fuel Generation

Solar energy is recognized as the primary source of renewable energy due to its applications in several fields, such as degradation of pollutant compounds, water splitting, and energy conversion by using photocatalysts (Archana et al., 2017; Goutam et al., 2018; Romeiro et al., 2021). Nanoparticles obtained by green synthesis are increasingly applied for photocatalytic applications, because of their high reactivity, large surface area, and size-dependent properties. The application of green nanoparticles (obtained from plant extracts, microbes, and enzymes) for contaminated water treatment, for instance, can be an excellent strategy to be combined with pollution problems. Figure 5 shows a schematic representation of different photocatalytic applications of nanomaterials.

Photodegradation of water pollutants has been studied by researchers worldwide. This process significantly reduces the toxicity of undesirable chemical compounds from industrial effluents and intensive agriculture. During photocatalysis, the absorption of ultraviolet-visible (UV) light promotes the formation of electron-hole pairs, which promote redox reactions and the formation of reactive oxygen species (ROS). These ROS are strong oxidants and can degrade chemical contaminants in water. Goutam et al. (2018) reported the synthesis of TiO_2 nanoparticles using leaf extract of *Jatropha curcas L.* In their work, the TiO_2 nanoparticles achieved chemical oxygen demand of 82.26% and 76.48% removal of Cr from tannery wastewater. In a recent study, hairy shaped ZnO was obtained by ureolytic bacteria mediated synthesis, in which ZnO exhibited a great photocatalytic degradation of textile dye (Dhandapani et al., 2020).

Water splitting is a chemical process that involves proton reduction and water oxidation (Romeiro et al., 2017). In this case, the band-gap of the semiconductor must exceed the oxidation potential of water (~1.23 eV). Considering this, Archana et al. synthesized ZnO nanoparticles with an average size of 100–200 nm using *Moringa oleífera* plant extract. The authors claimed that the smaller nanoparticles promoted higher H_2 yields up to 360 mmol hg^{-1} (Archana et al., 2017). Regarding the use of the microwave-assisted hydrothermal method, $BiVO_4$ was prepared by Chen et al. (2020). The authors observed that $BiVO_4$ presented a high O_2 evolution rate (2622 μmol g^{-1} h^{-1}) and also an overall water splitting process with a high H_2 generation rate (340 μmol g^{-1} h^{-1}) and O_2 formation rate of 172 μmol g^{-1} h^{-1}.

The conversion of CO_2 to value-added compounds or fuels is a promising approach to help with the energy crisis. The photocatalytic reduction of CO_2 can produce several high-added value products such as methanol (CH_3OH), formic acid (HCOOH), methane (CH_4), and carbon monoxide (CO) (Chang et al., 2016; Gao et al., 2016; Romeiro et al., 2021). Regarding this, Yang, et al. (2019) proposed that $ZnIn_2S_4$ flower-like microparticles decorated with CeO_2, synthesized through microwave-assisted hydrothermal method, presented the highest CH_3OH production rate of 0.542 μmol g^{-1} h^{-1} when the concertation of CeO_2 was 5 wt%. The authors ascribed this activity to the oxygen vacancies of CeO_2, which efficiently trapped the photogenerated electrons and improved the photogenerated charge separation and to the $ZnIn_2S_4$ structure, which contributed to a great specific area and reactive sites. Over the past few years, photocatalytic ammonia (NH_3) production using photocatalysts has attracted much attention. Industrially, NH_3 is synthesized *via* the Haber-Bosch process at high temperature (400–500 °C) and pressure (100–200 atm). However, semiconductor photocatalysis represents one interesting approach to explore the low-temperature and low-pressure NH_3 synthesis. Even considering the importance of this process, researches on the green synthesis of nanomaterials related to NH_3 production is scarce, making it a promising field of study.

Biohydrogen Production From Dark and Photo-Fermentations

Biodiesel is a non-toxic, biodegradable, and renewable fuel, therefore an alternative for replacing fossil fuels. Known as transesterification of glycerides, the process of biodiesel production involves reactions between the vegetable oils or animal fats, short-chain alcohol, i.e. methanol or ethanol, and sodium hydroxide as the catalyst. Each 100 kg of biodiesel can produce 10 kg of crude glycerol. Moreover, the waste cooking oils have been used by-products of biodiesel production due to lower costs by decreasing the water treatment costs. Specific microorganisms of a bacterial culture enable the consumption of ethanol,

Figure 5. Representation of photocatalytic applications

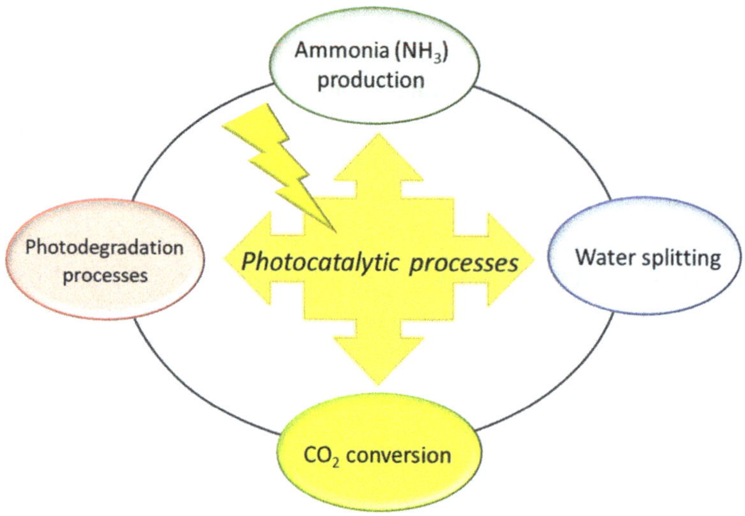

acetic acid, butyric acid, and mainly methanol by using the crude glycerol from biodiesel productions of waste cooking oils. The conversion of crude glycerol to high value-added products supports a sustainable action for biodiesel production (C. V. Rodrigues, Alcaraz, et al., 2020; C. V. Rodrigues, Santana, et al., 2020). Thus, new destinations are given to the crude glycerol generated by the transesterification process, converting the generated products to value-added ones (C. V. Rodrigues et al., 2019). The process of H_2 production is depicted in Figure 6.

The production of H_2 from dark-fermentation is a cost-effective and environmentally friendly process, which uses strictly anaerobic bacteria, employing, as substrates, pure sugars and organic wastes. Under these conditions, the bacteria enable to degrade them from oxidation processes, increasing their biomass and metabolic energy, while electrons are generated from reductions of protons to H_2. This dark-fermentation mostly occurs at a high rate, and it does not depend on light. Organic wastes act as the carbon source, generating the alcohols and acids accumulation as byproducts, that inhibit the H_2 production. The photo-fermentation uses bacteria and organic acids under anaerobic and light conditions, producing H_2 and CO_2. It produces a high volume of hydrogen from high efficiencies of substrate conversion and substrates that include effluents from dark-fermentation (Rai & Singh, 2016).

FUTURE RESEARCH DIRECTIONS

Although progress has been made in the eco-friendly methodologies to obtain nanomaterials for sustainable applications in the last years, there is still much researches to do associated with the use of non-toxic precursors, no release of contaminants into nature, and the harmful effects already caused on the environment by the human beings. The impacts have been studied; however, a more integrated approach is needed to establish a sustainable and non-harmful relationship between industry and the environment, in which the non-toxic and biological materials by eco-friendly approaches for sustainable applications could be a solution.

Figure 6. Production of H_2 from fermentation
Source: Adapted from (C. V. Rodrigues, Santana, et al., 2020)

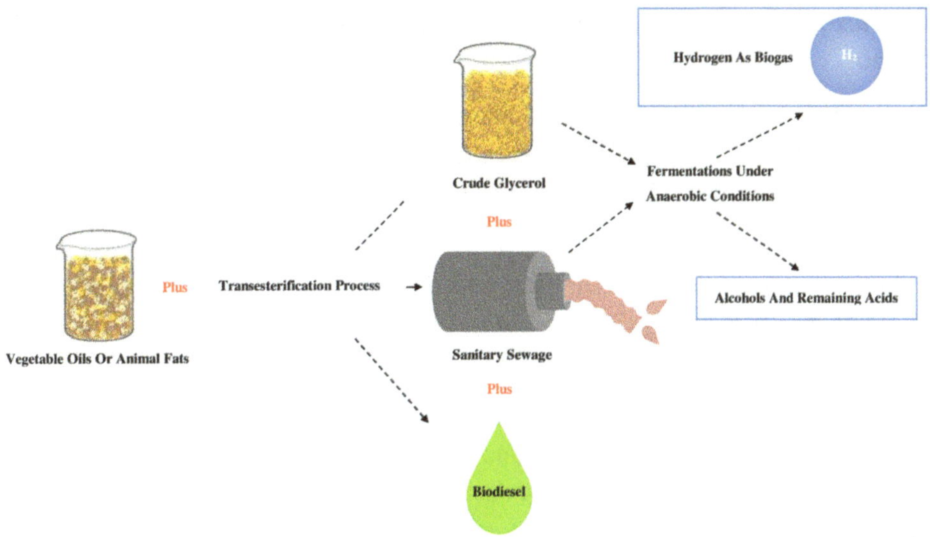

CONCLUSION

In conclusion, this book chapter summarizes eco-friendly technologies, such as green chemical and biological routes for the sustainable preparation of nanoparticles. Researchers have an enormous interest in green synthesis due to its economic prospects and feasibility. The presence of hazardous chemicals during synthesis procedures can be mitigated by using non-toxic solvents, natural materials, and micro-organisms. The sustainable use of green technologies and their applications can be paved by some of the environmental remediation areas, such as photocatalytic degradation, solar cells, and other energy conversions in general. Generated products to the value-added ones are also achievable, thus enhancing their applications.

ACKNOWLEDGMENT

The authors acknowledge the São Paulo State University-UNESP.

REFERENCES

Aditha, S. K., Kurdekar, A. D., Chunduri, L. A. A., Patnaik, S., & Kamisetti, V. (2016). Aqueous based reflux method for green synthesis of nanostructures: Application in CZTS synthesis. *MethodsX*, *3*, 35–42. doi:10.1016/j.mex.2015.12.003 PMID:27408826

Ahmad, N., Sharma, S., Singh, V. N., Shamsi, S. F., Fatma, A., & Mehta, B. R. (2011). Biosynthesis of Silver Nanoparticles from Desmodium triflorum: A Novel Approach Towards Weed Utilization. *Biotechnology Research International, 454090*, 1–8. Advance online publication. doi:10.4061/2011/454090 PMID:21350660

Ajitha, B., Ashok Kumar Reddy, Y., & Sreedhara Reddy, P. (2015). Green synthesis and characterization of silver nanoparticles using Lantana camara leaf extract. *Materials Science and Engineering C, 49*, 373–381. doi:10.1016/j.msec.2015.01.035 PMID:25686962

Ansari, M. A., Khan, H. M., Alzohairy, M. A., Jalal, M., Ali, S. G., Pal, R., & Musarrat, J. (2015). Green synthesis of Al2O3 nanoparticles and their bactericidal potential against clinical isolates of multi-drug resistant Pseudomonas aeruginosa. *World Journal of Microbiology & Biotechnology, 31*(1), 153–164. doi:10.100711274-014-1757-2 PMID:25304025

Archana, B., Manjunath, K., Nagaraju, G., Chandra Sekhar, K. B., & Kottam, N. (2017). Enhanced photocatalytic hydrogen generation and photostability of ZnO nanoparticles obtained via green synthesis. *International Journal of Hydrogen Energy, 42*(8), 5125–5131. doi:10.1016/j.ijhydene.2016.11.099

Bai, H.-J., & Zhang, Z.-M. (2009). Microbial synthesis of semiconductor lead sulfide nanoparticles using immobilized Rhodobacter sphaeroides. *Materials Letters, 63*(9), 764–766. doi:10.1016/j.matlet.2008.12.050

Bolat, G., Mareci, D., Chelariu, R., Izquierdo, J., González, S., & Souto, R. M. (2013). Investigation of the electrochemical behaviour of TiMo alloys in simulated physiological solutions. *Electrochimica Acta, 113*, 470–480. doi:10.1016/j.electacta.2013.09.116

Braithwaite, E. R. (1994). Molybdenum: an outline of its chemistry and users. In Occurrence, Extraction, Production and Uses of Molybdenum. Elsevier Science.

Chang, X., Wang, T., & Gong, J. (2016). CO2 photo-reduction: Insights into CO2 activation and reaction on surfaces of photocatalysts. *Energy & Environmental Science, 9*(7), 2177–2196. doi:10.1039/C6EE00383D

Chen, S.-H., Jiang, Y.-S., & Lin, H. (2020). Easy Synthesis of BiVO4 for Photocatalytic Overall Water Splitting. *ACS Omega, 5*(15), 8927–8933. doi:10.1021/acsomega.0c00699 PMID:32337456

da Rosa, M. B., Oreste, E. Q., Bönemann, D. H., & Rodrigues, A. A., Vendruscolo, C. T., Moreira, A. S., Ribeiro, A. S., & Nunes, A. M. (2016). Evaluation of the use of a reflux system for sample preparation of xanthan gum and subsequent determination of Ca, Cu, K, Mg, Na and Zn by atomic spectrometry techniques. *Journal of the Brazilian Chemical Society, 27*(5), 919–924. doi:10.1016/j.foodchem.2017.07.112

Dallinger, D., & Kappe, C. O. (2007). Microwave-Assisted Synthesis in Water as Solvent. *Chemical Reviews, 107*(6), 2563–2591. doi:10.1021/cr0509410 PMID:17451275

Day, J., Senthilarasu, S., & Mallick, T. K. (2019). Improving spectral modification for applications in solar cells: A review. *Renewable Energy, 132*, 186–205. doi:10.1016/j.renene.2018.07.101

de Medeiros, T. V., Manioudakis, J., Noun, F., Macairan, J.-R., Victoria, F., & Naccache, R. (2019). Microwave-assisted synthesis of carbon dots and their applications. *Journal of Materials Chemistry. C, Materials for Optical and Electronic Devices*, *7*(24), 7175–7195. doi:10.1039/C9TC01640F

Dhandapani, P., Prakash, A. A., AlSalhi, M. S., Maruthamuthu, S., Devanesan, S., & Rajasekar, A. (2020). Ureolytic bacteria mediated synthesis of hairy ZnO nanostructure as photocatalyst for decolorization of dyes. *Materials Chemistry and Physics*, *243*, 122619. doi:10.1016/j.matchemphys.2020.122619

Eisenbarth, E., Velten, D., Müller, M., Thull, R., & Breme, J. (2004). Biocompatibility of β-stabilizing elements of titanium alloys. *Biomaterials*, *25*(26), 5705–5713. doi:10.1016/j.biomaterials.2004.01.021 PMID:15147816

El-Rafie, H. M., El-Rafie, M. H., & Zahran, M. K. (2013). Green synthesis of silver nanoparticles using polysaccharides extracted from marine macro algae. *Carbohydrate Polymers*, *96*(2), 403–410. doi:10.1016/j.carbpol.2013.03.071 PMID:23768580

El-Sayed, E. I., Al-Ghamdi, A. A., Al-Heniti, S., Al-Marzouki, F., & El-Tantawy, F. (2011). Synthesis of ultrafine β-Ga2O3 nanopowder via hydrothermal approach: A strong UV "excimer-like" emission. *Materials Letters*, *65*(2), 317–321. doi:10.1016/j.matlet.2010.10.007

Erickson, C. S., Bradshaw, L. R., McDowall, S., Gilbertson, J. D., Gamelin, D. R., & Patrick, D. L. (2014). Zero-reabsorption doped-nanocrystal luminescent solar concentrators. *ACS Nano*, *8*(4), 3461–3467. doi:10.1021/nn406360w PMID:24621014

Fatima, R., Priya, M., Indurthi, L., Radhakrishnan, V., & Sudhakaran, R. (2020). Biosynthesis of silver nanoparticles using red algae Portieria hornemannii and its antibacterial activity against fish pathogens. *Microbial Pathogenesis*, *138*, 103780. doi:10.1016/j.micpath.2019.103780 PMID:31622663

Ferreira, A., Kunh, S. S., Fagnani, K. C., De Souza, T. A., Tonezer, C., Dos Santos, G. R., & Coimbra-Araújo, C. H. (2018). Economic overview of the use and production of photovoltaic solar energy in brazil. *Renewable & Sustainable Energy Reviews*, *81*, 181–191. doi:10.1016/j.rser.2017.06.102

Gabriel, S. B., Dille, J., Rezende, M. C., Mei, P., de Almeida, L. H., Baldan, R., & Nunes, C. A. (2015). Mechanical characterization of Ti-12Mo-13Nb alloy for biomedical application hot swaged and aged. *Materials Research*, *18*(Suppl 2), 8–12. doi:10.1590/1516-1439.329614

Gao, C., Meng, Q., Zhao, K., Yin, H., Wang, D., Guo, J., Zhao, S., Chang, L., He, M., Li, Q., Zhao, H., Huang, X., Gao, Y., & Tang, Z. (2016). Co3O4 Hexagonal Platelets with Controllable Facets Enabling Highly Efficient Visible-Light Photocatalytic Reduction of CO2. *Advanced Materials*, *28*(30), 6485–6490. doi:10.1002/adma.201601387 PMID:27171564

George, M. J., Sichilongo, K. F., Ramabulana, T., Madala, N. E., & Dubery, I. A. (2019). Comparison of Soxhlet and reflux techniques for extraction and characterisation of potential endocrine-disrupting compounds from solid waste dumpsite soil. *Environmental Monitoring and Assessment*, *191*(3), 1–11. doi:10.100710661-019-7294-6 PMID:30739205

Girija, K., Thirumalairajan, S., Avadhani, G. S., Mangalaraj, D., Ponpandian, N., & Viswanathan, C. (2013). Synthesis, morphology, optical and photocatalytic performance of nanostructured β-Ga2O3. *Materials Research Bulletin*, *48*(6), 2296–2303. doi:10.1016/j.materresbull.2013.02.047

Gonçalves, R. A., Toledo, R. P., Joshi, N., & Berengue, O. M. (2021). Green Synthesis and Applications of ZnO and TiO2 Nanostructures. *Molecules (Basel, Switzerland)*, *26*(8), 2236. Advance online publication. doi:10.3390/molecules26082236 PMID:33924397

González, J. E. G., & Mirza-Rosca, J. C. (1999). Study of the corrosion behavior of titanium and some of its alloys for biomedical and dental implant applications. *Journal of Electroanalytical Chemistry (Lausanne, Switzerland)*, *471*(2), 109–115. doi:10.1016/S0022-0728(99)00260-0

Goodwin, H., Jellicoe, T. C., Davis, N. J. L. K., & Böhm, M. L. (2018). Multiple exciton generation in quantum dot-based solar cells. *Nanophotonics*, *7*(1), 111–126. doi:10.1515/nanoph-2017-0034

Gour, A., & Jain, N. K. (2019). Advances in green synthesis of nanoparticles. *Artificial Cells, Nanomedicine, and Biotechnology*, *47*(1), 844–851. doi:10.1080/21691401.2019.1577878 PMID:30879351

Goutam, S. P., Saxena, G., Singh, V., Yadav, A. K., Bharagava, R. N., & Thapa, K. B. (2018). Green synthesis of TiO2 nanoparticles using leaf extract of Jatropha curcas L. for photocatalytic degradation of tannery wastewater. *Chemical Engineering Journal*, *336*, 386–396. doi:10.1016/j.cej.2017.12.029

Greer, H. F., Yu, F., & Zhou, W. (2011). Early stages of non-classic crystal growth. *Science China. Chemistry*, *54*(12), 1867–1876. doi:10.100711426-011-4441-5

Gul, M., Kotak, Y., & Muneer, T. (2016). Review on recent trend of solar photovoltaic technology. *Energy Exploration and Exploitation*, *34*(4), 485–526. doi:10.1177/0144598716650552

Guo, T. H., Liu, Y., Zhang, Y. C., & Zhang, M. (2011). Green hydrothermal synthesis and optical absorption properties of ZnO2 nanocrystals and ZnO nanorods. *Materials Letters*, *65*(4), 639–641. doi:10.1016/j.matlet.2010.11.032

Hua, S., de Matos, M. B. C., Metselaar, J. M., & Storm, G. (2018). Current Trends and Challenges in the Clinical Translation of Nanoparticulate Nanomedicines: Pathways for Translational Development and Commercialization. *Frontiers in Pharmacology*, *9*, 790. Advance online publication. doi:10.3389/fphar.2018.00790 PMID:30065653

Hulla, J. E., Sahu, S. C., & Hayes, A. W. (2015). Nanotechnology: History and future. *Human and Experimental Toxicology*, *34*(12), 1318–1321. doi:10.1177/0960327115603588 PMID:26614822

Jadoun, S., Arif, R., Jangid, N. K., & Meena, R. K. (2021). Green synthesis of nanoparticles using plant extracts: A review. *Environmental Chemistry Letters*, *19*(1), 355–374. doi:10.100710311-020-01074-x

Jebril, S., Khanfir Ben Jenana, R., & Dridi, C. (2020). Green synthesis of silver nanoparticles using Melia azedarach leaf extract and their antifungal activities: In vitro and in vivo. *Materials Chemistry and Physics*, *248*, 122898. doi:10.1016/j.matchemphys.2020.122898

Kabcum, S., Channei, D., Tuantranont, A., Wisitsoraat, A., Liewhiran, C., & Phanichphant, S. (2016). Ultra-responsive hydrogen gas sensors based on PdO nanoparticle-decorated WO3 nanorods synthesized by precipitation and impregnation methods. *Sensors and Actuators. B, Chemical*, *226*, 76–89. doi:10.1016/j.snb.2015.11.120

Kamble, G., Kashale, A., Rasal, A., Dengale, S., Kolekar, S., Chang, J. Y., Han, S.-H., & Ghule, A. (2021). Investigating the Influence of Reflux Condensation Reaction Temperature on the Growth of FeCo2O4 Thin Film for Flexible Supercapacitor. *ChemistrySelect*, *6*(8), 1838–1844. doi:10.1002lct.202004544

Kappe, C. O. (2004). Controlled Microwave Heating in Modern Organic Synthesis. *Angewandte Chemie International Edition*, *43*(46), 6250–6284. doi:10.1002/anie.200400655 PMID:15558676

Khalil, M. M. H., Ismail, E. H., El-Baghdady, K. Z., & Mohamed, D. (2014). Green synthesis of silver nanoparticles using olive leaf extract and its antibacterial activity. *Arabian Journal of Chemistry*, *7*(6), 1131–1139. doi:10.1016/j.arabjc.2013.04.007

Komarneni, S., Roy, R., & Li, Q. H. (1992). Microwave-hydrothermal synthesis of ceramic powders. *Materials Research Bulletin*, *27*(12), 1393–1405. doi:10.1016/0025-5408(92)90004-J

Kronka, M. S., Cordeiro-Junior, P. J. M., Mira, L., dos Santos, A. J., Fortunato, G. V., & Lanza, M. R. V. (2021). Sustainable microwave-assisted hydrothermal synthesis of carbon-supported ZrO2 nanoparticles for H2O2 electrogeneration. *Materials Chemistry and Physics*, *267*, 124575. doi:10.1016/j.matchemphys.2021.124575

Kumar, A., Kuang, Y., Liang, Z., & Sun, X. (2020). Microwave chemistry, recent advancements, and eco-friendly microwave-assisted synthesis of nanoarchitectures and their applications: A review. *Materials Today Nano, 11*.

Lee, K.-T., Park, J.-H., Kwon, S. J., Kwon, H.-K., Kyhm, J., Kwak, K.-W., Jang, H. S., Kim, S. Y., Han, J. S., Lee, S.-H., Shin, D.-H., Ko, H., Han, I.-K., Ju, B.-K., Kwon, S.-H., & Ko, D.-H. (2015). Simultaneous enhancement of upconversion and downshifting luminescence via plasmonic structure. *Nano Letters*, *15*(4), 2491–2497. doi:10.1021/nl5049803 PMID:25756859

Lemos, S. C. S., Nossol, E., Ferrari, J. L., Gomes, E. O., Andres, J., Gracia, L., Sorribes, I., & Lima, R. C. (2019). Joint Theoretical and Experimental Study on the La Doping Process in In2O3: Phase Transition and Electrocatalytic Activity. *Inorganic Chemistry*, *58*(17), 11738–11750. doi:10.1021/acs.inorgchem.9b01728 PMID:31415162

Li, L., Wei, W., & Behrens, M. (2012). Synthesis and characterization of α-, β-, and γ-Ga 2o 3 prepared from aqueous solutions by controlled precipitation. *Solid State Sciences*, *14*(7), 971–981. doi:10.1016/j.solidstatesciences.2012.04.037

Liu, Y. C., Li, J., Ahn, J., Pu, J., Rupa, E. J., Huo, Y., & Yang, D. C. (2020). Biosynthesis of zinc oxide nanoparticles by one-pot green synthesis using fruit extract of Amomum longiligulare and its activity as a photocatalyst. *Optik (Stuttgart)*, *218*, 165245. doi:10.1016/j.ijleo.2020.165245

Mareci, D., Chelariu, R., Gordin, D.-M., Ungureanu, G., & Gloriant, T. (2009). Comparative corrosion study of Ti-Ta alloys for dental applications. *Acta Biomaterialia*, *5*(9), 3625–3639. doi:10.1016/j.actbio.2009.05.037 PMID:19508903

Marinho, J. Z., Romeiro, F. C., Lemos, S. C. S., Motta, F. V., Riccardi, C. S., Li, M. S., Longo, E., & Lima, R. C. (2012). Urea-Based Synthesis of Zinc Oxide Nanostructures at Low Temperature. *Journal of Nanomaterials*, *2012*, 1–7. doi:10.1155/2012/427172

Mehwish, H. M., Rajoka, M. S. R., Xiong, Y., Cai, H., Aadil, R. M., Mahmood, Q., He, Z., & Zhu, Q. (2021). Green synthesis of a silver nanoparticle using Moringa oleifera seed and its applications for antimicrobial and sun-light mediated photocatalytic water detoxification. *Journal of Environmental Chemical Engineering*, *9*(4), 105290. doi:10.1016/j.jece.2021.105290

Milewicz, B., Bogacka, M., & Pikoń, K. (2021). Influence of solar concentrator in the form of luminescent pmma on the performance of a silicon cell. *Sustainability*, *13*(4), 1–14. doi:10.3390u13042086

Mirzaei, A., & Neri, G. (2016). Microwave-assisted synthesis of metal oxide nanostructures for gas sensing application: A review. *Sensors and Actuators. B, Chemical*, *237*, 749–775. doi:10.1016/j.snb.2016.06.114

Mittal, A. K., Chisti, Y., & Banerjee, U. C. (2013). Synthesis of metallic nanoparticles using plant extracts. *Biotechnology Advances*, *31*(2), 346–356. doi:10.1016/j.biotechadv.2013.01.003 PMID:23318667

Mulvihill, M. J., Beach, E. S., Zimmerman, J. B., & Anastas, P. T. (2011). Green Chemistry and Green Engineering: A Framework for Sustainable Technology Development. *Annual Review of Environment and Resources*, *36*(1), 271–293. doi:10.1146/annurev-environ-032009-095500

Muthu Kumara Pandian, A., Rajasimman, M., Rajamohan, N., Varjani, S., & Karthikeyan, C. (2021). Anaerobic mixed consortium (AMC) mediated enhanced biosynthesis of silver nano particles (AgNPs) and its application for the removal of phenol. *Journal of Hazardous Materials*, *416*, 125717. doi:10.1016/j.jhazmat.2021.125717 PMID:33819646

Navab, A. A., Nemati, A., Navab, A. A., & Abad, H. M. M. (2018). *Hydrothermal synthesis of TiO2 nanorod for using as an electron transport material in perovskite solar cells*. doi:10.1063/1.5018947

Ngom, I., Ngom, B. D., Sackey, J., & Khamlich, S. (2021). Biosynthesis of zinc oxide nanoparticles using extracts of Moringa Oleifera: Structural & optical properties. *Materials Today: Proceedings*, *36*, 526–533. doi:10.1016/j.matpr.2020.05.323

Orduz, R., Solórzano, J., Egido, M. Á., & Román, E. (2013). Analytical study and evaluation results of power optimizers for distributed power conditioning in photovoltaic arrays. *Progress in Photovoltaics: Research and Applications*, *21*(3), 359–373. doi:10.1002/pip.1188

Oufella, L. S., & Benchettara, A. (2016). Influence of fluoride on the electrochemical behvior of a new synthesized Ti-10Ta-2Mo alloy for biomedical applications. *Journal of Fundamental and Applied Sciences*, *8*(3), 731–752. doi:10.4314/jfas.v8i3.5

Parida, B., Iniyan, S., & Goic, R. (2011). A review of solar photovoltaic technologies. *Renewable & Sustainable Energy Reviews*, *15*(3), 1625–1636. doi:10.1016/j.rser.2010.11.032

Rai, P. K., & Singh, S. P. (2016). Integrated dark- and photo-fermentation: Recent advances and provisions for improvement. *International Journal of Hydrogen Energy*, *41*(44), 19957–19971. doi:10.1016/j.ijhydene.2016.08.084

Razmjoo, A., Qolipour, M., Shirmohammadi, R., Heibati, S. M., & Iraj, F. (2017). Techno-economic evaluation of standalone hybrid solar-wind systems for small residential districts in the central desert of Iran. *Environmental Progress & Sustainable Energy*, *36*(4), 1194–1207. doi:10.1002/ep.12554

Rodrigues, A. V., & Guastaldi, A. C. (2018). Ti-Mo alloys: Corrosion study in solutions simulating commercial gels. *Journal of Polymer Science and Engineering*, *1*(4), 1–7. doi:10.24294/jpse.v1i4.897

Rodrigues, A. V., Oliveira, N. T. C., dos Santos, M. L., & Guastaldi, A. C. (2015). Electrochemical behavior and corrosion resistance of Ti–15Mo alloy in naturally-aerated solutions, containing chloride and fluoride ions. *Journal of Materials Science. Materials in Medicine*, *26*(1), 1–9. doi:10.100710856-014-5323-0 PMID:25577207

Rodrigues, A. V., & Orlandi, M. O. (2019). Study of intense photoluminescence from monodispersed β-Ga2O3 ellipsoidal structures. *Ceramics International*, *45*(4), 5023–5029. doi:10.1016/j.ceramint.2018.11.203

Rodrigues, A. V., & Sabino, N. L. (2019a). Gallium oxide: A perspective of features in photoluminescence and gas sensor applications. *International Journal of Recent Scientific Research*, *10*(11(B)), 35816–35827. doi:10.24327/IJRSR

Rodrigues, A. V., & Sabino, N. L. (2019b). Synthesis of photoluminescent β-Ga2O3 nanostructures using electrospinning method, and control of length-diameter ratio by calcination heating rates. *Journal of Materials Science Materials in Electronics*, *30*(18), 16910–16916. doi:10.100710854-019-01631-3

Rodrigues, B. S., Branco, C. M., Corio, P., & Souza, J. S. (2020). Controlling Bismuth Vanadate Morphology and Crystalline Structure through Optimization of Microwave-Assisted Synthesis Conditions. *Crystal Growth & Design*, *20*(6), 3673–3685. doi:10.1021/acs.cgd.9b01517

Rodrigues, C. V., Alcaraz, F. A. R., Nespeca, M. G., Rodrigues, A. V., Motteran, F., Adorno, M. A. T., Varesche, M. B. A., & Maintinguer, S. I. (2020). Biohydrogen production in an integrated biosystem using crude glycerol from waste cooking oils. *Renewable Energy*, *162*, 701–711. doi:10.1016/j.renene.2020.08.061

Rodrigues, C. V., Rodrigues, A. V., Antoniosi, L., & Maintinguer, S. I. (2019). Applications of biodiesel and crude glycerol: the products generated by transesterification process. In Transesterification: process, mechanism and applications (pp. 83–122). Academic Press.

Rodrigues, C. V., Santana, K. O., Nespeca, M. G., Rodrigues, A. V., Pires, L. O., & Maintinguer, S. I. (2020). Energy valorization of crude glycerol and sanitary sewage in hydrogen generation by biological processes. *International Journal of Hydrogen Energy*, *45*(21), 11943–11953. doi:10.1016/j.ijhydene.2020.02.168

Romeiro, F. C., Marinho, J. Z., Silva, A. C. A., Cano, N. F., Dantas, N. O., & Lima, R. C. (2013). Photoluminescence and magnetism in Mn2+-doped ZnO nanostructures grown rapidly by the microwave hydrothermal method. *The Journal of Physical Chemistry C*, *117*(49), 26222–26227. doi:10.1021/jp408993y

Romeiro, F. C., Rodrigues, M. A., Silva, L. A. J., Catto, A. C., da Silva, L. F., Longo, E., Nossol, E., & Lima, R. C. (2017). rGO-ZnO nanocomposites for high electrocatalytic effect on water oxidation obtained by microwave-hydrothermal method. *Applied Surface Science*, *423*, 743–751. doi:10.1016/j.apsusc.2017.06.221

Romeiro, F. C., Silva, B. C., Martins, A. S., Zanoni, M. V. B., & Orlandi, M. O. (2021). Superior performance of rGO-tin oxide nanocomposite for selective reduction of CO2 to methanol. *Journal of CO2 Utilization, 46*.

Romeiro, F. C., Silva, S. C., Nossol, E., & Lima, R. C. (2020). One step microwave-hydrothermal synthesis of rGO–TiO2 nanocomposites for enhanced electrochemical oxygen evolution reaction. *New Journal of Chemistry, 44*(17), 6825–6832. doi:10.1039/D0NJ01475C

Sampaio, P. G. V., & González, M. O. A. (2017). Photovoltaic solar energy: Conceptual framework. *Renewable & Sustainable Energy Reviews, 74*, 590–601. doi:10.1016/j.rser.2017.02.081

Shan, J.-J., Li, C.-H., Wu, J.-M., Liu, J.-A., & Shi, Y.-S. (2017). Shape-controlled synthesis of monodispersed beta-gallium oxide crystals by a simple precipitation technique. *Ceramics International, 43*(8), 6430–6436. doi:10.1016/j.ceramint.2017.02.056

Shou, Q., Guo, C., Yang, L., Jia, L., Liu, C., & Liu, H. (2011). Effect of pH on the single-step synthesis of gold nanoparticles using PEO–PPO–PEO triblock copolymers in aqueous media. *Journal of Colloid and Interface Science, 363*(2), 481–489. doi:10.1016/j.jcis.2011.07.021 PMID:21855892

Simas, A. B. C., Pereira, V. L. P., Barreto, C. B. Jr, de Sales, D. L., & de Carvalho, L. L. (2009). An expeditious and consistent procedure for tetrahydrofuran (THF) drying and deoxygenation by the still apparatus. *Quimica Nova, 32*(9), 2473–2475. doi:10.1590/S0100-40422009000900042

Singh, J., Dutta, T., Kim, K.-H., Rawat, M., Samddar, P., & Kumar, P. (2018). 'Green' synthesis of metals and their oxide nanoparticles: Applications for environmental remediation. *Journal of Nanobiotechnology, 16*(1), 84. doi:10.118612951-018-0408-4 PMID:30373622

Sivaraj, A., Kumar, V., Sunder, R., Parthasarathy, K., & Kasivelu, G. (2020). Commercial Yeast Extracts Mediated Green Synthesis of Silver Chloride Nanoparticles and their Anti-mycobacterial Activity. *Journal of Cluster Science, 31*(1), 287–291. doi:10.100710876-019-01626-4

Sonbol, H., Ameen, F., AlYahya, S., Almansob, A., & Alwakeel, S. (2021). Padina boryana mediated green synthesis of crystalline palladium nanoparticles as potential nanodrug against multidrug resistant bacteria and cancer cells. *Scientific Reports, 11*(1), 5444. doi:10.103841598-021-84794-6 PMID:33686169

Terraschke, H., Olchowka, J., Geringer, E., Rodrigues, A. V., & Wickleder, C. (2018). Facile Ionic Liquid-Assisted Strategy for Direct Precipitation of Eu2+-Activated Nanophosphors under Ambient Conditions. *Small, 14*(17), 1–6. doi:10.1002mll.201703707 PMID:29573331

Velsankar, K., R.M., A. K., R., P., V., M., & Sudhahar, S. (2020). Green synthesis of CuO nanoparticles via Allium sativum extract and its characterizations on antimicrobial, antioxidant, antilarvicidal activities. *Journal of Environmental Chemical Engineering, 8*(5).

Wang, C., Gao, Y., Wang, L., & Li, P. (2017). Morphology regulation, structural, and photocatalytic properties of ZnO hierarchical microstructures synthesized by a simple hydrothermal method. *Physica Status Solidi. A, Applications and Materials Science, 214*(6), 1600876. doi:10.1002/pssa.201600876

Wang, L., Hou, Z., Quan, Z., Lian, H., Yang, P., & Lin, J. (2009). Preparation and luminescence properties of Mn2+-doped ZnGa2O4 nanofibers via electrospinning process. *Materials Research Bulletin*, *44*(10), 1978–1983. doi:10.1016/j.materresbull.2009.06.008

Yang, C., Li, Q., Xia, Y., Lv, K., & Li, M. (2019). Enhanced visible-light photocatalytic CO2 reduction performance of Znln2S4 microspheres by using CeO2 as cocatalyst. *Applied Surface Science*, *464*, 388–395. doi:10.1016/j.apsusc.2018.09.099

Yang, G., & Park, S.-J. (2019). Conventional and Microwave Hydrothermal Synthesis and Application of Functional Materials: A Review. *Materials (Basel, Switzerland)*, *12*(7), 1177. doi:10.3390/ma12071177 PMID:30978917

Yoon, Y., Han, K. I., Kim, B. H., Lee, I. G., Kim, Y., Kim, J. P., & Hwang, W. S. (2018). Formation of β-Ga2O3 nanofibers of sub-50 nm diameter synthesized by electrospinning method. *Thin Solid Films*, *645*, 358–362. doi:10.1016/j.tsf.2017.11.010

Zhang, D., Ma, X., Gu, Y., Huang, H., & Zhang, G. (2020). Green Synthesis of Metallic Nanoparticles and Their Potential Applications to Treat Cancer. *Frontiers in Chemistry*, *8*, 799. doi:10.3389/fchem.2020.00799 PMID:33195027

Zhang, Y., Yang, J., Li, Q., & Cao, X. (2007). Preparation of Ga2O3 nanoribbons and tubes by electrospinning. *Journal of Crystal Growth*, *308*(1), 180–184. doi:10.1016/j.jcrysgro.2007.07.036

Zhao, J., Zhang, W., Xie, E., Ma, Z., Zhao, A., & Liu, Z. (2011). Structure and photoluminescence of β-Ga2O3 :Eu3+ nanofibers prepared by electrospinning. *Applied Surface Science*, *257*(11), 4968–4972. doi:10.1016/j.apsusc.2010.12.157

Zhao, J.-G., Zhang, Z.-X., Ma, Z.-W., Duan, H.-G., Guo, X.-S., & Xie, E.-Q. (2008). Structural and photoluminescence properties of β-Ga2O3 nanofibers fabricated by electrospinning method. *Chinese Physics Letters*, *25*(10), 3787–3789. doi:10.1088/0256-307X/25/10/073

ADDITIONAL READING

Dresselhaus, M. S., & Thomas, I. L. (2001). Alternative energy technologies. *Nature*, *414*(6861), 332–337. doi:10.1038/35104599 PMID:11713539

Lefebvre, O., & Moletta, R. (2006). Treatment of organic pollution in industrial saline wastewater: A literature review. *Water Research*, *40*(20), 3671–3682. doi:10.1016/j.watres.2006.08.027 PMID:17070895

Mishra, B., Varjani, S., Kumar, G., Awasthi, M. K., Awasthi, S. K., Sindhu, R., Binod, P., Rene, E. R., & Zhang, Z. (2021). Microbial approaches for remediation of pollutants: Innovations, future outlook, and challenges. *Energy & Environment*, *32*(6), 1029–1058. doi:10.1177/0958305X19896781

KEY TERMS AND DEFINITIONS

Antimicrobial Activity: Inhibition process of bacteria growth or prevention of microbial colonies formation.

Dark-Fermentation: The process of H2 production which uses pure sugars, bacteria, and organic acids in the absence of light under anaerobic conditions.

Environmentally Friendly: Not harmful to the environment.

Luminescent Solar Concentrator: A module which is connected with the photovoltaic cell in order to decrease losses of energy from self-absorption processes, improving the device efficiency.

Morphology: The study of the form or shape of a particle observed with the help of a microscope (scientific study).

Nanomaterial: A material composed of particles or constituents in nanoscale dimensions, or a material produced by nanotechnology.

Photo-Fermentation: The process of H2 and CO2 production which uses bacteria and organic acids under anaerobic and light conditions.

Photocatalysis: Interaction process between a light source and the surface of semiconductor (catalyst) material.

Photovoltaic Cell: A system which converts sunlight to electricity, either storing or not the energy in order to use it later.

Semiconductor: A solid substance whose conductivity is between the respective insulator and metal.

Chapter 3
Phyto-Nanofabrication:
Plant-Mediated Synthesis of Metal and Metal Oxide Nanoparticles

Israt Jahan
Dicle University, Turkey

ABSTRACT

Green synthesis of metallic nanoparticles through natural entities (i.e., bacteria, actinomycetes, yeast, fungus, microalgae, seaweed, plants, and plant-derived materials) has become an advantageous and ecofriendly approach. However, phytocompounds of plant extract have achieved huge attention since by utilizing them high yield NPs with desirable size and shape, which can be produced through single-step synthesis scheme. Plants retain diverse biochemicals that exhibit strong hyper-accumulating potential, crucial for metallic ion reduction to metallic NPs, like platinum, gold, silver, titanium oxide, iron oxide, copper oxide, zinc oxide, palladium, etc. Here, previously published studies were reviewed for providing the latest scientific evidence on biosynthesis of metal and metal oxide NPs using different plant materials, especially medicinal plants and food and agro-wastes.

INTRODUCTION AND BACKGROUND

Utilization of metal and metal oxide nanoparticles (NPs) in various areas such as cosmetic, pharmaceutical, health care, drug delivery, food processing and packaging, photonics, surface coatings, biosensors, agricultural sectors, fuel cells, piezoelectric components, microelectronic fabrication, and toxic elements elimination and many other environmental applications has become a foreseeable tendency over the last decade (Hussain et al., 2016; Iravani, 2011; M. Rai & Ingle, 2012; Vijayaraghavan & Ashokkumar, 2017). This is due to the fact that they possess nano-scale structure with a high proportion of surface area as well as exhibit enhanced catalytic, morphological, physiochemical, electrical, magnetic, optical, mechanical, and biological characteristics, which differ from those of bulk materials (Ishak et al., 2019; Kamat, 2002). Because of possessing these unique and fascinating properties, metal and metal oxide

DOI: 10.4018/978-1-7998-8936-6.ch003

nanoparticles have become the focus of intensive studies over the last few years, and many studies have been accompanied synthesizing these nanoparticles applying different approaches.

Two routes are mainly followed for the production of nanoparticles, i.e., the 'top-down' approach where NPs are formed by size reduction and decomposing of larger macro particles into fine particles (Ahmed et al., 2016), and 'bottom-up' approach that involves the assembly of small substances to synthesize nanoparticles through the oxidation of metallic ions into zero-valent metal particles (Zheng et al., 2016). Top-down process consists of physical and chemical methods whereas bottom-up approach can be achieved through wet chemical methods (Figure 1).

Nevertheless, these approaches require a significant amount of harmful and toxic ingredients, prolonged reaction times, enormous consumption of energy, expensive instrumentations, and also produce non-eco-friendly toxic by-products, which increase the cost of the experiment as well as intensify environmental toxicity (Shah et al., 2015; Soundarrajan et al., 2012; B. Zheng et al., 2013). Therefore, green chemistry by using environment-friendly materials for nanoparticle production has become more acceptable and has been established as a superior, advantageous, and alternative synthesis approach over other conventional techniques (Chandra et al., 2020). Green synthesis of metallic nanoparticles is in the category of bottom-up approach, and the nanoparticle formation mechanisms are similar to the chemical reduction process (reduction of metal ions to nanoparticles). Major emphases of this approach are i) utilizing harmless solvents/reagents for synthesis, ii) modeling, adapting, and development of an inexpensive, facile, and energy-saving synthesis protocol, iii) involving diverse biological entities available in natural surroundings, and iv) fabrications of nontoxic and biocompatible nanoparticles for further applications (Jahan et al., 2019; Malik et al., 2014). To evaluate and differentiate the biocompatibility of synthesized silver nanoparticles, comparisons were conducted using both chemical reduction (conventional wet-chemistry) and plant-mediated green synthesis of AgNPs, and comparatively lower phytotoxicity and cytotoxicity were observed in AgNPs from green synthesis (Kummara et al., 2016).

Naturally occurring biomaterials mainly derived from unicellular and multicellular bacteria and actinomycetes, yeast, fungus, microalgae, seaweed, plants have been exploited in green chemistry-based techniques, especially for bio-fabrication of various metal and metal oxide nanoparticles (Figure 1) (Ahmad et al., 2003; Amooaghaie et al., 2015; Du et al., 2007; Gong et al., 2011; Kowshik et al., 2003, p. 3; P. Mukherjee et al., 2001). However, the maintenance for synthesis procedures using a lower group of organisms rather than plants is time-consuming because the extra complex steps are needed to follow through the process, such as microbial maintenance, sampling, isolation, and culturing. Therefore, the utilization of raw materials from plants to fabricate metallic NPs has become an immense and significant option that is inexpensive, facile, rapid, reasonable, practicable, and feasible (Vijayaraghavan & Ashokkumar, 2017). Additionally, high yield NPs with desirable size and shape can be produced by utilizing plant extracts through a controlled single-step synthesis process (Ahmed et al., 2016). Besides, plant-based industrial wastes and agricultural residues are suitable for extracting significant phytocompounds which could be potential reagents to fabricate metallic nanoparticles (Gan et al., 2012; Harish et al., 2015). Hence, this chapter highlighted previous studies and the latest scientific evidence on plant extract mediated biosynthesis approaches of metal and metal oxide nanoparticles using different plant materials, especially medicinal plants and agro-wastes.

Figure 1. Different available routes followed for the fabrication of metallic nanoparticles (Singh et al., 2018)

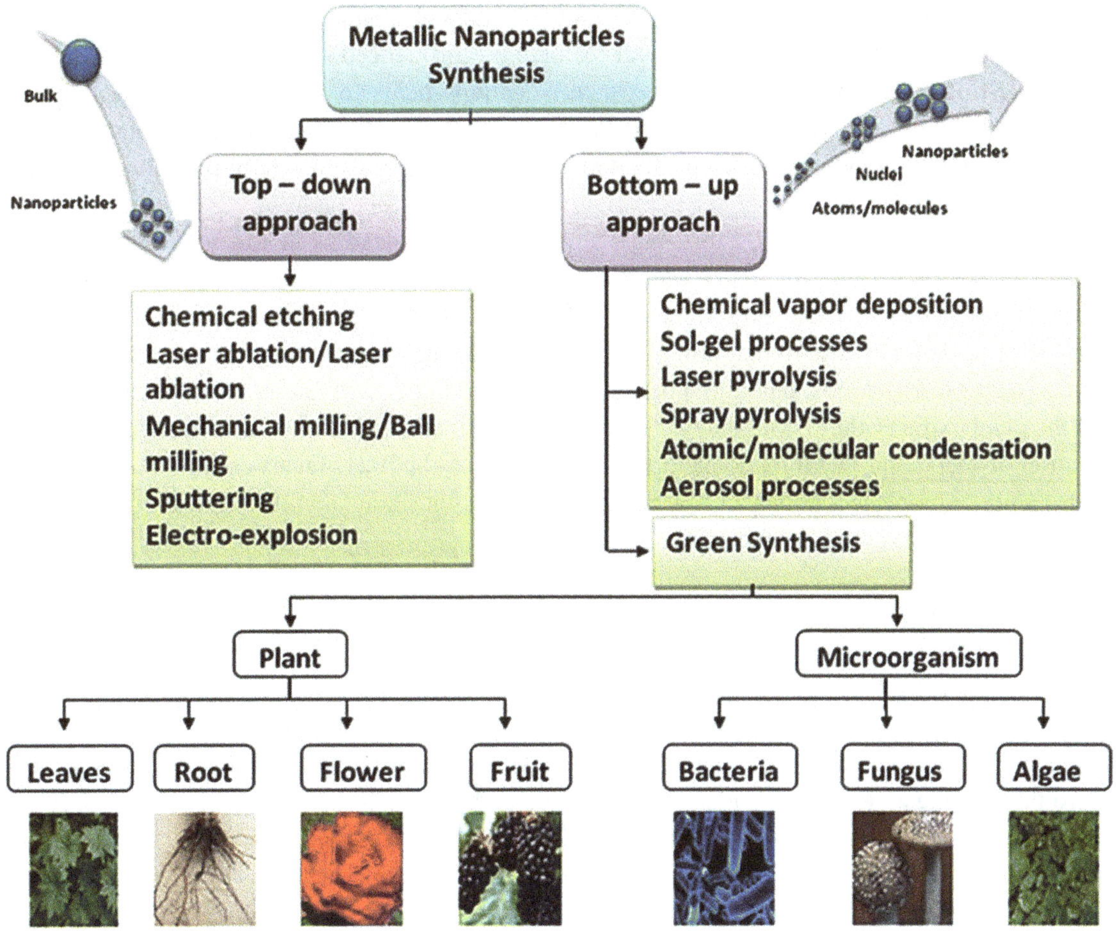

PLANT-BASED BIOSYNTHESIS OF METAL AND METAL OXIDE NANOPARTICLES

Since plants possess a wide range of phytocompounds, they exhibit strong hyper-accumulating potential which is crucial for metallic ion reduction. Besides, they are considered as the ultimate ecofriendly, non-toxic, and inexpensive green-process of metal and metal oxide nanoparticle production because of their abundances, highly rich genetic variability, and diversity (Kulkarni & Muddapur, 2014). In the process of biosynthesis, extracts from different plant parts i.e., whole plant, aerial parts, root, bulb, leaf, stem, bud, flower, fruit, seed, isolated phytocompounds, etc. can straightforwardly reduce metal salts to metal or metal oxide nanoparticles at ambient temperature and pressure (Malik et al., 2014). Like the chemical reduction process, phytonanofabrication in the reaction medium occurs by assembling metal atoms into clusters, which subsequently form nanoparticles. At beginning of the synthesis, metal salt solution and suitable sample of plant extract are mixed together where bioreduction of the salts starts instantly by re-leasing metallic ions in the medium; afterwards, the color change in the reaction medium indicates metal or metal oxide nanoparticles formation by the reduction of their ions (Figure 2). In the initial activation period, mono or divalent oxidation states of metals transformed into ionic state, which later reduced to zero-valent metal atoms (Malik et al., 2014). After that reduction, metal atoms start nucleation to form thermodynamically stable larger nanoparticles by integrating with smaller surrounding particles (Akhtar et al., 2013). Shapes and other morphological features are influenced by this nucleation process and depending on this progression stage, diverse structured nanoparticles i.e., spheres, pentagons, hexagons, triangles, rods, cubes, and wires can be formed (Akhtar et al., 2013), which finally are stabilized with the help of stabilizing phytochemicals of plant extracts for accomplishing more functional and stable morphology (Figure 2).

Solvent medium, reducing, and stabilizing agents are the three leading factors involved in the green synthesis of metallic nanoparticles, in which diverse phytochemicals of plant extract in an aqueous medium work for the reduction and stabilization of nano-sized materials (Jahan & Işildak, 2020; Kuppusamy et al., 2016). Because of retaining various phytochemicals such as i.e., vitamins, organic acids, phenolics, terpenoids, tannins, saponins, polysaccharides, proteins, enzymes, amino acids, and alkaloids, etc. in the plant extracts, they could play a vital role as reducing as well as a stabilizing agent during the fabrication of metal and metal oxide nanoparticles (Jahan et al., 2021; Ovais et al., 2018). Moreover, the role of potential phytochemicals of plant extracts for synthesizing, stabilizing, and functionalizing metallic nanoparticles are presented below.

Role of Flavonoids

Flavonoids are considered as a large group of secondary aromatic compounds (more than 7000 flavonoid compounds are reported until now), which are generally responsible for plant pigmentations (de Souza Farias et al., 2021). Chemically, they comprise a 15-carbon skeleton (C6-C3-C6) i.e., a heterocyclic ring and two phenyl rings joined by a bridge (C) with three carbons (de Souza Farias et al., 2021). In plant extracts, these diverse secondary metabolites are believed to be the main bioreducing ingredients for nanoparticle synthesis due to retaining hydrogen atoms or electron-donating capability (Zhou et al., 2010). Previously published literature indicated that the electron and hydrogen releasing capacity of flavonoids in a liquid medium is the significant mechanism that can be utilized for metallic nanoparticle synthesis (Lee et al., 2011; Mittal et al., 2014). For assessing untouched plants with the purpose of metal and metal oxide NPs synthesis, therefore, quality and quantity of flavonoids in the plant extracts are considered as the primary and crucial tools (Raghunandan et al., 2009). It was observed that during the bioreduction of metal ions, –OH groups of myricetin and quercetin (two important flavonoids) work as reducing agents since they are capable of oxidizing to carbonyl groups by releasing electrons (Ghoreishi et al., 2011). Both in purified and unprocessed (crude) forms, these polyphenolic secondary metabolites are widely used for metallic NPs synthesis. Silver nanoparticles at the range of 5 to 80 nm were individually and effectively fabricated by using pure aqueous extract of three flavonoids namely, naringin, hesperidin, and diosmin, where these three poly-hydroxylated bio compounds were utilized for the bioreduction of $Ag+$ ions (Sahu et al., 2016). Moreover, they also take part in the capping process of NPs not only for stabilizing but also for increasing the antimicrobial activities of AgNPs (Sahu et al., 2016). Likewise, in the case of a successful biosynthesis of AgNPs by *Psidium guajava* leaf extract, it was proposed that the bioreduction of ionic silver was performed flavonoids available in leaf extract (Wang et al., 2018).

Figure 2. Mechanism for plant-mediated metallic nanoparticle synthesis
(Malik et al., 2014)

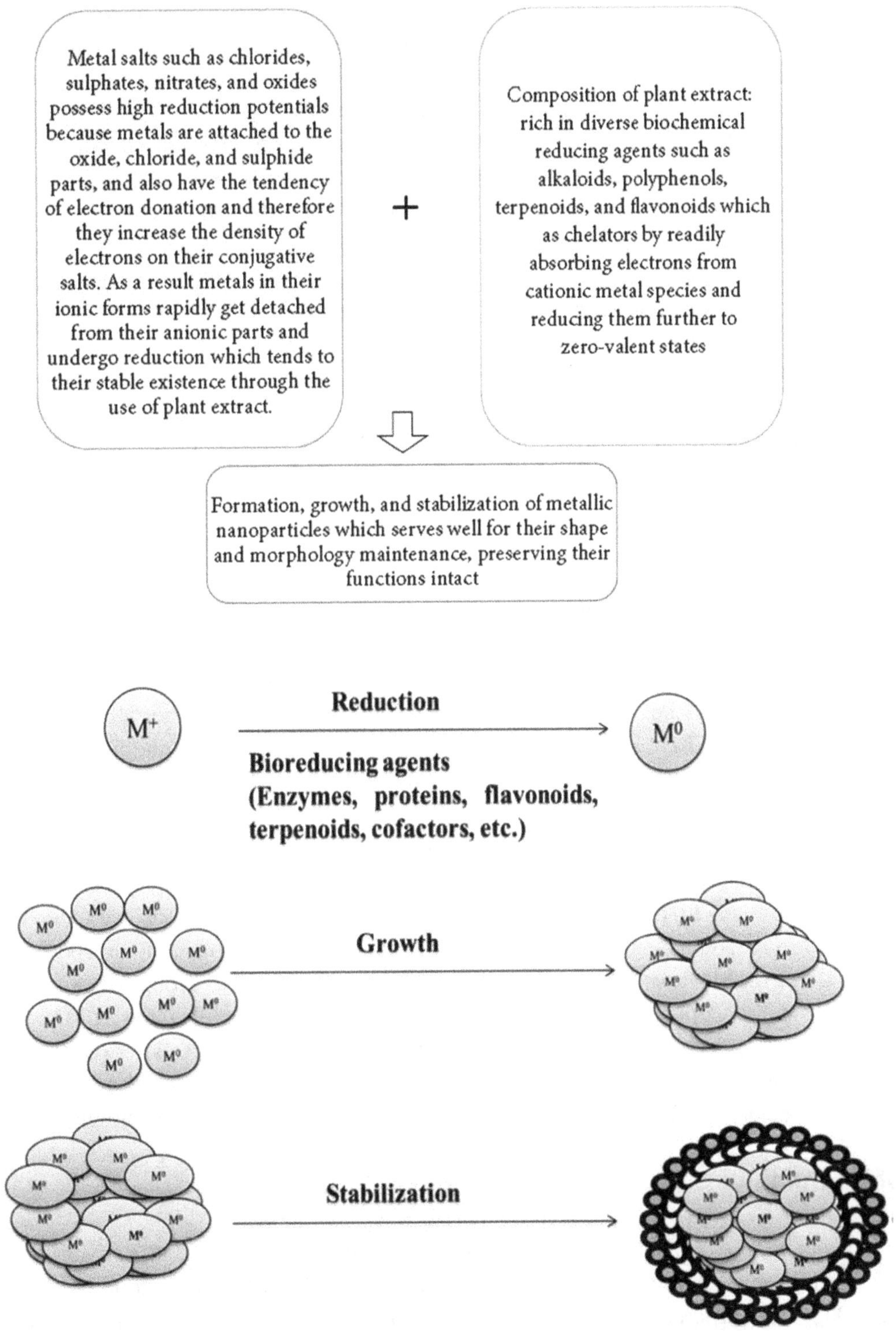

Role of Phenolic Acids

Phenolic acids are vital plant secondary metabolites, derivatives of benzoic and cinnamic acids, and consist of an aromatic phenolic ring and functional group (carboxylic acid; C6-C1 skeleton), in which the ring exhibits antioxidant potential as well as the metal-chelating ability (Kumar & Goel, 2019). As reducing agents, phenolic acids like ellagic acid, protocatechuic acid, gallic acid, and caffeic acid have been found to play a vital role in reducing metallic nanoparticles from their ionic form (Aromal et al., 2012; Edison & Sethuraman, 2012). It has been reported that phenolic hydroxyl groups of gallic acid are capable of forming a transitional complex of Ag+ ions during the synthesis of AgNPs synthesis, which consequently modifies to quinine through the oxidation process to generate silver nanoparticles (Edison & Sethuraman, 2012; S. Mukherjee & Patra, 2017). Similarly, phytochemically synthesized gold nanoparticles showed extended stability with excellent uniformity (size of ~20 nm) by developing interaction with isoflavone, protocatechuic acids, and gallic acid (Lee et al., 2011). Leaf extracts of *Eucalyptus tereticornis, Rosmarinus officinalis,* and *Melaleuca nesophil* were used for synthesizing Iron nanoparticles at the range of 50 to 90 nm, in which polyphenol work as the key factor for synthesizing NPs (organic character containing surface; iron–polyphenol nanoparticles) by reacting with positively charged Fe ions (Z. Wang et al., 2014).

Role of Terpenoids

Isoprenoids or terpenoids are naturally occurring and structurally diverse classes of essential oils that comprise different types and number of cyclic structures and oxygen-containing additional functional groups. Essential oils mainly consist of hydrocarbons such as terpenoids, terpenes, or sesquiterpenes, and oxygenated compounds i.e., phenol ethers, lactones, phenols, ketones, ether alcohols, esters, and aldehydes (Ashour et al., 2010; Edris, 2007). Many aforementioned pieces of the literature suggested that low molecular aliphatic and aromatic terpenoids, sesquiterpenoids, and monoterpenoids could be the leading plant-derived secondary metabolites for playing a potential role for metal and metal oxide nanoparticle biosynthesis, where their functional lactones, aldehydes, ketones, alcohols act not only as reducing agent but also capping and complexing agents by involving –C=O in a redox reaction (Ajitha et al., 2015; Ganaie et al., 2016; Song et al., 2009). For instance, sesquiterpenoids and monoterpenoids of essential oils are vital donors in the fabrication of silver nanoparticles (Song & Kim, 2009). In a study it was reported that the bioreduction of Ag ions was influenced by the –OH groups of geraniol and citronellol (terpenoids) available in *Pelargonium graveolens* leaf extract (Shankar et al., 2003). Moreover, three terpenoids i.e., methyl chavicol, eugenol, and linalool present in the bark extract of *Cinnamomum zeylanicum*, and eugenol in clove extract of *Syzygium aromaticum* L. demonstrate excellent metal reduction potential, especially gold and silver nanoparticles (Sathishkumar et al., 2009; Singh et al., 2010).

Role of Vitamins

Using only water instead of insoluble templates, ascorbic acid (vitamin C) and its derivatives (Jahan et al., 2021b; Nadagouda & Varma, 2007a, 2007b), vitamin B_1, and riboflavin (vitamin B_2) (Nadagouda & Varma, 2006) from different plant extract can be used as both reducing and capping agents, which provide facile and eco-friendly one-pot fabrication of metallic nanoparticles, nanoballs, nanowires, nanorods, nanospheres, aligned nanobelts, and nanospheres (Nadagouda & Varma, 2006; Varma, 2012).

Vitamin B_2 or riboflavin is a water-soluble natural compound, widely available in food products, especially in dark-colored vegetables (Powers, 2003). It consists of two important structures, the isoalloxazine ring (characteristic of flavin family) and four hydroxyl groups containing ribitol side chain (Yamashita et al., 2002). The ring structure takes part in the formation of all three redox states i.e., fully reduced, one-electron reduced, and fully oxidized, which are capable of transferring electrons (Jaiswal et al., 2002). The reduction potential (RP) of riboflavin is -0.3 V versus SCE which is adequate to reduce silver (Ag, RP 0.80 V versus SCE) and palladium (Pd, 0.915 V versus SCE) (Massey, 2000). The possible mechanism for the reduction of noble nanostructures by riboflavin (vitamin B_2) from their salt solution could be followed by the sequential steps: i) formation of a complex structure of noble metal salts and riboflavin; ii) immediate reduction of zero-valent metals from their ionic form and formation of capping with oxidized vitamin structure for stabilization; iii) self-assembly of nanoparticles to form different nanostructure (Nadagouda & Varma, 2008b, 2008a).

Role of Low and High Molecular Proteins

By donating electrons to the reaction medium, proteins act a potential role in the phytogenic synthesis of nanoparticles, especially AgNPs where the donated electrons can easily react with silver ions, and subsequently also play role in the stabilization process during the formation of AgNPs (Li et al., 2007). In a study for the formation of AgNPs, phosphoprotein e.g., casein was successfully utilized in the process of reduction and stabilization of NPs (Ashraf et al., 2013). SDS-PAGE analysis was utilized for confirming the significant role of plant-based proteins in NPs fabrication by evaluating concentrated supernatant of *Olax scandens*-mediated AgNPs and leaf extracts of *O. scandens*. The outcomes revealed that some low molecular weight protein bands of leaf extracts were absent in nanoparticle supernatant, conforming the utilization of these proteins during the formation of AgNPs (S. Mukherjee et al., 2014). Similarly, involvement of proteins in metallic NPs biosynthesis was also observed for gold nanoparticles synthesis using *Glycine max* seed extract as a phytochemical reservoir (Shukla et al., 2008).

Role of Carbohydrates, Sugars, Ketones, and Acids

Several studies also confirmed that carbohydrates play a vital role in NPs synthesis. Biosyntheses of metallic nanoparticles were successfully accomplished by using starch, hemicellulose, sugar, and glucose as reducing and stabilizing agents (Raveendran et al., 2003). In recent times, silver nanoparticles by using fructose and glucose, and their specific role in the synthesis process have been evaluated where it was observed that fructose exhibited limited reduction potential although glucose displayed excellent reducing potential for AgNPs production (Shah et al., 2015).

Role of Polyphenols From Agricultural Residues

Polyphenolic antioxidants from plant extracts such as coffee and tea extracts (Nadagouda & Varma, 2008b), different natural renewable resources such as carboxymethyl cellulose or other biodegradable polymers (Nadagouda & Varma, 2007a), reducing sugars (García-Serrano et al., 2008) from agronomic residues such as red grape residual waste during wine preparation (Baruwati & Varma, 2009) are capable of synthesizing and stabilizing metal and metal oxide nanoparticles, and polymeric nanostructures.

Table 1. List of metal and metal oxide nanoparticles synthesized using medicinal plant

Plant	Optimized Reaction Conditions		Plant parts Used for the Synthesis	Synthesized Nanoparticles	Morphology of NPs		References
	Time	Temperature (C°)			Size (nm)	Shapes	
Acalypha indica	1 h	30°C	whole plant	ZnS NPs	17.87 nm	spherical	(M. S. Kumar et al., 2019)
Acantholimon serotinum	6 h	60°C	aerial parts	ZnO NPs	20-80 nm	irregular & spherical	(Karimzadeh et al., 2020)
Achillea wilhelmsii	few minutes	room temp.	fresh flowers	CdO NPs	35 nm	irregular & spherical	(Andeani & Mohsenzadeh, 2013)
Aloe vera	5 h	room temp.	fresh leaves	Au NPs	50 - 350 nm	triangular to hexagonal	(Chandran et al., 2006)
Aloe vera	several hours	60°C	leaves	Indium oxide (In_2O_3) NPs	5-50 nm	spherical	(Maensiri et al., 2008)
Artemisia marshalliana Sprengel	-	room temp.	aerial parts	Ag NPs	5-50 nm	mostly spherical	(Ardestani et al., 2016)
Artemisia persica	15 s	micro. heating (850 W)	aerial parts	Cd NPs	11.2-18.6 nm	hexagonal	(Shakibaie et al., 2021)
Artemisia quttensis Podlech	2 min	room temp.	aerial parts	Ag NPs	5-25 nm	spherical	(Ghanbar et al., 2017)
Azadirachta indica	4 h	room temp.	fresh leaves	Ag NPs	20 nm	spherical	(Prathna et al., 2011)
Azadirachta indica	2 h	room temp.	fresh leaves	Au NPs	50-100 nm	mostly triangular, hexagonal at low percentage	(Shankar et al., 2004)
Beta vulgaris	10 min	80°C	whole plants	Ag NPs	~20 nm	spherical, circular and triangular	(Venugopal et al., 2017)
Cacumen Platycladi	10 min	90°C	fresh leaves	Pt NPs	2.4±0.8 nm	spherical	(Zheng et al., 2013)
Cassia fistula	30 min	70°C	fresh leaves	ZnO NPs	68.10 nm		
Catha edulis	-	80°C	fresh leaves	CuO NPs	18.20-28.10 nm	spherical	(Andualem et al., 2020)
Cinnamomum camphora	1 h	30°C	dried leaf powder	Ag NPs	55- 80 nm	mostly spherical; few triangular	(Huang et al., 2007)
Citrullus colocynthis	24 h	35°C	stem derived dried callus	Ag NPs	31 nm	spherical	(Satyavani et al., 2011)
Coleus aromaticus	24 h	room temp.	fresh leaves	TiO_2 NPs	16.5 nm	hexagonal	(Narayanan et al., 2021)
Cuminum cyminum	15 min	90 °C (microwave heat)	dried seeds	Ag NPs	1.84-20.57 nm	spherical	(Jahan, Erci, & Isildak, 2021a)
Curcuma longa	72 h	30°C	dried tuber powder	Palladium nanocrystal (Pd NPs)	15-20 nm	spherical	(Sathishkumar, Sneha, & Yun, 2009)
Cymbopogon flexuosus	12 h	room temp.	fresh leaves	Au NPs	25-80 nm	triangular along with spherical particles	(Rai et al., 2006)
Gardenia jasminoides Ellis	24 h	70°C	fruits	Pd NPs	3-5 nm	spherical, rod & polyhedral	(Jia et al., 2009)
Gloriosa superba	5 h	100°C	fresh tubers	Pt NPs	0.8-3 nm	spherical	(Rokade et al., 2018)
Gloriosa superba	5 h	100°C	fresh tubers	Pd NPs	5-8 nm	spherical	(Rokade et al., 2018)
Lippia adoensis	2 h	room temp.	aerial parts	ZnO NPs	19.78 nm	spherical & hexagonal	(Demissie et al., 2020)
Macrotyloma uniflorum	3 h	room temp.	seeds	Au NPs	14–17 nm	spherical	(Aromal et al., 2012)
Manihot esculenta	12 h	room temp.	fresh leaves	MgO NPs	36.7 nm	hexagonal	(Essien et al., 2020)

continued on following page

Table 1. Continued

Plant	Optimized Reaction Conditions		Plant parts Used for the Synthesis	Synthesized Nanoparticles	Morphology of NPs		References
	Time	Temperature (C°)			Size (nm)	Shapes	
Medicago sativa	0.5 h	21°C	whole plant	Fe_2O_3 NPs	2-10 nm	crystal-like structure	(Herrera-Becerra et al., 2008)
Mentha pulegium	-	room temp.	dried leaves	Ag NPs	5-50 nm	spherical to irregular	(Kelkawi et al., 2017)
Mukia maderaspatana (L.) M.	30 min	70°C	fresh leaves	Ag NPs	20-50 nm	irregular shaped	(Devi & Sathishkumar, 2017)
Mukia maderaspatana (L.) M.	4 hours	80°C	fresh leaves	Au NPs	20-50 nm	sphere-shaped, circular and triangular	(Devi & Sathishkumar, 2017)
Plumbago zeylanica	over night	room temp.	root bark	Ag NPs	80-98 nm	spherical	(Prasannaraj et al., 2016)
Psidium guajava L.	2 h	room temp.	fresh leaves	Ag NPs	20-35 nm	spherical	(Wang et al., 2018)
Rosmarinus officinalis L.	4 h	70°C	fresh flowers	MgO nano-flowers	≥20 nm	round-shaped	(Abdallah et al., 2019)
Salvia hispanica	12 h	80°C	seeds	NiO NPs	30 nm	spherical	(Sabouri et al., 2021)
Semecarpus anacardium	over night	room temp.	nuts	Ag NPs	60-95 nm	cuboid	(Prasannaraj et al., 2016)
Sesbania grandiflora	24 h	37°C	fresh leaves	Ag NPs	22 nm	spherical	(Jeyaraj et al., 2013)
Tanacetum vulgare	10 min	room temp.	fruits	Ag NPs, Au NPs	10-40 nm	mostly spherical & triangular, some hexagonal	(Dubey et al., 2010)
Terminalia arjuna	over night	room temp.	stem bark	Ag NPs	34-70 nm	spherical	(Prasannaraj et al., 2016)
Terminalia chebula Retz.	10 h	room temp.	fruits	Ag NPs	25 nm	irregular	(Edison & Sethuraman, 2012)
Tridax procumbens	2 h	room temp.	fresh leaves	ZnS NPs	~ 40 nm	spherical, rod & hexagonal	(Sathishkumar et al., 2017)

To utilize diverse plant secondary metabolites for metallic nanoparticles synthesis, exploiting medicinal plants in plant extract preparation is a more proficient, reliable, and profitable approach since they exhibit many therapeutics potentials due to having the capacity of secreting diverse active secondary phytocompounds (Chandra et al., 2020). Until now, many useful plants with medicinal values have been utilized for synthesizing metal and metal oxide NPs such as platinum, gold, silver, metallic copper and copper oxide, zinc oxide, heavy metal oxide, etc. (Table 1).

For consuming and managing plant wastes and biomass effectively, the biosynthesis of metal and metal oxide nanoparticle by exploiting plant-based biological wastes and agro-waste resources is a sustainable and environmentally friendly approach. This is due to fact that these waste products are regularly discharged into the environment from large-scale food production and agro-processing industry, crop harvesting procedures, and domestic usages. These wastes have their limitation to be used as animal feeds since they are poor in nutritional quality or might comprise anti-nutritional factors (Lateef et al., 2008). However, studies have indicated that these biodegradable wastes are rich in diverse biocompounds i.e., vitamins, phenolics, terpenoids, tannins, saponins, polysaccharides, proteins, enzymes, amino acids, and alkaloids, which could be a great source in green nanotechnology for nanoparticle synthesis (Adelere &

Table 2. List of metal and metal oxide nanoparticles synthesized using plant-based biological wastes

Waste Materials of Plants	Plant	Synthesized Nanoparticles	Optimized Reaction Conditions		Morphology of NPs		References
			Time	Temperature (C°)	Size (nm)	Shapes	
Agroforestry waste flower petals	*Moringa oleifera* Lam.	Au NPs	1 h	room temp.	3–5 nm	mostly spherical, few hexagonal & triangular	(Anand et al., 2015)
Air dried banana peels	*Musa paradisiaca*	Pd NPs	3 min	80°C	50 nm	irregular crystalline	(Bankar et al., 2010)
Air dried banana peels	*Musa paradisiaca*	Mn_3O_4 NPs	5 h	room temp.	20-50 nm	spherical	(Yan et al., 2014)
Bilberry wastes	*Vaccinium myrtillus L.*	Ag NPs	3-5 h	25°C	16 nm	spherical	(Baiocco et al., 2016)
Coir powder (mesocarp layer of coconut)	*Cocos nucifera*	Ag NPs	1 h	60°C	23 ± 2 nm	spherical	(Roopan et al., 2013)
De-oiled jatropha waste (DJW).	*Jatropha curcas* L.	Au NPs	1 h	room temp.	~14 nm	spherical, hexagonal & triangle	(Kanchi et al., 2018)
Dried cocoa pod sheel/ husk powder	*Theobroma cacao*	Ag NPs	Few minutes	30±2°C	4-32 nm	spherical	(Lateef et al., 2016)
Dried rinds (Watermelon)	*Citrullus lanatus* (Thunb.) Matsum. & Nakai	Pd NPs	24 h	30°C	~96.4 nm	spherical	(Lakshmipathy et al., 2015)
Dried rinds (Watermelon)	*Citrullus lanatus* (Thunb.) Matsum. & Nakai	Fe_3O_4 magnetic NPs	3 h	80°C	2-20 nm	spherical	(Prasad et al., 2016)
Flower petals	*Rosa santana*	Ag NPs	25 min	90 °C (microwave heat)	6.52-25.24nm	spherical	(Jahan et al., 2019)
Flower petals	*Rosa damascena*	Au NPs	2 min	room temp.	15.3 nm	quasi-spherical	(Ghoreishi et al., 2011)
Flower petals	*Rosa damascena*	Ag NPs	5 min	room temp.	18.3 nm	spherical	(Ghoreishi et al., 2011)
Fresh fruit peels (Pomegranate)	*Punica granatum*	Pt NPs	30 min	90°C	16-23 nm	spherical	(Dauthal & Mukhopadhyay, 2015)
Fresh mango peels	*Mangifera indica*	Au NPs	15 min	100°C	6-18 nm	spherical	(Yang et al., 2014)
Fruit peel powder (Pomegranate)	*Punica granatum*	Ag NPs	-	room temp.	3-13 nm	spherical	(Nasiriboroumand et al., 2018)
Fruit peels	*Citrus sinensis*	Ag NPs	75 min	25°C	35 nm	spherical	(Kaviya et al., 2011)
Fruit peels	*Citrus sinensis*	Ag NPs	45 min	60°C	10 nm	spherical	(Kaviya et al., 2011)
Fruit peels	*Citrus unshiu*	Ag NPs	10 min	30°C	5-20 nm	spherical	(Basavegowda & Rok Lee, 2013)
Fruit peels	*Annona squamosa*		4 h	25°C & 60°C	35 ± 5 nm	irregular	(Kumar et al., 2012)
Fruit peels (Rambutan)	*Nephelium lappaceum L.*	Ag NPs	3 h	room temp.	132.6 ± 42 nm	hexagonal & triangle	(B. Kumar et al., 2015)
Fruit peels (Rambutan)	*Nephelium lappaceum L.*	NiO nanocrystals	2 h	80°C	50 nm	-	(Yuvakkumar et al., 2014)
Kola seed shell	*Cola nitida*	Ag NPs	2 h	30±2°C	5-40 nm	spherical	(Lateef et al., 2015)
Oak fruit hull (Jaft)	*Quercus infectoria*	Ag NPs	4 h	45°C	40nm	spherical	(Heydari & Rashidipour, 2015)
Palm oil mill effluent (POME)	*Elaeis guineensis*	Au NPs	48 h	room temp.	18.75±5.96 nm	mostly spherical, few hexagonal & triangular	(Gan et al., 2012)
Peanut pods	*Arachis hypogaea* L.	Ag NPs	1 h	28°C	20-50 nm	spherical	(Velu et al., 2015)
Pomace or grape marc (stalk, skin, & seed of grape)	*Vitis vinifera*	Au NPs	5 min	40°C	20-25 nm	spherical to quasi-spherical	(Krishnaswamy et al., 2014)
Rice bran	*Oryza sativa* L.	Au NPs	1 h	45°C	50-100 nm	spherical	(Malhotra et al., 2014)

continued on following page

Table 2. Continued

Waste Materials of Plants	Plant	Synthesized Nanoparticles	Optimized Reaction Conditions		Morphology of NPs		References
			Time	Temperature (C°)	Size (nm)	Shapes	
Spent coffee grounds	*Coffea arabica* L.	Ag NPs	3-5 h	25°C	10-20 nm	spherical	(Baiocco et al., 2016)
Wheat bran xylan	*Triticum aestivum* L.	Ag NPs	Cycle 1: 5 min at 30°C Cycle 2: 30 min at 100°C		20-45 nm	spherical	(Harish et al., 2015)

Lateef, 2016). Successful fabrication of a wide variety of metallic NPs has been completed through the utilization of agro-wastes, where extracts from these biological entities played as reducing and stabilizing agents (Table 2). For instance, grape marc, palm oil mill effluent, rice bran, wheat bran, fruit peels, and seeds, corn cob, *Cocos nucifera* coir have been utilized as plant extract for complete reduction and stabilization of metallic NPs (Basavegowda & Rok Lee, 2013; Gan et al., 2012; Harish et al., 2015; Krishnaswamy et al., 2014; Malhotra et al., 2014; Roopan et al., 2013).

FACTORS AFFECTING GREEN SYNTHESIS OF NANOPARTICLES

The production of metallic nanoparticles is an extremely sensitive procedure since diverse environmental factors influence the fabrication processes. Quite a few factors can affect the creation, categorization, and appliance of nanoparticles while manufacturing the metallic nanoparticles. Among them, the most important physical and chemical parameters are the pH of the reaction mixture, reaction temperature, duration of reaction and agitation, pressure, reactant concentration, the content of the reducing agent, environment, proximity, and above all the protocols that are followed for the synthesis process (Baker et al., 2013). The sections following present a glimpse of some of those influential factors that play an important role in nanoparticles synthesis.

Particular Method or Technique

Different schemes have been intended for the fabrication of metal NPs. This could include the physical techniques of using mechanical and photolytic procedures, the chemical protocols with the help of various organic or inorganic chemicals as well as the biological protocols using different kinds of living organisms and bio-based raw materials. However, each of these procedures has its certain advantages and disadvantages. Providing that the biological protocols mainly use nontoxic and environmentally benign materials, they are economically cheap, eco-friendly, safe, have fewer side effects as well as more reliable and convenient than the protocols used in traditional methods (Faried et al., 2016). Additionally, the preference of technique used to purify the fabricated metallic nanoparticles can influence nanoparticle quantity and quality.

Type of Solvent Used for the Synthesis

In the case of the green synthesis of NPs, the nature of the solvent may influence the size and shapes of synthesized NPs. For instance, dried and powdered leaves extracts of squaw mint (*Mentha pulegium*) in three different solvents i.e., absolute methanol, ethanol solution (70% v/v), and distilled water were used to synthesize silver nanoparticles, where the aqueous extract synthesized the smallest AgNPs ($\geq$ 20 nm) (Kelkawi et al., 2017).

Chemical Constitute of Reducing and Stabilizing Agent

A variety of living systems (e.g. plants) are being exercised as reducing, synthesizing, and stabilizing mediators, meant for the nanoparticle syntheses. These living systems are rich in secondary metabolites. However, the substances of these secondary metabolites diverge according to their types as well as the procedures, which have been followed for their extraction (Park et al., 2011).

Influence and Effect of pH

The pH value of the redaction solution is very influential in the course of nanoparticle fabrications. Studies have shown that the pH of the solution medium has the propensity to construct variability in the morphology of synthesized nanomaterials (Nayaka, 2020). On the contrary of the high pH values, larger particles have a propensity to be fabricated particularly at lower acidic pH values (Dubey et al., 2010). There is an example, conversely, of rod-shaped gold nanoparticles fabrication by using the common oat (*Avena sativa*). In this case, it has been found that larger-sized gold nanoparticles (25 nm to 85 nm) were synthesized at pH 2, whereas comparatively smaller-sized gold nanoparticles (5 to 20 nm) were synthesized at pH 3 to 4 (Armendariz et al., 2004). Therefore, it is strongly evident that the dimension of NPs may possibly be monitored by altering the pH of the reaction solution. On the other hand, some studies also present the effect of pH on the structure and morphology of the synthesized silver nanoparticles (Heinrich, 2011).

Influence of Reaction Temperature

In all kinds of methods, the temperature is also a vital cause that influences the morphology and yield of fabricated nanoparticles. The highest temperature (>350°C) is needed in physical processes, whereas less than 350°C requires chemical schemes (Prathna et al., 2011). Generally, using green technology for the synthesis of nanoparticles requires ambient temperature or temperatures less than 100°C, and the temperature during synthesis established the nature of the nanoparticles (Rai et al., 2006). For instance, by using the sweet orange peel extract (*Citrus sinensis*), Ag nanoparticles with a mean dimension of just about 35 nm have been produced at a reaction temperature of 25°C. However, the average particle size in this synthesis was reduced to 10 nm while the reaction heat was amplified to 60°C (Kaviya et al., 2011). On the other hand, it has been illustrated that higher temperatures moreover, influence the structural form of gold nanoparticles. For example, rod-like and plate-like AuNPs have been formed at higher temperature whereas spherical-shaped NPs have been produced predominantly at lower temperatures (Gericke & Pinches, 2006).

Influence of Reaction Time

While using green technology, the reaction condition like time also comprises an immense effect upon the quality and type of synthesized metallic NPs. The variations in the time may affect the synthesis process in many ways. For example, particles might shrivel or grow during long storage; the aggregation of particles might also occur due to long-time storage (Darroudi, 2011). There is also an example that a quick color change occurred within 2 minutes while synthesizing spherical silver nanoparticles (AgNPs) using pineapple (*Ananas comosus*) extract. The reaction sustained up to 5 min and however, subsequently only an insignificant difference in color may possibly be monitored (Shah et al., 2015).

Influence of Pressure

Pressure also becomes an important factor while synthesizing nanoparticles. The shape and size can be affected by the pressure applied to the reaction medium. It has been found that the metal ions reduction rate occurs much faster at ambient pressure conditions using biological agents (Tran et al., 2013).

Effects of Reactant Concentration

In green chemistry, the formation of metallic nanoparticles has extensively been influenced by the concentration and density of bio-particles that occur in plant extracts. A recent study has demonstrated the difference in concentration of sundried camphor leaf extract (*Cinnamomum camphora*) in the synthesis medium; possibly possess a significant influence on the morphology of Ag and Au synthesized nanoparticles (Huang et al., 2007). Similarly, another study shows that the ratio of gold triangular plates to spherical nanoparticles was influenced depending on the amount of chloroaurate ions containing the extract from the juicy leaves of *Aloe vera* (Asphodelaceae) in the synthesis medium of the fabrication process (Chandran et al., 2006)

Environment

The surrounding environment especially humidity also operates as an important factor for the construction of synthesized NPs. Additionally; the environment also influences the chemical and physical structure of the synthesized nanoparticles. In the biological system, a single nanoparticle would quickly become core-shell nanoparticles by reacting with other materials. Quick changes might also occur when it absorbs materials by the oxidation or corrosion process in its environment (Sarathy et al., 2008). For instance, the crystalline nature of the zinc sulfide nanoparticles changed immediately when their surrounding has been changed from a wet environment into a dry circumstance (Khalkhali et al., 2015).

FUTURE RESEARCH DIRECTIONS

The identification and categorization of phytocompounds responsible for reducing, stabilizing, and capping of nanoparticles during the synthesis process are still unclear though naturally occurring biomaterials predominantly derived from plants and plant-based residues have been explored for metallic nanoparticle fabrication; thus, further detail inquiries are required to determine the exact roles of different biomolecules as well as to understand more specific mechanism of the photosynthesis process. Since no additional chemicals are used as a stabilizer during photosynthesis of NPs, stability of these synthesized NPs for a long period of storage time without any morphological changes is a concerning issue and it is necessary to resolve this concern. Regarding the safety concern of public health and environment impacts, biocompatibility and the toxicity of bio fabricated NPs are additional circumstances that are still not explored properly, and therefore, more detailed *in vitro* and *in vivo* cytotoxicity studies for toxicity profiling of biosynthesized NPs, and multidisciplinary practical applications are necessary to reveal practical accuracy of these nanomaterials based on trial, test and performance.

Therefore, more comprehensive, unified, efficient, and systemic assessments and investigations are required for large-scale commercial phytonanofabrication of well-defined and functionalized metal and metal oxide nanoparticles by considering existing problems and issues, particularly health and environmental risks, which would be more stable, cost-effective, and non-toxic approach, fit for human or societal benefit.

CONCLUSION

Over the last decade, the production of metallic nanoparticles through green chemistry has become an immensely striking and promising research area because a wide range of accessible, natural, non-toxic, and eco-friendly organic compounds from bacteria, fungi, yeast, and plant parts are capable of utilizing as proficient resources for fabricating nanostructured materials. So far, plant-based extracts as reducing, capping, and stabilizing agents have been established to retain extraordinary potentials among them for fabricating metal and metal oxide nanoparticles with desirable features (i.e., controlled shapes, sizes, structures, and other specific characteristics). Hence, this chapter was designed for comprehending the potential of plants, especially medicinal herbs, and plant-based biological wastes, and agro-waste as well as exploring the 'state of the art' of the 'green' synthesis of metal/metal oxide nanoparticles by employing these plant materials. For encountering the significance of phytonanofabrication of metallic NPs, a detailed discussion including the mechanisms of plant-mediated nano-synthesis process and the latest status of this study have been reviewed thoroughly by adding the information from updated and recently published relevant literature.

REFERENCES

Abdallah, Y., Ogunyemi, S. O., Abdelazez, A., Zhang, M., Hong, X., Ibrahim, E., Hossain, A., Fouad, H., Li, B., & Chen, J. (2019). The green synthesis of MgO nano-flowers using Rosmarinus officinalis L. (Rosemary) and the antibacterial activities against Xanthomonas oryzae pv. Oryzae. *BioMed Research International*, *2019*, 1–8. doi:10.1155/2019/9395146 PMID:30906776

Adelere, I. A., & Lateef, A. (2016). A novel approach to the green synthesis of metallic nanoparticles: The use of agro-wastes, enzymes, and pigments. *Nanotechnology Reviews*, *5*(6). Advance online publication. doi:10.1515/ntrev-2016-0024

Ahmad, A., Senapati, S., Khan, M. I., Kumar, R., & Sastry, M. (2003). Extracellular biosynthesis of monodisperse gold nanoparticles by a novel extremophilic actinomycete, Thermomonospora sp. *Langmuir, 19*(8), 3550–3553. doi:10.1021/la0267721

Ahmed, S., Ahmad, M., Swami, B. L., & Ikram, S. (2016). A review on plants extract mediated synthesis of silver nanoparticles for antimicrobial applications: A green expertise. *Journal of Advanced Research, 7*(1), 17–28. doi:10.1016/j.jare.2015.02.007 PMID:26843966

Ajitha, B., Ashok Kumar Reddy, Y., & Sreedhara Reddy, P. (2015). Green synthesis and characterization of silver nanoparticles using Lantana camara leaf extract. *Materials Science and Engineering C, 49*, 373–381. doi:10.1016/j.msec.2015.01.035 PMID:25686962

Akhtar, M. S., Panwar, J., & Yun, Y.-S. (2013). Biogenic synthesis of metallic nanoparticles by plant extracts. *ACS Sustainable Chemistry & Engineering, 1*(6), 591–602. doi:10.1021c300118u

Amooaghaie, R., Saeri, M. R., & Azizi, M. (2015). Synthesis, characterization and biocompatibility of silver nanoparticles synthesized from Nigella sativa leaf extract in comparison with chemical silver nanoparticles. *Ecotoxicology and Environmental Safety, 120*, 400–408. doi:10.1016/j.ecoenv.2015.06.025 PMID:26122733

Anand, K., Gengan, R. M., Phulukdaree, A., & Chuturgoon, A. (2015). Agroforestry waste Moringa oleifera petals mediated green synthesis of gold nanoparticles and their anti-cancer and catalytic activity. *Journal of Industrial and Engineering Chemistry, 21*, 1105–1111. doi:10.1016/j.jiec.2014.05.021

Andeani, K. (2013). Javad, & Mohsenzadeh, S. (2013). Phytosynthesis of cadmium oxide nanoparticles from Achillea wilhelmsii flowers. *Journal of Chemistry, 1–4*, 1–4. Advance online publication. doi:10.1155/2013/147613

Andualem, W. W., Sabir, F. K., Mohammed, E. T., Belay, H. H., & Gonfa, B. A. (2020). Synthesis of copper oxide nanoparticles using plant leaf extract of Catha edulis and its antibacterial activity. *Journal of Nanotechnology, 2020*, 1–10. doi:10.1155/2020/2932434

Ardestani, M. S., Sadat Shandiz, S. A., Salehi, S., Ghanbar, F., Darvish, M. R., Mirzaie, A., & Jafari, M. (2016). Phytosynthesis of silver nanoparticles using Artemisia marschalliana Sprengel aerial part extract and assessment of their antioxidant, anticancer, and antibacterial properties. *International Journal of Nanomedicine, 1835*, 1835. Advance online publication. doi:10.2147/IJN.S99882

Armendariz, V., Herrera, I., peralta-videa, J. R., Jose-yacaman, M., Troiani, H., Santiago, P., & Gardea-Torresdey, J. L. (2004). Size controlled gold nanoparticle formation by Avena sativa biomass: Use of plants in nanobiotechnology. *Journal of Nanoparticle Research, 6*(4), 377–382. doi:10.100711051-004-0741-4

Aromal, S. A., Vidhu, V. K., & Philip, D. (2012). Green synthesis of well-dispersed gold nanoparticles using Macrotyloma uniflorum. *Spectrochimica Acta. Part A: Molecular and Biomolecular Spectroscopy, 85*(1), 99–104. doi:10.1016/j.saa.2011.09.035 PMID:22018585

Ashour, M., Wink, M., & Gershenzon, J. (2010). Biochemistry of terpenoids: Monoterpenes, sesquiterpenes and diterpenes. In M. Wink (Ed.), *Biochemistry of Plant Secondary Metabolism* (pp. 258–303). Wiley-Blackwell. doi:10.1002/9781444320503.ch5

Ashraf, S., Abbasi, A. Z., Pfeiffer, C., Hussain, S. Z., Khalid, Z. M., Gil, P. R., Parak, W. J., & Hussain, I. (2013). Protein-mediated synthesis, pH-induced reversible agglomeration, toxicity and cellular interaction of silver nanoparticles. *Colloids and Surfaces. B, Biointerfaces*, *102*, 511–518. doi:10.1016/j.colsurfb.2012.09.032 PMID:23107938

Baiocco, D., Lavecchia, R., Natali, S., & Zuorro, A. (2016). Production of metal nanoparticles by agro-industrial wastes: A green opportunity for nanotechnology. *Chemical Engineering Transactions*, *47*, 67–72. doi:10.3303/CET1647012

Baker, S., Rakshith, D., Shanthamma Kavitha, K., Santosh, P., Umapathy Kavitha, Y., Rao, Y., & Satish, S. (2013). Plants: Emerging as nanofactories towards facile route in synthesis of nanoparticles. *BioImpacts*. Advance online publication. doi:10.5681/BI.2013.012 PMID:24163802

Bankar, A., Joshi, B., Kumar, A. R., & Zinjarde, S. (2010). Banana peel extract mediated novel route for the synthesis of palladium nanoparticles. *Materials Letters*, *64*(18), 1951–1953. doi:10.1016/j.matlet.2010.06.021

Baruwati, B., & Varma, R. S. (2009). High value products from waste: Grape pomace extract—A three-in-One package for the synthesis of metal nanoparticles. *ChemSusChem*, *2*(11), 1041–1044. doi:10.1002/cssc.200900220 PMID:19842157

Basavegowda, N., & Rok Lee, Y. (2013). Synthesis of silver nanoparticles using Satsuma mandarin (Citrus unshiu) peel extract: A novel approach towards waste utilization. *Materials Letters*, *109*, 31–33. doi:10.1016/j.matlet.2013.07.039

Chandra, H., Kumari, P., Bontempi, E., & Yadav, S. (2020). Medicinal plants: Treasure trove for green synthesis of metallic nanoparticles and their biomedical applications. *Biocatalysis and Agricultural Biotechnology*, *24*, 101518. doi:10.1016/j.bcab.2020.101518

Chandran, S. P., Chaudhary, M., Pasricha, R., Ahmad, A., & Sastry, M. (2006). Synthesis of gold nanotriangles and silver nanoparticles using Aloe vera plant extract. *Biotechnology Progress*, *22*(2), 577–583. doi:10.1021/bp0501423 PMID:16599579

Darroudi, M. (2011). Time-dependent effect in green synthesis of silver nanoparticles. *International Journal of Nanomedicine*, *677*, 677. Advance online publication. doi:10.2147/IJN.S17669 PMID:21556342

Dauthal, P., & Mukhopadhyay, M. (2015). Biofabrication, characterization, and possible bio-reduction mechanism of platinum nanoparticles mediated by agro-industrial waste and their catalytic activity. *Journal of Industrial and Engineering Chemistry*, *22*, 185–191. doi:10.1016/j.jiec.2014.07.009

de Souza Farias, S. A., da Costa, K. S., & Martins, J. B. L. (2021). Analysis of conformational, structural, magnetic, and electronic properties related to antioxidant activity: Revisiting flavan, anthocyanidin, flavanone, flavonol,isoflavone, flavone, and flavan-3-ol. *ACS Omega*, *6*(13), 8908–8918. doi:10.1021/acsomega.0c06156 PMID:33842761

Demissie, M. G., Sabir, F. K., Edossa, G. D., & Gonfa, B. A. (2020). Synthesis of zinc oxide nanoparticles using leaf extract of Lippia adoensis (Koseret) and evaluation of its antibacterial activity. *Journal of Chemistry*, *2020*, 1–9. doi:10.1155/2020/7459042

Devi, G. K., & Sathishkumar, K. (2017). Synthesis of gold and silver nanoparticles using Mukia maderaspatna plant extract and its anticancer activity. *IET Nanobiotechnology / IET*, *11*(2), 143–151. doi:10.1049/iet-nbt.2015.0054 PMID:28476996

Du, L., Jiang, H., Liu, X., & Wang, E. (2007). Biosynthesis of gold nanoparticles assisted by Escherichia coli DH5α and its application on direct electrochemistry of hemoglobin. *Electrochemistry Communications*, *9*(5), 1165–1170. doi:10.1016/j.elecom.2007.01.007

Dubey, S. P., Lahtinen, M., & Sillanpää, M. (2010). Tansy fruit mediated greener synthesis of silver and gold nanoparticles. *Process Biochemistry*, *45*(7), 1065–1071. doi:10.1016/j.procbio.2010.03.024

Edison, T. J. I., & Sethuraman, M. G. (2012). Instant green synthesis of silver nanoparticles using Terminalia chebula fruit extract and evaluation of their catalytic activity on reduction of methylene blue. *Process Biochemistry*, *47*(9), 1351–1357. doi:10.1016/j.procbio.2012.04.025

Edris, A. E. (2007). Pharmaceutical and therapeutic Potentials of essential oils and their individual volatile constituents: A review. *Phytotherapy Research*, *21*(4), 308–323. doi:10.1002/ptr.2072 PMID:17199238

Essien, E. R., Atasie, V. N., Okeafor, A. O., & Nwude, D. O. (2020). Biogenic synthesis of magnesium oxide nanoparticles using Manihot esculenta (Crantz) leaf extract. *International Nano Letters*, *10*(1), 43–48. doi:10.100740089-019-00290-w

Faried, M., Shameli, K., Miyake, M., Hajalilou, A., Zamanian, A., Zakaria, Z., Abouzari-lotf, E., Hara, H., Ahmad Khairudin, N. B. B., & Binti Mad Nordin, M. F. (2016). A green approach for the synthesis of silver nanoparticles using ultrasonic radiation's times in sodium alginate media: Characterization and antibacterial evaluation. *Journal of Nanomaterials*, *2016*, 1–11. doi:10.1155/2016/4941231

Gan, P. P., Ng, S. H., Huang, Y., & Li, S. F. Y. (2012). Green synthesis of gold nanoparticles using palm oil mill effluent (POME): A low-cost and eco-friendly viable approach. *Bioresource Technology*, *113*, 132–135. doi:10.1016/j.biortech.2012.01.015 PMID:22297042

Ganaie, S. U., Abbasi, T., & Abbasi, S. A. (2016). Rapid and green synthesis of bimetallic Au–Ag nanoparticles using an otherwise worthless weed Antigonon leptopus. *Journal of Experimental Nanoscience*, *11*(6), 395–417. doi:10.1080/17458080.2015.1070311

García-Serrano, J., Pal, U., Herrera, A. M., Salas, P., & Ángeles-Chávez, C. (2008). One-step "Green" synthesis and stabilization of Au and Ag nanoparticles using ionic polymers. *Chemistry of Materials*, *20*(16), 5146–5153. doi:10.1021/cm703201d

Gericke, M., & Pinches, A. (2006). Biological synthesis of metal nanoparticles. *Hydrometallurgy*, *83*(1–4), 132–140. doi:10.1016/j.hydromet.2006.03.019

Ghanbar, F., Mirzaie, A., Ashrafi, F., Noorbazargan, H., Dalirsaber Jalali, M., Salehi, S., & Sadat Shandiz, S. A. (2017). Antioxidant, antibacterial and anticancer properties of phytosynthesised Artemisia quttensis Podlech extract mediated AgNPs. *IET Nanobiotechnology / IET*, *11*(4), 485–492. doi:10.1049/iet-nbt.2016.0101 PMID:28530200

Ghoreishi, S. M., Behpour, M., & Khayatkashani, M. (2011). Green synthesis of silver and gold nanoparticles using Rosa damascena and its primary application in electrochemistry. *Physica E, Low-Dimensional Systems and Nanostructures*, *44*(1), 97–104. doi:10.1016/j.physe.2011.07.008

Gong, N., Shao, K., Feng, W., Lin, Z., Liang, C., & Sun, Y. (2011). Biotoxicity of nickel oxide nanoparticles and bio-remediation by microalgae Chlorella vulgaris. *Chemosphere*, *83*(4), 510–516. doi:10.1016/j.chemosphere.2010.12.059 PMID:21216429

Harish, B. S., Uppuluri, K. B., & Anbazhagan, V. (2015). Synthesis of fibrinolytic active silver nanoparticle using wheat bran xylan as a reducing and stabilizing agent. *Carbohydrate Polymers*, *132*, 104–110. doi:10.1016/j.carbpol.2015.06.069 PMID:26256330

Heinrich. (2011). Factors affecting the geometry of silver nanoparticles synthesis in Chrysosporium tropicum and Fusarium oxysporum. *American Journal of Nanotechnology*, *2*(1), 112–121. doi:10.3844/ajnsp.2011.112.121

Herrera-Becerra, R., Zorrilla, C., Rius, J. L., & Ascencio, J. A. (2008). Electron microscopy characterization of biosynthesized iron oxide nanoparticles. *Applied Physics. A, Materials Science & Processing*, *91*(2), 241–246. doi:10.100700339-008-4420-7

Heydari, R., & Rashidipour, M. (2015). Green synthesis of silver nanoparticles using extract of oak fruit hull (Jaft): Synthesis and in vitro cytotoxic effect on MCF-7 cells. *International Journal of Breast Cancer*, *2015*, 1–6. doi:10.1155/2015/846743 PMID:25685560

Huang, J., Li, Q., Sun, D., Lu, Y., Su, Y., Yang, X., Wang, H., Wang, Y., Shao, W., He, N., Hong, J., & Chen, C. (2007). Biosynthesis of silver and gold nanoparticles by novel sundried Cinnamomum camphora leaf. *Nanotechnology*, *18*(10), 105104. doi:10.1088/0957-4484/18/10/105104

Hussain, I., Singh, N. B., Singh, A., Singh, H., & Singh, S. C. (2016). Green synthesis of nanoparticles and its potential application. *Biotechnology Letters*, *38*(4), 545–560. doi:10.100710529-015-2026-7 PMID:26721237

Iravani, S. (2011). Green synthesis of metal nanoparticles using plants. *Green Chemistry*, *13*(10), 2638. doi:10.1039/c1gc15386b

Ishak, N. A. I., Kamarudin, S. K., & Timmiati, S. N. (2019). Green synthesis of metal and metal oxide nanoparticles via plant extracts: An overview. *Materials Research Express*, *6*(11), 112004. doi:10.1088/2053-1591/ab4458

Jahan, I., Erci, F., Cakir-Koc, R., & Isildak, I. (2021). Microwave-irradiated green synthesis of metallic silver and copper nanoparticles using fresh ginger (Zingiber officinale) rhizome extract and evaluation of their antibacterial potentials and cytotoxicity. *Inorganic and Nano-Metal Chemistry*, *51*(5), 722–732. doi:10.1080/24701556.2020.1808017

Jahan, I., Erci, F., & Isildak, I. (2019). Microwave-assisted green synthesis of non-cytotoxic silver nanoparticles using the aqueous extract of Rosa santana (rose) petals and their antimicrobial activity. *Analytical Letters*, *52*(12), 1860–1873. doi:10.1080/00032719.2019.1572179

Jahan, I., Erci, F., & Isildak, I. (2021a). Rapid green synthesis of non-cytotoxic silver nanoparticles using aqueous extracts of 'Golden Delicious' apple pulp and cumin seeds with antibacterial and antioxidant activity. *SN Applied Sciences*, *3*(1), 94. doi:10.100742452-020-04046-6

Jahan, I., Erci, F., & Isildak, I. (2021b). Facile microwave-mediated green synthesis of non-toxic copper nanoparticles using Citrus sinensis aqueous fruit extract and their antibacterial potentials. *Journal of Drug Delivery Science and Technology*, *61*, 102172. doi:10.1016/j.jddst.2020.102172

Jaiswal, P. V., Ijeri, V. S., & Srivastava, A. K. (2002). *Voltammetric behavior of certain vitamins and their determination in surfactant media*. The Japan Society for Analytical Chemistry. doi:10.14891/analscisp.17icas.0.i741.0

Jeyaraj, M., Sathishkumar, G., Sivanandhan, G., MubarakAli, D., Rajesh, M., Arun, R., Kapildev, G., Manickavasagam, M., Thajuddin, N., Premkumar, K., & Ganapathi, A. (2013). Biogenic silver nanoparticles for cancer treatment: An experimental report. *Colloids and Surfaces. B, Biointerfaces*, *106*, 86–92. doi:10.1016/j.colsurfb.2013.01.027 PMID:23434696

Jia, L., Zhang, Q., Li, Q., & Song, H. (2009). The biosynthesis of palladium nanoparticles by antioxidants in Gardenia jasminoides Ellis: Long lifetime nanocatalysts for p-nitrotoluene hydrogenation. *Nanotechnology*, *20*(38), 385601. doi:10.1088/0957-4484/20/38/385601 PMID:19713585

Kamat, P. V. (2002). Photophysical, photochemical and photocatalytic aspects of metal nanoparticles. *The Journal of Physical Chemistry B*, *106*(32), 7729–7744. doi:10.1021/jp0209289

Kanchi, S., Kumar, G., Lo, A.-Y., Tseng, C.-M., Chen, S.-K., Lin, C.-Y., & Chin, T.-S. (2018). Exploitation of de-oiled jatropha waste for gold nanoparticles synthesis: A green approach. *Arabian Journal of Chemistry*, *11*(2), 247–255. doi:10.1016/j.arabjc.2014.08.006

Karimzadeh, M. R., Soltanian, S., Sheikhbahaei, M., & Mohamadi, N. (2020). Characterization and biological activities of synthesized zinc oxide nanoparticles using the extract of Acantholimon serotinum. *Green Processing and Synthesis*, *9*(1), 722–733. doi:10.1515/gps-2020-0058

Kaviya, S., Santhanalakshmi, J., Viswanathan, B., Muthumary, J., & Srinivasan, K. (2011). Biosynthesis of silver nanoparticles using citrus sinensis peel extract and its antibacterial activity. *Spectrochimica Acta. Part A: Molecular and Biomolecular Spectroscopy*, *79*(3), 594–598. doi:10.1016/j.saa.2011.03.040 PMID:21536485

Kelkawi, A. H. A., Abbasi Kajani, A., & Bordbar, A. (2017). Green synthesis of silver nanoparticles using Mentha pulegium and investigation of their antibacterial, antifungal and anticancer activity. *IET Nanobiotechnology / IET*, *11*(4), 370–376. doi:10.1049/iet-nbt.2016.0103 PMID:28530184

Khalkhali, M., Liu, Q., Zeng, H., & Zhang, H. (2015). A size-dependent structural evolution of ZnS nanoparticles. *Scientific Reports*, *5*(1), 14267. doi:10.1038rep14267 PMID:26381583

Kowshik, M., Ashtaputre, S., Kharrazi, S., Vogel, W., Urban, J., Kulkarni, S. K., & Paknikar, K. M. (2003). Extracellular synthesis of silver nanoparticles by a silver-tolerant yeast strain MKY3. *Nanotechnology*, *14*(1), 95–100. doi:10.1088/0957-4484/14/1/321

Krishnaswamy, K., Vali, H., & Orsat, V. (2014). Value-adding to grape waste: Green synthesis of gold nanoparticles. *Journal of Food Engineering*, *142*, 210–220. doi:10.1016/j.jfoodeng.2014.06.014

Kulkarni, N., & Muddapur, U. (2014). Biosynthesis of metal nanoparticles: A review. *Journal of Nanotechnology*, *2014*, 1–8. doi:10.1155/2014/510246

Kumar, B., Smita, K., Cumbal, L., & Angulo, Y. (2015). Fabrication of silver nanoplates using Nephelium lappaceum (Rambutan) peel: A sustainable approach. *Journal of Molecular Liquids*, *211*, 476–480. doi:10.1016/j.molliq.2015.07.067

Kumar, M. S., Saroja, M., & Venkatachalam, M. (2019). Acalypha indica and Curcuma longa plant extracts mediated ZnS nanoparticles. *Material Science Research India*, *16*(2), 174–182. doi:10.13005/msri/160210

Kumar, N., & Goel, N. (2019). Phenolic acids: Natural versatile molecules with promising therapeutic applications. *Biotechnology Reports (Amsterdam, Netherlands)*, *24*, e00370. doi:10.1016/j.btre.2019.e00370 PMID:31516850

Kumar, R., Roopan, S. M., Prabhakarn, A., Khanna, V. G., & Chakroborty, S. (2012). Agricultural waste Annona squamosa peel extract: Biosynthesis of silver nanoparticles. *Spectrochimica Acta. Part A: Molecular and Biomolecular Spectroscopy*, *90*, 173–176. doi:10.1016/j.saa.2012.01.029 PMID:22336049

Kummara, S., Patil, M. B., & Uriah, T. (2016). Synthesis, characterization, biocompatible and anticancer activity of green and chemically synthesized silver nanoparticles – A comparative study. *Biomedicine and Pharmacotherapy*, *84*, 10–21. doi:10.1016/j.biopha.2016.09.003 PMID:27621034

Kuppusamy, P., Yusoff, M. M., Maniam, G. P., & Govindan, N. (2016). Biosynthesis of metallic nanoparticles using plant derivatives and their new avenues in pharmacological applications – An updated report. *Saudi Pharmaceutical Journal*, *24*(4), 473–484. doi:10.1016/j.jsps.2014.11.013 PMID:27330378

Lakshmipathy, R., Palakshi Reddy, B., Sarada, N. C., Chidambaram, K., & Khadeer Pasha, S. (2015). Watermelon rind-mediated green synthesis of noble palladium nanoparticles: Catalytic application. *Applied Nanoscience*, *5*(2), 223–228. doi:10.100713204-014-0309-2

Lateef, A., Azeez, M. A., Asafa, T. B., Yekeen, T. A., Akinboro, A., Oladipo, I. C., Ajetomobi, F. E., Gueguim-Kana, E. B., & Beukes, L. S. (2015). Cola nitida-mediated biogenic synthesis of silver nanoparticles using seed and seed shell extracts and evaluation of antibacterial activities. *BioNanoScience*, *5*(4), 196–205. doi:10.100712668-015-0181-x

Lateef, A., Azeez, M. A., Asafa, T. B., Yekeen, T. A., Akinboro, A., Oladipo, I. C., Azeez, L., Ojo, S. A., Gueguim-Kana, E. B., & Beukes, L. S. (2016). Cocoa pod husk extract-mediated biosynthesis of silver nanoparticles: Its antimicrobial, antioxidant and larvicidal activities. *Journal of Nanostructure in Chemistry*, *6*(2), 159–169. doi:10.100740097-016-0191-4

Lateef, A., Oloke, J. K., Gueguim Kana, E. B., Oyeniyi, S. O., Onifade, O. R., Oyeleye, A. O., Oladosu, O. C., & Oyelami, A. O. (2008). Improving the quality of agro-wastes by solid-state fermentation: Enhanced antioxidant activities and nutritional qualities. *World Journal of Microbiology & Biotechnology*, *24*(10), 2369–2374. doi:10.100711274-008-9749-8

Lee, J., Kim, H. Y., Zhou, H., Hwang, S., Koh, K., Han, D.-W., & Lee, J. (2011). Green synthesis of phytochemical-stabilized Au nanoparticles under ambient conditions and their biocompatibility and antioxidative activity. *Journal of Materials Chemistry*, *21*(35), 13316. doi:10.1039/c1jm11592h

Li, S., Shen, Y., Xie, A., Yu, X., Qiu, L., Zhang, L., & Zhang, Q. (2007). Green synthesis of silver nanoparticles using Capsicum annuum L. extract. *Green Chemistry*, *9*(8), 852. doi:10.1039/b615357g

Maensiri, S., Laokul, P., Klinkaewnarong, J., Phokha, S., & Seraphin, S. (2008). Indium oxide (In2O3) nanoparticles using Aloe vera plant extract: Synthesis and optical properties. *Optoelectronics and Advanced Materials Rapid Communications*, *2*, 161–165.

Malhotra, A., Sharma, N., Fnu, N., Kumar, N., Dolma, K., Sharma, D., Nandanwar, H., & Roy Choudhury, A. (2014). Multi-analytical approach to understand biomineralization of gold using rice bran: A novel and economical route. *RSC Advances*. Advance online publication. doi:10.1039.C4RA05404K

Malik, P., Shankar, R., Malik, V., Sharma, N., & Mukherjee, T. K. (2014). Green chemistry based benign routes for nanoparticle synthesis. *Journal of Nanoparticles*, *2014*, 1–14. doi:10.1155/2014/302429

Massey, V. (2000). The chemical and biological versatility of riboflavin. *Biochemical Society Transactions*, *28*(4), 283–296. doi:10.1042/bst0280283 PMID:10961912

Mittal, A. K., Kumar, S., & Banerjee, U. C. (2014). Quercetin and gallic acid mediated synthesis of bimetallic (silver and selenium) nanoparticles and their antitumor and antimicrobial potential. *Journal of Colloid and Interface Science*, *431*, 194–199. doi:10.1016/j.jcis.2014.06.030 PMID:25000181

Mukherjee, P., Ahmad, A., Mandal, D., Senapati, S., Sainkar, S. R., Khan, M. I., Parishcha, R., Ajaykumar, P. V., Alam, M., Kumar, R., & Sastry, M. (2001). Fungus-mediated synthesis of silver nanoparticles and their immobilization in the mycelial matrix: A novel biological approach to nanoparticle synthesis. *Nano Letters*, *1*(10), 515–519. doi:10.1021/nl0155274

Mukherjee, S., Chowdhury, D., Kotcherlakota, R., Patra, S. B. V., Bhadra, M. P., Sreedhar, B., & Patra, C. R. (2014). Potential theranostics application of aio-synthesized silver nanoparticles (4-in-1 system). *Theranostics*, *4*(3), 316–335. doi:10.7150/thno.7819 PMID:24505239

Mukherjee, S., & Patra, C. R. (2017). Biologically synthesized metal nanoparticles: Recent advancement and future perspectives in cancer theranostics. *Future Science OA*, *3*(3), FSO203. Advance online publication. doi:10.4155/fsoa-2017-0035 PMID:28884002

Nadagouda, M. N., & Varma, R. S. (2006). Green and controlled synthesis of gold and platinum nanomaterials using vitamin B2: Density-assisted self-assembly of nanospheres, wires and rods. *Green Chemistry*, *8*(6), 516. doi:10.1039/b601271j

Nadagouda, M. N., & Varma, R. S. (2007a). Synthesis of thermally stable carboxymethyl cellulose/metal biodegradable nanocomposites for potential biological applications. *Biomacromolecules*, *8*(9), 2762–2767. doi:10.1021/bm700446p PMID:17665946

Nadagouda, M. N., & Varma, R. S. (2007b). A greener synthesis of core (Fe, Cu)-shell (Au, Pt, Pd, and Ag) nanocrystals using aqueous vitamin C. *Crystal Growth & Design*, *7*(12), 2582–2587. doi:10.1021/cg070554e

Nadagouda, M. N., & Varma, R. S. (2008a). Green synthesis of Ag and Pd nanospheres, nanowires, and nanorods using vitamin B 2: Catalytic polymerisation of aniline and pyrrole. *Journal of Nanomaterials, 2008*, 1–8. doi:10.1155/2008/782358

Nadagouda, M. N., & Varma, R. S. (2008b). Green synthesis of silver and palladium nanoparticles at room temperature using coffee and tea extract. *Green Chemistry, 10*(8), 859. doi:10.1039/b804703k

Nasiriboroumand, M., Montazer, M., & Barani, H. (2018). Preparation and characterization of biocompatible silver nanoparticles using pomegranate peel extract. *Journal of Photochemistry and Photobiology. B, Biology, 179*, 98–104. doi:10.1016/j.jphotobiol.2018.01.006 PMID:29351880

Nayaka, S. (2020). Extracellular fabrication of bio-nanostructures from Ralstonia sp. strain NS-7: Characterizations and their microbiological evaluation. *Indian Journal of Science and Technology, 13*(31), 3128–3140. doi:10.17485/IJST/v13i31.289

Ovais, M., Khalil, A. T., Raza, A., Islam, N. U., Ayaz, M., Saravanan, M., Ali, M., Ahmad, I., Shahid, M., & Shinwari, Z. K. (2018). Multifunctional theranostic applications of biocompatible green-synthesized colloidal nanoparticles. *Applied Microbiology and Biotechnology, 102*(10), 4393–4408. doi:10.100700253-018-8928-2 PMID:29594356

Park, Y., Hong, Y. N., Weyers, A., Kim, Y. S., & Linhardt, R. J. (2011). Polysaccharides and phytochemicals: A natural reservoir for the green synthesis of gold and silver nanoparticles. *IET Nanobiotechnology / IET, 5*(3), 69. doi:10.1049/iet-nbt.2010.0033 PMID:21913788

Powers, H. J. (2003). Riboflavin (vitamin B-2) and health. *The American Journal of Clinical Nutrition, 77*(6), 1352–1360. doi:10.1093/ajcn/77.6.1352 PMID:12791609

Prasad, Ch., Gangadhara, S., & Venkateswarlu, P. (2016). Bio-inspired green synthesis of Fe3O4 magnetic nanoparticles using watermelon rinds and their catalytic activity. *Applied Nanoscience, 6*(6), 797–802. doi:10.100713204-015-0485-8

Prasannaraj, G., Sahi, S. V., Ravikumar, S., & Venkatachalam, P. (2016). Enhanced cytotoxicity of biomolecules loaded metallic silver nanoparticles against human liver (HepG2) and prostate (PC3) cancer cell lines. *Journal of Nanoscience and Nanotechnology, 16*(5), 4948–4959. doi:10.1166/jnn.2016.12336 PMID:27483851

Prathna, T. C., Chandrasekaran, N., Raichur, A. M., & Mukherjee, A. (2011). Kinetic evolution studies of silver nanoparticles in a bio-based green synthesis process. *Colloids and Surfaces. A, Physicochemical and Engineering Aspects, 377*(1–3), 212–216. doi:10.1016/j.colsurfa.2010.12.047

Raghunandan, D., Basavaraja, S., Mahesh, B., Balaji, S., Manjunath, S. Y., & Venkataraman, A. (2009). Biosynthesis of stable polyshaped gold nanoparticles from microwave-exposed aqueous extracellular anti-malignant guava (Psidium guajava) leaf extract. *NanoBiotechnology, 5*(1–4), 34–41. doi:10.100712030-009-9030-8

Rai, A., Singh, A., Ahmad, A., & Sastry, M. (2006). Role of halide ions and temperature on the morphology of biologically synthesized gold nanotriangles. *Langmuir, 22*(2), 736–741. doi:10.1021/la052055q PMID:16401125

Rai, M., & Ingle, A. (2012). Role of nanotechnology in agriculture with special reference to management of insect pests. *Applied Microbiology and Biotechnology*, *94*(2), 287–293. doi:10.100700253-012-3969-4 PMID:22388570

Raveendran, P., Fu, J., & Wallen, S. L. (2003). Completely "green" synthesis and stabilization of metal nanoparticles. *Journal of the American Chemical Society*, *125*(46), 13940–13941. doi:10.1021/ja029267j PMID:14611213

Rokade, S. S., Joshi, K. A., Mahajan, K., Patil, S., Tomar, G., Dubal, D. S., Parihar, V. S., Kitture, R., Bellare, J. R., & Ghosh, S. (2018). Gloriosa superba mediated synthesis of platinum and palladium nanoparticles for induction of apoptosis in breast cancer. *Bioinorganic Chemistry and Applications*, *2018*, 1–9. doi:10.1155/2018/4924186 PMID:30057593

Roopan, S. M., Rohit, Madhumitha, G., Rahuman, A. A., Kamaraj, C., Bharathi, A., & Surendra, T. V. (2013). Low-cost and eco-friendly phyto-synthesis of silver nanoparticles using Cocos nucifera coir extract and its larvicidal activity. *Industrial Crops and Products*, *43*, 631–635. doi:10.1016/j.indcrop.2012.08.013

Sabouri, Z., Rangrazi, A., Amiri, M. S., Khatami, M., & Darroudi, M. (2021). Green synthesis of nickel oxide nanoparticles using Salvia hispanica L. (chia) seeds extract and studies of their photocatalytic activity and cytotoxicity effects. *Bioprocess and Biosystems Engineering*, *44*(11), 2407–2415. Advance online publication. doi:10.100700449-021-02613-8 PMID:34272979

Sahu, N., Soni, D., Chandrashekhar, B., Satpute, D. B., Saravanadevi, S., Sarangi, B. K., & Pandey, R. A. (2016). Synthesis of silver nanoparticles using flavonoids: Hesperidin, naringin and diosmin, and their antibacterial effects and cytotoxicity. *International Nano Letters*, *6*(3), 173–181. doi:10.100740089-016-0184-9

Sarathy, V., Tratnyek, P. G., Nurmi, J. T., Baer, D. R., Amonette, J. E., Chun, C. L., Penn, R. L., & Reardon, E. J. (2008). Aging of iron nanoparticles in aqueous solution: Effects on structure and reactivity. *The Journal of Physical Chemistry C*, *112*(7), 2286–2293. doi:10.1021/jp0777418

Sathishkumar, M., Saroja, M., Venkatachalam, M., Parthasarathy, G., & Rajamanickam, A. T. (2017). Biosynthesis and characterization of zinc sulphide nanoparticles using leaf extracts of Tridax procumbens. *Oriental Journal of Chemistry*, *33*(2), 903–909. doi:10.13005/ojc/330240

Sathishkumar, M., Sneha, K., Won, S. W., Cho, C.-W., Kim, S., & Yun, Y.-S. (2009). Cinnamon zeylanicum bark extract and powder mediated green synthesis of nano-crystalline silver particles and its bactericidal activity. *Colloids and Surfaces. B, Biointerfaces*, *73*(2), 332–338. doi:10.1016/j.colsurfb.2009.06.005 PMID:19576733

Sathishkumar, M., Sneha, K., & Yun, Y. S. (2009). Palladium nanocrystals synthesis using Curcuma longa tuber extract. *International Journal of Material Science*, *4*(1), 11–17.

Satyavani, K., Ramanathan, T., & Gurudeekan. (2011). Green synthesis of silver nanoparticles using stem dried callus extract of bitter apple (Citrullus colocynthis). *Digest Journal of Nanomaterials and Biostructures*, *6*(3), 1019–1024.

Shah, M., Fawcett, D., Sharma, S., Tripathy, S., & Poinern, G. (2015). Green synthesis of metallic nanoparticles via biological entities. *Materials (Basel), 8*(11), 7278–7308. doi:10.3390/ma8115377 PMID:28793638

Shakibaie, M., Riahi-Madvar, S., Ameri, A., Amiri-Moghadam, P., Adeli-Sardou, M., & Forootanfar, H. (2021). Microwave assisted biosynthesis of cadmium nanoparticles: Characterization, antioxidant and cytotoxicity studies. *Journal of Cluster Science*. Advance online publication. doi:10.100710876-021-02107-3

Shankar, S. S., Ahmad, A., & Sastry, M. (2003). Geranium leaf assisted biosynthesis of silver nanoparticles. *Biotechnology Progress, 19*(6), 1627–1631. doi:10.1021/bp034070w PMID:14656132

Shankar, S. S., Rai, A., Ahmad, A., & Sastry, M. (2004). Rapid synthesis of Au, Ag, and bimetallic Au core–Ag shell nanoparticles using Neem (Azadirachta indica) leaf broth. *Journal of Colloid and Interface Science, 275*(2), 496–502. doi:10.1016/j.jcis.2004.03.003 PMID:15178278

Shukla, R., Nune, S. K., Chanda, N., Katti, K., Mekapothula, S., Kulkarni, R. R., Welshons, W. V., Kannan, R., & Katti, K. V. (2008). Soybeans as a phytochemical reservoir for the production and stabilization of biocompatible gold nanoparticles. *Small, 4*(9), 1425–1436. doi:10.1002mll.200800525 PMID:18642250

Singh, A. K., Talat, M., Singh, D. P., & Srivastava, O. N. (2010). Biosynthesis of gold and silver nanoparticles by natural precursor clove and their functionalization with amine group. *Journal of Nanoparticle Research, 12*(5), 1667–1675. doi:10.100711051-009-9835-3

Singh, J., Dutta, T., Kim, K.-H., Rawat, M., Samddar, P., & Kumar, P. (2018). 'Green' synthesis of metals and their oxide nanoparticles: Applications for environmental remediation. *Journal of Nanobiotechnology, 16*(1), 84. doi:10.118612951-018-0408-4 PMID:30373622

Song, J. Y., Jang, H.-K., & Kim, B. S. (2009). Biological synthesis of gold nanoparticles using Magnolia kobus and Diopyros kaki leaf extracts. *Process Biochemistry, 44*(10), 1133–1138. doi:10.1016/j.procbio.2009.06.005

Song, J. Y., & Kim, B. S. (2009). Rapid biological synthesis of silver nanoparticles using plant leaf extracts. *Bioprocess and Biosystems Engineering, 32*(1), 79–84. doi:10.100700449-008-0224-6 PMID:18438688

Soundarrajan, C., Sankari, A., Dhandapani, P., Maruthamuthu, S., Ravichandran, S., Sozhan, G., & Palaniswamy, N. (2012). Rapid biological synthesis of platinum nanoparticles using Ocimum sanctum for water electrolysis applications. *Bioprocess and Biosystems Engineering, 35*(5), 827–833. doi:10.100700449-011-0666-0 PMID:22167464

Tran, Q. H., Nguyen, V. Q., & Le, A.-T. (2013). Silver nanoparticles: Synthesis, properties, toxicology, applications and perspectives. *Advances in Natural Sciences: Nanoscience and Nanotechnology, 4*(3), 033001. doi:10.1088/2043-6262/4/3/033001

Varma, R. S. (2012). Greener approach to nanomaterials and their sustainable applications. *Current Opinion in Chemical Engineering, 1*(2), 123–128. doi:10.1016/j.coche.2011.12.002

Velu, K., Elumalai, D., Hemalatha, P., Janaki, A., Babu, M., Hemavathi, M., & Kaleena, P. K. (2015). Evaluation of silver nanoparticles toxicity of Arachis hypogaea peel extracts and its larvicidal activity against malaria and dengue vectors. *Environmental Science and Pollution Research International*, *22*(22), 17769–17779. doi:10.100711356-015-4919-3 PMID:26154036

Venugopal, K., Ahmad, H., Manikandan, E., Thanigai Arul, K., Kavitha, K., Moodley, M. K., Rajagopal, K., Balabhaskar, R., & Bhaskar, M. (2017). The impact of anticancer activity upon Beta vulgaris extract mediated biosynthesized silver nanoparticles (ag-NPs) against human breast (MCF-7), lung (A549) and pharynx (Hep-2) cancer cell lines. *Journal of Photochemistry and Photobiology. B, Biology*, *173*, 99–107. doi:10.1016/j.jphotobiol.2017.05.031 PMID:28570910

Vijayaraghavan, K., & Ashokkumar, T. (2017). Plant-mediated biosynthesis of metallic nanoparticles: A review of literature, factors affecting synthesis, characterization techniques and applications. *Journal of Environmental Chemical Engineering*, *5*(5), 4866–4883. doi:10.1016/j.jece.2017.09.026

Wang, L., Wu, Y., Xie, J., Wu, S., & Wu, Z. (2018). Characterization, antioxidant and antimicrobial activities of green synthesized silver nanoparticles from Psidium guajava L. leaf aqueous extracts. *Materials Science and Engineering C*, *86*, 1–8. doi:10.1016/j.msec.2018.01.003 PMID:29525084

Wang, Z., Fang, C., & Megharaj, M. (2014). Characterization of iron–polyphenol nanoparticles synthesized by three plant extracts and their fenton oxidation of azo dye. *ACS Sustainable Chemistry & Engineering*, *2*(4), 1022–1025. doi:10.1021c500021n

Yamashita, M., Rosatto, S. S., & Kubota, L. T. (2002). Electrochemical comparative study of riboflavin, FMN and FAD immobilized on the silica gel modified with zirconium oxide. *Journal of the Brazilian Chemical Society*, *13*(5), 635–641. doi:10.1590/S0103-50532002000500015

Yan, D., Zhang, H., Chen, L., Zhu, G., Wang, Z., Xu, H., & Yu, A. (2014). Supercapacitive properties of Mn3O4 nanoparticles bio-synthesized from banana peel extract. *RSC Advances*, *4*(45), 23649. doi:10.1039/c4ra02603a

Yang, N., WeiHong, L., & Hao, L. (2014). Biosynthesis of Au nanoparticles using agricultural waste mango peel extract and its in vitro cytotoxic effect on two normal cells. *Materials Letters*, *134*, 67–70. doi:10.1016/j.matlet.2014.07.025

Yuvakkumar, R., Suresh, J., Nathanael, A. J., Sundrarajan, M., & Hong, S. I. (2014). Rambutan (Nephelium lappaceum L.) peel extract assisted biomimetic synthesis of nickel oxide nanocrystals. *Materials Letters*, *128*, 170–174. doi:10.1016/j.matlet.2014.04.112

Zheng, B., Kong, T., Jing, X., Odoom-Wubah, T., Li, X., Sun, D., Lu, F., Zheng, Y., Huang, J., & Li, Q. (2013). Plant-mediated synthesis of platinum nanoparticles and its bioreductive mechanism. *Journal of Colloid and Interface Science*, *396*, 138–145. doi:10.1016/j.jcis.2013.01.021 PMID:23403109

Zheng, F., Luk, S.-Y., Kwong, T.-L., & Yung, K.-F. (2016). Synthesis of hollow PtAg alloy nanospheres with excellent electrocatalytic performances towards methanol and formic acid oxidations. *RSC Advances*, *6*(50), 44902–44907. doi:10.1039/C6RA06398E

Zhou, Y., Lin, W., Huang, J., Wang, W., Gao, Y., Lin, L., Li, Q., Lin, L., & Du, M. (2010). Biosynthesis of gold nanoparticles by foliar broths: Roles of biocompounds and other attributes of the extracts. *Nanoscale Research Letters*, *5*(8), 1351–1359. doi:10.100711671-010-9652-8 PMID:20676207

ADDITIONAL READING

Abdullaeva, Z. (2017). *Synthesis of Nanoparticles and Nanomaterials*. Springer International Publishing. doi:10.1007/978-3-319-54075-7

Ahmed, S., & Ali, W. (Eds.). (2020). *Green Nanomaterials: Processing, Properties, and Applications* (Vol. 126). Springer Singapore. doi:10.1007/978-981-15-3560-4

Egorova, E. M., Kubatiev, A. A., & Schvets, V. I. (2016). *Biological Effects of Metal Nanoparticles*. Springer International Publishing. doi:10.1007/978-3-319-30906-4

Pal, G., Rai, P., & Pandey, A. (2019). Green synthesis of nanoparticles: A greener approach for a cleaner future. In Green Synthesis, Characterization and Applications of Nanoparticles (pp. 1–26). Elsevier. doi:10.1016/B978-0-08-102579-6.00001-0

Patra, J. K., Fraceto, L. F., Das, G., & Campos, E. V. R. (Eds.). (2020). *Green Nanoparticles: Synthesis and Biomedical Applications*. Springer International Publishing. doi:10.1007/978-3-030-39246-8

Saquib, Q., Faisal, M., Al-Khedhairy, A. A., & Alatar, A. A. (Eds.). (2020). *Green Synthesis of Nanoparticles: Applications and Prospects*. Springer Singapore. doi:10.1007/978-981-15-5179-6

Wu, J., Cao, J., Han, W.-Q., Janotti, A., & Kim, H.-C. (Eds.). (2012). *Functional Metal Oxide Nanostructures* (Vol. 149). Springer. doi:10.1007/978-1-4419-9931-3

KEY TERMS AND DEFINITIONS

Capping Agent/Stabilizing Agent: A substance that provides colloidal stability along with preventing agglomeration and stopping the uncontrolled growth of synthesized nanoparticles.

Green Chemistry: An area of chemistry and chemical engineering focused on the design of products and processes that minimize or eliminate the use and generation of hazardous substances.

Metal Nanoparticles: Are the particles of metal atoms with diameters between 1 nm and about a few hundreds of nanometers.

Nanostructured Materials: Materials whose structural elements - clusters, crystallites, or molecules - have dimensions in the 1 to 100 nm range.

Nanotechnology: The branch of technology that deals with dimensions and tolerances of less than 100 nanometres, especially the manipulation of individual atoms and molecules.

Reducing Agent: A substance that reduces a chemical compound usually by donating electrons.

Thermodynamics: Is the study of the relations between heat, work, temperature, and energy.

Chapter 4
Green Synthesis of Nanocomposites:
A Greener Approach for a Cleaner Future

Meenal D. Patil
https://orcid.org/0000-0002-1529-4094
Shivaji University, Kolhapur, India

Suprimkumar D. Dhas
Shivaji University, Kolhapur, India

Annasaheb V. Moholkar
https://orcid.org/0000-0002-5564-3957
Shivaji University, Kolhapur, India

ABSTRACT

Nanotechnology has been a dynamic research area over the past few decades because it assures the resolution to the problems that hamper progress. Currently, a new era of 'green synthesis' is an emerging multidisciplinary field in nanotechnology which employs reliable, sustainable, low-cost, non-hazardous, and eco-friendly techniques. Green synthesis is considered a vital tool to reduce the negative impacts accompanying the traditional methods of synthesis for NPs commonly employed in industry and laboratory. This chapter unveils a comprehensive overview of the recent research on available green techniques for the synthesis of various nanocomposites in order to solve future generation challenges. This chapter also focuses on the green synthesis of various nanocomposites, synthesis parameters, potential applications, merits/demerits, and future prospects.

INTRODUCTION

Nanotechnology has garnered attention in a modern scientific field as it plays a dominant role in various fields of natural and agricultural sciences and technology concerning healthcare, food security, medi-

DOI: 10.4018/978-1-7998-8936-6.ch004

cal, and environmental monitoring. The continuously growing chemical, metallurgical, pharmaceutical, and other industries, urbanization and population explosion are implementing a negative impact on the earth's atmosphere by releasing a huge amount of hazardous and unwanted substances. To tackle this problem green synthesis is an emerging hope in nanotechnology, which employs reliable, sustainable, low-cost, non-hazardous, and eco-friendly protocols for synthesizing nanomaterials (NMs) (Saratale et al., 2018). Green synthesis is the predictable method for the production of different Nanomaterials such as metal nanoparticles, quantum dots (QDs), carbon nanotubes (CNTs), graphene, and their composites useful in the day-to-day life of human beings. New trends to overcome the problems such as toxicity and non-environmental being, researchers have been focused on the precise green routes, i.e., the natural materials and their derivatives that can be used for the synthesis of Nanomaterials. The nanomaterials are minuscule structures having at least one of its dimensions within the 1-100 nm range. These nanomaterials attribute characteristic properties in contrast to bulk material, due to their significantly high surface area to volume ratio (Bajpai et al., 2018). The idea behind nanocomposite is to embed carbon-based, ceramic, metal, semiconductor, and polymeric nanoparticles (NPs) as basic building blocks and create new materials with enhanced physico-chemical and opto-electronic properties. The nanocomposites are multiphase nanomaterials with one of the phases in the nanoscale regime. The combination of Nanoparticles with other Nanoparticles or bulky material forms the desired nanocomposite for the particular application of the interest.

BACKGROUND

In material science, 'Green' synthesis of nanocomposite has gained extensive research interest in the scientific community and is believed as an important tool to show synergetic effects of nanocomposites. The most widely used chemical synthesis route like the sol-gel (Ahir et al., 2021), pyrolysis (Mane & Moholkar, 2018), sonochemical method (Digraskar et al., 2021), SILAR (A. S. Patil et al., 2017), hydrothermal (Dhas et al., 2021), thermal decomposition (M. D. Patil et al., 2021), and many more. Whereas physical methods include sputtering (Agawane et al., 2015), microwave irradiation (Z. Liu et al., 2021), and ball-milling (Bhadane et al., 2020), etc. These chemical and physical synthesis methods are based on fundamental principles of synthesis (i.e., top-down, and bottom-up approaches). The traditional methods are toxic, less eco-friendly as compared with green synthesis (bottom-up approach). Therefore, green synthesis methods using benign reagents have been proposed, which will minimize the production of unnecessary and harmful waste, reduce pollution, use non-toxic solvents, and can be used for large-scale production. As external experimental conditions such as high pressure, energy, etc. are not required, the green synthesis methods are energy-saving progression. The synthesis of nanoparticles by employing diverse methods and controlling the process parameters to achieve desired morphology is reckoned for the specific application is of paramount importance to obtain desired nanoparticles with specific sizes and shapes. There are vivid varieties of appropriate and most suitable candidates for enormous industrial marketable and local applications. Using the green synthesis route, metal oxide, carbon-based materials, polymer-based materials, and their nanocomposites can be synthesized with different characteristic morphologies such as nanosheets, nanorods, nanowires, nanoplates, flower-like, sea-urchin like, etc (S.A. et al., 2016).

The different physicochemical, optoelectrical, and chemical compositional properties of green route synthesized nanocomposites have been exploited using modern up-to-date techniques viz. FE-SEM, TEM,

Figure 1. Illustration of nanocomposite synthesis strategies

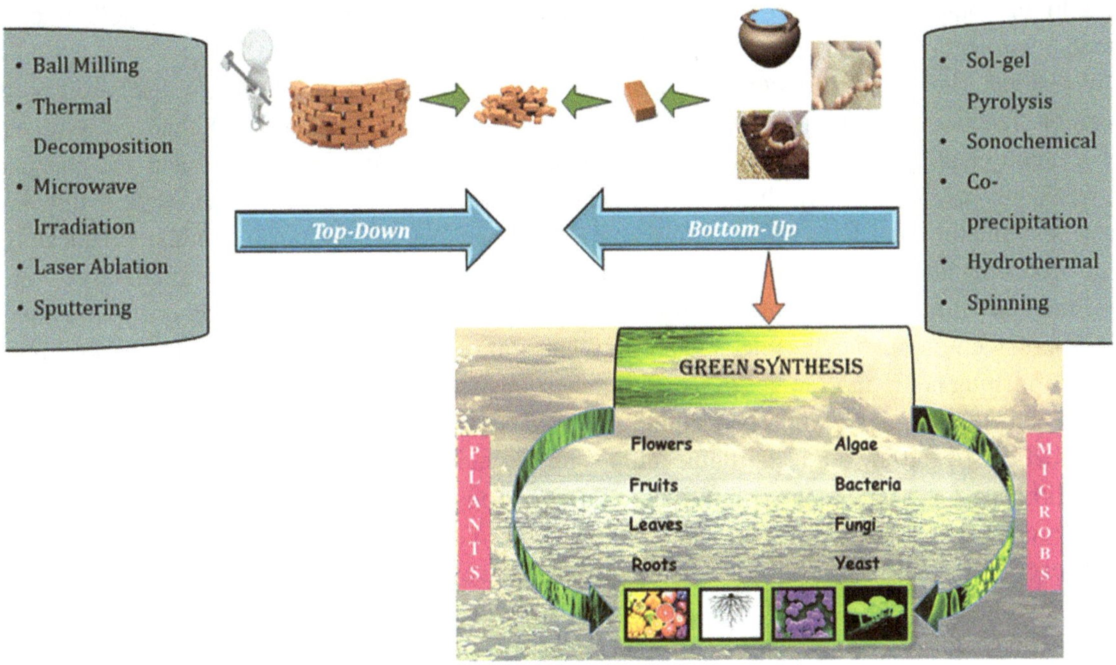

AFM, FT-IR, XRD, NMR, XPS (Jagwani & Hari Krishna, 2021). These functionalized nanocomposites are used in various applications such as environmental remediation, water remediation, solar cells, gas sensors, biosensing, surface coating, clothing, cosmetics, smart agriculture, etc. in order to uplift the eco-friendliness (Jagwani & Hari Krishna, 2021; S.A. et al., 2016). Green nanocomposites reduce the exposure of hazardous chemicals to the environment which enables special recognition owing to their advanced properties.

Our main focus in the present work is to study the different green synthesis strategies for nanocomposites, synthesis parameters, and their applications.

GREEN SYNTHESIS STRATEGIES

An efficient methodology for the synthesis of Nanomaterials has two basic approaches: top-down and bottom-up. (Illustrated in Figure 1) In top-down approach, Nanomaterials are prepared using suitable bulk material, which systematically degrade into fine particles by size reduction with divergent physical synthesis techniques like lithographic, ball milling, thermal decomposition, microwave irradiation, pulse laser ablation, and sputtering, etc. This approach is based on destructive technique. The major drawback of the top–down approach is the limitation of surface structure and thus not appropriate for synthesizing shaped materials. The limitations of this route have adverse effects on surface chemistry and physical properties of Nanomaterials (Jagwani & Hari Krishna, 2021; I. Khan et al., 2019; S.A. et al., 2016; Saratale et al., 2018). The bottom-up approach or building up approach is employed in reverse as Nanomaterials are grown by self-assembling simpler atoms, molecules, or clusters. The most common method of this approach is chemical reduction via chemical and biological synthesis techniques such as sol-gel method, pyrolysis, sonochemical method, SILAR, hydrothermal, green synthesis, and many more. In this route, by controlling the synthesis parameters, one can get the desired morphology for the potential application of Nanomaterials. (Rafique et al., 2017; Saquib et al., 2020).

In general, Nanomaterial's synthesis methods can be categorized into three groups based on the operation, reaction condition, and adopted protocols – (a) physical synthesis, (b) chemical synthesis, and (c) biological synthesis methods. The enormous traditional methods viz. physical and chemical, are frequently used for the synthesis of Nanomaterials with the preferred characteristics. But these physical synthetic procedures are generally complex, expensive, labor-intensive, while chemical synthetic procedures have a pernicious effect on the environment and living organisms. In the true words, these methods limit the mass-production of Nanomaterials due to high production cost, high energy consumption, environmental hostile, and produce hazardous waste by-products. However, these limitations are opening new opportunities to search for alternative pathways to synthesize Nanomaterials. The biological method or green synthesis method is the need of the hour as it provides a feasible competent, cost-effective, clean, safe, and environmentally sound technique for Nanomaterial production. It gives a new direction for synthesis procedure using non-hazardous chemicals as reducing agents and wiping out the production of hazardous substances. For the adopted biosynthesis method, nature acts as a "Bio-laboratory" for the production of well-functionalized nanoparticles accomplished via either a wide range of viable biological entities such as fungi, bacteria, algae, and plant extracts or water as green (Mondal et al., 2020; Saquib et al., 2020). From the literature study, researchers across the globe successfully have bio-synthesizes nanoparticles such as Fe, Ag, Au, Cu, etc. from plant extract and microbes (Alavi & Varma, 2021; Mondal et al., 2020; Sajjad et al., 2018).

Currently, significant advances in the bio-fabrication of nanocomposites have been achieved owing to their ease of synthesis, industrial-scale production, and multi-applications. Nanocomposites are multiphase nanomaterials made by the combination of two or more Nanomaterials in the nanoscale regime. The nanocomposite materials show the merging properties from the individual components, in addition to drastic enhancement in novel mechanical, chemical and electric properties. Various green routes such as hydrothermal, sol-gel, microwave, and many others were employed for synthesizing the nanocomposites, in similar manner to nanoparticles, under the control of biological entities as reducing agents. Different bio-synthesized nanocomposites are summarized in this work.

GREEN SYNTHESIS OF NANOCOMPOSITE

Plant Mediated Synthesis

Since ancient times the homemade remedies prepared from the above plants-flower and fruits have been recommended and used in houses as a primary aid, which is known as Grandma's purse. They are very much useful in rural and urban areas treating some normal as well as hard diseases. The extracts of different flavors made from the fruits, flowers, leaves, and other wheatgrass; twisted cereals are very much operative as far the energy throughout the day. In the Hindu Sanskriti at the beginning of the year i. e. the month of Chaitra, the neem (*Azadirachta indica: Meliaceae*) and its leaves, its flowering, coconut, jaggery, or sugar are not forcibly eaten by everyone. The neem is an immunity booster, which incorporates healthy practices right from the first day of the newly begun year. These make people feel fresh and dynamic, due to which they have become masters in their routine work and assignments. The

Figure 2. Plant mediated green synthesis of nanocomposites

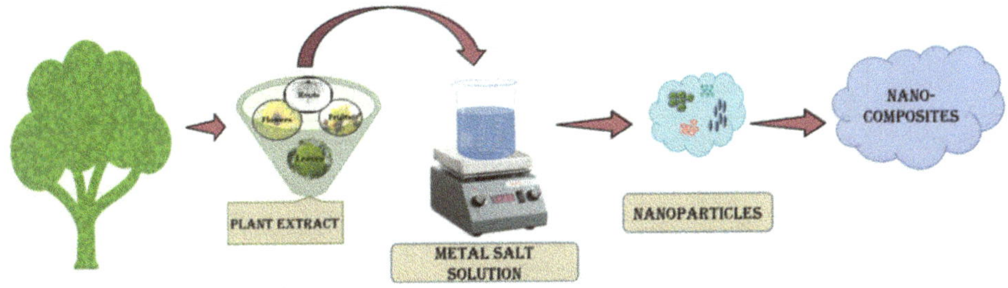

overall effect is that they are fulfilled with a lot of positive thinking and acting styles. It has become a very popular medicine for asthma, respiratory disorders, and shortness of breath.

In the past couple of years, the plant-mediated green synthesis of nanocomposite has emphasized focus due to their advantage in diminishing environmental risk, economic viability, cost-effective and simple process. Plants can capture light energy from the sun and convert it into chemical energy in the form of carbohydrates, proteins, peptides, sugars, enzymes, and bioactive alkaloids that act as a driving force to reduce metal ions (N. T. Khan & Khan, 2020; Saquib et al., 2020; Saratale et al., 2018). All parts of a plant, namely flowers, fruits, leaves, bark, seeds, and roots, resulted in favorable contenders for greener pathways in the fabrication of noble nanocomposites (Figure 2).

Various researchers have successfully reported the green synthesis of diverse nanocomposites using different plant extracts like tulsi, aloe vera, lemon, neem, *Averrhoa carambola, fenugreek,* grapes, and pomegranate, etc. Let us take a brief outlook on the biosynthesis of these individual nanoparticles or nanocomposites and their applications.

Flowers

From longer time, flowers were used by human beings to bring beauty to their environment. Now in modern times, the flower extract of various shrubs, herbs, trees, climbers, etc. is used in the synthesis of the nanocomposite. The eco-friendly biosynthesis approach uses flower extract for the purpose of either reducing metal ions or as a stabilizer or capping agent. Navya Rani et al., 2019 used *Tagetes erecta* extract as a reducing agent for the reduction of Graphene Oxide (GO) to Reduced Graphene Oxide (RGO), in order to prepare RGO decorated Cu_2O nanocomposite. Kolya et al., 2019 used an aqueous extract of mango (*Mangifera indica*) flower as mild stabilizing and reducing agent. Valian et al., 2020 reported the synthesis of $DyMnO_3$-ZnO ceramic nanocomposites using *Hibiscus sabdariffa* extract as a chelating agent.

The synthesis of nanocomposites using flower extract *via* various chemical methods such as co-precipitation (Rostami-Vartooni et al., 2018), hydrothermal (Vinay et al., 2021), sonication assisted (Sarangapany & Mohanty, 2020) and many other methods have been reported. These methods are simple, flexible, reduce energy consumption and the time used to obtain desired materials. Literature studies explore the demonstration of synthesized nanocomposite in various eco-friendly and biomedical applications. The compilation of the various nanocomposites synthesized using flower extract and their applications are presented in Table 1.

Table 1. The various nanocomposites synthesized using flower extract and their applications

Nanocomposite	Precursor	Method	Application	Ref.
RGO-Cu_2O	Marigold *(Tagetes erecta)*	-	antibacterial activity and cytotoxicity	(Navya Rani et al., 2019)
Ag/rGO	mango (*Mangifera indica*) flower	-	catalytic reduction and energy storage material	(Kolya et al., 2019)
$DyMnO_3$-ZnO ceramic	*Hibiscus sabdariffa*	Modified Pechini	Detection of atenolol	(Valian et al., 2020)
Ag/C/Fe_3O_4	*Caesalpinia gilliesii*	co-precipitation	Catalytic reduction of organic pollutants	(Rostami-Vartooni et al., 2018)
Ag/rGO@CTFE	chamomile tea flower	Hydrothermal	photocatalyst for degradation	(Vinay et al., 2021)
Au-Pt@TiO_2	dahlia	Sonication assisted	catalytic reduction of organic pollutants and selective oxidation	(Sarangapany & Mohanty, 2020)
Ag–Cu, Cu–Ag, Ag–TiO2 and TiO2–Ag	*Artemisia haussknechtii*	One pot	Antiplanktonic, antibiofilm, antiswarming motility and antiquorum sensing	(Alavi & Karimi, 2018)
Ag/MgO	*Acalypha hispida*	-	Catalytic reduction of pollutants	(Nasrollahzadeh et al., 2020)

Fruits

The bioactive fleshy seed-associated structures of plants- fruits- plays a key role in the green synthesis. Researchers have found it suitable to exploit the fruits extract as well as fruit seeds, as a source of green reducing agent for developing nanocomposites. The pollutant removal Fe@GPHC nanocomposite synthesized using Grape (*Vitis vinifera L.*) pulp reported by Hasan (Sayğılı, 2019). While Dharmalingam et al. (Dharmalingam & Anandalakshmi, 2020) and Koneru et al. (Koneru et al., 2020), exploited grape seed for sodium carboxy methyl cellulose/ hydroxy-propyl methyl cellulose nanocomposite in wound healing applications. Further, an intensive research study has been executed utilizing fruit peel extract. Fruit peels are considered a wastage in society. However, Akshatha et. al., (Akshatha et al., 2021) used orange peel as a green source for synthesizing cobalt decorated mesoporous carbon. These nanocomposites are synthesized via sol-gel (Gnanasekaran et al., 2021), sonochemical (Naghizadeh et al., 2020; Orooji et al., 2020), Solution casting (Basumatary et al., 2018; Koneru et al., 2020), and many others. Also, the various applications of fruits mediated nanocomposites in photocatalytic degradation (Akshatha et al., 2021; Gnanasekaran et al., 2021), antimicrobial food packaging (Basumatary et al., 2018), wastewater treatment (Baladi et al., 2021), flame retardation (Joulaei et al., 2019), etc. Table 2 displays the various nanocomposites synthesized using fruits, seeds, peel extract and their significant applications.

Leaves

In recent years, the synthesis of nanocomposite by greener route exhibits better applications in medical and other sectors compared to other traditional methods. The principal lateral appendage of the vascular plant stem is the leaf, promoting photosynthesis in plants as they are directly exposed to the sunlight. Also, they store the chemical energy and water and modify the functions according to need. Thus, vari-

Table 2. The various nanocomposites synthesized using fruits, seeds, peels extract and their significant applications.

Nanocomposite	Precursor	Method	Application	Ref.
Fe@GPHC	Grape (*Vitis vinifera L.*) pulp	Hydrothermal carbonization	Pollutant removal	(Sayğılı, 2019)
Sodium carboxy methyl cellulose/ hydroxy-propyl methyl cellulose	grapefruit seed	Solution casting method	wound healing applications	(Koneru et al., 2020)
zinc oxide complex NaCMC-HPMC hydrogel	grapefruit seed	-	wound healing applications	(Dharmalingam & Anandalakshmi, 2020)
Co_3O_4@Mesoporous carbon spheres	Orange peel	Microwave assisted	Simultaneous degradation of dyes and photocatalytic hydrogen Evolution reaction	(Akshatha et al., 2021)
TiO_2–CeO_2	Lemon juice	Sol-gel and Precipitation	Photocatalytic degradation	(Gnanasekaran et al., 2021)
Fe_3O_4@SiO_2-hydroxyapatite	strawberry	Sonochemical	Sulfasalazine delivery and Cytotoxicity	(Orooji et al., 2020)
$CoFe_2O_4$@ZnO-CeO_2	*Crataegus microphylla*	Sonochemical	Catalytic and antibacterial activities	(Naghizadeh et al., 2020)
Agar-Ag	*Lagerstroemia speciosa*	Solution casting	Antimicrobial Food packaging	(Basumatary et al., 2018)
$Sr_7Mn_7O_{19.62}$-Dy_2O_3	Ficus carica Latex, Tomato juice, Blackberry juice	Sol-gel combustion	Wastewater treatment	(Baladi et al., 2021)
ZnO/MgO	Ricinus communis L. plant seedless fruit	-	Photocatalytic and Antibacterial properties	(Panchal et al., 2019)
Ni/Ag@rGO	Pomegranate (Punica granatum) juice	Hydrothermal	Electrochemical detection of ascorbic acid	(Das & Sharma, 2020)
SnO_2-bentonite	jujube	-	Photodegradation	(Honarmand et al., 2020)
$DyMnO_3$-ZnO ceramic	Vitis vinifera, rhus	Modified Pechini method	Detection of atenolol	(Valian et al., 2020)
Ag-Starch	pomegranate seeds	-	Plastic packaging materials/ Plastisizer	(Mohseni et al., 2020)
$MgFe_2O_4$	Lime and orange	Sol-gel auto-combustion	Flame retardant	(Joulaei et al., 2019)
Ferric oxide/ activated carbon	date seed	Hydrothermal	Biohydrogen Production	(Rambabu et al., 2021)
Au/TiO_2	*Averrhoa bilimbi*	-	Photocatalytic activity under sodium light irradiation	(Yulizar et al., 2019)
Ag-F-rGO	*Capparis spinosa*	-	Catalytic reduction of organic dyes	(Zarei et al., 2021)

ous plant leaves extract has been widely used in the synthesis of nanocomposites. Hassanin et al. 2021 reported the synthesis of ZnO–NiO and Ag/Co_3O_4 nanocomposites respectively, using Neem leaves extract. Neem is a traditional immunity booster medicinal plant containing terpenoids and flavanones, which have the ability to reduce and stabilize nanoparticles. The leaves of fenugreek have antimicrobial,

Table 3. The various nanocomposites synthesized using leaves extract and their applications.

Nanocomposite	Precursor	Method	Application	Ref.
ZnO–NiO	neem	hydrothermal	-	(Hessien et al., 2021)
Ag/Co_3O_4	neem	-	Catalytic and microbial behavior	(Hassanin et al., 2021)
Ag-ZnO	*Trigonella foenum-graecum*	-	antibacterial, antifungal, antioxidant and photocatalytic properties	(Noohpisheh et al., 2020)
Ag/GO	green tea leaf (*Camellia Sinensis*)	-	anticancer and antibacterial efficacy	(Emima Jeronsia et al., 2020)
Ni/montmorillonite	*Allium jesdianum*	Sonication assisted ion exchange	electrocatalytic oxidation of methanol	(Sheikh-Mohseni et al., 2020)
RGO/Ag	*Avicennia marina*	-	Antifouling coating	(Soleimani et al., 2021)
ZnO/CuO	*Mentha longifolia*	-	Antibacterial agents	(Mohammadi-Aloucheh et al., 2018)
Fe_2O_3/Ag	Aloe vera gel	one pot microwave-assisted	photocatalyst, antibacterial and anticancer agent	(A et al., 2020)
Au@g-C_3N_4	*Averrhoa carambola*	-	photocatalytic degradation of methyl orange	(Manjari Mishra et al., 2021)
PPy/ Fe_3O_4	*Callistemon viminalis*	-	Electrochemical detection of serotonin in banana	(Uwaya & Fayemi, 2020)
Copper-Chitosan	*Ficus carica*	-	antibacterial properties	(Syame et al., 2017)

antitumor, antivirus, anti-inflammatory, and antioxidant properties, and are used in the synthesis of Ag-ZnO nanocomposites (Noohpisheh et al., 2020). Emima Jeronsia et al., 2020 reported the Ag/GO nanocomposite by exploiting the green tea leaf (*Camellia Sinensis*) extract as a reducing and stabilizing agent. The biomolecules present in the tea structure are useful in the diagnosis of cardiovascular diseases, diabetes, inflammatory diseases, bacterial and fungal infections, etc. Table 3 depicts the various nanocomposites synthesized using leaves extract and their applications.

Root

The root is an organ of the plant majorly functioning to absorb water and nutrition and providing anchorage for the plant. Once photosynthesis occurs in leaves, synthesized sugars are again transported to the root for active growth. The outermost layers of stems and roots of woody plants i. e. bark consist of various biopolymers, tannins, lignin, and polysaccharides (Vane et al., 2006). Thus, A recent study suggests that the part of plants i. e. roots, bark can be utilized in the synthesis of nanocomposites.

The traditionally used oral hygiene tool Miswak Roots (*Salvadora persica L.*) contain various bioactive compounds, organic and inorganic chemicals. The excellent reducing ability of the miswak root extract has been exploited for synthesizing numerous nanocomposites (Ahmadi et al., 2019; Al-Marri et al., 2016; Fadlallah et al., 2020). A promising pH indicator for the packaging industry is anthocyanins, which are stable in an acidic medium. The black carrot is a rich source of anthocyanins and is used in synthesis of active chitosan/PVA/bentonite films (Koosha & Hamedi, 2019). The commonly used 'Cinnamon' is noted as a robust antibacterial and antioxidant agent. The aromatic compound rich cinnamon reported

Table 4. The various nanocomposites synthesized using roots, bark extract, and their applications.

Nanocomposite	Precursor	Method	Application	Ref.
Ag/ Titania	*Salvadora persica L.* Roots (Miswak)	-	Dental applications	(Fadlallah et al., 2020)
Pd@graphene	*Salvadora persica L.* Roots (Miswak)	-	Selective oxidation of alcohols	(Al-Marri et al., 2016)
CMC/CNF/SPE	*Salvadora persica L.* Roots (Miswak)	Casting	Antibacterial properties	(Ahmadi et al., 2019)
$Ca_{10}(PO_4)_6(OH)_2$/Li-BioMOFs resin	*Glycyrrhiza glabra* (licorice) root	hydrothermal	Mechanical properties	(Asadi et al., 2020)
PLGA-CS-PEG	Moringa oleifera Root	-	Anti-cancer properties	(Abd-Rabou et al., 2017)
chitosan/PVA/bentonite	black carrot	-	Mechanical, thermal & antibacterial properties	(Koosha & Hamedi, 2019)
ZnO/CuO	*Theobroma cacao* seed bark	one-pot synthesis	-	(Yulizar et al., 2018)
cellulose/TiO_2/Cinnamon bark	Cinnamon bark	Sonication assisted	Catalytic reduction	(Althomali et al., 2021)
$CoFe_2O_4$-SiO_2	*Salix alba* bark	sol-gel	Removal of malachite green from water	(Amiri et al., 2017)
Au-rGO	Syzygium cumini seed	green reducing route	Anti-cancer and antibacterial activity	(Kadiyala et al., 2018)
Ag/GO	*Catharanthus roseus* dried bark	solar irradiation	Adsorption of methylene blue dye	(Rohaizad et al., 2020)
ZnO-RGO	garlic clove	-	Anticancer efficacy	(Ahamed et al., 2021)

for the synthesis of cellulose/TiO_2/Cinnamon bark nanocomposite (Althomali et al., 2021). We compiled the various nanocomposites synthesized using roots, bark extract, and their applications in Table 4.

Microorganism Mediated Synthesis

Enormous scientific efforts were made to develop stable and well-functionalized nanocomposites exploiting various microorganisms like algae, bacteria, fungi, yeast, leading to major contributions in nanotechnology (Figure 3).

These nano-factories i. e. microorganisms play the role of environment-friendly, sustainable precursors and provide robust controlled, energy-efficient, scalable biosynthesis processes (Gangadhar et al., 2020). The recent studies found that microbial synthesis occurs mainly through either the intracellular or extracellular route. The intracellular approach involves the diffusion of metal ions into the microbial cell in presence of enzymes to form Nanomaterials, while the extracellular mechanism involves the electrostatic attraction of metal ions on the surface of the cell wall and carrying out the enzymatic reduction of metal ions (Mondal et al., 2020). Table 5 represents the various microorganisms (algae, bacteria, fungi, yeast) mediated synthesis of nanocomposites.

Figure 3. Microorganism mediated green synthesis of nanocomposites

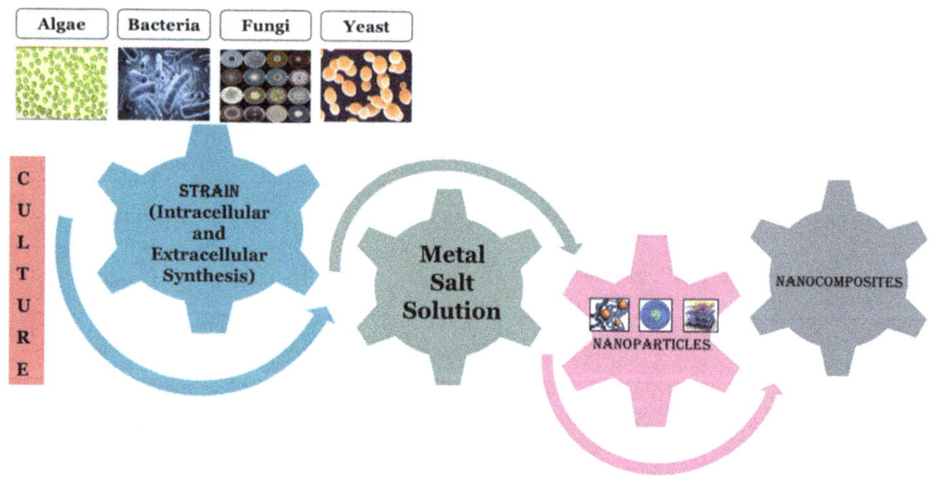

Algae

Algae are a large and diverse group of unicellular or multicellular organisms. In the vogue of nanotechnology, various types of algal species synthesize different types of nanocomposites. Phyconanotechnology is considered a branch of algae-mediated biosynthesis of nanometals (Li et al., 2011). In recent years, a number of publications have been published on the green synthesis of nanoparticles using algae as precursors (Dhand et al., 2015; Gangadhar et al., 2020). However, in recent years there are only a few publications on synthesizing the nanocomposites using algae.

Bacteria

For the preparation of metallic and other novel nanocomposites, bacteria are considered as potential bio-factory, as they possess the ability to reduce the metallic ions (Alavi & Rai, 2019; Iravani, 2014). Numerous momentous bacterial species such as *Lactobacillus paracasei, Escherichia coli, Alcaligenes aquatilis, Actinobacter sp.* have been widely adopted for synthesis of bio-reduced nanocomposites. Pomastowski et. al (Pomastowski et al., 2020), post-cultured ZnO nanocomposites using *Lactobacillus paracasei* isolated from sweet whey, which showed antibacterial activity against pathogenic bacteria like *E. coli, Pseudomonas aeruginosa, Salmonella typhi,* and *Klebsiella pneumoniae*.

3D network structured nanofiber of Bacterial cellulose pinched momentous attraction in the modern research area. Bacterial cellulose holds high mechanical stability, high crystallinity (84–89%), high degree of polymerization, and excellent water holding capacity (Wasim et al., 2020). Bacterial cellulose developed by static growth of bacterial strain (most commonly used *Gluconacetobacter*- highest yield) (Wasim et al., 2020). The time-dependent chemical deposition of CuNps on the bacterial cellulose membranes has been successfully proposed by Araújo et. al (Araújo et al., 2018).

Table 5. The various microorganisms (algae, bacteria, fungi, yeast) mediated synthesis of nanocomposites.

Algae				
Nanocomposite	**Precursor**	**Method**	**Application**	**Ref.**
Bacterial cellulose-polyaniline-titanium-dioxide	*Shewanella xiamenensis*	Sonication assisted	Microbial fuel cell	(Truong et al., 2021)
algae/$CoFe_2O_4$ algae/Ag-Fe-O	*Chlorella vulgaris*	Auto combustion	Biomedical applications	(Satheeshkumar et al., 2020)
Bacteria				
Zinc Oxide	*Lactobacillus paracasei*	Extracellular	Antibacterial	(Pomastowski et al., 2020)
Ag_2O/AgO-TiO_2	*Alcaligenes aquatilis*	Extracellular bacterial mediated	Pollution abatement	(Kulal & Shetty Kodialbail, 2021)
silicon/silica	*Actinobacter sp.*	-	Cytotoxicity	(Singh et al., 2008)
bacterial cellulose–copper oxide	*Gluconacetobacter hansenii*	Hydrothermal	antimicrobial activity	(Araújo et al., 2018)
Fungi				
ZnO/GQD	*Protoparmeliopsis muralis*	-	Antibacterial and antibiofilm activity	(Rashki et al., 2021)
ZnO-MoO_3	mushroom	Thermal calcination	Electrochemical sensor	(C. Liu et al., 2021)
non-animal mushroom carboxymethyl chitosan-ZnO	mushroom	Ultrasonic-assisted	Wound dressing	(Rao et al., 2020)
Prussian blue nanocubes/ RGO	mushroom	One-pot synthesis	4-nitrophenol reduction	(Chen et al., 2015)
chitosan nanocomposite	*Nomuraea rileyi*	-	Insect pestsizer	(Karthick Raja Namasivayam et al., 2018)
Ag-MgO	*Protoparmeliopsis muralis*	-	Antibacterial activities	(Alavi & Karimi, 2020)
LrGO-AgAu	*Cetraria Islandica (L.)*	One pot approach	Catalytic reduction	(Çıplak, Getiren, et al., 2020)
GO- au@polyaniline	*Cetraria Islandica (L.)*	-	Supercapacitor	(Çıplak, Yıldız, et al., 2020)
Yeast				
N-TiO_2/MnO_2	Commercial baker's yeast powder (*Saccharomyces cerevisiae*)	Hydrothermal carbonization	Gaseous formaldehyde degradation	(Mohamed & Awad, 2021)
Yeast-titania	*Saccharomyces cerevisae*	Immobilization	Removal of toxic Cr (vi) ion	(Choudhury et al., 2017)
magnetic silica-mannan	*Saccharomyces cerevisae*	-	Antigen delivery	(Rodponthukwaji et al., 2021)
polycarbazole- TiO_2	Yeast alcohol dehydrogenase	-	In vivo cytotoxicity	(Shakir et al., 2016)

Fungi

Fungi are eukaryotic microorganisms that include organisms such as mushroom, lichen, etc. Fungal-mediated green approach for the synthesising nanocomposite holds immense potential as they are rich source of bioactive compounds like flavonoids, phenolics, terpenoids, alkaloids, quinones, etc., having antitumour, antibacterial, antiviral and many other properties. A well-known bio-carbon rich material i. e. mushrooms widely used in the synthesis of nanocomposites. Due to its fleshy, porous, and eco-friendly nature, it is used in various applications such as electrochemical sensors (C. Liu et al., 2021), wound dressing (Rao et al., 2020), nitrophenol reduction (Chen et al., 2015).

Yeast

The single-celled Yeasts microorganisms are classified as members of the fungus kingdom. The yeast species *Saccharomyces cerevisiae* undergo fermentation process and it converts carbohydrates to carbon dioxide and alcohols and been used in many fields such as food, medicine, fuel industry, and cosmetics for thousands of years. They are majorly utilized in the finding of new, simple, eco-friendly green techniques. Mohamed & Awad (Mohamed & Awad, 2021), reported the use of Commercial baker's yeast powder (*Saccharomyces cerevisiae*) to fabricate N-doped/TiO_2/MnO_2 hollow nanocomposite spheres. Choudhury et. al. (Choudhury et al., 2017) prepared Yeast-titania nanocomposite by immobilizing dead yeast biomass for sorption of toxic heavy metal from contaminated water.

APPLICATION

The synergetic bio-synthesized nanocomposites are contributing to diverse environmental benign applications. The diverse applications of nanocomposites are displayed in Figure 4.

Numerous nanocomposites show applicability in the biomedical prospects such as antimicrobial (Araújo et al., 2018; Noohpisheh et al., 2020), antioxidant (Althomali et al., 2021; Noohpisheh et al., 2020), antifungal (Noohpisheh et al., 2020), Antiplanktonic, antibiofilm, antiswarming motility, and antiquorum sensing (Alavi & Karimi, 2018) activity. Several nanocomposites are effective in cancer treatment (A et al., 2020; Abd-Rabou et al., 2017; Ahamed et al., 2021; Emima Jeronsia et al., 2020; Kadiyala et al., 2018). The nanocomposites are used in wound dressings because of their antiseptic property (Dharmalingam & Anandalakshmi, 2020; Koneru et al., 2020; Rao et al., 2020). Functionalized nanocomposites are utilized in environmental remediation and waste treatment. The aspects such as catalytic reduction, photocatalysis, environmental catalysis for remediation, and biocatalysis are acknowledged in the application of nanocomposites (Akshatha et al., 2021; Çıplak, Getiren, et al., 2020; Gnanasekaran et al., 2021; Hassanin et al., 2021; Sarangapany & Mohanty, 2020). Excessively used toxic heavy metals in the industry are soluble in water, which can contaminate the water resources. Therefore, from an environmental point of view, these heavy metals can be removed with the help of nanocomposite (Choudhury et al., 2017; Prajapati & Mondal, 2021). The contribution of nanocomposites in nanosensors, biosensors, and sensors for heavy metals detection is on (Das & Sharma, 2020; Uwaya & Fayemi, 2020; Valian et al., 2020). Also, they find their way in the field of energy storage devices (Çıplak, Yıldız, et al., 2020; Kolya et al., 2019), Biohydrogen production (Rambabu et al., 2021). Some nanocomposites introduced their resistance behaviour against micro and macro-fouling in natural seawater environments (Soleimani

Figure 4. Diverse applications of nanocomposites

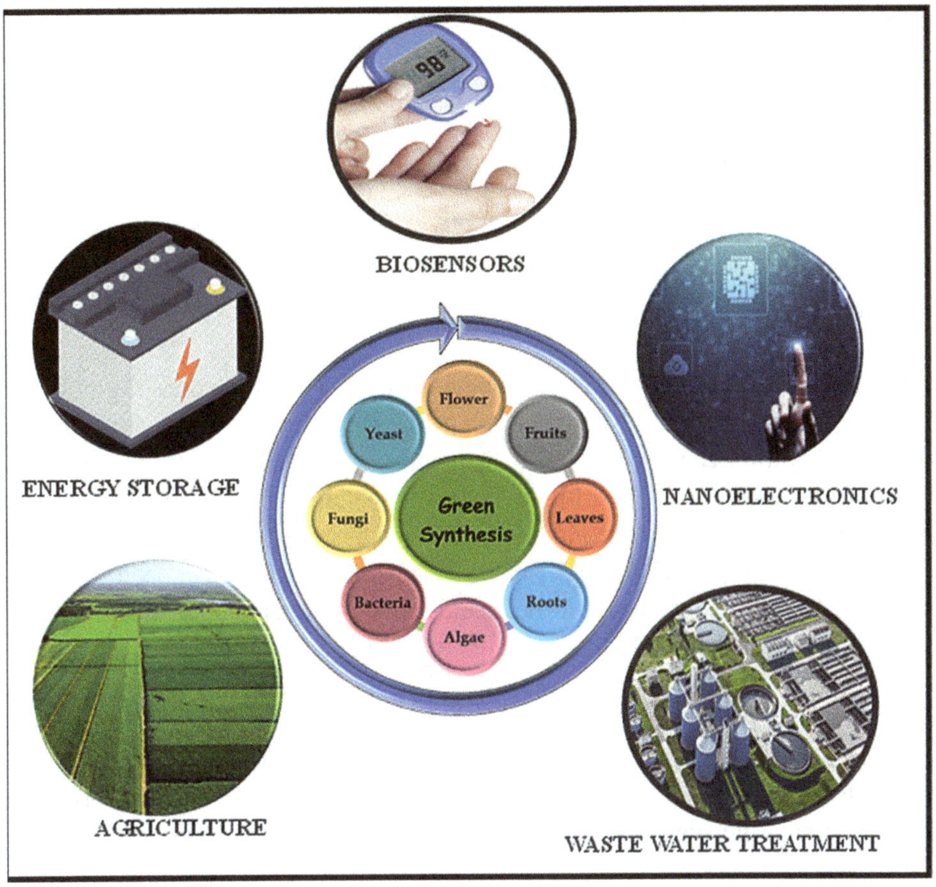

et al., 2021). The suitable constituent nanoparticle enhances the flame retardancy, thermal and mechanical properties of the nanocomposite (Asadi et al., 2020; Joulaei et al., 2019; Koosha & Hamedi, 2019). Also, they act as plasticizers in food packaging applications (Basumatary et al., 2018; Mohseni et al., 2020)

FUTURE RESEARCH DIRECTIONS

This study emphasized the recent years' research results on the biosynthesis of nanocomposites from renewable, eco-friendly biological entities such as plants and microorganisms, which is a significant step in the field of applied nanotechnology. The extract/ strain of green sources like flowers, fruits, leaves, roots, algae, bacteria, fungi, yeast are acting as stabilizing, reducing, or capping agents. The green synthesis is a reliable, clean, biologically appropriate, and environmental-friendly approach and demonstrates its versatility in many fields. This technique includes several physical and chemical methods in combination with natural precursors as reducing agents. Nature acts as a green source for the synthesis of nanocomposites. However, factors like local availability of resources, social adaptability, and economic feasibility heavily influence the overall synthesis process. Although this synthetic approach reduces negative impact on the environment, it cannot completely prevent the use of hazardous precursors.

Also, considerable effort has been utilized to synthesize different nanoparticles using the green route. Future prospect in the area of green synthesis is an improvement of reliable and eco-friendly processes for biosynthesized nanocomposites, as they show the synergetic effect of constituent Nanoparticles with enhanced physical, mechanical, thermal properties. This research area is under the developmental stage and opens challenges for future budding researchers.

Finally, we hope this study will contribute to accepting the challenges in the development of ever-growing interesting research area.

CONCLUSION

Over the last two decades, the constantly expanding field of 'Green synthesis' grabbed noteworthy attention, which allows synthesizing Nanomaterials with biocompatible, environmentally benevolent, non-toxic matrix and synthesized through non-toxic precursors. The advocacy for multiphase nanocomposites showing the synergic properties of individual nanomaterial is on the rise. Nowadays numerous nanocomposites are bio-synthesized by exploiting the extract of various plant parts (flowers, fruits, leaves, Roots) and microbes (algae, bacteria, fungi, yeast). Fascinating efforts have been made to categorize these biological entities as raw materials for the green synthesis of nanocomposites. The analysis made here brings forward minute details of green precursor, synthetic strategy, and application of different nanocomposites. There are wide opportunities and challenges in the green synthesis of nanocomposites, as it has applications in the field of environmental remediation, water remediation, solar cells, gas sensors, and many more. Hence, from this perspective, nanocomposite synthesis using 'Nature' as bio-laboratory can open new horizons in the future. The intensified research in this area could transform the large-scale production of nanocomposite synthesis on both the laboratory and industrial scale.

REFERENCES

A, S., Thamer, A., K, R., A, P., V, R., K, M., Murad, A., & P, M. (2020). Facile one pot microwave-assisted green synthesis of Fe_2O_3/Ag nanocomposites by phytoreduction: Potential application as sunlight-driven photocatalyst, antibacterial and anticancer agent. *Journal of Photochemistry and Photobiology B: Biology, 207*(April), 111885. doi:10.1016/j.jphotobiol.2020.111885

Abd-Rabou, A. A., Abdalla, A. M., Ali, N. A., & Zoheir, K. M. A. (2017). Moringa oleifera root induces cancer apoptosis more effectively than leave nanocomposites and its free counterpart. *Asian Pacific Journal of Cancer Prevention, 18*(8), 2141–2149. doi:10.22034/APJCP.2017.18.8.2141 PMID:28843248

Agawane, G. L., Shin, S. W., Vanalakar, S. A., Suryawanshi, M. P., Moholkar, A. V., & Kim, J. H. (2015). Investigations on chemo-mechano stabilities of the molybdenum thin films deposited by DC-sputter technique. *Zeitschrift für Physikalische Chemie, 229*(3), 377–393. doi:10.1515/zpch-2014-0624

Ahamed, M., Javed Akhtar, M., Majeed Khan, M. A., & Alhadlaq, H. A. (2021). Facile green synthesis of ZnO-RGO nanocomposites with enhanced anticancer efficacy. *Methods (San Diego, Calif.)*, (March). Advance online publication. doi:10.1016/j.ymeth.2021.04.020 PMID:33930572

Ahir, N. A., Takaloo, A. V., Nirmal, K. A., Kundale, S. S., Chougale, M. Y., Bae, J., Kim, D., & Dongale, T. D. (2021). Capacitive coupled non-zero I–V and type-II memristive properties of the $NiFe_2O_4$–TiO_2 nanocomposite. *Materials Science in Semiconductor Processing*, *125*, 105646. doi:10.1016/j.mssp.2020.105646

Ahmadi, R., Ghanbarzadeh, B., Ayaseh, A., Kafil, H. S., Özyurt, H., Katourani, A., & Ostadrahimi, A. (2019). The antimicrobial bio-nanocomposite containing non-hydrolyzed cellulose nanofiber (CNF) and Miswak (Salvadora persica L.) extract. *Carbohydrate Polymers*, *214*(March), 15–25. doi:10.1016/j.carbpol.2019.03.010 PMID:30925983

Akshatha, S., Sreenivasa, S., Parashuram, L., Udaya kumar, V., Alharthi, F. A., Chakrapani Rao, T. M., & Kumar, S. (2020, January). Microwave assisted green synthesis of p-type Co_3O_4@Mesoporous carbon spheres for simultaneous degradation of dyes and photocatalytic hydrogen evolution reaction. *Materials Science in Semiconductor Processing*, *121*, 105432. doi:10.1016/j.mssp.2020.105432

Al-Marri, A. H., Khan, M., Shaik, M. R., Mohri, N., Adil, S. F., Kuniyil, M., Alkhathlan, H. Z., Al-Warthan, A., Tremel, W., Tahir, M. N., Khan, M., & Siddiqui, M. R. H. (2016). Green synthesis of Pd@ graphene nanocomposite: Catalyst for the selective oxidation of alcohols. *Arabian Journal of Chemistry*, *9*(6), 835–845. doi:10.1016/j.arabjc.2015.12.007

Alavi, M., & Karimi, N. (2018). Antiplanktonic, antibiofilm, antiswarming motility and antiquorum sensing activities of green synthesized Ag–TiO_2, TiO_2–Ag, Ag–Cu and Cu–Ag nanocomposites against multi-drug-resistant bacteria. *Artificial Cells, Nanomedicine and Biotechnology, 46*(sup3), S399–S413. doi:10.1080/21691401.2018.1496923

Alavi, M., & Karimi, N. (2020). Hemoglobin self-assembly and antibacterial activities of bio-modified Ag-MgO nanocomposites by different concentrations of Artemisia haussknechtii and Protoparmeliopsis muralis extracts. *International Journal of Biological Macromolecules*, *152*, 1174–1185. doi:10.1016/j.ijbiomac.2019.10.207 PMID:31760003

Alavi, M., & Rai, M. (2019). Recent advances in antibacterial applications of metal nanoparticles (MNPs) and metal nanocomposites (MNCs) against multidrug-resistant (MDR) bacteria. *Expert Review of Anti-Infective Therapy*, *17*(6), 419–428. doi:10.1080/14787210.2019.1614914 PMID:31046483

Alavi, M., & Varma, R. S. (2021). Phytosynthesis and modification of metal and metal oxide nanoparticles/nanocomposites for antibacterial and anticancer activities: Recent advances. *Sustainable Chemistry and Pharmacy*, *21*(March), 100412. doi:10.1016/j.scp.2021.100412

Althomali, R. H., Alamry, K. A., Hussein, M. A., Khan, A., & Alosaimi, A. M. (2021). A green nanocomposite based modified cellulose/TiO_2/Cinnamon bark for the reduction of toxic organic compounds using spectrophotometric technique. *Journal of Materials Research and Technology*, *12*, 947–966. doi:10.1016/j.jmrt.2021.03.002

Amiri, M., Salavati-Niasari, M., Akbari, A., & Gholami, T. (2017). Removal of malachite green (a toxic dye) from water by cobalt ferrite silica magnetic nanocomposite: Herbal and green sol-gel autocombustion synthesis. *International Journal of Hydrogen Energy*, *42*(39), 24846–24860. doi:10.1016/j.ijhydene.2017.08.077

Araújo, I. M. S., Silva, R. R., Pacheco, G., Lustri, W. R., Tercjak, A., Gutierrez, J., Júnior, J. R. S., Azevedo, F. H. C., Figuêredo, G. S., Vega, M. L., Ribeiro, S. J. L., & Barud, H. S. (2018). Hydrothermal synthesis of bacterial cellulose–copper oxide nanocomposites and evaluation of their antimicrobial activity. *Carbohydrate Polymers, 179*(September), 341–349. doi:10.1016/j.carbpol.2017.09.081

Asadi, F., Forootanfar, H., & Ranjbar, M. (2020). A facile one-step preparation of $Ca_{10}(PO_4)_6(OH)_2$/ Li-BioMOFs resin nanocomposites with Glycyrrhiza glabra (licorice) root juice as green capping agent and mechanical properties study. *Artificial Cells, Nanomedicine, and Biotechnology, 48*(1), 1331–1339. doi:10.1080/21691401.2020.1842748 PMID:33170039

Bajpai, V. K., Kamle, M., Shukla, S., Mahato, D. K., Chandra, P., Hwang, S. K., Kumar, P., Huh, Y. S., & Han, Y.-K. (2018). Prospects of using nanotechnology for food preservation, safety, and security. *Yao Wu Shi Pin Fen Xi, 26*(4), 1201–1214. doi:10.1016/j.jfda.2018.06.011 PMID:30249319

Baladi, M., Hajizadeh-Oghaz, M., Amiri, O., Valian, M., & Salavati-Niasari, M. (2021). Enhanced photocatalytic activity of $Sr_7Mn_7O_{19.62}$-Dy_2O_3 nanocomposite synthesized via a green method. *International Journal of Hydrogen Energy, 46*(5), 3763–3779. doi:10.1016/j.ijhydene.2020.10.167

Basumatary, K., Daimary, P., Das, S. K., Thapa, M., Singh, M., Mukherjee, A., & Kumar, S. (2018). Lagerstroemia speciosa fruit-mediated synthesis of silver nanoparticles and its application as filler in agar-based nanocomposite films for antimicrobial food packaging. *Food Packaging and Shelf Life, 17*(June), 99–106. doi:10.1016/j.fpsl.2018.06.003

Bhadane, M. S., Gavhane, K. H., Kulkarni, P. P., Dahiwale, S. S., Bhoraskar, V. N., More, M. A., Patil, P. S., & Dhole, S. D. (2020). Particle size dependent TL response of CaF_2: Dy phosphor for gamma dosimetry. *Journal of Luminescence, 223*, 117168. doi:10.1016/j.jlumin.2020.117168

Chen, R., Zhang, Q., Gu, Y., Tang, L., Li, C., & Zhang, Z. (2015). One-pot green synthesis of Prussian blue nanocubes decorated reduced graphene oxide using mushroom extract for efficient 4-nitrophenol reduction. *Analytica Chimica Acta, 853*(1), 579–587. doi:10.1016/j.aca.2014.10.049 PMID:25467506

Choudhury, P. R., Bhattacharya, P., Ghosh, S., Majumdar, S., Saha, S., & Sahoo, G. C. (2017). Removal of Cr (VI) by synthesized titania embedded dead yeast nanocomposite: Optimization and modeling by response surface methodology. *Journal of Environmental Chemical Engineering, 5*(1), 214–221. doi:10.1016/j.jece.2016.11.041

Çıplak, Z., Getiren, B., Gökalp, C., Yıldız, A., & Yıldız, N. (2020). Green synthesis of reduced graphene oxide-AgAu bimetallic nanocomposite: Catalytic performance. *Chemical Engineering Communications, 207*(4), 559–573. doi:10.1080/00986445.2019.1613227

Çıplak, Z., Yıldız, A., & Yıldız, N. (2020). Green preparation of ternary reduced graphene oxide-au@ polyaniline nanocomposite for supercapacitor application. *Journal of Energy Storage, 32*(August), 101846. doi:10.1016/j.est.2020.101846

Das, T. R., & Sharma, P. K. (2020). Hydrothermal-assisted green synthesis of Ni/Ag@rGO nanocomposite using Punica granatum juice and electrochemical detection of ascorbic acid. *Microchemical Journal, 156*(March), 104850. doi:10.1016/j.microc.2020.104850

Dhand, C., Dwivedi, N., Loh, X. J., Jie Ying, A. N., Verma, N. K., Beuerman, R. W., Lakshminarayanan, R., & Ramakrishna, S. (2015). Methods and strategies for the synthesis of diverse nanoparticles and their applications: A comprehensive overview. *RSC Advances*, *5*(127), 105003–105037. doi:10.1039/C5RA19388E

Dharmalingam, K., & Anandalakshmi, R. (2020). Functionalization of cellulose-based nanocomposite hydrogel films with zinc oxide complex and grapefruit seed extract for potential applications in treating chronic wounds. *Polymer*, *202*(April), 122620. doi:10.1016/j.polymer.2020.122620

Dhas, S. D., Maldar, P. S., Patil, M. D., Hubali, K. M., Shembade, U. V., Abitkar, S. B., Waikar, M. R., Sonkawade, R. G., Agawane, G. L., & Moholkar, A. V. (2021). Hydrothermal synthesis of mesoporous $NiMnO_3$ nanostructures for supercapacitor application: Effect of electrolyte. *Journal of Energy Storage*, *35*, 102277.

Digraskar, R., Sapner, V. S., Ghule, A. V., & Sathe, B. R. (2021). CZTS/MoS_2-rGO Heterostructures: An efficient and highly stable electrocatalyst for enhanced hydrogen generation reactions. *Journal of Electroanalytical Chemistry (Lausanne, Switzerland)*, *882*, 114983. doi:10.1016/j.jelechem.2021.114983

Emima Jeronsia, J., Ragu, R., Sowmya, R., Mary, A. J., & Jerome Das, S. (2020). Comparative investigation on Camellia Sinensis mediated green synthesis of Ag and Ag/GO nanocomposites for its anticancer and antibacterial efficacy. *Surfaces and Interfaces*, *21*(November), 100787. doi:10.1016/j.surfin.2020.100787

Fadlallah, S., Gad El-Rab, S. M. F., & Halwani, E. M. (2020). Innovative Nanoporous Titania Surface with Stabilized Antimicrobial Ag-Nanoparticles via Salvadora persica L. Roots (Miswak) Extract for Dental Applications. *BioNanoScience*, *10*(4), 998–1009. doi:10.100712668-020-00765-7

Gangadhar, L., Reddy, K. B., Garg, A. P., & Sana, S. S. (2020). Green Synthesis of Bio-polymer Composites of Iron for Pharmaceutical Applications. *Journal of Nanomedicine & Nanotechnology*, *11*, 551. doi:10.35248/2157-7439.20.11.551

Gnanasekaran, L., Rajendran, S., Priya, A. K., Durgalakshmi, D., Vo, D. V. N., Cornejo-Ponce, L., Gracia, F., & Soto-Moscoso, M. (2021). Photocatalytic degradation of 2,4-dichlorophenol using bio-green assisted TiO_2–CeO_2 nanocomposite system. *Environmental Research*, *195*(February), 110852. Advance online publication. doi:10.1016/j.envres.2021.110852 PMID:33556356

Hassanin, H. A., Taha, A., & Afkar, E. (2021). Novel bio-mediated Ag/Co_3O_4 nanocomposites of different weight ratios using aqueous neem leaf extract: Catalytic and microbial behaviour. *Ceramics International*, *47*(3), 3099–3107. doi:10.1016/j.ceramint.2020.09.146

Hessien, M., Da'na, E., & Taha, A. (2021). Phytoextract assisted hydrothermal synthesis of ZnO–NiO nanocomposites using neem leaves extract. *Ceramics International*, *47*(1), 811–816. doi:10.1016/j.ceramint.2020.08.192

Honarmand, M., Golmohammadi, M., & Naeimi, A. (2020). Green synthesis of SnO_2-bentonite nanocomposites for the efficient photodegradation of methylene blue and eriochrome black-T. *Materials Chemistry and Physics*, *241*(May), 122416. doi:10.1016/j.matchemphys.2019.122416

Indira, J., & Sreeja, V. (2020). Synthesis of silver/hydroxyapatite/tryptophan nanocomposite particles by biological method. *Materials Today: Proceedings*. Advance online publication. doi:10.1016/j.matpr.2020.09.579

Iravani, S. (2014). Bacteria in Nanoparticle Synthesis: Current Status and Future Prospects. *International Scholarly Research Notices*, *2014*, 1–18. doi:10.1155/2014/359316 PMID:27355054

Jagwani, D., & Hari Krishna, P. (2021). Nature's nano-assets: Green synthesis, characterization techniques and applications – A graphical review. *Materials Today: Proceedings*, 2307–2317. Advance online publication. doi:10.1016/j.matpr.2021.04.185

Joulaei, M., Hedayati, K., & Ghanbari, D. (2019). Investigation of magnetic, mechanical and flame retardant properties of polymeric nanocomposites: Green synthesis of $MgFe_2O_4$ by lime and orange extracts. *Composites. Part B, Engineering*, *176*(July), 107345. doi:10.1016/j.compositesb.2019.107345

Kadiyala, N. K., Mandal, B. K., Ranjan, S., & Dasgupta, N. (2018). Bioinspired gold nanoparticles decorated reduced graphene oxide nanocomposite using Syzygium cumini seed extract: Evaluation of its biological applications. *Materials Science and Engineering C, 93*(October), 191–205. doi:10.1016/j.msec.2018.07.075

Karthick Raja Namasivayam, S., Arvind Bharani, R. S., & Karunamoorthy, K. (2018). Insecticidal fungal metabolites fabricated chitosan nanocomposite (IM-CNC) preparation for the enhanced larvicidal activity - An effective strategy for green pesticide against economic important insect pests. *International Journal of Biological Macromolecules*, *120*, 921–944. doi:10.1016/j.ijbiomac.2018.08.130 PMID:30172810

Khan, I., Saeed, K., & Khan, I. (2019). Nanoparticles: Properties, applications and toxicities. *Arabian Journal of Chemistry*, *12*(7), 908–931. doi:10.1016/j.arabjc.2017.05.011

Khan, N. T., & Khan, M. J. (2020). Metallic Nanoparticles Fabrication Methods– A Brief Overview. *SunKrist Nanotechnology and Nanoscience Journal*, *2*(1), 1–6. doi:10.46940nnj.02.1002

Kolya, H., Kuila, T., Kim, N. H., & Lee, J. H. (2019). Bioinspired silver nanoparticles/reduced graphene oxide nanocomposites for catalytic reduction of 4-nitrophenol, organic dyes and act as energy storage electrode material. *Composites. Part B, Engineering*, *173*(May), 106924. doi:10.1016/j.compositesb.2019.106924

Koneru, A., Dharmalingam, K., & Anandalakshmi, R. (2020). Cellulose based nanocomposite hydrogel films consisting of sodium carboxymethylcellulose–grapefruit seed extract nanoparticles for potential wound healing applications. *International Journal of Biological Macromolecules*, *148*, 833–842. doi:10.1016/j.ijbiomac.2020.01.018 PMID:31953172

Koosha, M., & Hamedi, S. (2019). Intelligent Chitosan/PVA nanocomposite films containing black carrot anthocyanin and bentonite nanoclays with improved mechanical, thermal and antibacterial properties. *Progress in Organic Coatings, 127*(May), 338–347. doi:10.1016/j.porgcoat.2018.11.028

Kulal, D., & Shetty Kodialbail, V. (2021). Visible light mediated photocatalytic dye degradation using $Ag_2O/AgO\text{-}TiO_2$ nanocomposite synthesized by extracellular bacterial mediated synthesis - An eco-friendly approach for pollution abatement. *Journal of Environmental Chemical Engineering*, *9*(4), 105389. doi:10.1016/j.jece.2021.105389

Li, X., Xu, H., Chen, Z. S., & Chen, G. (2011). Biosynthesis of nanoparticles by microorganisms and their applications. *Journal ofNanomaterials, 2011*, 1–16. Advance online publication. doi:10.1155/2011/270974

Liu, C., Zhang, N., Huang, X., Wang, Q., Wang, X., & Wang, S. (2021). Fabrication of a novel nanocomposite electrode with ZnO-MoO_3 and biochar derived from mushroom biomaterials for the detection of acetaminophen in the presence of DA. *Microchemical Journal, 161*(November), 105719. doi:10.1016/j.microc.2020.105719

Mane, A. A., & Moholkar, A. V. (2018). Effect of solution concentration on physicochemical and NO2 gas sensing properties of sprayed MoO_3 nanobelts. *Thin Solid Films, 648*(2), 50–61. doi:10.1016/j.tsf.2018.01.008

Manjari, G., Saran, S., Radhakrishanan, S., Rameshkumar, P., Pandikumar, A., & Devipriya, S. P. (2020). Facile green synthesis of Ag–Cu decorated ZnO nanocomposite for effective removal of toxic organic compounds and an efficient detection of nitrite ions. *Journal of Environmental Management, 262*(January), 110282. doi:10.1016/j.jenvman.2020.110282 PMID:32090885

Manjari Mishra, P., Pattnaik, S., & Prabha Devi, A. (2021). Green synthesis of bio-based Au@g-C_3N_4 nanocomposite for photocatalytic degradation of methyl orange. *Materials Today: Proceedings, xxxx*, 1218–1223. Advance online publication. doi:10.1016/j.matpr.2021.03.712

Mohamed, E. F., & Awad, G. (2021). Hollow N-TiO_2/MnO_2 nanocomposite-based yeast biomass for gaseous formaldehyde degradation under visible light. *Journal of Industrial and Engineering Chemistry, 98*, 366–374. doi:10.1016/j.jiec.2021.03.028

Mohammadi-Aloucheh, R., Habibi-Yangjeh, A., Bayrami, A., Latifi-Navid, S., & Asadi, A. (2018). Green synthesis of ZnO and ZnO/CuO nanocomposites in Mentha longifolia leaf extract: Characterization and their application as anti-bacterial agents. *Journal of Materials Science Materials in Electronics, 29*(16), 13596–13605. doi:10.100710854-018-9487-0

Mohseni, M. S., Khalilzadeh, M. A., Mohseni, M., Hargalani, F. Z., Getso, M. I., Raissi, V., & Raiesi, O. (2020). Green synthesis of Ag nanoparticles from pomegranate seeds extract and synthesis of Ag-Starch nanocomposite and characterization of mechanical properties of the films. *Biocatalysis and Agricultural Biotechnology, 25*(March), 101569. doi:10.1016/j.bcab.2020.101569

Mondal, P., Anweshan, A., & Purkait, M. K. (2020). Green synthesis and environmental application of iron-based nanomaterials and nanocomposite: A review. *Chemosphere, 259*, 127509. Advance online publication. doi:10.1016/j.chemosphere.2020.127509 PMID:32645598

Naghizadeh, A., Mohammadi-Aghdam, S., & Mortazavi-Derazkola, S. (2020). Novel CoFe2O4@ ZnO-CeO2 ternary nanocomposite: Sonochemical green synthesis using Crataegus microphylla extract, characterization and their application in catalytic and antibacterial activities. *Bioorganic Chemistry, 103*(August), 104194. doi:10.1016/j.bioorg.2020.104194 PMID:32890997

Nasrollahzadeh, M., Akbari, R., Issaabadi, Z., & Sajadi, S. M. (2020). Biosynthesis and characterization of Ag/MgO nanocomposite and its catalytic performance in the rapid treatment of environmental contaminants. *Ceramics International, 46*(2), 2093–2101. doi:10.1016/j.ceramint.2019.09.191

Navya Rani, M., Murthy, M., Shyla Shree, N., Ananda, S., Yogesh, S., & Dinesh, R. (2019). Cuprous oxide anchored Reduced Graphene oxide ceramic nanocomposite using Tagetes erecta flower extract and evaluation of its antibacterial activity and cytotoxicity. *Ceramics International*, *45*(18), 25020–25026. doi:10.1016/j.ceramint.2019.04.195

Noohpisheh, Z., Amiri, H., Farhadi, S., & Mohammadi-gholami, A. (2020). Green synthesis of Ag-ZnO nanocomposites using Trigonella foenum-graecum leaf extract and their antibacterial, antifungal, antioxidant and photocatalytic properties. *Spectrochimica Acta. Part A: Molecular and Biomolecular Spectroscopy*, *240*, 118595. doi:10.1016/j.saa.2020.118595 PMID:32599480

Orooji, Y., Mortazavi-Derazkola, S., Ghoreishi, S. M., Amiri, M., & Salavati-Niasari, M. (2020). Mesopourous Fe_3O_4@SiO_2-hydroxyapatite nanocomposite: Green sonochemical synthesis using strawberry fruit extract as a capping agent, characterization and their application in sulfasalazine delivery and cytotoxicity. *Journal of Hazardous Materials*, *400*(April), 123140. doi:10.1016/j.jhazmat.2020.123140 PMID:32563904

Panchal, P., Paul, D. R., Sharma, A., Hooda, D., Yadav, R., Meena, P., & Nehra, S. P. (2019). Phytoextract mediated ZnO/MgO nanocomposites for photocatalytic and antibacterial activities. *Journal of Photochemistry and Photobiology A Chemistry*, *385*(June), 112049. doi:10.1016/j.jphotochem.2019.112049

Patil, A. S., Patil, M. D., Lohar, G. M., Jadhav, S. T., & Fulari, V. J. (2017). Supercapacitive properties of CuO thin films using modified SILAR method. *Ionics*, *23*(5), 1259–1266. Advance online publication. doi:10.100711581-016-1921-9

Patil, M. D., Dhas, S. D., Mane, A. A., & Moholkar, A. V. (2021). Clinker-like V2O5 nanostructures anchored on 3D Ni-foam for supercapacitor application. *Materials Science in Semiconductor Processing*, *133*, 105978. doi:10.1016/j.mssp.2021.105978

Pomastowski, P., Kr, A., & Railean-plugaru, V. (2020). *Zinc Oxide Nanocomposites — Extracellular Synthesis*. Academic Press.

Prajapati, A. K., & Mondal, M. K. (2021). Novel green strategy for CuO–ZnO–C nanocomposites fabrication using marigold (Tagetes spp.) flower petals extract with and without CTAB treatment for adsorption of Cr(VI) and Congo red dye. *Journal of Environmental Management, 290*(November), 112615. doi:10.1016/j.jenvman.2021.112615

Rafique, M., Sadaf, I., Rafique, M. S., & Tahir, M. B. (2017). A review on green synthesis of silver nanoparticles and their applications. *Artificial Cells, Nanomedicine, and Biotechnology*, *45*(7), 1272–1291. doi:10.1080/21691401.2016.1241792 PMID:27825269

Rambabu, K., Bharath, G., Banat, F., Hai, A., Show, P. L., & Phong Nguyen, T. H. (2021). Ferric oxide/date seed activated carbon nanocomposites mediated dark fermentation of date fruit wastes for enriched biohydrogen production. *International Journal of Hydrogen Energy*, *46*(31), 16631–16643. doi:10.1016/j.ijhydene.2020.06.108

Rao, K. M., Suneetha, M., Park, G. T., Babu, A. G., & Han, S. S. (2020). Hemostatic, biocompatible, and antibacterial non-animal fungal mushroom-based carboxymethyl chitosan-ZnO nanocomposite for wound-healing applications. *International Journal of Biological Macromolecules, 155*, 71–80. doi:10.1016/j.ijbiomac.2020.03.170 PMID:32217121

Rashki, S., Abbas Alshamsi, H., Amiri, O., Safardoust-Hojaghan, H., Salavati-Niasari, M., Nazari-Alam, A., & Khaledi, A. (2021). Eco-friendly green synthesis of ZnO/GQD nanocomposites using Protoparmeliopsis muralis extract for their antibacterial and antibiofilm activity. *Journal of Molecular Liquids, 335*, 116195. doi:10.1016/j.molliq.2021.116195

Rodponthukwaji, K., Saengruengrit, C., Tummamunkong, P., Leelahavanichakul, A., Ritprajak, P., & Insin, N. (2021). Facile synthesis of magnetic silica-mannan nanocomposites for enhancement in internalization and immune response of dendritic cells. *Materials Today. Chemistry, 20*, 100417. doi:10.1016/j.mtchem.2020.100417

Rohaizad, A., Shahabuddin, S., Shahid, M. M., Rashid, N. M., Hir, Z. A. M., Ramly, M. M., Awang, K., Siong, C. W., & Aspanut, Z. (2020). Green synthesis of silver nanoparticles from Catharanthus roseus dried bark extract deposited on graphene oxide for effective adsorption of methylene blue dye. *Journal of Environmental Chemical Engineering, 8*(4), 103955. Advance online publication. doi:10.1016/j.jece.2020.103955

Rostami-Vartooni, A., Moradi-Saadatmand, A., & Mahdavi, M. (2018). Catalytic reduction of organic pollutants using biosynthesized Ag/C/Fe_3O_4 nanocomposite by red water and Caesalpinia gilliesii flower extract. *Materials Chemistry and Physics, 219*(June), 328–339. doi:10.1016/j.matchemphys.2018.08.026

S.A., A. R., & Tamsilian, Y. (2016). Synthesis of Nanomaterials. In Nanocomposite Materials. CRC Press. doi:10.1201/9781315372310-3

Sajjad, S., Leghari, S. A. K., Ryma, N.-U.-A., & Farooqi, S. A. (2018). Green Synthesis of Metal-Based Nanoparticles and Their Applications. In *Green Metal Nanoparticles* (pp. 23–77). John Wiley & Sons, Ltd. doi:10.1002/9781119418900.ch2

Saquib, Q., Faisal, M., & Abdulrahman, A. A. A. (2020). Green Synthesis of Nanoparticles: Applications and Prospects. In Green Synthesis of Nanoparticles: Applications and Prospects. doi:10.1007/978-981-15-5179-6

Sarangapany, S., & Mohanty, K. (2020). Facile green synthesis of magnetically separable Au–Pt@TiO_2 nanocomposite for efficient catalytic reduction of organic pollutants and selective oxidation of glycerol. *Journal of Alloys and Compounds, 830*, 154636. doi:10.1016/j.jallcom.2020.154636

Saratale, R. G., Saratale, G. D., Shin, H. S., Jacob, J. M., Pugazhendhi, A., Bhaisare, M., & Kumar, G. (2018). New insights on the green synthesis of metallic nanoparticles using plant and waste biomaterials: Current knowledge, their agricultural and environmental applications. *Environmental Science and Pollution Research International, 25*(11), 10164–10183. doi:10.100711356-017-9912-6 PMID:28815433

Satheeshkumar, M. K., Kumar, E. R., Indhumathi, P., Srinivas, C., Deepty, M., Sathiyaraj, S., Suriyanarayanan, N., & Sastry, D. L. (2020). Structural, morphological and magnetic properties of algae/ $CoFe_2O_4$ and algae/Ag-Fe-O nanocomposites and their biomedical applications. *Inorganic Chemistry Communications, 111*(August), 107578. doi:10.1016/j.inoche.2019.107578

Sayğılı, H. (2019). Hydrothermal synthesis of magnetic nanocomposite from biowaste matrix by a green and one-step route: Characterization and pollutant removal ability. *Bioresource Technology, 278*(November), 242–247. doi:10.1016/j.biortech.2019.01.103

Shakir, M., Khan, M. S., Baig, U., Alam, M. F., Younus, H., & Alam, M. (2016). In vivo cytotoxicity, molecular docking and study of yeast alcohol dehydrogenase on polycarbazole-titanium dioxide nanocomposite. *Journal of Molecular Catalysis. B, Enzymatic, 134*, 79–88. doi:10.1016/j.molcatb.2016.09.018

Sheikh-Mohseni, M. H., Sedaghat, S., Derakhshi, P., & Safekordi, A. (2020). Green bio-synthesis of Ni/ montmorillonite nanocomposite using extract of Allium jesdianum as the nano-catalyst for electrocatalytic oxidation of methanol. *Chinese Journal of Chemical Engineering, 28*(10), 2555–2565. doi:10.1016/j.cjche.2020.04.017

Singh, S., Bhatta, U. M., Satyam, P., Dhawan, A., Sastry, M., & Prasad, B. L. V. (2008). Bacterial synthesis of silicon/silica nanocompositesc. *Journal of Materials Chemistry, 18*(22), 2601–2606. doi:10.1039/ b719528a

Soleimani, S., Jannesari, A., Yousefzadi, M., Ghaderi, A., & Shahdadi, A. (2021). Eco-friendly foul release coatings based on a novel reduced graphene oxide/Ag nanocomposite prepared by a green synthesis approach. *Progress in Organic Coatings, 151*(August), 106107. doi:10.1016/j.porgcoat.2020.106107

Syame, S. M., Mohamed, W. S., Mahmoud, R. K., & Omara, S. T. (2017). Synthesis of copper-chitosan nanocomposites and their applications in treatment of local pathogenic isolates bacteria. *Oriental Journal of Chemistry, 33*(6), 2959–2969. doi:10.13005/ojc/330632

Truong, D. H., Dam, M. S., Bujna, E., Rezessy-Szabo, J., Farkas, C., Vi, V. N. H., Csernus, O., Nguyen, V. D., Gathergood, N., Friedrich, L., Hafidi, M., Gupta, V. K., & Nguyen, Q. D. (2021). In situ fabrication of electrically conducting bacterial cellulose-polyaniline-titanium-dioxide composites with the immobilization of Shewanella xiamenensis and its application as bioanode in microbial fuel cell. *Fuel, 285*(August). doi:10.1016/j.fuel.2020.119259

Uwaya, G. E., & Fayemi, O. E. (2020). Electrochemical detection of serotonin in banana at green mediated PPy/Fe_3O_4NPs nanocomposites modified electrodes. *Sensing and Bio-Sensing Research, 28*(April), 100338. doi:10.1016/j.sbsr.2020.100338

Valian, M., Khoobi, A., & Salavati-Niasari, M. (2020). Green synthesis and characterization of $DyMnO_3$-ZnO ceramic nanocomposites for the electrochemical ultratrace detection of atenolol. *Materials Science and Engineering C, 111*(March), 110854. doi:10.1016/j.msec.2020.110854 PMID:32279805

Vane, C. H., Drage, T. C., & Snape, C. E. (2006). Bark decay by the white-rot fungus Lentinula edodes: Polysaccharide loss, lignin resistance and the unmasking of suberin. *International Biodeterioration & Biodegradation, 57*(1), 14–23. doi:10.1016/j.ibiod.2005.10.004

Vinay, S. P., Alharthi, F. A., Udayabhanu, Alsalme, A., & Nagaraju, G. (2021). Hydrothermal synthesis of Ag/rGO@CTFE nanocomposite as a promising photocatalyst for degradation action. *Journal of Molecular Structure*, *1228*, 129722. doi:10.1016/j.molstruc.2020.129722

Wasim, M., Mushtaq, M., Khan, S. U., Farooq, A., Naeem, M. A., Khan, M. R., Salam, A., & Wei, Q. (2020). Development of bacterial cellulose nanocomposites: An overview of the synthesis of bacterial cellulose nanocomposites with metallic and metallic-oxide nanoparticles by different methods and techniques for biomedical applications. *Journal of Industrial Textiles*. Advance online publication. doi:10.1177/1528083720977201

Yulizar, Y., Bakri, R., Apriandanu, D. O. B., & Hidayat, T. (2018). ZnO/CuO nanocomposite prepared in one-pot green synthesis using seed bark extract of Theobroma cacao. *Nano-Structures and Nano-Objects*, *16*, 300–305. doi:10.1016/j.nanoso.2018.09.003

Yulizar, Y., Sudirman, Apriandanu, D. O. B., & Wibowo, A. P. (2019). Plant extract mediated synthesis of Au/TiO_2 nanocomposite and its photocatalytic activity under sodium light irradiation. *Composites Communications*, *16*(May), 50–56. doi:10.1016/j.coco.2019.08.006

Zarei, M., Seyedi, N., Maghsoudi, S., Nejad, M. S., & Sheibani, H. (2021). Green synthesis of Ag nanoparticles on the modified graphene oxide using Capparis spinosa fruit extract for catalytic reduction of organic dyes. *Inorganic Chemistry Communications, 123*(November), 108327. doi:10.1016/j.inoche.2020.108327

ADDITIONAL READING

Ameta, R., Rai, A. K., Vyas, S., Bhatt, J. P., & Ameta, S. C. (2021). Green synthesis of nanocomposites using plant extracts and their applications. Handbook of Greener Synthesis of Nanomaterials and Compounds, 663–682. doi:10.1016/B978-0-12-821938-6.00020-7

Chokkareddy, R., & Redhi, G. G. (2018). Green Synthesis of Metal Nanoparticles and its Reaction Mechanisms. In *Green Metal Nanoparticles* (pp. 113–139). Academic Press.

Liou, T. H., Tseng, Y. K., Liu, S. M., Lin, Y. T., Wang, S. Y., & Liu, R. T. (2021). Green synthesis of mesoporous graphene oxide/silica nanocomposites from rich husk ash: Characterization and adsorption performance. *Environmental Technology & Innovation*, *22*, 101424. doi:10.1016/j.eti.2021.101424

Singh, A., Kaur, N., Parmar, A., & Chopra, H. K. (2021). The Fundamental perspectives of greener synthesis. Handbook of Greener Synthesis of Nanomaterials and Compounds, 3–36. doi:10.1016/B978-0-12-821938-6.00001-3

KEY TERMS AND DEFINITIONS

Bottom-Up Approach: It refers to the build-up of a material from the bottom: atom-by-atom, molecule-by-molecule, or cluster-by cluster.

Extract: The concentrated preparation containing the active ingredient of a substance.

Green Synthesis: The use of environmentally compatible materials such as microorganisms and plants for the synthesis of nanoparticles/nanocomposites.

Nanocomposite: A multiphase solid material where one of the phases has dimensions of less than 100 nanometres (nm) or structures having nano-scale repeat distances between the different phases that make up the material.

Nanotechnology: An emerging area of research which deal with the matter with at least one dimension sized from 1 to 100 nanometres.

Synergetic Effect: When atleast two processes interact together and produce an effect which is greater than the cumulative effect that those processes produce when used individually.

Top-Down Approach: It involves the breaking down of the bulk material into nanosized structures or particles.

Chapter 5
New Advances in Graphene-Based Three-Dimensional Structures:
Synthesis and Applications

Javier Cencerrero Fernández del Moral
University of Castilla la Mancha, Spain

Amaya Romero Izquierdo
University of Castilla la Mancha, Spain

Paula Sánchez Paredes
University of Castilla la Mancha, Spain

Osmín Avilés-García
Universidad Michoacana de San Nicolás de Hidalgo, Mexico

Israel Fernandez-Reina
University of Castilla la Mancha, Spain

ABSTRACT

Graphene is defined as a two-dimensional network of carbon atoms with a single atom thickness and a hexagonal crystalline structure with sp2 hybridization compacted by covalent bonds. Due to its structure and geometry, graphene has unique properties, including a large specific surface area, rapid mobility of load carriers, and high thermal and electrical conductivity. However, these characteristics are limited due to the restructuring of graphene sheets. For this reason, there are many studies devoted to the synthesis of three-dimensional structures that prevent the agglomeration of the sheets and the loss of properties of the graphene structure. These three-dimensional structures have low density, high porosity and surface area, stable mechanical properties, and good mass and electron transfer properties. This chapter aims to summarize the synthesis methods of the different three-dimensional structures derived from graphene as well as their wide range of applications in environmental remediation, sensors, biomedical and energy-related applications, among many others.

DOI: 10.4018/978-1-7998-8936-6.ch005

INTRODUCTION

Since its discovery, graphene has become one of the most interesting materials in the 21st century. Its chemical bond and structure were first described in the 1930s, but at that time, it was viewed as having little importance as it was assumed to be unstable when undergoing thermal changes. However, in 2004, a research group, led by physicist Andre Gueim, managed to isolate graphite sheets by means of mechanical exfoliation of highly oriented pyrolytic graphite (HOPG). HOPG is a form of high order and high purity synthetic graphite. Thanks to their discovery, this research group was awarded the Nobel Prize in Physics in 2010. (W. Gao, 2012)

Graphene is composed of a series of carbon atoms organized in a regular honeycomb pattern, which is generated by covalent bonds from the superposition of sp^2 hybrids from bonded carbon. As a result of this structure, graphene acquires unique properties such as it is extremely hard and resistant, it has high thermal and electrical conductivity, being very light, flexible, and elastic, and it has a low Joule effect. Despite all these characteristics, this material has had few applications, mainly as the graphene sheets agglomerate and restructure, by which some of their properties are removed (Y. Wu et al., 2019). In order to maintain these unique properties, one alternative developed was to manipulate these sheets in order to build three-dimensional structures derived from graphene, thus preventing them from restructuring.

Three-dimensional structures derived from graphene not only retain their intrinsic properties but also acquire other important characteristics such as high porosity, low density, large specific surface area, excellent mechanical properties, etc. (Foster et al., 2017; Xia et al., 2017). Figure 1 shows the upward trend in articles concerning the main three-dimensional structures derived from graphene, such as spheres, adsorbents, foams, and graphene aerogels, showing the growing interest of research into these structures. In addition, note that 2014 showed the greatest year-on-year difference in the number of articles published. That year studies began to focus more on this type of structure, its synthesis, and ways of using it. This book chapter summarises synthesis methods and the main applications of 3D graphene structures. Graphene spheres, aerogels, adsorbents, and porous graphene structures are those considered in this study.

BACKGROUND

As mentioned above, three-dimensional structures derived from graphene prevent some of the properties characteristic of two-dimensional structures from being suppressed, by preventing the sheets from restructuring and agglomerating. However, three-dimensional structures have different geometry depending on the synthesis method used and its application.

As a summary, Figure 2 presents the different graphene-derived three-dimensional structures discussed in this book chapter (spheres, aerogels, foams, and adsorbents), as well as their synthesis methods and main applications.

Figure 1. Number of articles published per year related to different graphene structures. Source: Scopus. Key Words Searched: Three-Dimensional graphene, graphene spheres, Graphene Adsorbent Porous Structure, Graphene Foams, and Graphene Aerogels.

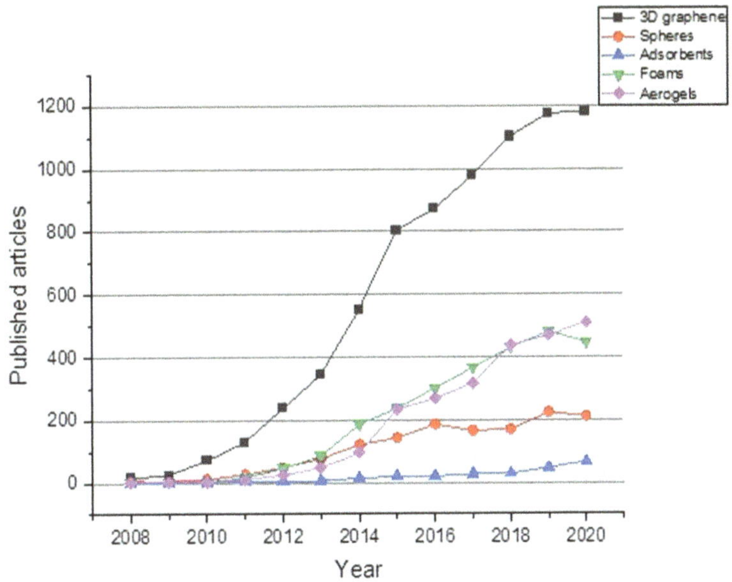

Figure 2. Synthesis and applications of three-dimensional graphene structures.

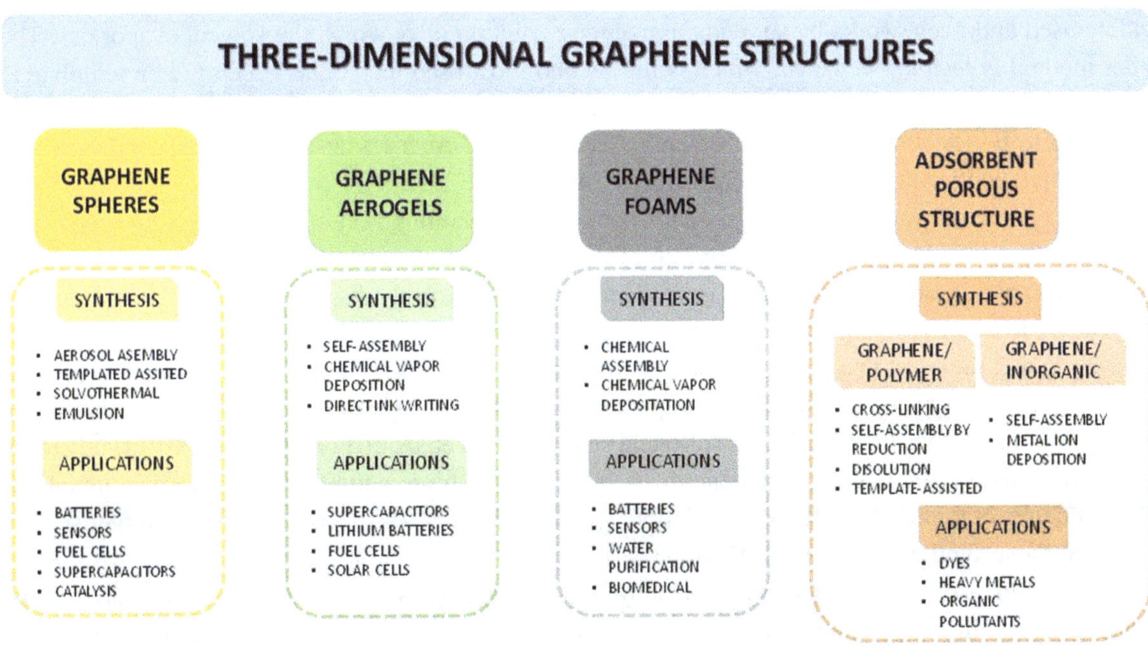

Figure 3. Synthesis route of graphene spheres by the aerosol assembly (Ha et al., 2019).

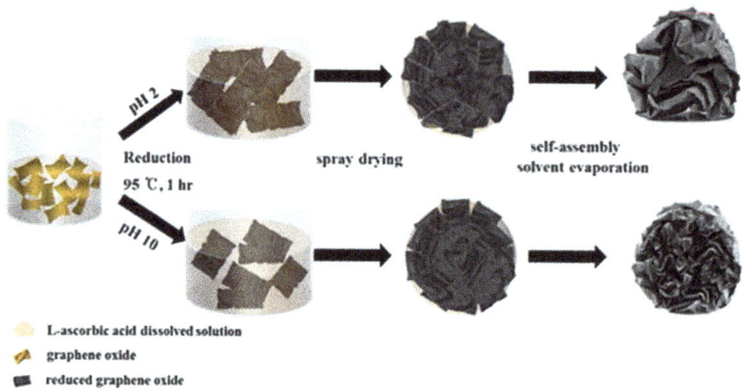

GRAPHENE SPHERES

These materials are among the most promising options for solving the problems associated with loss of properties in two-dimensional structures by preventing sheets from restructuring. These structures have many porous channels, which provide more free space between the spheres, thereby improving conductivity and structural stability (Yuan et al., 2021).

Synthesis Methods

There are two main methods of synthesizing graphene spheres. A*erosol assembly*, which is the most widely used approach, works by spraying a precursor solution after which the solvent evaporates. The other method is *template-assisted*, which is the second most used technique which uses a template to promote the spherical structure of the graphene sheets. Both techniques enable these three-dimensional structures to be formed simply and efficiently. Additionally, other lesser-used synthesis methods are *solvothermal* and *water-oil emulsion* (Nazarian-Samani et al., 2016).

From a precursor solution, *aerosol assembly* consists in obtaining very small particles by aerosolization and subsequently forms spherical structures as the solvent evaporates at a high temperature. In a typical synthesis, graphene oxide (GO) nanosheets are first uniformly dispersed in water. They are then transported to a spray system that generates droplets. Finally, they are sent to a reactor using an inert gas. By evaporating the water inside the reactor, the micro-droplets shrink, and the GO sheets are compressed to form spherical structures as a result of capillary force. In this way, the particle is formed within the space of a droplet, thus maintaining its spherical shape (J. Luo et al., 2011). Ha et al., 2019 synthesized graphene spheres from reduced graphene oxide in acidic and basic conditions with *aerosol assembly*. The structures obtained at pH 10 exhibited more spherical and denser shapes, as well as a larger surface area than those synthesized at pH 2 (Figure 3).

The *template-assisted* method involves using different materials, which are considered as core and shell during synthesis. The core must be a hard material such as silica (M. Li et al., 2019) (see Figure 4), poly(methyl methacrylate) (Haghighi Poudeh et al., 2015), polystyrene (Shao et al., 2013), or even nanoparticles of a metal or metal oxide (Z. Zhang, Xiao, Xi, et al., 2014). Also, the shell is made of

Figure 4. Synthesis of three-dimensional reduced GO spheres by the template-assisted method using silica (M. Li et al., 2019).

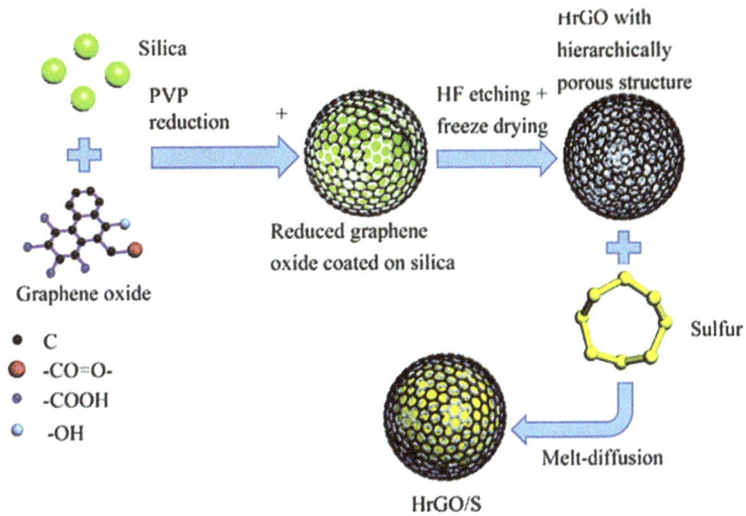

graphene-derived materials such as GO or reduced graphene oxide (rGO). The structure is initially prepared by electrostatic interactions between the core and the shell, which arise as a result of the attraction between the negatively charged graphene sheets and the positively charged compounds (template). Another potential interaction is through chemical vapour deposition (CVD) (Y. Zhang et al., 2013), which consists of exposing a hydrocarbon source to a metallic substrate (generally copper or nickel), on which the graphite will be deposited. Finally, the material that constitutes the core is chemically removed to obtain the spherical structure.

The manufacture of three-dimensional graphene structures by the *template-assisted* method has been extensively studied and different alternative templates have been proposed to enhance the characteristics and properties of graphene. Trung et al. (Trung et al., 2014) developed spherical structures made of poly(methyl methacrylate) (PMMA) as the core and reduced graphene oxide (rGO) as the shell. Synthesis was based on the difference in charges between both compounds: PMMA (positively charged) bonds to the GO sheets (negatively charged) by electrostatic attractions in aqueous media. Subsequently, the template is removed and the spherical structure remains. Furthermore, Wu et al. (P. Wu et al., 2014) described the development of spherical, three-dimensional, porous micro/nanostructures of graphene and silicon. SiO_2 spheres were used as a template, which was coated with a layer-by-layer assembly with GO due to electrostatic attractions between both species. Finally, the material was thermally reduced to obtain a three-dimensional Si-Graphene structure.

The *solvothermal method* involves a solvent and a precursor species in a sealed system above the boiling point of the former. Thermal energy produces an increase in pressure, which promotes efficient precursor-solvent interaction, by which spherical structures are formed. Finally, the spherical structures of graphene can be synthesized under atmospheric temperature and pressure by self-assembly with an *emulsion (oil-water)* and without surfactants. This process is possible because the functional groups and basal planes of the graphene oxide sheets are hydrophilic and hydrophobic, respectively, so the sheets can be grouped around drops of water (Yi Ding et al., 2012).

Figure 5. (a) CV curves for rGO, pure PANI, G@PANI, and G@PANI@G. (b) Power density vs energy density at different charge-discharge current densities for G@PANI@G, G@PANI, pure PANI, and rGO. (X. Liu et al., 2015)

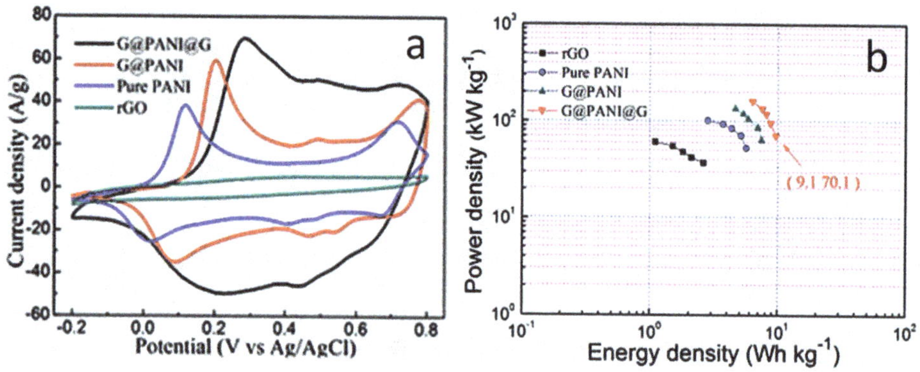

Applications

Three-dimensional graphene spheres can be used for a wide variety of applications due to their excellent properties, which are obtained during the different synthesis methods. Some properties they have are large surface area, excellent thermal conductivity and stability, high flexibility and hardness, and good electrical conductivity. Their main applications are related to energy storage in supercapacitors and batteries since there is a great deal of concern today about the harmful effects fossil fuels have on the environment. However, these spherical structures can also be used in sensors, catalysis, and energy conversion in fuel cells, among other applications (Nazarian-Samani et al., 2016).

Within the energy sector (specifically in developing devices that require energy storage to operate them), there has been a marked focus on rechargeable *batteries*. Using lithium metal ions in batteries is ideal for carrying out charge transfers. Likewise, using spherical graphene structures instead of graphite helps diffuse these ions and reduces the effects of polarisation. This occurs on account of their physical properties: porosity and large surface area (Mei et al., 2015).

Composite materials with graphene spheres have been considered as potential materials for use in *sensors*, due to their available surface area that enhances the adsorption capacity of the species to be detected (J. Zhang et al., 2021).

Three-dimensional graphene-based structures are also used as anode materials in microbial *fuel cells* with *aerosol assembly*. Among the graphene anodes that have been tested in biofuel cells are nitrogen-doped graphene spheres capable of forming hybrid structures with cobalt oxide nanoparticles (S. Mao et al., 2014). These structures exhibited high catalytic activity in oxygen reduction reactions (ORR) and oxygen evolution reactions (OER) when compared to commercial platinum catalysts, which are also highly expensive. Moreover, in some research, it has been observed that when heteroatoms act as the doping species inside the graphene structure, metal-free electroactivity is enhanced. This is due to an increase in active sites for oxygen adsorption, as well as to changes in electronegativities between the two species (carbon and dopants) (Dou et al., 2016; L. Tao et al., 2016).

Supercapacitors are devices that combine the high energy supply and storage capacity of conventional capacitors and batteries, respectively. There are double-layer capacitors built with carbon electrodes for energy storage. Liu et al. (X. Liu et al., 2015) designed hollow spheres with two layers of graphene and

Figure 6. Mechanism of PAS-GO preparation for heavy metal adsorption
(S. Luo et al., 2014).

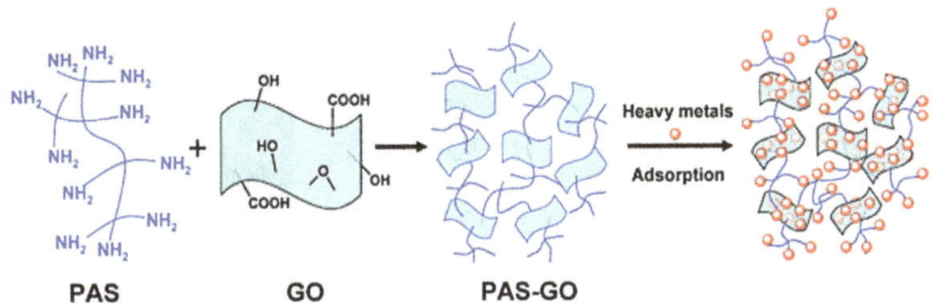

polyaniline to be used as electrodes in supercapacitors. Figure 5a shows how the hierarchical graphene@ polyaniline@graphene (G@PANI@G) has larger areas of CV (Cyclic voltammetry) than the pure components (pure PANI and the reduced graphene oxide rGO), which means it has a higher specific capacity. Figure 5b demonstrates the wide range of applications for these materials because power density and energy density (two important parameters in supercapacitors) in G@PANI@G are higher than in pure PANI, G@PANI, or rGO.

ADSORBENT POROUS STRUCTURES

These resulting materials are highly adsorbent and recyclable, which can then be used to remove heavy metals with minimal environmental impact. There can be two types of these materials, depending on the synthesis pathway: materials based on the graphene/polymer bond and those based on the graphene/ nanomaterial inorganic bond. Owing to the interconnections of the crystalline network and their well-defined porous structure, as well as their mechanical stability, these compounds help transport matter through their internal structure, thus making them excellent adsorbent materials (Yao et al., 2021).

Synthesis of Graphene/polymer Adsorbent Porous Structures

The bonding of graphene and polymer structures has a synergistic effect because GO can interact with the positively charged polymers through electrostatic interactions, hydrogen bonds, and π-π interactions. This leads to the layer-by-layer assembly of the GO sheets with polymers between them. This usually begins with graphene oxide (GO) due to the ease with which three-dimensional structures are assembled. The most used strategies for synthesizing graphene/polymer adsorbent structures may be *cross-linking*, *self-assembly by reduction*, *template-assisted,* or *synthesis by dissolution*. Each of these leads to materials with different microstructures, morphologies, and properties (K. Yang et al., 2018).

By using *cross-linking*, GO films and polymers are mixed in an organic or aqueous medium to give rise to three-dimensional adsorbent structures through their interaction by covalent bonds, hydrogen bonding, or other bonds created from the forces of electrostatic attraction. Luo et al. (S. Luo et al., 2014) succeeded in synthesizing adsorbent products by means of covalent bonds such as those formed by graphene oxide sheets and poly 3-aminopropyltriethoxysilane (PAS) as a cross-linking agent. The polymer structure of PAS prevents graphene films from aggregating and helps gain access to external molecules and this gives it adsorbent characteristics (Figure 6).

Figure 7. Modified scheme for the synthesis route of r-EPGM
(Y. Tao et al., 2014).

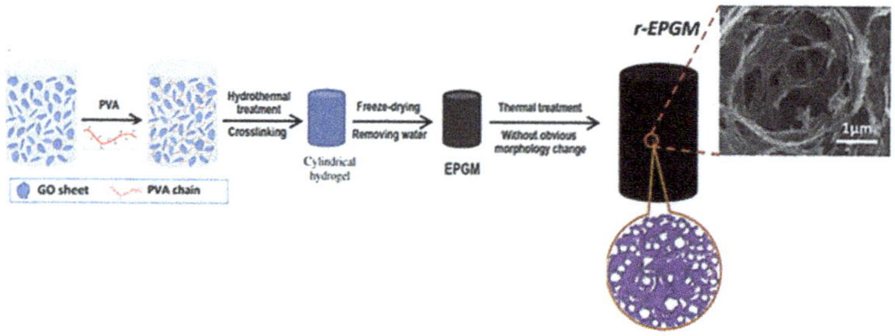

Self-assembly by reduction is considered to be one of the simplest, most attractive and effective methods for synthesizing three-dimensional graphene/polymer adsorbent structures. This is because the polymers inserted during this procedure help reduce agglomeration more efficiently and prevents the graphene sheets from restructuring. This method consists in manufacturing the three-dimensional graphene-based structures by mixing graphene oxide with the polymer to be used. This is followed by synthesis by means of GO reduction . To reduce graphite oxide strong chemicals reducers such as $NaHSO_3$, Na_2S or HI can be used. However, these agents are unstable or harmful to the environment, so, ethylenediaminetetraacetic acid (EDTA) was used as an environmentally-friendly reducing agent (Huang et al., 2014). Tao et al. (Y. Tao et al., 2014) used polyvinyl alcohol (PVA) as a polymer for synthesizing a reduced expanded porous graphene macroform (r-EPHM) graphene-based adsorbent structure by this technique. This polymer when joined to graphene sheets creates an expanded structure once the mixture is reduced, characterised by its high porosity. This is obtained by gradually removing the water contained in the internal structure of the compound by which macropores are created that help transport the ions rapidly and conduct electrons (see Figure 7).

The *template-assisted method* consists of coating a three-dimensional porous polymer template with graphene oxide sheets. Using this technique, Zhu et al. (H. Zhu et al., 2016), used a polyurethane (PU) template coated with rGO films (PU@rGO), which gave rise to a porous and spongy structure with high mechanical resistance, hydrophobicity, and high electrical conductivity which could be used to separate oil from the water after spills. The procedure works by adding $CaCO_3$ and graphite oxide nanoparticles to the polyurethane. The porous structure of the polyurethane is obtained due to $CaCO_3$ action. Finally, by reduction, graphite oxide coats the porous PU structure, thus forming the graphene-based adsorbent compound.

The *synthesis by dissolution* approach was applied by Wang et al. (Y. Wang et al., 2018) who synthesized rGO/polycarbonate (PC) structures with a solution-based method. For this purpose, tetrahydrofuran was used to obtain a homogeneous mixture between PCs and reduced graphene oxide. Deionized water was added to this mixture as a solvent by vigorously agitating it, resulting in a mixture that underwent phase separation. A three-dimensional rGO/PC structure was obtained from this process with a high hydrophobicity for absorbing oils and organic solvents from water (Figure 8).

Figure 8. Preparation route for rGO/PC porous structures
(Y. Wang et al., 2018).

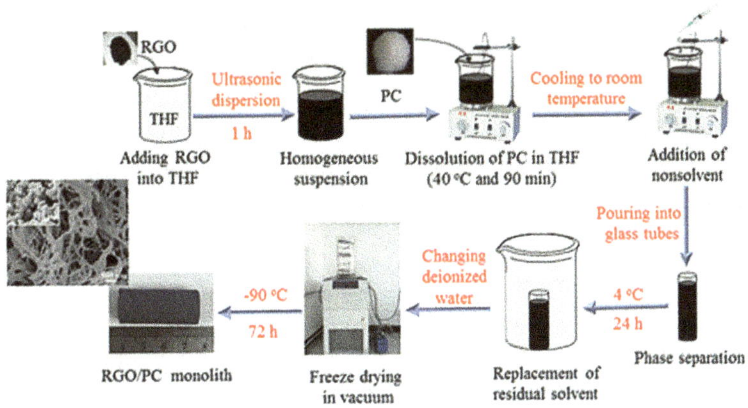

Synthesis of Graphene/nanomaterial Inorganic Adsorbent Structures

Numerous studies have been conducted in order to obtain adsorbent structures based on binding graphene to an inorganic nanomaterial. These structures are highly efficient at preventing the graphite sheets from restructuring and creating three-dimensional structures. In addition, they have a large creased surface area of graphene films, which helps to deposit inorganic nanomaterials in small volume spaces. These structures may mainly be synthesized by *self-assembly*, but other approaches are also possible.

Self-assembly is most widely used for manufacturing three-dimensional adsorbent structures based

Figure 9. Preparation scheme for MrGO/ZIF material
(Lin & Lee, 2016).

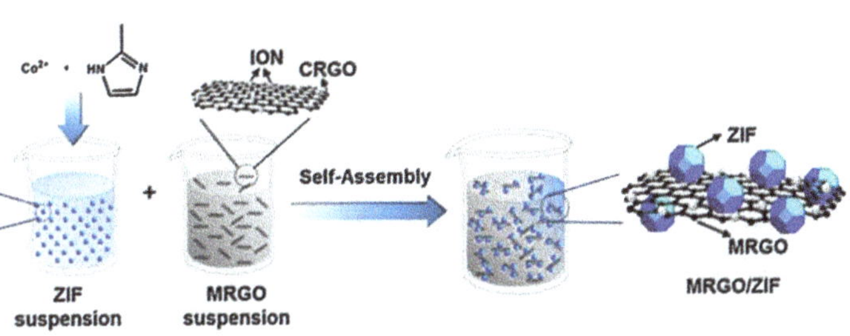

on graphene and inorganic nanomaterials. Here, synthesis is based on the difference in charge between graphene sheets (negative) and inorganic particles (positive). In this way, a homogeneous solution is obtained, and both elements are finally assembled by reduction or a hydrothermal process. Lin et al. (Lin & Lee, 2016) achieved self-assembly of magnetic reduced graphene oxide (MrGO) with a zeolitic adsorbent structure (ZIF), due to the electrostatic attraction between both materials. The resulting material (MrGO/ZIF) is outstanding in terms of its excellent performance as an adsorbent, for enhancing the individual properties of the zeolitic structure, and for its recyclability and stability (see Figure 9).

Metal ion deposition is another possible way to synthesise three-dimensional adsorbent structures based on graphene/inorganic nanomaterial. Using this technique, Yang et al. (Y. Yang et al., 2015) successfully synthesized three-dimensional porous structures through Fe_3O_4 and GO by means of a method which works on coordination reactions (interactions with the ligands from GO sheets). However, the surface of graphite oxide only contains a small number of functional groups that act as ligands. Therefore, nucleophilic substitution carboxylation was performed in order to increase the number and size of ligands in the graphite oxide. This process helps coordination reactions to progress, and, consequently, to achieve a three-dimensional graphene-based structure with good adsorptive properties.

Applications

As briefly mentioned in previous sections, most graphene-based adsorbent compounds, whether they are graphene/polymer structures or inorganic graphene/nanomaterials, are used to remove contaminants from certain media in environmental treatment processes. This adsorption works by transferring matter from the contaminants in an aqueous medium to the surface of the adsorbent material. Therefore, there must be a strong interaction between the contaminants and the adsorbent. The properties graphene-based adsorbent structures have which help such adsorption are: the large specific surface area that helps diffuse the contaminants inside the adsorbent and a porous structure that helps diffuse these contaminants, which leads to high adsorption rates. Moreover, they are easy to separate from the aqueous medium in which they have been used, thus making it easy to reuse them (L. Chen et al., 2018).

Interactions between the pollutants and the graphene-based adsorbent structures can be electrostatic, interactions involving π links, complexation, hydrophobic, van der Waals forces, or bonds by hydrogen bridges. Electrostatic interactions are very useful for the adsorption of cationic contaminants, such as heavy metals. This is due to the electronegativity of these graphene structures. Interactions by π links can also play an important role here, provided that the contaminant has an aromatic ring in its structure to form this type of link with graphene. If the contaminant has at least one hydrophobic chain in its structure, hydrophobic interaction takes place between it and the graphene structure.

In view of the above, three-dimensional adsorbent structures based on graphene could be used for removing dyes, heavy metals, and organic contaminants. *Heavy metals* (mainly Cu^{2+}, Ag^{+}, Pb^{2+}, Hg^{2+}, Cd^{2+}, and Cr^{6+})are highly significant pollutants found in aqueous media due to their toxic effect on plants and animals. The great adsorption effect heavy metals have is mainly due to the negative charges of graphene compounds as a result of the oxygen-derived functional groups found in their structure. These interact with the positive charges of most heavy metals. Therefore, the content of oxygen functional groups in the graphene structure will determine how negative charges are distributed and hence the ability these structures have to adsorb heavy metals. In addition, various functional groups can be incorporated into these structures in order to enhance the adsorptive performance of the compounds.

Furthermore, Andjelkovic et al. (Andjelkovic et al., 2015) synthesized a structure consisting of graphene and α-FeOOH for the adsorption of arsenic. The material exhibited excellent adsorption capacity in real contaminated water, because of its porous structure and three-dimensional network. It could reduce the concentration of arsenic from 150 μg/L to 10 μg/L in 5 minutes (Figure 10).

Currently, there has been a great focus on treating wastewater containing *dyes*, which can be cationic or anionic in nature. Due to the electronegativity of graphene-based adsorbent structures, their adsorption capacity is practically limited to treating cationic dyes. This can be seen in the study carried out by Yan et al. (X. Yan et al., 2016), in which the adsorption efficiency of graphene and montmorillonite (a

Figure 10. Removal of arsenic from contaminated drinking water vs time with two different adsorbents. (Concentration of arsenic = 150 μg/L)
(Andjelkovic et al., 2015)

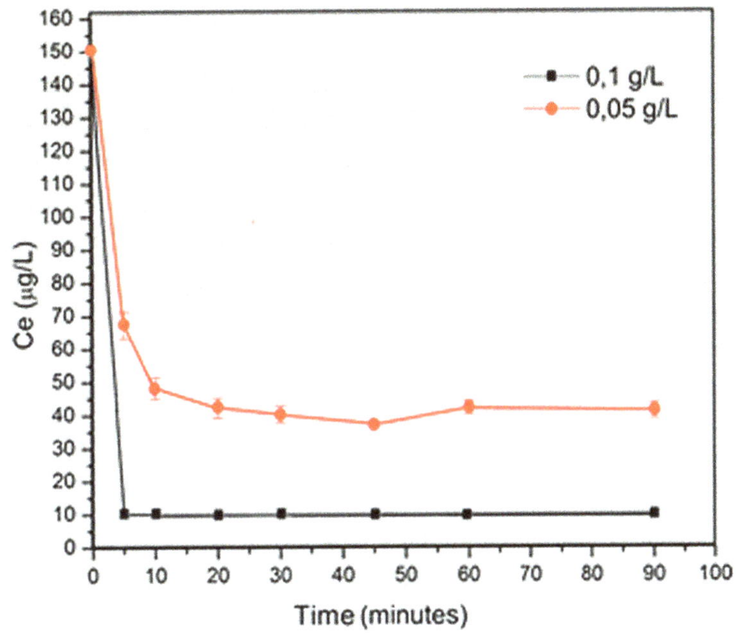

Figure 11. The adsorption capacity of different compounds from aqueous media by 3D graphene/poly-pyrrole structures
(H. Li et al., 2013).

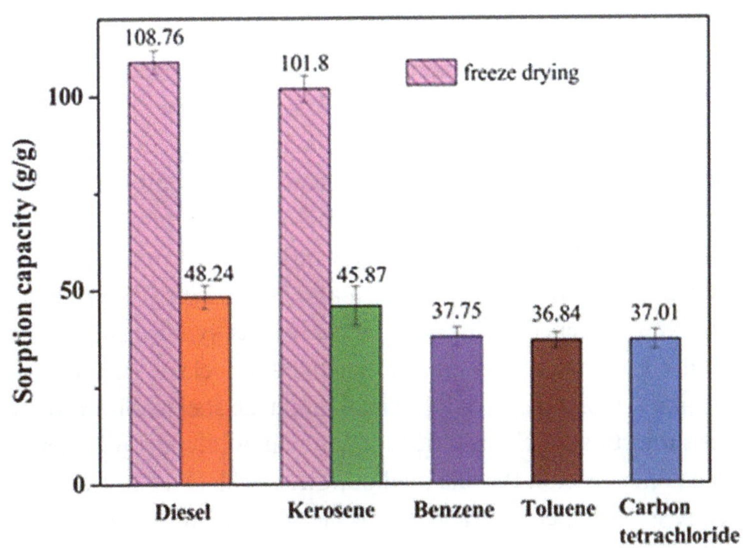

mineral of the silicate group that acts as a cross-linking agent) based structure was evaluated. Adsorbent performance was compared with methyl orange (anionic) and methylene blue (cationic). Under certain pH conditions, the amount of methylene blue (MB) removed from contaminated water (reaching over 90%), is greater than that adsorbed from methyl orange (MO), (barely 60%).

Several studies on the need to remove *organic contaminants* have increased in recent decades because of industrial discharges in aquatic environments. Graphene structures low in oxygen are more hydrophobic than those high in oxygen. Therefore, the former will be more useful in the adsorption of oils and other organic substances. Aromatic compounds such as toluene or benzene can also be removed from aqueous media by interactions through π links. Moreover, others such as diesel can be adsorbed through van der Waals forces. In order to remove these substances, Li et al. (H. Li et al., 2013) successfully obtained a graphene/polymer structure by self-assembly, using polypyrrole (PP) as the polymeric agent. The results obtained for the adsorption capacity of different compounds (diesel, kerosene, benzene, toluene, and CCl_4) are summarised in Figure 11. Furthermore, by subjecting the structure to freeze-drying, its adsorption properties are considerably enhanced.

Graphene Aerogels

The porosity of these materials leads to an increase in surface area which, apart from flexibility, low density, and the good chemical and electronic properties found in most graphene-derived structures, makes aerogels a structure with great potential for use in energy storage and exchange devices.

Like most structures derived from graphene, aerogels have emerged in response to growing environmental concerns, mainly due to a large number of toxic substances (SOx, NOx, VOCs, etc.) given off by fossil fuels. Therefore, research and development into materials that can be used in batteries, supercapacitors, or fuel/solar cells have led to the production of three-dimensional structures such as graphene aerogels (J. Mao et al., 2018a).

Synthesis

There are several methods by which graphene aerogels can be synthesized. Among these is *self-assembly*, in which graphene nanosheets act as precursor elements for assembly for obtaining the three-dimensional structures. *Self-assembly* can be carried out due to the abundance of oxygen-containing functional groups in the graphene oxide structure. This enables interaction among several two-dimensional sheets. This procedure can be carried out by hydrothermal or chemical reduction processes. Jia et al. (J. Zhu et al., 2021) successfully obtained a graphene-derived aerogel with high conductivity, high strength, and thermal stability by a two-step environment-friendly hydrothermal *self-assembly*. To obtain the desired structure, a cobalt derivate, thiourea, and ethylenediamine were mixed with as-synthesized ultrasonicated GO sheets. This was then treated at high temperatures in a teflon-lined autoclave. In this procedure, the aerogel forms due to the interactions among the initially treated graphene sheets. Moreover, the chemical reduction technique involves subjecting graphene oxide to treatment with reducing agents under atmospheric pressure conditions in an open reactor. By this method, Chen et al. (W. Chen & Yan, 2011) synthesized an aerogel at a relatively low temperature and atmospheric pressure, without agitation, cross-linking agents, or high pressures. The reducing agents used were $NaHSO_3$, vitamin C, hydroquinone, Na_2S, and HI. In addition, different structural shapes could be obtained with respect to the type of reactor and the

Figure 12. Synthesis procedure for three-dimensional carbon-graphene aerogel (H. L. Gao et al., 2016).

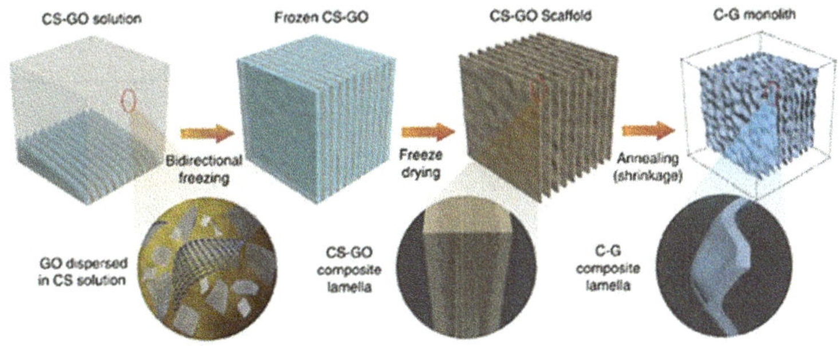

operating conditions used. The main characteristics of the aerogels obtained by this method were: low density, high electrical conductivity, mechanical resistance, and thermal stability.

Gao et al. (H. L. Gao et al., 2016) synthesized 3D graphene aerogels using biopolymers as cross-linking agents, which bind the graphene sheets by *self-assembly*. The biopolymer used was chitosan (CS), a linear polysaccharide, which was incorporated into the graphene oxide by freeze drying to obtain plat sheets and finally shrunk in a wavy morphology. With this procedure, very elastic, compressible, and fatigue-resistant three-dimensional lamellar structures(properties that are generally very difficult to obtain simultaneously) were synthesized (Figure 12).

The *chemical vapour deposition (CVD)* growth technique is an alternative method for obtaining graphene structures with high conductivity, due to the formation of porous structures that help to transport

Figure 13. Fabrication process for 3D graphene aerogels by direct ink writing (C. Zhu et al., 2015).

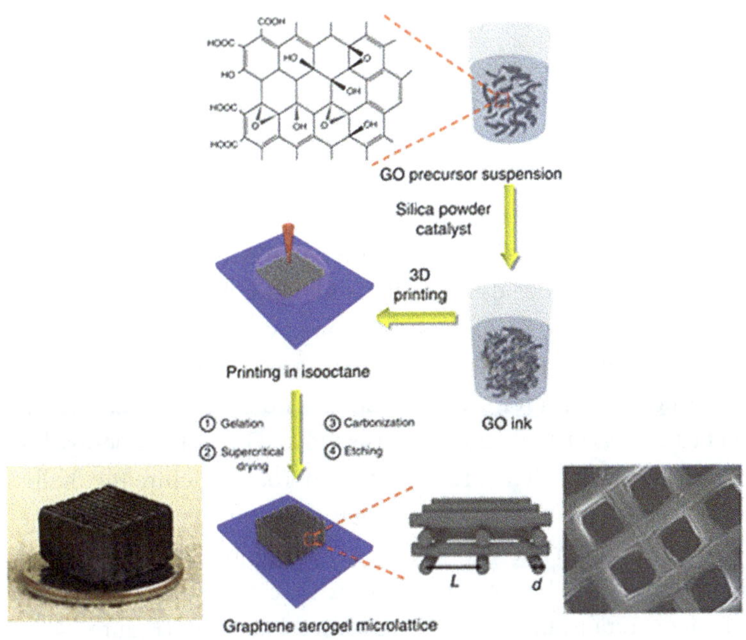

Figure 14. (a) specific capacitance versus current density and (b) capacitive retention versus current density
(C. Zhu et al., 2016).

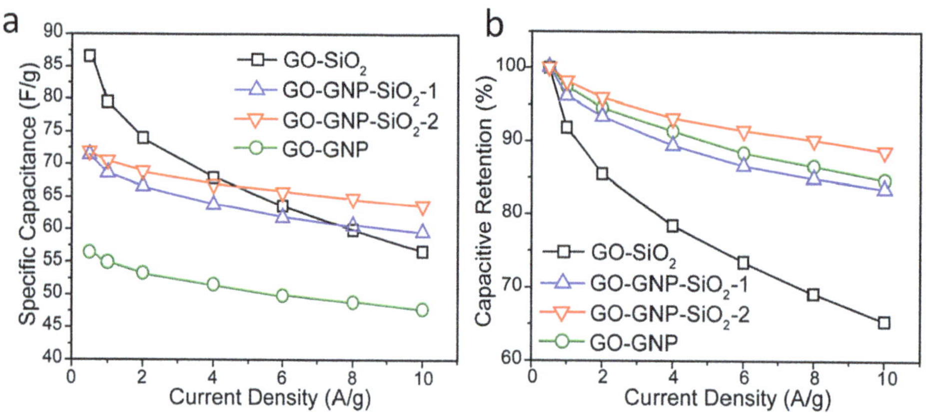

electrons. This process can be carried out both with and without using a material as a template (X. Wang et al., 2013). However, using templates limits large-scale manufacturing due to the high production cost involved.

Recently, Zhu et al. (C. Zhu et al., 2015) built three-dimensional graphene structures using *3D printing*. Manufacturing these graphene aerogels with macroscopic architectures involved modifying the initial graphene oxide solution, as well as the synthesis process to prevent the structure from drying rapidly. The obtained material had a broad surface area, mechanical strength, compressibility, and electrical conductivity (see Figure 13).

Applications

Graphene aerogels can be used both in energy storage devices (supercapacitors and batteries), and energy transformation processes (fuel and solar cells).

Supercapacitors are efficient capacitators with higher energy and power densities than conventional batteries and capacitors. Double-layer *supercapacitors* are characterised by an interface with two electrical layers formed by the electrodes and the electrolytes that enable the charge to accumulate. By applying an electric field to a supercapacitor, the electrodes accumulate the charges (positive and negative) and the potential difference with the electrolytes causes ion migration which balances the electric field, thereby storing the energy. The type of electrode used and its available active surface are essential characteristics in determining the optimal performance of the capacitors. For this reason, graphene-derived materials (with high electrical conductivity and a large surface area due to their porous structure of interconnected channels) have been considered ideal for use in supercapacitors (Xiong Zhang et al., 2015).

Additionally, Zhu et al. (C. Zhu et al., 2016) fabricated a 3D graphene aerogel by *direct-ink writing*. The graphene aerogel electrodes, with an ordered macroporous structure that helps transport ions, had very low weight and high conductivity. This meant they were excellent devices for energy storage. Note in Figure 14a, where the material made with GO graphene nanopellets (GNP) and SiO_2 with a higher amount of GNP (GO-GNP-SiO_2-2) has the highest specific capacitance at a current density of 10 A/g. Moreover, this is what most retains this capacitance with this current (Figure 14b).

Figure 15. (a) LSV before and after accelerated degradation test in ORR (b) LSV curves for OER before and after stability test for N-GRW and Ir/C
(H. Bin Yang et al., 2016).

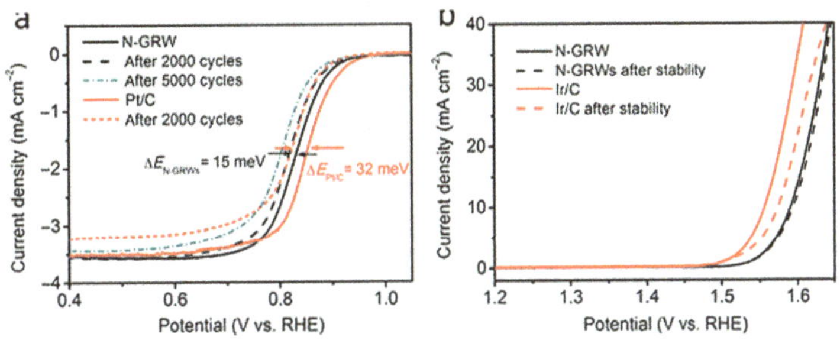

Energy storage systems, like batteries, become more powerful and efficient as technology advances. *Lithium batteries* are the most widely used due to their long charge/discharge cycles, high energy, and the fact that they do not self-discharge in numerous electronic devices. These batteries are composed of an electrolyte, anode, and cathode. They work by transporting lithium ions to and from the positive and negative electrodes through the electrolyte, which depends on whether the battery is charged or not. The materials used as cathodes must help store energy efficiently. Those most commonly used are $LiMO_2$ (where M is usually cobalt, nickel, or manganese), $LiMn_2O_4$, and $LiFePO_4$ structures; these have good properties but have poor electrical conductivity (Kuppan et al., 2017; P. Yan et al., 2017). For this reason, three-dimensional structures derived from graphene have been synthesized to be used as cathodes. Le et al. (He et al., 2016) prepared a three-dimensional graphene aerogel with carbon nanotubes and selenium, which were kept within the periodic graphene sheets. The material had a mesoporous structure with interconnected pore channels for moving ions and electrons, thereby making the cathode highly effective in lithium-selenium batteries. In addition, graphene-based aerogels have also been used to manufacture anodes with excellent results in lithium batteries (Fang et al., 2016; Sun et al., 2017).

Fuel cells are two electrodes that convert chemical energy into electrical energy. The electrochemical processes that take place inside (depending on the type of fuel cell used) are the oxygen evolution reaction (OER), oxygen reduction reaction (ORR), and hydrogen evolution reaction (HER) (J. Mao et al., 2018b). Graphene aerogels provide high contact, catalytic activity, and stability which are good properties for oxygen reduction reactions. Likewise, adding metallic and non-metallic species to the graphene structures provides more active sites in these types of reactions with enhanced activity (Fu et al., 2016; Z. Luo et al., 2016).

ORR and OER are considered critical in fuel cells and batteries for energy storage and conversion systems. In this respect, different research has focused on the synthesis of attractive materials for such purposes (W. Yan et al., 2016; H. Bin Yang et al., 2016). Yang et al. (H. Bin Yang et al., 2016) reported an interconnected 3D architecture N-doped graphene nanoribbon (N-GRW), which had a greater density of active sites for oxygen evolution and oxidation reactions. This is on account of its three-dimensional network that helps transport electrons. N-GRW was tested with linear sweep voltammetry (LSV) to compare it with a Pt/C catalyst whose activity was similar but more stable after 2000 cycles (Figure 15a). This material was more durable than the catalyst Ir/C in OER and it only fell by 10% over 24h of continuous operation (Figure 15b).

Figure 16. Synthesis of three-dimensional graphene foams by electrochemical reduction (Z. Zhang, Xiao, Qian, et al., 2014).

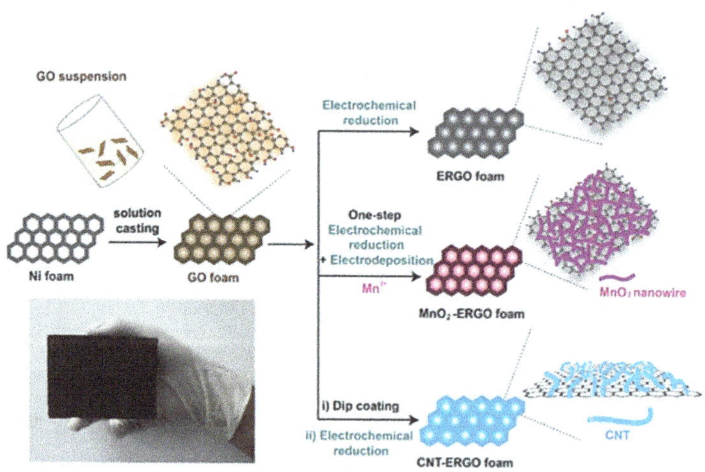

Hydrogen generation is another example of how fuel cells can be used for splitting water, and are one option for reducing our reliance on fossil fuels. Several catalysts have been developed for HER (Z. Li et al., 2016; Xueping Zhang et al., 2016). Li et al. (Z. Li et al., 2016) developed a graphene-based aerogel with reduced graphene oxide/carbon nanotubes (rGO/O-MWCNT) functionalised with phenylenediamine (PPD) and MoS_2 by a green hydrothermal method, with excellent activity and stability in hydrogen evolution reactions (HER).

Direct conversion of solar energy into electricity is a widely used and well-known renewable energy source. Dye-sensitized *solar cells* (in which the performance of the cathode is crucial as it must transfer charges quickly and be a good reducing agent) are one of the most commonly used types of cells in these processes. Therefore, the potential for using structures based on graphene aerogels as solar cell cathodes has been of great interest (Ergen et al., 2017; H. Wang et al., 2013).

Xue et al. (Xue et al., 2012) manufactured nitrogen-doped graphene aerogels using the chemical vapour deposition (CVD) growth method. This doping not only improved the electrical conductivity of the graphene structures, but also increased the hydrophilicity of these structures, thus promoting charge transfer and electrolyte-electrode interactions. In this way, a cathode with enhanced performance was obtained which was more economical than the most used platinum-based cathodes.

Graphene Foams

Graphene foams are another type of three-dimensional structures with different applications due to their large surface area (Kim et al., 2021; Xu et al., 2015) which can be synthesized from two-dimensional graphene using specific templates (Du et al., 2016).

Synthesis

Synthesis of 3D graphene-based foams can be carried out by two methods: *CVD* (Chemical Vapour Deposition) and *chemical assembly*. Synthesis by the latter goes through several stages in which the

Figure 17. Procedure for synthesis of 3D graphene foams by CVD
(Sha et al., 2016).

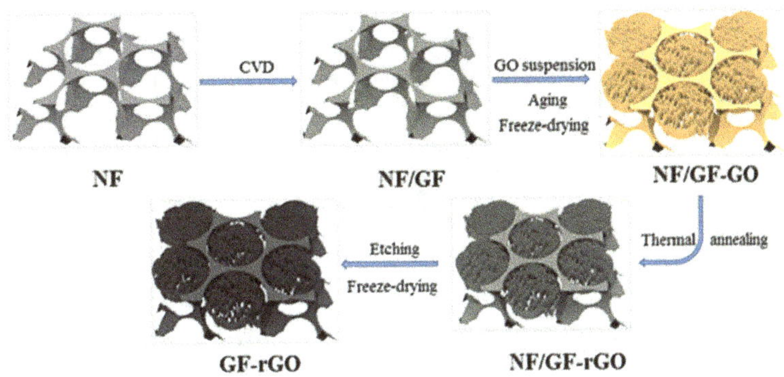

graphene sheets are obtained from graphite exfoliation. Afterwards, the template surface is coated and the template removed in order to obtain the three-dimensional structure. Zhang et al. (Z. Zhang, Xiao, Qian, et al., 2014) synthesized graphene foams by combining a deposition with an electrochemical method. The template used was nickel to which graphene oxide was superficially added. Later a potential was applied to reduce the graphene oxide and thus obtain an electrochemically reduced GO (ERGO) three-dimensional structure. Likewise, other species were incorporated into the MnO_2-carbon nanotubes (CNT) structure, providing them with unique properties such as high flexibility, large surface area, and chemical stability (see Figure 16).

Furthermore, synthesis of 3D graphene foams by *chemical vapour deposition* uses substrates for growing graphene and (depending on the catalytic properties of the substrate); these processes may occur freely or with a catalyst. There has been different research on catalysed CVD (Yujie Ding et al., 2016; Dursun et al., 2020; Kumar et al., 2021; B. Liu et al., 2021). Hu et al. (Hu et al., 2016) obtained 3D graphene foams by CVD. They first grew graphene foams using Ni foam as a template. Then they added graphene oxide by immersing and drying it. Subsequently, the graphene oxide was reduced by heat treatment. Finally, the template was removed and the material was dried again (see Figure 17). Thus, a three-dimensional hybrid structure, which exhibited electrochemical properties, was obtained. Sha et al. (Sha et al., 2016) synthesized graphene foams by combining CVD and the traditional powder metal-

Figure 18. Synthesis route for 3D graphene foams by CVD using seashells as template
(Shi et al., 2016).

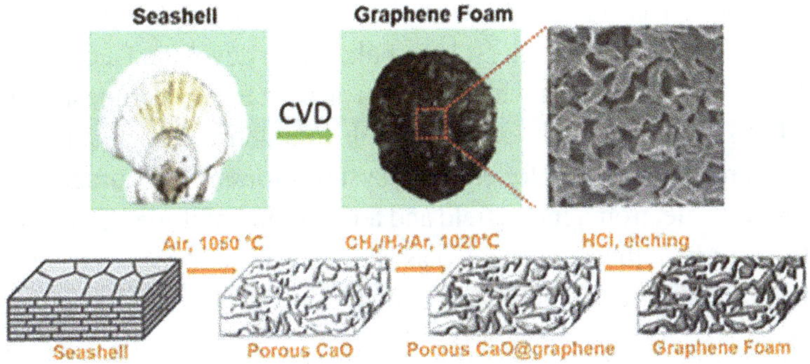

Figure 19. 3D graphene foam for biomedical purposes in the stimulation of neuronal progenitor cells (Nguyen et al., 2018).

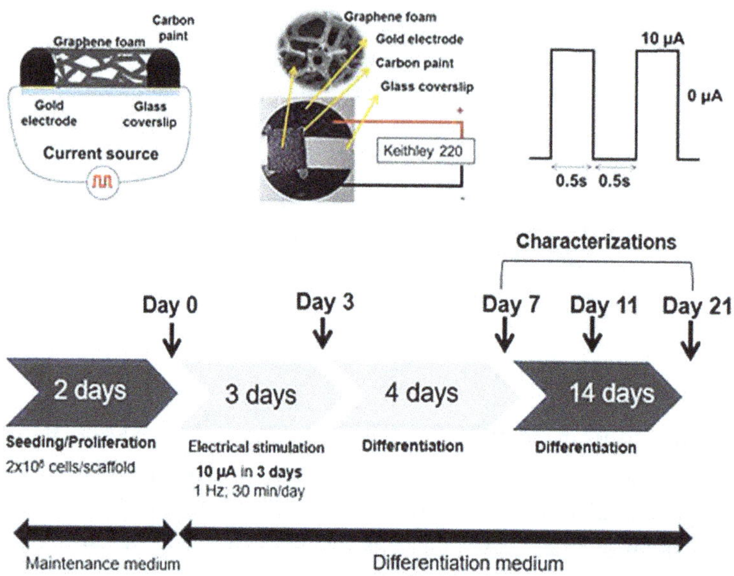

lurgy method. A wet mixture of nickel powder and sucrose (NF, template) and sucrose (carbon source) was dried, ground, and pressed. Then CVD was applied, for depositing the graphite foam (GF) onto the surface of the pellets. Finally, the nickel was removed to obtain a 3D structure with a large surface area.

In order to reduce costs and develop a growth system without metals, seashells or cuttlebones were used as templates. These materials are abundant and can be used in the synthesis of 3D graphene foams (K. Chen et al., n.d.; Shi et al., 2016). Shi et al. (Shi et al., 2016) developed a graphene foam grown by CVD on seashells. These natural materials (composed of calcium carbonates) were first thermally treated to obtain calcium oxides. In this case, instead of using metallic foams, porous calcium oxides were used as templates to grow the graphene structures. Next, the template was removed and the graphene foam in the form of a shell was obtained (Figure 18). These structures had very low density and high porosity, as well as sound electrical conductivity and mechanical flexibility.

Applications

Some different ways graphene foams can be used are batteries, sensors, water purification, and biomedical applications. Using graphene-based foams in *batteries* has been addressed by different research (Hussain et al., 2018; J. Li et al., 2021). Liu et al. (J. Liu et al., 2014) developed a rechargeable battery using Fe_2O_3 nanorods and $Ni(OH)_2$ nanosheets which were grown on a film composed of carbon nanotubes, and graphene foam. The synthesized cells exhibited high volumetric energy density, as well as stability during cycles. It has promising potential for creating systems for energy storage.

Due to the changes in the electromagnetic field and temperature of three-dimensional graphene foams when they are locally disturbed, they can be used as *sensors* or *biosensors* (Yue et al., 2020). Fan et al. (Fan et al., 2017) developed a glucose biosensor, which was constructed of graphene foam doped with

Table 1. Summary of the synthesis methods and applications for the 3D graphene structures.

Three-dimensional Graphene Structure	Synthesis	Applications
Spheres	Aerosol assembly Templated assisted Solvothermal Emulsion	Batteries (Mei et al., 2015). Sensors (J. Zhang et al., 2021). Fuel Cells (L. Tao et al., 2016) Supercapacitors (X. Liu et al., 2015) Catalysis (Dou et al., 2016)
Aerogels	Self-assembly Chemical vapour deposition Direct ink writing	Supercapacitors (Xiong Zhang et al., 2015) Lithium batteries (Kuppan et al., 2017 Fuel Cells (J. Mao et al., 2018b) Solar Cells (Ergen et al., 2017)
Foams	Chemical assembly Chemical vapour deposition	Batteries (J. Li et al., 2021) Sensors (Yue et al., 2020) Water purification (H. Zhu et al., 2015) Biomedical (Nguyen et al., 2018)
Adsorbent porous structure	Graphene/polymer Cross-linking Self-assembly by reduction Dissolution Template-assisted Graphene/inorganic Self-assembly Metal ion deposition	Dyes (X. Yan et al., 2016) Heavy metals (Andjelkovic et al., 2015) Organic pollutants (H. Li et al., 2013)

nitrogen and functionalised with carbon nanotubes. The biosensor displayed good stability and high selectivity.

Biomedical applications using materials of graphene-based three-dimensional foams have been described in different research. Nguyen et al. (Nguyen et al., 2018) developed metal-free graphene foams, which were used as conductors in the electrical stimulation of neuronal progenitor cells in humans. The 3D structures were seen to form a neural network during the first week. Also, the electrically stimulated cells presented a greater soma than those that were not stimulated. This electrical stimulation led to an increase in neuron maturation and differentiation (Figure 19). Akhavan et al. (Akhavan et al., 2016) fabricated 3D graphene foams to regenerate neuronal tissue from electrical stimuli in stem cells.

Graphene foams have also been used in *environmental remediation*; here, Zhu et al. (H. Zhu et al., 2015) synthesized functionalised graphene foams with an amphiphilic copolymer (polyhexadecyl acrylate, P2VP-*b*-PHA) for removing organic solvents and oil from contaminated water. The material showed a synergistic effect between the high adsorption capacity of the graphene foam and the pH-responsive surface property in the P2VP-*b*-PHA. This foam could be used in 10 cycles for the absorption/desorption of chloroform without losing absorptive capacity. Hence, it shows it has a good recyclability property.

Finally, Table 1 summarises all the synthesis methods and applications for the 3D graphene structures mentioned in this chapter.

FUTURE RESEARCH DIRECTIONS

Nowadays, graphene derivatives are used in a wide range of different fields, but new fields of research are emerging. One of the most innovative recent studies is for using these materials as muscle movement sensors. Graphene-derived compounds are not usually cytotoxic, so their interaction with tissues/cells is being researched This is achieved by taking care of their synthesis route, because toxic reagents/solvents (e.g., hydrazine) may remain, which would lead to problems of biocompatibility and biodegradability. This entails developing environmentally friendly synthesis that would be compatible with human tissue. In addition, due to their good electrical properties, biosensors for biomolecules such as DNA, enzymes, and hormones are also being developed. The search for graphene derivatives with good electrochemical properties that are fast, non-destructive, selective, and highly sensitive for detecting all these biomolecules is still in progress. For this type of application, the most promising technique is unzipping graphene nanotubes, because many imperfections that enhance electrochemical activity are generated and biosensors of a well-known size can be created. This method must be optimized using different oxidants and reducing reagents to generate the appropriate graphene oxide derivate.

For synthesizing metal-free graphene derivatives new nitrogen, phosphorous, etc., dopants that would be toxic or corrosive must still be found. At present, research is being undertaken on this, such as on algae as N dopants. However, an available industrial method for synthesis must be found.

CONCLUSION

This chapter presents recent breakthroughs in the synthesis and applications of graphene-based three-dimensional materials. The thickness of graphene is due to a series of chemical bonds that protrude above and below the 2D plane of carbon atoms. It can therefore be considered as a 3D material, albeit it is very thin. Recently, numerous studies showed that assembling graphene sheets in 3D architectures is one of the most effective ways of using and maintaining the properties of the two-dimensional material.

In conclusion, *aerosol assembly* was observed to be the optimal synthesis for graphene spheres due to their large specific surface area and good electrical conductivity. They are mostly used in the synthesis of supercapacitors. To synthesize these materials, cross-linking is first used, with which a great range of materials can be made by mixing different derivatives of grains with polymers, which yields many different structures with great potential for researching and improving them. Secondly, with self-assembly, graphene/nanomaterial inorganic materials can be developed.

Aerogels are mostly synthesized by self-assembly whose main use is as supercapacitors or electrocatalysts. Finally, we have graphene foams, which are mainly synthesized by using templates with CVD and they are largely used to make batteries.

REFERENCES

Akhavan, O., Ghaderi, E., Shirazian, S. A., & Rahighi, R. (2016). Rolled graphene oxide foams as three-dimensional scaffolds for growth of neural fibers using electrical stimulation of stem cells. *Carbon*, *97*, 71–77. doi:10.1016/j.carbon.2015.06.079

Andjelkovic, I., Tran, D. N. H., Kabiri, S., Azari, S., Markovic, M., & Losic, D. (2015). Graphene aerogels decorated with α-FeOOH nanoparticles for efficient adsorption of arsenic from contaminated waters. *ACS Applied Materials & Interfaces*, *7*(18), 9758–9766. doi:10.1021/acsami.5b01624 PMID:25871444

Chen, K., Li, C., Chen, Z., Shi, L., Reddy, S., Meng, H., Ji, Q., Zhang, Y., & Liu, Z. (n.d.). *Bioinspired synthesis of CVD graphene flakes and graphene-supported molybdenum sulfide catalysts for hydrogen evolution reaction*. doi:10.1007/s12274-016-1013-1

Chen, L., Han, Q., Li, W., Zhou, Z., Fang, Z., Xu, Z., Wang, Z., & Qian, X. (2018). Three-dimensional graphene-based adsorbents in sewage disposal: a review. In Environmental Science and Pollution Research (Vol. 25, Issue 26, pp. 25840–25861). Springer Verlag. doi:10.100711356-018-2767-7

Chen, W., & Yan, L. (2011). In situ self-assembly of mild chemical reduction graphene for three-dimensional architectures. *Nanoscale*, *3*(8), 3132–3137. doi:10.1039/c1nr10355e PMID:21698339

Ding, Y., Cheng, H., Zhou, C., Fan, Y., Zhu, J., Shao, H., & Qu, L. (2012). Functional microspheres of graphene quantum dots. *Nanotechnology*, *23*(25), 255605. doi:10.1088/0957-4484/23/25/255605 PMID:22653222

Ding, Y., Zhu, J., Wang, C., Dai, B., Li, Y., Qin, Y., Xu, F., Peng, Q., Yang, Z., Bai, J., Cao, W., Yuan, Y., & Li, Y. (2016). Multifunctional three-dimensional graphene nanoribbons composite sponge. *Carbon*, *104*, 133–140. doi:10.1016/j.carbon.2016.03.058

Dou, S., Tao, L., Huo, J., Wang, S., & Dai, L. (2016). Etched and doped Co9S8/graphene hybrid for oxygen electrocatalysis. *Energy & Environmental Science*, *9*(4), 1320–1326. doi:10.1039/C6EE00054A

Du, X., Liu, H. Y., & Mai, Y. W. (2016). Ultrafast synthesis of multifunctional n-doped graphene foam in an ethanol flame. *ACS Nano*, *10*(1), 453–462. doi:10.1021/acsnano.5b05373 PMID:26635121

Dursun, S., Akay, R. G., & Yazici, M. S. (2020). CVD graphene supported cobalt (II) phthalocyanine as cathode electrocatalyst for PEM fuel cells. *International Journal of Hydrogen Energy*, *45*(60), 34837–34844. doi:10.1016/j.ijhydene.2020.04.168

Ergen, O., Gilbert, S. M., Pham, T., Turner, S. J., Tan, M. T. Z., Worsley, M. A., & Zettl, A. (2017). Graded bandgap perovskite solar cells. In Nature Materials (Vol. 16, Issue 5, pp. 522–525). Nature Publishing Group. doi:10.1038/nmat4795

Fan, P., Liu, L., Guo, Q., Wang, J., Yang, J., Guan, X., Chen, S., & Hou, H. (2017). Three-dimensional N-doped carbon nanotube@carbon foam hybrid: An effective carrier of enzymes for glucose biosensors. *RSC Advances*, *7*(43), 26574–26582. doi:10.1039/C7RA02592K

Fang, W., Zhang, N., Fan, L., & Sun, K. (2016). Bi2O3 nanoparticles encapsulated by three-dimensional porous nitrogen-doped graphene for high-rate lithium ion batteries. *Journal of Power Sources*, *333*, 30–36. doi:10.1016/j.jpowsour.2016.09.155

Foster, C. W., Down, M. P., Zhang, Y., Ji, X., Rowley-Neale, S. J., Smith, G. C., Kelly, P. J., & Banks, C. E. (2017). 3D Printed Graphene Based Energy Storage Devices. *Scientific Reports*, *7*(1), 1–11. doi:10.1038rep42233 PMID:28256602

Fu, X., Choi, J. Y., Zamani, P., Jiang, G., Hoque, M. A., Hassan, F. M., & Chen, Z. (2016). Co-N Decorated Hierarchically Porous Graphene Aerogel for Efficient Oxygen Reduction Reaction in Acid. *ACS Applied Materials & Interfaces*, *8*(10), 6488–6495. doi:10.1021/acsami.5b12746 PMID:26937737

Gao, H. L., Zhu, Y. B., Mao, L. B., Wang, F. C., Luo, X. S., Liu, Y. Y., Lu, Y., Pan, Z., Ge, J., Shen, W., Zheng, Y. R., Xu, L., Wang, L. J., Xu, W. H., Wu, H. A., & Yu, S. H. (2016). Super-elastic and fatigue resistant carbon material with lamellar multi-arch microstructure. *Nature Communications*, *7*(1), 1–8. doi:10.1038/ncomms12920 PMID:27676215

Gao, W. (2012). *Graphite Oxide: Structure, Reduction and Applications*. https://scholarship.rice.edu/handle/1911/64614

Ha, T., Kim, S. K., Choi, J. W., Chang, H., & Jang, H. D. (2019). pH controlled synthesis of porous graphene sphere and application to supercapacitors. *Advanced Powder Technology*, *30*(1), 18–22. doi:10.1016/j.apt.2018.10.002

Haghighi Poudeh, L., Saner Okan, B., Seyyed Monfared Zanjani, J., Yildiz, M., & Menceloglu, Y. (2015). Design and fabrication of hollow and filled graphene-based polymeric spheres via core-shell electrospraying. *RSC Advances*, *5*(111), 91147–91157. doi:10.1039/C5RA19581K

He, J., Chen, Y., Lv, W., Wen, K., Li, P., Wang, Z., Zhang, W., Qin, W., & He, W. (2016). Three-Dimensional Hierarchical Graphene-CNT@Se: A Highly Efficient Freestanding Cathode for Li-Se Batteries. *ACS Energy Letters*, *1*(1), 16–20. doi:10.1021/acsenergylett.6b00015

Hu, G., Xu, C., Sun, Z., Wang, S., Cheng, H. M., Li, F., & Ren, W. (2016). 3D Graphene-Foam-Reduced-Graphene-Oxide Hybrid Nested Hierarchical Networks for High-Performance Li-S Batteries. *Advanced Materials*, *28*(8), 1603–1609. doi:10.1002/adma.201504765 PMID:26677000

Huang, Y., Li, C., & Lin, Z. (2014). EDTA-induced self-assembly of 3D graphene and its superior adsorption ability for paraquat using a teabag. *ACS Applied Materials & Interfaces*, *6*(22), 19766–19773. doi:10.1021/am504922v PMID:25359004

Hussain, A., Abidi, I. H., Tso, C. Y., Chan, K. C., Luo, Z., & Chao, C. Y. H. (2018). Thermal management of lithium ion batteries using graphene coated nickel foam saturated with phase change materials. *International Journal of Thermal Sciences*, *124*, 23–35. doi:10.1016/j.ijthermalsci.2017.09.019

Kim, J., Eum, J.-H., Kang, J., Kwon, O., Kim, H., & Kim, D. W. (2021). Tuning the hierarchical pore structure of graphene oxide through dual thermal activation for high-performance supercapacitor. *Scientific Reports*, *11*(1), 2063. doi:10.103841598-021-81759-7 PMID:33483594

Kumar, S., Rehman, M. A., Lee, S., Kim, M., Hong, H., Park, J. Y., & Seo, Y. (2021). Supercapacitors based on Ti3C2Tx MXene extracted from supernatant and current collectors passivated by CVD-graphene. *Scientific Reports*, *11*(1), 649. Advance online publication. doi:10.103841598-020-80799-9 PMID:33436987

Kuppan, S., Xu, Y., Liu, Y., & Chen, G. (2017). Phase transformation mechanism in lithium manganese nickel oxide revealed by single-crystal hard X-ray microscopy. *Nature Communications*, *8*(1), 1–10. doi:10.1038/ncomms14309 PMID:28145406

Li, H., Liu, L., & Yang, F. (2013). Covalent assembly of 3D graphene/polypyrrole foams for oil spill cleanup. *Journal of Materials Chemistry. A, Materials for Energy and Sustainability*, *1*(10), 3446–3453. doi:10.1039/c3ta00166k

Li, J., Du, S., Tao, H., & Yang, X. (2021). Self-supporting and 3D MoS2/MoO2/CNT/graphene foam as high-performance anode for lithium ion batteries. *Ionics*, *27*(1), 75–84. doi:10.100711581-020-03815-y

Li, M., Dai, Y., Pei, X., & Chen, W. (2019). Three-dimensional hollow reduced graphene oxide spheres with a hierarchically porous structure for high-performance lithium-sulfur batteries. *Inorganic Chemistry Frontiers*, *6*(9), 2528–2538. doi:10.1039/C9QI00656G

Li, Z., Dai, X., Du, K., Ma, Y., Liu, M., Sun, H., Ma, X., & Zhang, X. (2016). Reduced graphene oxide/O-MWCNT hybrids functionalized with p-phenylenediamine as high-performance MoS2 electrocatalyst support for hydrogen evolution reaction. *The Journal of Physical Chemistry C*, *120*(3), 1478–1487. doi:10.1021/acs.jpcc.5b09523

Lin, K. Y. A., & Der Lee, W. (2016). Highly efficient removal of Malachite green from water by a magnetic reduced graphene oxide/zeolitic imidazolate framework self-assembled nanocomposite. *Applied Surface Science*, *361*, 114–121. doi:10.1016/j.apsusc.2015.11.108

Liu, B., Pavlou, C., Wang, Z., Cang, Y., Galiotis, C., & Fytas, G. (2021). Determination of the elastic moduli of CVD graphene by probing graphene/polymer Bragg stacks. *2D Materials, 8*(3), 035040. doi:10.1088/2053-1583/abfedb

Liu, J., Chen, M., Zhang, L., Jiang, J., Yan, J., Huang, Y., Lin, J., Fan, H. J., & Shen, Z. X. (2014). A flexible alkaline rechargeable Ni/Fe battery based on graphene foam/carbon nanotubes hybrid film. *Nano Letters*, *14*(12), 7180–7187. doi:10.1021/nl503852m PMID:25402965

Liu, X., Wen, N., Wang, X., & Zheng, Y. (2015). A High-performance hierarchical graphene@polyaniline@graphene sandwich containing hollow structures for supercapacitor electrodes. *ACS Sustainable Chemistry & Engineering*, *3*(3), 475–482. doi:10.1021c5006999

Luo, J., Jang, H. D., Sun, T., Xiao, L., He, Z., Katsoulidis, A. P., Kanatzidis, M. G., Gibson, J. M., & Huang, J. (2011). Compression and aggregation-resistant particles of crumpled soft sheets. *ACS Nano*, *5*(11), 8943–8949. doi:10.1021/nn203115u PMID:21995602

Luo, S., Xu, X., Zhou, G., Liu, C., Tang, Y., & Liu, Y. (2014). Amino siloxane oligomer-linked graphene oxide as an efficient adsorbent for removal of Pb(II) from wastewater. *Journal of Hazardous Materials*, *274*, 145–155. doi:10.1016/j.jhazmat.2014.03.062 PMID:24780856

Luo, Z., Tan, C., Zhang, X., Chen, J., Cao, X., Li, B., Zong, Y., Huang, L., Huang, X., Wang, L., Huang, W., & Zhang, H. (2016). Preparation of Cobalt Sulfide Nanoparticle-Decorated Nitrogen and Sulfur Co-Doped Reduced Graphene Oxide Aerogel Used as a Highly Efficient Electrocatalyst for Oxygen Reduction Reaction. *Small*, *12*(43), 5920–5926. doi:10.1002mll.201602615 PMID:27592615

Mao, J., Iocozzia, J., Huang, J., Meng, K., Lai, Y., & Lin, Z. (2018). Graphene aerogels for efficient energy storage and conversion. In Energy and Environmental Science (Vol. 11, Issue 4, pp. 772–799). Royal Society of Chemistry. doi:10.1039/C7EE03031B

Mao, S., Wen, Z., Huang, T., Hou, Y., & Chen, J. (2014). *Electronic Supplementary Information (ESI) High-performance Bi-functional Electrocatalysts of 3D Crumpled Graphene-Cobalt Oxide Nanohybrids for Oxygen Reduction and Evolution Reactions*. Academic Press.

Mei, R., Song, X., Hu, Y., Yang, Y., & Zhang, J. (2015). Hollow reduced graphene oxide microspheres as a high-performance anode material for Li-ion batteries. *Electrochimica Acta, 153*, 540–545. doi:10.1016/j.electacta.2014.05.154

Nazarian-Samani, M., Kim, H. K., Park, S. H., Youn, H. C., Mhamane, D., Lee, S. W., Kim, M. S., Jeong, J. H., Haghighat-Shishavan, S., Roh, K. C., Kashani-Bozorg, S. F., & Kim, K. B. (2016). Three-dimensional graphene-based spheres and crumpled balls: Micro- and nano-structures, synthesis strategies, properties and applications. In RSC Advances (Vol. 6, Issue 56, pp. 50941–50967). Royal Society of Chemistry. doi:10.1039/C6RA07485E

Nguyen, A. T., Mattiassi, S., Loeblein, M., Chin, E., Ma, D., Coquet, P., Viasnoff, V., Teo, E. H. T., Goh, E. L., & Yim, E. K. F. (2018). Human Rett-derived neuronal progenitor cells in 3D graphene scaffold as an in vitro platform to study the effect of electrical stimulation on neuronal differentiation. *Biomedical Materials (Bristol), 13*(3), 034111. doi:10.1088/1748-605X/aaaf2b PMID:29442069

Sha, J., Gao, C., Lee, S. K., Li, Y., Zhao, N., & Tour, J. M. (2016). Preparation of three-dimensional graphene foams using powder metallurgy templates. *ACS Nano, 10*(1), 1411–1416. doi:10.1021/acsnano.5b06857 PMID:26678869

Shao, Q., Tang, J., Lin, Y., Zhang, F., Yuan, J., Zhang, H., Shinya, N., & Qin, L. C. (2013). Synthesis and characterization of graphene hollow spheres for application in supercapacitors. *Journal of Materials Chemistry. A, Materials for Energy and Sustainability, 1*(48), 15423–15428. doi:10.1039/c3ta12789c

Shi, L., Chen, K., Du, R., Bachmatiuk, A., Rümmeli, M. H., Xie, K., Huang, Y., Zhang, Y., & Liu, Z. (2016). Scalable Seashell-Based Chemical Vapor Deposition Growth of Three-Dimensional Graphene Foams for Oil-Water Separation. *Journal of the American Chemical Society, 138*(20), 6360–6363. doi:10.1021/jacs.6b02262 PMID:27157548

Sun, H., Mei, L., Liang, J., Zhao, Z., Lee, C., Fei, H., Ding, M., Lau, J., Li, M., Wang, C., Xu, X., Hao, G., Papandrea, B., Shakir, I., Dunn, B., Huang, Y., & Duan, X. (2017). Three-dimensional holey-graphene/niobia composite architectures for ultrahigh-rate energy storage. *Science, 356*(6338), 599–604. doi:10.1126cience.aam5852 PMID:28495745

Tao, L., Wang, Q., Dou, S., Ma, Z., Huo, J., Wang, S., & Dai, L. (2016). Edge-rich and dopant-free graphene as a highly efficient metal-free electrocatalyst for the oxygen reduction reaction. *Chemical Communications, 52*(13), 2764–2767. doi:10.1039/C5CC09173J PMID:26757794

Tao, Y., Kong, D., Zhang, C., Lv, W., Wang, M., Li, B., Huang, Z. H., Kang, F., & Yang, Q. H. (2014). Monolithic carbons with spheroidal and hierarchical pores produced by the linkage of functionalized graphene sheets. *Carbon, 69*, 169–177. doi:10.1016/j.carbon.2013.12.003

Trung, N. B., Van Tam, T., Kim, H. R., Hur, S. H., Kim, E. J., & Choi, W. M. (2014). Three-dimensional hollow balls of graphene-polyaniline hybrids for supercapacitor applications. *Chemical Engineering Journal, 255*, 89–96. doi:10.1016/j.cej.2014.06.028

Wang, H., Sun, K., Tao, F., Stacchiola, D. J., & Hu, Y. H. (2013). 3D honeycomb-like structured graphene and its high efficiency as a counter-electrode catalyst for dye-sensitized solar cells. *Angewandte Chemie International Edition*, *52*(35), 9210–9214. doi:10.1002/anie.201303497 PMID:23897636

Wang, X., Zhang, Y., Zhi, C., Wang, X., Tang, D., Xu, Y., Weng, Q., Jiang, X., Mitome, M., Golberg, D., & Bando, Y. (2013). Three-dimensional strutted graphene grown by substrate-free sugar blowing for high-power-density supercapacitors. *Nature Communications*, *4*(1), 1–8. doi:10.1038/ncomms3905 PMID:24336225

Wang, Y., Wang, B., Wang, J., Ren, Y., Xuan, C., Liu, C., & Shen, C. (2018). Superhydrophobic and superoleophilic porous reduced graphene oxide/polycarbonate monoliths for high-efficiency oil/water separation. *Journal of Hazardous Materials*, *344*, 849–856. doi:10.1016/j.jhazmat.2017.11.040 PMID:29190582

Wu, P., Wang, H., Tang, Y., Zhou, Y., & Lu, T. (2014). Three-dimensional interconnected network of graphene-wrapped porous silicon spheres: In situ magnesiothermic-reduction synthesis and enhanced lithium-storage capabilities. *ACS Applied Materials & Interfaces*, *6*(5), 3546–3552. doi:10.1021/am405725u PMID:24506494

Wu, Y., Zhu, J., & Huang, L. (2019). A review of three-dimensional graphene-based materials: Synthesis and applications to energy conversion/storage and environment. In *Carbon* (Vol. 143, pp. 610–640). Elsevier Ltd. doi:10.1016/j.carbon.2018.11.053

Xia, K., Li, Q., Zheng, L., You, K., Tian, X., Han, B., Gao, Q., Huang, Z., Chen, G., & Zhou, C. (2017). Controllable fabrication of 2D and 3D porous graphene architectures using identical thermally exfoliated graphene oxides as precursors and their application as supercapacitor electrodes. *Microporous and Mesoporous Materials*, *237*, 228–236. doi:10.1016/j.micromeso.2016.09.015

Xu, Y., Shi, G., & Duan, X. (2015). Self-Assembled Three-Dimensional Graphene Macrostructures: Synthesis and Applications in Supercapacitors. *Accounts of Chemical Research*, *48*(6), 1666–1675. doi:10.1021/acs.accounts.5b00117 PMID:26042764

Xue, Y., Liu, J., Chen, H., Wang, R., Li, D., Qu, J., & Dai, L. (2012). Nitrogen-doped graphene foams as metal-free counter electrodes in high-performance dye-sensitized solar cells. *Angewandte Chemie International Edition*, *51*(48), 12124–12127. doi:10.1002/anie.201207277 PMID:23124846

Yan, P., Zheng, J., Gu, M., Xiao, J., Zhang, J. G., & Wang, C. M. (2017). Intragranular cracking as a critical barrier for high-voltage usage of layer-structured cathode for lithium-ion batteries. *Nature Communications*, *8*(1), 1–9. doi:10.1038/ncomms14101 PMID:28091602

Yan, W., Cao, X., Tian, J., Jin, C., Ke, K., & Yang, R. (2016). Nitrogen/sulfur dual-doped 3D reduced graphene oxide networks-supported CoFe2O4 with enhanced electrocatalytic activities for oxygen reduction and evolution reactions. *Carbon*, *99*, 195–202. doi:10.1016/j.carbon.2015.12.011

Yan, X., Hu, W., Guo, J., Cai, X., Huang, L., Xiong, Y., & Tan, S. (2016). Easily Separated a Novel rGO-MMT Three-Dimensional Aerogel with Good Adsorption and Recyclable Property. *ChemistrySelect*, *1*(18), 5828–5837. doi:10.1002lct.201601028

Yang, H., Miao, J., Hung, S.-F., Chen, J., Tao, H. B., Wang, X., Zhang, L., Chen, R., Gao, J., Chen, H. M., Dai, L., & Liu, B. (2016). Identification of catalytic sites for oxygen reduction and oxygen evolution in N-doped graphene materials: Development of highly efficient metal-free bifunctional electrocatalyst. *Science Advances*, *2*(4), e1501122. doi:10.1126ciadv.1501122 PMID:27152333

Yang, K., Wang, J., Chen, X., Zhao, Q., Ghaffar, A., & Chen, B. (2018). Application of graphene-based materials in water purification: From the nanoscale to specific devices. In Environmental Science: Nano (Vol. 5, Issue 6, pp. 1264–1297). Royal Society of Chemistry. doi:10.1039/C8EN00194D

Yang, Y., Hu, G., Chen, F., Liu, J., Liu, W., Zhang, H., & Wang, B. (2015). An atom-scale interfacial coordination strategy to prepare hierarchically porous Fe3O4-graphene frameworks and their application in charge and size selective dye removal. *Chemical Communications*, *51*(76), 14405–14408. doi:10.1039/C5CC06257H PMID:26271754

Yao, G., Liu, X., Zhang, G., Han, Z., & Liu, H. (2021). Green synthesis of tannic acid functionalized graphene hydrogel to efficiently adsorb methylene blue. *Colloids and Surfaces. A, Physicochemical and Engineering Aspects*, *625*, 126972. doi:10.1016/j.colsurfa.2021.126972

Yuan, L., Xin, N., Liu, Y., & Shi, W. (2021). In situ construction of multi-dimensional Co3O4/NiCo2O4 hierarchical flakes on self-supporting carbon substrate with ultra-high capacitance for hybrid supercapacitors. *Journal of Colloid and Interface Science*, *599*, 158–167. doi:10.1016/j.jcis.2021.04.036 PMID:33933790

Yue, H. Y., Zhang, H. J., Huang, S., Lu, X. X., Gao, X., Song, S. S., Wang, Z., Wang, W. Q., & Guan, E. H. (2020). Highly sensitive and selective dopamine biosensor using Au nanoparticles-ZnO nanocone arrays/graphene foam electrode. *Materials Science and Engineering C*, *108*, 110490. Advance online publication. doi:10.1016/j.msec.2019.110490 PMID:31923936

Zhang, J., Zhou, Q., Cao, J., Wu, W., Zhang, H., Shi, Y., Mao, Q., & Ma, H. (2021). Flexible textile ion sensors based on reduced graphene oxide/fullerene and their potential applications of sweat characterization. *Cellulose (London, England)*, *28*(5), 3123–3133. doi:10.100710570-021-03730-z

Zhang, X., Han, Y., Huang, L., & Dong, S. (2016). 3D Graphene Aerogels Decorated with Cobalt Phosphide Nanoparticles as Electrocatalysts for the Hydrogen Evolution Reaction. *ChemSusChem*, *9*(21), 3049–3053. doi:10.1002/cssc.201600904

Zhang, X., Lai, Y., Ge, M., Zheng, Y., Zhang, K.-Q., & Lin, Z. (2015). Fibrous and flexible supercapacitors comprising hierarchical nanostructures with carbon spheres and graphene oxide nanosheets. *Journal of Materials Chemistry. A, Materials for Energy and Sustainability*, *3*(24), 12761–12768. doi:10.1039/C5TA03252K

Zhang, Y., Zhang, L., & Zhou, C. (2013). Review of chemical vapor deposition of graphene and related applications. *Accounts of Chemical Research*, *46*(10), 2329–2339. doi:10.1021/ar300203n PMID:23480816

Zhang, Z., Xiao, F., Qian, L., Xiao, J., Wang, S., & Liu, Y. (2014). Facile synthesis of 3D MnO2-graphene and carbon nanotube-graphene composite networks for high-performance, flexible, all-solid-state asymmetric supercapacitors. *Advanced Energy Materials*, *4*(10), 1400064. doi:10.1002/aenm.201400064

Zhang, Z., Xiao, F., Xi, J., Sun, T., Xiao, S., Wang, H., Wang, S., & Liu, Y. (2014). Encapsulating Pd nanoparticles in double-shelled graphene@carbon hollow Spheres for excellent chemical catalytic property. *Scientific Reports*, *4*(1), 1–5. doi:10.1038rep04053 PMID:24514577

Zhu, C., Han, T. Y. J., Duoss, E. B., Golobic, A. M., Kuntz, J. D., Spadaccini, C. M., & Worsley, M. A. (2015). Highly compressible 3D periodic graphene aerogel microlattices. *Nature Communications*, *6*(1), 1–8. doi:10.1038/ncomms7962 PMID:25902277

Zhu, C., Liu, T., Qian, F., Han, T. Y. J., Duoss, E. B., Kuntz, J. D., Spadaccini, C. M., Worsley, M. A., & Li, Y. (2016). Supercapacitors Based on Three-Dimensional Hierarchical Graphene Aerogels with Periodic Macropores. *Nano Letters*, *16*(6), 3448–3456. doi:10.1021/acs.nanolett.5b04965 PMID:26789202

Zhu, H., Chen, D., Li, N., Xu, Q., Li, H., He, J., & Lu, J. (2015). Graphene foam with switchable oil wettability for oil and organic solvents recovery. *Advanced Functional Materials*, *25*(4), 597–605. doi:10.1002/adfm.201403864

Zhu, H., Chen, D., Yang, S., Li, N., Xu, Q., Li, H., Wang, L., He, J., Jiang, J., & Lu, J. (2016). A versatile and cost-effective reduced graphene oxide-crosslinked polyurethane sponge for highly effective wastewater treatment. *RSC Advances*, *6*(44), 38350–38355. doi:10.1039/C6RA05450A

Zhu, J., Zhao, Y., Ning, P., & Tang, S. (2021). Cobalt sulphide/graphene aerogel nanocomposite with enhanced anode performance for lithium energy storage. *Bulletin of Materials Science*, *44*(1), 1–8. doi:10.100712034-020-02275-4

ADDITIONAL READING

Ali, I., Basheer, A. A., Mbianda, X. Y., Burakov, A., Galunin, E., Burakova, I., Mkrtchyan, E., Tkachev, A., & Grachev, V. (2019). Graphene based adsorbents for remediation of noxious pollutants from wastewater. *Environment International*, *127*, 160–180. doi:10.1016/j.envint.2019.03.029 PMID:30921668

Fang, R., Chen, K., Yin, L., Sun, Z., Li, F., & Cheng, H.-M. (2019). The Regulating Role of Carbon Nanotubes and Graphene in Lithium-Ion and Lithium–Sulfur Batteries. *Advanced Materials*, *31*(9), 1800863. doi:10.1002/adma.201800863 PMID:29984484

Gao, X. G., Cheng, L. X., Jiang, W. S., Li, X. K., & Xing, F. (2021). Graphene and its Derivatives-Based Optical Sensors. *Frontiers in Chemistry*, *9*, 615164. Advance online publication. doi:10.3389/fchem.2021.615164 PMID:33614600

Kemp, K. C., Seema, H., Saleh, M., Le, N. H., Mahesh, K., Chandra, V., & Kim, K. S. (2013). Environmental applications using graphene composites: Water remediation and gas adsorption. *Nanoscale*, *5*(8), 3149–3171. doi:10.1039/c3nr33708a PMID:23487161

Kim, J. E., Oh, J. H., Kotal, M., Koratkar, N., & Oh, I. K. (2017). Self-assembly and morphological control of three-dimensional macroporous architectures built of two-dimensional materials. *Nano Today*, *14*, 100–123. doi:10.1016/j.nantod.2017.04.008

Lefebvre, O., & Moletta, R. (2006). Treatment of organic pollution in industrial saline wastewater: A literature review. *Water Research*, *40*(20), 3671–3682. doi:10.1016/j.watres.2006.08.027 PMID:17070895

Mamaghani, A. H., Haghighat, F., & Lee, C. S. (2019). Hydrothermal/solvothermal synthesis and treatment of TiO2 for photocatalytic degradation of air pollutants: Preparation, characterization, properties, and performance. *Chemosphere*, *219*, 804–825. doi:10.1016/j.chemosphere.2018.12.029 PMID:30572234

Shi, X., Zheng, S., Wu, Z. S., & Bao, X. (2018). Recent advances of graphene-based materials for high-performance and new-concept supercapacitors. *Journal of Energy Chemistry*, *27*(1), 25–42. doi:10.1016/j.jechem.2017.09.034

Song, F., Li, G., Zhu, Y., Wu, Z., Xie, X., & Zhang, N. (2020). Rising from the horizon: Three-dimensional functional architectures assembled with MXene nanosheets. *Journal of Materials Chemistry. A, Materials for Energy and Sustainability*, *8*(36), 18538–18559. doi:10.1039/D0TA06222G

Song, S., Shen, H., Wang, Y., Chu, X., Xie, J., Zhou, N., & Shen, J. (2020). Biomedical application of graphene: From drug delivery, tumor therapy, to theranostics. *Colloids and Surfaces. B, Biointerfaces*, *185*, 110596. doi:10.1016/j.colsurfb.2019.110596 PMID:31707226

Yin, Z., Zhu, J., He, Q., Cao, X., Tan, C., Chen, H., Yan, Q., & Zhang, H. (2014). Graphene-Based Materials for Solar Cell Applications. *Advanced Energy Materials*, *4*(1), 1300574. doi:10.1002/aenm.201300574

Zhao, J., Zhang, G. Y., & Shi, D. X. (2013). Review of graphene-based strain sensors. *Chinese Physics B*, *22*(5), 057701. Advance online publication. doi:10.1088/1674-1056/22/5/057701

Zheng, S., Wu, Z. S., Wang, S., Xiao, H., Zhou, F., Sun, C., Bao, X., & Cheng, H. M. (2017). Graphene-based materials for high-voltage and high-energy asymmetric supercapacitors. *Energy Storage Materials*, *6*, 70–97. doi:10.1016/j.ensm.2016.10.003

KEY TERMS AND DEFINITIONS

Antimicrobial Activity: Inhibition process of bacteria growth or prevention of microbial colonies formation.

Graphene Aerogels: These materials are three-dimensional structures derived from graphene, synthesized by assembling nanosheets, thus forming a structure with interconnected micro, meso and macro pores, which help transport molecules rapidly.

Graphene Oxide (GO): A graphene derivative obtained from a partial oxidation of the graphene.

Graphene Spheres: Three-dimensional structures are formed by assembling the graphene sheets within circular structures.

Graphene-Based Adsorbent Structures: These materials consist of a three-dimensional porous network that (in addition to maintaining the properties of two-dimensional carbon structures) have high porosity, low density and a large surface area.

Chapter 6
Polymers and Graphene-Based Materials as Barrier Coatings:
A Quick Review on Their Properties and Synthesis Techniques

Shamma Al Hashmi
Technology Innovation Institute, UAE

Shroq Al Zadjali
Technology Innovation Institute, UAE

Nitul S. Rajput
Technology Innovation Institute, UAE

Meriam Mohammedture
Technology Innovation Institute, UAE

Monserrat Gutierrez
Technology Innovation Institute, UAE

Amal M. K. Esawi
The American University in Cairo, Egypt

ABSTRACT

Currently, a wide range of materials are being used as barrier coatings for different applications. Among them, polymers and graphene have been the focus of many studies. Polymers are used in numerous industries due to their remarkable properties, including resistance to thermal degradation, resistance to chemical permeation, and good mechanical properties. On the other hand, graphene, a one-atom-thick and two-dimensional material, does not allow the permeation of gases or liquid molecules through its plane; thus, it has been considered one of the promising nanomaterials used in gas and liquid barrier applications and corrosion inhibition coatings.

DOI: 10.4018/978-1-7998-8936-6.ch006

INTRODUCTION

The impermeable nature of several materials is important to different industrial sectors. If a component in a system is not accurately coated it can result in the catastrophic failure of the entire system. Just the worldwide amount spent on fixing corrosion failure is approximately US$ 2.5 trillion, indicating that effective barriers are vital to the required systems. The most important requirements for materials to be used in barrier applications are good impermeable nature to water vapour and oxygen, high transparency, and good mechanical properties (Sangroniz et al., 2019).

Barrier properties of a material correspond to the specific permeable objects or species such as gases, water vapour, or even light that it is acting as a wall against. Coatings do not only provide barrier properties but can also improve the efficiency of the system, such as the thermal barrier coatings used in turbine engines (Xu et al., 2008). Thermal barrier coatings are a family of coatings specifically used in high temperature applications such as energy generation and storage, and nuclear industries. Aerospace turbine engines use low thermal conducting ceramics or its composites to coat the superalloy, with the first type of coatings used was frit enamels in 1950. This enables a temperature reduction in the surface enabling the engine to operate above the melting temperature of the superalloy, thus improving performance and energy efficiency. Furthermore, coatings are usually multi-layers of different types of physical and chemical properties, with each layer providing its specific function to the system. This includes functions such as corrosion, erosion, or oxidation inhibitors, balancing thermal mismatch between the layers of coatings, or preventing interdiffusion of elements.

Polymers are well known for their barrier properties. Some of the advantageous properties of polymers that make them favorable to be used in multiple industries, which also include, coating for electronics, solar cells, and in the medical industry, are their excellent mechanical properties, and resistance to chemicals and heat. For example, food packaging needs barrier coatings that will prevent the mixing of moisture and oils from the interior and exterior of the packaging as it will compromise the sensory and hygienic integrity of the packed food (Tyagi et al., 2021). The good barrier properties of polymers result from the molecular packing and orientation within the polymer and are influenced by the degree of crystallization of the semicrystalline polymers. The packing and orientation of the molecules determine the fractional free volume (FFV) (Klopffer & Flaconneche, 2001), where a rise in the amorphous density leads to a reduction in the FFV. FFV refers to the volume within the polymer that is available for the penetrant to diffuse into. It does not encompass the volume taken up by the polymer molecules and the volume that is inaccessible by the penetrant (Dhoot et al., 2002). Whereas the presence of crystals within the polymer determines the tortuous pathway that the penetrant must follow, therefore increasing the degree of crystallinity leads to an increase in the barrier properties (Klopffer & Flaconneche, 2001).

Additionally, among the existing methods for corrosion protection, protective coatings are gaining popularity due to their scalable and ease of application. Graphene and its derivatives have caught attention in developing new types of protective coatings due to their thermal stability, good mechanical properties, impermeability to different types of gases, and great chemical resistance (Nine et al., 2015). Graphene provides sufficient protective coatings due to the formation of dense electron clouds that restrict the gaps in the aromatic rings (Sreeprasad & Berry, 2013). Graphene is a nanomaterial, so its thickness is ideal for protecting and maintaining or even enhancing the substrate's properties. To protect against corrosion, the requirement is to either minimize or entirely stop the electrochemical process by blocking access of corrosive ions to the substrate surface (Kumar et al., 2021). Graphene-based materials properties including chemical inertness, small geometric pore, enhanced adsorption capacity, and high surface area, contribute to its capability to be used as effective corrosion inhibitors (Othman et al., 2019a).

This review discusses and introduces current research and challenges found in advanced materials used as barrier and coating applications to accomplish different functionalities such as oil, grease, water, and gas resistance. Detailed advances of polymer and graphene-based materials used for barrier coating applications are concentrated on to give a clearer insight on the developments needed to achieve the highest level of barrier properties for these materials.

POLYMER-BASED MATERIALS

Polymers have been used in a multitude of applications, but barrier polymers are commonly used in many protective and packaging applications (DeLassus, 2002). Some of the notable ones are polylactic acid (PLA), poly(ethylene isophthalate) (PEI), Ethylene Vinyl Alcohol Copolymer (EVOH), poly(ethylene-2,2'-bifuran-5,5'-dicarboxylate) (PEBF), poly(caprolactone) (PCL) blends and poly(butylene furanedicarboxylate-co-glycolate). Although the polymers can be used without changing their composition, it is possible to improve the barrier properties by adding nanofillers.

Polylactic Acid (PLA)

Polylactic acid (PLA), also known as polylactide, is a thermoplastic polyester composed of repeating lactic acid. There are several methods to form PLA, the most common is lactide molecule undergoes the ring-opening polymerization, lactide molecules contain two lactic acid structures; another method is the condensation polymerization of lactic acid (Koppolu et al., 2019). Polymerization of lactide to PLA results in three different configurations of PLA due to the asymmetry of lactic acid structures in lactide; lactide molecules often exist in 2 different configurations. The lactide configurations are L-lactide, D-lactide, and meso-lactide, which are dimers of L-lactic acid, D-lactic acid, and both L-lactic acid and D-lactic acid respectively (Madhavan Nampoothiri et al., 2010). The crystallinity of PLA depends on the optical purity of the monomer. For example, PLA composed of 93% L-lactic acid is considered to be semi-crystalline. Duan and Thomas investigated how the degree of crystallinity would affect the water vapor permeability of PLA (Duan & Thomas, 2014). In order to produce a range of different degrees of crystallinity, samples were annealed at different temperatures for different periods of time. Water vapor transmission rate (WVTR) was measured for the PLA films produced using the MOCON water vapor transmission rate tester model. The results were that as the crystallinity of the samples increases, the WVTR decreases linearly.

Poly(ethylene isophthalate) (PEI)

Poly(ethylene isophthalate) (PEI) is a polyester consisting of a repeating unit of $C10H8O4$ containing phenyl rings as a backbone and 2 carboxyl groups in the 1,3-meta position. It is a structural isomer of poly(ethylene terephthalate) (PET). PET can be used in an extensive range of applications due to its properties, including sufficient thermal and chemical resistance and excellent mechanical properties. Nevertheless, the range of applications of PET can be extended if the barrier properties of PET are improved. These applications could be packaging, e.g. for carbonated beverages or oxygen-sensitive

Figure 1. The structure of the phenyl ring backbone of (a) PET and (b)PEI

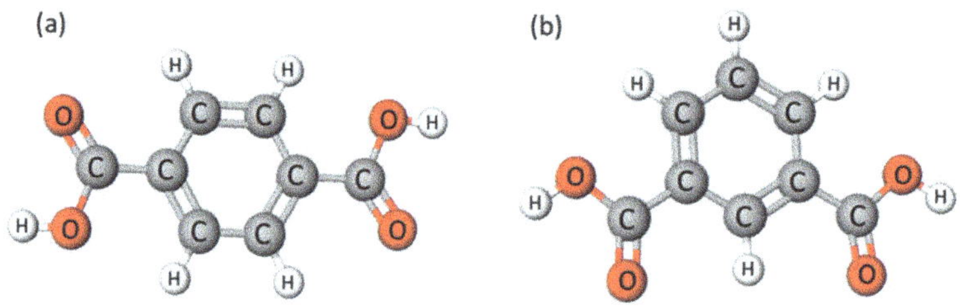

products. PET also consists of a phenyl ring as a backbone with 2 carboxyl groups; however, the position of the carboxyl groups is different and is located in the 1,4-para position, as shown in Figure 1. As a result, the phenyl ring of PET can flip, acting like a valve or trapdoor for gas penetration. Conversely, the non-colinear position of the carboxyl groups in the phenyl ring in PEI, the phenyl rings cannot flip. The process for the preparation of PEI consists of the polymerization of the monomer bis-(2- hydroxyethyl) isophthalate (BHEI), which is prepared from Isophthalic acid (IPA) and ethylene glycol.

Samples of PEI and PET were tested in a Differential Scanning Calorimeter (DSC) to study the thermal properties of the polymers. The DSC results showed the glass transition of both PEI and PET as well as PET's crystallization peak and its melting peak. PEI showed neither a crystallization peak nor a melting peak, indicating that PEI remained amorphous and did not crystallize. The permeability of PEI and PET was evaluated using an electrobalance apparatus. The analysis showed that PET allowed 12 times more CO2 to permeate than PEI. This difference is attributed to the structural difference between the two. The folding over of the phenyl ring in PET creates a diffusion pathway that allows gasses to pass through the film (Kotek et al., 2004).

Ethylene Vinyl Alcohol Copolymer (EVOH)

EVOH is currently used as an oxygen barrier in various applications, including food packaging, pharmaceutical packaging, and floor tubes to prevent oxygen from corroding the reboiler. EVOH has also been used as a barrier to organic material, such as hydrocarbons and flavors. It has been successfully used to prevent fuel from leaking into the surroundings in fuel tanks, to increase retention of herbicide as agricultural films, and as geomembrane films to prevent methane leakage from landfills. EVOH forms excellent barriers due to the structures of its components. The ethylene units show good water resistance but have poor gas barrier properties. In contrast, the vinyl alcohol unit exhibits good gas barrier properties because the hydroxyl group forms inter-and intra-molecular bonds. However, vinyl alcohol is soluble in water which causes it to be challenging to process. Therefore, the combination of these components results in a processable copolymer with a good gas barrier that remains sensitive to water. EVOH is synthesized in 2 stages, mainly because vinyl alcohol alone is unstable. In the first stage, the ethylene-vinyl acetate random copolymer is produced by copolymerizing vinyl acetate and ethylene. The second stage is the transesterification of polyethylene vinyl acetate in the presence of methanol, resulting in EVOH with methyl acetate as a by-product (Maes et al., 2018). Since the barrier properties of EVOH are good, EVOH has been used in polymer blends to increase the barrier properties(J. Kim et al., 2020).

Poly (ethylene 2,2'-bifuran-5,5'dicarboxylate) (PEBF)

Poly (ethylene-2,2'-bifuran-5,5'-dicarboxylate) (PEBF) is a biologically based polyester. It has lately been tested to have significantly better higher tensile modulus and oxygen barrier properties compared to PET. Due to its structure, PEBF can absorb UV rays, which can extend its application range to UV-sensitive packaging. However, these properties depend on the polymer synthesis; for example, if high amounts of catalyst were used in the synthesis, the PEBF films produced will be extremely brittle. The catalysts used in the production of PEBF were manganese and zinc catalysts, which are known to drive the transesterification reaction that produces hydroxyethoxy end groups from the methyl ester ends. The production of PEBF can be achieved by 2 methods. The first requires the transesterification of 2,2'-bifuran-5,5'-dicarboxylate (BFE) prior to melt polycondensation. This is done by removing methanol molecules and replacing them with ethylene glycol (EG). This reaction is monitored periodically with proton nuclear magnetic resonance (1H NMR) to observe the methyl ends and determine the time required. The other method uses bis(2-hydroxyethoxy) 2,2'-bifuran-5,5'-dicarboxylate (HBEBF) and does not require the transesterification reaction structure, which already contains the EG moiety instead of the methyl ester. The polyesters prepared with HBEBF have fewer diethylene glycol (DEG) concentrations in the produced polymer; this occurs because DEG tends to form more rapidly at elevated temperatures and in the absence of excess EG in the polycondensation melt. The permeability of the polymer is lower than polyesters that are commercially available which are known for their low permeability, such as poly (ethylene 2,6-naphthalene dicarboxylate) (PEN) and PET, with the exception of poly (ethylene furanoate) (PEF), which still has a lower oxygen permeability than PEBF (Edling et al., 2020). PEBF showed low oxygen permeability due to its molecular structure, mainly the nonlinear structure of bifuran, which decreased the segmental mobility of the polymer chain and hindered the molecular movements, and consequently lowered the permeability of PEBF (Edling et al., 2020).

Poly(caprolactone) (PCL) Blends

Poly (Hydroxy Amino Ether) (PHAE)

Poly(caprolactone) (PCL) is a biodegradable polyester. It has many advantageous properties such as low viscosity, resistant to solvents, and easy to process. Although PCL has many advantageous properties, it does not have good barrier properties, which leads to limitations in its applications. To improve its barrier properties, PCL is blended with poly(hydroxyamino ether) (PHAE). PHAE is a polyamide with better barrier properties than PCL and has already been shown to be miscible with several polymers (Yang et al., 2010), such as polyamide 6, poly(butylene terephthalate), poly(ethylene terephthalate), and PCL. The PCL/PHEA was previously investigated and found to have good compatibility and sufficient mechanical properties. The adsorption of water vapor of the polymer blends was researched by Eceolaza et al.(Eceolaza et al., 2019). The result shows that PCL had absorbed less water vapor than PHAE, indicating that PCL is more hydrophobic than PHEA. This is supported by the structures of PCL and PHEA, where PHEA contains groups such as ether, amine, and hydroxyl groups which are capable of forming hydrogen bonds with water and promoting the absorption of water. While PCL contains only ester groups that can interact with water molecules. The diffusion coefficient of blend polymers was measured and the analysis showed that PCL has the highest diffusion coefficient which decreases with increasing weight fraction of PHEA. This occurrence can be explained by the presence of small free-volume holes within PHAE. Although the free volume holes are abundant, they are small enough to prevent the permeation of a larger penetrant, thus hindering it. Secondly, the measurement temperature plays an important role. The glass transition temperature of PHAE was not reached, unlike PCL, whose glass transition temperature was exceeded, resulting in the polymer chains in PCL having more mobility and giving a higher diffusion coefficient than that of PHEA.

Poly (Hydroxy Ether of Bisphenol A) (PH)

Another polymer utilized to increase the barrier properties of PCL is the blend of PCL with poly(hydroxyether of bisphenol A) (PH). PH has excellent barrier properties because it includes a hydroxyl group that is capable of interacting with the penetrant. Blends of PCL and PH have been prepared, and their thermal and barrier properties have been reported by Corres et al. (Corres et al., 2020). The thermal tests showed that with an increasing weight percentage of amorphous polymer (PH) in PCL/PH blends, the crystallinity decreases. It has been shown that as the crystallinity increases, the permeability of the polymer decreases; crystals within the polymer could cause the path of the penetrant to become more tortuous. The free volume of the blends was analyzed and it was found that as the weight percentage of PH in the blends increases, the size of the holes in the free volume decreases. It was also shown that PCL had a higher free volume fraction than PH; this can be explained by the low glass transition temperature of PCL, which leads to inefficient packing of the polymer chains compared to PH. Thus, the free volume analysis of the blends leads to the assumption that the permeability to gasses and vapors decreases with increasing PH's content in the blends. The permeation of carbon dioxide, water vapor, and oxygen in PCL/PH polymer was studied and showed the same trend. The permeation trend decreases with the increasing weight percentage of PH in the blend polymers, which leads to an increase in barrier properties. This improvement is attributed to the reduction of free volume hole. It can also be credited to the forces between the two polymers leading to a miscible polymer blend which hinders diffusion and consequently reduces permeability.

Poly(butylene furandicarboxylate-co-glycolate)

Polyglycolic acid (PGA) (Figure 2) is a polymer of glycolic acid (GA), which is a type of hydroxy acid (HA), and has gained interest because its' monomer can be produced from a renewable source (Salusjärvi et al., 2019). However, the properties of PGA are limited by the method of synthesis. Melt polycondensation cannot produce high molecular weight PGA because the conditions of melt polycondensation promote depolymerization more than the growth of polymer (Ayyoob et al., 2017). This greatly affects the properties of PGA and limits its potential applications. To produce usable polymers from HA, it is incorporated into aromatic polyesters to produce a copolymer. The HA would improve biodegradability, facilitate processing and change the physical properties of the copolymer. PGA has shown higher gas barrier properties than PLA and PCL; this is due to how close the stacked segments of PGA are compared to other polymers, which could be due to the regular structure GA compared to the other 2 monomers. Biodegradable polymers prepared from an aromatic polyester and GA are theoretically feasible; however, in reality, it has rarely been possible to prepare a polymer with good properties due to the thermal instability of PGA, which decomposes at temperatures below those required for melt polycondensation (Hu et al., 2019)

Figure 2. The structure of Polyglycolic acid (PGA)

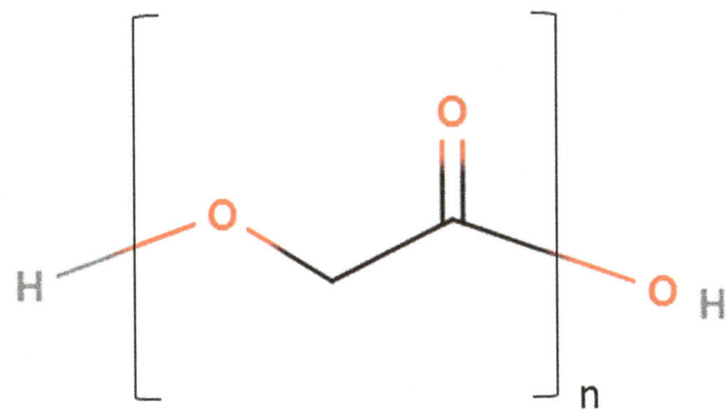

Several studies (Pajoumshariati et al., 2018; Yang et al., 2013) have attempted to prepare a high molecular weight copolymer containing GA. Recent work investigated different methods of preparation. It was found that the temperature of melt polycondensation has a considerable effect on the molecular weight of the copolymer, where low temperature does not lead to polycondensation and high temperature would lead to the degradation of PGA. It was found that the temperature of melt polycondensation has a great effect on the molecular weight of the copolymer, where a low temperature would not lead to polycondensation and a high temperature would cause the PGA to degrade.

GRAPHENE-BASED MATERIALS

Graphene

Graphene is the thinnest carbon allotrope consisting of hexagonal crystalline structure with sp^2 hybridized carbon atoms densely packed in aromatic rings; the gaps within graphene's aromatic rings are blocked due to the dense π-orbitals forming a cloud of delocalized electrons which does not allow the permeation of molecules and gases (Berry, 2013).

The synthesis process plays a vital role in determining the properties of graphene layers, purity, and structural defect percentage. Thus, the production of defect-free and large-area graphene is still challenging. Many synthesis techniques have been reported, such as chemical vapor deposition (CVD), plasma-enhanced chemical vapor deposition (PECVD) on different substrates, epitaxial growth on metal substrates, unzipping CNTs to produce graphene nanoribbons, and mechanical exfoliation of graphite producing graphene material (Rangel et al., 2009). It is well known that graphene is impermeable to any gaseous molecules. The impermeability property comes from the localized electron density around the aromatic rings that repel atoms and molecules' penetration. Thus, single-crystal graphene can act as a great barrier against gas diffusion (Yoo et al., 2014).

Graphene as a Corrosion Inhibiting Coating

The use of reactive metals in different applications is widely spread. Thus, an intense effort in the production of corrosion inhibitor coatings based on 2D materials such as graphene and its derivatives has been under research for applications in which highly corrosive fluids are carried and high corrosion resistance is required for long periods.

Graphene prepared by CVD can be transferred to metal and employed as a corrosion inhibiting coating. The properties and the structure of graphene sheet produced are influenced by the substrate of the prepared graphene films (Babichev et al., 2015). Different types of metals can be used as a substrate to produce graphene sheets, for instance, titanium alloys, copper alloys, aluminium, and steel. Oxidation resistance Cu and Cu/Ni alloy coated with atomically thin graphene layers grown by CVD has been investigated both in air and at high temperature. It was demonstrated that graphene sheets prevented the formation of oxides on the metal surfaces even after heating the samples to 200°C (S. Chen et al., 2011). This is due to the thermal and chemical stability of graphene at high temperatures without undergoing any chemical reaction (De Heer et al., 2007; Shivaraman et al., 2009).

CVD graphene grown on Cu has been transferred to pristine Nickel to study the corrosion protection properties of graphene. The flexible C-C bonds within the graphene structure allow it to adhere to the Ni substrate and to form a smooth coating (Prasai et al., 2012). However, CVD graphene introduces many intrinsic defects that range between 1-15 nm in size, allowing the diffusion of unfavorable molecules. Thus, it is challenging to prepare a defect-free graphene sheet (Koenig et al., 2012). Kirkland et al. studied the effect of graphene-coated on Ni on corrosion resistance performance. Both cathodic and anodic reaction rates and the current densities decreased, indicating that protection against corrosion was achieved on Ni (Kirkland et al., 2012). The electrochemical degradation of Cu coated with graphene in the aggressive chloride environment was investigated by Raman et al. It was found that the anodic and current densities of graphene-Cu coated decreased 1-2 orders of magnitude, indicating a decrease in the Cu dissolution. It was also demonstrated that the corrosion potential was measured to be 40 mV compared to uncoated Cu samples, which were associated with the corrosion resistance due to graphene. Compared with Kirkland et al. and Parasai et al., it has been shown that graphene coating produced by Raman et al. was more uniform and continuous (Singh Raman et al., 2012). Scanning electron microscopy revealed that graphene grown on stainless steel by thermal CVD retains high surface roughness leading to an increase in the surface area of the graphene sheet. Electrochemical tests also reveal that corrosion resistance can be increased by growing graphene on stainless steel due to the decrease in the corrosion current by 23% (Ruammaitree et al., 2018).

Graphene Oxide (GO)

The synthesis of large surface area, monocrystalline, and defect-free graphene is still challenging. Thus, pristine graphene is rarely used due to its limited scale-up processes. GO is a type of functionalised graphene used in different applications due to its availability, mass production, and ease of fabrication. The chemical structure of the GO sheet includes hydroxyl, alkoxy, carbonyl, and carboxylic acid with other oxygen functional groups (Smith et al., 2019). A schematic picture indicating the structural difference among graphene, GO and rGO is presented in Figure 3. The densely packed sp2 carbon atoms act as a gas barrier that can be used in packaging materials, protection of electronic devices, and corrosion inhibition coatings. Graphene oxide is usually prepared by two processes including top-down and bottom-up. Bottom-up processes involving chemical vapour deposition (CVD), pyrolysis, epitaxial growth, and plasma synthesis (Fogarassy et al., 2014; Liu et al., 2017). Nevertheless, top-down processes including mechanical exfoliation, chemical exfoliation, and chemical fabrication where GO is extracted from graphite are more favorable. Those processes are easy to scale up and have higher processability

Figure 3. A schematic representation of (a) graphene, (b) GO, and (c) r-GO.

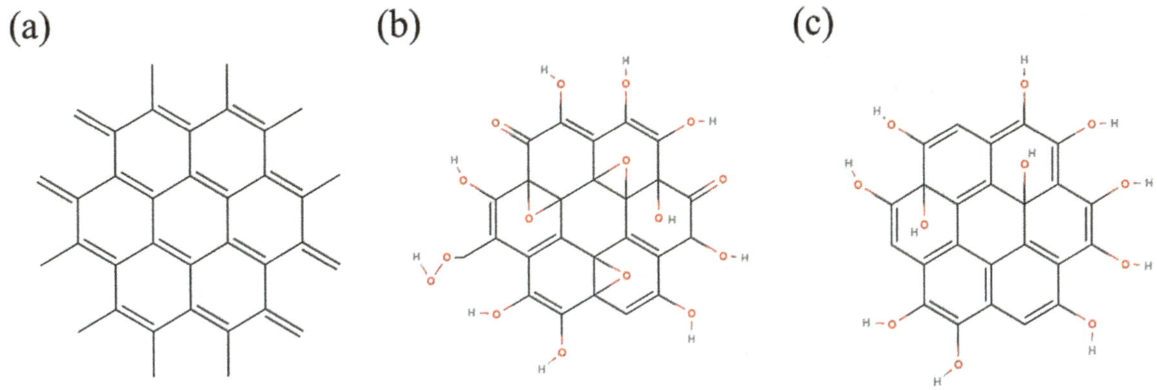

(Kiang Chua & Pumera, 2014). The advantages of using graphene oxide over graphene in different applications is the higher solubility and the possibility to tune the surface structure with different functional groups (Smith et al., 2019).

The first synthesis technique reported by Brodie to produce GO was by oxidizing graphite using nitric acid and potassium chlorate (Brodie, 1860). Later, concentrated sulfuric acid and potassium permanganate were used as an oxidizing agent by Hummers and Offeman (Hummers & Offeman, 1958). Both scientists produced GO by oxidising graphite. However, Hummers and Offeman improved the oxidation process by using $KMnO_4$ as an oxidizer due to evolution of toxic ClO_2 gas when using $KClO_3$, and the addition of sodium nitrate rather than solvent nitric acid that will lead to the formation of nitric acid in situ (Smith et al., 2019). The GO's structure consists of forming functional groups of oxygen on the basal plane, such as carbonyl, hydroxyl, and carboxylic groups at the edges which results in structural defects (Pei et al., 2010). However, carboxylic groups attached at the edges of the plane promote the stabilization of GO colloidal suspension in water without surfactants (Compton & Nguyen, 2010).

Graphene Oxide (GO) as a Barrier Material

Graphene oxide was used in barriers as a filler material within poly(vinyl alcohol) polymer matrix. Full dispersion, exfoliation, and orientation of the GO nanosheets were achieved due to the attachment of chemical functional groups that promote the interfacial bonding between GO nanosheets and PVA polymer matrix. Huang et al. investigated the barrier properties of graphene oxide (GONs) nanosheets/ poly(vinyl alcohol) (PVA) nanocomposite films. It was demonstrated that an increase in the barrier properties has been accomplished because of the full dispersion and exfoliation of the GONs filler associated with the oxygen functional groups that are responsible for decreasing the oxygen water vapour permeability coefficient. It was proved that 98% and 68% reduction in PO_2 and PH_2O respectively were achieved with only 72 vol% of GONs. The reduction in oxygen and water permeability is attributed to the tortuous path effect, which is explained by the increase in the molecule adsorption/diffusion path length through the nanocomposite film (H. D. Huang et al., 2012).

The Tortuosity effect is vital for the impermeability nature of the nanocomposite. It is influenced by the distribution of the filler within the matrix and the volume fraction. Nanoparticles or nanoplatelets act as obstructions to the molecules' diffusivity forcing them to go along a more tortuous pathway (Wolf et al., 2018).

Nielsen demonstrated the gas permeability model for a regular arrangement of GO platelets where gas molecules diffuse in a perpendicular direction. The impermeability of GO film is associated with the orientation and shape of GO platelets. Perfectly impermeable GO film against gas diffusion consists of rectangular GO platelets oriented perpendicular to the direction of gas diffusion. GO films possess high potential energy barriers which block small gas molecules from penetrating through the film. Thus, the only penetration path will be through the top grain boundaries layer and pass through the interlayer spacing, creating a tortuous path. It was revealed that GO platelet size is critical in the permeability nature of the film. The diffusion pathways depend on the size of the GO platelets, by using large GO platelets the gas diffusion pathway will be longer compared to small GO platelets. Moreover, the kinetic energy of gas molecules at low and high applied pressure affect the impermeable nature of the film. The kinetic energy of gas molecules at low applied pressure is less than the potential energy barrier of the GO film; thus, it will not permeate through the film. As the applied pressure increases, the gas molecules will have sufficient energy to overcome the film's potential energy barrier and thus penetrate through it. Nevertheless, as the GO platelets size increases, there is also an increase in the potential energy barrier, and thus higher applied pressure is required for the gas molecules to overcome the potential energy barrier (Yoo et al., 2014).

Reduced Graphene Oxide (r-GO)

In the previous section it was explained that graphene oxide consists of different oxygen functional groups on the basal plane and at the edges. The reduction of GO consists of eliminating the oxygen functional groups to produce reduced graphene oxide, eliminating the surface defects in GO, improving the mechanical and chemical properties. Thus, rGO will have a lesser number of functional groups attached to it as compared to GO (Figure 2). Due to the difference in the structural composition of rGO, GO and graphene, they would behave in different physical, chemical, and mechanical properties. A comparison among the derivatives is given in Table 1. The reduction of graphene oxide can be obtained in different ways such as chemical, electrochemical, and thermal. Each reduction process gives a different microstructure, properties, and performance of the reduced graphene oxide (Smith et al., 2019). Nowadays, there is a great interest in producing reduced graphene oxide by using green reduction agents such as ascorbic acid, sugars, amino acids, and microorganisms (De Silva et al., 2017). Elias et al., 2009, reported a reversible hydrogenation process where the hydrogen atom can be removed and attached without the destruction of the crystal lattice (Elias et al., 2009). The use of hydroiodic acid (HI) as a reducing agent proved to be effective at low temperatures. It was reported that GO reduced by HI leads to an improvement in strength, ductility, flexibility (Pei et al., 2010). Pei et al. demonstrated the reduction of graphene oxide using HI, which increases electrical conductivity up to 298 S/cm. It was also demonstrated that the rGO sheet retained good flexibility, structural integrity, and a 21% increase in tensile strength (Pei et al., 2010). It was proved later in 2014 by Su et al. that GO reduced by HI maintains its structural shape by producing fewer structural defects, which in return increases the mechanical strength (Su et al., 2014).

Y. Su et al. report the chemical reduction of GO by hydroiodic acid (HI) and ascorbic acids (VC). HI-rGO and VC-rGO films were fabricated; it demonstrated transparent and flexible rGO films with no detectable permeation of hydrogen, water, and salt. The material was also tested against HF acid, and it demonstrated full protection against corrosion (Su et al., 2014).

Table 1. Synthesis processes and different properties of graphene, graphene oxide (GO), and reduced graphene oxide (rGO)

Properties	Graphene	Graphene Oxide	Reduced Graphene Oxide
Synthesis	Chemical vapor deposition, thermal decomposition of silicon carbide (SiC), graphite exfoliation	Oxidation and exfoliation of graphite	Reduction of graphene oxide
C:O ratio	No oxygen	2-4	8-246
Interlayer distance	0.33 nm	0.80-1.09 nm	0.35-0.45 nm
Young's modulus (TPa)	1	0.2	0.25
Electron mobility ($cm^2V^{-1}s^{-1}$)	10000-50000	Insulator	0.05-200
Production cost	High	Low	Low

(Rajput et al., 2021)

As stated earlier, the oxygen functional groups on the GO have as a result of structural defects. It can be eliminated by various reduction processes. Reduction processes occurring at a temperature greater than 200°C are more favorable because oxygen functional groups are decomposed (Stankovich et al., 2007). However, rGO also possesses defective basal plane structure due to the presence of vacancies produced by the evolution of carbon such as CO or CO_2 (Schniepp et al., 2006).

Reduced Graphene Oxide (r-GO) as a Barrier Material

Humidity effect on the HI rGO was investigated at 45°C and 100% humidity, and it was demonstrated that no degradation occurs for the moisture barrier. Secondly, the anticorrosion performance of reduced GO by ascorbic acid was investigated. rGO reduced by ascorbic acid was coated on SiO_2 wafer and it fully protected the substrate, and no sign of etching or degradation was detected in presence of hydrofluoric acid (Su et al., 2014). The oxidation resistance of Copper (Cu) and Iron (Fe) coated with five bilayer thermally reduced GO was examined by Kang et al. Five bilayer rGO proved to be sufficient in protecting the Fe and Cu from corrosion as the colour and the morphology of the metals didn't change after oxidation (Kang et al., 2012).

Nano-clays

Nano clay is a layered mineral silicate with desirable mechanical and thermal properties for barrier applications within polymer systems (W. Huang, 2018; Thrivikraman et al., 2017). Fundamentally, when the nano clay material is dispersed within the polymer solution, the mechanical and barrier properties of the composite can be unusually enhanced which is highly due to the high value of surface to volume ratio of the nano clays as well as due to the increased tortuosity of the diffusion path (Ko, 2014). The tortuosity phenomenon due to the implementation of nanoclay can be realized from the schematic Figure 4. Different nanocomposites based on nano clay fillers have been synthesized to investigate their barrier performance. Examples of such nano clays are montmorillonite, bentonite, hectorite, and kaolinite (Gupta & Roy, 2019). Natural clays are hydrophilic in nature and they can easily interact with polar polymers; thus, the polymer or the clay is modified. For instance, montmorillonite has been modified by using alkyl and quaternary ammonium halides to enhance the exfoliation level when combined with nylon

(Fernes et al., 2004). Nanocomposite of polymer and clay was prepared using a simple ultrasonication method (SUN et al., 2007). It is found that the nanocomposite has excellent barrier properties against water or water vapor. Rhim et al. demonstrated the synthesis of Agar-clay nanocomposite films and investigated the effect of the clay content on the mechanical and water barrier properties (Rhim, 2011). The study reveals that the prepared composite film would become more hydrophilic with the addition of hydrophilic nano clay and this reflects in the increase in the swelling ratio. Thus, the film properties could be tuned by controlling the concentration of the nano clay. The group also investigated water vapor barrier properties of PLA / nanoclay composite films. With the implementation of the nanoclays, water vapor permeability (WVP) was found to decrease by 6-33%. The study also showed that the WVP highly depends on the nanoclay material as well as the concentration of the nanoclay.

Xylan-alginate films with nanoclays of bentonite or hallosite were tested by Naidu et al. (Naidu & John, 2020). The composite was prepared using the solvent casting technique. It was found that the addition of clay materials resulted in a decrease in the water vapor permeability, eventually increasing the humidity barrier properties. Mohan et al. synthesized a composite of epoxy-sisal fibre and nano clays (T. P. Mohan & Kanny, 2011). The impact of the nanoclays in the water barrier properties was investigated and further compared with the composited filled with micro clay. The study found that water intake gradually decreases with the nanoclay content; thus, proving a promising clue of using nanoclay material for water barrier applications. Composite of unsaturated polyester and nanoclay was developed by Merah et al. using a combined process of high shear mixing (HSM) and ultrasonication (Merah & Mohamed, 2019).

Slavutsky et al. prepared a composite of corn starch (a polymer) and montmorillonite (nanoclay) using two different processes and films were then prepared using casting method (Slavutsky et al., 2012). Various functional properties of the film were modified, which was due to the interaction between the nanoclay and the starch chains. The composite film has lower solubility and opacity and improved water barrier properties. The barrier properties of nanoclay composite materials could also depend on the mixing process. In a study conducted by Müller et al., it was found that the composite film (starch-nanoclay) with the nanoclay material previously dispersed in glycerol indicates enhanced mechanical, barrier, and hygroscopicity properties (Müller et al., 2011).

Polymer clay nanocomposite and graphene nanocomposite demonstrate similar barrier properties, which could be decided by the aspect ratio of the filler. Well dispersed fillers with a large aspect ratio create a tortuous path within the matrix for molecules and gases to travel through (Joshi et al., 2006; Yoo et al., 2014). A reduction in water vapour permeability of PCL clay nanocomposite was demonstrated by Messersmith and Giannelis as the aspect ratio of the platelets increased (Messersmith & Giannelis, 1995). Moreover, water permeability was also investigated in the case of epoxy clay nanocomposite and it demonstrated a decrease with increasing nano clay concentration until 10 wt% (Madhup et al., 2017).

Comparing Graphene-based, Nano-clays, and Polymer Nanofillers

Nanocomposite coatings such as, Graphene based polymer are also used as a barrier material. It can be produced by physical or chemical methods through the mixing of graphene, GO, or rGO with polymer. The graphene nanofiller will act as the physical barrier that can reduce the permeability of the nanocomposite by increasing the tortuous path (Othman et al., 2019b). It was indicated that with the addition of 0.001% GO nanofiller within a polyamide matrix, an 83% decrease in water absorption rate was achieved (Tseng et al., 2012). Moreover, modified graphene oxide and polydimethylsiloxane nanocomposite film

Figure 4. The schematic figure shows the impact of nanoclay-polymer composite material in the water diffusion path

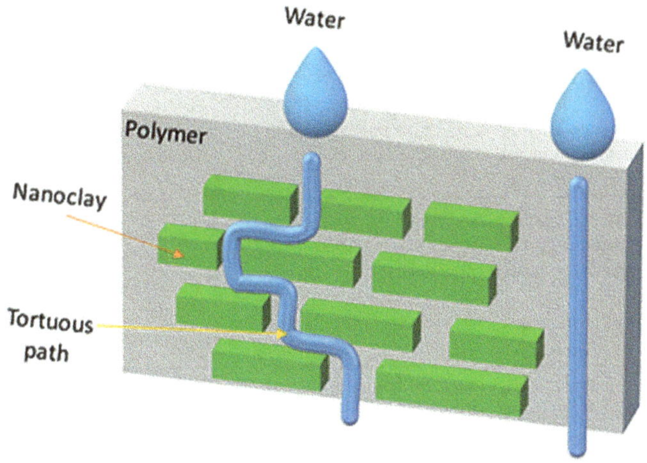

possessed a contact angle equal to 112° larger than the contact angle of polydimethylsiloxane film, which was equal to 87.23° indicating an improvement in surface hydrophobicity (Sharif et al., 2020). Table 2 compares the advantages and disadvantages of adding graphene-based, nano clays, and polymer as a filer within a polymer matrix.

FUTURE RESEARCH DIRECTIONS

This paper reviewed recent studies which focused on the use of polymers, graphene, and its derivatives in barriers, membranes, and coatings applications. In spite of the numerous advantageous properties of polymers, they have limited barrier properties (Yoo et al., 2014). In order to improve the barrier properties, polymer blends are made and studied. However, the polymer blends' barrier behavior was shown not to follow common models used to predict the change in polymer blends' properties compared to the individual polymers (Corres et al., 2020). This would lead to prolonged research before polymer blends could be used as barrier films.

Another technique used to improve the barrier properties is fabricating copolymers. Each monomer contributes to the overall properties of the copolymers. Nevertheless, the weakness of each monomer is still translated into the copolymer. For example, although the copolymer Ethylene vinyl alcohol copolymer (EVOH) has excellent barrier properties, it does not perform effectively in humid conditions due to vinyl alcohol monomer 's high water solubility (Maes et al., 2018).

The limitations with the use of polymers, polymer blends, and copolymers as barriers coatings instigated the study of 2D materials such as graphene and its derivatives. However, 2D materials still face some challenges when used for barrier application. Such challenges fall into three categories. Firstly, the deposition process of 2D materials into large, curved, and rough surfaces is not straightforward. Secondly, poor adhesion, damages due to friction, and galvanic corrosion leads to a failure of ultrathin coatings. Thirdly, challenges in upscaling of 2D materials based coating (Camilli et al., 2019).

Table 2. Advantages and disadvantages of graphene-based, nano clays, and polymer as a nanofillers

	Advantages	Disadvantages
Graphene and its derivatives	Increase in the thermal diffusivity of the nanocomposite (Prolongo et al., 2014). Improvement in the hydrophobicity (Prolongo et al., 2014). Electrical conductivity increases as the content of graphene nanofiller increases above the percolation threshold (W. Chen et al., 2018).	Presence of holes attributed to the weak interface exist in between the polymer matrix and nanofiller (Prolongo et al., 2014). Poor dispersion of graphene within the rubber matrix when fabricated by melt compounding (Malas, 2017).
Nano clays	Improved thermal stability (Rao et al., 2020). High exfoliation and intercalation (Saikia, 2019). Tensile strength increases with percentage of nano clay (Saikia, 2019). Improvement in water and oxygen permeability (Albdiry et al., 2013).	Agglomeration of nano clays due to low shear force when using mechanical mixing process (Albdiry et al., 2013). Limited loading level (Rothon & DeArmitt, 2017). High impurity level (Rothon & DeArmitt, 2017).
Polymers	Enhancement of the barrier properties compared to the pure polymer (Sadiku et al., 2016). Desired properties can be achieved by controlling the morphology and phase interaction of the polymer blend (Lagaron et al., 2004) . Incompatible polymers can be blended together using a compatibilizer to produce a polymer blend that has an acceptable barrier properties (Hamad et al., 2018).	The use of a compatibilizer can result in the increase in the viscosity of the polymer blend as well as loss of flexural strength and softening of the material (Hale et al., 1999) The degradation of one of the polymer during the fabrication method due to the thermal stability of the polymer (Hamad et al., 2018).

The mass production of high-quality monolayer graphene is still challenging, which needs to be overcome to allow the industrial use of graphene in practical applications. The main aspect of graphene commercialization relies on the production process as well as the cost-effectiveness of the produced graphene material (Zhong et al., 2015). Moreover, graphene production is lower than other carbon nanomaterials such as carbon fibers and carbon nanotubes (Lowe & Zhong, 2016). However, further reduction and oxidizing techniques have been applied to chemically functionalize graphene's surface to achieve good interfacial interaction with different surfaces that improve its applicability within different sectors (V. B. Mohan et al., 2018). Thus, the main features to take into consideration for mass production of graphene are the scalability, production cost, reproducibility, and the overall quality of the graphene (Zhong et al., 2015).

Different approaches have been utilized to produce graphene. The exfoliation of graphite by wet chemical processes is low in cost but produces low quality graphene (Zhong et al., 2015). Furthermore, graphene nanoplatelets have the advantage of mass production compared to graphene by the intercalation expansion exfoliation process. However, a concern of whether the mechanical, electrical, and thermal properties of graphene nanoplatelets exhibit the same performance as monolayer graphene. Nanocomposites of GO and polymers are widely used in applications where defect-free graphene is not important. Nevertheless, GO properties and structure should be altered further depending on the application. For instance, to achieve high electrical conductivity, GO should be reduced using thermal or chemical reduction methods (Lowe & Zhong, 2016).

Hummer's method used to produce graphene oxide is associated with two types of challenges. Firstly, the evolution of hazardous gases such as, the NO_2 and the N_2O_4. Secondly, the explosion of manganese heptoxide due to the chemical reaction between potassium permanganate and sulfuric acid. However, the use of sodium nitrate can be eliminated due to the intercalation of sulfuric acid and potassium permanganate. Moreover, managing and controlling the surrounding environment can reduce the explosion risk of manganese heptoxide which can easily react with organic compounds at temperature above 55°C (Lowe & Zhong, 2016). Nevertheless, the reduction of graphene oxide to produce reduced graphene oxide suffers of challenges associated with the chemical inhomogeneity, and the evolution of unavoidable holes and defects that will reduce the conductivity of the rGO (Eigler et al., 2013).

CONCLUSION

In conclusion, coating and barriers are vital in protecting and preserving a system. Different types of coatings are used for different applications and environments, as it increases the shelf life and the protection performance of the materials. The most common kinds of coatings used in industry are polymer and graphene-based coatings. A possible method of improving polymer-based coating is adding nano-sized filler and forming nanocomposite. While the main types of carbon-based coatings include graphene, GO, and r-GO. Graphene based materials used as a filler within polymer matrix demonstrated improved water permeability due to the increase of the tortuous pathway for the permeance particles. It was also revealed that scaling up of graphene oxide is limited due to the evolution of toxic gases, nevertheless mass production of graphene-based materials is still a challenge that is essential to overcome to allow the use of graphene in different sectors.

REFERENCES

Al-Hagri, A., Li, R., Rajput, N. S., Lu, J.-Y., Cong, S., Sloyan, K., Almahri, M. A., Tamalampudi, S. R., Chiesa, M., & Al Ghaferi, A. (2019). Direct growth of single-layer terminated vertical graphene array on germanium by plasma enhanced chemical vapor deposition. *Carbon*, *155*, 320–325. doi:10.1016/j.carbon.2019.08.069

Albdiry, M. T., Yousif, B. F., Ku, H., & Lau, K. T. (2013). A critical review on the manufacturing processes in relation to the properties of nanoclay/polymer composites. *Journal of Composite Materials*, *47*(9), 1093–1115. doi:10.1177/0021998312445592

Ayyoob, M., Lee, D. H., Kim, J. H., Nam, S. W., & Kim, Y. J. (2017). Synthesis of poly(glycolic acids) via solution polycondensation and investigation of their thermal degradation behaviors. *Fibers and Polymers*, *18*(3), 407–415. doi:10.100712221-017-6889-1

Babichev, A. V, Rykov, S. A., Tchernycheva, M., Smirnov, A. N., Davydov, V. Y., Kumzerov, Y. A., & Butko, V. Y. (2015). *Influence of Substrate Microstructure on the Transport Properties of CVD-Graphene*. doi:10.1021/acsami.5b08479

Berry, V. (2013). Impermeability of graphene and its applications. Carbon. doi:10.1016/j.carbon.2013.05.052

Brodie, B. C. (1860). Researches on the atomic weight of graphite. *Quarterly Journal of the Chemical Society of London*, *12*(1), 261–268. Advance online publication. doi:10.1039/QJ8601200261

Camilli, L., Yu, F., Cassidy, A., Hornekær, L., & Bøggild, P. (2019). Challenges for continuous graphene as a corrosion barrier. In 2D Materials (Vol. 6, Issue 2, p. 022002). IOP Publishing Ltd. doi:10.1088/2053-1583/ab04d4

Chen, S., Brown, L., Levendorf, M., Cai, W., Ju, S. Y., Edgeworth, J., Li, X., Magnuson, C. W., Velamakanni, A., Piner, R. D., Kang, J., Park, J., & Ruoff, R. S. (2011). Oxidation resistance of graphene-coated Cu and Cu/Ni alloy. *ACS Nano*, *5*(2), 1321–1327. doi:10.1021/nn103028d PMID:21275384

Chen, W., Weimin, H., Li, D., Chen, S., & Dai, Z. (2018). A critical review on the development and performance of polymer/graphene nanocomposites. *Science and Engineering of Composite Materials*, *25*(6), 1059–1073. doi:10.1515ecm-2017-0199

Compton, O. C., & Nguyen, S. T. (2010). Graphene oxide, highly reduced graphene oxide, and graphene: Versatile building blocks for carbon-based materials. Small. doi:10.1002mll.200901934

Corres, M. Á., Mayor, Á., Sangroniz, A., del Río, J., Iriarte, M., & Etxeberria, A. (2020). Blends based on biodegradable poly(caprolactone) with outstanding barrier properties for packaging applications: The role of free volume and interactions. *European Polymer Journal*, *135*(July), 109869. doi:10.1016/j.eurpolymj.2020.109869

De Heer, W. A., Berger, C., Wu, X., First, P. N., Conrad, E. H., Li, X., Li, T., Sprinkle, M., Hass, J., Sadowski, M. L., Potemski, M., Martinez, G., & Macdonald, H. (2007). Epitaxial graphene. *Solid State Communications*, *143*(1-2), 92–100. doi:10.1016/j.ssc.2007.04.023

De Silva, K. K. H., Huang, H. H., Joshi, R. K., & Yoshimura, M. (2017). Chemical reduction of graphene oxide using green reductants. In *Carbon* (Vol. 119, pp. 190–199). Elsevier Ltd. doi:10.1016/j.carbon.2017.04.025

DeLassus, P. (2002). Barrier polymers. In *Encyclopedia of Chemical Technology* (pp. 375–407). John Wiley & Sons, Inc.

Dhoot, S. N., Freeman, B. D., & Stewart, M. E. (2002). Barrier Polymers. In *Encyclopedia of Polymer Science and Technology*. John Wiley & Sons, Inc. doi:10.1002/0471440264.pst025

Duan, Z., & Thomas, N. L. (2014). Water vapour permeability of poly(lactic acid): Crystallinity and the tortuous path model. *Journal of Applied Physics*, *115*(6), 064903. Advance online publication. doi:10.1063/1.4865168

Eceolaza, S., Sangroniz, A., del Río, J., Iriarte, M., & Etxeberria, A. (2019). Improving the barrier character of poly(caprolactone): Transport properties and free volume of immiscible blends. *Journal of Applied Polymer Science*, *136*(40), 1–11. doi:10.1002/app.48018

Edling, H. E., Sun, H., Paschke, E., Schiraldi, D. A., Tanko, J. M., Paradzinsky, M., & Turner, S. R. (2020). High barrier biosourced polyester from dimethyl [2,2′-bifuran]-5,5′-dicarboxylate. *Polymer*, *191*(December). doi:10.1016/j.polymer.2020.122258

Eigler, S., Grimm, S., Hof, F., & Hirsch, A. (2013). Graphene oxide: A stable carbon framework for functionalization. *Journal of Materials Chemistry. A, Materials for Energy and Sustainability*, *1*(38), 11559–11562. doi:10.1039/c3ta12975f

Elias, D. C., Nair, R. R., Mohiuddin, T. M. G., Morozov, S. V., Blake, P., Halsall, M. P., Ferrari, A. C., Boukhvalov, D. W., Katsnelson, M. I., Geim, A. K., & Novoselov, K. S. (2009). Control of graphene's properties by reversible hydrogenation: Evidence for graphane. *Science, 323*(5914), 610–613. Advance online publication. doi:10.1126cience.1167130 PMID:19179524

Fernes, T. D., Hunter, D. L., & Paul, D. R. (2004). Nylon-6 Nanocomposites from Alkylammonium-Modified Clay: The Role of Alky1 Tails on Exfoliation. *Macromolecules, 37*(5), 1793–1798. doi:10.1021/ma0305481

Fogarassy, Z., Rümmeli, M. H., Gorantla, S., Bachmatiuk, A., Dobrik, G., Kamarás, K., Biró, L. P., Havancsák, K., & Lábár, J. L. (2014). Dominantly epitaxial growth of graphene on Ni (1 1 1) substrate. *Applied Surface Science, 314*, 490–499. doi:10.1016/j.apsusc.2014.06.197

Gupta, O., & Roy, S. (2019). Recent progress in the development of nanocomposite membranes. In *Nanocomposite Membranes for Water and Gas Separation* (pp. 29–67). Elsevier. doi:10.1016/B978-0-12-816710-6.00002-X

Hale, W., Lee, J., Keskkula, H., & Paul, D. R. (1999). Effect of PBT melt viscosity on the morphology and mechanical properties of compatibilized and uncompatibilized blends with ABS. *Polymer, 40*(13), 3621–3629. doi:10.1016/S0032-3861(98)00583-7

Hamad, K., Kaseem, M., Ayyoob, M., Joo, J., & Deri, F. (2018). Polylactic acid blends : The future of green, light and tough. *Progress in Polymer Science, 85*, 83–127. doi:10.1016/j.progpolymsci.2018.07.001

Hu, H., Zhang, R., Wang, J., Ying, W., Shi, L., Yao, C., Kong, Z., Wang, K., & Zhu, J. (2019). A mild method to prepare high molecular weight poly(butylene furandicarboxylate-co-glycolate) copolyesters: Effects of the glycolate content on thermal, mechanical, and barrier properties and biodegradability. *Green Chemistry, 21*(11), 3013–3022. doi:10.1039/C9GC00668K

Huang, H. D., Ren, P. G., Chen, J., Zhang, W. Q., Ji, X., & Li, Z. M. (2012). High barrier graphene oxide nanosheet/poly(vinyl alcohol) nanocomposite films. *Journal of Membrane Science, 409-410*, 156–163. Advance online publication. doi:10.1016/j.memsci.2012.03.051

Huang, W. (2018). Clay Nanopapers. In Nanopapers: From Nanochemistry and Nanomanufacturing to Advanced Applications (pp. 59–86). Elsevier Inc. doi:10.1016/B978-0-323-48019-2.00003-7

Hummers, W. S. Jr, & Offeman, R. E. (1958). Preparation of Graphitic Oxide. *Journal of the American Chemical Society, 80*(6), 1339. Advance online publication. doi:10.1021/ja01539a017

Joshi, M., Banerjee, K., & Thakare, V. (2006). Polymer/clay nanocomposite based coatings for enhanced gas barrier property. Indian Journal of Fibre & Textile Research, 31.

Kang, D., Kwon, J. Y., Cho, H., Sim, J. H., Hwang, H. S., Kim, C. S., Kim, Y. J., Ruoff, R. S., & Shin, H. S. (2012). Oxidation resistance of iron and copper foils coated with reduced graphene oxide multilayers. *ACS Nano, 6*(9), 7763–7769. doi:10.1021/nn3017316 PMID:22881084

Kiang Chua, C., & Pumera, M. (2014). Chemical reduction of graphene oxide: A synthetic chemistry viewpoint. *Chemical Society Reviews, 43*(1), 291–312. doi:10.1039/C3CS60303B PMID:24121318

Kim, D.-H., & Kim, H.-S. (2013). Waterproof characteristics of nanoclay/epoxy nanocomposite in adhesively bonded joints. *Composites. Part B, Engineering*, *55*, 86–95. doi:10.1016/j.compositesb.2013.05.041

Kim, J., Oh, S., Cho, S. M., Jun, J., & Kwak, S. (2020). Oxygen barrier properties of polyketone/EVOH blend films and their resistance to moisture. *Journal of Applied Polymer Science*, *137*(47), 1–10. doi:10.1002/app.49537

Kirkland, N. T., Schiller, T., Medhekar, N., & Birbilis, N. (2012). Exploring graphene as a corrosion protection barrier. Corrosion Science. doi:10.1016/j.corsci.2011.12.003

Klopffer, M. H., & Flaconneche, B. (2001). Transport Properdines of Gases in Polymers: Bibliographic Review. *Oil & Gas Science and Technology*, *56*(3), 223–244. doi:10.2516/ogst:2001021

Ko, G. K. S. (2014). Overview on in Polymer-Nano Clay Composite Paper Coating for Packaging Application. *Journal of Marine Science and Engineering*, *4*(1). Advance online publication. doi:10.4172/2169-0022.1000151

Koenig, S. P., Wang, L., Pellegrino, J., & Bunch, J. S. (2012). Selective molecular sieving through porous graphene. *Nature Nanotechnology*, *7*(11), 728–732. Advance online publication. doi:10.1038/nnano.2012.162 PMID:23042491

Koppolu, R., Lahti, J., Abitbol, T., Swerin, A., Kuusipalo, J., & Toivakka, M. (2019). Continuous Processing of Nanocellulose and Polylactic Acid into Multilayer Barrier Coatings. *ACS Applied Materials & Interfaces*, *11*(12), 11920–11927. doi:10.1021/acsami.9b00922 PMID:30829474

Kotek, R., Pang, K., Schmidt, B., & Tonelli, A. (2004). Synthesis and gas barrier characterization of poly(ethylene isophthalate). *Journal of Polymer Science. Part B, Polymer Physics*, *42*(23), 4247–4254. doi:10.1002/polb.20284

Kumar, S. S. A., Bashir, S., Ramesh, K., & Ramesh, S. (2021). New perspectives on Graphene/Graphene oxide based polymer nanocomposites for corrosion applications: The relevance of the Graphene/Polymer barrier coatings. *Progress in Organic Coatings*, *154*, 106215. doi:10.1016/j.porgcoat.2021.106215

Lagaron, J. M., Catalá, R., & Gavara, R. (2004). *Structural characteristics defining high barrier properties in polymeric materials Structural characteristics de ning high barrier properties in polymeric materials*. Academic Press.

Liu, M. Q., Wang, C., & Kim, N. Y. (2017). High-sensitivity and low-hysteresis porous mim-type capacitive humidity sensor using functional polymer mixed with tio2 microparticles. *Sensors (Switzerland)*, *17*(2), 0284. Advance online publication. doi:10.339017020284 PMID:28157167

Lowe, S. E., & Zhong, Y. L. (2016). Challenges of Industrial-Scale Graphene Oxide Production. In *Graphene Oxide*. Fundamentals and Applications. doi:10.1002/9781119069447.ch13

Madhavan Nampoothiri, K., Nair, N. R., & John, R. P. (2010). An overview of the recent developments in polylactide (PLA) research. *Bioresource Technology*, *101*(22), 8493–8501. doi:10.1016/j.biortech.2010.05.092 PMID:20630747

Madhup, M. K., Shah, N. K., & Parekh, N. R. (2017). Investigation and improvement of abrasion resistance, water vapor barrier and anticorrosion properties of mixed clay epoxy nanocomposite coating. *Progress in Organic Coatings, 102*, 186–193. doi:10.1016/j.porgcoat.2016.10.012

Maes, C., Luyten, W., Herremans, G., Peeters, R., Carleer, R., & Buntinx, M. (2018). Recent Updates on the Barrier Properties of Ethylene Vinyl Alcohol Copolymer (EVOH): A Review. *Polymer Reviews (Philadelphia, Pa.), 58*(2), 209–246. doi:10.1080/15583724.2017.1394323

Malas, A. (2017). Rubber nanocomposites with graphene as the nanofiller. In *Progress in Rubber Nanocomposites* (pp. 179–229). Elsevier Inc. doi:10.1016/B978-0-08-100409-8.00006-1

Merah, N., & Mohamed, O. (2019). Nanoclay and Water Uptake Effects on Mechanical Properties of Unsaturated Polyester. *Journal of Nanomaterials, 2019*, 1–11. doi:10.1155/2019/8130419

Messersmith, P. B., & Giannelis, E. P. (1995). Synthesis and barrier properties of poly(ε-caprolactone)-layered silicate nanocomposites. *Journal of Polymer Science. Part A, Polymer Chemistry, 33*(7), 1047–1057. doi:10.1002/pola.1995.080330707

Mohan, T. P., & Kanny, K. (2011). Water barrier properties of nanoclay filled sisal fibre reinforced epoxy composites. *Composites. Part A, Applied Science and Manufacturing, 42*(4), 385–393. doi:10.1016/j.compositesa.2010.12.010

Mohan, V. B., Lau, K. T., Hui, D., & Bhattacharyya, D. (2018). Graphene-based materials and their composites: A review on production, applications and product limitations. Composites Part B: Engineering. doi:10.1016/j.compositesb.2018.01.013

Müller, C. M. O., Laurindo, J. B., & Yamashita, F. (2011). Effect of nanoclay incorporation method on mechanical and water vapor barrier properties of starch-based films. *Industrial Crops and Products, 33*(3), 605–610. doi:10.1016/j.indcrop.2010.12.021

Naidu, D. S., & John, M. J. (2020). Effect of Clay Nanofillers on the Mechanical and Water Vapor Permeability Properties of Xylan–Alginate Films. *Polymers, 12*(10), 2279. doi:10.3390/polym12102279 PMID:33020377

Nine, M. J., Cole, M. A., Tran, D. N. H., & Losic, D. (2015). Graphene: A multipurpose material for protective coatings. *Journal of Materials Chemistry. A, Materials for Energy and Sustainability, 3*(24), 12580–12602. doi:10.1039/C5TA01010A

Othman, N. H., Che Ismail, M., Mustapha, M., Sallih, N., Kee, K. E., & Ahmad Jaal, R. (2019a). Graphene-based polymer nanocomposites as barrier coatings for corrosion protection. *Progress in Organic Coatings, 135*, 82–99. doi:10.1016/j.porgcoat.2019.05.030

Othman, N. H., Che Ismail, M., Mustapha, M., Sallih, N., Kee, K. E., & Ahmad Jaal, R. (2019b). Graphene-based polymer nanocomposites as barrier coatings for corrosion protection. In *Progress in Organic Coatings* (Vol. 135, pp. 82–99). Elsevier B.V. doi:10.1016/j.porgcoat.2019.05.030

Pajoumshariati, S., Shirali, H., Yavari, S. K., Sheikholeslami, S. N., Lotfi, G., Mashhadi Abbas, F., & Abbaspourrad, A. (2018). GBR membrane of novel poly (butylene succinate-co-glycolate) co-polyester co-polymer for periodontal application. *Scientific Reports*, *8*(1), 1–16. doi:10.103841598-018-25952-1 PMID:29760507

Pei, S., Zhao, J., Du, J., Ren, W., & Cheng, H. M. (2010). Direct reduction of graphene oxide films into highly conductive and flexible graphene films by hydrohalic acids. *Carbon*, *48*(15), 4466–4474. Advance online publication. doi:10.1016/j.carbon.2010.08.006

Prasai, D., Tuberquia, J. C., Harl, R. R., Jennings, G. K., & Bolotin, K. I. (2012). Graphene: Corrosion-inhibiting coating. *ACS Nano*, *6*(2), 1102–1108. Advance online publication. doi:10.1021/nn203507y PMID:22299572

Prolongo, S. G., Moriche, R., Jiménez-Suárez, A., Sánchez, M., & Ureña, A. (2014). Advantages and disadvantages of the addition of graphene nanoplatelets to epoxy resins. *European Polymer Journal*, *61*, 206–214. doi:10.1016/j.eurpolymj.2014.09.022

Rajput, N. S., Al Zadjali, S., Gutierrez, M., Esawi, A. M. K., & Al Teneiji, M. (2021). Synthesis of holey graphene for advanced nanotechnological applications. *RSC Advances*, *11*(44), 27381–27405. doi:10.1039/D1RA05157A

Rangel, N. L., Sotelo, J. C., & Seminario, J. M. (2009). Mechanism of carbon nanotubes unzipping into graphene ribbons. *The Journal of Chemical Physics*, *131*(3), 031105. doi:10.1063/1.3170926 PMID:19624173

Rao, G. R., Srikanth, I., & Reddy, K. L. (2020). Effect of organo-modified montmorillonite nanoclay on mechanical, thermal and ablation behavior of carbon fiber/phenolic resin composites. *Defence Technology*, *17*(3), 812–820. doi:10.1016/j.dt.2020.05.012

Rhim, J.-W. (2011). Effect of clay contents on mechanical and water vapor barrier properties of agar-based nanocomposite films. *Carbohydrate Polymers*, *86*(2), 691–699. doi:10.1016/j.carbpol.2011.05.010

Rothon, R., & DeArmitt, C. (2017). Fillers (Including Fiber Reinforcements). In Brydson's Plastics Materials: Eighth Edition (pp. 169–204). Elsevier Inc. doi:10.1016/B978-0-323-35824-8.00008-6

Ruammaitree, A., Phokharatkul, D., Nuntawong, N., & Wisitsoraat, A. (2018). Improvement in corrosion resistance of stainless steel foil by graphene coating using thermal chemical vapor deposition. *Surface Review and Letters*, *25*(Supp01), 1840003. Advance online publication. doi:10.1142/S0218625X18400036

Sadiku, E. R., Babul Reddy, A., Gnanasekarana, D., Oboirien, B., Aderibigbe, B. A., Varaprasad, K., & Goddeti, S. M. R. (2016). Nanostructured Polymer Blends for Gas/Vapor Barrier and Dielectric Applications. In *Design and Applications of Nanostructured Polymer Blends and Nanocomposite Systems* (pp. 239–259). Elsevier. doi:10.1016/B978-0-323-39408-6.00011-X

Saikia, P. (2019). Clay nanostructures for biomedical applications. In *Two Dimensional Nanostructures for Biomedical Technology: A Bridge between Material Science and Bioengineering* (pp. 137–172). Elsevier. doi:10.1016/B978-0-12-817650-4.00005-X

Salusjärvi, L., Havukainen, S., Koivistoinen, O., & Toivari, M. (2019). Biotechnological production of glycolic acid and ethylene glycol: current state and perspectives. In Applied Microbiology and Biotechnology (Vol. 103, Issue 6, pp. 2525–2535). Applied Microbiology and Biotechnology. doi:10.100700253-019-09640-2

Sangroniz, A., Zhu, J.-B., Tang, X., Etxeberria, A., Chen, E. Y.-X., & Sardon, H. (2019). Packaging materials with desired mechanical and barrier properties and full chemical recyclability. *Nature Communications, 10*(1), 3559. doi:10.103841467-019-11525-x PMID:31395871

Schniepp, H. C., Li, J. L., McAllister, M. J., Sai, H., Herrera-Alonson, M., Adamson, D. H., Prud'homme, R. K., Car, R., Seville, D. A., & Aksay, I. A. (2006). Functionalized single graphene sheets derived from splitting graphite oxide. *The Journal of Physical Chemistry B, 110*(17), 8535–8539. Advance online publication. doi:10.1021/jp060936f PMID:16640401

Sharif, N. A., Ehsani, M., Zaarei, D., Kalaee, M. R., & Khajavi, R. (2020). Nanocomposite Coatings Based on Modified Graphene Oxide and Polydimethylsiloxane: Characterization and Thermal Properties. *Russian Journal of Applied Chemistry, 93*(11), 1765–1773. doi:10.1134/S1070427220110191

Shivaraman, S., Chandrashekhar, M. V. S., Boeckl, J. J., & Spencer, M. G. (2009). Thickness estimation of epitaxial graphene on sic using attenuation of substrate raman intensity. *Journal of Electronic Materials, 38*(6), 725–730. doi:10.100711664-009-0803-6

Singh Raman, R. K., Chakraborty Banerjee, P., Lobo, D. E., Gullapalli, H., Sumandasa, M., Kumar, A., Choudhary, L., Tkacz, R., Ajayan, P. M., & Majumder, M. (2012). Protecting copper from electrochemical degradation by graphene coating. *Carbon, 50*(11), 4040–4045. Advance online publication. doi:10.1016/j.carbon.2012.04.048

Slavutsky, A. M., Bertuzzi, M. A., & Armada, M. (2012). Water barrier properties of starch-clay nanocomposite films. *Brazilian Journal of Food Technology, 15*(3), 208–218. doi:10.1590/S1981-67232012005000014

Smith, A. T., LaChance, A. M., Zeng, S., Liu, B., & Sun, L. (2019). Synthesis, properties, and applications of graphene oxide/reduced graphene oxide and their nanocomposites. *Nano Materials Science, 1*(1), 31–47. doi:10.1016/j.nanoms.2019.02.004

Sreeprasad, T. S., & Berry, V. (2013). How Do the Electrical Properties of Graphene Change with its Functionalization? *Small, 9*(3), 341–350. doi:10.1002mll.201202196 PMID:23169614

Stankovich, S., Dikin, D. A., Piner, R. D., Kohlhaas, K. A., Kleinhammes, A., Jia, Y., Wu, Y., Nguyen, S. B. T., & Ruoff, R. S. (2007). Synthesis of graphene-based nanosheets via chemical reduction of exfoliated graphite oxide. *Carbon, 45*(7), 1558–1565. Advance online publication. doi:10.1016/j.carbon.2007.02.034

Su, Y., Kravets, V. G., Wong, S. L., Waters, J., Geim, A. K., & Nair, R. R. (2014). Impermeable barrier films and protective coatings based on reduced graphene oxide. *Nature Communications, 5*(1), 4843. Advance online publication. doi:10.1038/ncomms5843 PMID:25208890

Sun, Q., Schork, F., & Deng, Y. (2007). Water-based polymer/clay nanocomposite suspension for improving water and moisture barrier in coating. *Composites Science and Technology, 67*(9), 1823–1829. doi:10.1016/j.compscitech.2006.10.022

Thrivikraman, G., França, C. M., Athirasala, A., Tahayeri, A., & Bertassoni, L. E. (2017). 3D printing: Prospects and challenges: Nanomaterials in 3D bioprinting: Current approaches and future possibilities. In *Nanotechnologies in Preventive and Regenerative Medicine: An Emerging Big Picture* (pp. 349–379). Elsevier Inc. doi:10.1016/B978-0-323-48063-5.00004-6

Tseng, I. H., Liao, Y. F., Chiang, J. C., & Tsai, M. H. (2012). Transparent polyimide/graphene oxide nanocomposite with improved moisture barrier property. *Materials Chemistry and Physics*, *136*(1), 247–253. doi:10.1016/j.matchemphys.2012.06.061

Tyagi, P., Salem, K. S., Hubbe, M. A., & Pal, L. (2021). Advances in barrier coatings and film technologies for achieving sustainable packaging of food products – A review. *Trends in Food Science & Technology*, *115*, 461–485. doi:10.1016/j.tifs.2021.06.036

Wolf, C., Angellier-Coussy, H., Gontard, N., Doghieri, F., & Guillard, V. (2018). How the shape of fillers affects the barrier properties of polymer/non-porous particles nanocomposites: A review. *Journal of Membrane Science*, *556*(April), 393–418. doi:10.1016/j.memsci.2018.03.085

Xu, H., Guo, H., & Gong, S. (2008). Thermal barrier coatings. In *Developments in High Temperature Corrosion and Protection of Materials* (pp. 476–491). Elsevier. doi:10.1533/9781845694258.2.476

Yang, Z., Peng, H., Wang, W., & Liu, T. (2010). Crystallization behavior of poly(ε-caprolactone)/layered double hydroxide nanocomposites. *Journal of Applied Polymer Science*, *116*(5), 2658–2667. doi:10.1002/app.31787

Yang, Z., Zhang, X., Luo, X., Jiang, Q., Liu, J., & Jiang, Z. (2013). Enzymatic synthesis of poly(butylene-co-sebacate-co-glycolate) copolyesters and evaluation of the copolymer nanoparticles as biodegradable carriers for doxorubicin delivery. *Macromolecules*, *46*(5), 1743–1753. doi:10.1021/ma302433x

Yoo, B. M., Shin, H. J., Yoon, H. W., & Park, H. B. (2014). Graphene and graphene oxide and their uses in barrier polymers. *Journal of Applied Polymer Science*, *131*(1). Advance online publication. doi:10.1002/app.39628

Zhong, Y. L., Tian, Z., Simon, G. P., & Li, D. (2015). Scalable production of graphene via wet chemistry: Progress and challenges. In Materials Today (Vol. 18, Issue 2, pp. 73–78). Elsevier. doi:10.1016/j.mattod.2014.08.019

ADDITIONAL READING

Camilli, L., Yu, F., Cassidy, A., Hornekær, L., & Bøggild, P. (2019). Challenges for continuous graphene as a corrosion barrier. In 2D Materials (Vol. 6, Issue 2, p. 022002). IOP Publishing Ltd. doi:10.1088/2053-1583/ab04d4

Kim, J., Oh, S., Cho, S. M., Jun, J., & Kwak, S. (2020). Oxygen barrier properties of polyketone/EVOH blend films and their resistance to moisture. *Journal of Applied Polymer Science*, *137*(47), 1–10. doi:10.1002/app.49537

Maes, C., Luyten, W., Herremans, G., Peeters, R., Carleer, R., & Buntinx, M. (2018). Recent Updates on the Barrier Properties of Ethylene Vinyl Alcohol Copolymer (EVOH): A Review. *Polymer Reviews (Philadelphia, Pa.)*, *58*(2), 209–246. doi:10.1080/15583724.2017.1394323

Othman, N. H., Che Ismail, M., Mustapha, M., Sallih, N., Kee, K. E., & Ahmad Jaal, R. (2019). Graphene-based polymer nanocomposites as barrier coatings for corrosion protection. In *Progress in Organic Coatings* (Vol. 135, pp. 82–99). Elsevier B.V. doi:10.1016/j.porgcoat.2019.05.030

Wolf, C., Angellier-Coussy, H., Gontard, N., Doghieri, F., & Guillard, V. (2018). How the shape of fillers affects the barrier properties of polymer/non-porous particles nanocomposites: A review. *Journal of Membrane Science*, *556*(April), 393–418. doi:10.1016/j.memsci.2018.03.085

KEY TERMS AND DEFINITIONS

Aromatic Rings: A highly stable ring molecular structure consisting of sp^2 hybridized atoms.

Crystallization: The degree of structural order of molecular chains in polymers.

Graphene: A carbon allotrope consisting of hexagonal crystalline structure with sp^2 hybridized carbon atoms forming aromatic rings.

Hydrophobicity: The ability of the surface of the material to repel water.

Nanocomposite: A class of composite material composed of a polymer matrix and a filler within the nanorange.

Permeability: The ability of a material to allow gases or liquid to pass through.

Thermoplastic: A class of polymers that can be melted and processed by applying heat.

Tortuosity: An intrinsic property of a porous material defined as the ratio of actual flow path length to the straight distance between the ends of the flow path.

Chapter 7
Characterization Tools for Nanomaterials

Tripti Ahuja
Indian Institute of Technology Delhi, Delhi, India

Deeksha Satyabola
Arizona State University, USA

Sujan Manna
Indian Institute of Technology Madras, Chennai, India

ABSTRACT

Nanomaterials are nanostructures that have gained massive interest in the scientific community due to their exciting functional and physicochemical properties. They offer high stability, advanced optical-electronic properties, and tuneable surface functionalization. Such properties have made them a material of everyday use in the pharmaceutical industries, cosmetics, electronics, and many more. Considering their increasing usage, there is a demand for reliable and better quantitative and qualitative characterization tools. This chapter is dedicated to such characterization methodologies for the analysis of nanomaterials. Numerous distinct and integrated correlated techniques are discussed in detail. Such tools are arranged from the basic to advanced methods, each mentioning the unique property of nanomaterial such as elemental composition, morphology, crystal structure, magnetism, and strength. Lastly, some recent advancements in the characterization methodologies of nanostructures are provided. Understanding the properties of nanomaterials will enable us to expand their applications in society.

INTRODUCTION

"Why cannot we write the entire 24 volumes of the Encyclopaedia Britannica on the head of a pin?" Richard Feynman from the historical lecture, 'There's Plenty of Room at the Bottom' in 1959 (Geoff, 1960).

DOI: 10.4018/978-1-7998-8936-6.ch007

Richard Feynman, the pioneer of science and technology, painted the first canvas of the nanoscale. Sixty-two years from then, today, we have not only maneuvered things at the length scale of an atom but have developed a whole new discipline of science called Nanoscience and Nanotechnology. Although the first paper in this area was published in 1981 by Eric Drexler, the real breakthrough came only in 1989 after the development of a scanning tunneling microscope (STM) by Binnig, Gerber, and Quate at IBM Zurich, Switzerland. Soon after, nanotechnology got spread throughout the world, encompassing every sphere of our life, from medicine to industries. Nanomaterials refer to functional materials developed on the concept of nanotechnology, whose dimensions range from 1 to 100 nm. They have gained considerable interest among scientists because of their exclusive optical, magnetic, electronic, and physicochemical properties. They have also become an integral part of nanoscience and nanotechnology not only because of their functional properties but also due to their future technological applications in various fields such as sensors, bioimaging, catalysis, optoelectronics, photothermal therapy, water purification, drug-gene delivery, cosmetics, lab-on-chip devices, energy harvesting, etc. (Stancheva, 2012; Toropov & Vartanyan, 2019; Vajtai, 2013). The considerable demand for novel nanomaterials opens the challenge for their accurate and reliable characterization techniques.

Recently, various characterization tools and methodologies have been developed to identify and specify the elemental composition, size, shape, crystal structure, surface charge, and other physicochemical properties of nanomaterials. An individual characterization tool is often insufficient to describe the complete nature of nanomaterials. In that case, a combinatorial approach of techniques and methodologies is required for the detailed analysis of such functional nanomaterials. However, a complete characterization of nanomaterials is still a challenging task to accomplish. Some of the limitations associated with the incomplete characterization of nanomaterials are; 1) The size, shape, and morphology of nanomaterials generally vary from bulk materials to the single-dimension level; 2) The inaccessibility of every research lab to have quick and easy access to diverse characterization facilities. In the study of nanomaterials, analytical targets and characterization tools are employed right from the stage of synthesis to the determination of morphology, understanding surface functionalization, and other physicochemical properties. In this chapter, the focus will roam around different standard, complementary, and specific techniques such as absorption spectroscopy, transmission electron microscopy, X-ray crystallography, etc. for the analysis of the distinct nanomaterials. Besides the characterization tools, the weaknesses, and the strengths associated with the analysis are also mentioned in brief.

Towards the end, few recent advancements and the present state-of-art in characterization methodologies are reflected. Cryo-transmission electron microscopy (cryo-TEM), electron tomography (ET), and tip-enhanced Raman spectroscopy (TERS) are some of the cutting-edge techniques developed lately, which have gained immense popularity due to their enhanced resolution and improved sampling conditions. The timeline of characterization techniques is illustrated as a flowchart in Figure 1.

Accounting for the importance of nanomaterials in society, this book chapter is well-organized focusing on their characterization tools. Hence, different aspects have been focused on 2D materials, nanoparticles (NPs), quantum dots, and so forth with their futuristic applications in diverse fields.

BACKGROUND

Analytical tools with a high spatial and temporal resolution are crucial for understanding the synthetic aspects, architecture, various physicochemical properties, and surface chemistry of nanomaterials. Several

Figure 1. A timeline of the various characterization tools developed in the last few decades. Blue, yellow, pink, and orange colors categorize the characterization tools into spectroscopic, microscopic, mass spectrometric, and chromatographic methods, respectively.

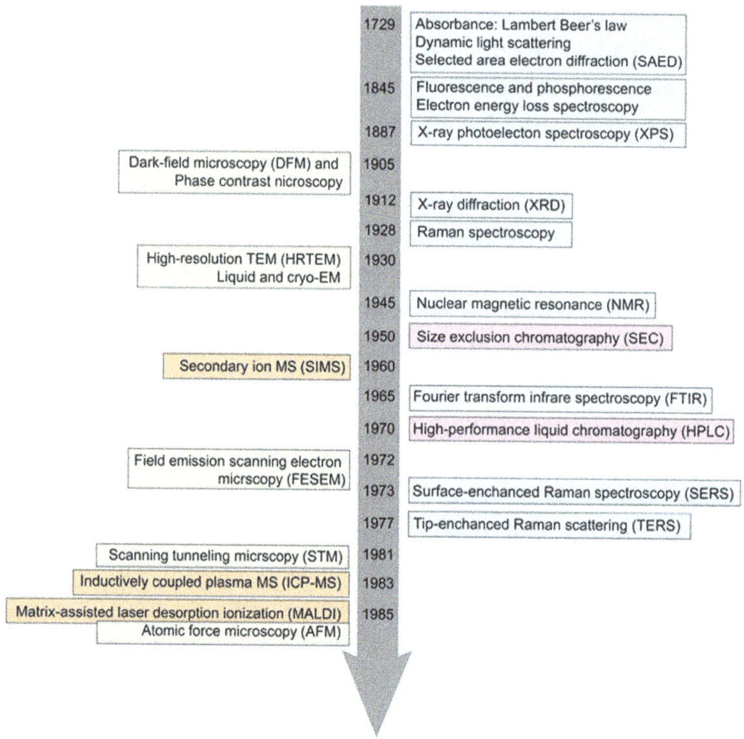

characterization tools such as absorbance, fluorescence, dynamic light scattering (DLS), surface-enhanced Raman spectroscopy (SERS), mass spectrometry (MS), and so forth have been routinely used for the analysis of nanomaterials. These experimental characterization techniques can be classified broadly on the following aspects:

1. Type of the probes or sources utilized,
2. Physical and chemical properties of nanomaterials,
3. Operational principle of tools – standard and *in-situ* operando techniques,
4. Mode of availability, cost, and sensitivity of the instrument,
5. Effect of the probe on sample after analysis.

These tools can also be categorized based on ease of operation, simplicity, robustness, and efficiency. But the five categories as mentioned above efficiently enfold the standard and frequently used analytical tools for nanomaterials. The most common identifiable physicochemical properties of nanomaterials include a) size and its distribution, b) degree of aggregation, c) surface charge, d) surface area, e) surface chemistry, f) crystal structure, g) chemical composition, h) optical and electronic properties, i) morphology and so forth. Such analytical tools characterize the nanomaterials and aid in identifying, investigating, and evaluating the physicochemical properties and stability associated with them. Table 1 presents the comprehensive overview of numerous distinct characterization tools mentioning the type of sources or probes used, physicochemical properties of nanomaterials detected, the principle of operation of the tool, mode of availability, cost, sensitivity, and non-destructive nature of the analytical techniques.

SPECTROSCOPY AND CRYSTALLOGRAPHY

The physicochemical properties of nanomaterials are firmly influenced by various factors such as shape, size, free electron density, composition, interparticle interactions, and the surrounding medium. Such properties help to obtain information about the electron injection and aggregation of nanomaterials that are determined using various spectroscopic techniques. The basic instrumentation in spectroscopy includes a light source to irradiate the sample, a monochromator to converge and produce single-wavelength radiation, a sample holder, and a detector to detect output radiations (Figure 2).

Nanomaterials display characteristic *UV-vis* extinction spectra in the visible and near-IR regions of the spectrum. It has been determined by Haiss *et al.* that *UV-vis spectroscopy* can be used as a primary method to determine the size and concentration of gold NPs (Haiss *et al.*, 2007). Furthermore, UV-Vis can be combined with the particle instability parameter (PIP) to quantitatively determine the stability and nature of plasmonic NPs (Ray *et al.*, 2015). As UV-vis spectra cannot distinguish the particles based on their size, it is not an excellent tool to determine polydispersity.

Similarly, *photoluminescence (PL)* spectroscopy is a useful characterization tool for fluorescent materials and metallic nanoclusters. It is an alternative technique for fluorescent molecules for optical labeling applications such as sensing and imaging. In addition, PL of NPs does not suffer photobleaching and photoblinking (T. Zhang *et al.*, 2014).

Dynamic light scattering (DLS) is an analytical technique that involves the scattering of light by nanoarchitectures using a low concentration of material to avoid multiple scattering effects. The size parameter is affected by the surface coating, colloidal stability, and the morphology of nanomaterials. *DLS* can provide real-time screening of a nanomaterial, producing essential insights into their aggregation (Kato, 2012). Its approach to detect particles is based on ensemble averaging, which produces sensitive and reproducible results for monodisperse and homogeneous colloidal solutions, unlike polydisperse and heterogeneous samples.

X-ray diffraction (XRD) is an excellent tool for the analysis of nanomaterials. Principally, it gives information about the crystalline structure, lattice parameters, crystallite grain size, and the nature of the phase of nanostructures. Using XRD, one can determine the composition of a sample by comparing the intensity and peak positions with the reference patterns available at the International Centre for Diffraction Data (ICDD). The 3D structure of samples can be obtained via crystallography. The basic instrumentation involved in the diffraction methodology is shown in Figure 3. A drawback of this technique is that it cannot be used accurately for amorphous materials.

Selected area electron diffraction (SAED) is a technique used inside the *TEM* system to study the crystal structure and defects of nanomaterials. Majorly, it helps to evaluate multiply-twinned structures and kinetics in nanomaterials. Schamp and Jesser have determined that *SAED* can calculate lattice parameters and interplanar spacing, which can be further evaluated for precise determination of the strains in the anisotropic gold NPs (Schamp & Jesser, 2005).

X-ray photoelectron spectroscopy (XPS) is widely used for the surface analysis of nanomaterials. Nag *et al.* have demonstrated the role of XPS to study internal heterostructures of nanomaterials (Sarma *et al.*, 2013), and Tunc *et al.* have shown a simple method for direct detection, location, and identification

Table 1. Overview of the standard and essential experimental techniques used for the characterization of nanomaterials.

Sl. No.	Analytical Tools	Source/Probe Used	Property of Nanomaterials Detected	Principle of Operation	Mode of Availability, Cost, and Sensitivity	Sample Damage
Spectroscopy and Crystallography						
1	Absorbance (UV-visible spectroscopy), emission, and excitation (Photoluminescence/ PL spectroscopy)	UV-visible electromagnetic radiations (optical)	Optical and electronic properties, size and shape, concentration, agglomeration state	UV-visible determines the intensity of the light absorbed from a specimen compared to a reference, and PL tracks the intensity and the wavelength of the emitted light from molecules that have absorbed photons.	Facile, low-cost, moderate sensitivity, and quick	Non-destructive (ND) samples can be reused
2	Dynamic light scattering (DLS)	Electromagnetic radiations (EMRs)	Hydrodynamic size, agglomeration of NPs	It assesses the scattering of the light as a function of time.	Facile, low-cost, high sensitivity, and quick	ND
3	Selected area electron diffraction (SAED)	Backscattered electrons	Structure and defects of crystalline nanomaterials	It predictably exploits electron backscattered diffraction.	Not easily accessible, high-cost, high-sensitivity, and time-consuming	Destructive, sample cannot be reused
4	X-ray diffraction (XRD) spectroscopy	X-rays	Crystal structure, crystal defects, crystallite grain size, composition	It measures an interference pattern of the X-rays caused under the uniform spacing in the NPs crystal structure.	Easily accessible, high-cost, high-sensitivity, and time-consuming	Destructive
5	X-ray photoelectron spectroscopy (XPS)	X-rays	Nature of ligand interactions with nanomaterials, the oxidation state of nanoarchitectures, electronic structure, elemental composition	It uses the principle of the photoelectric effect.	Not easily accessible, high-cost, high-sensitivity, and time-consuming	Destructive
6	Fourier transform infrared (FTIR) spectroscopy	Infrared radiations	Ligand vibrational binding, surface composition, the structure of ligands adsorbed on NPs	It exploits EMRs in the IR region to state the changes in the vibrational energy of the molecules.	Facile, low-cost, high sensitivity, and quick	ND
7	Raman spectroscopy	UV, visible and near IR radiations	Vibrational chemical bonds, stress-strain, crystal defects, plasmon resonance, structure of ligands adsorbed on NPs	It determines the vibrational modes of molecules, although rotational and other low-frequency modes of systems using inelastic scattering of photons.	Facile, low-cost, high sensitivity, and quick	ND
8	Surface-enhanced Raman scattering (SERS) spectroscopy	UV, visible and near IR radiations	Different types of anisotropic NPs, vibrational features, structure of ligands adsorbed on NPs	An advanced Raman spectroscopy tool that uses LSPR nature of NPs to increase vibrational signals of ligand bound with the NPs.	Facile, low-cost, high sensitivity, and quick	ND
9	Electron energy loss spectroscopy (EELS)	Beam of electrons	Elemental composition, bonding and oxidation state, distribution of near neighboring atoms, valence/ conduction band electron density	It exploits the energy loss due to the inelastic interactions of electrons (possessing kinetic energy of well-known and narrow range) with NPs	Not easily accessible, high-cost, high-sensitivity, and time-consuming	Destructive
10	Nuclear magnetic resonance (NMR)	Nuclear spin	Effect of ligands on NP shape, size, electronic core structure, atomic composition	An active nuclear spin interacts with another spin to create two energy states in presence of an external magnetic field	Not easily accessible, average-cost, high sensitivity, and quick	ND
Microscopy						
11	Tranmission electron microscopy (TEM)/ High-resolution transmission electron microscopy (HRTEM)	Electrons	Particles morphology, size, shape, polydispersity, phase distribution within crystal structure using FFT, aggregation state, studying growth kinetics	TEM makes use of the interaction between the current density of the electron beam and a thin sample in the transmission mode. Furthermore, HRTEM exploits both transmitted and scattered electrons to create the phase-contrast images of the sample.	Not easily accessible, high-cost, high-sensitivity, and quick	Destructive

continued on following page

Table 1. Continued

Sl. No.	Analytical Tools	Source/Probe Used	Property of Nanomaterials Detected	Principle of Operation	Mode of Availability, Cost, and Sensitivity	Sample Damage
12	Field emission scanning electron microscopy (SEM/ FESEM)	Backscattered and secondary electrons	NPs morphology, size, shape, surface roughness, elemental composition, dispersion of NPs in cells and other matrices/supports	It exploits backscattered and secondary electrons for imaging of nanoscale surfaces	Not easily accessible, high-cost, high-sensitivity, and quick	Based on sample conditions, it can be destructive or ND
13	Atomic force microscopy (AFM)	Cantilever and feedback laser	Size distribution for polydisperse particles, height, and thickness of NPs	It measures the interaction between a fine probe (cantilever) and the sample.	Not easily accessible, high-cost, high-sensitivity, and time-consuming	ND in tapping and non-contact mode while destructive in contact mode.
14	Dark field microscopy (DFM) and phase contrast microscopy	Electromagnetic visible radiations	Imaging, visualization of NPs, localization of cells inside NPs	It utilizes scattered photons and excludes unscattered beams resulting in a dark image around the specimen.	Facile, low-cost, low sensitivity, and quick	ND
Mass Spectrometry (MS)						
15	Matrix-assisted laser desorption ionization (MALDI)	laser	Size distribution, elemental composition, NP concentration	The laser hits the sample contained in the matrix, ionizes it, triggers ablation and desorption of sample as ions go into spectrometer	Not easily accessible, high-cost, high-sensitivity, and quick	Destructive for samples
16	Inductively coupled plasma MS (ICP-MS)	Inductively coupled plasma or ionized matter	Size distribution, elemental composition, NP concentration	Uses an inductively coupled plasma to ionize the sample which generates ions that are detected	Not easily accessible, high-cost, high-sensitivity, and quick	Destructive for samples
17	Secondary Ion MS (SIMS)	Focused ion beam by ionization of noble gases	Surface analysis	Focused ion beam hits the sample, ionize it, and analyzed with detector	Not easily accessible, high-cost, high-sensitivity, and quick	Destructive for samples
Chromatography						
18	High-performance liquid chromatography (HPLC)	Silica monolith column is used	Separate, identify and quantify nanomaterials	A high-pressure liquid solvent is passed through a column filled with solid adsorbent, each component elutes out of the column based on the size	Accessible, moderate-cost, timely process	Samples can be recovered in pure state
19	Size exclusion chromatography (SEC)	Nucleosil or silica microspheres column	Separate, identify and quantify nanomaterials	The chromatography column is packed with fine, porous beads of agarose or other polymers through which the sample passes and is adsorbed at different times.	Accessible, moderate-cost, timely process	Samples can be recovered in pure state
Gravimetric analysis						
20	Thermogravimetric analysis (TGA)	Temperature and micro-precision balance	Composition and mass of stabilizers	It monitors the changes in weight of the NPs as a function of time and temperature.	Facile, low-cost, moderate sensitivity, and quick	Destructive
Electroanalytical methods						
21	Electrochemical detection and tangential flow filtration (EC-TFF)	Poly(amic) acid membrane filter electrode (PMFE)	Molecular weight and charge	It tracks the molecular weight of the sample by using various electrodes.	Accessible, low-cost process	ND

of charges developed on the surface of a core-shell nanoarchitecture in a non-contact mode (Tunc *et al.*, 2005). Thus, *XPS* can provide information on both the chemical identity, oxidation state, and dielectric properties of nanomaterials.

Apart from the spectroscopy techniques mentioned above, several other techniques such as FTIR and SERS can determine the structure, size, composition, and other characteristic features of nanomaterials. These techniques monitor the changes in the vibrational modes of the molecule and evaluate the binding of ligands pinned on the surface of nanostructures. They also serve as the quantitative tools for the determination of unknown concentrations of molecules, stress-strain, and crystal defects in nanostructures.

Table 2. Recent advanced experimental techniques for the characterization of nanomaterials.

SI. No.	Recent Analytical Tools	Advancement to Previous Tools Mentioned in Table 1	Additional Information Gained Than Previous Tools of Table 1	Principle of Operation of Tools	Mode of Availability, Cost, and Sensitivity of Tools	Sample Damage
1	Liquid TEM and cryo TEM	TEM	Reaction kinetics on nanostructures, synthetic formation of nanomaterials	Same as TEM but the major change in the holder of sample	Highly inaccessible, high-cost, high-sensitivity, and quick	Destructive
2	Scanning tunneling microscopy (STM)	AFM	Surface roughness, NPs polydispersity, and composition	It uses tunneling quantum current to image the surface of NPs	Highly inaccessible, high-cost, high-sensitivity, and time-consuming	ND
3	Correlated SEM-DFM-SERS	SEM, DFM, and SERS	Both morphology and spectroscopy aspects of nanomaterials	It combines the principle of SEM, DFM, and SERS.	Facile, high-cost, high-sensitivity, and quick	ND
4	Tip-enhanced Raman scattering (TERS)	SERS	Chemical mapping of ligands on the surface of NPs, ligand binding, and ligand-NP interactions	It utilizes the principle of AFM and SERS	Highly inaccessible, high-cost, high-sensitivity, and quick	ND
5	Particle dynamics induced-plasmon enhanced Raman scattering (PDI-PERS)	SERS	Simultaneously visualizing plasmonic particle and monitoring plamonic and Raman scattering spectra at a single-particle level.	It combines the principle of DFM and SERS.	Inaccessible, high-cost, high-sensitivity, and quick	ND
6	HADDF-STEM	HRTEM	3D reconstruction obtained from 2D images of nanostructures	It involves slicing a 3D object plane-by-plane at different angles by using a penetrating wave.	Highly inaccessible, high-cost, high-sensitivity, and quick	Destructive/ ND based on type of source and probe used

Figure 2. Diagrammatic representation of a basic spectroscopic instrument. A light source is directed on a collimated lens which when passed through a prism/monochromator produces a single wavelength of light interacting with the sample. The remaining light is detected by a detector placed after the sample, which produces a signal in the form of a spectrum.

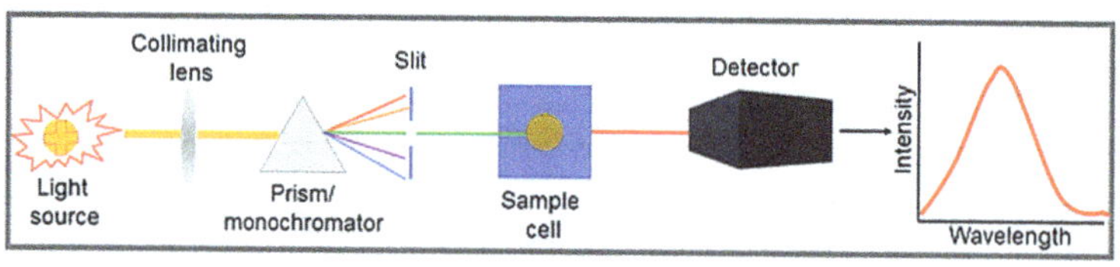

Figure 3. A schematic representation of XRD. X-rays are directed onto the sample resulting in a characteristic diffraction pattern, which upon analysis gives a 3D structure of the sample.

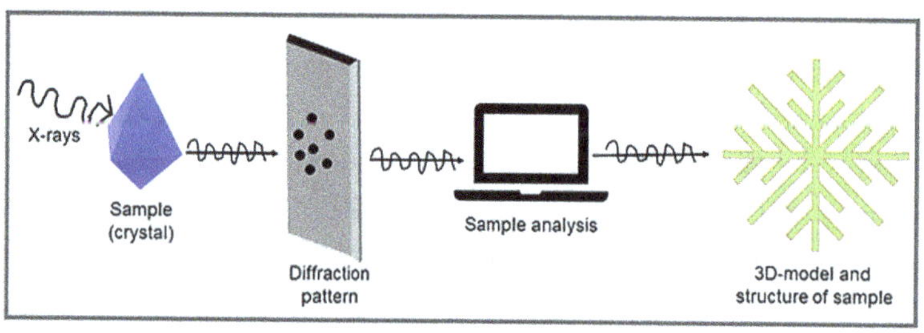

Figure 4. A comparative diagrammatic representation of electron (a) and optical microscopy (b). An optical microscope has a light source, unlike a transmission electron microscope that has a source as an electron beam.

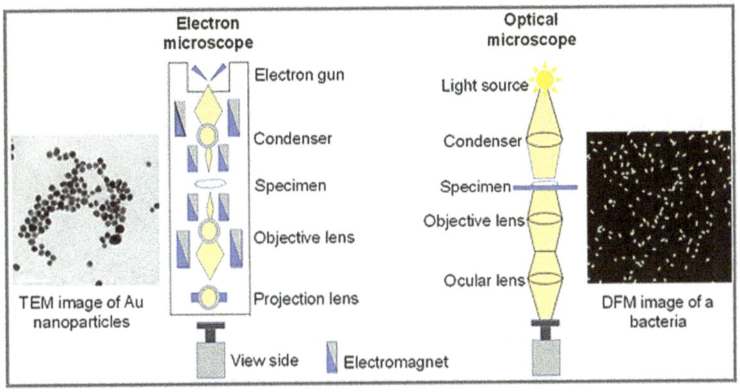

Fourier transform infrared spectroscopy (FTIR) deals with the vibrational transitions of a molecule, which in turn provides molecular information. Nanoparticles having their intrinsic vibrations corresponding to any functional groups at their surface can be characterized by IR spectroscopy (López-lorente & Mizaikoff., 2016). Fourier transformation is a mathematical tool that helps to decode the raw data to the original spectrum. A presence of a covalent bond between some functional molecules and carbon nanotubes is established by grafting some molecules consisting of silicon and tin using the FTIR spectrum (Blake *et al.*, 2004).

Raman spectroscopy (RS) is also a non-destructive vibrational spectroscopic technique that is utilized to extract molecular information (structure and composition) (Jones *et al.*, 2019). It provides the vibrational, rotational, and other states of the molecule. The last few years have observed a huge improvement from the technical aspects of RS as problems regarding fluorescence, reproducible data, and poor sensibility are addressed by researchers around the globe. *Stimulated Raman scattering* is a coherent technique that provides stronger signals than *spontaneous Raman spectroscopy*. The additional advantage is the power to time-resolve the corresponding vibrational motions (Brown *et al.*, 2008).

SERS is a robust and advanced Raman spectroscopy methodology for the ultrasensitive detection of analytes (M. Fleischmann, 1974). The major event being the excitation of the localized surface plasmons (LSP) which help in the amplification of the electromagnetic field. LSP can be defined as a sea of oscillating electrons localized on a NP surface. SERS detection can successfully reach a single-molecule level using the bianalyte SERS method (Blackie *et al.*, 2009).

Several other analytical tools exist, such as NMR, TGA, and ICP-MS, to characterize different aspects of nanomaterials such as size, chemical nature, elemental composition, and so on.

Electron energy loss spectroscopy (EELS) involves the incidence of an electron beam with varying kinetic energies to hit the sample. The inelastically scattered electrons are captured by an electron detector and grouped based on the amount of energy loss. It can be used in conjunction with STEM or TEM to provide the spatial distribution of the components present in NPs. For instance, the components of immuno-functionalized core-shell superparamagnetic magnetite NPs were determined with spatially resolved EELS with STEM. This allowed the researchers to locate the biomacromolecules on specific areas on the surface of NP (Arenal *et al.*, 2013).

Nuclear magnetic resonance (NMR) involves the presence of an active nuclear spin. The most common nuclei that are studied in NMR are hydrogen and carbon. It is a boon for researchers as it provides information on the interaction of nanomaterials with a functional group (Devience *et al.*, 2015). This enables an understanding of their surface chemistry and physical properties. Guo *et al.* used a solid-state NMR technique called cross-polarization magic-angle spinning (CP-MAS) to understand the adsorption of alanine on the interface of fumed silica NPs. Several other NMR techniques have aided researchers in characterizing and investigating the adsorption behavior of substances on nanomaterials. It is also beneficial due to its non-destructive nature and sample recovery.

MICROSCOPY

Transmission electron microscopy (TEM/High-resolution HRTEM) is a widely used technique to determine the morphology of nanomaterials. It not only provides direct images of the sample but also predicts the sample polydispersity. In this technique, the electron beam interaction with the sample depends on sample density, size, and elemental composition of nanoarchitectures. It has also been used to analyze the NPs aggregation for several biomedical applications such as sensing, diagnosis, therapy, and imaging (Nam *et al.*, 2009; Perrault & Chan, 2010; Thanh & Rosenzweig, 2002; Vilela *et al.*, 2012). In addition, it has been used to image the growth of CuNPs (van den Berg *et al.*, 2016). Substantial control over NP-assembly leading to superlattices has also been achieved (Courty *et al.*, 2005). Apart from several such advantages, *TEM* lacks quantifying many particles, which can generate ambiguous images due to orientation effects. Moreover, additional care must be taken regarding sample preparation because inappropriate samples can create artifacts. A comparative study between the basic methodologies involved in optical versus electron microscopy has been illustrated in Figure 4.

Scanning electron microscopy (SEM/Field emission FESEM), another electron microscopy like *TEM,* is capable of imaging surfaces of nanoarchitectures down to a 1 nm resolution. Goldstein *et al.* has used high-resolution *SEM* (*HRSEM*) to track the location of AuNPs in the cells and tissues (Goldstein *et al.*, 2014). This technique ensures the straightforward visualization of nanomaterials with fast and quick sample preparation. Though it provides surface information, precise height determination of samples cannot be achieved.

Atomic force microscopy (AFM) is a surface probe technique that creates three-dimensional images of nanomaterials. It also measures the height and thickness of nanostructures with sub-nanometer accuracy. The complete set-up occupies lesser space in the laboratory than *SEM, TEM, XRD, XPS, NMR,* etc. It also determines the grain size and NP coverage on the surface. However, it is limited with the cantilever resolution, sample preparation, and requires a skillful operation.

Darkfield microscopy (DFM) is optical microscopy for visualization and localization of plasmonic NPs and bioorganisms. When DFM is coupled with a spectrograph, it provides high sensitivity, high S/N ratio, and consumes less time in terms of the studies involving morphology, composition, and surrounding medium of the NPs. It is commonly used for monitoring chemical reactions in real-time, biosensing, and single-molecule detection. (H. Z. Zhang *et al.*, 2017; Zhou *et al.*, 2016, 2017)

Figure 5. The principle of operation of a mass spectrometer. Sample loaded in the vaporized state passed through the electromagnets get separated into various species based on the charge by mass ratio. An amplifier strengthens the signals detected by the detector, which gives a mass spectrum.

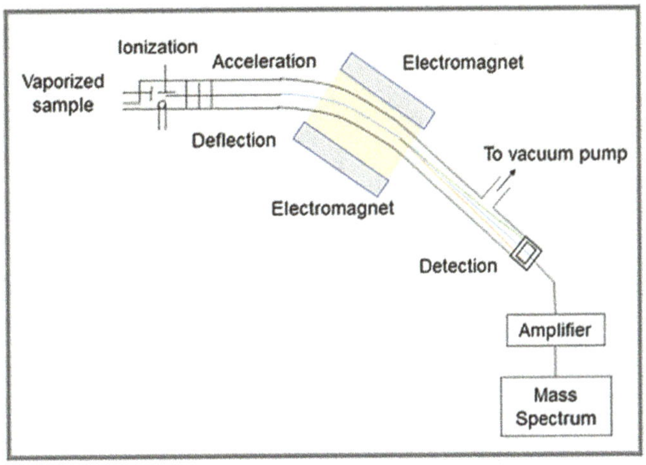

MASS SPECTROMETRY

Mass spectrometry (MS) provides an accurate method for measuring mass and therefore, the size and size distribution of nanomaterials (Mclafferty, 2011). It provides valuable insights into the polymer, biological, and nanoscale materials. The instrumentation of MS is presented in Figure 5. Matrix-assisted laser desorption/ionization (MALDI) inductively coupled plasma mass spectrometry (ICP-MS) and secondary ion mass spectrometry (SIMS) are a few commonly used MS techniques for nanomaterials characterization.

MALDI is an ionization technique in which a laser source is directed onto the sample embedded with the matrix (Pan *et al.*, 2007). This triggers ablation and desorption of the sample and matrix. Ultimately, the analyte molecules are ionized by protonation or deprotonation in the hot plume of gases and passed on to the spectrometer. Khitrov *et al.* have used MALDI-TOF (time of flight) MS for the determination of mass and its distribution of a series of 2.5–3.7 nm by assuming ZnS NPs to be spherical in morphology. The results were correlated with the observed TEM and optical data, to reaffirm the masses (Khitrov & Strouse, 2003).

ICP-MS involves vaporization, atomization, and ionization of the substance under consideration. It is a powerful and highly sensitive technique with high selectivity (Sadik *et al.*, 2014). A calibration standard is essential here for which isotope dilution techniques are established for accurate elemental quantifications (Montoro Bustos *et al.*, 2009). Multi-elemental trace impurities can be determined in high-purity NPs after microwave-assisted acid digestion using high-resolution ICP-MS (Yue *et al.*, 2001). *ICP-MS* coupled with other analytical techniques provides additional information. For instance, *size exclusion chromatography ICP-MS (SEC-ICP-MS):* It is an integrated technique, generally used to get the idea of the size distribution of NPs. *Single-particle ICP-MS (SP-ICP-MS)* is used to determine the size and concentration of NPs at the single-particle level. The dissolved metal and metal NPs such as Ag and Au under single particle detection will behave differently, resulting in the successful distinction between the different NP samples (Mitrano *et al.*, 2012).

Figure 6. The diagrammatic representation of the methodology involved in chromatography. A solvent is systematically injected into the pump that takes the sample with it through the column. The sample is separated into its components based on the affinity to the column and a chromatogram is displayed on the screen.

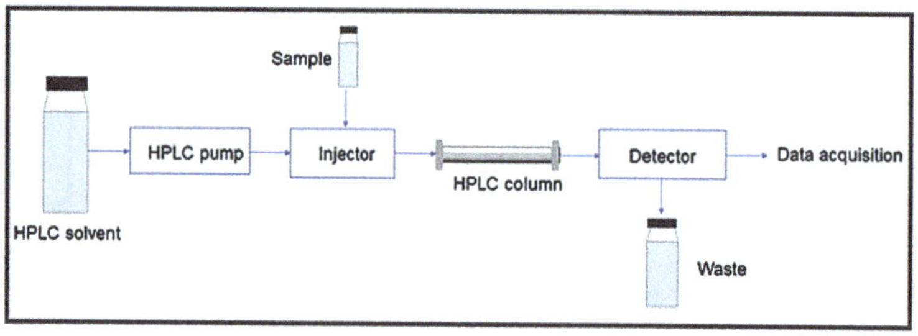

Secondary ion mass spectrometry (SIMS) has a great potential to characterize films. It analyses the surface of a specific sample. The procedure involves sputtering with a focused primary ion beam, which ejects the secondary ions that are collected and analyzed. It provides a detailed picture of the molecular, elemental, and isotopic information at the surfaces of the desired NPs (Yang *et al.*, 2013). A major drawback of SIMS is that it is destructive, which can be reduced by setting up an upper primary ion dose limit. Despite these, it is used for the characterization of nanomaterials because of its robust and straightforward analysis (Seebauer & Barlaz, 2016).

CHROMATOGRAPHY

In analytical chemistry, chromatography is used for the separation of components present in a mixture. It is thus used for the purification and identification of individual substances in a mixture.

High-performance liquid chromatography (HPLC), a chromatography method, is used for the quantitative analysis of nanomaterials. It is suitable for the separation of NPs from a mixer and aims at understanding the interaction of the NP with the stationary phase (Itoh *et al.*, 2017). The basic instrumentation involved in HPLC is shown in Figure 6. Generally, larger NPs (~100 nm) move slower than smaller NPs (< 10 nm) through the packed column and are thus separated based on size. Different columns are used depending on the molecular weight. N. Itoh *et al.* have developed a three polymer-coated column made of glycidyl methacrylate (GMA), vinylpyrrolidone (VP), and 2-hydroxyethyl methacrylate (HEMA) for the effectual separation of 100 nm liposomal NPs (Itoh *et al.*, 2017). These special columns do not allow strong adsorption of large NPs, and they are easily eluted through the column enabling their separation.

Size Exclusion Chromatography (SEC) is a type of partition chromatography. The column is made of a porous matrix which allows the partition of particles based on their size (Brusotti *et al.*, 2018). Smaller molecules have higher penetration compared to larger molecules. The column length and mobile phase flow rate also matters. Appropriate eluent and pore size of the stationary phase is crucial for the proper separation of NPs. For instance, SEC was combined with the information obtained from DLS and cryo-TEM to evaluate the lipid NPs with the size range of 60-140 nm effectively. Although the particle assembly process was similar for various groups of particles, high-resolution SEC enabled to get a firm idea of the LNP size heterogeneity in the native solution conditions (J. Zhang *et al.*, 2012).

GRAVIMETRIC ANALYSIS

Thermogravimetric analysis (TGA) is a thermal analysis method that monitors the change in the weight of nanomaterials as a function of time or temperature. It is used for the determination of the purity of materials, primary standards, the composition of alloys and metals. It is also used to study polymers and the oxidation of alloys.

ELECTROANALYTICAL METHODS

The charges corresponding to the nanoparticles bring the concept of electrochemistry into the picture. *Electrochemical techniques* involve filtration process entangled with the electrochemical approach is a distinctive tool to characterize nanomaterials based on their sizes and charges. Recently, a portable nanoparticle analyzer has been designed which relies on the principles of electrochemical detection and tangential flow filtration (EC-TFF) (Andreescu *et al.*, 2005).

As a step further to the recent advanced techniques, many correlative techniques have been developed to provide multi-signals simultaneously. An example of AFM combined with Kelvin probe microscopy to provide 3D maps of surface potential distributions of nanostructures. Many other advanced and correlative techniques have been established to find useful information and inputs of nanoparticles at a single-particle and single-molecule level. The details about some of such advanced techniques are described in Table 2 and the next section.

RECENT ADVANCEMENTS

Since challenges prevailed in the scientific community for the development of the accurate and high-resolution analysis of nanomaterials at the single-particle level. Several correlative and integrated techniques were established to detect multifunctional properties simultaneously, such as SEM-Raman, which has been combined to predict the morphology and spectroscopy characteristics of nanomaterials. Some of these recent advanced techniques are elaborated to get the essence of the cutting-edge technologies deployed to decode the complex nanostructures.

Cryo-transmission electron microscopy (cryo-TEM) was evolved in the early 1980s to visualize the biological specimens in nearly their native states (frozen-hydrated states) with the similar resolution of a transmission electron microscope (0.1 to 2 nm) (Dubochet et al., 1988; Lyumkis, 2019). The technique deals with instant flash-freezing of the solutions containing biomolecules at first, followed by the bombardment of the electrons on them to get the microscopic images of the individual molecules. The presence of polydispersity within synthesized nanoparticle samples can also be visualized by two-dimensional (2D) cryo-TEM. The use of Fourier transform helps to obtain the structural periodicity in a sample. 3D structures can be acquired for both homogenous and heterogeneous structures by following two separate protocols. It can also be integrated with focused ion beam milling. Cryo-FIBSEM is often used for serial block-face imaging which generates the volume representation in 3D geometry (Schertel *et al.*, 2013).

SEM-DFM-SERS is another correlative technique, which combines the functions of SEM, DFM, and SERS. It has enabled the study of the structure-activity relationships down to a single-particle level. Some of the optical and electron microscopy correlated instrumentations have been evolved in the past

Figure 7. (a) Schematic representation of integrated Ambient Electrospray Deposition Raman spectroscopy (AERS) set-up, (b) Optical image of the actual set-up consisting of ambient electrospray coupled with a miniature Raman set-up, and (c) zoomed-in region of B marked with a red-colored ellipse. (d) SERS spectra of p-MBA of various concentrations ranging from 100 μM to 1 nM, inset shows the pictorial representation of p-MBA molecules adsorbed on the AgNP surface and (e) quantification of the analyte: calibration curve of SERS intensity of the 1377 cm^{-1} peak vs. concentration of p-MBA. Data are fitted with a straight line.
(Adapted with the permission of Tripti Ahuja et al., 2019).

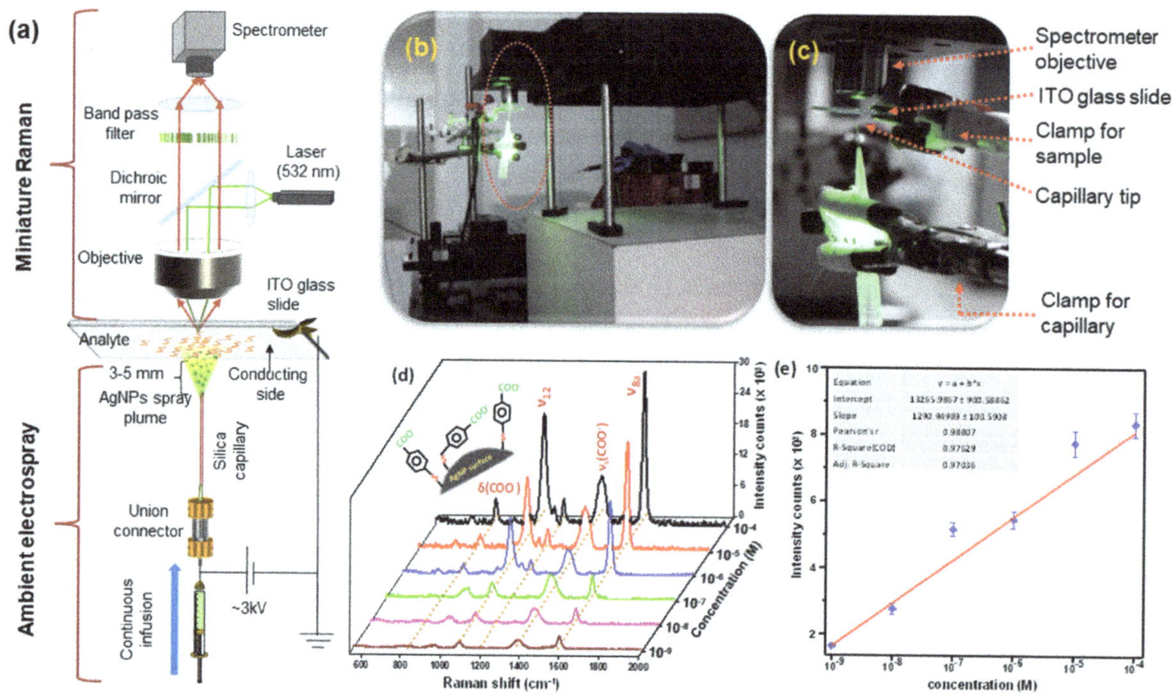

for different applications (Perassi *et al.*, 2014; Rosman *et al.*, 2012; Song *et al.*, 2010; Tran *et al.*, 2018; Z. Zhang *et al.*, 2019). This technique helps us to determine the fate of NPs before and after laser irradiation with the potential of SEM.

Tip-enhanced Raman spectroscopy (TERS) is a derived imaging technique of SERS for nanoscale resolution imaging (Yeo *et al.*, 2009). It's a classic example of the coupling of two well-known techniques. A scanning probe microscope (which is used in scanning tunnelling and atomic force microscopy) is entangled with a confocal Raman spectrometer by an optomechanical amalgamation (Sonntag *et al.*, 2014). A metallic tip (usually gold or silver) is used to direct the field of the incident light at the apex of the system. It helps to enhance the local field, in turn increasing the sensitivity (the factor can vary between 10^3-10^7). The paramount importance of this technique lies in spatial resolution. The diffraction limit of the light is nearly 100 times larger than the spatial resolution obtained for the chemical surfaces in TERS imaging (Pozzi *et al.*, 2013).

Particle Dynamics Induced-Plasmon Enhanced Raman Scattering (PDI-PERS) deals with the enhancement of the SERS activity of nanoparticles. It has been explored to tinker the SERS activity (Cardinal *et al.*, 2017; Kim *et al.*, 2011; Rao & Radhakrishnan, 2015; Sreeprasad & Pradeep, 2011). The underlying principle to get this enhancement is based upon the successful manipulation of the plasmonic nanocavity. The possible factors being the gap between the nanoparticles in an aggregated state (R. Zhang *et al.*, 2013) and/or the use of linker molecules of various lengths for the smooth transfer of the electrons (Sreeprasad & Pradeep, 2011). PDI-PERS provides a convenient path for the successful determination of SERS activity in such cases. Reshaping or reorienting of NPs can induce on-resonance situations which favors the SERS. Laser-induced changes on individual silver nanoparticles help to switch between various binding modes of citrate ligand (T. Ahuja *et al.*, 2021).

Ambient electrospray deposition-Raman spectroscopy (AESD-RS) is a Raman spectroscopy technique to determine the vibrational and rotational modes of a molecule, irrespective of its bond polarity. AESD is a technique to form charged micro or nano-droplets by using direct current (DC) voltages, ranging in few kilovolts on the liquid present in a microcapillary. RS helps to detect the vibrational features of the electrosprayed particles. A combined schematic representation of AESD-RS is shown in Figure 7(a–c). This combination enables the detection of molecules up to nanomolar concentrations (Figure 7(d–e)) such as the detection of p-mercaptobenzoic acid (*p*-MBA) and 2,4-dinitrotoluene (2,4-DNT) using preformed silver nanoparticles (Ahuja *et al.*, 2019). Using the same line of approach, precise changes in the different molecular orientations of various thiol ligands on the interaction with the soft-landed silver nanoparticles have also been reported with the AESD-RS set-up (Ghosh *et al.*, 2020).

High-annular dark-field scanning transmission electron microscopy (*HADDF-STEM)* is a standard procedure used to investigate the atomic structure of nanocrystals. It gathers two-dimensional information about the crystal by taking images along several projections. In 2003, it was combined with tomography by Midgley *et al.*, which allowed researchers to obtain three-dimensional information about the material. It can also be used for *in-situ* imaging. Some limitations of this method are that it takes more time than the conventional method (~ one hour slower than continuous mode). Also, it is more challenging to perform due to a stepwise increment in the tilt angle. Large tilt increments can be performed but it may complicate the tracking of the particles (Vanrompay *et al.*, 2021). In totality, it involves image correction, angle assignment, alignment, image selection, reconstruction, and evaluation.

Thus, the characterization tools have enabled us to utilize the properties of nanomaterials to create materials of importance and design tools of relevance such as biosensors, diagnostic and targeting tools.

FUTURE RESEARCH DIRECTIONS

The chapter has highlighted existing and advanced characterization tools needed for nanomaterials. While the content may require more expertise to execute, but the information presented is of broad interest. This can help the researchers to choose the most suitable technique for the characterization of a desired property of nanomaterial. As the discovery and understanding at the nanoscale are augmenting, the demand for compact, faster, and improved technologies is gaining interest. And for which, sensitive and advanced techniques are arising every year with a promise to improve the way we look at nanomaterials in the future.

CONCLUSION

The present surge in the use of nanomaterials has increased the demand for quick and reliable characterization techniques. Such relevant characterization tools and methodologies were highlighted throughout the chapter. They were categorized based on various parameters such as properties of nanomaterials, source or probe, availability, cost, non-destructive nature, and sensitivity of tools. The strengths and limitations associated with these tools were also mentioned so that they can be effectively combined and complement each other to provide a detailed picture of nanomaterials. The type and role of each technique have been tabulated comparatively that can act as a guide to help the scientific community to get a firm grip on the discussed topic.

REFERENCES

Ahuja, T., Chaudhari, K., Paramasivam, G., Ragupathy, G., Mohanty, J. S., & Pradeep, T. (2021). Toward Vibrational Tomography of Citrate on Dynamically Changing Individual Silver Nanoparticles. *The Journal of Physical Chemistry C*, *125*(6), 3553–3566. doi:10.1021/acs.jpcc.0c09981

Ahuja, T., Ghosh, A., Mondal, S., Basuri, P., Jenifer, S. K., Srikrishnarka, P., Mohanty, J. S., Bose, S., & Pradeep, T. (2019). Ambient electrospray deposition Raman spectroscopy (AESD RS) using soft-landed preformed silver nanoparticles for rapid and sensitive analysis. *Analyst (London)*, *144*(24), 7412–7420. doi:10.1039/C9AN01700C PMID:31674602

Andreescu, D., Wanekaya, A. K., Sadik, O. A., & Wang, J. (2005). Nanostructured polyamic acid membranes as novel electrode materials. *Langmuir*, *21*(15), 6891–6899. doi:10.1021/la050141k PMID:16008401

Blackie, E. J., Le Ru, E. C., & Etchegoin, P. G. (2009). Single-Molecule Surface-Enhanced Raman Spectroscopy of Nonresonant Molecules. *JACS*, *131*(40), 14466–14472. doi:10.1021/ja905319w PMID:19807188

Blake, R., Gun, Y. K., Coleman, J., Cadek, M., Fonseca, A., Nagy, J. B., & Blau, W. J. (2004). A Generic Organometallic Approach toward Ultra-Strong Carbon Nanotube Polymer Composites. *JACS*, *126*(33), 10226–10227. doi:10.1021/ja0474805 PMID:15315418

Brown, P. O., Botstein, D., Lapidot, M., Pilpel, Y., Mazo, A., Hodgson, J. W., Petruk, S., Sedkov, Y., Brock, H. W., Timmons, J. A., Good, L., Kapranov, P., Willingham, A. T., Gingeras, T. R., Yazgan, O., Krebs, J. E., Fahey, M. E., Moore, T. F., Higgins, D. G., ... Schadt, E. E. (2008). Label-Free Biomedical Imaging with. *Science*, *322*(5909), 1857–1862. doi:10.1126cience.1165758 PMID:19095943

Brusotti, G., Calleri, E., Colombo, R., Massolini, G., Rinaldi, F., & Temporini, C. (2018). Advances on Size Exclusion Chromatography and Applications on the Analysis of Protein Biopharmaceuticals and Protein Aggregates : A Mini Review. *Chromatographia*, *81*(1), 3–23. doi:10.100710337-017-3380-5

Cardinal, M. F., Vander Ende, E., Hackler, R. A., McAnally, M. O., Stair, P. C., Schatz, G. C., & Van Duyne, R. P. (2017). Expanding applications of SERS through versatile nanomaterials engineering. *Chemical Society Reviews*, *46*(13), 3886–3903. doi:10.1039/C7CS00207F PMID:28640313

Courty, A., Mermet, A., Albouy, P. A., Duval, E., & Pileni, M. P. (2005). Vibrational coherence of self-organized silver nanocrystals in f.c.c. supra-crystals. *Nature Materials*, *4*(5), 395–398. doi:10.1038/nmat1366 PMID:15834414

Devience, S. J., Pham, L. M., Lovchinsky, I., Sushkov, A. O., Bar-gill, N., Belthangady, C., Casola, F., Corbett, M., Zhang, H., Lukin, M., Park, H., Yacoby, A., & Walsworth, R. L. (2015). Nanoscale NMR spectroscopy and imaging of multiple nuclear species. *Nature Nanotechnology*, *10*(2), 129–134. doi:10.1038/nnano.2014.313 PMID:25559712

Dubochet, J., Adrian, M., Chang, J., Homo, J., Lepault, J., Mcdowall, A. W., & Schultz, P. (1988). Cryo-electron microscopy of vitrified specimens. *Quarterly Reviews of Biophysics*, *2*(2), 129–228. doi:10.1017/S0033583500004297 PMID:3043536

Fleischmann, M., Hendra, P. J., & McQuillan, A. J. (1974). Raman spectra of pyridine adsorbed at a silver electrode. *Chemical Physics Letters*, *26*(2), 2–5. doi:10.1016/0009-2614(74)85388-1

Geoff, W. (n.d.). *Richard Fenyman*. Academic Press.

Ghosh, A., Ahuja, T., Chaudhari, K., & Pradeep, T. (2020). Probing Subtle Changes in Molecular Orientations Using Ambient Electrospray Deposition Raman Spectroscopy (AESD RS). *The Journal of Physical Chemistry C*, *124*(30), 16644–16651. doi:10.1021/acs.jpcc.0c04116

Goldstein, A., Soroka, Y., Frušić-Zlotkin, M., Popov, I., & Kohen, R. (2014). High resolution SEM imaging of gold nanoparticles in cells and tissues. *Journal of Microscopy*, *256*(3), 237–247.

Haiss, W., Thanh, N. T. K., Aveyard, J., & Fernig, D. G. (2007). Determination of size and concentration of gold nanoparticles from UV-Vis spectra. *Analytical Chemistry*, *79*(11), 4215–4221. doi:10.1021/ac0702084 PMID:17458937

Itoh, N., Kimoto, A., Yamamoto, E., Higashi, T., Santa, T., Funatsu, T., & Kato, M. (2017). High performance liquid chromatography analysis of 100-nm liposomal nanoparticles using polymer-coated, silica monolithic columns with aqueous mobile phase. *Journal of Chromatography. A*, *1484*, 34–40. doi:10.1016/j.chroma.2016.12.080 PMID:28089273

Jones, R. R., Hooper, D. C., Zhang, L., Wolverson, D., & Valev, V. K. (2019). Raman Techniques : Fundamentals and Frontiers. *Nanoscale Research Letters*, *14*(231). PMID:31300945

Kato, H. (2012). Size Determination of Nanoparticles by Dynamic Light Scattering. *Nanomaterials*, 535–554.

Khitrov, G. A., & Strouse, G. F. (2003). ZnS nanomaterial characterization by MALDI-TOF mass spectrometry. *Journal of the American Chemical Society*, *125*(34), 10465–10469. doi:10.1021/ja035508r PMID:12926972

Kim, N. H., Lee, S. J., & Moskovits, M. (2011). Reversible tuning of SERS hot spots with aptamers. *Advanced Materials*, *23*(36), 4152–4156. doi:10.1002/adma.201101847 PMID:21823179

López-lorente, Á. I., & Mizaikoff, B. (2016). Trends in Analytical Chemistry Recent advances on the characterization of nanoparticles using infrared spectroscopy. *Trends in Analytical Chemistry*, *84*, 97–106. doi:10.1016/j.trac.2016.01.012

Lyumkis, D. (2019). Challenges and Opportunities in Cryo-EM Single-Particle Analysis Challenges and Opportunities in Cryo-EM. *The Journal of Biological Chemistry*, *294*(13), 5181–5197. doi:10.1074/jbc.REV118.005602 PMID:30804214

Mclafferty, F. W. (2011). A Century of Progress in Molecular Mass Spectrometry. *Annual Review of Analytical Chemistry (Palo Alto, Calif.)*, *4*(1), 1–22. doi:10.1146/annurev-anchem-061010-114018 PMID:21351881

Mitrano, D. M., Barber, A., Bednar, A., Westerhoff, P., Higgins, C. P., & Ranville, J. F. (2012). Silver nanoparticle characterization using single particle ICP-MS (SP-ICP-MS) and asymmetrical flow field flow fractionation ICP-MS (AF4-ICP-MS). *Journal of Analytical Atomic Spectrometry*, *27*(7), 1131–1142. doi:10.1039/c2ja30021d

Montoro Bustos, A. R., Encinar, J. R., Fernández-Argüelles, M. T., Costa-Fernández, J. M., & Sanz-Medel, A. (2009). Elemental mass spectrometry: A powerful tool for an accurate characterisation at elemental level of quantum dots. *Chemical Communications*, *03*(21), 3107–3109. doi:10.1039/b901493d PMID:19462102

Nam, J., Won, N., Jin, H., Chung, H., & Kim, S. (2009). pH-Induced Aggregation of Gold Nanoparticles for Photothermal Cancer Therapy. *Journal of the American Chemical Society*, *131*(38), 13639–13645. doi:10.1021/ja902062j PMID:19772360

Pan, C., Xu, S., Zhou, H., Fu, Y., Ye, M., & Zou, H. (2007). Recent developments in methods and technology for analysis of biological samples by MALDI-TOF-MS. *Analytical and Bioanalytical Chemistry*, *387*(1), 193–204. doi:10.100700216-006-0905-4 PMID:17086385

Perassi, E. M., Hrelescu, C., Wisnet, A., Döblinger, M., Scheu, C., Jäckel, F., Coronado, E. A., & Feldmann, J. (2014). Quantitative understanding of the optical properties of a single, complex-shaped gold nanoparticle from experiment and theory. *ACS Nano*, *8*(5), 4395–4402. doi:10.1021/nn406270z PMID:24787120

Perrault, S. D., & Chan, W. C. W. (2010). In vivo assembly of nanoparticle components to improve targeted cancer imaging. *Proceedings of the National Academy of Sciences of the United States of America*, *107*(25), 11194–11199. doi:10.1073/pnas.1001367107 PMID:20534561

Pozzi, E. A., Sonntag, M. D., Jiang, N., Klingsporn, J. M., Hersam, M. C., & Van Duyne, R. P. (2013). Tip-Enhanced Raman Imaging : An Emergent Tool for Probing Biology at the Nanoscale. *ACS Nano*, *7*(2), 885–888. doi:10.1021/nn400560t PMID:23441673

Rao, V. K., & Radhakrishnan, T. P. (2015). Tuning the SERS Response with Ag-Au Nanoparticle-Embedded Polymer Thin Film Substrates. *ACS Applied Materials & Interfaces*, *7*(23), 12767–12773. doi:10.1021/acsami.5b04180 PMID:26035249

Ray, T. R., Lettiere, B., de Rutte, J., & Pennathur, S. (2015). Quantitative Characterization of the Colloidal Stability of Metallic Nanoparticles Using UV–vis Absorbance Spectroscopy. *Langmuir*, *31*(12), 3577–3586. doi:10.1021/la504511j PMID:25730093

Rosman, C., Pierrat, S., Henkel, A., Tarantola, M., Schneider, D., Sunnick, E., Janshoff, A., & Sönnichsen, C. (2012). A new approach to assess gold nanoparticle uptake by mammalian cells: Combining optical dark-field and transmission electron microscopy. *Small*, *8*(23), 3683–3690. doi:10.1002mll.201200853 PMID:22888068

Sadik, O. A., Du, N., Kariuki, V., Okello, V., & Bushlyar, V. (2014). Current and emerging technologies for the characterization of nanomaterials. *ACS Sustainable Chemistry & Engineering*, *2*(7), 1707–1716. doi:10.1021c500175v

Sarma, D. D., Santra, P. K., Mukherjee, S., & Nag, A. (2013). X-ray Photoelectron Spectroscopy: A Unique Tool To Determine the Internal Heterostructure of Nanoparticles. *Chemistry of Materials*, *25*(8), 1222–1232. doi:10.1021/cm303567d

Schamp, C. T., & Jesser, W. A. (2005). On the measurement of lattice parameters in a collection of nanoparticles by transmission electron diffraction. *Ultramicroscopy*, *103*(2), 165–172. doi:10.1016/j.ultramic.2004.11.007 PMID:15774277

Schertel, A., Snaidero, N., Han, H., Ruhwedel, T., Laue, M., Grabenbauer, M., & Möbius, W. (2013). Cryo FIB-SEM : Volume imaging of cellular ultrastructure in native frozen specimens. *Journal of Structural Biology*, *184*(2), 355–360. doi:10.1016/j.jsb.2013.09.024 PMID:24121039

Seebauer, E. G., & Barlaz, D. E. (2016). SIMS for analysis of nanostructures. *Current Opinion in Chemical Engineering*, *12*, 8–13. doi:10.1016/j.coche.2016.01.007

Song, Y., Nallathamby, P. D., Huang, T., Elsayed-Ali, H. E., & Xu, X.-H. N. (2010). Correlation and Characterization of Three-Dimensional Morphologically Dependent Localized Surface Plasmon Resonance Spectra of Single Silver Nanoparticles Using Dark-Field Optical Microscopy and Spectroscopy and Atomic Force Microscopy. *The Journal of Physical Chemistry C*, *114*(1), 74–81. doi:10.1021/jp9083019 PMID:20190865

Sonntag, M. D., Pozzi, E. A., Jiang, N., Hersam, M. C., & Van Duyne, R. P. (2014). Recent Advances in Tip-Enhanced Raman Spectroscopy. *The Journal of Physical Chemistry Letters*, *5*(18), 3125–3130. doi:10.1021/jz5015746 PMID:26276323

Sreeprasad, T. S., & Pradeep, T. (2011). Reversible assembly and disassembly of gold nanorods induced by EDTA and its application in SERS tuning. *Langmuir*, *27*(7), 3381–3390. doi:10.1021/la104828e PMID:21366283

Stancheva, K. A. (2012). Noble metal nanoparticles. In Oxidation Communications (Vol. 35, Issue 3). doi:10.1002/9781119120940.ch3

Thanh, N. T. K., & Rosenzweig, Z. (2002). Development of an Aggregation-Based Immunoassay for Anti-Protein A Using Gold Nanoparticles. *Analytical Chemistry*, *74*(7), 1624–1628. doi:10.1021/ac011127p PMID:12033254

Toropov, N., & Vartanyan, T. (2019). Noble metal nanoparticles: Synthesis and optical properties. In Comprehensive Nanoscience and Nanotechnology (Vols. 1–5, Issue March 2015). Elsevier Ltd. doi:10.1016/B978-0-12-803581-8.00585-3

Tran, V., Thiel, C., Svejda, J. T., Jalali, M., Walkenfort, B., Erni, D., & Schlücker, S. (2018). Probing the SERS brightness of individual Au nanoparticles, hollow Au/Ag nanoshells, Au nanostars and Au core/Au satellite particles: Single-particle experiments and computer simulations. *Nanoscale, 10*(46), 21721–21731. doi:10.1039/C8NR06028B PMID:30431039

Tunc, I., Suzer, S., Correa-Duarte, M. A., & Liz-Marzán, L. M. (2005). XPS characterization of Au (Core)/ SiO2 (shell) nanoparticles. *The Journal of Physical Chemistry B, 109*(16), 7597–7600. doi:10.1021/ jp050767j PMID:16851877

Vajtai, R. (2013). Springer handbook of nanomaterials. In Springer Handbook of Nanomaterials. doi:10.1007/978-3-642-20595-8

van den Berg, R., Elkjaer, C. F., Gommes, C. J., Chorkendorff, I., Sehested, J., de Jongh, P. E., de Jong, K. P., & Helveg, S. (2016). Revealing the Formation of Copper Nanoparticles from a Homogeneous Solid Precursor by Electron Microscopy. *Journal of the American Chemical Society, 138*(10), 3433–3442. doi:10.1021/jacs.5b12800 PMID:26891132

Vanrompay, H., Skorikov, A., Bladt, E., Béché, A., Freitag, B., Verbeeck, J., & Bals, S. (2021). Fast versus conventional HAADF-STEM tomography of nanoparticles: advantages and challenges. *Ultramicroscopy, 221*(October). doi:10.1016/j.ultramic.2020.113191

Vilela, D., González, M. C., & Escarpa, A. (2012). Sensing colorimetric approaches based on gold and silver nanoparticles aggregation: Chemical creativity behind the assay. A review. *Analytica Chimica Acta, 751*, 24–43. doi:10.1016/j.aca.2012.08.043 PMID:23084049

Yang, L., Seah, M. P., Gilmore, I. S., Morris, R. J. H., Dowsett, M. G., Boarino, L., Sparnacci, K., & Laus, M. (2013). Depth profiling and melting of nanoparticles in secondary ion mass spectrometry (SIMS). *The Journal of Physical Chemistry C, 117*(31), 16042–16052. doi:10.1021/jp4048538

Yeo, B., Stadler, J., Schmid, T., Zenobi, R., & Zhang, W. (2009). Tip-enhanced Raman Spectroscopy – Its status, challenges and future directions. *Chemical Physics Letters, 472*(1–3), 1–13. doi:10.1016/j. cplett.2009.02.023 PMID:20161522

Yue, X., Cai, S., Shao, R., Liu, Y., & Lu, Y. (2001). Determination of trace impurities in high purity gallium by inductively coupled plasma mass spectrometry. *Fenxi Huaxue, 29*(11), 1309–1310.

Zhang, H. Z., Li, R. S., Gao, P. F., Wang, N., Lei, G., Huang, C. Z., & Wang, J. (2017). Real-time dark-field light scattering imaging to monitor the coupling reaction with gold nanorods as an optical probe. *Nanoscale, 9*(10), 3568–3575. doi:10.1039/C6NR09453H PMID:28244517

Zhang, J., Haas, R. M., & Leone, A. M. (2012). Polydispersity characterization of lipid nanoparticles for siRNA delivery using multiple detection size-exclusion chromatography. *Analytical Chemistry, 84*(14), 6088–6096. doi:10.1021/ac3007768 PMID:22816783

Zhang, R., Zhang, Y., Dong, Z. C., Jiang, S., Zhang, C., Chen, L. G., Zhang, L., Liao, Y., Aizpurua, J., Luo, Y., Yang, J. L., & Hou, J. G. (2013). Chemical mapping of a single molecule by plasmon-enhanced Raman scattering. *Nature, 498*(7452), 82–86. doi:10.1038/nature12151 PMID:23739426

Zhang, T., Lu, G., Shen, H., Shi, K., Jiang, Y., Xu, D., & Gong, Q. (2014). Photoluminescence of a single complex plasmonic nanoparticle. *Scientific Reports*, *4*(1), 3867. doi:10.1038rep03867 PMID:24463794

Zhang, Z., Fujita, K., Kawata, S., Bando, K., Mochizuki, K., & Taguchi, A. (2019). Quantitative Evaluation of SERS Nanoparticles for Intracellular pH Sensing at a Single Particle Level. *Analytical Chemistry*, 1–25. doi:10.1021/acs.analchem.8b03276 PMID:30698014

Zhou, J., Gao, P. F., Zhang, H. Z., Lei, G., Zheng, L. L., Liu, H., & Huang, C. Z. (2017). Color resolution improvement of the dark-field microscopy imaging of single light scattering plasmonic nanoprobes for microRNA visual detection. *Nanoscale*, *9*(13), 4593–4600. doi:10.1039/C6NR09452J PMID:28322387

Zhou, J., Lei, G., Zheng, L. L., Gao, P. F., & Huang, C. Z. (2016). HSI colour-coded analysis of scattered light of single plasmonic nanoparticles. *Nanoscale*, *8*(22), 11467–11471. doi:10.1039/C6NR01089J PMID:27194457

KEY TERMS AND DEFINITIONS

Crystallography: Crystallography is the branch of science that study the arrangement of atoms and molecules in the crystalline solids. It is a diffraction based technique.

Electron Microscopy: Electron microscopy is a technique that utilizes the beam of electrons to enlarge the image of the sample.

Localized Surface Plasmon Resonance (LSPR): LSPR is an optical phenomenon produced by light when it interacts with the conductive nanoparticles smaller than the incident wavelength.

Nanotechnology: Nanotechnology is the synthesis, manipulation, and utilization of nanosized (1 nm = 10^{-9} m) materials (atomic, molecular, and supramolecular scale) to harvest the property in terms of technological applications.

Optical Microscopy: Optical microscopy is a technique that uses visible light and a set of lens to magnify the image of the original sample.

Spectroscopy: Spectroscopy is the study of the light-matter interaction as a function of any of the properties of light.

Tomography: Tomography is a unique imaging technique that involves slicing a 3D object plane-by-plane at different angles by using a penetrating wave.

Section 2

Applications of Nanomaterials

Chapter 8
Applications of Nanomaterials in the Oil and Gas Industry

Kamel Fahmi Bou-Hamdan
https://orcid.org/0000-0003-0826-3473
Beirut Arab University, Lebanon

ABSTRACT

The petroleum industry has been an ever-growing industry. New technologies are always being introduced to encompass the challenges that are encountered. Nanomaterials are being included in these technologies to improve the operation of different processes. Their distinctive physical and chemical characteristics encourage their use in different sectors such as the upstream, midstream, and downstream of the oil and gas industry. In this chapter, the nanomaterials that are utilized in the oil and gas industries are highlighted. Their implementation in various applications is also provided. These applications include hydrocarbon exploration, well drilling and completion, production operations, enhanced oil recovery mechanisms, transportation, and refining operations. There is also a discussion about existing problems and possibilities for future uses.

INTRODUCTION

Over the years, the energy demand has increased because of growth in population and limitations in fossil fuels. This encouraged researchers to seek new ways of providing energy under more economical conditions. In the oil and gas industry, the studies focused on all its different sectors: upstream, midstream, and downstream. Figure 1 shows a schematic representation of the main oil and gas sectors. The main objective of these studies is to improve the hydrocarbon extraction, transportation, and refining processes (Khalil et al., 2017). As a result, new technologies were invented to match the needs of the industry such as drilling extended reach wells to cover an additional surface area in the hydrocarbon reservoir (Bou Hamdan et al., 2015); Utilizing enhanced oil recovery methods to improve the recovery factors (Agista et al., 2018); Hydraulically fracturing troublesome or low permeability reservoirs to produce the trapped hydrocarbons within the micro- or nano-pores of the rock (Bou-Hamdan, 2020). In addition to that, new

DOI: 10.4018/978-1-7998-8936-6.ch008

Figure 1. Schematic representation of the oil and gas industry operations

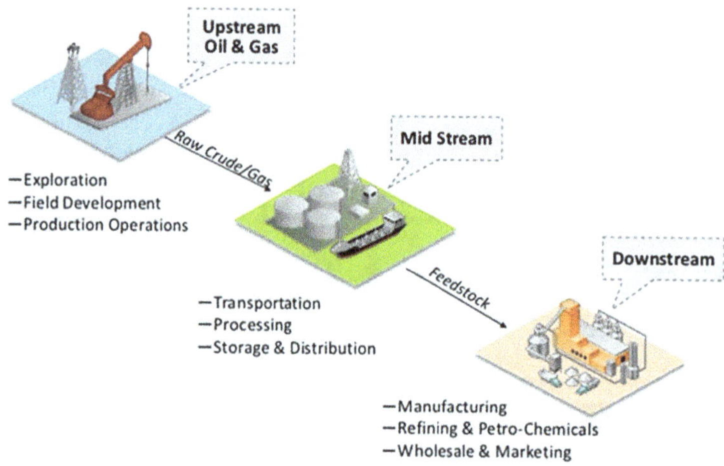

approaches attempt to make use of nanotechnology to optimize different processes within this industry (Hassani et al., 2020; Ko & Huh, 2019).

The discovery of nanomaterials encouraged engineers and researchers to vastly study their properties and use them in manufacturing unique tools. These materials are characterized by their very small size which ranges between 1 – 100 nanometers. Due to their nanosized dimensions, the chemical and physical properties of these materials behave differently than the larger size. Many properties are altered, including their magnetic characteristics, thermal behavior, internal pressure, and others (B. H. Kim et al., 2014; Kong & Ohadi, 2010). They were incorporated in different fields including, but not limited to, medical applications, aerospace, photography, electronics, and petroleum (Mout et al., 2012; Otsuka et al., 2012; Shen et al., 2014; S. Xu et al., 2015; Yadav et al., 2015; You et al., 2013).

Nanotechnology also found its way into the oil and gas industry. Many hydrocarbon reservoirs pose a great challenge to produce and henceforth require more advanced methods. Nanoparticles were utilized in upstream sectors such as seismic exploration, drilling systems, production, and enhanced oil recovery (EOR). It was observed that applying nanotechnology in the upstream greatly improved the capacity of extracted hydrocarbons. This benefit was also observed in unconventional reservoirs, which are harder to produce than conventional ones (Yekeen et al., 2019). Furthermore, nanomaterials were also used in midstream and downstream sectors such as transportation and storage of hydrocarbons, and other refining and processing applications. These applications will be covered later in this chapter.

Nanomaterials show a promising future in different applications because of their exceptional chemical and physical characteristics. Therefore, the goal of this chapter is to highlight the different uses and contributions of nanotechnology in the petroleum industry. The next section covers the relevant types of nanomaterials. It is then followed by another section that covers the different applications of nanotechnology in the petroleum industry, in all the three main sectors: upstream, midstream, and downstream. The fourth section gives a brief discussion on the challenges that this technology might face before being fully utilized in real field projects. The last section is the conclusion where it summarizes the main takeaway from this chapter.

Figure 2. Schematic representation of the formation of nanoparticles

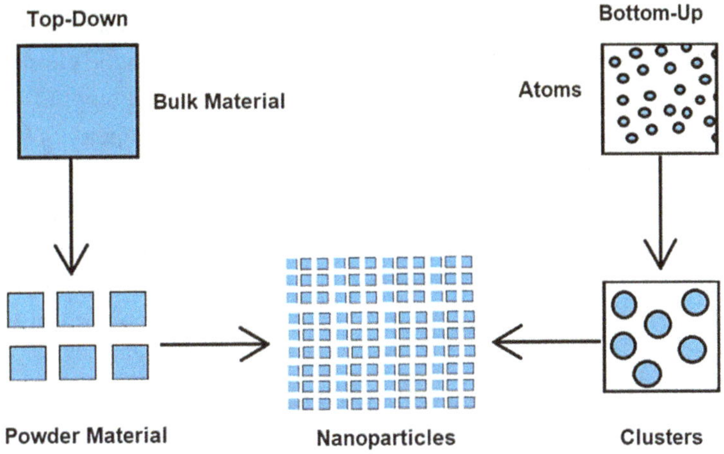

BACKGROUND

Significant attempts have been made to synthesize nanomaterials with precise control over their shape, size, and other properties. Nanoparticles may be manufactured using techniques such as top-down and bottom-up methods as shown in Figure 2. The top-down synthesis technique involves breaking down large particles into smaller pieces using different physical or chemical processes (Mittal et al., 2013). In the bottom-up technique, smaller nanoparticles are mixed to produce the required nanomaterial using either biological or chemical processes (Thakkar et al., 2010). The petroleum sector uses a wide range of nanomaterials, including carbon nanomaterials, metallic and metal oxide nanoparticles, magnetic nanoparticles, and nanocomposites (Maity et al., 2010).

Carbon nanomaterials such as activated carbon, carbon nanotubes, graphenes, and others attracted considerable attention in the petroleum industry. This is because of their electrical and mechanical properties (high electrical conductivity, tensile strength, flexibility, and thermal conductivity (H, 2015). Carbon nanotubes (CNTs) are cylinder-shaped carbon structures that are produced by rolling graphene sheets. Their diameters range from a few nanometers to 50 nanometers, while their lengths range from a few micrometers to a few centimeters (Laurent et al., 2010). Arc discharge, chemical vapor deposition (CVD), electrolytic techniques, hydrothermal, Laser ablation, and sonochemical methods are all common ways for CNT synthesis, and each has its own set of benefits and drawbacks (Prasek et al., 2011). Afterwards, the CNTs that have been produced must be purified using the appropriate method (Duque et al., 2010).

Metallic nanoparticles, for example, have been utilized in a broad range of oil and gas applications, depending on the metal type, oxidation states, form, functionality, and other characteristics, and may be produced by using physical and chemical methods (Jamkhande et al., 2019). Various nanoparticles, including silver (Ag), gold (Au), copper (Cu), iridium (Ir), osmium (Os), palladium (Pd), platinum (Pt), and rhodium (Rh), were successfully produced using the metal salt reduction technique (Alsaba et al., 2020).

Metal oxide nanoparticles are regarded as ideal candidates in many sectors of the oil and gas industry because of their unique thermal conductivity and heat transmission characteristics. They are being used in different applications in the exploration and production phases as additives for different fluids. Some examples of these nanoparticles are aluminum oxide Al_2O_3, ceria CeO_2, ferric oxide Fe_2O_3, magnesia MgO, silica SiO_2, titania TiO_2, zirconia ZrO_2, and zinc oxide ZnO (Fedele et al., 2011). These nanoparticles along with the metallic ones were used in different upstream and midstream applications.

Magnetic nanoparticles are a type of nanomaterial made up of magnetic elements, including cobalt, iron, and nickel, as well as their oxides (Kandpal et al., 2014). Their potential applications are being researched and have been tested in sensors for characterizing the heterogeneity of hydrocarbon reservoirs (Araujo et al., 2020; Cheng et al., 2015; Rahmani, Bryant, Huh, Ahmadian, et al., 2015).

Co-precipitation, hydrothermal, microemulsion, sol-gel, solvothermal, sonochemical, and CVD techniques are used to synthesize both the magnetic and the metal oxide nanoparticles (Besenhard et al., 2021; Kandpal et al., 2014; Kumar & Sangwan, 2018; Malhotra et al., 2020; Song et al., 2015). Additional physical, chemical, and biological methods can be used to synthesize these nanoparticles (Archana S., 2021).

Furthermore, the oil and gas sector has intensively investigated nanocomposites, which are materials in which nanoparticles are integrated into a medium to improve certain characteristics. Many applications of nanocomposites in oil and gas drilling are discussed in this chapter, such as in drilling mud, enhanced oil recovery, fracturing fluids, and others.

APPLICATIONS OF NANOMATERIALS IN THE PETROLEUM INDUSTRY

Nanotechnologies can transform the fossil fuel industry in all its sectors: upstream, midstream, and downstream. This includes different activities such as exploration processes, extraction operations, Enhanced Oil Recovery (EOR) techniques, transportation methods, and refining systems (Kong & Ohadi, 2010; Vivien et al., 2019; Zou et al., 2019). It offers a diverse variety of options for the use of materials and technologies in the hydrocarbon sector. Nanosized substances, like solid nanocomposites, nonlinear fluid, functional nanoparticles, and fluid configurations, are critical to future advancements in technology (C. Li et al., 2021). In recent years, there has been a fast increase in the amount of research on nanoscale materials connected to the oil industry. The number of scientific publications produced on this subject demonstrates a prominent level of research activity even during the downtime of this industry. Engineered nanoparticles have triggered the interest of researchers because of their possible applications. Table 1 contains thirteen different studies on the use of nanoparticles in various disciplines, part of which is described in the following sections.

Hydrocarbon Exploration

Hydrocarbon exploration is the initial step in the search for hydrocarbons. It comprises huge risks and is very costly. The goal of this procedure is to locate hydrocarbon deposits under the earth's surface. However, it frequently offers unique obstacles, such as unexpected geology and unforeseeable hazards, which can significantly raise the entire cost of extraction. Numerous traditional sensing approaches are said to be limited in their ability to offer details about the reservoir since they are limited to just a few centimeters from the borehole as shown in Figure 3 (Kong & Ohadi, 2010). Modern geophysical techniques are incapable of producing subsurface images with high resolutions. They are also unable to reach far into reservoirs to gather critical reservoir information. Moreover, sensing methods, such as traditional electrical sensing devices, are frequently unable to give meaningful information when reser-

Table 1. Summary of studies that have been published in the literature

Topic	Nanomaterial	Objectives
Hydrocarbon Exploration	Nano-optical Fiber (Tokonami, 2021)	Determine reservoir pressure, temperature, and detect oil microbes.
	Hyperpolarized Porous Silicon Nanoparticles (Seo et al., 2018)	Oil imaging sensors in a reservoir.
	Magnetic Nanoparticle (Noqta et al., 2020)	Flood front, fluid contact, fracture, and hydrocarbon bypass detection.
Drilling and Completion	Silica nanoparticles between 1 and 100 nm (Lau et al., 2017)	Introduced to the drilling mud and slurry to reduce water intrusion and enhance the drilling process.
	Calcined clay (Murtaza et al., 2020)	Lower the cement's porosity and permeability and improve its compressive strength properties.
Production Operations	Polymer-nanoparticle mixture (Yekeen et al., 2019)	Increases the fracturing fluid viscosity for better performance. Then polymer separates from the nanoparticles to obtain a fluid with a lower viscosity which allows easier fluid extraction from the fracture.
	Calcium Carbonate $CaCO_3$ nanoparticles (F. Huang et al., 2019)	To create a highly stable emulsion that can then be emulsified in water using another type of nanoparticles or surfactants as a stabilizer for better matrix acidizing.
Enhanced Oil Recovery	Silica nanoparticles containing Al2O3 (Habibi et al., 2020)	Improve wettability and oil recovery factors.
	Metal oxides based (L. Li et al., 2016)	Enhance the viscosity and stability of fracturing fluid at elevated temperatures.
Midstream Processes	Zinc oxide nanoparticles (Kathuria et al., 2020)	Inhibits corrosion of equipment.
	Graphite nanofluids (Z. Deng et al., 2021)	Promotes hydrate formation for storing and transporting natural gas.
Downstream Processes	Nanocatalysts such as metal oxides (Kosari et al., 2017)	Promotes thermal cracking of asphaltenes.
	Ceramic Nanofiltration membranes (Yang et al., 2019)	Used in water purification. They can separate oils, emulsions, and silts.

voir conditions are severe. Despite recent advances in exploration technologies like three-dimensional and four-dimensional seismic surveys, innovative technologies that may reliably detect hydrocarbons and are simple to use, affordable to obtain, and non-destructive remain desirable.

Many scholars studied the unique advantage of nanoparticles as nanosensors. The notion of utilizing hyperpolarized silicon nanoparticles in hydrocarbon exploration was first proposed in 2007. These nanoparticles were previously investigated in biomedical applications as a possible replacement for magnetic resonance imaging (MRI) (Pudakalakatti et al., 2021).

The use of nano-optical fibers to detect microbes in subsurface reservoirs was proposed based on the idea of Resonance Raman Spectroscopy (RRS) (Tokonami, 2021). Such a method could be managed to partly assess reservoir characteristics such as temperatures, pressures, and water salinity because microorganisms can only exist in certain conditions.

The use of magnetic nanomaterials in electromagnetic reservoir monitoring (Noqta et al., 2020; Ryoo et al., 2012) is one of the most frequently explored technologies. Magnetic nanoparticles are thought to be capable of providing high resolutions during measurements by altering the magnetic properties of the rock formations and reservoir fluids at extremely low frequency (Rahmani, Bryant, Huh, Athey, et al.,

Figure 3. Comparison of vertical resolution and radial depth of investigation for different well logs (Diniz-Ferreira & Torres-Verdín, 2012)

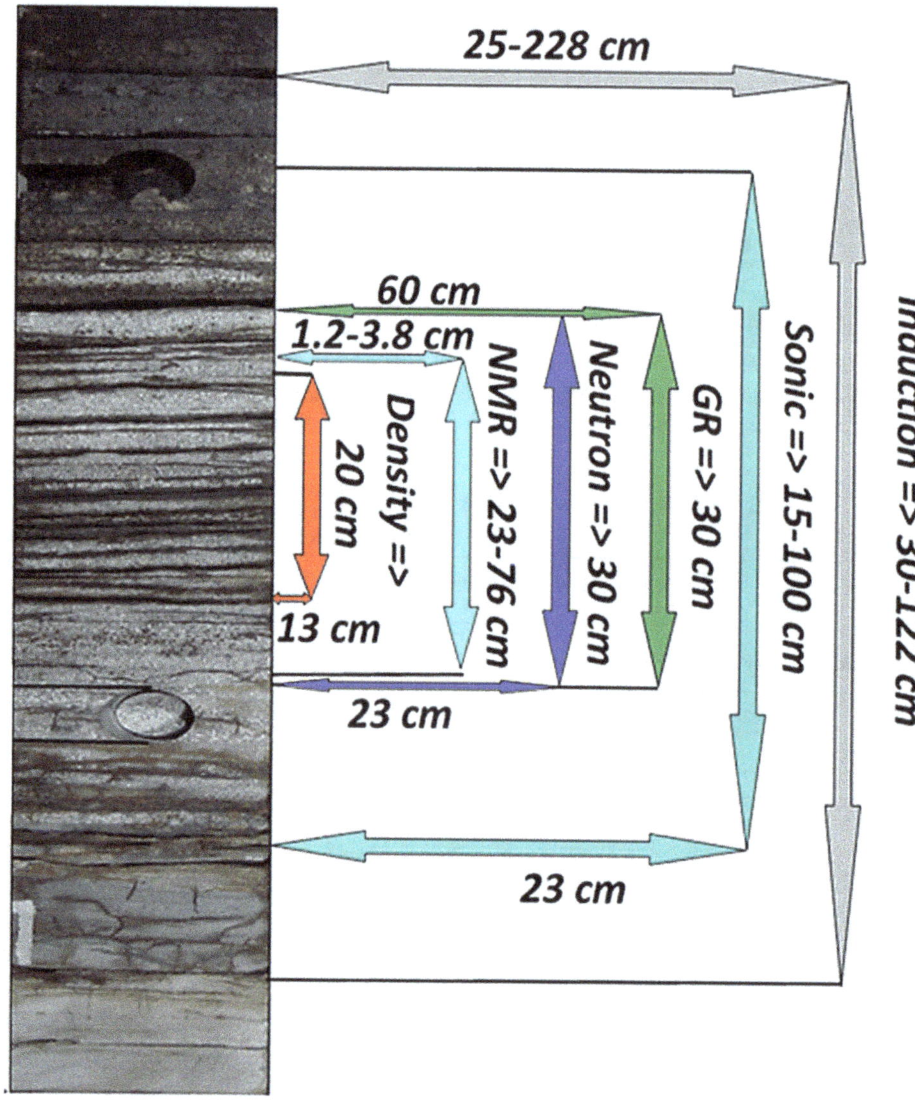

2015). Magnetic nanoparticles have been shown in recent research to be capable of probing the spread of immiscible fluids because they can be simply manufactured to be predominantly adsorbed at the oil and water contact (Al-Shehri et al., 2013; Ryoo et al., 2012). The displacement of nanoparticles that are adsorbed at the oil-water contact or gas-water contact and are subjected to additional magnetic fields causes interfacial fluctuations and pressure waves, which are used to detect the presence of nanoparticles. This technique is known as magneto-acoustic tomography.

Similarly, superparamagnetic iron oxide nanomaterials could be employed as a contrast material in nuclear magnetic resonance (NMR) imaging for characterizing rocks and detecting flood fronts (Rahmani, Bryant, Huh, Ahmadian, et al., 2015). In recent research, superparamagnetic iron oxide nanoparticles were shown to improve NMR logging contrast by reducing the NMR T2 relaxation time (Samrot et al.,

2021). The nanoparticles are intended to function as NMR nanosensors in this technique, with NMR relaxometry performed immediately within the formations at the borehole or on fluid samples retrieved from a well. The nanomaterials that are pumped inside the reservoir are then adsorbed at the boundary between the immiscible fluids. Therefore, it is possible to evaluate variations in fluid characteristics as the well is put on production. Magnetic nanomaterials have also been used in other studies as contrast agents for underground magnetic sensing surveys and reservoir classification (Aderibigbe et al., 2016; Fleming, 2014).

One of the recent advancements in the application of nanotechnology in the petroleum industry is the use of reservoir nano-robots such as nano-detecting devices (LIU et al., 2016). These devices integrate a reservoir sensor, a micro-dynamic mechanism, and micro-signal transmission. Patents were awarded for different inventions of nano-robot systems in well logging and other measurement operations.

Drilling and Completion

For the last several years, nanotechnology has been extensively researched for use in drilling and completion operations, including its use in drilling mud, well cement, and drilling equipment. Some nanoparticles were offered as a solution to several drilling difficulties, including unconsolidated formations, shallow waters, borehole instability, and lost circulation (Amanullah & Al-Tahini, 2009). Drilling muds that assist boost borehole stability and enhance their rheological characteristics have been the most effective application of nanotechnology in drilling operations to date. A schematic representation of the main drilling equipment is shown in Figure 4.

Drilling Mud

Drilling mud is used in field operations for multiple reasons related to well control and drill bit cooling. While drilling a shale layer, it is difficult to utilize a water-based drilling mud of low cost. This is due to numerous risks of losing the borehole integrity as shale swells when water invasion takes place into rock voids. Since the salinity in shales has a significantly higher ionic strength compared to the drilling mud, a substantial osmotic pressure builds, forcing the mud to be spontaneously sucked into the shale pores (HUANG et al., 2021; Rafieepour et al., 2020). Colloidal particles are introduced into drilling mud to establish a mud cake which prevents water movement towards the region surrounding the wellbore. This process works for rocks having micron-sized pore diameter, but not for shale rock formations having nanosized pores. This is a result of excessive osmotic force, where physically clogging the pores is not sufficient to overcome water seepage (Lau et al., 2017). Silica nanoparticles having a size range between 1 – 100 nm were introduced to the drilling mud and slurry to reduce water intrusion more efficiently (Lau et al., 2017; Yadav et al., 2015). The physical blocking of the considerably smaller pores by trapped nanoparticles is thought to be the mechanism for preventing water penetration. The silica Nanoparticles reduced shale invasion by 10 up to 100 times, resulting in enhanced mud stability at high pressures and temperatures for a range of rheological properties (Medhi et al., 2020). Metallic and metal oxide nanoparticles, especially those that are iron-based and calcium-based (Contreras et al., 2014), and MgO and ZnO (Gurluk et al., 2013), have also been explored for improving the characteristics of drilling fluids (Aderibigbe et al., 2016; Gurluk et al., 2013). Adding Nanoparticles to drilling fluids was discovered to increase thermal stability and reduce filtration loss from the reservoir. This method was tested in the field with great success by MI-Swaco and Baker-Hughes (El Sherbeny et al., 2014;

Hoelscher et al., 2012). However, the gradual seepage of water with low-ionic strength deep into the shales continues. Additional research into developing higher efficiency nanoparticle surface coatings is required to decrease the osmotic pressures even more.

The use of nanoparticles as an emulsifier for drilling mud emulsions is another application where drilling fluids might be improved. In many applications, oil-based mud is employed to prevent shale formation from expanding when in contact with water-based mud. However, the oil-wet borehole and oil-coated filter cake can cause additional well-completion problems (Das & Al-Enezi, 2014). This can be overcome using reversible drilling mud emulsions, which are fluids that contain both, the oil characteristics needed for drilling and water qualities needed for completions. The reversible drilling mud was created to overcome the well-completion problems and has been utilized successfully in different applications (Binks, 2002; Kokal, 2018; Liu & Lu, 2019). The loss of drilling mud, alteration in rheological properties, lost circulation, and wellbore instability are all typical difficulties when drilling in severe environments, including high-temperature and high-pressure wells (HTHP), deep and ultra-deep water, and unconventional settings. Adding nanoparticles to drilling fluid, such as silica and iron oxides, decreased fluid loss, maintained optimal rheological characteristics, and increased wellbore strength (Carpenter, 2017; Contreras et al., 2014).

Oil Well-cement

Cement is used in field operations to seal off the zone between the casing and formation. Nanoparticles have demonstrated great benefits as cement additives. Many kinds of nanoparticles have been examined as cement additives, including alumina, magnesium oxide, nano-clay, and silica (Adesina, 2020; McElroy et al., 2019; Metaxa et al., 2021; Murtaza et al., 2019; Muzenski et al., 2019; Záleská et al., 2020). Functioning as an accelerator and enhancing the mechanical properties, silica nanoparticle has been shown to increase cement stability, rheological properties, and hydration degree (Adesina, 2020; Muzenski et al., 2019). The nanoparticle can also increase cement flexibility by lowering permeability and compressive strength (Murtaza et al., 2020; Záleská et al., 2020). In high-pressure high-temperature (HPHT) settings, a carbon nanotube (CNT) has been demonstrated to enhance cement compressive strength substantially (W. A. Khan et al., 2016).

Drilling Tools

Nanotechnology has a profound impact on drilling equipment through nano-modifications. Drilling operations can be improved by coating the equipment, such as drill bits, with nanoceramic particles, which improves and extends their life (Sengupta & Kumar, 2013). Another option is to increase drilling efficiency in tough and demanding environments by using nano-diamond (Wilson, 2012). Numerous patents for the application of nanomaterials in drilling operations were published, and it is expected that many others will be published as more innovative technologies are being discovered. Forthcoming technologies in drilling will be directed by advancements in nanotechnology (Baig et al., 2021; Khalil et al., 2017).

Figure 4. Schematic representation of the main drilling systems

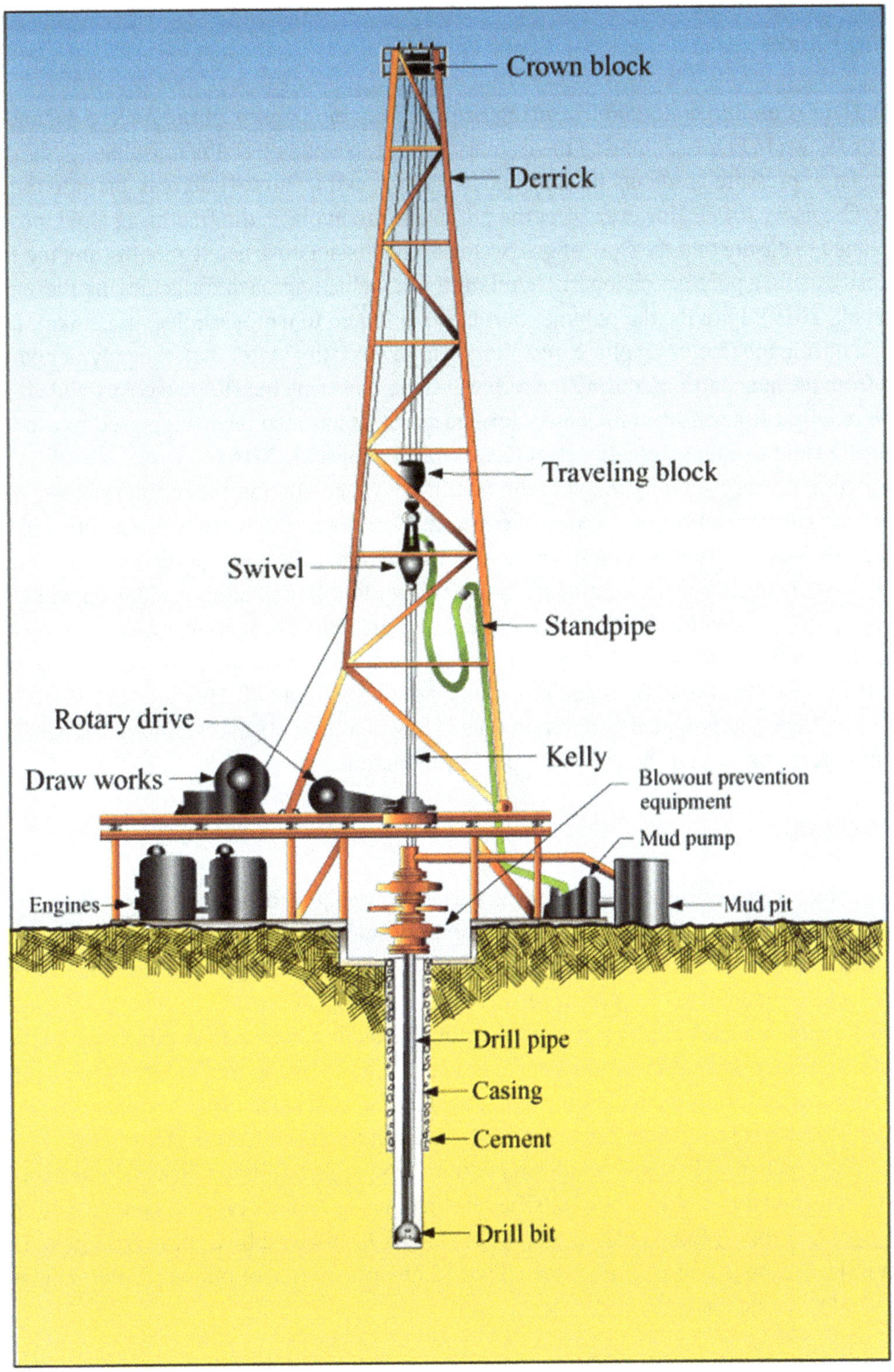

Production Operations

Fracturing Fluids

Fracturing fluid is used in hydraulic fracturing operations to break parts of the rock to enhance its permeability as shown in Figure 5. During the fracturing stage, fracturing fluids must have a high viscosity to produce large pressure gradients in the spreading crack and to limit fluid leak-off into the adjacent formation (Osiptsov, 2017). However, after the proppants are in place, the fracturing fluid must be thoroughly cleaned to ensure that the flow of gas, water, and oil is not obstructed. Controlling the fracturing fluid viscosity using a polymer nanoparticle mixture is a viable approach for achieving these dual aims (Yekeen et al., 2019). Initially, the polymer particles are linked to nanoparticles, in network form or as dendrimers, during the fracturing phase, resulting in high viscosity. After that, the polymer particles are separated from the nanoparticles during the removal stage, lowering the fluid viscosity and allowing the fluid to be readily extracted from the newly formed crack. It has also been suggested to use ferrofluid as a fracturing fluid to gain external control (Zawadzki & Bogacki, 2016).

The high use of water volumes in hydraulic fracturing, especially for unconventional resources such as shale oil and gas production, is a source of concern. Therefore, CO_2 foam has been utilized as a fracturing fluid in few applications to minimize water consumption. Silica nanoparticles were used to create stable CO_2 foams using low water volumes (around 2%) while still maintaining a high apparent viscosity around 100 cp (Rognmo et al., 2018; Xue et al., 2016). Furthermore, nanomaterials were used in solving additional problems in hydraulic fracturing operations. Proppants were coated with nanoparticles to control the migration of rock fines and proppants into the wellbore (T. Huang et al., 2010; Jatykov et al., 2021). The fines and proppants migration could cause sanding off the wellbore and hence the well would require remedial actions before being put in production.

Matrix Acidizing

Matrix acidizing is a process used to enhance the permeability of the formation by using acids to dissolve mud solids and sediments within the rock. The transport of acid into the appropriate reservoir layer, while avoiding any damage to the wellbore tools or surface equipment, is a key challenge in matrix acidizing. A powerful acid is required to create high permeability networks in the matrix zones, but it will interact with surface equipment and wellbore materials prior to reaching the target depth. Utilizing two emulsions in the acid, such as acid-in-oil-in-water emulsion, is a promising method to avoid the issue. The surface equipment and wellbore materials do not contact the acid since the water phase contacts them. Surfactant-based double emulsions are notoriously difficult to create. On the other hand, nanoparticles may be used to create highly stable emulsions with nearly uniform droplet sizes (acid-in-oil), and the aforementioned emulsion can then be emulsified in water using another type of nanoparticles or surfactants as a stabilizer (N. N. Deng et al., 2013; He et al., 2019). It would be a very efficient well stimulation agent if double emulsions could be created with a modest volume of oil, where the fluid stability is sustained until reaching the targeted depths.

Figure 5. Schematic representation of hydraulic fracturing

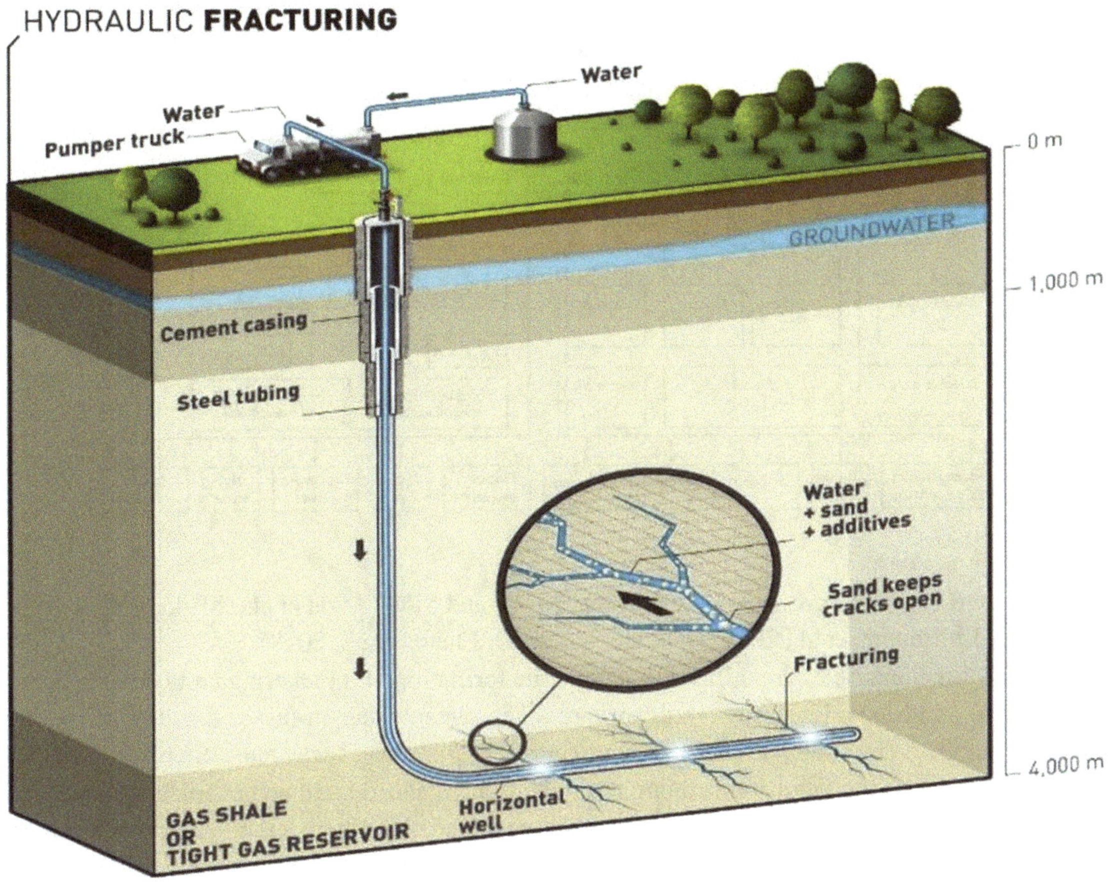

Enhanced Oil Recovery

Enhanced Oil Recovery (EOR) refers to advanced operations that can be used to enhance the production of partially depleted reservoirs. The main branches of EOR can be divided into four parts, gaseous EOR (Fahandezhsaadi et al., 2019; Hamouda & Chughtai, 2018), EOR using chemicals (Abbas et al., 2021; M. N. Khan et al., 2021), thermal EOR (Medina et al., 2019; Menad et al., 2019), and other advanced EOR methods (Ghaseminezhad & Sahraei, 2020; Suleimanov et al., 2020). Figure 6 shows a comprehensive list of different EOR methods.

Nanomaterials like nanofluids might be utilized to speed up hydrocarbon recovery during secondary or tertiary recovery. The use of nanoparticles for enhanced oil recovery is one of the most prevalent uses of nanotechnology in this field. Nanoparticles are thought to provide a better and more efficient approach to regulate and increase oil recovery than any other current or past technology (Kong & Ohadi, 2010). From core-displacement analysis, it was discovered that most nanomaterials performed better than surfactants and polymers in altering the rock's wettability. Compared to other EOR approaches like microbiological and thermal recoveries, the increasing use of nanoparticles in chemical EOR has received great interest. Nanoparticles are widely used to increase oil sweep efficiency and driving fluid stability (Cheraghian et al., 2020; Panchal et al., 2021). Several nanoparticle types were found to have

Figure 6. Enhanced Oil Recovery methods
(Nazar et al., 2011)

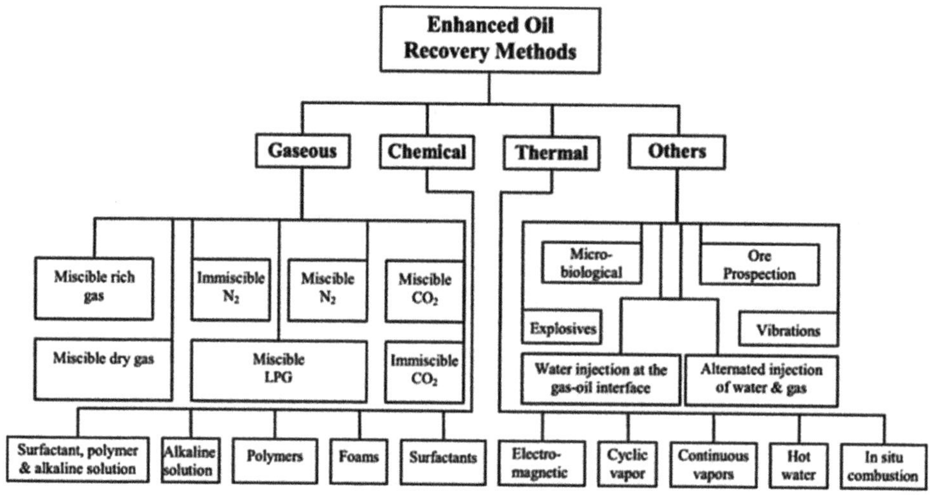

a promising EOR application such as Al_2O_3, Fe_2O_3, MgO, and SiO_2 (Afzal et al., 2014; Arab & Pouraf-shary, 2013; El-hoshoudy et al., 2019; Habibi et al., 2020; Llanos et al., 2018).

Additionally, nanoparticles are utilized to help in the formation of a Pickering emulsion. In contrast to conventional emulsions, Pickering emulsions are oil-in-water, water-in-oil, or a combination of the two, and are stabilized by solid particles like nanoparticles rather than surfactants. Stabilizing the solid nanomaterials in Pickering emulsion is thought to result in enhanced emulsion characteristics, such as strong resistance to agglomeration, high-temperature resilience, and high pH tolerance (Chevalier & Bolzinger, 2013; Dieng et al., 2019). Several organic and inorganic solids, such as barium sulfate, calcium carbonate, carbon nanotubes (CNT), cationic nanocrystals, clay, laponite, latex, magnetic nanoparticles, and protein, were employed as Pickering emulsion stabilizers in numerous studies (Angkuratipakorn et al., 2020; Ashby & Binks, 2000; Bao et al., 2019; Briggs et al., 2018; Gálvez-Vergara et al., 2020; F. Huang et al., 2019; Keogh et al., 2017; Yan et al., 2020). However, silica and magnetic nanomaterials are among the most popular particles employed to make a Pickering emulsion for EOR (Berezovska et al., 2021; Sharma et al., 2015). It is claimed that using nanoparticles in a Pickering emulsion improves emulsion stability and gives more control over viscous fingering effects during the EOR application. Moreover, nanoparticles can be utilized to stabilize CO_2-generated Pickering emulsions for CO_2-foam EOR flooding. Surface-modified silica nanoparticles were shown to considerably enhance the production of stable CO_2 foams (Yu et al., 2014).

Midstream Processes

Transportation of Oil and Gas

As oil is extracted from an offshore reservoir, it will be exposed to very cold temperatures close to the seafloor. When the temperature drops suddenly, waxes or asphaltenes precipitate out of the oil and may deposit on the inner walls of the pipes. This can severely interrupt the production process (Moncayo-Riascos et al., 2019). Likewise, methane hydrate buildup at the inner surface of natural gas-carrying pipelines is a major issue. To address the issue, electrical resistive heating is used with a heating jacket wrapped around the pipe, or the oil pipe is placed into another pipe to provide a secure insulating gap. When exposed to a magnetic field oscillation with a frequency ranging from 500 to 1000 MHz, nanoparticles may efficiently create high localized heat, which is known as "hyperthermia" (J. wook Kim et al., 2021). Researchers studied the feasibility of embedding nanoparticles in the paint that covers the inside of the subsea pipeline since hyperthermia heating may be well localized (Wang et al., 2019). Asphaltenes, hydrates, or wax can be formed at the boundary of the pipe's inner walls. By implementing magnetic field oscillations to the "nano-paint," the heated layer causes the deposits to dissociate from the wall of the pipe and be transported downstream by the flowing hydrocarbon.

Corrosion of equipment is another critical issue in the oil and gas industry, with millions of dollars spent each year to repair the damage it causes. Most reservoirs contain hydrogen sulfide, a highly poisonous and corrosive gas that might damage drilling equipment if it leaks into the drilling fluid. It can also affect the pipelines if not dealt with prior to transportation. Hydrogen sulfide may be effectively removed from drilling fluids using zinc oxide nanoparticles with a size of 14 – 25 nm. This nanoparticle is useful for avoiding equipment corrosion and for cleaning the environment (Kathuria et al., 2020). Other nanofluids showed an enhanced reduction in corrosion such as fluids containing ferromagnetic nanoparticles and iron oxide nanoparticles.

Additionally, recent studies assessed graphite nanofluids as means of promoting methane hydrate formation. The aim is to store and transport natural gas as solidified natural gas (SNG). By liquifying and solidifying the gas, larger quantities of hydrocarbons can be transported from one location to another using specially designed ships (without pipelines). This eliminates many issues related to pipeline transportation of natural gas such as the formation of methane hydrates that can block the flow inside the pipe (Z. Deng et al., 2021).

Downstream Processes

Nanocatalysts

Unlike the other sectors, the downstream has used nanotechnology for over a decade, mostly in the refining and treating processes. Sulfur and CO_2 emissions cause serious problems for the downstream business. This problem worsens as the need for energy upsurges to satisfy the rapidly growing demand (Kong & Ohadi, 2010). These difficulties encourage the industry to pursue more cost-effective, energy-efficient, and technologically oriented development (Khalil et al., 2017). The use of nanoparticle-based catalysts is one of the most frequent applications of nanomaterials in the petrochemical industries. Nanotechnology has made significant contributions to the development of more efficient and effective refining processes for transforming crude hydrocarbons into beneficial products such as diesel, gasoline, jet fuel, kerosene, liquefied petroleum gas (LPG), and other valuable chemical feedstocks.

As nanocatalysts provide a vast surface area for the required catalytic processes, the use of nanomaterials in the transformation and improvement of heavy crude oil and its derivatives has recently garnered great interest. Because of their excellent asphaltenes adsorption, oxidation, high oxygen storage, and release capability, several types of nanocatalysts, such as metal oxides, have been employed in crude oil hydroprocessing. Some studies looked into the impact of nickel oxide (NiO) and palladium oxide (PdO) supported on pyrogenic silica nanoparticles catalysts in thermal cracking of asphaltenes (Kosari et al.,

2017; Montoya et al., 2016). According to their findings, the addition of NiO or PdO enhanced catalytic activity compared to pyrogenic silica (fume) alone.

Nanofiltration

Nanofiltration (NF) is a membrane technique, commonly used to extract multi-valent cations and organic materials with a submicron size (Ko & Huh, 2019). NF filters out nanomaterials of sizes ranging between the method of reverse osmosis (RO), which may eliminate mono-valent ions, and the traditional filtration processes, which typically eliminate colloidal particles. As a pre-processing step for reverse osmosis, nanofiltration is often utilized along with RO for water desalination (Yang et al., 2019). NF membranes that are made up of ceramic are novel materials used in the purification of water. Their demand is growing because of their superior mechanical and technological capabilities. They can separate oils, emulsions, and silts, but they are susceptible to fouling by waxes and asphaltenes.

In the petroleum industry, nanofiltration is used in three primary areas. The first use is the removal of colloidal particles and organic debris from water that is produced from hydrocarbon reservoirs. This allows the water to be re-used for applications such as waterflooding, injection with chemicals, or proper dumping. The separation of asphaltenes from natural gas condensates is the second usage of nanofiltration in the industry. The existence of large-sized asphaltene particles reduces the value of condensate, making nanofiltration removal a more cost-effective alternative (L. Xu et al., 2018). Finally, several studies are looking at utilizing nanofiltration to remove H_2S and CO_2 from the natural gas stream.

FUTURE RESEARCH DIRECTIONS

Nanotechnology's implementation in the oil and gas sector has the potential to transform the existing technique of exploring and producing hydrocarbons from the subsurface. The synthesis of smart and unique collections of nanoparticles is steadily reaching every area of the oil and gas industry, thanks to the fast growth of nanotechnology in numerous applications.

Many existing exploration and production processes may benefit from the implementation of nanotechnology. However, there are still numerous obstacles in the oil and gas sector when it comes to applying these technologies. In particular, finding a simple and low-cost manufacturing process is one of the key problems in the production of nanomaterials for a variety of industrial applications, including the oil industry. Although several studies were conducted to discover synthetic methods for nanosized materials, exact control of nanoparticle's properties such as shape, size, and structure remains a challenge. Furthermore, most of the existing techniques continue to rely on inefficient equipment and environmentally unfriendly materials.

The cost-benefit of several types of nanomaterials compared to the existing technology is still quite low, especially in oil and gas applications. The desired outcomes in oilfield tests may only be attained by using a substantial dose of nanomaterials. However, using too many nanomaterials might raise the cost of operating substantially. As a result, the ability to produce massive quantities of nanoparticles simply and cost-effectively will become crucial for the widespread use of nanomaterials.

Additional obstacles that might be encountered in the future deployment of nanomaterials could be due to extreme subsurface environments. The ability of nanoparticles to cluster under severe circumstances such as severe temperatures and water salinity is well recognized. Therefore, the particle grain size will dramatically increase, and the particles' distinctive characteristics will be lost.

To solve this issue, modifying the surface of the nanoparticle or coating it can help. This can be accomplished by adding additional molecules to the nanoparticle surface or coating it with polymers to minimize the attraction force between the particles. However, many of these procedures will weaken or even eradicate numerous unique characteristics of nanoparticles. As a result, it appears that another strategy is required.

Another big challenge for the implementation of nanotechnology in the petroleum industry is the lack of knowledge of the migration and transportation behavior of nanoparticles through porous media. There are currently just a few publications focusing on the transportation of nanofluids in rocks. Some studies showed a lack of consistency in the results of different research and a lack of understanding of the fundamental theories. As a result, future studies should focus on the basic principles of both science and engineering, rather than just on how to enhance hydrocarbon production.

CONCLUSION

Nanotechnology offers a wide range of applications in a variety of fields. Nanotechnology applications have provided tremendous benefits to the oil and gas sector in recent years. The following are the three areas of nanotechnology that are most important to the upstream oil industry: Nanofluids in drilling mud, fracturing fluids, and enhanced oil recovery; nanosensors in different hydrocarbon exploration methods; structural nanomaterials as in strengthening the tools used in the industry. The oil and gas sector uses a variety of nanoparticles, including carbon nanomaterials, magnetic nanoparticles, metallic nanoparticles, and metal oxide nanoparticles.

Many obstacles must be solved in the development of these intriguing applications. For instance, nanomaterials are still relatively costly for several applications. However, those reservoir operations need a huge amount of nanomaterials, which might decrease the production cost. Another obstacle is the environmental risk of depositing nanoparticles in reservoirs. All nanomaterials that will be injected into the reservoir should be made environmentally friendly such as iron oxide and silica nanoparticles. It seems that the problems posed by nanomaterials still require further investigations before they can be properly implemented in field operations.

REFERENCES

Abbas, A. H., Elhag, H. H., Sulaiman, W. R. W., Gbadamosi, A., Pourafshary, P., Ebrahimi, S. S., Alqohaly, O. Y., & Agi, A. (2021). Modelling of continuous surfactant flooding application for marginal oilfields: A case study of Bentiu reservoir. *Journal of Petroleum Exploration and Production Technology*. Advance online publication. doi:10.100713202-020-01077-y

Aderibigbe, A., Cheng, K., Heidari, Z., Killough, J., & Fuss-Dezelic, T. (2016). Application of magnetic nanoparticles mixed with propping agents in enhancing near-wellbore fracture detection. *Journal of Petroleum Science Engineering*, *141*, 133–143. Advance online publication. doi:10.1016/j.petrol.2016.01.027

Adesina, A. (2020). Overview of Workability and Mechanical Performance of Cement-Based Composites Incorporating Nanomaterials. Silicon. doi:10.100712633-020-00800-y

Afzal, S., Nikookar, M., Ehsani, M. R., & Roayaei, E. (2014). An Experimental Investigation of the Catalytic Effect of Fe2O3 Nanoparticle on Steam Injection Process of an Iranian Reservoir. *Iranian Journal of Oil & Gas Science and Technology*, *3*(2). Advance online publication. doi:10.22050/ijogst.2014.6033

Agista, M. N., Guo, K., & Yu, Z. (2018). A state-of-the-art review of nanoparticles application in petroleum with a focus on enhanced oil recovery. Applied Sciences, 8(6). doi:10.3390/app8060871

Al-Shehri, A. A., Ellis, E. S., Servin, J. M. F., Kosynkin, D. V., Kanj, M. Y., & Schmidt, H. K. (2013). Illuminating the reservoir: Magnetic nanomappers. *SPE Middle East Oil and Gas Show and Conference, MEOS, Proceedings, 3*. 10.2118/164461-MS

Alsaba, M. T., Al Dushaishi, M. F., & Abbas, A. K. (2020). A comprehensive review of nanoparticles applications in the oil and gas industry. Journal of Petroleum Exploration and Production Technology, 10(4). doi:10.100713202-019-00825-z

Amanullah, M., & Al-Tahini, A. M. (2009). Nano-technology- its significance in smart fluid development for oil and gas field application. *Society of Petroleum Engineers - SPE Saudi Arabia Section Technical Symposium 2009*. 10.2118/126102-MS

Angkuratipakorn, T., Chung, C., Koo, C. K. W., Mundo, J. L. M., McClements, D. J., Decker, E. A., & Singkhonrat, J. (2020). Development of food-grade Pickering oil-in-water emulsions: Tailoring functionality using mixtures of cellulose nanocrystals and lauric arginate. *Food Chemistry*, *327*, 127039. Advance online publication. doi:10.1016/j.foodchem.2020.127039 PMID:32454273

Arab, D., & Pourafshary, P. (2013). Nanoparticles-assisted surface charge modification of the porous medium to treat colloidal particles migration induced by low salinity water flooding. *Colloids and Surfaces. A, Physicochemical and Engineering Aspects*, *436*, 803–814. Advance online publication. doi:10.1016/j.colsurfa.2013.08.022

Araujo, J. F. D. F., Tahir, Arsalani, S., Freire, F. L. Jr, Mariotto, G., Cremona, M., Mendoza, L. A. F., Luz-Lima, C., Zaman, Q., Del Rosso, T., Baffa, O., & Bruno, A. C. (2020). Novel scanning magnetic microscopy method for the characterization of magnetic nanoparticles. *Journal of Magnetism and Magnetic Materials*, *499*, 166300. Advance online publication. doi:10.1016/j.jmmm.2019.166300

Archana, S. (2021). A Comparative Study of Iron Oxide Nanoparticles Surface Modified Using Carboxylic Acids. *International Journal for Research in Applied Sciences and Biotechnology*, *8*(1), 116–125. Advance online publication. doi:10.31033/ijrasb.8.1.13

Ashby, N. P., & Binks, B. P. (2000). Pickering emulsions stabilised by Laponite clay particles. *Physical Chemistry Chemical Physics*, *2*(24), 5640–5646. Advance online publication. doi:10.1039/b007098j

Baig, N., Kammakakam, I., Falath, W., & Kammakakam, I. (2021). Nanomaterials: A review of synthesis methods, properties, recent progress, and challenges. Materials Advances, 2(6). doi:10.1039/D0MA00807A

Bao, Y., Zhang, Y., Liu, P., Ma, J., Zhang, W., Liu, C., & Simion, D. (2019). Novel fabrication of stable Pickering emulsion and latex by hollow silica nanoparticles. *Journal of Colloid and Interface Science, 553*, 83–90. Advance online publication. doi:10.1016/j.jcis.2019.06.008 PMID:31195217

Berezovska, I., Kapilov, K., Dhavalikar, P., Cosgriff-Hernandez, E., & Silverstein, M. S. (2021). Reactive Surfactants for Achieving Open-Cell PolyHIPE Foams from Pickering Emulsions. *Macromolecular Materials and Engineering, 306*(6), 2000825. Advance online publication. doi:10.1002/mame.202000825

Besenhard, M. O., Panariello, L., Kiefer, C., Lagrow, A. P., Storozhuk, L., Perton, F., Begin, S., Mertz, D., Thanh, N. T. K., & Gavriilidis, A. (2021). Small iron oxide nanoparticles as MRI T 1contrast agent: Scalable inexpensive water-based synthesis using a flow reactor. *Nanoscale, 13*(19), 8795–8805. Advance online publication. doi:10.1039/D1NR00877C PMID:34014243

Binks, B. P. (2002). Particles as surfactants - Similarities and differences. *Current Opinion in Colloid & Interface Science, 7*(1–2), 21–41. Advance online publication. doi:10.1016/S1359-0294(02)00008-0

Bou-Hamdan, K. (2020). *Key Design Considerations for Maximizing the Recovery Rate of Unconventional Reservoirs. The Way Ahead.* SPE.

Bou Hamdan, K. F., Harkouss, R., & Chakra, H. A. (2015, November 16). An overview of Extended Reach Drilling: Focus on design considerations and drag analysis. *Mediterranean Gas and Oil International Conference, MedGO 2015 - Conference Proceedings.* 10.1109/MedGO.2015.7330328

Briggs, N., Raman, A. K. Y., Barrett, L., Brown, C., Li, B., Leavitt, D., Aichele, C. P., & Crossley, S. (2018). Stable pickering emulsions using multi-walled carbon nanotubes of varying wettability. *Colloids and Surfaces. A, Physicochemical and Engineering Aspects, 537*, 227–235. Advance online publication. doi:10.1016/j.colsurfa.2017.10.010

Carpenter, C. (2017). New Magnetite Nanoparticles Allow Smart Drilling Fluids with Superior Properties. *Journal of Petroleum Technology, 69*(11), 41–42. Advance online publication. doi:10.2118/1117-0041-JPT

Cheng, K. K., Chan, P. S., Fan, S., Kwan, S. M., Yeung, K. L., Wáng, Y. X. J., Chow, A. H. L., Wu, E. X., & Baum, L. (2015). Curcumin-conjugated magnetic nanoparticles for detecting amyloid plaques in Alzheimer's disease mice using magnetic resonance imaging (MRI). *Biomaterials, 44*, 155–172. Advance online publication. doi:10.1016/j.biomaterials.2014.12.005 PMID:25617135

Cheraghian, G., Rostami, S., & Afrand, M. (2020). Nanotechnology in enhanced oil recovery. Processes, 8(9). doi:10.3390/pr8091073

Chevalier, Y., & Bolzinger, M. A. (2013). Emulsions stabilized with solid nanoparticles: Pickering emulsions. *Colloids and Surfaces. A, Physicochemical and Engineering Aspects, 439*, 23–34. Advance online publication. doi:10.1016/j.colsurfa.2013.02.054

Contreras, O., Hareland, G., Husein, M., Nygaard, R., & Alsaba, M. (2014). Application of in-house prepared nanoparticles as filtration control additive to reduce formation damage. *SPE - European Formation Damage Conference, Proceedings, EFDC*, 1. 10.2118/168116-MS

Das, O. P., & Al-Enezi, K. (2014). Successful application of innovative reversible invert emulsion reservoir drilling fluid and improved filter-cake cleanup agent increased sustained well productivity in smart multilateral wells with inflow mart completion in high mobility reservoir, west Kuwait. *Society of Petroleum Engineers - 30th Abu Dhabi International Petroleum Exhibition and Conference, ADIPEC 2014: Challenges and Opportunities for the Next 30 Years, 5*. 10.2118/172030-MS

Deng, N. N., Wang, W., Ju, X. J., Xie, R., Weitz, D. A., & Chu, L. Y. (2013). Wetting-induced formation of controllable monodisperse multiple emulsions in microfluidics. *Lab on a Chip*, *13*(20), 4047–4052. Advance online publication. doi:10.1039/C3LC50638J PMID:23948718

Deng, Z., Wang, Y., Yu, C., Li, G., Lang, X., Wang, S., & Fan, S. (2021). Promoting methane hydrate formation with expanded graphite additives: Application to solidified natural gas storage. *Fuel*, *299*, 120867. Advance online publication. doi:10.1016/j.fuel.2021.120867

Dieng, S. M., Anton, N., Bouriat, P., Thioune, O., Sy, P. M., Massaddeq, N., Enharrar, S., Diarra, M., & Vandamme, T. (2019). Pickering nano-emulsions stabilized by solid lipid nanoparticles as a temperature sensitive drug delivery system. *Soft Matter*, *15*(40), 8164–8174. Advance online publication. doi:10.1039/C9SM01283D PMID:31593197

Diniz-Ferreira, E. L., & Torres-Verdín, C. (2012). Improved estimation of pore connectivity and permeability in deepwater carbonates with the construction of multi-layer static and dynamic petrophysical models. *SPWLA 53rd Annual Logging Symposium 2012*.

Duque, J. G., Parra-Vasquez, A. N. G., Behabtu, N., Green, M. J., Higginbotham, A. L., Price, B. K., Leonard, A. D., Schmidt, H. K., Lounis, B., Tour, J. M., Doorn, S. K., Cognet, L., & Pasquali, M. (2010). Diameter-dependent solubility of single-walled carbon nanotubes. *ACS Nano*, *4*(6), 3063–3072. Advance online publication. doi:10.1021/nn100170f PMID:20521799

El-hoshoudy, A. N., Gomaa, S., & Selim, O. A. (2019). Application of acrylamide polymer grafted with SiO2 nanoparticles in enhanced oil recovery-design project. *Petroleum and Coal*, *61*(6).

El Sherbeny, W., Al-Baddaly, H., Rahal, A., Said, M., Hardman, D., & Henry, T. (2014). Nanotechnology applications to minimize geo-mechanical related challenges while drilling intercalated sediments, Western Desert, Egypt. *Society of Petroleum Engineers - SPE Saudi Arabia Section Technical Symposium and Exhibition*. 10.2118/172242-MS

Fahandezhsaadi, M., Amooie, M. A., Hemmati-Sarapardeh, A., Ayatollahi, S., Schaffie, M., & Ranjbar, M. (2019). Laboratory evaluation of nitrogen injection for enhanced oil recovery: Effects of pressure and induced fractures. *Fuel*, *253*, 607–614. Advance online publication. doi:10.1016/j.fuel.2019.05.039

Fedele, L., Colla, L., Bobbo, S., Barison, S., & Agresti, F. (2011). Experimental stability analysis of different water-based nanofluids. *Nanoscale Research Letters*, *6*(1), 300. Advance online publication. doi:10.1186/1556-276X-6-300 PMID:21711817

Fleming, N. (2014). Technology Focus: Nanotechnology (February 2014). *Journal of Petroleum Technology*, *66*(02), 116. Advance online publication. doi:10.2118/0214-0116-JPT

Gálvez-Vergara, A., Fresco-Cala, B., & Cárdenas, S. (2020). Switchable Pickering emulsions stabilized by polystyrene-modified magnetic nanoparticles. *Colloids and Surfaces. A, Physicochemical and Engineering Aspects, 606*, 125462. Advance online publication. doi:10.1016/j.colsurfa.2020.125462

Ghaseminezhad, H., & Sahraei, E. (2020). Experimental investigation of foam assisted water alternate gas injection in relation to enhanced oil recovery applications. *Petroleum Science and Technology, 38*(17), 859–869. Advance online publication. doi:10.1080/10916466.2020.1790599

Gurluk, M. R., Wang, G., Nasr-Ei-din, H. A., & Crews, J. (2013). The effect of different brine solutions on the viscosity of VES micelles. *SPE - European Formation Damage Conference, Proceedings, EFDC*, 2. 10.2118/165164-MS

H, S. (2015). Synthesis of Carbon Nanotubes for Oil-water Interfacial Tension Reduction. *Oil & Gas Research, 1*(1). doi:10.4172/2472-0518.1000104

Habibi, S., Jafari, A., & Fakhroueian, Z. (2020). Application of novel functionalized Al2O3/silica by organosiloxane and amine reagents for enhanced oil recovery. *Applied Nanoscience (Switzerland), 10*(7), 2085–2100. Advance online publication. doi:10.100713204-020-01337-7

Hamouda, A. A., & Chughtai, S. (2018). Miscible CO2 flooding for EOR in the presence of natural gas components in displacing and displaced fluids. *Energies, 11*(2), 391. Advance online publication. doi:10.3390/en11020391

Hassani, S. S., Daraee, M., & Sobat, Z. (2020). Advanced development in upstream of petroleum industry using nanotechnology. Chinese Journal of Chemical Engineering, 28(6). doi:10.1016/j.cjche.2020.02.030

He, F., Zhang, M. J., Wang, W., Cai, Q. W., Su, Y. Y., Liu, Z., Faraj, Y., Ju, X. J., Xie, R., & Chu, L. Y. (2019). Designable Polymeric Microparticles from Droplet Microfluidics for Controlled Drug Release. Advanced Materials Technologies, 4(6). doi:10.1002/admt.201800687

Hoelscher, K. P., De Stefano, G., Riley, M., & Young, S. (2012). Application of nanotechnology in drilling fluids. *Society of Petroleum Engineers - SPE International Oilfield Nanotechnology Conference 2012*. 10.2118/157031-MS

Huang, B., Li, C., Zhang, Y., Ding, W., Yang, M., Yang, Y., Zhai, H., Xu, X., Wang, D., Debnath, S., Jamil, M., Li, H. N., Ali, H. M., Gupta, M. K., & Said, Z. (2021). Advances in fabrication of ceramic corundum abrasives based on sol–gel process. *Chinese Journal of Aeronautics, 34*(6), 1–17. Advance online publication. doi:10.1016/j.cja.2020.07.004

Huang, F., Liang, Y., & He, Y. (2019). On the Pickering emulsions stabilized by calcium carbonate particles with various morphologies. *Colloids and Surfaces. A, Physicochemical and Engineering Aspects, 580*, 123722. Advance online publication. doi:10.1016/j.colsurfa.2019.123722

Huang, T., Evans, B. A., Crews, J. B., & Belcher, C. K. (2010). Field case study on formation fines control with nanoparticles in offshore wells. *Proceedings - SPE Annual Technical Conference and Exhibition, 5*. 10.2118/135088-MS

Jamkhande, P. G., Ghule, N. W., Bamer, A. H., & Kalaskar, M. G. (2019). Metal nanoparticles synthesis: An overview on methods of preparation, advantages and disadvantages, and applications. Journal of Drug Delivery Science and Technology, 53. doi:10.1016/j.jddst.2019.101174

Jatykov, T., Sagindykov, B., & Bimuratkyzy, K. (2021). Hydraulic fracturing design improvements by proppant flowback mitigation method application during hydrocarbon production. *Rudarsko Geolosko Naftni Zbornik*, *36*(1), 99–109. Advance online publication. doi:10.17794/rgn.2021.1.8

Kandpal, N. D., Sah, N., Loshali, R., Joshi, R., & Prasad, J. (2014). Co-precipitation method of synthesis and characterization of iron oxide nanoparticles. *Journal of Scientific and Industrial Research*, *73*(2).

Kathuria, L., Din Reshi, N. U., & Samuelson, A. G. (2020). N-Heterocyclic Carbene (NHC)-Stabilized Ru0 Nanoparticles: In Situ Generation of an Efficient Transfer Hydrogenation Catalyst. *Chemistry (Weinheim an der Bergstrasse, Germany)*, *26*(34), 7622–7630. Advance online publication. doi:10.1002/chem.202000142 PMID:32048353

Keogh, W., Neville, A., Huggan, M., Eroini, V., Olsen, J. H., Nielsen, F. M., Baraka-Lokmane, S., Bourdelet, E., Ellingsen, J. A., Bache, O., & Charpentier, T. (2017). Deposition of Inorganic Carbonate, Sulfate, and Sulfide Scales on Antifouling Surfaces in Multiphase Flow. *Energy & Fuels*, *31*(11), 11838–11851. Advance online publication. doi:10.1021/acs.energyfuels.7b02165

Khalil, M., Jan, B. M., Tong, C. W., & Berawi, M. A. (2017). Advanced nanomaterials in oil and gas industry: Design, application and challenges. Applied Energy, 191. doi:10.1016/j.apenergy.2017.01.074

Khan, M. N., Wan Sulaiman, W. R., & Abbas, A. H. (2021). Study of Sulfosuccinate and Extended Sulfated Sodium Surfactants on the Malaysian Crude/Water Properties for ASP Application in Limestone. *Arabian Journal for Science and Engineering*, *46*(7), 6915–6924. Advance online publication. doi:10.100713369-020-05252-5

Khan, W. A., Rahman, M. K., Mahmoud, M. A., & Sarmah, P. (2016). MWCNT for enhancing mechanical properties of oil well cement for HPHT applications. *Proceedings of the SPE/IADC Middle East Drilling Technology Conference and Exhibition*. 10.2118/178175-MS

Kim, B. H., Hackett, M. J., Park, J., & Hyeon, T. (2014). Synthesis, characterization, and application of ultrasmall nanoparticles. Chemistry of Materials, 26(1). doi:10.1021/cm402225z

Kim, J., Wang, J., Kim, H., & Bae, S. (2021). Concentration-dependent oscillation of specific loss power in magnetic nanofluid hyperthermia. *Scientific Reports*, *11*(1), 733. Advance online publication. doi:10.103841598-020-79871-1 PMID:33436765

Ko, S., & Huh, C. (2019). Use of nanoparticles for oil production applications. Journal of Petroleum Science and Engineering, 172. doi:10.1016/j.petrol.2018.09.051

Kokal, S. (2018). Technology Focus: CO2 Applications (July 2018). *Journal of Petroleum Technology*, *70*(07), 69. Advance online publication. doi:10.2118/0718-0069-JPT

Kong, X., & Ohadi, M. M. (2010). Applications of micro and nano technologies in the oil and gas industry-an overview of the recent progress. *Society of Petroleum Engineers - 14th Abu Dhabi International Petroleum Exhibition and Conference 2010, ADIPEC 2010, 3*. 10.2118/138241-MS

Kosari, M., Golmohammadi, M., Ahmadi, S. J., Towfighi, J., & Chenari, A. H. (2017). On the catalysis capability of transition metal oxide nanoparticles in upgrading of heavy petroleum residue by supercritical water. *The Journal of Supercritical Fluids, 126*, 14–24. Advance online publication. doi:10.1016/j.supflu.2017.02.021

Kumar, H., & Sangwan, P. (2018). Synthesis and Characterization of Cobalt Oxide Nanoparticles by Sol-Gel Method. *International Journal of Advanced Research in Science and Engineering, 7.*

Lau, H. C., Yu, M., & Nguyen, Q. P. (2017). Nanotechnology for oilfield applications: Challenges and impact. *Journal of Petroleum Science Engineering, 157*, 1160–1169. Advance online publication. doi:10.1016/j.petrol.2017.07.062

Laurent, C., Flahaut, E., & Peigney, A. (2010). The weight and density of carbon nanotubes versus the number of walls and diameter. Carbon, 48(10). doi:10.1016/j.carbon.2010.04.010

Li, C., Hassan, A., Palmai, M., Xie, Y., Snee, P. T., Powell, B. A., Murdoch, L. C., & Darnault, C. J. G. (2021). Experimental measurements and numerical simulations of the transport and retention of nanocrystal CdSe/ZnS quantum dots in saturated porous media: Effects of pH, organic ligand, and natural organic matter. *Environmental Science and Pollution Research International, 28*(7), 8050–8073. Advance online publication. doi:10.100711356-020-11097-0 PMID:33051847

Li, L., Al-Muntasheri, G. A., & Liang, F. (2016). Nanomaterials-enhanced high-temperature fracturing fluids prepared with untreated seawater. *Proceedings - SPE Annual Technical Conference and Exhibition*. 10.2118/181283-MS

Liu, H., Jin, X., & Ding, B. (2016). Application of nanotechnology in petroleum exploration and development. *Petroleum Exploration and Development, 43*(6), 1107–1115. Advance online publication. doi:10.1016/S1876-3804(16)30129-X

Liu, X., & Lu, Y. (2019). Tailoring Emulsion Polymerization for High-Yield Synthesis of Tween 80 Stabilized Magnetic Cross-Linked Polystyrene Nanocomposite Particles. *Industrial & Engineering Chemistry Research, 58*(19), 8140–8147. Advance online publication. doi:10.1021/acs.iecr.9b01103

Llanos, S., Giraldo, L. J., Santamaria, O., Franco, C. A., & Cortés, F. B. (2018). Effect of Sodium Oleate Surfactant Concentration Grafted onto SiO2 Nanoparticles in Polymer Flooding Processes. *ACS Omega, 3*(12), 18673–18684. Advance online publication. doi:10.1021/acsomega.8b02944 PMID:31458433

Maity, S. K., Ancheyta, J., & Marroquín, G. (2010). Catalytic aquathermolysis used for viscosity reduction of heavy crude oils: A review. *Energy & Fuels, 24*(5), 2809–2816. Advance online publication. doi:10.1021/ef100230k

Malhotra, N., Lee, J. S., Liman, R. A. D., Ruallo, J. M. S., Villaflore, O. B., Ger, T. R., & Der Hsiao, C. (2020). Potential toxicity of iron oxide magnetic nanoparticles: A review. Molecules, 25(14). doi:10.3390/molecules25143159

McElroy, P., Emadi, H., Surowiec, K., & Casadonte, D. J. (2019). Mechanical, rheological, and stability performance of simulated in-situ cured oil well cement slurries reinforced with alumina nanofibers. *Journal of Petroleum Science Engineering, 183*, 106415. Advance online publication. doi:10.1016/j.petrol.2019.106415

Medhi, S., Chowdhury, S., Gupta, D. K., & Mazumdar, A. (2020). An investigation on the effects of silica and copper oxide nanoparticles on rheological and fluid loss property of drilling fluids. *Journal of Petroleum Exploration and Production Technology*, *10*(1), 91–101. Advance online publication. doi:10.100713202-019-0721-y

Medina, O. E., Olmos, C., Lopera, S. H., Cortés, F. B., & Franco, C. A. (2019). Nanotechnology applied to thermal enhanced oil recovery processes: A review. *Energies*, *12*(24), 4671. Advance online publication. doi:10.3390/en12244671

Menad, N. A., Noureddine, Z., Hemmati-Sarapardeh, A., Shamshirband, S., Mosavi, A., & Chau, K. (2019). Modeling temperature dependency of oil - water relative permeability in thermal enhanced oil recovery processes using group method of data handling and gene expression programming. *Engineering Applications of Computational Fluid Mechanics*, *13*(1), 724–743. Advance online publication. doi:10.1080/19942060.2019.1639549

Metaxa, Z. S., Tolkou, A. K., Efstathiou, S., Rahdar, A., Favvas, E. P., Mitropoulos, A. C., & Kyzas, G. Z. (2021). Nanomaterials in cementitious composites: An update. *Molecules (Basel, Switzerland)*, *26*(5), 1430. Advance online publication. doi:10.3390/molecules26051430 PMID:33800797

Mittal, A. K., Chisti, Y., & Banerjee, U. C. (2013). Synthesis of metallic nanoparticles using plant extracts. Biotechnology Advances, 31(2). doi:10.1016/j.biotechadv.2013.01.003

Moncayo-Riascos, I., De Leon, J., Garcia-Martinez, J. A., Garcia-Cruz, I., & Lira-Galeana, C. (2019). Multiscale simulation of asphaltene deposition in pipeline flows. *Journal of Petroleum Science Engineering*, *183*, 106376. Advance online publication. doi:10.1016/j.petrol.2019.106376

Montoya, T., Argel, B. L., Nassar, N. N., Franco, C. A., & Cortés, F. B. (2016). Kinetics and mechanisms of the catalytic thermal cracking of asphaltenes adsorbed on supported nanoparticles. *Petroleum Science*, *13*(3), 561–571. Advance online publication. doi:10.100712182-016-0100-y

Mout, R., Moyano, D. F., Rana, S., & Rotello, V. M. (2012). Surface functionalization of nanoparticles for nanomedicine. *Chemical Society Reviews*, *41*(7), 2539. Advance online publication. doi:10.1039/c2cs15294k PMID:22310807

Murtaza, M., Mahmoud, M., Elkatatny, S., Al Majed, A., Chen, W., & Jamaluddin, A. (2019). Experimental investigation of the impact of modified nano clay on the rheology of oil well cement slurry. *International Petroleum Technology Conference 2019, IPTC 2019*. 10.2523/IPTC-19456-MS

Murtaza, M., Rahman, M. K., & Al Majed, A. A. (2020). The application and comparative study of calcined clay and nanoclaymixed cement slurries at hpht conditions. *Proceedings of the Annual Offshore Technology Conference*. 10.4043/30872-MS

Muzenski, S., Flores-Vivian, I., & Sobolev, K. (2019). Ultra-high strength cement-based composites designed with aluminum oxide nano-fibers. *Construction & Building Materials*, *220*, 177–186. Advance online publication. doi:10.1016/j.conbuildmat.2019.05.175

Nazar, M. F., Shah, S. S., & Khosa, M. A. (2011). Microemulsions in enhanced oil recovery: A review. *Petroleum Science and Technology*, *29*(13), 1353–1365. Advance online publication. doi:10.1080/10916460903502514

Noqta, O. A., Sodipo, B. K., & Aziz, A. A. (2020). One-pot synthesis of highly magnetic and stable citrate coated superparamagnetic iron oxide nanoparticles by modified coprecipitation method. *Functional Composites and Structures*, *2*(4), 045005. Advance online publication. doi:10.1088/2631-6331/abcdbc

Osiptsov, A. A. (2017). Fluid Mechanics of Hydraulic Fracturing: A Review. *Journal of Petroleum Science Engineering*, *156*, 513–535. Advance online publication. doi:10.1016/j.petrol.2017.05.019

Otsuka, H., Nagasaki, Y., & Kataoka, K. (2012). PEGylated nanoparticles for biological and pharmaceutical applications. *Advanced Drug Delivery Reviews* , 64. doi:10.1002/9781118135440.ch47

Panchal, H., Patel, H., Patel, J., & Shah, M. (2021). A systematic review on nanotechnology in enhanced oil recovery. Petroleum Research. doi:10.1016/j.ptlrs.2021.03.003

Prasek, J., Drbohlavova, J., Chomoucka, J., Hubalek, J., Jasek, O., Adam, V., & Kizek, R. (2011). Methods for carbon nanotubes synthesis - Review. Journal of Materials Chemistry, 21(40). doi:10.1039/c1jm12254a

Pudakalakatti, S., Enriquez, J. S., McCowan, C., Ramezani, S., Davis, J. S., Zacharias, N. M., Bourgeois, D., Constantinou, P. E., Harrington, D. A., Carson, D., Farach-Carson, M. C., & Bhattacharya, P. K. (2021). Hyperpolarized MRI with silicon micro and nanoparticles: Principles and applications. In *Wiley Interdisciplinary Reviews*. Nanomedicine and Nanobiotechnology. doi:10.1002/wnan.1722

Rafieepour, S., Zamiran, S., & Ostadhassan, M. (2020). A cost-effective chemo-thermo-poroelastic wellbore stability model for mud weight design during drilling through shale formations. *Journal of Rock Mechanics and Geotechnical Engineering*, *12*(4), 768–779. Advance online publication. doi:10.1016/j.jrmge.2019.12.008

Rahmani, A. R., Bryant, S., Huh, C., Athey, A., Ahmadian, M., Chen, J., & Wilt, M. (2015). Crosswell magnetic sensing of superparamagnetic nanoparticles for subsurface applications. *SPE Journal*, *20*(5), 1067–1082. Advance online publication. doi:10.2118/166140-PA

Rahmani, A. R., Bryant, S. L., Huh, C., Ahmadian, M., Zhang, W., & Liu, Q. H. (2015). Characterizing reservoir heterogeneities using magnetic nanoparticles. *Society of Petroleum Engineers - SPE Reservoir Simulation Symposium 2015, 1*. 10.2118/173195-MS

Rognmo, A. U., Heldal, S., & Fernø, M. A. (2018). Silica nanoparticles to stabilize CO2-foam for improved CO2 utilization: Enhanced CO2 storage and oil recovery from mature oil reservoirs. *Fuel*, *216*, 621–626. Advance online publication. doi:10.1016/j.fuel.2017.11.144

Ryoo, S., Rahmani, A. R., Yoon, K. Y., Prodanović, M., Kotsmar, C., Milner, T. E., Johnston, K. P., Bryant, S. L., & Huh, C. (2012). Theoretical and experimental investigation of the motion of multiphase fluids containing paramagnetic nanoparticles in porous media. *Journal of Petroleum Science Engineering*, *81*, 129–144. Advance online publication. doi:10.1016/j.petrol.2011.11.008

Samrot, A. V., Ali, H. H., & Selvarani, A. (2021). Adsorption efficiency of chemically synthesized Superparamagnetic Iron Oxide Nanoparticles (SPIONs) on crystal violet dye. *Current Research in Green and Sustainable Chemistry*, *4*, 100066. Advance online publication. doi:10.1016/j.crgsc.2021.100066

Sengupta, S., & Kumar, A. (2013). Nano-ceramic coatings - A means of enhancing bit life and reducing drill string trips. *Society of Petroleum Engineers - International Petroleum Technology Conference 2013, IPTC 2013: Challenging Technology and Economic Limits to Meet the Global Energy Demand, 1*. 10.2523/IPTC-16474-MS

Seo, H., Choi, I., Whiting, N., Hu, J., Luu, Q. S., Pudakalakatti, S., McCowan, C., Kim, Y., Zacharias, N., Lee, S., Bhattacharya, P., & Lee, Y. (2018). Hyperpolarized Porous Silicon Nanoparticles: Potential Theragnostic Material for 29Si Magnetic Resonance Imaging. *ChemPhysChem, 19*(17), 2143–2147. Advance online publication. doi:10.1002/cphc.201800461 PMID:29779258

Sharma, T., Kumar, G. S., & Sangwai, J. S. (2015). Comparative effectiveness of production performance of Pickering emulsion stabilized by nanoparticle-surfactant-polymerover surfactant-polymer (SP) flooding for enhanced oil recoveryfor Brownfield reservoir. *Journal of Petroleum Science Engineering, 129*, 221–232. Advance online publication. doi:10.1016/j.petrol.2015.03.015

Shen, W., Zhang, X., Huang, Q., Xu, Q., & Song, W. (2014). Preparation of solid silver nanoparticles for inkjet printed flexible electronics with high conductivity. *Nanoscale, 6*(3), 1622–1628. Advance online publication. doi:10.1039/C3NR05479A PMID:24337051

Song, Y., Li, X., Wei, C., Fu, J., Xu, F., Tan, H., Tang, J., & Wang, L. (2015). A Green Strategy to Prepare Metal Oxide Superstructure from Metal-Organic Frameworks. *Scientific Reports, 5*(1), 8401. Advance online publication. doi:10.1038rep08401 PMID:25669731

Suleimanov, B. A., Rzayeva, S. J., & Keldibayeva, S. S. (2020). A new microbial enhanced oil recovery (MEOR) method for oil formations containing highly mineralized water. *Petroleum Science and Technology, 38*(23), 999–1006. Advance online publication. doi:10.1080/10916466.2020.1793777

Thakkar, K. N., Mhatre, S. S., & Parikh, R. Y. (2010). Biological synthesis of metallic nanoparticles. Nanomedicine: Nanotechnology, Biology, and Medicine, 6(2). doi:10.1016/j.nano.2009.07.002

Tokonami, S. (2021). External-field-induced Assembly for Biological Analytical Chemistry. *Analytical Sciences, 37*(3), 395–396. Advance online publication. doi:10.2116/analsci.highlights2103 PMID:33692263

Vivien, A., Guillaumont, M., Meziane, L., Salzemann, C., Aubert, C., Halbert, S., Gérard, H., Petit, M., & Petit, C. (2019). Role of Oleylamine Revisited: An Original Disproportionation Route to Monodispersed Cobalt and Nickel Nanocrystals. *Chemistry of Materials, 31*(3), 960–968. Advance online publication. doi:10.1021/acs.chemmater.8b04435

Wang, N., Prodanović, M., & Daigle, H. (2019). Nanopaint application for flow assurance with electromagnetic pig. *Journal of Petroleum Science Engineering, 180*, 320–329. Advance online publication. doi:10.1016/j.petrol.2019.05.028

Wilson, A. (2012). With Functionalization, Nanodiamonds May Increase Durability of PDC Cutters. *Journal of Petroleum Technology, 64*(12), 134–137. Advance online publication. doi:10.2118/1212-0134-JPT

Xu, L., Abedini, A., Qi, Z. B., Kim, M., Guerrero, A., & Sinton, D. (2018). Pore-scale analysis of steam-solvent coinjection: Azeotropic temperature, dilution and asphaltene deposition. *Fuel, 220*, 151–158. Advance online publication. doi:10.1016/j.fuel.2018.01.119

Xu, S., Habib, A. H., Pickel, A. D., & McHenry, M. E. (2015). Magnetic nanoparticle-based solder composites for electronic packaging applications. Progress in Materials Science, 67. doi:10.1016/j.pmatsci.2014.08.001

Xue, Z., Worthen, A., Qajar, A., Robert, I., Bryant, S. L., Huh, C., Prodanović, M., & Johnston, K. P. (2016). Viscosity and stability of ultra-high internal phase CO2-in-water foams stabilized with surfactants and nanoparticles with or without polyelectrolytes. *Journal of Colloid and Interface Science, 461*, 383–395. Advance online publication. doi:10.1016/j.jcis.2015.08.031 PMID:26414421

Yadav, P., Kosandar, B. A., Jadhav, P. B., Ishaq, L. G., Addagalla, A. K. V., & Mohammad, A. R. (2015). Customized high-performance, water-based mud for unconventional reservoir drilling. *SPE Middle East Oil and Gas Show and Conference, MEOS, Proceedings*. 10.2118/172603-MS

Yan, X., Ma, C., Cui, F., McClements, D. J., Liu, X., & Liu, F. (2020). Protein-stabilized Pickering emulsions: Formation, stability, properties, and applications in foods. Trends in Food Science and Technology, 103. doi:10.1016/j.tifs.2020.07.005

Yang, Z., Zhou, Y., Feng, Z., Rui, X., Zhang, T., & Zhang, Z. (2019). A review on reverse osmosis and nanofiltration membranes for water purification. Polymers, 11(8). doi:10.3390/polym11081252

Yekeen, N., Padmanabhan, E., Idris, A. K., & Chauhan, P. S. (2019). Nanoparticles applications for hydraulic fracturing of unconventional reservoirs: A comprehensive review of recent advances and prospects. Journal of Petroleum Science and Engineering, 178. doi:10.1016/j.petrol.2019.02.067

You, H., Yang, S., Ding, B., & Yang, H. (2013). Synthesis of colloidal metal and metal alloy nanoparticles for electrochemical energy applications. *Chemical Society Reviews, 42*(7), 2880–2904. Advance online publication. doi:10.1039/C2CS35319A PMID:23152097

Yu, J., Khalil, M., Liu, N., & Lee, R. (2014). Effect of particle hydrophobicity on CO2 foam generation and foam flow behavior in porous media. *Fuel, 126*, 104–108. Advance online publication. doi:10.1016/j.fuel.2014.02.053

Záleská, M., Pivák, A., Pavlíková, M., Jankovský, O., & Pavlík, Z. (2020). Influence of graphite oxide addition on the properties of magnesium oxychloride cement composites. *IOP Conference Series. Materials Science and Engineering, 960*(2), 022080. Advance online publication. doi:10.1088/1757-899X/960/2/022080

Zawadzki, J., & Bogacki, J. (2016). Smart magnetic markers use in hydraulic fracturing. Chemosphere, 162. doi:10.1016/j.chemosphere.2016.07.058

Zou, R., Xu, J., Kuffner, S., Becker, J., Li, T., Guan, X., Zhang, X., Li, L., Cohen Stuart, M. A., & Guo, X. (2019). Spherical Poly(vinyl imidazole) Brushes Loading Nickel Cations as Nanocatalysts for Aquathermolysis of Heavy Crude Oil. *Energy & Fuels, 33*(2), 998–1006. Advance online publication. doi:10.1021/acs.energyfuels.8b03964

ADDITIONAL READING

Bou-Hamdan, K. (2020). *Key Design Considerations for Maximizing the Recovery Rate of Unconventional Reservoirs. The Way Ahead.* SPE.

KEY TERMS AND DEFINITIONS

Downstream Operations: This includes all the processes needed to transport hydrocarbons into a final product.

Drilling Fluids: A fluid that is used in drilling operations and provides many benefits such as cooling the drill bit and cleaning the hole of cuttings.

Enhanced Oil Recovery: This represents a group of methods that are used to enhance the oil recovery once the primary and secondary methods can no longer provide efficient production.

Hydraulic Fracturing: A well stimulation technique that involves pumping fluids into the reservoir rock at high rates to break the rock and create permeable pathways for the hydrocarbons.

Midstream Operations: This includes all the processes involved in transporting and storing hydrocarbons. It serves as a connection between Upstream and Downstream operations.

Oil Well Cement: This cement is used in drilling operations to seal the surrounding wellbore area. This cement is pumped behind the casing to prevent fluids to pass from or into the formations.

Upstream Operations: This includes all the processes involved in exploring and producing the hydrocarbons from the subsurface.

Chapter 9
Application of Nanoparticles for Environmental Remediation:
Potential Impacts of Carbon- and Metal-Based Engineered Nanoparticles

Gamze Ersan
Clemson University, USA

Mahmut Selim Ersan
Arizona State University, USA

ABSTRACT

In the last decades, numerous research studies have been focused on the mitigation of different classes of contaminants by varying types of NPs. Treatment mechanisms were controlled by properties of contaminants (e.g., organic vs inorganic, molecular size, and hydrophobicity, etc.), NPs (e.g., surface charge, area, pore size, and surface functional groups, etc.), and aqueous water (e.g., organic matter types, pH, ionic strength, surfactants, and temperature). Therefore, this chapter includes (1) a literature summary for the removal of contaminants by carbon- and metal-based NPs, (2) a discussion on the controlling mechanisms for the mitigation of contaminants by carbon- and metal-based NPs, and (3) an evaluation on the application and implication of carbon- and metal-based NPs in environmental studies. This chapter also identifies future research needs and challenges on the application of carbon- and metal-based NPs for environmental remediation.

INTRODUCTION

Nanoparticles (NPs) are known to exhibit novel properties that range between 1 to 100 nanometers (Amin et al., 2014). They have different classifications based on their physicochemical properties including backbone material, shapes, and sizes. Up to date, carbon- and metal-based NPs are the next generation nano-adsorbents and nano-catalysts in water treatment technologies because of their physical and chemical features and affinity towards contaminants in aqueous environments (Ersan et al., 2016; 2019;

DOI: 10.4018/978-1-7998-8936-6.ch009

Khadir et al., 2020; Sagir and Tahir, 2021). Since they hold promise as high-performance nanomaterials, nanotechnology-based sorption and catalysts technologies for environmental remediation have attracted scientific interest worldwide. Therefore, NPs having high surface area can bring new prospects for pollution management in the aqueous environment.

Depending on the dimension, morphology, and chemical properties of NPs, they can be arranged into different classifications. According to their origin, physicochemical properties, and composition, several types of NPs have been commonly used for mitigation of contaminants (i.e., metal ions, dyes, pesticides, pharmaceuticals, other organics, etc.) in the aqueous environment (Aillon et al., 2009; Gottschalk and Nowack, 2011; Tiwari et al., 2015, Ahmad et al., 2020; Sabzehmeidani et al., 2021).

Carbon-based NPs including carbon nanotubes (CNTs) can provide potential advantages over the existing state-of-the-art adsorbents (i.e., activated carbons (ACs) and carbon fibers). The unique features of CNTs may increase the adsorption capacity and initial sorption rates and mitigate fouling by natural organic matters (NOMs). Compared with CNTs, graphene is two-dimensional nanostructure material, which has attracted great attention from many researchers (Novoselov et al., 2004). Since graphene has a high theoretical specific surface area (>2000 m^2g^{-1}), it can be used as an adsorbent for the adsorption of organic contaminants (OCs) (Zhu et al., 2010; Ersan and Ersan, 2021).

Metal-based NPs are made from metals precursors including silver, copper, gold, zinc, titanium, iron elements. In the last two decades, metal-based NPs have been widely used as highly efficient catalysts for mitigating water and air pollution (Taghavi et al., 2017; Ahmad et al., 2020). Their unique properties are highly related to size, shape, chemical composition, agglomeration state, and fractal dimension (OECD et al., 2010). In the literature, the use of metal-based NPs as a nano-catalyst in the chemical oxidation process offers several advantages including compounds removal in short contact time, and the ability to transform some recalcitrant compounds into their by-products when compared with conventional treatment technologies (Karki et al., 2018; Zhang et al., 2021).

This chapter summarizes up-to-date literature information on the mitigation of contaminants by most common NPs (carbon- and metal-based NPs) and provides insights about mechanisms and controlling factors for the removal of contaminants by NPs. Besides, the application and implication of carbon- and metal-based NPs examine in environmental studies. Lastly, this chapter also identifies gaps and future research needs on the application of carbon- and metal-based NPs for environmental remediation.

BACKGROUND

Trace amounts of organic and inorganic contaminants such as pesticides, pharmaceuticals, perfluorooctanoic acid, arsenic, and lead in sources of drinking water have posed significant health risks to humans in recent years. These micro-based organic and inorganic contaminants are not completely removed by conventional treatment processes. Therefore, alternative innovative treatment technologies have been developed for new applications in wastewater and water treatment. Engineered NPs are safely taking advantage of their unique properties to treat contaminated water using fewer chemicals, electricity, and smaller reactors than current treatment technologies. In addition to this, carbon- and metal-based NPs may enable reuse and regenerate to decline treatment costs. Therefore, more research is needed for the regeneration of carbon- and metal-based NPs in different background water chemistries.

Figure 1. The structure of SWNT bundles (left) and MWNT (right) by capturing from Transmission Electron Microscopy.
**The distance between the fringes (layers of the tube)*

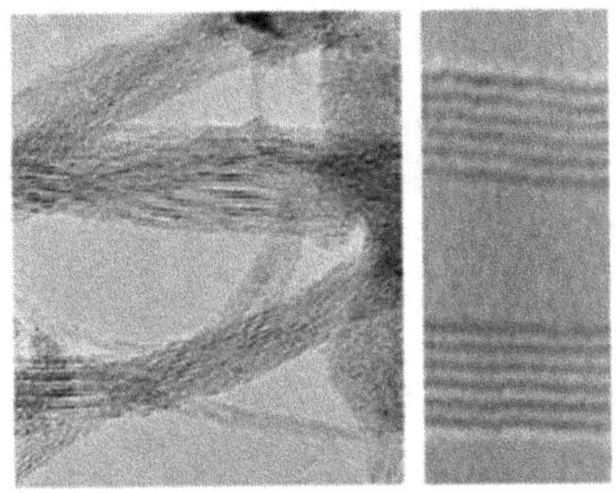

CARBON-BASED NPs

Carbon Nanotubes (CNTs)

With increasingly stringent water quality standards for both drinking waters and treated wastewaters, recalcitrant contaminants (such as per- and polyfluoroalkyl substances, pharmaceuticals, and endocrine-disrupting compounds, etc.) have stimulated the research for novel adsorbent materials. Ideally, CNTs have a hydrophobic surface with a unique electronic structure that is expected to interact strongly with OCs having electrons, and they have a morphology that provides high-energy sorption sites in addition to those found in micropores. Therefore, CNTs hold promise as high-performance adsorbents for removing OCs from the aquatic environment. CNTs have attracted scientific interest worldwide due to their peerless structure and electronic features (Iijima, 1991; Dresselhaus et al., 2001; Apul et al., 2015; Ersan et al., 2017; Sabzehmeidani et al., 2021). Because of their distinctive structures, CNTs have been described as matter in one dimension (1-D), which is a honeycomb lattice rolled on to itself, with a diameter of the order of nanometers. In general, two main types of CNTs exist depending upon whether the tubes are made of only one graphene sheet (single-walled carbon nanotubes, SWNTs) or more than one graphene sheet (multi-walled carbon nanotubes, MWNTs), as seen in Table 1. Besides, Figure 1 presents transmission electron micrographs of SWNTs and MWNTs materials. SWNTs have a wall thickness of one atom, and a circumference equal to tens of atoms (typical diameter of 1.4 nm).

SWNTs self-assemble into bundles during their production process, evident in Figure 1 and depicted schematically in Figure 2. The Van der Waals interactions among tubes cause them to order in a hexagonal array. The formation of bundles has important consequences for the adsorption characteristics of these materials; various sites are formed which are energetically favorable for adsorption. As shown in Figure 2, for CNT bundles, these adsorption sites are the outermost surface, interstitial channels, inner cavities, and external grooves (Agnihotri et al., 2008; Zhang et al., 2010, 2011; Tang et al., 2017; Sabzehmeidani et al., 2021; Ersan et al., 2017; Ersan and Ersan, 2021) (Figure 2). In addition, when SWNTs are cut open, sites may be formed in the hollow interior of the open tubes. These sites are most like classical micropores and could only be accessed by adsorbates having sufficiently small dimensions. SWNTs have a wide range of application potential as sorbents for gaseous materials, composite material

Figure 2. Potential adsorption sites (bundled and rolled) on SWNT bundles and (open-up layered) on graphene nanosheets

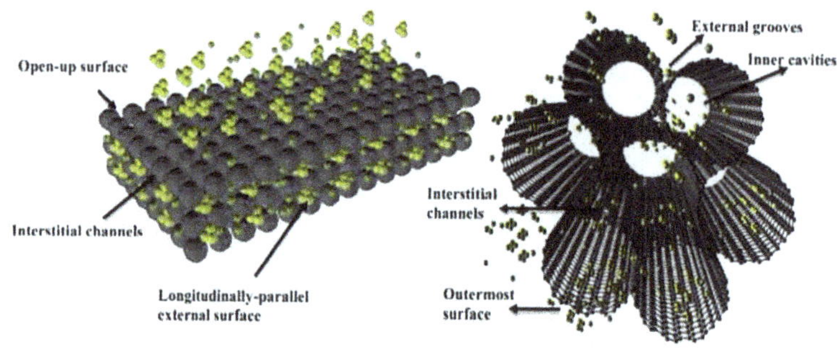

fillers, medical materials, and electronic devices (Ajayan and Ebbsen, 1997; Jia and We, 2017). In addition, recent research results are indicating that SWNTs are promising adsorbents for removing organic and inorganic contaminants in the single and multi-solute systems, and may possess superior sorption capability over conventional sorbents, including granular activated carbons (GACs).

CNTs have a comparatively large specific surface area and favorable adsorption characteristics (Dillon et al., 1997; Teizer et al., 1999; Wang et al., 1999; Kuznetsova et al., 2000; Talapatra and Migone, 2001; Calbi et al., 2001; Migone and Talapatra, 2004; Yang et al., 2008; Zhang et al., 2013; Gupta et al., 2016; Tang et al., 2017; Yang et al., 2018; Hasnain and Nayak, 2019; Sabzehmeidani et al., 2021). The specific surface area of MWNTs generally ranges from 30 – 200 m^2/g and the surface area of purified SWNTs can approach that of activated carbons (~1000 m^2/g). Even when the surface area is not as high as many activated carbons, it is expected that the surface area is more accessible, which contributes to the potential of CNTs as highly efficient adsorbents. For example, Long and Yang (2001) reported that CNTs were superior to activated carbon in the removal of dioxin because of a strong interaction between dioxin and CNT. They showed the temperatures and activation energy for desorption and the Langmuir isotherm constant were higher than that of activated carbon. Additionally, it was seen that the amount of dioxin absorbed onto CNTs was higher than that of ACs in the low concentration – Henry's Law region, suggesting higher adsorption efficiency than that of activated carbon. Peng et al. (2003) measured the rate and uptake of 1, 2-dichlorobenzene by CNTs. They found rapid sorption kinetics (equilibrium reached in < 40 min) and an increase in uptake with temperature. Therefore, sorption was endothermic and the free energy (ΔG) of adsorption was dominated by a positive entropy change ($+\Delta S$). Moreover, they found near constant adsorption capacity over a wide pH 3 to pH 10. Lu et al. (2005) studied the adsorption of trihalomethanes (THMs) by MWNTs. Pore size distributions suggested that the presence of aggregation and the existence of pores in the aggregated network. The kinetics of adsorption was fast (< 180 min) but not as fast as that observed by Peng et al. (2003). Uptake decreased with increasing adsorbate size and degree of bromine substitution, indicating either a size exclusion, steric or halogen substituent (polarity) effect. Uptake decreased with pH above pH 7. The uptake and kinetics of the MWNT were compared to that of a commercial GAC (unspecified particle size). The adsorption kinetics of contaminants onto CNTs were faster. The uptake of chlorinated THMs (i.e., chloroform) by MWNTs was higher than GAC, while uptake of bromine substituted THMs (i.e., bromoform) by GAC was much higher than MWNTs.

Lu et al. (2006) observed the adsorption uptake of THMs onto CNTs was decreased with increasing temperature but was still two to three times the amount adsorbed onto PAC. Cheng et al. (2005) studied the adsorption (and desorption) of naphthalene and 1,2-dichlorobenzene by C60 nano-scale aggregates dispersed in water. Mean aggregate sizes ranged from about 80 to 170 nm, depending on the method of preparation. Linear isotherms were noted for aqueous phase concentrations ranging from about 1 to 4 mg/L. Higher and more non-linear uptake by smaller aggregates (1 to 3 nm) was reported in an earlier study (Cheng et al., 2004). The adsorption of three polycyclic aromatic hydrocarbons (PAHs) compounds by fullerenes, SWNTs and MWNTs were examined by Yang et al. (2006a). According to their finding, the adsorption affinities for phenanthrene, naphthalene, and pyrene were increased with increasing octanol-water partition coefficient values, K_{ow}. The uptake of phenanthrene onto CNTs was up to 4 orders of magnitudes faster than fullerene, and sorption kinetics of SWNT was higher than MWNTs. Sorption isotherms were well described by the 3-parameter Polanyi-Manes model, and isotherm fitting findings suggested the adsorption of PAHs by carbon nanomaterials was monolayer sorption on a homogeneous surface or multilayer formations on a heterogeneous surface; linear partitioning did not significantly contribute to the total uptake. Gotovac et al. (2006) showed higher purity CNTs exhibited greater PAH uptake. More specifically, they found that phenanthrene favored the smoother surface of a purified nanotube, which was shown to be clean from functional groups and amorphous phases. They proposed, based on XPS analysis, that phenanthrene multilayered physisorbed film on the surface of the CNTs. Additionally, they argued π-π coupling interaction between the π electrons of the aromatic rings on phenanthrene and CNT surface (Ersan et al., 2016). Moreover, when the phenanthrene was compared to tetracene, the adsorption behavior of phenanthrene was remarkably different; the amount of phenanthrene adsorbed was much smaller than that of tetracene. This difference was attributed to the molecular size and geometry of the two chemicals allowing for tetracene to have π-π interactions with CNT`s axis while phenanthrene cannot (Gotovac et al. 2007).

Pan et al. (2008) reported adsorption interactions between 17α-ethinyl estradiol and bisphenol A and CNTs. Their findings emphasized that there was a positive correlation between surface area and single point adsorption coefficient K, suggesting that surface area plays a major role in the adsorption of these two compounds. Additionally, they concluded that the strong interaction between the two endocrine disruptors and the CNTs is because of π-π electron-donor-acceptor relationships (EDA). This relationship is possible because the two compounds contain phenol groups, which are π-electron donors, and CNTs can act as either electron donors or acceptors. Ji et al. (2009a) studied the adsorption of tetracycline and naphthalene on four different carbonaceous sorbents (SWNT, MWNT, AC), and graphite). Their results showed SWNTs exhibited higher adsorption capacities for tetracycline (SWNT > MWNT > AC > graphite) while activated carbon had the highest adsorption affinity for naphthalene (AC > SWNT > MWNT > graphite). They suggest the strong sorption interaction between tetracycline and SWNT and MWNT`s surface is because of the combination of π-π electron-donor-acceptor (EDA) relationship and cation-π bonding, this is supported by a decrease in K_d (single solute adsorption coefficient [L/kg]) values with increasing pH. However, for the case of naphthalene, there was no such change with pH, and the higher adsorption affinity to activated carbon was linked to the pore-filling mechanism since the molecular size of naphthalene is comparable in size with the pores. In another study performed by Ji et al. (2009b), it was shown that two sulfonamide antibiotics adsorbed strongly to MWNTs, which was contributed to the π-π electron coupling onto nanotube. The effects of both varying pH between pH values 3 to 8 and varying the ionic strength between values of 0.01 to 0.1 M (using NaCl and $CaCl_2$) showed minimal change in adsorption uptake, although adsorption was suppressed in the humic acid solution. Chen et al. (2007) presented that the adsorption mechanisms may also include π-electron polarizability in addition to the hydrophobic impacts. Moreover, adsorption affinity to CNTs followed the pattern of aliphatic <

aromatic < nonpolar aromatic < nitroaromatic and suggested that nitroaromatic had a larger affinity to CNTs than nonpolar aromatics was because specifically of the strong π-π EDA relationships between the nitro group and the CNT surface. Consistent with their work on hydroxyl and amino-substituted aromatics, they found that π-EDA interaction was the most dominant adsorption mechanism because of strong interaction between the π-electron-rich aromatic ring of the compound and the π-region onto CNT as well as the impacts hydroxyl functionalized aromatics being able to interact with the electron depleted sites on the surface of the CNTs (Hasnain and Nayak, 2019; Sabzehmeidani et al., 2021). They also emphasized that the pH parameter had a prominent impact on the adsorption capacity of OH-substituted aromatics at high pH values when the OH group was ionized. Similarly, Yang et al. (2008) reported nitro-substituted functional groups on aniline and phenol promoted the sorption capacities of CNT, which were also higher than chloride and methyl groups. Meanwhile, the adsorption affinity was increased with increasing degree of substitution. On the other side, the sorption affinity increased with increasing aromatic rings of compounds (Lin and Xing, 2008). However, the role of the aromatic ring system shape or size remains unclear. The planarity, ring conformations, and the size of compounds influence their sorption affinity because of the curvature of CNT or access to the preferential sorption sites. Zhang et al. (2010) investigated the adsorption affinities of different planarity (planar phenanthrene molecule, nonplanar biphenyl, and nonplanar 2- phenyl phenol) compounds. They suggest the interstitial channels and longitudinally parallel exterior surfaces of SWNT offer more homogenous adsorption sites than MWNTs for the more rigid planar molecules. Additionally, nonplanar OCs are more flexible and can easily absorb into the inner cavities of MWNTs. Zhang et al. (2010) also reported the influence of surface chemistry on these OCs. They concluded that π-π EDA relationships between phenanthrene and CNT surface were not a major adsorption mechanism due to a lack of pH dependence. They propose that if π-π EDA relationship played the major role in the adsorption of OCs, then the deprotonation of acidic functional groups of the CNTs would enhance the π-π EDA relationship. They emphasized that electrostatic interactions and water cluster formation on the CNTs have occurred, since the highest adsorption affinity existed at a neutral pH the major mechanisms affecting OC adsorption onto CNTs is a combination of hydrophobic interactions. They argue that pH and ionic strength values in solution within the range found in natural sources should not play an important role in altering the adsorption affinities of OCs on CNTs (Abdel et al., 2015; Saka et al., 2018; Ersan et al., 2017; Ersan and Ersan, 2021).

The tunable surface chemistry, non-corrosive property, high surface area, and hydrophobic surface of CNTs lead to improve adsorption capacities with respect to conventional adsorbents. In the last decade, CNTs as unique adsorbents have been specifically preferred for the adsorption of recalcitrant OCs and heavy metals in the aqueous environment. Duman et al., (2016) reported that subsequently synthesized MWCNT were successfully used for the mitigation of methylene blue in aqueous water. Besides, Shi et al., (2016) also reported that modified CNTs are used for the removal of several antibiotics such as tetracyclines, sulfonamides, macrolides, and quinolones. During the last few years, researchers have been focusing on the removal of unknown organic and inorganic contaminants by functionalized CNTs under different background chemistry (Hasnain and Nayak, 2019; Sabzehmeidani et al., 2021). These modified CNTs may also be integrated into the hybrid treatment technologies.

Graphene

Graphene consists of the two-dimensional planar nanosheets (Novoselov et al., 2004; Wang et al., 2014a-b; Wang et al., 2016; Ye et al., 2018; Liu et al., 2020; Ersan et al., 2017; Ersan and Ersan, 2021). Due to the peerless structural properties of graphene, they can be used in wide range of applications (Geim and Novoselov, 2007; Geim, 2009). Therefore, the industrial application of graphene has been exponentially growing very soon (Ersan et al., 2017; Velusamy et al., 2021). Highly hydrophobic graphene nanomaterials provide highly accessible surface area and therefore, they are anticipated as a beneficial sorbent for contaminants (Stoller et al., 2008). Graphene has several adsorption sites from external surfaces to interstitial channels (Figure 2). Besides, graphene can modify by covalent bonding with oxygen-containing functional groups of graphene oxides (GO), as seen in Table 1. The oxygen-containing functional groups can decline the hydrophobicity of graphene surface, while GO dispersion in water was quickly increased, and thus influence the adsorption mechanisms.

There are numerous peer-reviewed articles in the current literature, which have been focused on the adsorption of OCs onto pristine and functionalized graphene nanosheets (Ramesha et al., 2011; Carolin et al., 2017; Liu et al., 2020; Kommu et al., 2018; Wang et al., 2014a; 2018; 2021; Ersan et al., 2021). Due to the rapidly increasing interest on application of graphene in full scale systems, understanding the adsorption of OCs on graphene has raised some concerns (i) as adsorbents, the feasibility of graphene in water treatment systems, (ii) the fate of contaminants loaded graphene materials in the aqueous environment, and (iii) the toxicological impacts of graphene onto plants, animal, and environment when the contaminant loaded graphene materials are directly released into the environment.

In natural waters, graphene can directly interact with organic matter molecules, which is omnipresent in surface water and groundwater. NOM molecules carry a net negative charge in freshwaters, and it is a heterogeneous mixture of humic substances, hydrophilic acids, etc. The presence of organic matters (i.e., NOM, algal, or wastewater impacted waters) have two opposite impacts on OC adsorption onto graphene: (i) an increase in adsorption capacity because of better dispersion of adsorbent, and (ii) a decrease in adsorption capacity because of organic matter and contaminant competition onto graphene surface. Previously, the impact of NOM on the OC adsorption by GACs and CNTs has been examined (Ersan et al., 2016; Jiang et al., 2017), however, more study still needs to be evaluated for the adsorption behavior of OC on the algal and wastewater impacted water loaded carbon-based NPs.

METAL-BASED NPs

Silver NPs (AgNPs)

As same as other metal-based NP, AgNPs have a common size of 10–100 nm with a wide variety of shapes. In the last decades, AgNPs are most commonly used in several fields, including biomedical applications, food industry, and different fields of industrial purposes (Lee and Jun, 2019; Wei et al., 2018; Chugh et al., 2018; Gurunathan et al., 2015; Li et al., 2010; Mukherje et al., 2001). Among various metal-based NPs in Table 2, AgNPs have shown distinctive features, such as unique optical, electrical, and thermal properties, and therefore used in the field of water disinfection, AC filters, and water treatment processes (especially small-scale, point-of-use, or emergency response treatment systems). AgNPs can be manufactured with various properties according to the used recipe during the fabrication process. In general, silver spheres, wires, and plates can be generated by several manufacturing methods. AgNPs can be surface functionalized with different materials. Polyvinylpyrrolidone (PVP) and tannic acid are commonly preferred to use capping agents for AgNPs. Besides, AgNPs are used in biological applications, which are commonly coated with polyethylene, serum albumin, bovine glycol, and other

Table 1. Properties of the most used Carbon-based NPs

Properties	Graphene	Graphene Oxide (GO)	Single-Walled Carbon Nanotubes (SWCNT)	Multi-Walled Carbon Nanotubes (MWCNT)
Structure	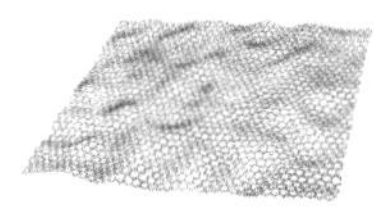	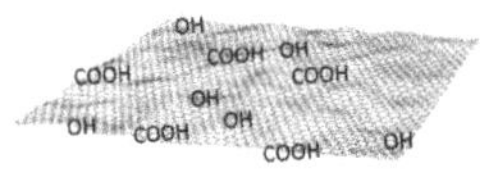		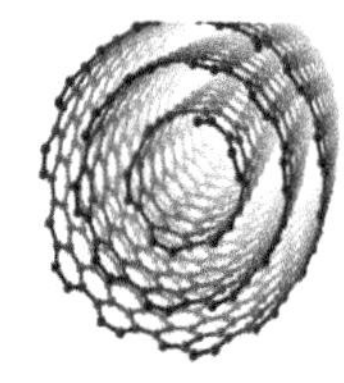
Surface Area (m^2/g)	132-1556	4-771	<1315	<500
Oxygen contents (%)	0.8-5	10.4-54.4	0-60	0-5
Synthesis	• Graphite exfoliation • Chemical vapor deposition	• Oxidation and exfoliation of graphite	• Chemical vapor deposition of hydrocarbons	• Chemical vapor deposition of acetylene
Cost	$10-240/1g	$97-250/1g	$40-150/g	$12-100/g
References	Liu et al., 2014; Ersan et al., 2016; Jiang et al., 2017; Liu et al., 2020	Cai and Casanova, 2014; Ersan et al., 2016; Jiang et al., 2017; Liu et al., 2020	Katep et al., 2010, Ersan et al., 2019, Liao · 2018; Sabzehmeidani et al., 2021	Apul et al., 2015, Bahgat et al., 2011; Ahmad et al., 2013; Hasnain and Nayak, 2019

proteins and peptides. Coated materials can be functionalized with molecules that surface charge of the negatively charged AgNPs switched to the positively surface charged.

The solubility of AgNPs is also impacted by their size, and thus the solubility of AgNPs increase with declining AgNPs diameter, which can be increased releasing of toxic Ag^{2+} ions. The effective solubility product of the crystal was determined by two coefficients, k_{sp}, which is typically understood to equal k_b. But these coefficients start to digress as the NPs enter into the nanoscale, due to the presence of a relatively high number of atoms at their surface, and thus, the surface tension can be more important for AgNPs (Auffan et al., 2009; Lee and Jun, 2019).

Up to now, a limited number of adsorption studies have been examined for AgNPs ability to remove contaminants from aqueous water including cadmium (Sadegh et al., 2017; Al-Qahtani, 2017) and rhodamine B (Azeez et al., 2018). However, the application of AgNPs in removing contaminants is still limited because of their high production costs. On the other side, since their unique physicochemical properties, metal-based NPs can be potential hazards in the natural environment (Kaegi et al., 2013). Therefore, AgNPs are gaining interest as an emerging contaminant in aqueous water (Boxal et al., 2012; Rasheed et al., 2019), because AgNPs have been enormously used in daily life. Since Serrão Sousa and Ribau Teixeira (2015) investigated the mitigation of AgNPs from synthetic NOM waters by different types of negatively charged nanofiltration (NF) (NF90 and NF270). Up to the 98% removal of AgNPs by NF90 and NF270 in the first 15 min. With increasing contact time (3h), the removal of AgNPs was almost removed from the waters. The results also emphasized that size exclusion mechanisms were the core mechanism for the AgNPs removal, and tighter membrane representing higher removal of AgNPs in aqueous water.

Gold NPs (AuNPs)

The intrinsic features (optics, electronics, and physicochemical characteristics) of gold NPs (AuNPs) can be changed depending on the characterization of the NPs (i.e., shape, size, and aspect ratio). Among all metal-based NPs, AuNPs are the most stable, due to their unique features. It has also been developed with different shapes and structures (i.e., nanoflowers, nanospheres, nanobranches, nanorods, and nanowires, etc.) by numerous synthetic approaches (O'Neal et al., 2004; Chen et al., 2008; Li et al., 2015; Ahmed et al., 2016; Chugh et al., 2018; Xiao et al., 2019; Khan et al., 2019). To date, AuNPs has been used for different type of medical applications, such as drug and gene delivery, computed tomography, photo-, dynamic-, and radiation-thermal therapy, X-ray imaging, and other biological activities. Meanwhile, recently, researchers in engineering fields have shown interest in gold AuNPs, due to their photothermal and photocatalytic properties. Photocatalytic properties of AuNPs can use light to activate a chemical process, which may have convicted effective in water treatment. However, some of the researchers emphasized that AuNPs, for water purification applications, are not promising sorbents because of the difficulty of dispersion of NPs in water without stabilizing agents (Khan et al., 2019). AuNPs can be expensive and energy-intensive for contaminants removal from aquatic waters. From another perspective, the AuNPs can be easily synthesized with plant-based approaches, which can be an alternative for pilot-scale systems (Zhong et al., 2015). After the synthesis, many AuNPs can be conducted and used straight in sequential analysis (MaríaJesús Lerma- García et al., 2014). According to the method, the one-step citrate method is used for reducing and stabilizer inhibitor in the AuNPs synthesis approach, and their diameter size change with several parameters such as pH, time, and one-step citrate concentration, which provide more stability and display high catalytic properties. The formation and occurrence of toxic by-products in the synthesis process of AuNPs during ecofriendly syntheses and their applications are needed to investigate soon.

On the other side, Koedrith et al., (2015) reported that the AuNPs as nanoprobes are started to use for the detection of many chemically differentiated contaminants, including heavy metals, organic contaminants, toxicants, and pathogens, which is attributed to their high selective capacity to bind to target contaminants (Tamer et al., 2013). Thus, AuNPs can offer new stable and dispersible Au nanostructure approaches with selective or SERS properties, after allowing highly sensitive for selective compound detection. AuNPs can provide good catalyzers for using in many chemical transformations, including hydrocarbon oxidation, C–C coupling, and other oxidation and reduction reactions. Pradeep and Anshup, (2009) investigated the oxidation of carbon monoxide at low temperature supported by coupled with AuNPs on numerous metal oxides and found catalytic properties of AuNPs may enhance the oxidation reactions.

Copper NPs (CuNPs)

Recently, Copper NPs (CuNPs) are particularly attractive to researchers due to the low cost, high natural abundance of Cu, and the practical and ease of synthesis. Specifically, as anti-bacterial and microbial agents, CuNPs exhibits uncommon behavior (Majzlik et al., 2011; Duan et al., 2015; Naseem and Waseem, 2021), although CuNPs is less effective than other metal-based NPs. CuNPs can also increase the porosity of the microbeads, and thus the dispersed AgNPs are relatively more accessible to microorganisms (Khare et al., 2014). When the CuNPs are compared with AgNPs, the antimicrobial properties of CuNPs may have a similar principle with AgNPs. Besides, CuNPs and CNTs have also shown strong antimicrobial properties. Dislike conventional chemical oxidants, these antimicrobial nanomaterials are weak oxidants and relatively inert in the aquatic environment. Therefore, these metal-based NPs may not expect to produce harmful disinfection byproducts, and thus they are suitably incorporated into treatment process application. As a result, CuNPs can potentially enhance the conventional oxidation process. Alsabagh et al., (2015) deeply investigated that the characterization of chitosan/AgNPs, chitosan/CuNPs, and chitosan/CNTs and mixed chitosan/NPs using different techniques. Among all NPs, Cu-based NPs with structural modifications can be more practically used for a water purification process as an active material.

The use of CuNPs is restricted by the inherent instability of Cu under atmospheric conditions since it can easily be prone to oxidation. To date, several studies have been developing new methods and supporting materials for increasing their stability by changing their sensitivity to polar compounds. On the other hand, due to their peerless physic-chemical properties, Cu and CuNPs retain an important place in many promising photocatalytic applications and implications such as hydrogen generation from water, photocatalytic self-cleaning, and CO_2 reduction, etc.

Iron NPs (FeNPs)

Micrometer nanoparticles of iron (Fe) are highly reactive metals because of their large surface area and backbone structures. In the presence of oxygen and water, FeNPs can rapidly oxidize and form free iron ions. FeNPs were first enormously applied for the remediation of polluted soil and groundwater. During these remediation processes, nano zero-valent Fe was oxidized to Fe^{2+} and Fe^{3+} ions, while the organics were reduced to inorganic complexes (Mukherjee et al., 2016; Shad et al., 2020). On the other side, FeNPs are also commonly used for membrane coating due to the enhanced hydrophilicity of the membrane. Therefore, nanosized and magnetic characteristics of iron NPs (FeNPs) are promising NPs for the removal of contaminants in the water environment. Gupta et al. 2015, reported that high removal affinity of several heavy metals onto modified FeNPs was observed from wastewater (Gupta et al., 2015). In recent studies, FeNPs can be promising adsorbents for the treatment of polluted soil and groundwater targeting multiple contaminations such as chlorinated contaminants (Muealler et al., 2012; Valiyeva et al., 2019) and reducing inorganic anions such as nitrate to ammonia (Valiyeva et al., 2019) or perchlorate to chloride (Cao et al., 2005).

The transformations of metal-based NPs can be exposed when they are released into the environments. Metal-based NPs transformations can lead to alters in NPs structure that is directly released upon into the environment. The metal-based NPs transformations can be classified as chemical, physical, and biological transformations, and interactions with macromolecules (Lowry et al., 2012; Shad et al., 2020). More specifically, chemical transformations are the redox reactions, dissolution, and adsorption processes. Several metal-based NPs can stimulate redox reactions due to the redox activity of metal elements (i.e., Fe, Cu, Ag, and Au). Besides, bacteria in the oxidizing/ reducing environments can pioneer to alter the oxidation process of metal-based NPs. As a result, these reactions can lead to exposing the surface functional groups of metal-based NPs due to the altering of charge of metal-based NPs and the precipitation of new phases (Bottero and Wiesner, 2010). Besides, these oxidation/reduction reactions can also alter the persistency of metal-based NPs in the environment (Lowry et al., 2012).

The adsorption and desorption process are another chemical reaction for metal-based NPs. As explained in the previous section, the adsorption behavior of metal-based NPs can occur via physisorption, ion exchange (electrostatic), and chemisorption with chemical bonding. Physisorption is a weaker interaction than electrostatic and chemisorption. Like aggregation, the adsorption process between metal-based NPs can result in the stabilization and destabilization of metal-based NPs. For instance, the destabilization and aggregation of metal-based NPs through bridging can occur, if the surface coverage of metal-based NPs is only partial. On the other side, if surface coverage of NPs is completed, metal-based NPs can stabilize or resist the aggregation forces. And therefore, desorption of metal-based NPs may be less common in the aqueous (Lowry et al., 2012; Ali et al., 2017). However, it is possible that the changes in aquatic water conditions can also impact the adsorption capacity of the adsorbed substance (Nowack et al., 2012; Ersan et al., 2016). Therefore, adsorption and desorption of metal-based NPs can change depending on their surface charge and chemistry, and reactivity, and toxic characteristics of metal-based NPs in the environment.

Table 2. Properties of the most used Metal-based NPs

	Metal-based NPs			
Properties	**Silver (Ag)**	**Gold (Au)**	**Copper (Cu)**	**Iron (Fe)**
Geometric Structure	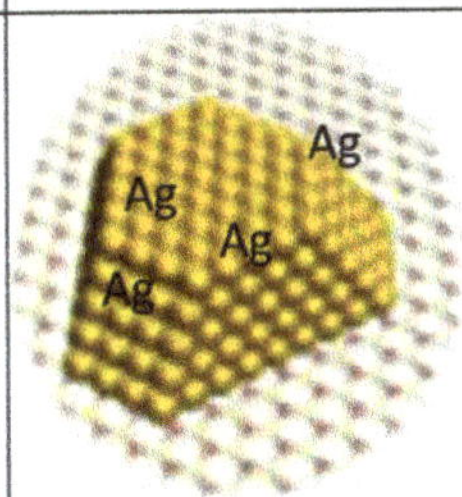	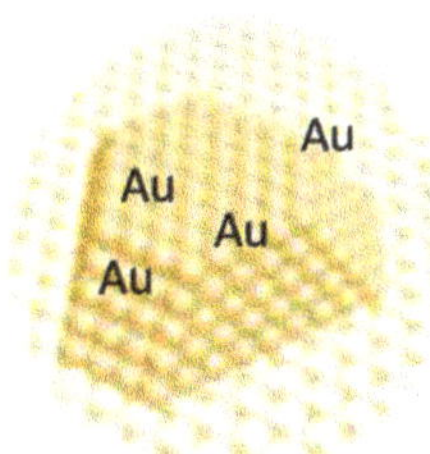	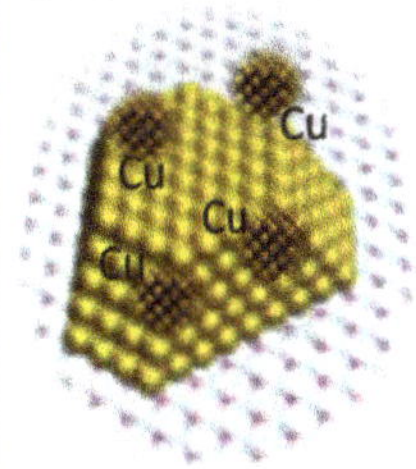	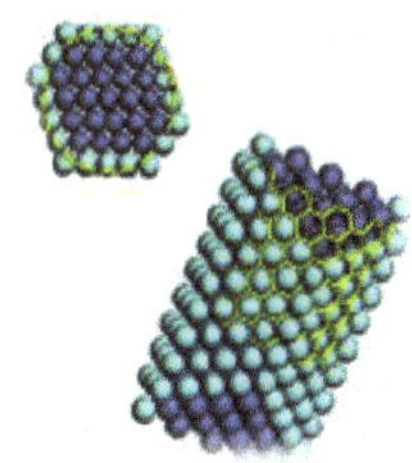
Surface Area (m^2/g)	>5.37	>62	>30	10-450
Synthesis	mixing the metallic salt solution, stabilization with a surfactant, and plant extract			
Cost*	$10/g	$135/ml	$1/gr	$1.58/g
References	Gottschalk et al., 2009; Fabrega et al., 2011; Benn and Westerhoff, 2008; Mueller et al., 2008; Huang et al., 2007; Schrang et al., 2010			

*Cost of NPs was collected from online marketing.

APPLICATION AND IMPLICATION OF NPs IN ENVIRONMENTAL REMEDIATION

Carbon-based NPs

The different types of NPs bring new prospects for the organic pollutants' management (Zhang et al., 2010). Compared with CNTs, graphene is more exciting. Graphene, a peerless 2-dimensional carbon-based nanostructure can possess atomic thickness, which was considerably attracted attention all around the world (Novoselov et al. 2004). This repercussion of NPs is highly related to their excellent physicochemical and mechanical features (Geim et al., 2007; Allen et al., 2010; Rao et al.,2009). Graphene has a large theoretical specific surface area (up to 3000 m^2g^{-1}), which shows its alternative adsorbents for the adsorption of contaminations (Zhu et al., 2010). The adsorption capacities of sulfonated graphene were one of the highest capabilities of today's NPs.

CNTs have attracted researchers' interest all around the world because of their peerless structure, physical and electronic properties. Since CNTs are promising alternative nano-based adsorbents, numerous studies have been focused on the evaluation of the fate of contaminations-loaded NPs in the water treatment industry. CNTs can offer several potential advantages over the traditional carbon-based adsorbents such as activated carbons and carbon fibers: i) surface of CNTs are highly hydrophobic, which exploits entropic free energy of contaminants; ii) electronic structure of CNTs is more favorable to support double-bonded carbon atoms interactions with aromatic structures of compounds; and, iii) shape and packing properties of CNTs can produce extra available sorption sites, which are not located in micropores. Overall, these peerless properties of CNTs can significantly increase the pollution uptake, initial sorption kinetics, and mitigate fouling by NOM, algal and wastewater effluent organic matter impacted waters, and also can improve the regeneration process.

Of particular importance is to compare the performance of CNT adsorbents with that of industry-standard materials such as granular activated carbons, over-concentration ranges that are relevant to water treatment conditions (e.g., extending from drinking water MCLs upwards several orders of magnitude). The exact types of adsorption sites and the mechanisms of adsorption need clarification and direct comparison with granular activated carbons.

Metal-based NPs

The metal-based NPs are implied into several industrial applications and implications, which can release from their wastewater including cosmetics, electronics, automotive industry, optics, coatings, semiconductors, paints, pigments, and energy applications to the environment (Keller et al., 2013; Duan et al., 2015; Naseem and Waseem, 2021). Unique intrinsic features of metal-based NPs such as large surface area to volume ratio (Brar et al., 2010). Size and crystallinity of metal-based NPs can highly influence their electric, magnetic, optical, and thermal properties. For instance, an order of magnitude reduction in diameter from 100 to 10 nm can decline the melting point values of two metal-based NPs by 80 and 120 ^{0}C, respectively (Auffan et al., 2009, 2010).

The catalytic activity of metal-based NPs presents a rich source for the chemical processes in industry and academic research. To date, metal-based NPs have applications and implications in diverse fields, such as chemical manufacturing, energy conversion and storage, biological and chemical applications, and environmental technologies. Some researchers have loaded metal-based NPs onto ACs to produce a new and cheap adsorbent for organic and inorganic contaminant removal. These ACs doped by metal-based NPs is revealed a high sorption capacity compared with ACs. Therefore, metal-based NPs have been found promising for cleaning water from contaminants because of their peerless optical and electrical properties.

Besides, metal-based NPs, in general, are stable without any special storage requirements for a long time. As seen in Figure 3, metal-based NPs can be applied in several treatment processes and approaches. Recent advances in catalysis or electrocatalysis process, metal-based NPs can be used to minimize mass transfer limitations and maximize contaminant removal by electrocatalytic processes, while providing quick and easy spent sorbent regeneration in situ. On the other hand, metal-based NPs can enhance the removal efficiency in the membrane process, while inhibiting membrane fouling. Meanwhile, for biological treatment technologies, metal-based NPs highly preferred to use for inducing the alteration of biological activities in the biological treatment (Bilal et al., 2019; Rasheed et al., 2019, 2018).

Figure 3. Application of carbon- and metal-based NPs

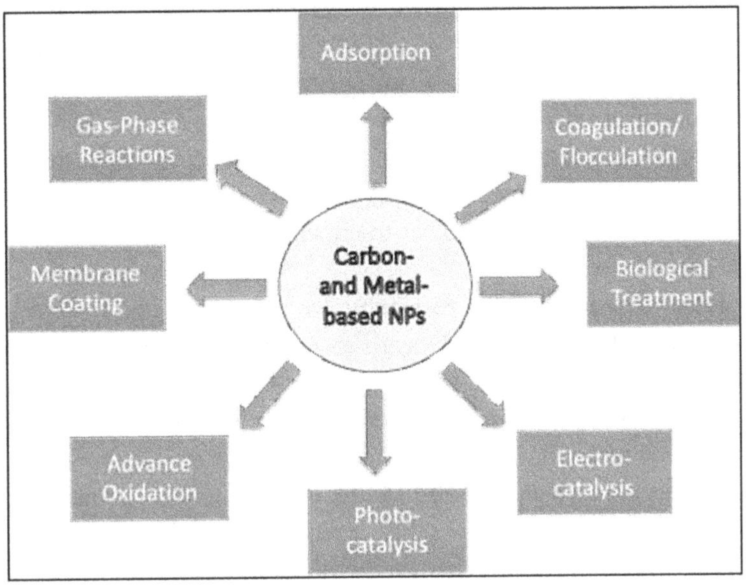

FUTURE RESEARCH DIRECTIONS

With the increasing utilization of NPs doped products, the release of NPs was highly increased into the environment. Since wastewater treatment plant, in general, contains activated sludge process, released NPs can easily trap in this stage. The possible impacts of NPs, such as metal- and metal oxide-based NPs (i.e., SiO_2, Al_2O_3, ZnO) on wastewater treatment performance were investigated. However, it was not deeply reported yet that the released carbon- and metal-based NPs were removed by activated sludge via adsorption, aggregation, and settling in wastewater treatment plants. Therefore, the main exiting pathway of engineered NPs is through the activated sludge process, which may eventually impact sludge digestion in the anaerobic reactor. Besides, the releasing of carbon- and metal-based NPs into the aquatic water can detect through different ways such as industrial and municipal wastewater discharges, stormwater runoff, and leaching from consumer products. Further research needs to be undertaken to understand the fate of released NPs in the different treatment processes.

The released NPs over the life cycle of the material were predicted to ultimately accumulate in landfills, and soils, while the predicted compartment eventually containing the lowest concentration of released NPs was in the air. This suggests that their chronical impacts and fats in aqueous waters should be carefully investigated in the future.

As described deeply in previous sections, aggregation of carbon- and metal-based NPs has been observed to decline the toxicity of NPs due to the limitations on chemical transformations and more hydrophobic surface of carbon- and metal-based NPs. In aqueous environments, NPs can be easily transported, aggregated, and settled out, since the highly dynamic nature of environmental conditions and particle characteristics can highly alter the aggregation structure of NPs. In the current literature, the researchers emphasized that aggregation of NPs is highly influenced by changing pH, ionic strength, and the presence of organic matters. The limited number of studies specifically reported the rate of aggregation and size of individual aggregates of NPs may change with higher concentrations of NPs. Further research concerning the interactions between concentration NPs and aqueous water will lead to better prediction for the NP interactions within the environmental systems.

Recently, the great interest in catalysis using metal-based NPs has prompted the synthesis and investigation of diverse highly functionalized NPs. According to the current literature, carbon-based catalysts (i.e., graphene) and metal-based NPs (i.e., Ag, Cu) have been widely investigated for the electrocatalysis process. With developing new materials, metal oxide-based NPs (i.e., TiO_2, SiO_2) and other mixture metal-based- NPs (i.e., magnetite-supported NPs, Ag/Cu mixed NPs) will be used for not only photo- and electrocatalysis process but also for the other treatment processes (advanced oxidation, membrane coating, and adsorption processes).

Overall, the laboratory and pilot scaled research have offered a good insight into adsorption and electrocatalysis mechanisms of contaminants by carbon- and metal-based NPs. The big challenge lies in understanding the removal mechanisms of different classes of fluorinated/chlorinated contaminants and heavy metals and/or competition among different contaminants on carbon- and metal-based NPs within complex natural, wastewater effluent, or algal organic matter impacted wasters. In addition, the metal- and metal oxide-based NPs adsorption database for the contaminants is much narrower than ACs and carbon-based NPs. Reporting more research in the current literature for adsorption of contaminants by carbon- and metal-based NPs with detailed characterization information can be useful in the near future.

CONCLUSION

In this study, the potential impacts of carbon- and metal-based engineered NPs on the mitigation of organic and inorganic in aquatic waters were comprehensively evaluated and compared side-by-side by their removal efficiency. The results showed that:

- The physicochemical properties of contaminants, NPs, and background water chemistry are significantly important variables on the contaminant uptake by carbon- and metal-based NPs.
- The characteristics of carbon- and metal-based NPs, can change the aggregation state of carbon- and metal-based NPs in water, and highly depend on background water characteristics including redox potential and quantity, pH, ionic strength, presence of organic matters (natural, algal or wastewater impacted), and temperature. Overall, all these individual parameters can influence the removal mechanisms.
- Considering the expected increase of carbon- and metal-based NPs production, understanding for the adsorption of contaminants by NPs has been known multiple important implications including assessing (i) the feasibility of using NPs as an adsorbent in engineered systems, (ii) the fate and transport of contaminants with NP materials in the aqueous water, and (iii) the potential toxicological impacts of NPs on the plants, animal, and environment when contaminants loaded NPs are escaped into the water.

- The surface of carbon-and metal-based NPs may be intentionally oxidized during the manufacturing process after its release to the environment. The oxygen on the NPs surface can inhibit compound adsorption due to the hydrophilic moieties or it can increase the dispersion of NPs, and thus expose more available adsorption sites on the NPs. Thus, the adsorption of contaminants by NPs with higher surface oxygen content and examining adsorption mechanisms are warranted.
- Overall, carbon-and metal-based NPs hold promise as high-performance nanomaterials for the nano-based sorption and catalysts technologies for environmental remediation, which have attracted scientific interest worldwide. Therefore, the high surface area of NPs brings new prospects for environmental pollutant management in aqueous water.

REFERENCES

Abd Ellah, N. H., & Abouelmagd, S. A. (2016). Surface functionalization of polymeric nanoparticles for tumor drug delivery: Approaches and challenges. *Expert Opinion on Drug Delivery*, 1–14. PMID:27426638

Abdel Ghani, N. T., El Chaghaby, G. A., & Helal, F. S. (2015). Individual and competitive adsorption of phenol and nickel onto multiwalled carbon nanotubes. *Journal of Advanced Research*, *6*(3), 405–415. doi:10.1016/j.jare.2014.06.001 PMID:26257938

Ahmad, M., Wang, J., Xu, J., Zhang, Q., & Zhang, B. (2020). Magnetic tubular carbon nanofibers as efficient Cu(II) ion adsorbent from wastewater. *Journal of Cleaner Production*, *252*, 119825. doi:10.1016/j.jclepro.2019.119825

Ahmed, S., Annu, S., & Yudha, S. S. (2016). Biosynthesis of gold nanoparticles: A green approach. *Journal of Photochemical Photobiology B: Biology*, *161*, 141–153. doi:10.1016/j.jphotobiol.2016.04.034 PMID:27236049

Aillon, K. L., Xie, Y., El-Gendy, N., Berkland, C. J., & Forrest, M. L. (2009). Effects of nanomaterial physicochemical properties on in-vivo toxicity. *Advanced Drug Delivery*, *61*(6), 457–466. doi:10.1016/j.addr.2009.03.010 PMID:19386275

Al-Qahtani, K. M. (2017). Cadmium removal from aqueous solution by green synthesis zero-valent silver nanoparticles with Benjamina leaves extract. *Egyptian Journal of Aquatic Research*, *43*(4), 269–274. doi:10.1016/j.ejar.2017.10.003

Ali, S., Khan, I., Khan, S. A., Sohail, M., Ahmed, R., Rehman, A., Ur Ansari, M. S., & Morsy, M. A. (2017). Electrocatalytic performance of Ni@Pt core–shell nanoparticles supported on carbon nanotubes for methanol oxidation reaction. *Electroanalytical Chemistry*, *795*, 17–25. doi:10.1016/j.jelechem.2017.04.040

Allen, M. J., Tung, V. C., & Kaner, R. B. (2010). Honeycomb Carbon: A Review of Graphene. *Chemical Reviews*, *110*(1), 132–145. doi:10.1021/cr900070d PMID:19610631

Alsabagh, A. M., Fathy, M., & Morsi, R.E. (2015). Preparation and characterization of chitosan/silver nanoparticle/copper nanoparticle/carbon nanotube multifunctional nanocomposite for water treatment: Heavy metals removal; kinetics, isotherms, and competitive studies. *Royal Science Chemistry Advance*, *5*, 55774–55783.

Amin, M.T., Alazba, A.A., & Manzoor, U. (2014). A review of removal of pollutants from water/wastewater using different types of nanomaterials. *Advance Material Science Engineering,* 825-910.

Apul, O. G., & Karanfil, T. (2015). Adsorption of synthetic organic contaminants by carbon nanotubes: A critical review. *Water Research, 68*, 34–55. doi:10.1016/j.watres.2014.09.032 PMID:25462715

Auffan, M., Bottero, J. Y., Chaneac, C., & Rose, J. (2010). Inorganic manufactured nanoparticles: How their physicochemical properties influence their biological effects in aqueous environments. *Nanomedicine (London), 5*(6), 999–1007. doi:10.2217/nnm.10.61 PMID:20735233

Auffan, M., Rose, J., Bottero, J. Y., Lowry, G. V., Jolivet, J. P., & Wiesner, M. R. (2009). Towards a definition of inorganic nanoparticles from an environmental, health and safety perspective. *Nature Nanotechnology, 4*(10), 634–641. doi:10.1038/nnano.2009.242 PMID:19809453

Azeez, L., Lateef, A., Adebisi, S. A., & Oyedeji, A. O. (2018). Novel biosynthesized silver nanoparticles from cobweb as adsorbent for Rhodamine B: Equilibrium isotherm, kinetic and thermodynamic studies. *Applied Water Science, 8*(1), 32. doi:10.100713201-018-0676-z

Benn, T. M., & Westerhoff, P. (2008). Nanoparticle silver released into water from commercially available sock fabrics. *Environmental Science & Technology, 42*(11), 4133–4139. doi:10.1021/es7032718 PMID:18589977

Bilal, M., Zhao, Y., Rasheed, T., Ahmed, I., Hassan, S. T., Nawaz, M. Z., & Iqbal, H. (2019). Biogenic nanoparticle–chitosan conjugates with antimicrobial, antibiofilm, and anticancer potentialities: Development and characterization. *International Journal of Environmental Research and Public Health, 16*(4), 598. doi:10.3390/ijerph16040598 PMID:30791374

Bottero, J.Y. & M.R. (2010). Wiesner Considerations in evaluating the physicochemical properties and transformations of inorganic nanoparticles in water. *Nanomedicine, 5*(6).

Boxall, A., Rudd, M. A., Brooks, B. W., Caldwell, D. J., Choi, K., Hickmann, S., Innes, E., Ostapyk, K., Staveley, J. P., Verslycke, T., Ankley, G. T., Beazley, K. F., Belanger, S. E., Berninger, J. P., Carriquiriborde, P., Coors, A., DeLeo, P. C., Dyer, S. D., Ericson, J. F., ... Van Der Kraak, G. (2012). Pharmaceuticals and Personal Care Products in the Environment: What Are the Big Questions? *Environmental Health Perspectives, 120*(9), 1221–1229. doi:10.1289/ehp.1104477 PMID:22647657

Brar, S. K., Verma, M., Tyagi, R. D., & Surampalli, R. Y. (2010). Engineered nanoparticles in wastewater and wastewater sludge - Evidence and impacts. *Waste Management (New York, N.Y.), 30*(3), 504–520. doi:10.1016/j.wasman.2009.10.012 PMID:19926463

Burdus, A.-C., Gherasim, O., Grumezescu, A.M., Mogoantă, L., Ficai, A., & Andronescu, E. (2018). Biomedical applications of silver nanoparticles: An up-to-date overview. *Nanomaterials, 8*, 681.

Cao, J., Elliott, D., & Zhang, W. (2005). Perchlorate reduction by nanoscale iron particles. *Journal of Nanoparticle Research, 7*(4 5), 499 506. doi:10.100711051 005 4412 x

Carolin, C. F., Kumar, P. S., Saravanan, A., Joshiba, G. J., & Naushad, M. (2017). Efficient techniques for the removal of toxic heavy metals from aquatic environment: A review. *Journal of Environmental Chemical Engineering, 5*(3), 2782–2799. doi:10.1016/j.jece.2017.05.029

Chen, H., Kou, X., Yang, Z., Ni, W., & Wang, J. (2008). Shape- and size dependent refractive index sensitivity of gold nanoparticles. *Langmuir, 24*(10), 5233–5237. doi:10.1021/la800305j PMID:18435552

Chugh, H., Sood, D., Chandra, I., Tomar, V., Dhawan, G., & Chandra, R. (2018). Role of gold and silver nanoparticles in cancer nanomedicine. *Artificial Cells, Nanomedicine, and Biotechnology, 46*(sup1), 1210–1220. doi:10.1080/21691401.2018.1449118 PMID:29533101

Duan, H., Wang, D., & Li, Y. (2015). Green chemistry for nanoparticle synthesis. *Chemical Society Reviews, 44*(16), 5778–5792. doi:10.1039/C4CS00363B PMID:25615873

Duman, O., Tun, S., Polat, T. G., & Bozoglan, B. K. I. (2016). Synthesis of magnetic oxidized multi-walled carbon nanotube- k-carrageenan-Fe_3O_4 nanocomposite adsorbent and its application in cationic Methylene Blue dye adsorption. *Carbohydrate Polymers, 147*, 79–88. doi:10.1016/j.carbpol.2016.03.099 PMID:27178911

Ersan, G., Apul, O. G., Perreault, F., & Karanfil, T. (2017). Adsorption of organic contaminants by graphene nanosheets: A review. *Water Research, 126*, 385–398. doi:10.1016/j.watres.2017.08.010 PMID:28987890

Ersan, G., & Ersan, M. S. (2021). Are carbon-based nanomaterials for the adsorption of organic contaminants perform better than nanoplastics (NPs) and microplastics (MPs)? *Journal of International Environmental Engineering and Science, 16*(2), 72–81.

Ersan, G., Kaya, Y., Apul, O. G., & Karanfil, T. (2016). Adsorption of organic contaminants by graphene nanosheets, carbon nanotubes and granular activated carbons under different natural organic matter preloading conditions. *The Science of the Total Environment, 565*, 811–817. doi:10.1016/j.scitotenv.2016.03.224 PMID:27107611

Ersan, G., Kaya, Y., Ersan, M. S., Apul, O. G., & Karanfil, T. (2019). Adsorption kinetics and aggregation for three classes of carbonaceous adsorbents in the presence of natural organic matter. *Chemosphere, 229*, 515–524. doi:10.1016/j.chemosphere.2019.05.014 PMID:31100622

Fabrega, J., Luoma, S. N., Tyler, C. R., Galloway, T. S., & Lead, J. R. (2011). Silver nanoparticles: Behavior and effects in the aquatic environment. *Environment International, 37*(2), 517–531. doi:10.1016/j.envint.2010.10.012 PMID:21159383

Geim, A. K., & Novoselov, K. S. (2007). The rise of graphene. *Nature Materials, 6*(3), 183–191. doi:10.1038/nmat1849 PMID:17330084

Gottschalk, F., & Nowack, B. (2011). The release of engineered nanomaterials to the environment. *Journal of Environmental Monitoring, 13*(5), 1145. doi:10.1039/c0em00547a PMID:21387066

Gottschalk, F., Sonderer, T., Scholz, R. W., & Nowack, B. (2009). Modeled environmental concentrations of engineered nanomaterials (TiO2, ZnO, Ag, CNT, Fullerenes) for different regions. *Environmental Science & Technology, 43*(24), 9216–9222. doi:10.1021/es9015553 PMID:20000512

Gupta, V. K., Moradi, O., Tyagi, I., Agarwal, S., Sadegh, H., Shahryari-Ghoshekandi, R., Makhlouf, A. S. H., Goodarzi, M., & Garshasbi, A. (2016). Study on the removal of heavy metal ions from industry waste by carbon nanotubes: Effect of the surface modification: A review. *Critical Reviews in Environmental Science and Technology*, *46*(2), 93–118. doi:10.1080/10643389.2015.1061874

Gupta, V. K., Tyagi, I., Sadegh, H., Shahryari-Ghoshekand, R., Makhlouf, A. S. H., & Maazinejad, B. (2015). Nanoparticles as adsorbent; a positive approach for removal of noxious metal ions, a review. *Science. Technology and Development*, *34*(3), 195–214. doi:10.3923td.2015.195.214

Gurunathan, S., Park, J. H., Han, J. W., & Kim, J. H. (2015). Comparative assessment of the apoptotic potential of silver nanoparticles synthesized by Bacillus tequilensis and Calocybe indica in MDA-MB-231 human breast cancer cells: Targeting p53 for anticancer therapy. *International Journal of Nanomedicine*, *10*, 4203–4222. doi:10.2147/IJN.S83953 PMID:26170659

Hasnain, M. S., & Nayak, A. K. (2019). Applications of carbon nanotubes. Springer Briefs Application Science Technology, 33–36.

Hering, J. G., Waite, T. D., Luthy, R. G., Drewes, J. E., & Sedlak, D. L. (2013). A changing framework for urban water systems. *Environmental Science & Technology*, *47*(19), 10721–10726. doi:10.1021/es4007096 PMID:23650975

Huang, X., Jain, P. K., El-Sayed, I. H., & El-Sayed, M. A. (2007). Gold nanoparticles: Interesting optical properties and recent applications in cancer diagnostics and therapy. *Nanomedicine (London)*, *2*(5), 681–693. doi:10.2217/17435889.2.5.681 PMID:17976030

Jain, P. K., Lee, K. S., El-Sayed, I. H., & El-Sayed, M. A. (2006). Calculated absorption and scattering properties of gold nanoparticles of different size, shape, and composition: Applications in biological imaging and biomedicine. *The Journal of Physical Chemistry B*, *110*(14), 7238–7248. doi:10.1021/jp057170o PMID:16599493

Jain, T. K., Roy, I., De, T. K., & Maitra, A. N. J. (1998). Nanometer Silica Particles Encapsulating Active Compounds: A Novel Ceramic Drug Carrier. *Journal of the American Chemical Society*, *120*(43), 11092–11095. doi:10.1021/ja973849x

Ji, L., Chen, W., Xu, Z., Zheng, S., & Zhu, D. (2013). Graphene nanosheets and graphite oxide as promising adsorbents for removal of organic contaminants from aqueous solution. *Journal of Environmental Quality*, *42*(1), 191–198. doi:10.2134/jeq2012.0172 PMID:23673754

Jia, X., & Wei, F. (2017). Advances in Production and Applications of Carbon Nanotubes. *Topics in Current Chemistry (Z)*, *375*(1), 18. doi:10.100741061-017-0102-2 PMID:28138948

Jiang, L., Liu, Y., Liu, S., Zeng, G., Hu, X., Hu, X., Guo, Z., Tan, X., Wang, L., & Wu, Z. (2017). Adsorption of Estrogen Contaminants by Graphene Nanomaterials under Natural Organic Matter Preloading: Comparison to Carbon Nanotube, Biochar, and Activated Carbon. *Environmental Science & Technology*, *51*(11), 6352–6359. doi:10.1021/acs.est.7b00073 PMID:28494154

Kaegi, R., Voagelin, A., Ort, C., Sinnet, B., Thalmann, B., Krismer, J., Hagendorfer, H., Elumelu, M., & Mueller, E. (2013). Fate and Transformation of Silver Nanoparticles in Urban Wastewater Systems. *Water Research*, *47*(12), 3866–3877. doi:10.1016/j.watres.2012.11.060 PMID:23571111

Karki, H. P., Ojha, D. P., Joshi, M. K., & Kim, H. J. (2018). Effective reduction of p-nitrophenol by silver nanoparticle loaded on magnetic Fe_3O_4/ATO nano-composite. *Applied Surface Science*, *435*(30), 599–608. doi:10.1016/j.apsusc.2017.11.166

Khadir, A., Motamedi, M., Negarestani, M., Sillanpaa, M., & Sasani, M. (2020). Preparation of a nano bio-composite based on cellulosic biomass and conducting polymeric nanoparticles for ibuprofen removal: Kinetics, isotherms, and energy site distribution. *International Journal of Biological Macromolecules*, *162*, 663–677. doi:10.1016/j.ijbiomac.2020.06.095 PMID:32544583

Khan, I., Saeed, K., & Khan, I. (2019). Nanoparticles: Properties, applications, and Toxicities. *Arabian Journal of Chemistry*, *12*(7), 908–931. doi:10.1016/j.arabjc.2017.05.011

Khare, P., Sharma, A., Verma, N. (2014). Synthesis of phenolic precursor-based porous carbon beads in situ dispersed with copper-silver bimetal nanoparticles for antibacterial applications. *Journal of Colloid Interface Science*, *418*, 216-224.

Koedrith, P., Thasiphu, T., Weon, J., Boonprasert, R., Tuitemwong, K., & Titemwo, P. (2015). Recent trends in rapid environmental monitoring of pathogens and toxicants: Potential of nanoparticle-based biosensor and applications. *TheScientificWorldJournal*, *2015*, 12. doi:10.1155/2015/510982 PMID:25884032

Kommu, A., & Singh, J. K. (2020). A review on graphene-based materials for removal of toxic pollutants from wastewater. *Soft Materials*, *18*(2-3), 297–322. doi:10.1080/1539445X.2020.1739710

Lee, S. H., & Jun, B. H. (2019). Silver Nanoparticles: Synthesis and Application for Nanomedicine. *International Journal of Molecular Sciences*, *20*(4), 865. doi:10.3390/ijms20040865 PMID:30781560

Lerma-García, M.Á., Simó-Alfonso, E.F., Ríos, Á., & Zougagh, M. (2014). Synthesis of gold nanoparticles using phenolic acids and its application in catalysis. *Journal of Material and Environmental Science*, (6), 1919–1926.

Li, S., Zhang, L., Wang, T., Li, L., Wang, C., & Su, Z. (2015). The facile synthesis of hollow Au nanoflowers for synergistic chemo-photothermal cancer therapy. *Chemical Communications*, *51*(76), 14338–14341. doi:10.1039/C5CC05676D PMID:26299901

Li, W. R., Xie, X. B., Shi, Q. S., Zeng, H. Y., Ou-Yang, Y. S., & Chen, Y. B. (2010). Antibacterial activity and mechanism of silver nanoparticles on Escherichia coli. *Applied Microbiology and Biotechnology*, *8*(4), 1115–1122. doi:10.100700253-009-2159-5 PMID:19669753

Liu, H., & Qiu, H. (2020). Recent advances of 3D graphene-based adsorbents for sample preparation of water pollutants: A review. *Chemical Engineering Journal*, *393*, 124691. doi:10.1016/j.cej.2020.124691

Lowry, G. V., Gregory, K. B., Apte, S. C., & Lead, J. R. (2012). Transformations of nanomaterials in the environment. *Environmental Science & Technology*, *46*(13), 6893–6899. doi:10.1021/es300839e PMID:22582927

Majzlik, P., Strasky, A., Adam, V., Nemec, M., Trnkova, L., & Zehnalek, J. (2011). Influence of zinc (II) and copper (II) Ions on Streptomyces bacteria revealed by electrochemistry. *International Journal of Electrochemical Science*, *6*, 2171-2191.

Mansha, M., Khan, I., Ullah, N., & Qurashi, A. (2017). Synthesis, characterization and visible-light-driven photoelectrochemical hydrogen evolution reaction of carbazole-containing conjugated polymers. *International Journal of Hydrogen Energy*, *42*(16), 10952–10961. doi:10.1016/j.ijhydene.2017.02.053

Mueller, N. C., Braun, J., Bruns, J., Černík, M., Rissing, P., Rickerby, D., & Nowack, B. (2012). Application of nanoscale zero valent iron (NZVI) for groundwater remediation in Europe. *Environmental Science and Pollution Research International*, *19*(2), 550–558. doi:10.100711356-011-0576-3 PMID:21850484

Mueller, N. C., & Nowack, B. (2008). Exposuremodeling of engineered nanoparticles in the environment. *Environmental Science & Technology*, *42*(12), 4447–4453. doi:10.1021/es7029637 PMID:18605569

Mukherjee, P., Ahmad, A., Mandal, D., Senapati, S., Sainkar, S. R., Khan, M. I., Renu, P., Ajaykumar, P. V., Alam, M., & Kumar, R. (2001). Fungus-mediated synthesis of silver nanoparticles and their immobilization in the mycelial matrix: A novel biological approach to nanoparticle synthesis. *Nano Letters*, *1*(10), 515–519. doi:10.1021/nl0155274

Mukherjee, R., Kumar, R., Sinha, A., Lama, Y., & Saha, A. K. (2016). A review on synthesis, characterization, and applications of nano zero valent iron (nZVI) for environmental remediation. *Critical Reviews in Environmental Science and Technology*, *35*(5), 360–384. doi:10.1080/10643389.2015.1103832

Naseem, T., & Waseem, M. (2021). A comprehensive review on the role of some important nanocomposites for antimicrobial and wastewater applications. *Chemical Reviews*, *107*, 2411–2502.

Novoselov, K. S., Geim, A. K., Morozov, S. V., Jiang, D., Zhang, Y., Dubonos, S. V., Grigorieva, I. V., & Firsov, A. A. (2004). Electric field effect in atomically thin carbon films. *Science*, *306*(5696), 666–669. doi:10.1126cience.1102896 PMID:15499015

Nowack, B., Ranville, J. F., Diamond, S., Gallego-Urrea, A., Metcalfe, C., Rose, J., Horne, N., Koelmans, A. A., & Klaine, S. J. (2012). Potential scenarios for nanomaterial release and subsequent alteration in the environment. *Environmental Toxicology and Chemistry*, *31*(1), 50–59. doi:10.1002/etc.726 PMID:22038832

O'Neal, D. P., Hirsch, L. R., Halas, N. J., Payne, J. D., & West, J. L. (2004). Photothermal tumor ablation in mice using near infrared-absorbing nanoparticles. *Cancer Letters*, *209*(2), 171–176. doi:10.1016/j.canlet.2004.02.004 PMID:15159019

OECD. (2010). List of Manufactured Nanomaterials and List of Endpoints for Phase One of the Sponsorship Program for the Testing of Manufactured Nanomaterials. Organization for Economic Co-operation and Development.

Rao, C. N. R., Sood, A. K., Subrahmanyam, K. S., & Govindaraj, A. (2009). Graphene: The new two-dimensional nanomaterial. *Angewandte Chemie International Edition*, *48*(42), 7752–7777. doi:10.1002/anie.200901678 PMID:19784976

Rao, J. P., & Geckeler, K. E. (2011). Polymer nanoparticles: Preparation techniques and size-control parameters. *Progress in Polymer Science*, *36*(7), 887–913. doi:10.1016/j.progpolymsci.2011.01.001

Rasheed, T., Bilal, M., Li, C., Nabeel, F., Khalid, M., & Iqbal, H. M. (2018). Catalytic potential of biosynthesized silver nanoparticles using Convolvulusarvensis extract for the degradation of environmental pollutants. *Journal of Photochemistry and Photobiology. B, Biology*, *181*, 44–52. doi:10.1016/j.jphotobiol.2018.02.024 PMID:29499463

Rasheed, T., Nabeel, F., Bilal, M., & Iqbal, H. M. (2019). Biogenic synthesis and characterization of cobalt oxide nanoparticles for catalytic reduction of direct yellow-142 and methyl orange dyes. *Biocatalysis and Agricultural Biotechnology*, *19*, 101154. doi:10.1016/j.bcab.2019.101154

Sabzehmeidani, M. M., Mahnaee, S., Ghaedi, M., Heidari, H., & Roy, W. L. (2021). Carbon based materials: A review of adsorbents for inorganic and organic compounds. *Material Advances*, *2*(2), 598–627. doi:10.1039/D0MA00087F

Sadegh, H., Ali, G. A. M., Gupta, V. K., Makhlouf, A. S. H., Shahryari-ghoshekandi, R., Nadagouda, M. N., Sillanpa, M., & Megiel, E. (2017). The role of nanomaterials as effective adsorbents and their applications in wastewater treatment. *Journal of Nanostructure in Chemistry*, *7*(1), 1–14. doi:10.100740097-017-0219-4

Sagir, M., & Tahir, M. B. (2021). Role of Nanocatalyst (Photocatalysts) for Wastewater Treatment. *Current Analytical Chemistry*, *17*(2), 138–149. doi:10.2174/1573411016666200226091404

Saka, C. (2018). Overview on the surface functionalization mechanism and determination of surface functional groups of plasma treated carbon nanotubes. *Critical Reviews in Analytical Chemistry*, *48*(1), 1–14. doi:10.1080/10408347.2017.1356699 PMID:28722465

Shad, S., Belinga-Desaunay-Nault, M.-F. A., Sohail, Bashir, N., & Lynch, I. (2020). Removal of contaminants from canal water using microwave synthesized zero valent iron nanoparticles. *Environmental Science. Water Research & Technology*, *6*(11), 3057–3065. doi:10.1039/D0EW00157K

Shi, Z., Wang, L., Deng, L., Peng, X., & Yuan, M. (2016). Adsorptive removal of Cr(VI) from aqueous solution using NiFe2O4/ ZnAl-LDH. *Chine Journal of Environmental Engineering*, *10*, 4635–4642.

Sigmund, W., Yuh, J., Park, H., Maneeratana, V., Pyrgiotakis, G., Daga, A., Taylor, J., & Nino, J. C. (2006). Processing and structure relationships in electrospinning of ceramic fiber systems. *Journal of the American Ceramic Society*, *89*(2), 395–407. doi:10.1111/j.1551-2916.2005.00807.x

Taghavi Fardood, S., Atrak, K., & Ramazani, A. (2017). Green synthesis using tragacanth gum and characterization of Ni-Cu-Zn ferrite nanoparticles as a magnetically separable photocatalyst for organic dyes degradation from aqueous solution under visible light. *Journal of Materials Science Materials in Electronics*, *28*(14), 10739–10746. doi:10.100710854-017-6850-5

Tamer, U., Cetin, D., Suludere, Z., Boyaci, I. H., Temiz, H. T., Yegenoglu, H., Daniel, P., Dinçer, I., & Elerman, Y. (2013). Gold-coated iron composite nanospheres targeted the detection of Escherichia coli. *International Journal of Molecular Sciences*, *14*(3), 6223–6240. doi:10.3390/ijms14036223 PMID:23507756

Tang, H., Zhao, Y., Yang, X., Liu, D., Shao, P., Zhu, Z., Shan, S., Cui, F., & Xing, B. (2017). New insight into the aggregation of graphene oxide using molecular dynamics simulations and extended Derjaguin-Landau- Verwey-Overbeek theory. *Environmental Science & Technology*, *51*(17), 9674–9682. doi:10.1021/acs.est.7b01668 PMID:28771343

Thomas, S. C., Harshita, B. S. P., Mishra, P., & Talegaonkar, S. (2015). Ceramic Nanoparticles: Fabrication Methods and Applications in Drug Delivery. *Current Pharmaceutical Design*, *21*(42), 6165–6188. doi:10.2174/1381612821666151027153246 PMID:26503144

Tiwari, J. N., Vij, V., Kemp, K. C., & Kim, K. S. (2015). Engineered carbon-nanomaterial-based electrochemical sensors for biomolecules. *ACS Nano*, *10*(1), 46–80. doi:10.1021/acsnano.5b05690 PMID:26579616

Valiyeva, G. G., Bavasso, I., Di Palma, L., Hajiyeva, S. R., Ramazanov, M. A., & Hajiyeva, F. V. (2019). Synthesis of Fe/ Ni Bimetallic Nanoparticles and Application to the Catalytic Removal of Nitrates from Water. *Nanomaterials (Basel, Switzerland)*, *9*(8), 1130. doi:10.3390/nano9081130 PMID:31390768

Velusamy, S., Roy, A., Sundaram, S., & Kumar Mallick, T. (2021). A review on heavy metal ions and containing dyes removal through graphene oxide-based adsorption strategies for textile wastewater treatment. *Chemical Record (New York, N.Y.)*, *21*(7), 1–42. doi:10.1002/tcr.202000153 PMID:33539046

Wang, B., Zhang, F., He, S., Huang, F., & Peng, Z. (2014). Adsorption behavior of reduced graphene oxide for removal of heavy metal ions. *Asian Journal of Chemistry*, *26*(15), 4901–4906. doi:10.14233/ajchem.2014.17024

Wang, J., Chen, B., & Xing, B. (2016). Wrinkles and folds of activated graphene nanosheets as fast and efficient adsorptive sites for hydrophobic organic contaminants. *Environmental Science & Technology*, *50*(7), 3798–3808. doi:10.1021/acs.est.5b04865 PMID:26938576

Wang, J., Zhang, J., Han, L., Wang, J., Zhu, L., & Zeng, H. (2021). Graphene-based materials for adsorptive removal of pollutants from water and underlying interaction mechanism. *Advances in Colloid and Interface Science*, *289*, 102360. doi:10.1016/j.cis.2021.102360 PMID:33540288

Wang, S., Li, X., Liu, Y., Zhang, C., Tan, X., Zeng, G., Song, B., & Jiang, L. (2018). Nitrogen-containing amino compounds functionalized graphene oxide: synthesis, characterization, and application for the removal of pollutants from wastewater: a review. *Journal of Hazardous Materials*, *342*, 177–191. doi:10.1016/j.jhazmat.2017.06.071 PMID:28829983

Wang, X., Liu, B., Lu, Q., & Qu, Q. (2014). Graphene-based materials: Fabrication and application for adsorption in analytical chemistry. *Journal of Chromatography. A*, *1362*, 1–15. doi:10.1016/j.chroma.2014.08.023 PMID:25160951

Wei, L., Lu, J., Xu, H., Patel, A., Chen, Z.-S., & Chen, G. (2015). Silver nanoparticles: Synthesis, properties, and therapeutic applications. *Drug Discovery Today*, *20*(5), 595–601. doi:10.1016/j.drudis.2014.11.014 PMID:25543008

Weinberg, H., Galyean, A., & Leopold, M. (2011). Evaluating engineered nanoparticles in natural waters. *Trends in Analytical Chemistry*, *30*(1), 72–83. doi:10.1016/j.trac.2010.09.006

Wigginton, N., Yeston, J., & Malakoff, D. (2012). More treasure than trash. *Science, 337*(6095), 662–663. doi:10.1126cience.337.6095.662 PMID:22879497

Xiao, T., Huang, J., Wang, D., Meng, T., & Yang, X. (2019). Au and Au-Based nanomaterials: Synthesis and recent progress in electrochemical sensor applications. *Talanta, 206*, 120210. doi:10.1016/j.talanta.2019.120210 PMID:31514855

Yang, K., Wu, W., Jing, Q., & Zhu, L. (2008). Aqueous adsorption of aniline, phenol, and their substitutes by multi-walled carbon nanotubes. *Environmental Science & Technology, 42*(21), 7931–7936. doi:10.1021/es801463v PMID:19031883

Yang, Z. F., Li, L. Y., Te Hsieh, C., & Juang, R. S. (2018). Coprecipitation of magnetic Fe3O4 nanoparticles onto carbon nanotubes for removal of copper ions from aqueous solution. *Journal of the Taiwan Institute of Chemical Engineers, 82*, 56–60. doi:10.1016/j.jtice.2017.11.009

Ye, N., Wang, Z., Wang, S., Fang, H., & Wang, D. (2018). Aqueous aggregation and stability of graphene nanoplatelets, graphene oxide, and reduced graphene oxide in simulated natural environmental conditions: Complex roles of surface and solution chemistry. *Environmental Science and Pollution Research International, 25*(11), 10956–10965. doi:10.100711356-018-1326-6 PMID:29399742

Zhang, L., Pan, F., Liu, X., Yang, L., Jiang, X., Yang, J., & Shi, W. (2013). Multi-walled carbon nanotubes as sorbent for recovery of endocrine disrupting compound-bisphenol F from wastewater. *Chemical Engineering Journal, 218*, 238–246. doi:10.1016/j.cej.2012.12.046

Zhang, N., Qiu, Y., Sun, H., Hao, J., Chen, J., Xi, J., Liu, J., He, B., & Bai, Z. W. (2021). Substrate-Assisted Encapsulation of Pd-Fe Bimetal Nanoparticles on Functionalized Silica Nanotubes for Catalytic Hydrogenation of Nitroarenes and Azo Dyes. *ACS Applied Nano Materials, 4*(6), 5854–5863. doi:10.1021/acsanm.1c00777

Zhang, S., Shao, T., & Karanfil, T. (2011). The effects of dissolved natural organic matter on the adsorption of synthetic organic chemicals by activated carbons and carbon nanotubes. *Water Research, 45*(3), 1378–1386. doi:10.1016/j.watres.2010.10.023 PMID:21093009

Zhang, S. J., Shao, T., Bekaroglu, S. S. K., & Karanfil, T. (2010). Adsorption of synthetic organic chemicals by carbon nanotubes: Effects of background solution chemistry. *Water Research, 44*(6), 2067–2074. doi:10.1016/j.watres.2009.12.017 PMID:20071001

Zhang, S. J., Shao, T., Kose, H. S., & Karanfil, T. (2010). Adsorption of aromatic compounds by carbonaceous adsorbents: A comparative study on granular activated carbon, activated carbon fiber, and carbon nanotubes. *Environmental Science & Technology, 44*(16), 6377–6383. doi:10.1021/es100874y PMID:20704238

Zhong, G., Liu, J., & Liu, X. (2015). A fast colorimetric assay for lead detection using label-free gold nanoparticles (AuNPs). *Micromachines, 6*(4), 462–472. doi:10.3390/mi6040462

Zhu, Y., Murali, S., Stoller, M. D., Ganesh, K. J., Cai, W., Ferreira, P. J., Pirkle, A., Wallace, R. M., Cychosz, K. A., Thommes, M., Su, D., Stach, E. A., & Ruoff, R. S. (2011). Carbon-Based Supercapacitors Produced by Activation of Graphene. *Science, 332*(6037), 1537–1541. doi:10.1126cience.1200770 PMID:21566159

Zuber, A., Purdey, M., Schartner, E., Forbes, C., Hoek, B., Giles, D., Abell, A., Monro, T., & Ebendorff, H. (2016). Detection of gold nanoparticles with different sizes using absorption and fluorescence-based method. *Sensors and Actuators. B, Chemical, 227,* 117–127. doi:10.1016/j.snb.2015.12.044

ADDITIONAL READING

Vladimir Torchilin. (2020). *Handbook of Materials for Nanomedicine: Metal-Based and Other Nanomaterials (Jenny Stanford Series on Biomedical Nanotechnology).* Jenny Stanford Publishing.

KEY TERMS AND DEFINITIONS

π-π Coupling Interaction: A type of non-covalent interaction that joins π systems.

π-π Electron-Donor-Acceptor (EDA) Relationships: Combine a cation-π force with a π–π EDA force resulting from charge polarization of the ring's quadrupole.

Carbon Nanomaterials: A growing family of nanosized material which formed from one or more carbon-based graphene sheets such as black nanoparticles, fullerenes, carbon nanotubes, graphene, fibers, and other related forms.

Electrostatic Interactions: Comprise the attractive and/or repulsive interactions between charged molecules.

Hydrophobic Interactions: The relationship between water and hydrophobes (low water-soluble) molecules.

Metal Nanoparticles: Submicron scale entities made from pure metals such as gold, platinum, silver, titanium, zinc, cerium, iron, and thallium.

Van Der Waals Interactions: Occur when adjacent molecules come close enough that their outer electron clouds just barely touch.

Chapter 10
Use of Nanoparticles for Environmental Remediation

Abu Barkat Md Gulzar
Assam University, Silchar, India

Pranab Behari Mazumder
Assam University, Silchar, India

ABSTRACT

Continuous release of contaminants into the environment and their toxic effects on the living world prompted public and scientific interests in the remediation of pollutants. The release of toxic pollutants not only hinders the soil environments but also disturbs the aquatic environments. The hindrance in soil and aquatic environments causes a serious warning to the productive environment and food security. Thus, excluding and counteracting the pollutants from the environment is necessary to develop a safe and productive environment. Currently, numerous techniques are used to remediate contaminants from the environment; moreover, nanoparticles are one of such most widely used techniques for environmental clean-up. This chapter briefly discusses the need for environmental remediation and nanoparticles for environmental clean-up, especially soil and water. Additionally, the complications and current research trends of using nanoparticles in environmental remediation are briefly discussed for determining future research prospects.

INTRODUCTION

Soil, water, and air are the most precious components of the living world. They perform the most significant functions like life supports, biodiversity conservation, carbon storage, food production, etc. To attain sustainable development requires a safe, healthy and productive environment. However, our environment is under serious threat of pollution. Pollution can be defined as the presence and persistence of toxic and hazardous substances into the environment. The concentration of the pollutants must be present in the environment below the threshold limit; beyond these limits, it can be harmful to the living world. Over the last decades, to fulfill the needs and demands of the growing populations, rapid industrializations,

DOI: 10.4018/978-1-7998-8936-6.ch010

various developmental processes, and synthetic chemicals in agriculture continuously release waste as a by-product and pollute the environment. The release of toxic contaminants hinders the soil environments and also disturbs the aquatic environments(Ukaogo et al., 2020). The toxic contaminants from the environment slowly make entry into the food chain and cause various complications in the living world. Excessive concentration of the pollutants in the environment, their interactions, and accumulation into the plants disturb plants' growth and developmental processes and ultimately reduce yield and productivity and sometimes cause death. The reduction in plant productivity and the exponential growth of the human population are some of the reasons for upcoming risks of food security(Rojas et al., 2016; Sun et al., 2017).

Thus, exclusion and counteracting the pollutants from the environment is necessary to develop a safe and productive environment. Currently, numerous techniques are used to remediate contaminants from the environment; but the techniques are not still enough for complete remediation of current and upcoming environmental contaminants. In parallel, the rapid advancement in nanotechnology has gained increased attention to the use of nanoparticles (NPs) for environmental applications. Nano-remediation is currently attempting to provide a novel and practical solution for the abatement of contaminants from the environment by employing an important role in source reduction, detection, monitoring, and remediation. Nano-remediation has the ability to conquer some challenges in environmental cleanup by increasing the overall efficiency of the remediation process in a cost-effective manner. However, the currently available wastewater and soil remediation techniques like surfactant co-solvent flushing, chemical oxidation, soil washing, thermal treatment, etc. are quite expensive, partially effective, and time-consuming; and hence, could be slowly get rid of over the time. In this case, nano-remediation technique has received a wide range of research acceptance as well as practical implications due to its effectiveness, rapidity, and eco-friendly as well as budget-friendly behaviour (Patil et al., 2016).

Nano-remediation techniques use nano sized particles like zeolite, nano-fibre, carbon nanotubes, metallic oxides, bimetallic NPs, etc. for the abatement of pollutants. US EPA 2004, reported 44 locations across the seven countries where nano-remediation techniques have been found successful for abatement of contaminants from contaminated sites. It has been found that the nano-remediation techniques reduce the overall cost as well as time frame as compared to the conventional remediation techniques (Patil et al., 2016; US EPA, 2004). In recent years, NPs are widely used to abate contaminants from the environment(Guerra et al., 2018). NPs possess high reactivity with large and specific surface areas and hence act as sensors, catalysts, and adsorbents. The large surface area to mass ratio enhances the adsorption capabilities of NPs. Similarly, the large radii of curvatures and the reduced size provide NPs a highly reactive surface. Additionally, the mobility of NPs is very high due to their small size and can be reached at any part of the medium. These unique properties of NPs make them the most efficient tool for environmental remediation(Khan et al., 2019).

Nanotechnology now becomes a known area of research providing diverse applications. Nanotechnology produces various nanoscale NPs that can be used in medicine, agriculture, environmental protection, etc. This chapter will provide a brief idea on the need for remediation and the use of NPs for environmental remediation to the students and also help the researchers as a thorough reference to understand the current research trends and the complications to find future research questions. Additionally, the complications and current research trends of using NPs in environmental remediation are briefly discussed for determining future research prospects.

BACKGROUND

The rapid growth of the human population, their needs, demands, changing lifestyle, and modernization continuously increase industries for rapid production/manufacturing in a short period. With enhanced production, the improper disposal of toxic wastes is constantly polluting the environment. Again, for improving the agricultural output, increased use of chemical fertilizers, pesticides, herbicides, and modern farming practices also has severe bearings on environments(Kalita & Baruah, 2020). The toxic contaminants beyond the threshold limit not only affect the environment but also causes a severe threat to the food security and living world, including humans(Ali et al., 2019). Contaminants are mainly categorized into organic and inorganic groups. The inorganic contaminants are considered as most toxic contaminants due to their non-destroyable and non-biodegradable nature. Radionuclides, heavy metals, metalloids, etc., are the most common form of inorganic pollutants. On the other hand, phenols, azo-dye, pesticides, polyaromatic hydrocarbons, etc., are examples of organic pollutants which are less toxic than inorganic pollutants(Bharagava et al., 2019). Among these organic pollutants, "persistent organic pollutants (POPs)" are of specific concern due to their long life span and toxicity. POPs are highly stable toxic chemical compounds that can persist long and move long distances through water and air. Due to high persistency and hydrophobic nature, these POPs regularly aggregate in fat tissues. Their concentrations get biomagnified; thus, their concentration increases in every higher level of the food chain. The anthropogenic processes generate POPs either intentionally as pesticides, dyes, plastics, electronic devices, etc., or unintentionally as by-products during the production of other valuable substances (Pham & Sillanpää, 2020). The continuous release of organic and inorganic toxic compounds and their persistence into the environment causes a severe threat to plants and animals, especially humans, as shown in ***Table 1.*** If the current trend of environmental pollution persists, the earth will be rendered unsuitable for life forms in the near future. To conquer this issue, scientists, researchers, and various governmental and non-governmental agencies are continuously working to develop techniques to remediate contaminants from the environment.

NANOPARTICLES IN ENVIRONMENTAL CLEAN-UP

Nanotechnology is a rising field of science and it has already shown its efficacy in bio-application such as medicine, pharmaceuticals, and agriculture. The advancement of nanotechnology and the production of different NPs provide new insight into the field of remediation. The use of nano-sized particles, i.e. NPs, to remediate contaminants from soil, water, and the air is termed nanoremediation. Nanotechnology provides the opportunity to recognize, monitor, prevent, reduce and treat the contaminants in soil, air, and water (Gamal Snousy & MF, 2017).

There are several explanations why NPs can be used to remove pollutants from the environment. The extended surface area of NPs enables to absorption of a large number of contaminants from the environment; chemical reactions are made quite possible by the quantum effects of NPs; NPs exhibit surface plasmon resonance, which aids in the detection of toxic pollutants. Aside from that, NPs can enter the polluted zone where other particles are unable to. With the rapid progress of nanotechnology, environmentalists are paying much more attention to nano-remediation for environmental clean-up (Rizwan et al., 2014). The use of NPs for polluted site remediation has been shown to be successful in many regions

Table 1. List of some commonly available environmental contaminants and their toxic effects on the living world

Name of the Pollutants	Toxic Effects		References
	On Plants	On Human	
Persistent Organic Pollutants (POPs)			
Aldrin	In plants, a low or moderate dose of aldrin exposure doesn't show toxic effects. But exposure to higher concentration disturbs plants' growth and developmental processes.	Moderate level exposure to aldrin causes headache, vomiting, chronic convulsion, irritability, dizziness. In contrast, a high dose of aldrin causes kidney damage and affects the nervous system.	(Mlynarczuk et al., 2020; Taiwo, 2019)
Endosulfan	Reduced seed viability, delayed germination, reduced root-shoot length, and a necrotic spot in leaves are the most common effects in plants.	It affects the nervous system as well as the reproductive organs.	(Sathishkumar et al., 2021)
Hexachlorobenzene	Moderate exposure doesn't show any toxic effects. The toxic effects might be observed only at higher doses than the recommended level.	Environmental Protection Agency (EPA) categorized hexachlorobenzene as a group B2 carcinogen. It also affects the liver, kidney, skin, bone, blood cells, thyroid, immune system, and nervous system.	(Starek-Świechowicz et al., 2017)
Polychlorinated biphenyl (PCB)	Reduced germination rate and altered plant growth are the most common phytotoxic effects of PCB.	Cancer, neurological disorder, and endocrine reproductive disorders are the most common effects of PCB.	(Subramanian et al., 2017; US EPA, 2021)
DDT	Reduced germination rate and plant growth are the direct effects of DDT toxicity in plants. It also alters the cellular metabolisms in plants.	Immune system suppression, cardiovascular diseases, cancer, liver, and reproductive organ damage are the most common toxic effects of DDT.	(Mlynarczuk et al., 2020)
Dioxin	Dioxin adversely affects the gene regulations in seeds that alter plants' developmental processes and ultimately affect plant productivity.	Diabetes, cardiovascular diseases, endometriosis, porphyria, cancer, hormonal imbalance, early menopause, reduced immunological responses, abnormalities in nails, skin, and tooth, altered metabolism, and growth factor signaling.	(Hanano et al., 2018; Wang et al., 2019)
Inorganic Pollutants			
Arsenic	Chlorosis, cellular necrosis, growth inhibition, retarded photosynthesis, respiration, and transpiration, reduced plant metabolism, oxidative stress leading to the damage of protein, DNA, nucleotides, etc.,	Cancer, skin diseases, diabetes, cardiovascular diseases, cellular dysfunctioning, autophagy, apoptosis. DNA damage, oxidative stress, chromosomal aberrations, hyperkeratosis, etc.	(Hasanuzzaman et al., 2015; Shankar et al., 2014)
Lead	Stunted growth, blackening root architecture, chlorosis, reduced photosynthesis, imbalance in nutrient and water uptake, disruption of membrane permeability and structure, alteration in hormonal status, etc.	Anemia, decreased immunity, cardiovascular diseases, male infertility, premature birth with low weight, alteration in enzyme and cellular processes, neurological disorders, etc.,	(Kumar et al., 2020; Küpper, 2017)
Mercury	Inactivation of proteins, enzymes, destruction of the cell membrane, displacement of important molecules from other functional compounds, reduced photosynthesis, water uptake, transpiration, chlorophyll biosynthesis, etc.,	Mercury exposure affects the immune system, nervous system, and digestive system, and the toxicity also affects the lung, kidney, skin, eye, and gastrointestinal tract, alteration in metabolic processes, etc.	(Mahbub et al., 2017)
Cadmium	Alteration in mineral uptake and transport, destruction of the cell membrane, cellular organelles, and cellular biomolecules, damage photosynthetic apparatus, etc.,	Cancer, respiratory diseases, reproductive, urinary illness, neurological disorders, etc.,	(Haider et al., 2021; Rafati Rahimzadeh et al., 2017)
Copper	Variation in nutrient uptake, transport, decreased photosynthesis, alteration in enzyme activities, and reduced chlorophyll biosynthesis, etc.,	Gastrointestinal disorder, liver and kidney damage, irritation in eyes and face, nausea, headache, diarrhea, etc.,	(Adrees et al., 2015; Taylor et al., 2020)
Chromium	Impaired metabolic processes, reduced and delayed germination, reduced biomass, and yield, oxidative damage through generation of ROS, etc.,	Cancer, ulcer, liver and kidney damage, respiratory disorder, nasal irritation, immunosuppression, hypersensitivity reactions, etc.,	(Rahman & Singh, 2019)
Selenium	Reduced growth and developmental processes, disruption of proteins, oxidative damage, reduced productivity, etc.,	Nausea, vomiting, fatigue, discoloration of the nail, hair loss, foul of breath, irritation in the mucus membrane, etc.,	(Gebreeyessus & Zewge, 2019)

Figure 1. An overview of mechanisms of NPs assisted environmental remediation

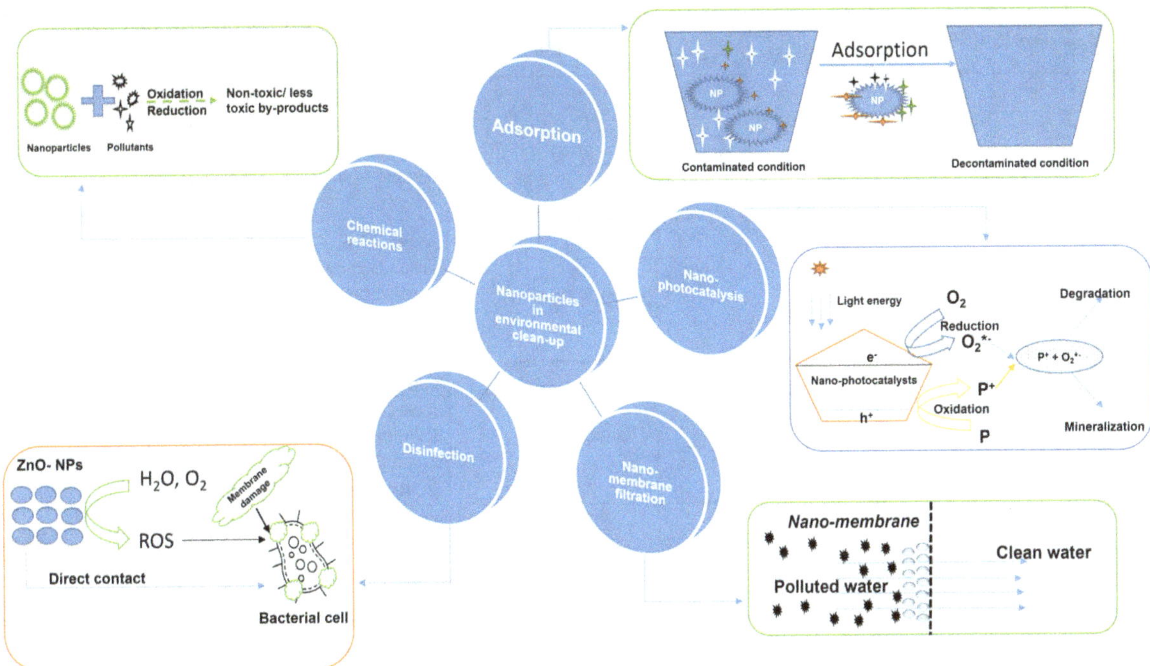

around the world. NPs are highly effective at removing pollutants from water and soil. Figure 1 gives an overview of mechanisms of NPs assisted environmental remediation.

Nanoparticles for Water Clean-Up

Water is the most vital life-supporting component, and it covers almost 70 percent of the earth's surface. Besides this, the bulk of it is inaccessible or unusable by people, and only 3 percent of the entire world's water is available as fresh water, and the rest of 97 percent is saltwater. In reality, just 0.5 percent of total available water is suitable for human consumption, and it comes from rainfall, streams, rivers, lakes, ponds, reservoirs, and aquifers. The remaining 2.5 percent of fresh water is stored as polar ice or groundwater (Beth Baker, Caleb Aldridge, Austin Omer, 2019). On the other hand, the rapid growth of the human population, changing their lifestyles, urbanization, industrialization, and economization continuously releases more and more waste as by-products and pollutes the water sources. Thus, the lack of pure water for the living world has become a worldwide concern in this twenty-first century. Therefore, wastewater treatment is necessary to make it suitable for living organisms. However, traditional contaminated water restoration procedures have several limitations, like inadequate pollutant removal with high energy consumption and the formation of hazardous sludge. The filtration process can eliminate impurities, but it generates concentrated forms of toxic contaminants that are problematic to dispose of. Thus, there is a legitimate need for the most effective and less energy-consuming techniques for treating contaminated water. Nanotechnology has shown remarkable potential for wastewater clean-up among the different developing technologies due to its high surface area and surface reactivity, enabling oxidation-reduction and degradation of contaminants. Due to the minute size, NPs also exhibit high mobility, and

Table 2. Some examples of commonly used nanoparticles (NPs) and their mechanisms in water clean-up

Nanoparticles(NPs)	Targeted Contaminants	Key Mechanism	Observations	References
CuO	Arsenic	Adsorption	CuO treatment showed selective adsorption and complete removal of arsenic from all ten water samples.	(McDonald et al., 2015)
Magnetic Fe_3O_4 amended graphene nanosheets	Rhodamine B	Adsorption	Magnetic Fe_3O_4 amended graphene nanosheets showed better rhodamine B adsorption from contaminated water.	(Wu et al., 2015)
FeO, CuFeO, MnFeO, FeO/ activated carbon	*E. coli* and *Staphylococcus aureus*	Disinfection	CuFeO and FeO are the most promising candidates for removing *E. coli* and *Staphylococcus aureus from contaminated water.*	(Pinto et al., 2020)
Fe_3O_4	Cr(VI)	Adsorption	Magnetic Fe_3O_4 NPs exhibit a high potential to remove Cr(VI) from water through adsorption.	(Aftabtalab A et al., 2014)
Iron oxide	Cd	Adsorption	Iron oxide NPs adsorbs Cd from an aqueous solution at pH 4.	(Ehrampoush et al., 2015)
Titanium oxide nanofiber membrane (Silver-doped)	Methylene blue and methylene red	Photocatalysis	The membrane can remove more than 92% methylene blue within 10 minutes, while 93% methylene red is decayed within 3 hours by photocatalysis.	(Kanjwal et al., 2009)
Multi-wall carbon nanotubes	Cr(VI)	Adsorption	Multi-wall carbon nanotubes show high Cr(VI) adsorption from an aqueous solution.	(Huang et al., 2015)
Magnetic zeolite polypyrrole composite	Vanadium	Adsorption	Magnetic zeolite polypyrrole composites adsorb upto 75% vanadium from the aqueous condition at pH 5.	(Mthombeni et al., 2015)

a small amount of NP can easily scan the large volume of contaminated water(Saikia et al., 2019). Some examples of commonly used NPs and their mechanisms in water clean-up are mentioned in Table 2.

Over the last few years, nanotechnology has been widely employed in wastewater clean-up. Based on the literature reports, the use of NPs in contaminated water treatment may fall under four major groups:

1. Water clean-up by nanotechnology via adsorption mechanisms
2. Water clean-up by nanotechnology via photocatalytic degradation
3. Water clean-up by nanotechnology via disinfection mechanisms
4. Water clean-up by nanotechnology via nano-membrane filtration (Saikia et al., 2019)

1. Water Clean-up by Nanotechnology via Adsorption Mechanisms

Adsorption is the most efficient strategy for clean-up contaminated water, in which the sorbate, i.e. the contaminant, adheres onto the surface of the adsorbent because of specific attractive forces. The adsorption of pollutants may involve both covalent and non-covalent interactions. But non-selective adsorption is a significant drawback of adsorption mechanisms in which non-targeted compounds may adhere to the adsorbent's surface; thus, the targeted contaminants remain in the medium. This problem could be overcome by using a surface modification of NPs which will allow targeted adherence of the contaminant onto the surface of NPs(L. Qian et al., 2019). Similarly, desorption is an essential factor in which the impurities are removed from the surface of NPs by altering the temperature, pH of the solution, which enhances the reusability of the NPs and reduces the cost. Based on the adsorption process, NPs are categorized into different groups like metallic and non-metallic oxides, magnetic NPs, nanostructured mixed oxides, carbonaceous NPs, silicon-based NPs, etc. (Saikia et al., 2019).

Metal and non-metal oxides of NPs are widely used in the removal of toxic contaminants from polluted water. Due to the high natural availability of iron and low-cost iron oxide NPs, it was extensively applicable in various fields. Iron-based NPs exhibit excellent redox potential, extensive surface charge, and reusable capacity and are widely used to remediation heavy metals and dyes from contaminated water. Factors like temperature, pH, adsorbent dose and incubation periods, etc., are responsible for enhancing the adsorption process (Saikia et al., 2019). Similarly, surface modification of Fe_2O_3 with additives like aminopropyltrimethoxysilane also enhances the adsorption onto the surface of Fe_2O_3 (Palimi et al., 2014). The NPs can also convert highly toxic forms of heavy metal ions into less toxic forms through reduction reactions. For example, nZVI catalyzes the reduction reaction of highly toxic Cr(VI) to less toxic Cr(III) and precipitates in ferric chromate (Farooqi et al., 2021). Again, magnetic NPs are also excellent candidates for removing contaminants from water due to their high reactivity, ease of separation, and reusability. Fato et al. (2019) reported that the magnetic NP Fe_2O_3 could remove multiple or co-existing heavy metal (Cu^{2+}, Cd^{2+}, Ni^{2+}, and Pb^{2+}) pollutants from river water(Fato et al., 2019). However, due to the strong reactivity under aqueous conditions, magnetic NPs are susceptible to corrosion, reducing their lifespan.

Besides the iron oxide NPs, other metallic NPs like magnesium oxide, manganese oxide, zinc oxide, etc., are also effective for efficiently eliminating heavy metals from water. For example, Tripathy & Kanungo (2005) studied the adsorption of heavy metal ions Co^{2+}, Cu^{2+}, Ni^{2+}, and Zn^{2+} from seawater and 0.5M NaCl solution onto the surface of δ-MnO_{2+} and FeOOH. They observed that both the metal oxides adsorb heavy metal onto their surface, but the rate of adsorption in δ-MnO2+ increases when it is present in a mixed oxide system than that of δ-MnO2+ alone (Tripathy & Kanungo, 2005). Similarly, Han et al. (2007) investigated the uranium(VI) removal from the aqueous solution through adsorption onto the surface of manganese oxide zeolite. They found that the highest uranium(VI) adsorption was 15mg g^{-1} at 20°C and pH 4.0(Han et al., 2007).

Carbonaceous NPs are also good adsorbers of organic pollutants through π-π interactions, hydrophobic interactions, electrostatic interactions, hydrogen bonding, and covalent interactions. Carbon nanotubes (CNTs) are the most widely studied carbonaceous NPs, which possess four different sites like peripheral groove, inner site, interstitial channel, and the external surface for the adsorption of contaminants. Thus, organic pollutants having benzene ring, polar aromatic molecules, polycyclic aromatic hydrocarbons (PAHs), and C=C bonds are quickly deposited onto the surface of CNTs. Similarly, organic compounds having functional groups –OH, -COOH, $-NH_2$ are also deposited onto CNTs through the formation of hydrogen bonds. Again, electron-rich CNTs and positively charged organic pollutants under suitable pH exhibit electrostatic interactions, and hence the organic contaminants get absorbed by CNTs(Y. Qian et al., 2020). However, the low dispersal capacity, the difficulty in separation, and the size of small particles are minor inconveniences to CNTs as adsorbents. As a result, the CNTs are regularly modified using different technologies to improve efficiency. Among these modifications, the development of multi-walled carbon nanotubes and the incorporation of magnetic properties are the most common, enhancing the adsorption efficiency (Saikia et al., 2019). Similarly, Yang et al. (2011) studied the Pb(II) and humic acid adsorption on polyacrylamide-grafted multi-walled carbon nanotubes. And observed selective adsorption of both Pb(II) and humic acid(Yang et al., 2011).

2. Water Clean-up by Nanotechnology via Photocatalytic Action

The term "photocatalysis" refers to the degradation of compounds in the presence of light. In nanophotocatalytic water treatment process, NPs capture light energy and utilize it to degrade or breakdown of a wide range of organic contaminants. The oxidation and reduction reactions are crucial to the photocatalytic water treatment process (Gupta et al., 2020). Photocatalysts produce highly reactive transitory species like OH*, H_2O_2, O_2^{*-}, O_3, etc., to degrade and mineralize organic contaminants (Chong et al., 2010). The degradation process indicates the splitting or decomposition of organic pollutants into several products, while mineralization indicates the total removal or conversion of organic pollutants into CO_2, H_2O, and other inorganic ions (Gupta et al., 2020).

Semiconductor metal oxide NPs like ZnO, TiO_2, ZnS, Fe_2O_3, CdS, GaP, etc., are the most promising candidates for photocatalysis-based water clean-up technology. These NPs possesses excellent band gap which upon the perception of an optimal wavelength of light produces hole pair of electron (e^-, h^+) that leads to the cascade of the waste degradation process (Saikia et al., 2019). The mechanism of photocatalytic water clean-up technology through a series of oxidation-reduction reactions in the presence of nano-photocatalysts are as follows (Chong et al., 2010) and in ***Figure 2***:

1. Photoexcitation through the perception of light

   Nano-photocatalysts + $h\vartheta \rightarrow e^- + h^+$

   Photoexcitation of nano-photocatalysts after perception of light and produces hole pair of electrons e^- and h^+.
2. Trapping of hole pair of electrons e^- and h^+
   The conduction band and valance band traps electron-hole pair e^- and h^+ generated after photoexcitation of nano-photocatalysts.

   Trapping of "e^- "on conduction band (CB): $e^-_{(CB)} \rightarrow e^-_{(TR)}$

   Trapping of "h^+" on valance band (VB): $h^+_{(VB)} \rightarrow h^+_{(TR)}$

3. Recombination of hole pair electrons e^- and h^+

   $e^-_{(TR)} + h^+_{(VB)}[\ h^+_{(TR)}] \rightarrow e^-_{(CB)}$ + heat

4. e^- scavenging by scavengers (reduction reaction)
   The O_2 molecule gets reduced by accepting electron (e^-) from the conduction band and forms O_2*.

   $$O_2 + e^- \rightarrow O_2^{*-}$$

5. oxidation reaction with h^+
   Pollutant (P) + $h^+ \rightarrow P^+$ (intermediate by-products)

   $$H_2O + h^+ \rightarrow H^+ + OH^*$$

   The organic pollutants (P) and surface trapped h+ holes get oxidized and form intermediate by-products P^+. Similarly, the H_2O molecules in the presence of h^+ holes get oxidized and form H^+, OH*.
6. Photo-degradation of contaminants

   $P^+ + OH^* \rightarrow$ ultimate degradation of contaminants

   $P^+ + O_2^{*-} \rightarrow$ ultimate degradation of contaminants

   R-H (intermediate products) + OH* $\rightarrow$ Ŕ* + H_2O

   The intermediate by-products (P^+) combine with OH* and O_2^{*-} radicals give rise to ultimate degraded products.
7. Other reactions
   i. Formation of HOO*
      The OH* and O_2^{*-} generated from oxidation and reduction reaction recombines and forms HOO*.

      $$O_2^{*-} + OH^* \rightarrow HOO^*$$

   ii. Scavenging of e^- by HOO*

      $$HOO^* + e^- \rightarrow HO_2^-$$

      The HOO* molecule accepts hole electrons (e^-) and forms HO_2^-
   iii. Generation of H_2O_2
      HO_2^- recombines with H^+ ions and forms H_2O_2

      $$HO_2^- + H^+ \rightarrow H_2O_2$$

Each nano-photocatalyst has its absorption range of light that corresponds to its' band gap. To excite the electron and initiate the photocatalytic reactions, the wavelength of light must equal to or greater than the band gap. In titanium dioxide, the electron excites only after the absorption of UV radiation and remains inactive in visible light, which reduces its' photocatalytic efficiency under solar irradiation. Therefore, to increase the absorption range for titanium dioxide or any other photocatalysts, different modification methods like doping, dye sensitization, composites or hybrid nano-photocatalysts, anions, etc., are used. All these methods are utilized to improve the photocatalytic efficiency of different nano-photo-catalysts (Saikia et al., 2019).

Figure 2. Photocatalytic mechanism of electron-hole pair formation and remediation of environmental contaminants.

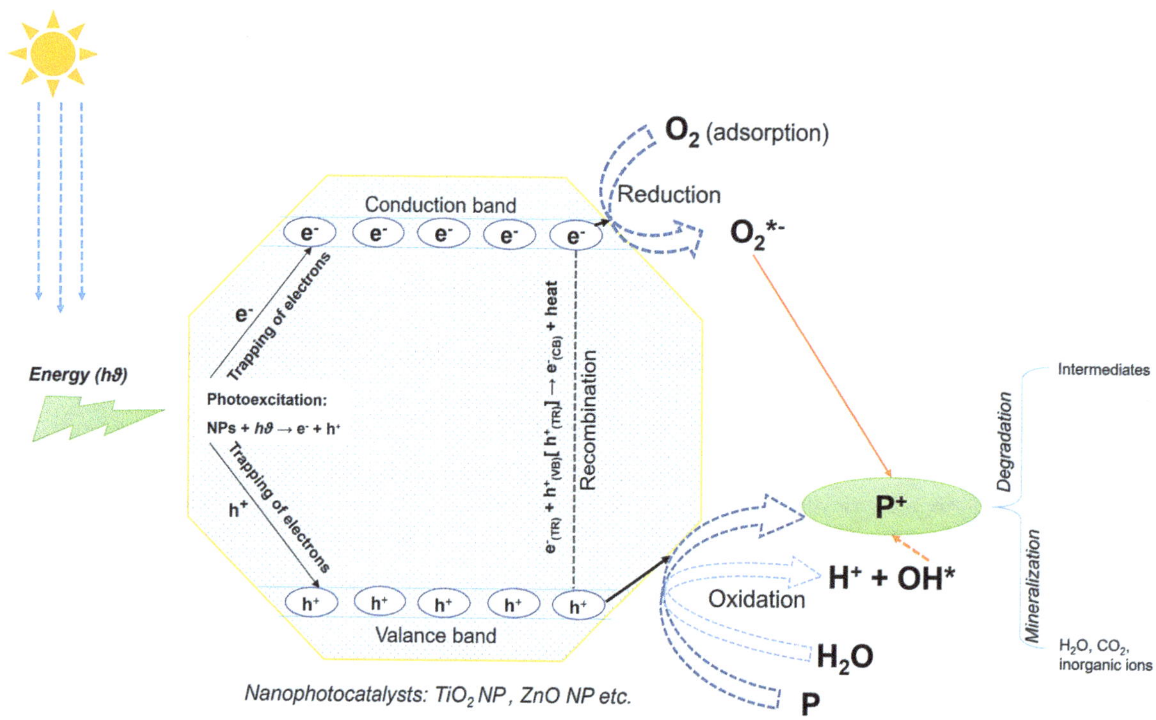

3. Water Clean-up by Nanotechnology via Disinfection Mechanisms

Water contamination with pathogenic microbes is a primary concern as it makes water unfit for consumption. Several chemical disinfectants like chloramines, chlorine, ozone, etc., are widely used as a conventional approach to water suitable for human use. However, these chemical disinfectants exhibit a high oxidation tendency, interact with other water components, and generate toxic disinfection by-products (DBPs)(Gupta et al., 2020). Different NPs like iron oxides, carbon CNTs, titanium dioxides, zinc oxide, silver NPs, etc., show great antimicrobial potential and exhibit excellent disinfection properties by lowering the DBPs production. However, in NPs assisted disinfection mechanisms, the NPs either interact with the cell membrane, disrupt it, or penetrate the cell membrane and disturb the electron transfer mechanisms or cause oxidative stress through the generation of reactive oxygen species (ROS) and causes cellular damage (Saikia et al., 2019).

Green synthesis of iron oxide (Fe_3O_4) NPs using two algal species, *Colpomenia sinuosa* and *Pterocladia capillacea,* exhibit excellent antimicrobial potential against gram-positive and gram-negative bacteria. Again, the Fe_3O_4 produced using *C. sinuosa* exhibited superior antifungal against *Aspergillus flavus* and

Fusarium oxysporum compared to Fe_3O_4 produced by using *P. capillacea* and controls (Salem et al., 2019). Furthermore, biosynthesis of ZnO NPs using *Pandanus odorifer* also exhibits great antimicrobial potency against *Bacillus subtilis* (Gram-positive) and *Escherichia coli* (Gram-negative) (Hussain et al., 2019). Hence, these biosynthetic NPs can be used as microbial disinfectant in wastewater treatment.

iv. Water Clean-up by Nanotechnology via Nano-membrane Filtration

The basic concept of membrane-based filtration uses a semipermeable membrane to separate or remove the unwanted compounds from water to make it suitable for use. Membrane technology is concerned with different materials and procedures used to increase membrane function, minimize membrane fouling, and reduce energy consumption to achieve better efficiency. Among the other existing membrane-based filtration processes like reverse osmosis, ultrafiltration, microfiltration etc., nano-membrane based filtration shows the advantage of functioning at relatively low osmotic pressure (Gehrke et al., 2015; Saikia et al., 2019). Nano-membranes are a particular type of barrier made up of NPs to separate or remove pollutants from aqueous solutions. Recently, nano-membrane-based filtration has become one of the most leading approaches in wastewater treatment. It exhibits high removal efficiency of heavy metals, various multivalent ions, pesticides, organic contaminants, and various pathogenic microbes. Rather than the size of pollutants, the electrostatic interactions between the charged nano-membrane and the charged ions of the pollutants significantly influence their penetration over the membrane. For example, the charged nano-membrane traps the divalent sulphate salts more efficiently than that monovalent chloride salts. Thus, nano-membrane-based filtration approach is beneficial for selective elimination of contaminants under aqueous conditions (Saikia et al., 2019).

The combination of NPs with inorganic or polymeric membranes make the membrane multifunctional. Various NPs are having hydrophilicity (TiO_2, Al_2O_3, zeolite etc.), antimicrobial activity (CNTs, Ag-NPs etc.), and photocatalytic activity (TiO_2, ZnO, bi-metallic NPs etc.) are widely used to increase membrane functioning (Arancibia-Miranda et al., 2016; Gehrke et al., 2015). For examples, the incorporation of TiO_2 NPs in inorganic membrane shows effective degradation of chlorinated contaminants in water. It has been reported that the combination of TiO_2 NPs with polyethylene membrane and the TiO_2 slurry with polymeric membranes are highly efficient in the degradation of 1,2-dichlorobenzene and chlorinated pharmaceutical wastes (Martínez et al., 2013). Similarly, TiO_2 with polyethersulfone composite membrane enhanced fluid flow and reduced the fouling properties (Guo & Kim, 2017). Again, the hydrophilic metal oxide NPs increase the membrane hydrophilicity and thus increases water permeability or reduces the fouling property. While the NPs having an antimicrobial activity like CNTs, Ag-NPs etc., are widely used to kill pathogenic microbes and reduce the membrane biofouling (Saikia et al., 2019). It has been reported that the silver-containing nanofiber membrane-like polyacrylonitrile (PAN) and polyvinyl alcohol (PVA) exhibit excellent antimicrobial activity. The Ag-PAN nanofiber can remove 100% bacteria, while Ag-PVA nanofiber has 91-99% from polluted water. Both PAN and PVA nanofiber don't leach silver NPs into the water, so it can be concluded that both Ag-PAN and Ag-PVA nanofiber can be used in the wastewater treatment process (Plessis & Marguerite, 2011).

Nanoparticles for Soil Clean-Up

Soil is one of the most crucial life-supporting components on the earth. However, the rapid growth of the human population and various anthropogenic activities continuously add various toxic components and deteriorate the quality of the soil. Soil remediation is simply a technique for removing contaminants from soil using a variety of optimized techniques. Several techniques, such as bio-remediation, soil washing, thermal remediation, air sparging, electrokinetic remediation, etc., are available for soil clean-up(Sarkar et al., 2019). However, such techniques show some limitations, which are mentioned in Table 3.

Table 3. List of available soil remediation techniques, key mechanisms, and their major drawbacks

Remediation Techniques	Key Mechanisms	Drawbacks/Limitations	References
Bioremediation	Use of living organisms like microbes, plants or their products to remediate contaminants from the environment.	Bioremediation is a slow process to remediate contaminants, and hence, it cannot be utilized as an instant remediation technique. Additionally, it is not easy to conduct continuous field monitoring to determine the pace or degree of bioremediation or evaluate performance.	(Gabriel-Alin Iosob, Maria Prisecaru, Ionuţ Stoica, Maria Călin, 2016)
Soil washing	Soil washing with water to eliminate pollutants.	Soil washing is an ex-situ technique, requiring ample space and a massive amount of water source. The contaminated soil is washed with water, and hence, it releases polluted water, which requires further treatments. Again, clay containing soils cannot be treated using this technique.	(Karthika et al., 2016)
Thermal remediation	Soil treated with high temperature at 650°F-900°F to eliminate contaminants in the form of vapour.	It is an ex-situ technique and requires high energy to raise the temperature; this increases the cost. In addition, this technique is not suitable for a wide range of contaminants as it treats only volatile, semi-volatile and hydrocarbon contaminants.	(Vidonish et al., 2016)
Air sparging	In the air sparging technique, injection of high pressurized air into the contaminated soil removes organic pollutants into vaporized form.	It is ineffective for stratified soil treatments. This technique is suitable for only volatile organic compounds. Additionally, the migration/movement of contaminants from one part to another is a significant problem created by the air sparging technique.	(EPA et al., 2017)
Electro-kinetic remediation	Direct current with low intensity is applied in the soil to remediate soil contaminants by generating electric potential.	It is a slow process and hence takes more time. In addition, the presence of non-targeted contaminants is another major limitation for electrokinetic remediation. Soil pH change due to the electrolysis reactions hinder the soil microbial community, which is disadvantageous.	(Koliyabandara et al., 2021)

Hence, the use of NPs is an alternate strategy with few limitations and high efficacy in soil remediation. The basic mechanisms behind the NP-mediated soil clean-up include immobilization, redox reduction, photocatalytic Degradation, Fenton or Fenton like reduction, and amalgamation of various abovementioned mechanisms(Y. Qian et al., 2020).

Recently, adsorption/immobilization phenomena gain much more attention as a most efficient and cost-effective technique to remove contaminants from soil. In this case, selecting suitable components based on the type of contaminants present and the soil condition is a key point for efficient remediation. To date, different types of NPs such as carbon NPs, metal oxide NPs, various engineered NPs, and nanocomposites are widely used in the adsorption/immobilization of organic and inorganic pollutants in the soil as shown in Table 4. Carbon NPs immobilize organic pollutants through π-π interactions and van der Waals forces. In contrast, metal oxide and other engineered NPs immobilize organic and inorganic pollutants through surface complexation, redox reduction, etc.(Y. Qian et al., 2020).

Carbon NPs are excellent components for the adsorption of organic contaminants from the soil due to their extensive surface area and hydrophobicity. To date, the adsorption and desorption mechanisms of carbon NPs towards organic pollutants have been widely studied. Carbon nanotubes (CNTs) have efficient adsorption ability of soil organic contaminants and thus, gain public attention for soil remediation. Moreover, CNTs have been found to act as good absorbers and sequester ionizable organic molecules like pesticides and pharmaceutical compounds through H-bonding and π-π interactions(Y. Qian et al., 2020). For example, the application of CNTs in PAHs (benzo[a]pyrene and phenanthrene) contaminated soil decreases the extractability and bioavailability of PAHs to soil microbes and plants by reducing their mineralization and mobilization ability(Towell et al., 2011). Similarly, S. Li et al. (2013) studied the PAHs (pyrene, fluorene, phenanthrene, naphthalene, acenaphthalene, etc.) sorption by CNTs. They reported that CNTs have efficient PAHs sorption ability under different environmental situations and thus can be used in PAHs remediation from soil(Li et al., 2013). Furthermore, CNTs have also been found to be an effective adsorbent for pyrene immobilization in soil(W. Zhang et al., 2019).

TiO_2-NP, iron oxide-NP, ZnO-NP, GO-NP, CuO-NP, etc., are the most common metal oxide NPs used in soil contamination remediation, which could have effectively decreased the bioavailability of the contaminants (Guerra et al., 2018). Shipley et al. (2011) reported that iron oxide NPs have strong arsenic (arsenate and arsenite) adsorption capacity from the soil. In their study report, Shipley et al. (2011) also stated that the iron oxide NPs also remove 12 more different types of metals from soil (Shipley et al., 2011). The result indicated that the iron oxide NPs are promising candidates for remediating metal-polluted soil. In a study, Ma et al. (2017) investigated the effects of TiO_2-NPs on tetracycline (TC) accumulation and toxicity in rice plants. They found that single exposure of TC reduced plant biomass and several essential nutrient accumulations but increased TC accumulation into the plants. But the application of TiO_2-NPs reduced TC accumulation and enhanced plant biomass by improving nutrient availability in plants. Therefore, it was suggested that the reduction of TC phytotoxicity is due to the sorption of TC onto the TiO_2-NPs, which makes TC unavailable for plants (Ma et al., 2017). Tan et al. (2007) studied the effects of fulvic acid, humic acid, ionic strength, and pH on Th(IV) sorption on TiO_2-NPs. It has been reported that both fulvic and humic acids favored the Th(IV) sorption on TiO_2-NPs, and a maximum of 94% sorption is found at pH 4(Tan et al., 2007).

Again, iron-containing NPs or nanocomposites are considered a promising catalyst for Fenton-like oxidation processes(Saikia et al., 2019). Usman et al. (2012) studied the impact of Fe_3O_4 incorporation in PAHs polluted soil. They reported that Fe_3O_4 acts as an efficient catalyst for Fenton reaction and remediate PAHs from polluted soil without generating harmful by-products and adjusting the pH(Usman et al., 2012). However, Fe_3O_4 mediated Fenton-like oxidation is a sluggish process; it's important to integrate it with additional techniques like sonication, properly scattering the catalysts in the matrix, and boosting overall degrading efficiency. Furthermore, ultrasonic irradiation might clean the catalysts' surface and allow them to be active, and increase overall reactivity throughout the Fenton like reactions (Hou et al., 2016).

The reduction reaction is another key mechanism utilized by NPs to remediate contaminants from the environment. It has been reported that NPs remediate heavy metals from the soil through a reduced reaction process; in this context, inorganic metallic NPs like nZVI, oxides of iron, oxides of sulfide, and iron-containing bimetallic NPs, etc. are widely studied(Farooqi et al., 2021). However, heavy metals are highly toxic and carcinogenic materials. For example, hexavalent chromium, i.e. Cr(VI), is highly poisonous, and several NPs have been identified, which tend to reduce highly toxic Cr(VI) to less toxic Cr(III). For example, nZVI shows high reduction ability and thus converts Cr(VI) to Cr(III) and forms complex chromium ion as ferrous chromate. But the iron NPs are highly susceptible to aggregation, which makes them inactive. Therefore, measures should be taken to keep them active for the efficient remediation of heavy metals. In that case, the incorporation of biochar, compost, and other monometallic NPs could keep them active and enhance the rate of remediation(Farooqi et al., 2021). For example, L. Qian et al. (2019) reported that incorporation of silicon-rich biochar and palladium along with nZVI boosted reduction and adsorption of Cr(VI)(L. Qian et al., 2019).

Recently, degradation of organic pollutants using nano-photocatalysts under sunlight or UV-IR radiation exposure has gained much and more research and public interest to remediate contaminants from the environment(Osin et al., 2018). This method requires the addition of nano-photocatalysts along with light penetration into the polluted soil. Due to the low light penetration into the soil, this strategy

Table 4. Some examples of commonly used nanoparticles (NPs) and their mechanisms in soil clean-up

Nanoparticles (NPs)	Targeted Contaminants	Key Mechanism	Observations	References
nZVI	As and Hg	Immobilization	nZVI reduced the exchangeable As and Hg into the soil through immobilization.	(Gil-Díaz et al., 2017)
CNTs	Pyrene	Adsorption	Dissolve organic matter amended CNTs removes pyrene from the soil through adsorption.	(W. Zhang et al., 2019)
Ni/Fe bimetallic NP	Decabromodiphenyl ether (BDE209)	Degradation	With ambient temperature and pH 5.6, the bimetallic NP removed upto 72% of BDE209 through Degradation.	(Xie et al., 2014)
TiO_2	PAHs	Photocatalysis	TiO_2 under UV radiation accelerates photodegradation of PAHs like benzo [α] pyrene, phenanthrene, and pyrene. The rate of degradation increased under acidic condition compared to neutral or basic conditions.	(L. Zhang et al., 2008)
nZVI	Pb/Zn	Immobilization	nZVI immobilizes Pb/Zn into the soil without affecting soil physicochemical and biological properties. And the rate of Pb immobilization is higher in Pb than Zn.	(Mar Gil-Díaz et al., 2014)
Fe_3O_4	PAHs	Fenton-like reaction	The Fe_3O_4 degrades 16 different types of PAHs through Fenton-like oxidation processes.	(Usman et al., 2012)
nZVI	DDT	Degradation	nZVI degrades DDT in spiked and aged soil; 1gm nZVI/kg soil is sufficient to degrade DDT in spiked soil, but a higher concentration is needed to treat aged soil.	(El-Temsah & Joner, 2013)
nZVI activated persulfate	PAHs	Degradation	nZVI activated persulfate degrades 82.21% PAHs into the soil. The persulfate releases H^+ ion and reduces the soil pH.	(Y. Song et al., 2019)

is severely limited and would necessitate regular ploughing of the contaminated land. Besides this, photocatalytic degradation efficiency was correlated to several factors like soil properties, organic content, and soil pH. These factors are responsible for the generation of the hydroxyl group by NPs and plays an efficient role in the sorption of pollutant from the contaminated soil(Y. Qian et al., 2020). TiO_2 is one of the most commonly used NPs in the photocatalytic degradation process. It has been reported that the addition of TiO_2 NPs under visible light exposure enhanced pyrene degradation rate in contaminated soil(Chang Chien et al., 2011).

FUTURE RESEARCH DIRECTIONS

The NPs possess superior physicochemical properties like size, surface area, and shape and are widely used in various fields. For example, using NPs in the remediation of contaminants is an emerging thought that appeared along with the advancement of nanotechnology and remediation technologies. Despite this, it faces numerous complications in terms of practical applications. Some of the key complications are aggregation, loss of activity and stability over time, retention, recovery and cost balance, real condition efficiency, and environmental and health risk of NPs(Saikia et al., 2019). Most of NPs show the tendency of aggregations when introduced into the water and soil. This aggregation leads to the settling of NPs and reduces the surface area, thus decreasing overall performance. Certain ions and salts in the medium also accelerate the aggregation process(Liu et al., 2015). Therefore, detailed investigations in this field are required to minimize the aggregation of NPs under contaminated conditions. The loss of activity and stability over time hamper the use of NPs in remediation processes. For example, the silver NPs initially show excellent antibacterial activity, but it loses Ag^+ ions and deteriorate their antibacterial properties with time. This issue can be resolved by inhibiting hole pair (e^-, h^+) electron recombination; still, the efficiency of NPs declines with time. Apart from this, most of the NPs assisted remediation studies are conducted in the laboratory under controlled condition. Whereas, in field condition, the scenario is even more complicated, containing a wide range of complexing ions and molecules. Most NPs perform better under laboratory conditions, but their performance deteriorates when applied under actual contaminated conditions (Saikia et al., 2019). However, further investigation at the field level is needed to understand the thorough interactions between NPs and host contaminants to synthesize and commercialize the most suitable NPs. Excluding a few NPs, the production and characterization of NPs are quite costlier(Gkika et al., 2017). Thus, the recovery and reusability of such NPs may reduce the excessive production of such NPs and ultimately reduce the production cost.

Although nanotechnology offers excellent potential for a sustainable environment, it has unintended effects on the environment and living organisms. The nanotoxicity against humans, plants, and beneficial microbes is a major concern. The intrinsic effects of NPs against living organisms are not apparent. However, the short-term NPs exposure may or may not be toxic, but long-term exposure can cause serious health issues like cancer, cardiovascular diseases, etc. Thus, a more detailed investigation is needed to understand the NP toxicity to achieve maximum advantages NPs assisted environmental clean-up by lowering environmental and health risks(B. Song et al., 2019).

CONCLUSION

Water, soil, and air are the most precious components of the living world. They perform the most significant functions like life supports, biodiversity conservation, carbon storage, food production, etc. To attain sustainable development requires a safe, healthy and productive environment. However, our environment is under serious threat of pollution. Various techniques have been developed to eliminate environmental contaminants. In this regard, nanotechnology has been established as the most promising approach where NPs are used to remediate contaminants from the environment. Nanoremediation is a state forward approach to remediate contaminants through adsorption, photocatalytic degradation, disinfection, Fenton-like reaction, nano-membrane separation, etc. This chapter highlighted the need for remediation and the role of NPs for environmental clean-up with their basic mechanisms. This chapter also highlighted the complications and current research trends of using NPs in environmental remediation, which will help researchers determine future research prospects. On the other hand, nanotoxicity is a major concern, and currently, the use of NPs for environmental clean-up at field conditions is in the phase of investigation and endeavor. Thus, more detailed research is needed to understand the NP toxicity to achieve maximum advantages NPs assisted environmental clean-up by lowering environmental and health risks.

FUNDING

This research received no specific grant from any funding agency in the public, commercial, or not-for-profit sectors.

ACKNOWLEDGMENT

We would like to acknowledge DBT- Electronic Library Consortium (DeLCON) of Ministry of Science and Technology, Government of India for providing access to research articles.

REFERENCES

Adrees, M., Ali, S., Rizwan, M., Ibrahim, M., Abbas, F., Farid, M., Zia-ur-Rehman, M., Irshad, M. K., & Bharwana, S. A. (2015). The effect of excess copper on growth and physiology of important food crops: A review. *Environmental Science and Pollution Research International*, *22*(11), 8148–8162. doi:10.100711356-015-4496-5 PMID:25874438

Aftabtalab, A., Sadabadi, H., Shilpa Chakra, C. H., & Venkateswara Rao, K. (2014). Magnetite nanoparticles (Fe3O4) synthesis for removal of Chromium (VI) from waste water. *International Journal of Scientific & Engineering Research, 5*(1). https://www.ijser.org/paper/Magnetite-nanoparticles-Fe3O4-synthesis-for-removal.html

Ali, H., Khan, E., & Ilahi, I. (2019). Environmental Chemistry and Ecotoxicology of Hazardous Heavy Metals: Environmental Persistence, Toxicity, and Bioaccumulation. *Journal of Chemistry, 2019*, 1–14. doi:10.1155/2019/6730305

Arancibia-Miranda, N., Baltazar, S. E., García, A., Muñoz-Lira, D., Sepúlveda, P., Rubio, M. A., & Altbir, D. (2016). Nanoscale zero valent supported by Zeolite and Montmorillonite: Template effect of the removal of lead ion from an aqueous solution. *Journal of Hazardous Materials, 301*, 371–380. doi:10.1016/j.jhazmat.2015.09.007 PMID:26384998

Baker, B., Aldridge, C., & Omer, A. M. H. (2019, November). *Water: Availability and Use*. Mississippi State University Extension Service. http://extension.msstate.edu/publications/water-availability-and-use

Bharagava, R. N., Purchase, D., Saxena, G., & Mulla, S. I. (2019). Applications of Metagenomics in Microbial Bioremediation of Pollutants. In *Microbial Diversity in the Genomic Era* (pp. 459–477). Elsevier. doi:10.1016/B978-0-12-814849-5.00026-5

Chang Chien, S. W., Chang, C. H., Chen, S. H., Wang, M. C., Madhava Rao, M., & Satya Veni, S. (2011). Effect of sunlight irradiation on photocatalytic pyrene degradation in contaminated soils by micro-nano size TiO2. *The Science of the Total Environment, 409*(19), 4101–4108. doi:10.1016/j.scitotenv.2011.06.050 PMID:21762957

Chong, M. N., Jin, B., Chow, C. W. K., & Saint, C. (2010). Recent developments in photocatalytic water treatment technology: A review. *Water Research, 44*(10), 2997–3027. doi:10.1016/j.watres.2010.02.039 PMID:20378145

Ehrampoush, M. H., Miria, M., Salmani, M. H., & Mahvi, A. H. (2015). Cadmium removal from aqueous solution by green synthesis iron oxide nanoparticles with tangerine peel extract. *Journal of Environmental Health Science & Engineering, 13*(1), 84. doi:10.118640201-015-0237-4 PMID:26682059

El-Temsah, Y. S., & Joner, E. J. (2013). Effects of nano-sized zero-valent iron (nZVI) on DDT degradation in soil and its toxicity to collembola and ostracods. *Chemosphere, 92*(1), 131–137. doi:10.1016/j.chemosphere.2013.02.039 PMID:23522781

EPA. (2017). *For Underground Storage Tank Sites A Guide For Corrective Action Plan Reviewers: Chapter V Landfarming*. EPA 510-B-17-003, October. https://www.epa.gov/sites/production/files/2014-03/documents/tum_ch9.pdf

Farooqi, Z. H., Akram, M. W., Begum, R., Wu, W., & Irfan, A. (2021). Inorganic nanoparticles for reduction of hexavalent chromium: Physicochemical aspects. *Journal of Hazardous Materials, 402*, 123535. doi:10.1016/j.jhazmat.2020.123535 PMID:33254738

Fato, F. P., Li, D.-W., Zhao, L.-J., Qiu, K., & Long, Y.-T. (2019). Simultaneous Removal of Multiple Heavy Metal Ions from River Water Using Ultrafine Mesoporous Magnetite Nanoparticles. *ACS Omega, 4*(4), 7543–7549. doi:10.1021/acsomega.9b00731 PMID:31459847

Gamal Snousy, M., & Mf, Z. (2017). Nanoparticles Restrictions in Environmental Cleanup. *Nano Research & Applications, 03*(01). Advance online publication. doi:10.21767/2471-9838.100018

Gebreeyessus, G. D., & Zewge, F. (2019). A review on environmental selenium issues. *SN Applied Sciences*, *1*(1), 55. doi:10.100742452-018-0032-9

Gehrke, I., Geiser, A., & Somborn-Schulz, A. (2015). Innovations in nanotechnology for water treatment. *Nanotechnology, Science and Applications*, *8*, 1–17. doi:10.2147/NSA.S43773 PMID:25609931

Gil-Díaz, M., Alonso, J., Rodríguez-Valdés, E., Gallego, J. R., & Lobo, M. C. (2017). Comparing different commercial zero valent iron nanoparticles to immobilize As and Hg in brownfield soil. *The Science of the Total Environment*, *584–585*, 1324–1332. doi:10.1016/j.scitotenv.2017.02.011 PMID:28190571

Gkika, D. A., Vordos, N., Nolan, J. W., Mitropoulos, A. C., Vansant, E. F., Cool, P., & Braet, J. (2017). Price tag in nanomaterials? *Journal of Nanoparticle Research*, *19*(5), 177. doi:10.100711051-017-3875-x

Guerra, F., Attia, M., Whitehead, D., & Alexis, F. (2018). Nanotechnology for Environmental Remediation: Materials and Applications. *Molecules (Basel, Switzerland)*, *23*(7), 1760. doi:10.3390/molecules23071760 PMID:30021974

Guo, J., & Kim, J. (2017). Modifications of polyethersulfone membrane by doping sulfated-TiO 2 nanoparticles for improving anti-fouling property in wastewater treatment. *RSC Advances*, *7*(54), 33822–33828. doi:10.1039/C7RA06406C

Gupta, A., Tandon, M., & Kaur, A. (2020). Role of metallic nanoparticles in water remediation with special emphasis on sustainable synthesis: A review. *Nanotechnology for Environmental Engineering*, *5*(3), 27. doi:10.100741204-020-00092-y

Haider, F. U., Liqun, C., Coulter, J. A., Cheema, S. A., Wu, J., Zhang, R., Wenjun, M., & Farooq, M. (2021). Cadmium toxicity in plants: Impacts and remediation strategies. *Ecotoxicology and Environmental Safety*, *211*, 111887. doi:10.1016/j.ecoenv.2020.111887 PMID:33450535

Han, R., Zou, W., Wang, Y., & Zhu, L. (2007). Removal of uranium(VI) from aqueous solutions by manganese oxide coated zeolite: Discussion of adsorption isotherms and pH effect. *Journal of Environmental Radioactivity*, *93*(3), 127–143. doi:10.1016/j.jenvrad.2006.12.003 PMID:17258360

Hanano, A., Almousally, I., Shaban, M., & Murphy, D. J. (2018). Arabidopsis plants exposed to dioxin result in a WRINKLED seed phenotype due to 20S proteasomal degradation of WRI1. *Journal of Experimental Botany*, *69*(7), 1781–1794. doi:10.1093/jxb/ery027 PMID:29394403

Hasanuzzaman, M., Nahar, K., Hakeem, K. R., Öztürk, M., & Fujita, M. (2015). Arsenic Toxicity in Plants and Possible Remediation. In *Soil Remediation and Plants* (pp. 433–501). Elsevier. doi:10.1016/B978-0-12-799937-1.00016-4

Hou, L., Wang, L., Royer, S., & Zhang, H. (2016). Ultrasound-assisted heterogeneous Fenton-like degradation of tetracycline over a magnetite catalyst. *Journal of Hazardous Materials*, *302*, 458–467. doi:10.1016/j.jhazmat.2015.09.033 PMID:26521091

Huang, Z., Wang, X., & Yang, D. (2015). Adsorption of Cr(VI) in wastewater using magnetic multiwall carbon nanotubes. *Water Science and Engineering*, *8*(3), 226–232. doi:10.1016/j.wse.2015.01.009

Hussain, A., Oves, M., Alajmi, M. F., Hussain, I., Amir, S., Ahmed, J., Rehman, M. T., El-Seedi, H. R., & Ali, I. (2019). Biogenesis of ZnO nanoparticles using Pandanus odorifer leaf extract: Anticancer and antimicrobial activities. *RSC Advances*, *9*(27), 15357–15369. doi:10.1039/C9RA01659G

Iosob, G.-A., Prisecaru, M., Stoica, I., & Călin, M. T. O. C. (2016). Biological Remediation of Soil Polluted With Oil Products : an Overview of Available Technologies. *Biological Remediation of Soil Polluted With Oil Products : An Overview of Available Technologie,* 98. https://www.researchgate.net/publication/309901995_BIOLOGICAL_REMEDIATION_OF_SOIL_POLLUTED_WITH_OIL_PRODUCTS_AN_OVERVIEW_OF_AVAILABLE_TECHNOLOGIES

Kalita, E., & Baruah, J. (2020). Environmental remediation. In *Colloidal Metal Oxide Nanoparticles*. Elsevier Inc., doi:10.1016/B978-0-12-813357-6.00014-0

Kanjwal, M. A., Barakat, N. A. M., Sheikh, F. A., Khil, M. S., & Kim, H. Y. (2009). Functionalization of Electrospun Titanium Oxide Nanofibers with Silver Nanoparticles: Strongly Effective Photocatalyst. *International Journal of Applied Ceramic Technology*, *7*, E54–E63. doi:10.1111/j.1744-7402.2009.02397.x

Karthika, N., Jananee, K., & Murugaiyan, V. (2016). Remediation of contaminated soil using soil washing-a review. *Journal of Engineering Research and Applications*, *6*(1–2), 13–18. http://www.ijera.com/papers/Vol6_issue1/Part - 2/C601021318.pdf

Khan, I., Saeed, K., & Khan, I. (2019). Nanoparticles: Properties, applications and toxicities. *Arabian Journal of Chemistry*, *12*(7), 908–931. doi:10.1016/j.arabjc.2017.05.011

Koliyabandara, S. M. P., Siriwardhana, C., De Silva, S. M., Walpita, J., & Cooray, A. T. (2021). Electrokinetic Remediation for the Removal of Organic Waste in Soil and Sediments. In *Electrokinetic Remediation for Environmental Security and Sustainability* (pp. 453–477). Wiley. doi:10.1002/9781119670186.ch19

Kumar, A., & Kumar, A., M.M.S., C.-P., Chaturvedi, A. K., Shabnam, A. A., Subrahmanyam, G., Mondal, R., Gupta, D. K., Malyan, S. K., Kumar, S. S., A. Khan, S., & Yadav, K. K. (2020). Lead Toxicity: Health Hazards, Influence on Food Chain, and Sustainable Remediation Approaches. *International Journal of Environmental Research and Public Health*, *17*(7), 2179. doi:10.3390/ijerph17072179 PMID:32218253

Küpper, H. (2017). Lead Toxicity in Plants. In Lead: Its Effects on Environment and Health (pp. 491–500). De Gruyter. doi:10.1515/9783110434330-015

Li, S., Turaga, U., Shrestha, B., Anderson, T. A., Ramkumar, S. S., Green, M. J., Das, S., & Cañas-Carrell, J. E. (2013). Mobility of polyaromatic hydrocarbons (PAHs) in soil in the presence of carbon nanotubes. *Ecotoxicology and Environmental Safety*, *96*, 168–174. doi:10.1016/j.ecoenv.2013.07.005 PMID:23896179

Liu, W., Tian, S., Zhao, X., Xie, W., Gong, Y., & Zhao, D. (2015). Application of Stabilized Nanoparticles for In Situ Remediation of Metal-Contaminated Soil and Groundwater: A Critical Review. *Current Pollution Reports*, *1*(4), 280–291. doi:10.100740726-015-0017-x

Ma, C., Liu, H., Chen, G., Zhao, Q., Eitzer, B., Wang, Z., Cai, W., Newman, L. A., White, J. C., Dhankher, O. P., & Xing, B. (2017). Effects of titanium oxide nanoparticles on tetracycline accumulation and toxicity in Oryza sativa (L.). *Environmental Science. Nano*, *4*(9), 1827–1839. doi:10.1039/C7EN00280G

Mahbub, K. R., Krishnan, K., Naidu, R., Andrews, S., & Megharaj, M. (2017). Mercury toxicity to terrestrial biota. *Ecological Indicators*, *74*, 451–462. doi:10.1016/j.ecolind.2016.12.004

Mar Gil-Díaz, M., Pérez-Sanz, A., Ángeles Vicente, M., & Carmen Lobo, M. (2014). Immobilisation of Pb and Zn in Soils Using Stabilised Zero-valent Iron Nanoparticles: Effects on Soil Properties. *CLEAN - Soil, Air. Water (Basel)*, *42*(12), 1776–1784. doi:10.1002/clen.201300730

Martínez, F., López-Muñoz, M. J., Aguado, J., Melero, J. A., Arsuaga, J., Sotto, A., Molina, R., Segura, Y., Pariente, M. I., Revilla, A., Cerro, L., & Carenas, G. (2013). Coupling membrane separation and photocatalytic oxidation processes for the degradation of pharmaceutical pollutants. *Water Research*, *47*(15), 5647–5658. doi:10.1016/j.watres.2013.06.045 PMID:23863375

McDonald, K. J., Reddy, K. J., Singh, N., Singh, R. P., & Mukherjee, S. (2015). Removal of arsenic from groundwater in West Bengal, India using CuO nanoparticle adsorbent. *Environmental Earth Sciences*, *73*(7), 3593–3601. doi:10.100712665-014-3645-3

Mlynarczuk, J., Górska, M., & Wrobel, M. H. (2020). Effects of DDT, DDE, aldrin and dieldrin on prostaglandin, oxytocin and steroid hormone release from smooth chorion explants of cattle. *Animal Reproduction Science*, *223*, 106623. doi:10.1016/j.anireprosci.2020.106623 PMID:33132122

Mthombeni, N. H., Mbakop, S., & Onyango, M. S. (2015). Magnetic Zeolite-Polymer Composite as an Adsorbent for the Remediation of Wastewaters Containing Vanadium. *International Journal of Environmental Sciences and Development*, *6*(8), 602–605. doi:10.7763/IJESD.2015.V6.665

Osin, O. A., Yu, T., Cai, X., Jiang, Y., Peng, G., Cheng, X., Li, R., Qin, Y., & Lin, S. (2018). Photocatalytic Degradation of 4-Nitrophenol by C, N-TiO2: Degradation Efficiency vs. Embryonic Toxicity of the Resulting Compounds. *Frontiers in Chemistry*, *6*, 192. Advance online publication. doi:10.3389/fchem.2018.00192 PMID:29915782

Palimi, M. J., Rostami, M., Mahdavian, M., & Ramezanzadeh, B. (2014). Surface modification of Fe 2 O 3 nanoparticles with 3-aminopropyltrimethoxysilane (APTMS): An attempt to investigate surface treatment on surface chemistry and mechanical properties of polyurethane/Fe 2 O 3 nanocomposites. *Applied Surface Science*, *320*, 60–72. doi:10.1016/j.apsusc.2014.09.026

Patil, S. S., Shedbalkar, U. U., Truskewycz, A., Chopade, B. A., & Ball, A. S. (2016). Nanoparticles for environmental clean-up: A review of potential risks and emerging solutions. *Environmental Technology & Innovation*, *5*, 10–21. doi:10.1016/j.eti.2015.11.001

Pham, T. D., & Sillanpää, M. (2020). Ultrasonic and electrokinetic remediation of low permeability soil contaminated with persistent organic pollutants. In *Advanced Water Treatment: Electrochemical Methods*. Elsevier Inc. doi:10.1016/B978-0-12-819227-6.00004-8

Pinto, M., Ramalho, P. S. F., Moreira, N. F. F., Gonçalves, A. G., Nunes, O. C., Pereira, M. F. R., & Soares, O. S. G. P. (2020). Application of magnetic nanoparticles for water purification. *Environmental Advances*, *2*, 100010. doi:10.1016/j.envadv.2020.100010

Plessis, D., & Marguerite, D. (2011). Fabrication and characterization of anti-microbial and biofouling resistant nanofibers with silver nanoparticles and immobilized enzymes for application in water filtration. *Materials Science*.

Qian, L., Liu, S., Zhang, W., Chen, Y., Ouyang, D., Han, L., Yan, J., & Chen, M. (2019). Enhanced reduction and adsorption of hexavalent chromium by palladium and silicon rich biochar supported nanoscale zero-valent iron. *Journal of Colloid and Interface Science*, *533*, 428–436. doi:10.1016/j.jcis.2018.08.075 PMID:30172153

Qian, Y., Qin, C., Chen, M., & Lin, S. (2020). Nanotechnology in soil remediation – applications vs. implications. *Ecotoxicology and Environmental Safety*, *201*, 110815. doi:10.1016/j.ecoenv.2020.110815 PMID:32559688

Rafati Rahimzadeh, M., Rafati Rahimzadeh, M., Kazemi, S., & Moghadamnia, A.-A. (2017). Cadmium toxicity and treatment: An update. *Caspian Journal of Internal Medicine*, *8*(3), 135–145. doi:10.22088/cjim.8.3.135 PMID:28932363

Rahman, Z., & Singh, V. P. (2019). The relative impact of toxic heavy metals (THMs) (arsenic (As), cadmium (Cd), chromium (Cr)(VI), mercury (Hg), and lead (Pb)) on the total environment: An overview. *Environmental Monitoring and Assessment*, *191*(7), 419. doi:10.100710661-019-7528-7 PMID:31177337

Rizwan, M., Singh, M., Mitra, C. K., & Morve, R. K. (2014). Ecofriendly Application of Nanomaterials: Nanobioremediation. *Journal of Nanoparticles*, *2014*, 1–7. doi:10.1155/2014/431787

Rojas, R. V., Achouri, M., Maroulis, J., & Caon, L. (2016). Healthy soils: A prerequisite for sustainable food security. *Environmental Earth Sciences*, *75*(3), 180. doi:10.100712665-015-5099-7

Saikia, J., Gogoi, A., & Baruah, S. (2019). Nanotechnology for Water Remediation. In Environmental Chemistry for a Sustainable World (pp. 195–211). doi:10.1007/978-3-319-98708-8_7

Salem, D. M. S. A., Ismail, M. M., & Aly-Eldeen, M. A. (2019). Biogenic synthesis and antimicrobial potency of iron oxide (Fe3O4) nanoparticles using algae harvested from the Mediterranean Sea, Egypt. *Egyptian Journal of Aquatic Research*, *45*(3), 197–204. doi:10.1016/j.ejar.2019.07.002

Sarkar, A., Sengupta, S., & Sen, S. (2019). Nanoparticles for Soil Remediation. In Environmental Chemistry for a Sustainable World (pp. 249–262). doi:10.1007/978-3-319-97922-9_9

Sathishkumar, P., Mohan, K., Ganesan, A. R., Govarthanan, M., Yusoff, A. R. M., & Gu, F. L. (2021). Persistence, toxicological effect and ecological issues of endosulfan – A review. *Journal of Hazardous Materials*, *416*, 125779. doi:10.1016/j.jhazmat.2021.125779

Shankar, S., Shanker, U., & Shikha. (2014). Arsenic Contamination of Groundwater: A Review of Sources, Prevalence, Health Risks, and Strategies for Mitigation. *TheScientificWorldJournal*, *2014*, 1–18. doi:10.1155/2014/304524 PMID:25374935

Shipley, H. J., Engates, K. E., & Guettner, A. M. (2011). Study of iron oxide nanoparticles in soil for remediation of arsenic. *Journal of Nanoparticle Research*, *13*(6), 2387–2397. doi:10.100711051-010-9999-x

Song, B., Xu, P., Chen, M., Tang, W., Zeng, G., Gong, J., Zhang, P., & Ye, S. (2019). Using nanomaterials to facilitate the phytoremediation of contaminated soil. *Critical Reviews in Environmental Science and Technology*, *49*(9), 791–824. doi:10.1080/10643389.2018.1558891

Song, Y., Fang, G., Zhu, C., Zhu, F., Wu, S., Chen, N., Wu, T., Wang, Y., Gao, J., & Zhou, D. (2019). Zero-valent iron activated persulfate remediation of polycyclic aromatic hydrocarbon-contaminated soils: An in situ pilot-scale study. *Chemical Engineering Journal, 355*, 65–75. doi:10.1016/j.cej.2018.08.126

Starek-Świechowicz, B., Budziszewska, B., & Starek, A. (2017). Hexachlorobenzene as a persistent organic pollutant: Toxicity and molecular mechanism of action. *Pharmacological Reports, 69*(6), 1232–1239. doi:10.1016/j.pharep.2017.06.013 PMID:29128804

Subramanian, S., Schnoor, J. L., & Van Aken, B. (2017). Effects of Polychlorinated Biphenyls (PCBs) and Their Hydroxylated Metabolites (OH-PCBs) on Arabidopsis thaliana. *Environmental Science & Technology, 51*(12), 7263–7270. doi:10.1021/acs.est.7b01538 PMID:28541669

Sun, F., Dai, Y., & Yu, X. (2017). Air pollution, food production and food security: A review from the perspective of food system. *Journal of Integrative Agriculture, 16*(12), 2945–2962. doi:10.1016/S2095-3119(17)61814-8

Taiwo, A. M. (2019). A review of environmental and health effects of organochlorine pesticide residues in Africa. *Chemosphere, 220*, 1126–1140. doi:10.1016/j.chemosphere.2019.01.001 PMID:33395800

Tan, X., Wang, X., Chen, C., & Sun, A. (2007). Effect of soil humic and fulvic acids, pH and ionic strength on Th(IV) sorption to TiO2 nanoparticles. *Applied Radiation and Isotopes, 65*(4), 375–381. doi:10.1016/j.apradiso.2006.10.014 PMID:17157512

Taylor, A. A., Tsuji, J. S., Garry, M. R., McArdle, M. E., Goodfellow, W. L. Jr, Adams, W. J., & Menzie, C. A. (2020). Critical Review of Exposure and Effects: Implications for Setting Regulatory Health Criteria for Ingested Copper. *Environmental Management, 65*(1), 131–159. doi:10.100700267-019-01234-y PMID:31832729

Towell, M. G., Browne, L. A., Paton, G. I., & Semple, K. T. (2011). Impact of carbon nanomaterials on the behaviour of 14C-phenanthrene and 14C-benzo-[a] pyrene in soil. *Environmental Pollution, 159*(3), 706–715. doi:10.1016/j.envpol.2010.11.040 PMID:21195517

Tripathy, S. S., & Kanungo, S. B. (2005). Adsorption of Co2+, Ni2+, Cu2+ and Zn2+ from 0.5 M NaCl and major ion sea water on a mixture of δ-MnO2 and amorphous FeOOH. *Journal of Colloid and Interface Science, 284*(1), 30–38. doi:10.1016/j.jcis.2004.09.054 PMID:15752781

Ukaogo, P. O., Ewuzie, U., & Onwuka, C. V. (2020). Environmental pollution: causes, effects, and the remedies. In *Microorganisms for Sustainable Environment and Health* (pp. 419–429). Elsevier., doi:10.1016/B978-0-12-819001-2.00021-8

US EPA. (2004). *Cleaning up the Nation's waste sites: Markets and technology trends.* https://nepis.epa.gov/Exe/ZyNET.exe/30006II3.TXT?ZyActionD=ZyDocument&Client=EPA&Index=2000+Thru+2005&Docs=&Query=&Time=&EndTime=&SearchMethod=1&TocRestrict=n&Toc=&TocEntry=&QField=&QFieldYear=&QFieldMonth=&QFieldDay=&IntQFieldOp=0&ExtQFieldOp=0&XmlQuery=&File=D%3A%5Czyfiles%5CIndex Data%5C00thru05%5CTxt%5C00000008%5C30006II3.txt&User=ANONYMOUS&Password=anonymous&SortMethod=h%7C-&MaximumDocuments=1&FuzzyDegree=0&ImageQuality=r75g8/r75g8/x150y150g16/i425&Display=hpfr&DefSeekPage=x&SearchBack=ZyActionL&Back=ZyActionS&BackDesc=Resultspage&MaximumPages=1&ZyEntry=1&SeekPage=x&ZyPURL

US EPA. (2021). *Learn about Polychlorinated Biphenyls (PCBs) | Polychlorinated Biphenyls (PCBs).* United States Environmental Protection Agency. https://www.epa.gov/pcbs/learn-about-polychlorinated-biphenyls-pcbs

Usman, M., Faure, P., Ruby, C., & Hanna, K. (2012). Remediation of PAH-contaminated soils by magnetite catalyzed Fenton-like oxidation. *Applied Catalysis B: Environmental, 117–118,* 10–17. doi:10.1016/j.apcatb.2012.01.007

Vidonish, J. E., Zygourakis, K., Masiello, C. A., Sabadell, G., & Alvarez, P. J. J. (2016). Thermal Treatment of Hydrocarbon-Impacted Soils: A Review of Technology Innovation for Sustainable Remediation. *Engineering, 2*(4), 426–437. doi:10.1016/J.ENG.2016.04.005

Wang, Z., Hang, J. G., Feng, H., Shi, L. L., Dong, J. J., Shen, B., Luo, T., Cai, R. M., Shen, L. J., Kido, T., & Sun, X. L. (2019). Effects of perinatal dioxin exposure on development of children: A 3-year follow-up study of China cohort. *Environmental Science and Pollution Research International, 26*(20), 20780–20786. doi:10.100711356-019-05362-0 PMID:31102233

Wu, Y., Li, Z., Chen, J., Yu, C., Huang, X., Zhao, C., Duan, L., Yang, Y., & Lü, W. (2015). Graphene nanosheets decorated with tunable magnetic nanoparticles and their efficiency of wastewater treatment. *Materials Research Bulletin, 68,* 234–239. doi:10.1016/j.materresbull.2015.03.073

Xie, Y., Fang, Z., Cheng, W., Tsang, P. E., & Zhao, D. (2014). Remediation of polybrominated diphenyl ethers in soil using Ni/Fe bimetallic nanoparticles: Influencing factors, kinetics and mechanism. *The Science of the Total Environment, 485–486,* 363–370. doi:10.1016/j.scitotenv.2014.03.039 PMID:24742544

Yang, S., Hu, J., Chen, C., Shao, D., & Wang, X. (2011). Mutual Effects of Pb(II) and Humic Acid Adsorption on Multiwalled Carbon Nanotubes/Polyacrylamide Composites from Aqueous Solutions. *Environmental Science & Technology, 45*(8), 3621–3627. doi:10.1021/es104047d PMID:21395259

Zhang, L., Li, P., Gong, Z., & Li, X. (2008). Photocatalytic degradation of polycyclic aromatic hydrocarbons on soil surfaces using TiO2 under UV light. *Journal of Hazardous Materials, 158*(2–3), 478–484. doi:10.1016/j.jhazmat.2008.01.119 PMID:18372106

Zhang, W., Lu, Y., Sun, H., Zhang, Y., Zhou, M., Song, Q., & Gao, Y. (2019). Effects of multi−walled carbon nanotubes on pyrene adsorption and desorption in soils: The role of soil constituents. *Chemosphere, 221,* 203–211. doi:10.1016/j.chemosphere.2019.01.030 PMID:30640002

ADDITIONAL READING

Njuguna, S. M., Onyango, J. A., Githaiga, K. B., Gituru, R. W., & Yan, X. (2020). Application of multivariate statistical analysis and water quality index in health risk assessment by domestic use of river water. Case study of Tana River in Kenya. *Process Safety and Environmental Protection*, *133*, 149–158. doi:10.1016/j.psep.2019.11.006

Sharma, S., & Bhattacharya, A. (2017). Drinking water contamination and treatment techniques. *Applied Water Science*, *7*(3), 1043–1067. doi:10.100713201-016-0455-7

Singh, S., Kumar, V., Romero, R., Sharma, K., & Singh, J. (2019). Applications of Nanoparticles in Wastewater Treatment. In Nanotechnology in the Life Sciences (pp. 395–418). doi:10.1007/978-3-030-17061-5_17

KEY TERMS AND DEFINITIONS

Degradation: The degradation process indicates the splitting or decomposition of organic pollutants into several intermediate products.

Mineralization: The term "mineralization" indicates the total removal or conversion of organic pollutants into CO_2, H_2O, and other inorganic ions.

Nanoparticles: The ultrafine particles having dimensions between 1-100 nm in size are termed nanoparticles.

Nanoremediation: The use of nano-sized particles, i.e. nanoparticles (NPs), to remediate contaminants from soil, water, and the air is termed as nanoremediation. Nanotechnology provides the opportunity to recognize, monitor, prevent, reduce and treat the contaminants in soil, air, and water.

Persistent Organic Compounds (POPs): The toxic organic molecules that resist biodegradation and persist for a long time in the environment are called Persistent Organic Compounds. For example, aldrin, DDT, chlordane, dieldrin, endrin etc.

Photocatalysis: The term "photocatalysis" refers to the degradation of compounds in the presence of light. In the nano-photocatalytic waste treatment process, NPs capture light energy and utilizes it to degrade or breakdown of a wide range of organic contaminants.

Remediation: The term "remediation" comes from the Latin word *remediātiō*, which means "process of healing". In this chapter, the word "remediation" refers to eliminating hazardous contaminants or pollutants from the environment.

Chapter 11
CaO Nanoparticles:
Synthesis and Applications in Water Remediation

Nnabuk Okon Eddy
https://orcid.org/0000-0001-7704-3082
Department of Pure and Industrial Chemistry, University of Nigeria, Nsukka, Nigeria

Rajni Garg
https://orcid.org/0000-0002-9629-8141
Rayat Bahra University, India

ABSTRACT

Adsorption is widely acknowledged as one of the best options that are available for the removal of contaminants from water. Contamination of water does not only create water scarcity, but it has the capacity to generate and transfer several environmental problems including threat to public health. This chapter reviewed calcium oxide nanoparticle (CaONP) as a noble metal oxide for the removal of contaminants from water. The review is concentrated in the general overview of water contamination, metal oxide nanoparticles, general application of CaONP, synthetic methods, characterization method, and applications. The chapter observed that little is done on the use of CaONP for the removal of contaminants from water except for dyes, some heavy metal ions, and few organic/inorganic compounds. It is also observed that CaONP can be applied as adsorbent and in photocatalytic degradation of dye. Suggestions are made on the possibility of utilizing local raw materials that are easily accessible, cheap, and environmental sources of raw materials for the synthesis of CaONP.

INTRODUCTION

According to a World Health Organization information sheet, by 2025, half of the world's population would be living in water-stressed regions. In the least developed nations, 22 percent of healthcare institutions do not have access to water, 21 percent do not have access to sanitation, and 22 percent do not have access to waste management (Boretti & Rosa, 2019). Contamination of different sources of water

DOI: 10.4018/978-1-7998-8936-6.ch011

has taken different dimensions and the final consequence is in the form of polluted water. The effect of water contamination on the ecosystem has prompted considerable attention for contamination of water by various forms of physicochemical parameters, heavy metals, emergent contaminants, nutrients, etc. (Njuguna et al., 2020). The presence of unwanted contaminants in water can affects aquatic life and alter the quality of water against its normal usage. Hence, water remediation involving the removal of unwanted and harmful contaminants is a must before its use for domestic, agricultural, and industrial use (Borgohain et al., 2020). Depending upon the nature of contaminants and feasibility of the process, several options are available for mitigating water pollution, involving reverse osmosis, filtration, oxidation, adsorption and biological treatment, etc. (Dave & Chopda, 2014). However, the use of the adsorption process has gained wider acceptance because it is cost-effective and can be implemented with biodegradable materials to remove contaminants from wastewater (Sarwar et al., 2021).

Adsorption involves the sticking of particles (i.e., adsorbates such as molecules, ions, atoms, etc) on a surface (called adsorbent). The adsorption process can be influenced by temperature, pressure, nature of adsorbent, the concentration of adsorbate, pH of the solution, the mass of adsorbent, and other factors (Danesh et al., 2021). Consequently, processes that control adsorption must be based on the control of these factors. Different adsorbents have been studied for their ability to remove pollutants from water including plant samples (Tomar et al., 2014), eggshells (Lunge et al., 2012), agricultural wastes (Bansal et al., 2009) such as rice husk (Chuah et al., 2005; Vijila et al., 2020), sawdust (Bansal et al., 2012), charcoal (Mishra & Patel, 2009), cow dung (Rajkumar et al., 2019), diatomite (Akafu et al., 2019), brick powder (Yadav et al., 2006), polymers (Solanki et al., 2021), chitosan (Kamble et al., 2007), metal oxides (Srivastav et al., 2013), etc. However, most of the known adsorbents suffer from factors such as renewability, formation of stable product with the adsorbate, thermal instability, solubility, and other factors that work to reduce their efficiency (Xia et al., 2013). On the other hand, nano-sized adsorbents or nano-adsorbents are highly effective not only for the removal of heavy metal ions and degradation of organic compounds but also for disinfecting harmful micro-organisms (Oprčkal et al., 2017). Therefore, the applications of nano-adsorbents for adsorption/removal of contaminants in water have garnered the interest of researchers across the world.

Nanomaterials with sizes ranging in nano-scale (1-100 nm) are extensively utilized in diverse fields, including environmental monitoring, biomedical and pharmaceutical sector, electronics and optoelectronics, textiles, and cosmetics (R. Sharma et al., 2020). The nano-scale causes a variety of changes in the physical characteristics of the nanomaterials, including an increase in the volume to surface area ratio and the impact of the quantum effect on the particle size (M. Kumar et al., 2021). The characteristics of nanoparticles, such as their magnetic, optical, and electrical properties, are substantially different from those of conventional bulk materials. Nanomaterials have unique characteristics such as high adsorption, catalytic activity, and reactivity, as a result of their vast surface area and compact size (Garg et al., 2020). Several studies have shown that nanoparticles are effective in the remediation of wastewater. Nanomaterials have been used for water remediation including various pollutants such as emerging pollutants, organic pollutants, and inorganic anions in addition to disinfection of water (Jaiswal et al., 2021). Literature reports the use of various nanomaterials with potential applications in wastewater remediation. These include carbon nanotubes, zerovalent nanoparticles, metal oxide nanoparticles, and nanocomposites, among others. Literature also reveals that among the various metal oxide nanoparticles (such as TiO, MgO, ZnO, etc), calcium oxide nanoparticles (CaO NP) are the most demanded because of their biocompatibility (Vanthana Sree et al., 2020).

Figure 1. Categorization of water contaminants

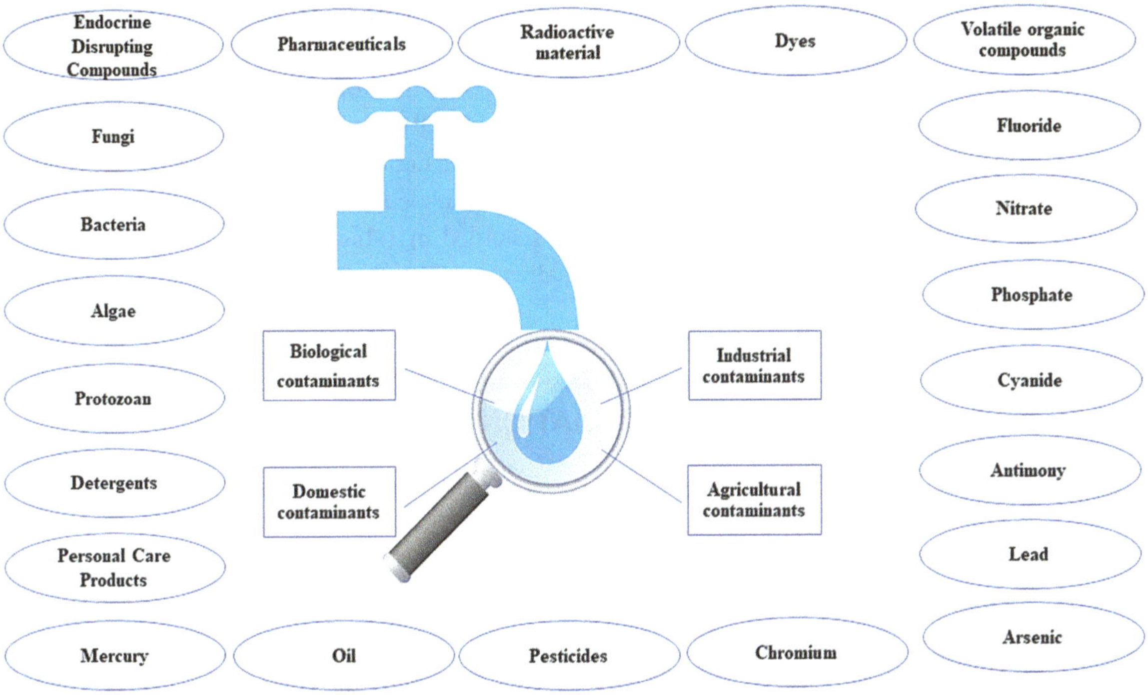

As a result of the above context, this study examines the most recent developments in the synthesis and use of CaO NP for the removal of different pollutants from wastewater. With the advent of green chemistry, greener approaches have fascinated the researchers for the synthesis of nanomaterials. Hence, the chapter also explores the greener techniques being used for the synthesis of CaO NP. Further, the challenges and the future aspects for green synthesis and application of CaO NP in wastewater remediation have also been discussed.

BACKGROUND

Water can be polluted by anthropogenic and industrial wastes. Therefore, the adoption of measures that can control the rate of contamination or treat contaminated water is necessary for the view of the importance of water in the ecosystem (Agasti, 2021). Some major classes of water contaminants include toxic organic contaminants (dyes, emerging organic contaminants, crude oil, phenol, etc.), toxic heavy metals, antibiotics, salts, physicochemical substances, microorganisms, and other organic matter as categorised in Figure 1.

Among all these contaminants, less attention is given to emergent pollutants (S. Sharma & Bhattacharya, 2017). Emergent pollutants are substances, either manufactured or naturally occurring, that are not routinely monitored yet have the potential to harm the environment if released into the ecosystem. Some emergent pollutants have been identified in some sources of water indicating that water quality assessment must include this class of pollutants. Available literature reveals that the presence of about 700 substances categorized into 20 classes of emergent pollutants (resulting from diffuse and point source pollutions) has been reported for the European aquatic environment (Gavrilescu et al., 2015). Pharmaceuticals (including antibiotics) are regarded as refractory organic micropollutants and their presence in the environment has generated major concern because of their adverse effect on aquatic phytoplankton and zooplankton (Cheriyamundath & Vavilala, 2021). Summary of water quality parameters and analytical methods is presented in Table 1 (Allard et al., 1992; DHV consultants & DELFT hydraulics with HALCROW, TAHAL, CES, 2003). Given the scarcity and high demand for potable water, several technologies have been developed to purify waste or contaminated water including methods such as flocculation, filtration, adsorption and those that are recorded in Table 2 (Bashambu et al., 2021; S. Sharma & Bhattacharya, 2017). The limitation and challenges associated with each of the listed methods are also listed in Table 2.

THE MECHANISM FOR WATER REMEDIATION BY NP

Photodegradation

NP can be applied in the treatment of contaminants from wastewater because of their large surface area to volume ratio (which give favorable enhancement on size-dependent properties such as high reactivity, high adsorption capacity, high dissolution capacity), high super-magneticity, and quantum confinement effect (Kamboj et al., 2020). Photocatalysis with NP is aimed at degrading organic pollutants to convert these to non-toxic products such as water and CO_2. It is a very useful method for the treatment of wastewater that is highly contaminated with non-degradable organic matter. NP can accelerate photoreaction because of their strong tendency to catalyze several reactions.

The process of photocatalysis can be analyzed by considering the transition of an electron from the valence band (VB) to the conduction band (CB) (which is capable of creating a hole in the VB) when the photon of energy is equal to or greater than the energy gap of the system under consideration. The liberated ejected electron is capable of creating an oxygen scavenger (when it reacts with the dissolved oxygen in the water) which can lead to the formation of active free radicals ($*O_2$). The *OH free radical can also be formed by the hole if its accepts an electron from the electron donor (H_2O) (Ajmal et al., 2014). The formation of these free radicals will lead to the attack of the organic molecule by the active species which may manifest by the opening of an aromatic ring and discoloration before final degradation as illustrated in Figure 2.

The likely mechanism may involve the following steps:

1. Absorption of a photon by the organic compound and subsequent excitation:

$$\text{Org.} + hv \rightarrow \text{org}^*$$

2. The formation of electron-hole pair formation

$$\text{NP} + hv \rightarrow \text{e}^- \left(\text{CB}\right) + \text{H}^+ \left(\text{CB}\right)$$

Table 1. Analytical methods for some water quality parameters.

Parameters	Analytical Method	Reagents
Phenolphthalein alkalinity	Titrimetric	Na_2CO_3, H_2SO_4, ethyl alcohol, phenolphthalein
Total alkalinity	Titrimetric	Na_2CO_3, H_2SO_4, ethyl alcohol, bromocresol indicator, methyl orange indicator
Aluminum	Spectrophotometric	Eriochrome cyanide R, HCl, $AlK(SO_4)_2$, sodium acetate, standard aluminum solution, ascorbic acid, acetic acid, solochrome cyanide, bromocresol green, EDTA, sodium hydroxide
Bicarbonate	Titrimetric	Na_2CO_3, H_2SO_4, ethyl alcohol, bromocresol indicator, methyl orange indicator, phenolphthalein
Biochemical oxygen demand	Bioassay	Phosphate buffer solution, KH_2PO_4, $Na_2HPO_4.7H_2O$, NH_4Cl, $MgSO_4.7H_2O$, $CaCl_2$, $FeCl_3.6H_2O$, H_2SO_4, NaOH, glutamic acid, glucose glutamic acid.
Boron	Spectrophotometric	HCl, ethyl alcohol boric acid, curcumin, ion exchange resin, boron solution, oxalic acid, and boric acid
Calcium	EDTA titrimetric	Standard calcium solution, EDTA, HCl, Murexide indictor, NaCl, NaOH, $CaCO_3$ and NH_4OH
Carbonate	Titrimetric	Na_2CO_3, H_2SO_4, ethyl alcohol, bromocresol indicator, methyl orange indicator, phenolphthalein
Chemical oxygen demand	Open reflux	Standard $K_2Cr_2O_7$, H_2SO_4, Ag_2SO_4, $FeSO_4.7H_2SO_4$, 1, 10-phenanthroline monohydrate and $Fe(NH_4)_2$ $(SO_4)_2.6H_2O$
Chloride	Titrimetric (Argentometric)	NaCl, $AgNO_3$, $AlK(SO_4)_2.12H_2SO_4$ $AlNH_4(SO_4)_2.12H_2O$, NH_4OH and $Al(OH)_3$
Chlorophyll-a	Spectrophotometric	Acetone, $Mg(CO_3)_2$
Faecal coliform	Elevated temperature fermentation	Tryptose, trypticase, bile salts, bile salts, dipotassium hydrogen phosphate, potassium dihydrogen phosphate, and sodium chloride
Total coliform	Standard multiple tube fermentation	Tryptose, lactose, dipotassium hydrogen phosphate, potassium dihydrogen phosphate, sodium chloride, and sodium lauryl sulfate
Colour	Visual comparison	K_2PtCl_6 crystallized cobaltous chloride and HCl
Dissolve oxygen	Titrimetric (Winkler azide modification)	$MnSO_4$ $.4H_2O$, $MnSO_4.2H_2O$, $MnSO_4.H_2O$,alkali-iodide-azide, NaOH, KOH, NaI, KI, NaN_3, sulphuric acid, starch, salicylic acid, standard sodium thiosulphate titrant, $Na_2S_2O_3.5H_2O$, NaOH, standardize with bi-iodate solution, $KH(IO_3)_2$
Conductivity	Potentiometric (Conductivity cell)	Chromic/sulphuric acid, chloroplatinic acid, $H_2PtCl_6.6H_2O$) and lead acetate
Flouride	Electrometric (Ion-selective electrode)	NaF, NaCl, NaOH, Glacial acetic acid, 1,2-cyclohexylenediaminetetraacetic acid
Total hardness	Titrimetric (EDTA)	NH_4Cl, NH_4OH, EDTA, eriochrome black, $CaCO_3$
Iron	Spectrophotometric	HCl, ammonium acetate, Hydroxylamine, sodium acetate, 1,10-phenanthroline monohydrate, H_2SO_4, ferrous ammonium sulfate, $KMnO_4$ and standard iron solution
Magnesium	Titrimetric	NH_4Cl, NH+OH, EDTA, eriochrome black, $CaCO_3$, Standard calcium solution, EDTA, HCl, Murexide indictor, NaCl, NaOH, $CaCO_3$ and NH_4OH
Manganese	Spectrophotometric	$HgSO_4$, HNO_3, $(NH_4)_2S_2O_8$, $KMnO_4$, $Na_2C_2O_4$, H_2SO_4 and $KMnO_4$
Nitrogen nitrate	Spectrophotometric	Stock Nitrate solution, KNO_3, $CHCl_3$ nitrate Solution, and HCl,
Nitrogen nitrite	Spectrophotometric	phosphoric acid, sulphanilamide, N-(1-naphthyl)-ethylenediamine dihydrochloride, sodium oxalate, $Na_2C_2O_4$, $NaNO_2$, $CHCl_3$ $KMnO_4$, H_2SO_4 and NO_2^- solution
Phosphorus as ortho phosphorus	Spectrophotometric	Sulphuric acid, Potassium antimonyl tartrate, ammonium molybdate solution, ascorbic acid, H_2SO_4, anhydrous KH_2PO_4 and standard phosphate solution
Total phosphorus	Spectrophotometric	Ammonium molybdate solution, persulphate, Sulphuric acid, Potassium antimonyl tartrate, ammonium molybdate solution, ascorbic acid, H_2SO_4, anhydrous KH_2PO_4. and standard phosphate solution
Potassium and sodium	Flame emission photometric	Standard solution of the metal
Sulfate	Nephelometry	magnesium chloride, sodium acetate, potassium nitrate, acetic acid, sodium sulfate, barium chloride, sodium carbonate, and sulphuric acid
Heavy metals (Co, Ni, Cr(VI), Pb, Cd, As, Se, V, Hg, Cu)	Spectrophotometric	Standard solution of the respective metals
Dyes (Sudan I dye, Methyl blue, Basic red 9 dye, crystal violet dye, dispersed red 1 dye, reactive and some azo dyes.	Spectrophotometric	Standard solution of the respective dyes
Antibiotics (tetracycline, ampicillin, amoxicillin, cloxacillin, ceftezole, penicillin, and few others)	Spectrophotometric	Standard solution of the respective antibiotics
Organic pollutants (phenol, crude oil, fertilizers)	GCMS	Standards

Table 2. Some water treatment processes and their limitations

Method	Description	Limitations
Distillation	Use of evaporation and condensation	The method is not cost-effective with the requirement of the high cost of operation and maintenance.
Chemical precipitation	Separation of the pollutants in the form of precipitates	The method is not environmentally friendly because of the employment of a large amount of chemicals and non-selective nature.
Flocculation/ Coagulation	Separation of the pollutants in the form of flocculates	The method is complex and not very inefficient with the requirement of basic reagents to attain optimum pH.
Flotation	Separation of the pollutants along with bubbles	The method is not cost-effective with the requirement of the high cost of operation and maintenance.
Ion exchange	Use of ion-exchange resins	The method is not cost-effective with the requirement of the high cost of operation and maintenance.
Chemical oxidation	Oxidizing pollutants employing oxidants such as Cl_2, O_3, $KMnO_4$, ClO_2, H_2O_2 etc.	The method needs large amounts of chemicals and the resistance of some contaminants to oxidation can limit its usefulness
Advanced oxidation process	Use of photocatalysis	The method is not economically viable for large-scale operations.
Nanofiltration	Use of membranc with nano-sized pores	The method consumes high energy and pre-treatment is a prerequisite.
Ultrafiltration	Involves the use of semi-permeable membrane	The method can't be used for the removal of soluble salts from the water and is energy-intensive.
Carbon filtration	Based on activated carbon filter	The method is highly selective and cannot remove chemicals that cannot be adsorbed by carbon.
Biological treatment	Use of micro-biological (mixed or pure) cultures	The method is not cost-effective and requires excessive time for growth, procurement, and processing of culture. A slight change in environmental conditions can result in inactivation.

Figure 2. Photocatalysis by NP

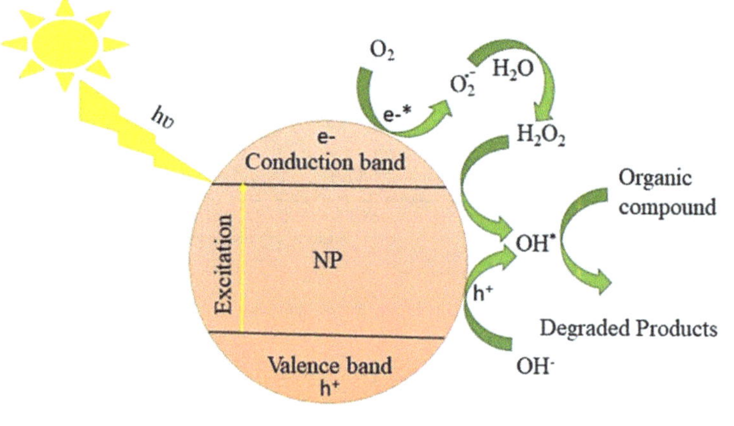

3. Oxygen ion sorption reaction

$$e^{-}(CB) + O_2 \rightarrow O_2^{*}$$

4. Neutralization by proton

$$2O_2 + 2H_2O \rightarrow H_2O_2 + O_2(O_2^{*})$$

5. Formation of OH* through the decomposition of H_2O_2

$$H_2O_2 + e^{-}(CB) \rightarrow OH^{-} + OH^{*}$$

6. Splitting of water by photo-hole to produce OH* radical

$$h^{+} + H_2O \rightarrow H^{+} + OH^{*}$$

7. Electrophilic attack of the organic compound, leading to the degradation of the organic compounds such as dye

$$O_2^{*} / OH^{*} + \text{org.} \rightarrow \text{Degradation product}$$

Adsorption

Adsorption is a surface process that involves the interaction of adsorbent with adsorbate. The mechanism associated with the adsorption of a contaminant to the surface of NP can be physical adsorption, chemical adsorption, or both physical and chemical adsorption (Okwunodulu & Eddy, 2014). Adsorption experiment can be implemented as batch adsorption or column adsorption. Batch adsorption is ideal for the determination of the concentration of a contaminant removed from a solution. The process is simple and involves little instrumentation as illustrated in Figure 3.

Models that can be used to analyze adsorption models can be generally grouped into the following:

1. Kinetic modelling: involving the influence of time on the adsorption. For example, pseudo-first-order and pseudo-second-order kinetics, intraparticle diffusion models, etc.
2. Thermodynamic models: Arrhenius model (activation energy evaluation), Transition state equation (entropy, enthalpy changes, and free energy of adsorption).
3. Adsorption isotherms models: unfolding the adsorption behavior of the adsorbent. For example, Langmuir Freundlich, Dubinin-Raduskevich isotherm, El awardy *et al* and other isotherms.
4. Computational chemistry modelling: Prediction of the mechanism of adsorption and computation of parameters and their correlation with the adsorption process. Simulation studies shall be carried

Figure 3. Adsorption of contaminants by NP

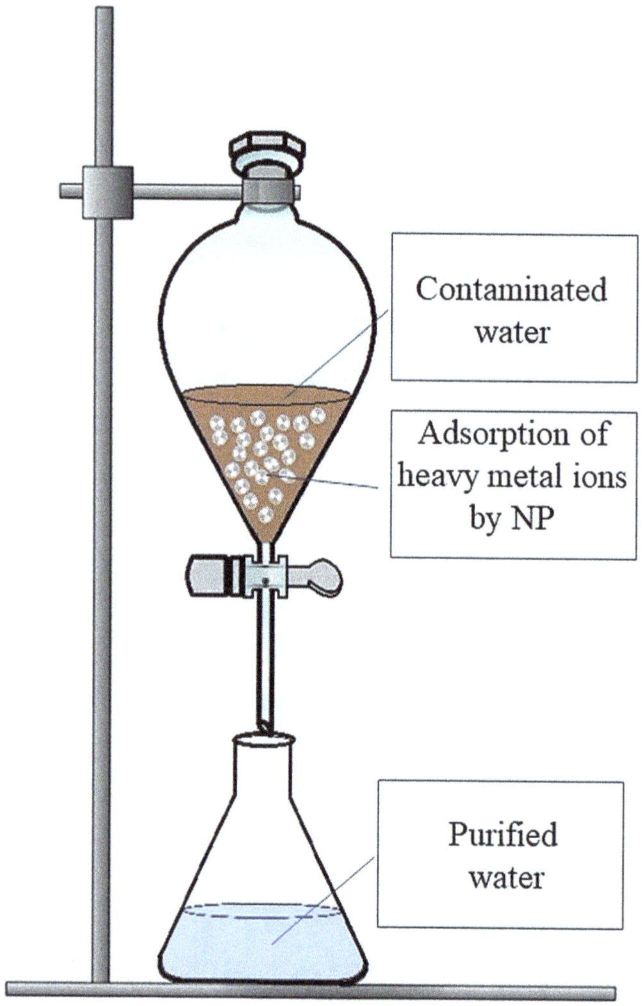

out to establish a quantitative structure-activity relationship (QSAR), which can serve as a model for predicting the adsorption efficiency of the MOF and nanomaterials for other pollutants that have not been studied.

METAL OXIDES AND WATER REMEDIATION

Literature reports many investigations on the usage of metal oxide nanoparticles for water and wastewater treatments. Studies conducted by Kumar et al., 2013 revealed the application of the precipitation method for the syntheses of nano ZnO and SnO (which had a surface area of 15.75 and 24.48 m^2/g respectively) as very effective adsorbents to remove hexavalent chromium and malachite green oxalate from aqueous solution. The effectiveness of the materials for adsorption was dependent upon the concentration, pH, temperature, adsorbent dosage, and time of contact. The adsorption equilibrium responded positively to the Langmuir and Freundlich adsorption models while its kinetics was consistent with second-order and particle diffusion models. Regenerated adsorbent (using CH_3COOH and NaOH) also showed good efficiency. Ghaedi et al., 2013 also synthesized, characterized, and applied palladium, silver, and zinc oxide nanoparticles for the removal of bromophenol red dye through the adsorption process. The kinetics of the adsorption process was investigated using pseudo-first, pseudo-second, Elovich, and the intraparticle diffusion models (Adedirin et al., 2011). However, the adsorption fitted the pseudo-second-order kinetics and the intraparticle diffusion model. The adsorption models that fitted the adsorption were the Langmuir, Freundlich, Tempkin, and Dubinin-Radushkevich models. Calculated thermodynamic parameters indicated that the various metal oxides nanoparticles showed no restrain in effectiveness towards the dye removal.

Bashambu et al., 2021 demonstrated the improvement of adsorption capacities of polymeric membranes (for impurities in water) concerning larger surface area, selectivity, permeability, and strength through a composite combination of nano TiO_2, Al_2O_3, ZrO, and others. Kapoor et al., 2018 reported that metal oxide nanoparticles are useful for the treatment of biological wastewater because these have the potential to initiate inhibition of nitrification. Odiongenyi & Afangide, 2020 published results from adsorption and thermodynamics studies employing alumina and nano-alumina to remove Congo red dye from an aqueous solution. They found better removal efficiency by the nano-alumina compared to that of the alumina. The effectiveness of dye removal was shown to be affected by time, initial concentration of dye, adsorbent dose, pH, and temperature. The adsorption data fitted the Langmuir adsorption isotherm. Nano-ZnO, nano-Ag, nano-Fe, and nano-TiO_2 have also been found to be excellent nano-adsorbents for the treatment of wastewater and anaerobic digestion by Yang et al., 2019. Singh et al., 2019 also reviewed the applications of nanoparticles (including some metal oxide nanoparticles) in wastewater treatment and stated that the presence of beneficial features such as a direct bandgap, a high optical absorption coefficient, a layered structure, adjustable band margins for optimal catalysis, minimal cost, and reduced toxicity are among the most significant benefits of these nanoparticles. The study also covered detailed characteristics and applications of these metal oxide nanoparticles including those of TiO_2, ZnO, and iron oxide nanomaterials. Nanoparticle suspensions of CeO_2, Fe_3O_4, and TiO_2 were developed and verified for their efficacy in the removal of lead ions from wastewater, and the results were encouraging. The adsorption capacities were determined to be 189 mg, 83 mg, and 159 mg of Pb/g of CeO_2, Fe_3O_4, and TiO_2 respectively (Recillas et al., 2011). Recently, with due consideration of the biocompatibility of CaO NP, many researchers have explored the synthesis of CaO NP using various methods and their applications in water remediation as discussed ahead.

PROGRESSES IN THE SYNTHESIS OF CaO NP

CaO NP have a strong industrial demand because of their significant applications. These are useful as nano-catalyst and are used in the treatment of toxic waste. These nanoparticles act as effective adsorbents for the adsorption of gases through the mechanism of chemical adsorption. These are also used as a calcitrant in paint, ceramics, and other products. There are several methods available for synthesizing nano metal oxides but synthetic methods are broadly classified into up and down methods (Eliseev et al., 2000). However, not all the available up and down methods can be applied for the synthesis of CaO NP. Useful methods for the synthesis of CaO are summarized in Table 3. The methods listed in the table

Table 3. CaO nanoparticle synthesis methods

Preparation Method	Precursor	Reference
Sol-gel	Eggshell	Habte et al., 2019
Microwave irradiation	$Ca(NO_3)_2.4H_2O$	Roy and Bhattacharya, 2011
Thermal decomposition	$Ca(OH)_2$	Bhavya et al., 2016
Green synthesis	Plant extract and $CaCl_2$	Ramola & Joshi, 2019
Precipitation synthesis	$Ca(NO_3)_2.6H_2O$	Madan & Wasewar, 2017

are sol-gel, microwave-assisted, thermal decomposition, precipitation, and green methods. The choice of method depends on the nature of the precursor, prevailing conditions, expected output, and other factors.

Researchers have outlined some of the advantages of the sol-gel method. The method is simple, cheap, and can be carried out under low temperature and pressure. It is an economical method because even a wide variety of wastes materials have been used to synthesize CaO NP. The synthetic process does not require elaborate equipment. However, the major disadvantage of the method is that it cannot be easily optimized for commercial-scale production because of low product yield. However, some successes have been recorded for the sol-gel method. For example, Habte et al., 2019 synthesized CaO NP using eggshells wastes as the precursor. The synthesized CaO NP were characterized and reported to be spherical in morphology while the particle sizes ranged from 50 to 198 nm. The rationale behind the use of eggshells in the synthesis of CaO NP is their $CaCO_3$ content. The theory behind the method involves four basic steps, namely, formulation of a homogenous solution, sol formation through hydrolysis, gel production via condensation, and drying of the produced gel. During the process, the $CaCO_3$ from the shell reacts with HCl to form $CaCl_2$ according to the following equation:

$$CaCO_3(s) + 2HCl(aq) \rightarrow CaCl_2(aq) + H_2O(l) + CO_2(g) \quad (1)$$

The next step involves the hydrolysis reaction to form a stable colloidal dispersion of precursor particles in presence of aq. NaOH. As a result, the homogeneous solution of $CaCl_2$ solution is converted to sol, and metal hydroxide is produced as the end product, at ambient temperature (Equation (2).

$$CaCl_2(aq) + 2NaOH(aq) \rightarrow Ca(OH)_2(aq) + 2NaCl(aq) \quad (2)$$

To accelerate the precipitation of $Ca(OH)_2$, the rate of addition of NaOH should be relatively slow such that a low rate of nucleation would be achieved. This solution containing $Ca(OH)_2$ gel is incubated for at least 24 hours at ambient temperature to obtain CaO NP (see equation 3):

$$Ca(OH)_2(aq) \xrightarrow{\Delta} CaO(s) + H_2O(l) \quad (3)$$

A solvothermal method is closely related to the sol-gel method involving the synthesis of nanoparticles in a solvent environment. If the solvent is water, the process is called the hydrothermal method. Imtiaz et al., 2013 defined the adoption of a solvothermal method for the synthesis of CaO NP using

Figure 4. Scheme for hydrothermal synthesis of CaO NP from $CaCl_2$

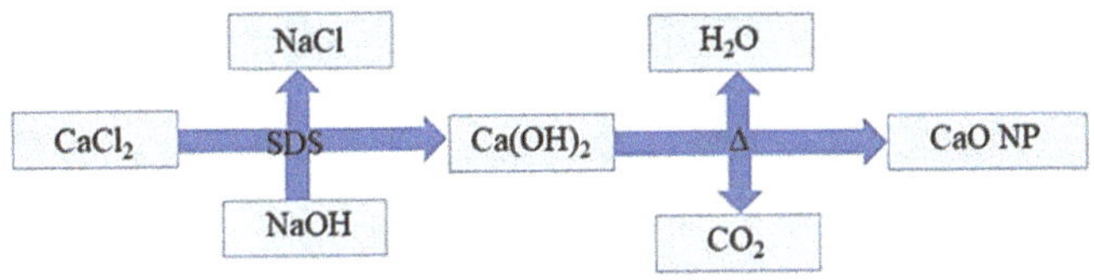

$CaCl_2$, NaOH, and sodium dodecyl sulfate (SDS). Generally, the scheme for a solvothermal method of this nature can be represented diagrammatically as shown in Figure 4.

The solvothermal method can also be achieved through a microwave-assisted process. A study on the usage of the microwave-assisted method to synthesize CaO NP using $Ca(NO)_3$ and NaOH as the precursors has been reported by Roy and Bhattacharya, 2011. The mechanism of the synthesis is closely similar to the sol-gel mechanism reported earlier. However, the necessary equations are as follows:

$$\mathrm{Ca}\left(\mathrm{NO}_3\right)_2\left(\mathrm{aq}\right)+2\mathrm{NaOH}\left(\mathrm{aq}\right)\xrightarrow{\Delta}\mathrm{Ca}\left(\mathrm{OH}_2\right)\left(\mathrm{aq}\right)+2\mathrm{NaNO}_3\left(\mathrm{aq}\right) \quad (4)$$

$$\mathrm{Ca}\left(\mathrm{OH}\right)_2\left(\mathrm{aq}\right)\xrightarrow{Microwave}\mathrm{CaO}\left(\mathrm{s}\right)+\mathrm{H_2O}\left(\mathrm{l}\right) \quad (5)$$

Poggi et al., 2016 reported the synthesis of CaO NP through a simple solvothermal method. Granular metallic calcium was hydrolyzed using alcohol in a high-pressure reactor to obtain alkoxide that was converted to hydroxide by dissolving in water. They also carried out optical characterization (using UV-visible spectrophotometer) and structural characterization with powder X-ray diffraction and TEM analysis. The particle size of the synthesized CaO NP was in the range of 170-180 nm. Bhavya et al., 2016 have also described the synthesis of CaO NP by solution combustion method in the presence of $Ca(NO_3)_2.4H_2O$ and NH_2CONH_2, which were dissolved in distilled water. The produced nanoparticles had an average size of 36 nm and were characterized by XRD, SEM, and UV visible spectroscopy. Sonication assisted synthesis of CaO NP (by dissolving $Ca(NO_3)_2$ (11.86 g) in ethylene glycol (25 ml) and then in 5 ml of NaOH) has been reported to yield different sizes of CaO NP that were a function of the sonication time (Tang et al., 2013). The scheme for such reaction can be presented as shown in Figure 5.

It can be deduced from Figure 5 that the initial reaction leads to the formation of two products, namely, calcium hydroxide and ethylene glycol nitrate. The former can be hydrolyzed with sodium hydroxide to free the solvent while the microwaves convert the $Ca(OH)_2$ to CaO NP.

In their study, Alavi & Morsali, 2010 observed that $Ca(OH)_2$ and CaO NP can be synthesized by the sonochemical method using calcium acetate and sodium hydroxide/tetramethylammonium hydroxide as precursors. Concentrations of the Ca^{2+}, aging time, and ultrasonic power were factors that affected the size, morphology, and growth process of the produced nanostructures. The $Ca(OH)_2$ formed from the reaction of calcium acetate with aq. NaOH (Figure 6) was calcined at 600 °C to obtain CaO NP, which were characterized by transmission electron microscopy, scanning electron microscopy, and X-ray powder diffraction instrumentation.

Figure 5. Scheme for the production of CaO NP by sonication using $Ca(NO_3)_2$, ethylene glycol, and NaOH.

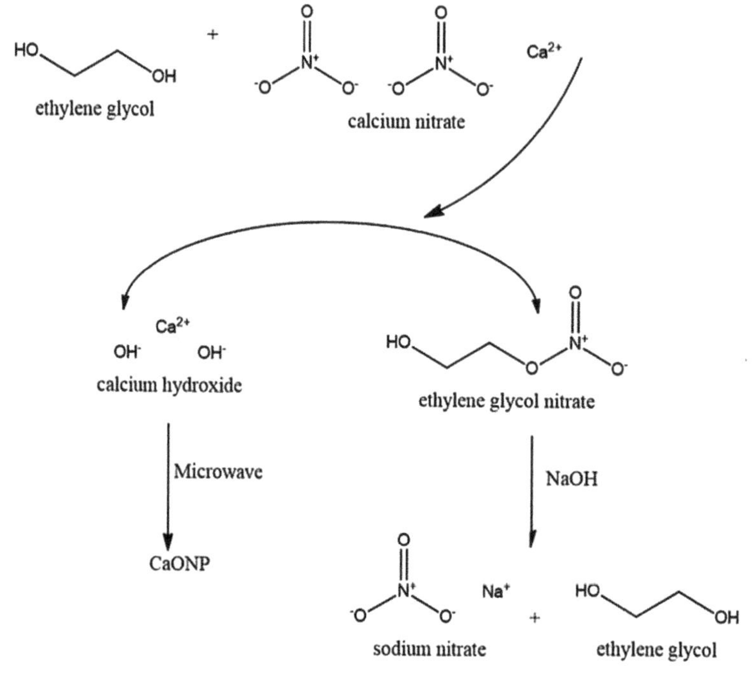

Figure 6. Scheme for the synthesis of CaO NP from calcium acetate and NaOH

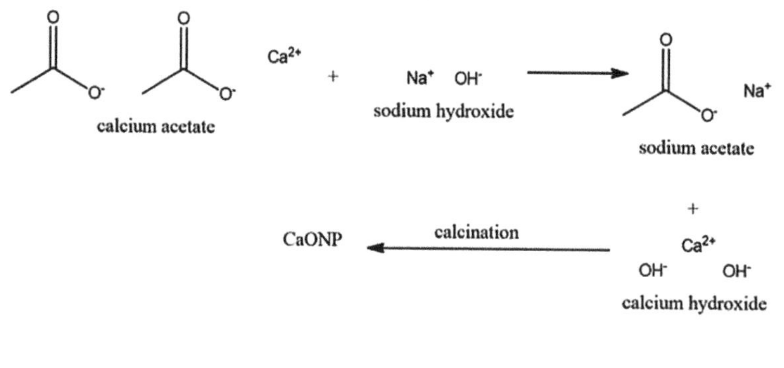

Mirghiasi et al., 2014 also reported a synthesis of CaO NP from $Ca(OH)_2$ using a wet chemical precursor obtained from the reaction of NaOH with $CaCl_2.4H_2O$. The synthesized CaO NP were characterized by XRD, TGA, FTIR, TEM, and BET. SEM revealed spherical shapes in the morphology, while the TEM imaging indicated that the particle sizes of the nanoparticles were 91 and 94 nm for those produced from 1 and 2 M NaOH concentrations.

Among all the methods, one of the methods that have attracted serious environmental concern is the green synthesis which can be implemented using aqueous extract of plants. For example, utilizing $CaCl_2$ as a precursor and aqueous broccoli extract as a reducing agent, Osuntokun et al. 2018 synthesized CaO NP. Different quantities of the aqueous broccoli extract were used to produce $Ca(OH)_2$, and further calcination resulted in the formation of CaO NP. Bano and Pillai, 2020 demonstrated that it is possible to effectively manufacture CaO NP with excellent optical characteristics using a green method with the employment of *Murraya Koenigii* leaf (curry leaves) extract as a capping agent. The optical absorbance spectrum revealed a large absorption peak between 400 and 500 nm. The FTIR analysis verified the presence of several functional groups that contributed to the stability of CaO NP. The PL emission spectrum revealed a high-intensity emission peak at 231 nm in addition to peaks at 375 and 400 nm. Additionally, they noticed an XRD spectrum that showed planes of CaO cubic phase, while the SEM pictures revealed a nonuniform distribution of spherical particles along with some clusters. Calcium and oxygen were verified by EDX spectra.

A Laboratory investigation carried out by Ramola & Joshi, 2019 on the interaction of $Ca(NO_3)_2$ with NaOH towards the production of CaO NP revealed that the nanoparticles can be synthesized from the leaf extract of *Rhododendron arboretum*. The synthesized nanoparticles were useful as adsorbents, catalyst, and antimicrobial agent. Also, the product was characterized by using FTIR, FESEM, XRD, and EDX. Gandhi et al., 2021 successfully synthesized CaO NP using the leaf extract of *Ocimum tenuiflorum* and $CaCl_2$ as precursors. The size distribution characteristics of the nanomaterial were analyzed using the dynamic scattering technique through a photon correlation spectrophotometer. Powder X-ray diffraction analysis was used for phase identification and in the determination of unit cell dimension, molecular and atomic structure of the nanocrystal. Also, a UV-visible spectrophotometer was used to ascertain the maximum wavelength of absorption while the surface morphology of the nanoparticle was analyzed through scanning electron microscopy. Maringgal et al., 2020 implemented a biosynthesis approach towards the synthesis of CaO NP through the deposition and precipitation of $CaCO_3$ in *Trigona sp.* honey. The image of the synthesized CaO NP obtained from the field emission scanning microscope (FESEM) indicated a spherical morphology with an average particle size of less than 100 nm. The obtained XRD pattern revealed sharper peaks that were typical for well crystalized CaO NP while the Debye-Scherrer calculated particle size was 51.64 nm. Absorption bands revealed by FTIR instrumentation showed intensities that corresponded to hydroxyl, carboxylic groups, amines, and amides.

Cissus quadrangularis (Marquis et al., 2016) and *Piper betel* (Yoonus et al., 2021) leaf extracts have also been reported as good mediators for the green synthesis of CaO NP by two research groups. Both groups adopted UV-visible spectroscopy, SEM, XRD, and FTIR to characterized their respective nanoparticles. The average size of the nanoparticles in both cases was less than 60 nm while the morphology still reflected spherical shapes. Closely similar results were obtained by Ijaz et al., 2017 who used *Mentha pipertia* extract for the synthesis of CaO NP. However, their SEM images reveal agglomeration of disc-shaped nanoparticles while the FTIR absorption bands were typical intensive peaks associated with calcium ions and oxygen. Employment of ball milling method for the synthesis of CaO NP is scantly reported. However, the method was adopted by Nuryantini et al., 2019 to synthesize CaO NP from duck eggshells. The product of the milling was calcinated at 700 °C for 7 hours and the resulting CaO NP were characterized using SEM. Variation in milling time was observed to generate different particle sizes of CaO NP. $Ca(OH)_2$ and $CaCO_3$ were also found to be constituents of the produced nanoparticles. This implies that the milling method is not very effective in removing unwanted components that are inherent in the precursor but can lead to the production of composite nanoparticles.

APPLICATION OF CaO NP IN WATER REMEDIATION

Removal of Heavy Metal Ions

In a demonstration of the effectiveness of CaO NP for environmental remediation of heavy metal contamination of water, Oladoja et al., 2012 synthesized the spongy/porous CaO NP with point zero charges of 11.80 and adopted the batch adsorption experiment to withdraw a toxic heavy metal, Cr(VI) from contaminated water. The adsorbent was very effective but was strongly influenced by pH, temperature, concentration of Cr(VI), and mass of the adsorbent. Langmuir adsorption model was supported by the suitability of a pseudo second-order kinetic model. The adsorption was exothermic and spontaneous. Khan et al., 2016 have reported the effectiveness of Ca-doped ZnO nanoparticles as an adsorbent for the removal of Pb^{2+}, Cd^{2+}, Cu^{2+}, Hg^{2+}, La^{2+}, Mn^{2+}, Pd^{2+}, and Y^{3+} and it was found to be selective for the removal of only Pb^{2+}. The equilibrium concentration of Pb^{2+} removed was 84.66 mg/g and the adsorption fitted the Langmuir isotherm while the kinetic obeyed a pseudo second-order model. The adsorption process was spontaneous.

Toamah & Fadhil, 2019 have described the syntheses of CaO NP by a sol-gel method using $CaCl_2$ as a precursor. Various calcination temperatures were used and the products were characterized using SEM, XRD, and TEM. The results obtained show that the products were efficient adsorbents for the removal of Cr^{2+} and Hg^{2+} from aqueous solution. The excellent removal efficiency was observed at optimum temperature, pH, stirring time, temperature, adsorbent dosage, concentration of the heavy metal ions, and contact time. Jalu et al., 2021 synthesized CaO NP from eggshell using a sol-gel method and successfully characterized the product through FTIR, XRD proximate analysis, determination of particle density, bulk density, porosity and point of zero charges. The synthesized CaO NP were found to be a good adsorbent for the removal of Pb^{2+} from an aqueous solution. The optimum percentage removal of lead from the aqueous solution was 99.07% and this was achieved at pH of 6.94, initial concentration of 75.46, adsorbent dosage of 0.838 g, and contact time of 101.97 minutes. The Langmuir adsorption model and the pseudo second-order model best described the adsorption process.

Dyes and Organic Compounds

Patel et al., 2009, synthesized CaO NP using the sol-gel method. The product was characterized using FTIR, TGA, DTA, and TEM. The synthesized CaO NP exhibited 98% removal efficiency for fluoride ions from water within 39 minutes period of contact and an optimum dosage of 0.6 g/L. Xia et al., 2013 were able to establish from their study that CaO NP has an excellent adsorption capacity for Congo red dye from aqueous solution. The maximum adsorption capacity for the nanomaterials was 357.14 mg/g in 10 minutes, which was found to be more than the maximum adsorption capacity for the non-nano CaO by 1194.48 mg/g and by significantly higher values for other metal oxides that have been reported as good adsorbent for Congo red dye in aqueous solution. Madhusudhana et al., 2012 reported the results of synthesis and application of CaO NP for adsorption removal of Coralene dark red 28 azo dye. The effectiveness of the adsorbent was observed by decoloration of the dye and was dependent on irradiation time, pH, and amount of CaO NP. Optimum removal was observed after 6 hours for all solutions at various pH values. Madan & Wasewar, 2017 have jointly reported the application of the chemical precipitation method for the synthesis of CaO and the utilization of powder XRD, EDX, SEM, TEM for characterization of the produced CaO NP. Adsorption removal of benzenacetic acid was successful when the CaO NP was employed as adsorbent. The extent of adsorption depended on contact time, adsorbent dosage, and concentration of the benzeneacetic acid. Optimum conditions were observed at an adsorbent dosage amount of 0.03 g, initial benzeneacetic concentration of 94.49%, and contact time of 30 minutes.

Vanthana Sree et al., 2020 synthesized CaO NP from eggshells using a sol-gel method. The product was characterized by XRD, SEM, EDAX, XPS, and BET analysis. The synthesized CaO NP were used to enhance photocatalysis of Toluidine blue and methylene blue dyes in an aqueous solution. The results obtained indicated that the CaO NP successfully degraded both dyes after 15 and 10-minutes exposure respectively. Optimum conditions were a pH of 7, dye concentration of 20 ppm, and CaO NP catalyst concentration of 50 mg. The observed kinetics for the adsorption of the dye degradation was in excellent agreement with the pseudo-first-order kinetics. Thakur et al., 2021 have reported a recent study on the use of CaO NP prepared from eggshells as the precursor, for the adsorption removal of cationic (Brilliant Green) and anionic (phenol Red) dyes from aqueous solution. The synthesized adsorbent was investigated for their surface area (using BET), hydrodynamic size (using DLS) and surface morphology (using SEM and high-resolution TEM), crystallographic profile (using XRD). The electrokinetic factor (Zeta potential). the lattice strain and crystallite size (using W-H analysis from powder XRD data) for the annealed NP were also determined. The CaO NP were very effective in the removal of the dye. However, in the binary solution of the two dyes, the removal efficiency towards the Brilliant Green was better than the removal efficiency for the Phenol Red dye. There were also some improvements in the chemical oxygen demand while the adsorption process followed the Langmuir adsorption model. Jaiswal et al., 2021 synthesized and characterized CaO NP using waste chicken eggshells. The synthesized nanomaterial was applied for the degradation of methylene blue from an aqueous solution and it was found to exhibit a degradation efficiency approximating 98% indicating the use of CaO NP as photo-biocatalyst.

FUTURE RESEARCH DIRECTIONS

Nanomembranes involve filtration at the nanoscale and are generally fabricated from organic polymer-based nanocomposites such that the resulting thickness should be less than 100 nm. The biocompatibility of CaO NP can be suitably explored in nanofiltration. CaO NP are stable and possess excellent structural characteristics. The role of CaO NP in removing microorganisms from water has been linked to their biological activities against certain microorganisms and also their dimension is sufficient to block the progress of microorganisms. However, the major challenges in solving problems through the engineered processes are the availability of raw materials, cost of production/instrumentation, and the required technology. Therefore, it is necessary to examine the prospects of CaO NP concerning raw materials availability, cost, and production technology. In most countries of the world, the chemical raw materials acting as precursors for CaO NP including $CaCl_2$ and $CaCO_3$, are cheap and easily available. Likewise, the use of plant extract does not require elaborate technology and the cost implication is very low. Recent development in the field of green synthesis has shown that several plant extracts can be used to reduce most CaO NP precursors to the nanoscale. The method is not only cheap but also can be implemented easily and does not require robust technology. Raw materials inventory for the production of CaO NP can also be sorted out from the biological waste containing calcium carbonate such as waste shells, bones, and other animal skeletal parts. Raw materials assay has also revealed that shells of some lower animal shells (such as shells of egg, periwinkle, oyster, snail, and others) have a significant amount of $CaCO_3$ which enables them to be suitable precursors for the production of CaO NP (Eddy &

Ekop, 2007; Odoemelam & Eddy, 2009). There are also natural deposits of calcium carbonate including limestone, calcite, etc. in several locations. Therefore, cost-benefit analysis positions CaO NP as economic nano-adsorbent for the treatment of wastewater.

CONCLUSION

The review opens a wider prospect for the utilization of CaO NP for the treatment of wastewater. CaO NP can be synthesized at ease from local raw materials and the process of synthesis is cheap and easily accessible. Even the use of the shells of some lower organisms can be explored. The treatment process can be achieved through adsorption (batch or fixed bed), photocatalysis, and nanomembrane. The adsorption process seems to be the most wide explored method and can be operated to follow the kinetics and thermodynamic of the adsorption process. However, at an industrial scale, most of the known synthetic methods may not meet the market demand and still need upgradation.

REFERENCES

Adedirin, O., Adamu, U., & Nnabuk, E. O. (2011). Removal of Cd(II) from aqueous solution using Bacillus subtilis and Escherichia coli immobilized in agarose gel: Equilibrium, kinetics and thermodynamic study. *Archives of Applied Science Research*, *3*(3), 59–76.

Agasti, N. (2021). Decontamination of heavy metal ions from water by composites prepared from waste. *Current Research in Green and Sustainable Chemistry, 4*(October), 100088. doi:10.1016/j.crgsc.2021.100088

Ajmal, A., Majeed, I., Malik, R. N., Idriss, H., & Nadeem, M. A. (2014). Principles and mechanisms of photocatalytic dye degradation on TiO 2 based photocatalysts: A comparative overview. *RSC Advances*, *4*(70), 37003–37026. doi:10.1039/C4RA06658H

Akafu, T., Chimdi, A., & Gomoro, K. (2019). Removal of Fluoride from Drinking Water by Sorption Using Diatomite Modified with Aluminum Hydroxide. *Journal of Analytical Methods in Chemistry*, *2019*, 1–11. Advance online publication. doi:10.1155/2019/4831926 PMID:31886021

Allard, M. (1992). GEMS/WATER operational guide / prepared by the WHO Collaborating Centre for Surface and Ground Water under the joint sponsorhsip of the United Nations Environment Programme, World Health Organization, United Nations Educational, Scientific and Cultural Organization, World Meteorological Organization (3rd ed.). World Health Organization.

Amin Alavi, M., & Morsali, A. (2010). Ultrasonic-assisted synthesis of Ca(OH) 2 and CaO nanostructures. *Journal of Experimental Nanoscience*, *5*(2), 93–105. doi:10.1080/17458080903305616

Bano, S., & Pillai, S. (2020). Green synthesis of calcium oxide nanoparticles at different calcination temperatures. *World Journal of Science, Technology and Sustainable Development*, *17*(3), 283–295. doi:10.1108/WJSTSD-12-2019-0087

Bansal, M., Singh, D., & Garg, V. K. (2009). A comparative study for the removal of hexavalent chromium from aqueous solution by agriculture wastes' carbons. *Journal of Hazardous Materials, 171*(1–3), 83–92. doi:10.1016/j.jhazmat.2009.05.124 PMID:19553015

Bansal, M., Singh, D., & Garg, V. K. (2012). Use of chemically treated sawdust for chromium removal from aqueous solutions: Effect of process parameters. *International Journal of Environment and Waste Management, 10*(2–3), 190–200. doi:10.1504/IJEWM.2012.048370

Bashambu, L., Singh, R., & Verma, J. (2021). Metal/metal oxide nanocomposite membranes for water purification. *Materials Today: Proceedings, 44*(January), 538–545. doi:10.1016/j.matpr.2020.10.213

Bhavya, C., Yogendra, K., Mahadevan, K. M., & Madhusudhana, N. (2016). Synthesis of Calcium Oxide Nanoparticles and Its Mortality Study on Fresh Water Fish Cyprinus Carpio. *IOSR Journal of Environmental Science, Toxicology and Food Technology, 10*(12), 55–60. doi:10.9790/2402-1012015560

Boretti, A., & Rosa, L. (2019). Reassessing the projections of the World Water Development Report. *Npj Clean Water, 2*(1), 15. doi:10.103841545-019-0039-9

Borgohain, X., Boruah, A., Sarma, G. K., & Rashid, M. H. (2020). Rapid and extremely high adsorption performance of porous MgO nanostructures for fluoride removal from water. *Journal of Molecular Liquids, 305*, 112799. doi:10.1016/j.molliq.2020.112799

Cheriyamundath, S., & Vavilala, S. L. (2021). Nanotechnology-based wastewater treatment. *Water and Environment Journal: the Journal / the Chartered Institution of Water and Environmental Management, 35*(1), 123–132. doi:10.1111/wej.12610

Chuah, T. G., Jumasiah, A., Azni, I., Katayon, S., & Thomas Choong, S. Y. (2005). Rice husk as a potentially low-cost biosorbent for heavy metal and dye removal: An overview. *Desalination, 175*(3), 305–316. doi:10.1016/j.desal.2004.10.014

DHV Consultants & DELFT Hydraulics. (2003). *Operational Manual-water quality analysis (SW)*. Author.

Danesh, N., Ghorbani, M., & Marjani, A. (2021). Separation of copper ions by nanocomposites using adsorption process. *Scientific Reports, 11*(1), 1–23. doi:10.103841598-020-80914-w PMID:33462314

Dave, P. N., & Chopda, L. V. (2014). Application of iron oxide nanomaterials for the removal of heavy metals. *Journal of Nanotechnology, 2014*, 1–14. Advance online publication. doi:10.1155/2014/398569

Eddy, N. O., & Ekop, A. S. (2007). Assessment of the quality of water treated and distributed by the Akwa Ibom State Water Company. *E-Journal of Chemistry, 4*(2), 180–186. https://doi.org/10.1155/2007/343042

Eliseev, A. A., Lukashin, A. V., Vertegel, A. A., Heifets, L. I., Zhirov, A. I., & Tretyakov, Y. D. (2000). Complexes of Cu(II) with polyvinyl alcohol as precursors for the preparation of CuO/SiO 2 nanocomposites. *Materials Research Innovations, 3*(5), 308–312. https://doi.org/10.1007/PL00010877

Gandhi, N., Shruthi, Y., Sirisha, G., & Anusha, C. R. (2021). Facile and Eco-Friendly Method for Synthesis of Calcium Oxide (CaO) Nanoparticles and its Potential Application in Agriculture. *Haya: The Saudi Journal of Life Sciences, 6*(5), 89–103. doi:10.36348/sjls.2021.v06i05.003

Garg, R., Kumari, M., Kumar, M., Dhiman, S., & Garg, R. (2020). Green synthesis of calcium carbonate nanoparticles using waste fruit peel extract. *Materials Today: Proceedings, 46*, 6665–6668. doi:10.1016/j.matpr.2021.04.124

Gavrilescu, M., Demnerová, K., Aamand, J., Agathos, S., & Fava, F. (2015). Emerging pollutants in the environment: Present and future challenges in biomonitoring, ecological risks and bioremediation. *New Biotechnology, 32*(1), 147–156. https://doi.org/10.1016/j.nbt.2014.01.001

Ghaedi, M., Ghayedi, M., Kokhdan, S. N., Sahraei, R., & Daneshfar, A. (2013). Palladium, silver, and zinc oxide nanoparticles loaded on activated carbon as adsorbent for removal of bromophenol red from aqueous solution. *Journal of Industrial and Engineering Chemistry, 19*(4), 1209–1217. https://doi.org/10.1016/j.jiec.2012.12.020

Habte, L., Shiferaw, N., Mulatu, D., Thenepalli, T., Chilakala, R., & Ahn, J. (2019). Synthesis of Nano-Calcium Oxide from Waste Eggshell by Sol-Gel Method. *Sustainability, 11*(11), 3196. https://doi.org/10.3390/su11113196

Ijaz, U., Bhatti, I. A., Mirza, S., & Ashar, A. (2017). Characterization and evaluation of antibacterial activity of plant mediated calcium oxide (CaO) nanoparticles by employing Mentha pipertia extract. *Materials Research Express, 4*(10), 105402. https://doi.org/10.1088/2053-1591/aa8603

Imtiaz, A., Farrukh, M. A., Khaleeq-ur-rahman, M., & Adnan, R. (2013). Micelle-Assisted Synthesis of Al 2 O 3 ·CaO Nanocatalyst: Optical Properties and Their Applications in Photodegradation of 2,4,6-Trinitrophenol. *TheScientificWorldJournal, 2013*, 1–11. https://doi.org/10.1155/2013/641420

Jaiswal, K. K., Dutta, S., Pohrmen, C. B., Verma, R., Kumar, A., & Ramaswamy, A. P. (2021). Bio-waste chicken eggshell-derived calcium oxide for photocatalytic application in methylene blue dye degradation under natural sunlight irradiation. *Inorganic and Nano-Metal Chemistry, 51*(7), 995–1004. doi:10.1080/24701556.2020.1813769

Jalu, R. G., Chamada, T. A., & Kasirajan, D. R. (2021). Calcium oxide nanoparticles synthesis from hen eggshells for removal of lead (Pb(II)) from aqueous solution. *Environmental Challenges, 4*(June), 100193. doi:10.1016/j.envc.2021.100193

Kamble, S. P., Jagtap, S., Labhsetwar, N. K., Thakare, D., Godfrey, S., Devotta, S., & Rayalu, S. S. (2007). Defluoridation of drinking water using chitin, chitosan and lanthanum-modified chitosan. *Chemical Engineering Journal, 129*(1–3), 173–180. https://doi.org/10.1016/j.cej.2006.10.032

Kamboj, A., Amjad, M., Ahmad, W., & Singh, A. (2020). A general survey on Green synthesis and application of calcium oxide nanoparticles. *International Journal of Health and Clinical Research, 3*(2), 41–48. https://ijhcr.com/index.php/ijhcr/article/view/63

Kapoor, V., Phan, D., & Pasha, A. B. M. T. (2018). Effects of metal oxide nanoparticles on nitrification in wastewater treatment systems: A systematic review. *Journal of Environmental Science and Health. Part A, Toxic/Hazardous Substances & Environmental Engineering, 53*(7), 659–668. https://doi.org/10.1080/10934529.2018.1438825

Khan, S. B., Marwani, H. M., Asiri, A. M., & Bakhsh, E. M. (2016). Exploration of calcium doped zinc oxide nanoparticles as selective adsorbent for extraction of lead ion. *Desalination and Water Treatment, 57*(41), 19311–19320. https://doi.org/10.1080/19443994.2015.1109560

Kumar, K. Y., Muralidhara, H. B., Nayaka, Y. A., Balasubramanyam, J., & Hanumanthappa, H. (2013). Low-cost synthesis of metal oxide nanoparticles and their application in adsorption of commercial dye and heavy metal ion in aqueous solution. *Powder Technology, 246*, 125–136. https://doi.org/10.1016/j.powtec.2013.05.017

Kumar, M., Bansal, M., & Garg, R. (2021). An overview of beneficiary aspects of zinc oxide nanoparticles on performance of cement composites. *Materials Today: Proceedings, 43*, 892–898. doi:10.1016/j.matpr.2020.07.215

Lunge, S., Thakre, D., Kamble, S., Labhsetwar, N., & Rayalu, S. (2012). Alumina supported carbon composite material with exceptionally high defluoridation property from eggshell waste. *Journal of Hazardous Materials, 237–238*, 161–169. doi:10.1016/j.jhazmat.2012.08.023

Madan, S. S., & Wasewar, K. L. (2017). Optimization for benzeneacetic acid removal from aqueous solution using CaO 2 nanoparticles based on Taguchi method. *Journal of Applied Research and Technology, 15*(4), 332–339. https://doi.org/10.1016/j.jart.2017.02.007

Madhusudhana, N., Yogendra, K., & Mahadevan, K. M. (2012). Decolorization of Coralene Dark Red 2B azo dye using Calcium Oxide Nanoparticle as an adsorbent. *International Journal of Research in Chemistry and Environment, 2*(2), 21–25. https://doi.org/10.13140/RG.2.2.31460.83849

Maringgal, B., Hashim, N., Tawakkal, I. S. M. A., Hamzah, M. H., & Mohamed, M. T. M. (2020). Biosynthesis of CaO nanoparticles using Trigona sp. Honey: Physicochemical characterization, antifungal activity, and cytotoxicity properties. *Journal of Materials Research and Technology, 9*(5), 11756–11768. https://doi.org/10.1016/j.jmrt.2020.08.054

Marquis, G., Ramasamy, B., Banwarilal, S., & Munusamy, A. P. (2016). Evaluation of antibacterial activity of plant mediated CaO nanoparticles using C issus quadrangularis extract. *Journal of Photochemistry and Photobiology. B, Biology, 155*, 28–33. https://doi.org/10.1016/j.jphotobiol.2015.12.013

Mirghiasi, Z., Bakhtiari, F., Darezereshki, E., & Esmaeilzadeh, E. (2014). Preparation and characterization of CaO nanoparticles from Ca(OH)2 by direct thermal decomposition method. *Journal of Industrial and Engineering Chemistry, 20*(1), 113–117. https://doi.org/10.1016/j.jiec.2013.04.018

Mishra, P. C., & Patel, R. K. (2009). Use of agricultural waste for the removal of nitrate-nitrogen from aqueous medium. *Journal of Environmental Management, 90*(1), 519–522. https://doi.org/10.1016/j.jenvman.2007.12.003

Njuguna, S. M., Onyango, J. A., Githaiga, K. B., Gituru, R. W., & Yan, X. (2020). Application of multivariate statistical analysis and water quality index in health risk assessment by domestic use of river water. Case study of Tana River in Kenya. *Process Safety and Environmental Protection, 133*, 149–158. https://doi.org/10.1016/j.psep.2019.11.006

Nuryantini, A. Y., Sundari, C. D. D., Halimahtussa'diah, H., & Nuryadin, B. W. (2019). Synthesis and Characterization of Calcium Oxide Nanoparticles from Duck Eggshells using Ball Milling Methods. *Jurnal Kimia Valensi, 5*(2), 231–135. doi:10.15408/jkv.v5i2.8779

Odiongenyi, A. O., & Afangide, U. N. (2020). Adsorption and Thermodynamic Studies on the Removal of Congo Red Dye from Aqueous Solution by Alumina and Nano-alumina. *Communication in Physical Sciences*, *4*(1), 1–7.

Odoemelam, S. A., & Eddy, N. O. (2009). Studies on the Use of Oyster, Snail and Periwinkle Shells as Adsorbents for the Removalof Pb 2+ Aqueous Solution. *E-Journal of Chemistry*, *6*(1), 213–222. https://doi.org/10.1155/2009/407549

Okash Toamah, W., & Kadhim Fadhil, A. (2019). Preparation of nanoparticles from CaO and use it for removal of chromium (II), and mercury (II) from aqueous solutions. *Journal of Physics: Conference Series*, *1234*(1), 012086. https://doi.org/10.1088/1742-6596/1234/1/012086

Okwunodulu, F. U., & Eddy, N. O. (2014). Equilibrium and Thermodynamic Consideration of Cd2+, Ni2+ and Pb2+ Removal From Aqueous Solution Onto Treated and Untreated Cola Nitida Waste Biomass. *International Journal of Scientific Research (Ahmedabad, India)*, *3*(3), 567–570.

Oladoja, N. A., Ololade, I. A., Olaseni, S., Olatujoye, V. O., Jegede, O. S., & Agunloye, A. O. (2012). Synthesis of Nano Calcium Oxide from a Gastropod Shell and the Performance Evaluation for Cr (VI) Removal from Aqua System. *Industrial & Engineering Chemistry Research*, *51*(2), 639–648. https://doi.org/10.1021/ie201189z

Oprčkal, P., Mladenovič, A., Vidmar, J., Mauko Pranjić, A., Milačič, R., & Ščančar, J. (2017). Critical evaluation of the use of different nanoscale zero-valent iron particles for the treatment of effluent water from a small biological wastewater treatment plant. *Chemical Engineering Journal*, *321*, 20–30. https://doi.org/10.1016/j.cej.2017.03.104

Osuntokun, J., Onwudiwe, D. C., & Ebenso, E. E. (2018). Aqueous extract of broccoli mediated synthesis of CaO nanoparticles and its application in the photocatalytic degradation of bromocrescol green. *IET Nanobiotechnology, 12*(7), 888–894. doi:10.1049/iet-nbt.2017.0277

Patel, G., Pal, U., & Menon, S. (2009). Removal of Fluoride from Aqueous Solution by CaO Nanoparticles. *Separation Science and Technology*, *44*(12), 2806–2826. https://doi.org/10.1080/01496390903014425

Poggi, G., Toccafondi, N., Chelazzi, D., Canton, P., Giorgi, R., & Baglioni, P. (2016). Calcium hydroxide nanoparticles from solvothermal reaction for the deacidification of degraded waterlogged wood. *Journal of Colloid and Interface Science*, *473*, 1–8. https://doi.org/10.1016/j.jcis.2016.03.038

Rajkumar, S., Murugesh, S., Sivasankar, V., Darchen, A., Msagati, T. A. M., & Chaabane, T. (2019). Low-cost fluoride adsorbents prepared from a renewable biowaste: Syntheses, characterization and modeling studies. *Arabian Journal of Chemistry*, *12*(8), 3004–3017. https://doi.org/10.1016/j.arabjc.2015.06.028

Ramola, B., & Joshi, N. C. (2019). Green Synthesis, Characterisations and Antimicrobial Activities of CaO Nanoparticles. *Oriental Journal of Chemistry*, *35*(3), 1154–1157. https://doi.org/10.13005/ojc/350333

Recillas, S., García, A., González, E., Casals, E., Puntes, V., Sánchez, A., & Font, X. (2011). Use of CeO2, TiO2 and Fe3O4 nanoparticles for the removal of lead from water. *Desalination*, *277*(1–3), 213–220. https://doi.org/10.1016/j.desal.2011.04.036

Roy, A., & Bhattacharya, J. (2011). Microwave assisted synthesis of CaO nanoparticles and use in waste water treatment. *TechConnect Briefs, 3,* 565–568.

Sarwar, A., Wang, J., Khan, M. S., Farooq, U., Riaz, N., Nazir, A., Mahmood, Q., Hashem, A., Al-Arjani, A.-B. F., & Alqarawi, A. A., & Abd-Allah, E. F. (2021). Iron Oxide (Fe3O4)-Supported SiO2 Magnetic Nanocomposites for Efficient Adsorption of Fluoride from Drinking Water: Synthesis, Characterization, and Adsorption Isotherm Analysis. *Water (Basel)*, *13*(11), 1514. https://doi.org/10.3390/w13111514

Sharma, R., Garg, R., & Kumari, A. (2020). A review on biogenic synthesis, applications and toxicity aspects of zinc oxide nanoparticles. *EXCLI Journal*, *19*, 1325–1340. https://doi.org/10.17179/excli2020-2842

Sharma, S., & Bhattacharya, A. (2017). Drinking water contamination and treatment techniques. *Applied Water Science*, *7*(3), 1043–1067. https://doi.org/10.1007/s13201-016-0455-7

Singh, S., Kumar, V., Romero, R., Sharma, K., & Singh, J. (2019). Applications of Nanoparticles in Wastewater Treatment. In *Nanotechnology in the Life Sciences* (pp. 395–418). doi:10.1007/978-3-030-17061-5_17

Solanki, Y. S., Agarwal, M., Maheshwari, K., Gupta, S., Shukla, P., & Gupta, A. B. (2021). Removal of fluoride from water by using a coagulant (inorganic polymeric coagulant). *Environmental Science and Pollution Research International*, *28*(4), 3897–3905. https://doi.org/10.1007/s11356-020-09579-2

Srivastav, A. L., Singh, P. K., Srivastava, V., & Sharma, Y. C. (2013). Application of a new adsorbent for fluoride removal from aqueous solutions. *Journal of Hazardous Materials*, *263*, 342–352. https://doi.org/10.1016/j.jhazmat.2013.04.017

Tang, Z.-X., Yu, Z., Zhang, Z.-L., Zhang, X.-Y., Pan, Q.-Q., & Shi, L.-E. (2013). Sonication-assisted preparation of CaO nanoparticles for antibacterial agents. *Quimica Nova*, *36*(7), 933–936. https://doi.org/10.1590/S0100-40422013000700002

Thakur, S., Singh, S., & Pal, B. (2021). Superior adsorptive removal of brilliant green and phenol red dyes mixture by CaO nanoparticles extracted from egg shells. *Journal of Nanostructure in Chemistry*. doi:10.1007/s40097-021-00412-x

Tomar, V., Prasad, S., & Kumar, D. (2014). Adsorptive removal of fluoride from aqueous media using Citrus limonum (lemon) leaf. *Microchemical Journal*, *112*, 97–103. https://doi.org/10.1016/j.microc.2013.09.010

Vanthana Sree, G., Nagaraaj, P., Kalanidhi, K., Aswathy, C. A., & Rajasekaran, P. (2020). Calcium oxide a sustainable photocatalyst derived from eggshell for efficient photo-degradation of organic pollutants. *Journal of Cleaner Production*, *270*, 122294. https://doi.org/10.1016/j.jclepro.2020.122294

Vijila, B., Edinsha Gladis, E. H., Michael Ahitha Jose, J., Sharmila, T. M., & Joseph, J. (2020). Removal of fluoride with rice husk derived adsorbent from agro waste materials. *Materials Today: Proceedings*. doi:10.1016/j.matpr.2020.09.729

Xia, H., Chen, L., & Fang, Y. (2013). Highly Efficient Removal of Congo red from Wastewater by Nano-Cao. *Separation Science and Technology*, *48*(17), 2681–2687. https://doi.org/10.1080/01496395.2013.805340

Yadav, A., Kaushik, C., Haritash, A., Kansal, A., & Rani, N. (2006). Defluoridation of groundwater using brick powder as an adsorbent. *Journal of Hazardous Materials*, *128*(2–3), 289–293. https://doi.org/10.1016/j.jhazmat.2005.08.006

Yang, J., Hou, B., Wang, J., Tian, B., Bi, J., Wang, N., Li, X., & Huang, X. (2019). Nanomaterials for the Removal of Heavy Metals from Wastewater. *Nanomaterials (Basel, Switzerland)*, *9*(3), 424. https://doi.org/10.3390/nano9030424

Yoonus, J., Resmi, R., & Beena, B. (2021). Greener nanoscience: Piper betel leaf extract mediated synthesis of CaO nanoparticles and evaluation of its antibacterial and anticancer activity. *Materials Today: Proceedings, 41*, 535–540. doi:10.1016/j.matpr.2020.05.246

ADDITIONAL READING

Naseem, T., & Durrani, T. (2021). The role of some important metal oxide nanoparticles for wastewater and antibacterial applications: A review. *Environmental Chemistry and Ecotoxicology*, *3*, 59–75. doi:10.1016/j.enceco.2020.12.001

Pandey, N., Shukla, S. K., & Singh, N. B. (2017). Water purification by polymer nanocomposites: An overview. *Nanocomposites*, *3*(2), 47–66. doi:10.1080/20550324.2017.1329983

Patel, H. (2019). Fixed-bed column adsorption study: A comprehensive review. *Applied Water Science*, *9*(3), 45. doi:10.100713201-019-0927-7

Qu, X., Alvarez, P. J. J., & Li, Q. (2013). Applications of nanotechnology in water and wastewater treatment. *Water Research*, *47*(12), 3931–3946. doi:10.1016/j.watres.2012.09.058 PMID:23571110

KEY TERMS AND DEFINITIONS

Adsorption: Adhesion of atoms, molecules, or other small particles from a dissolved state to an inert surface.

Biocompatibility: The property of a chemical or material that prevents the biological tissue that comes into touch with it from exhibiting harmful or immunological reactions.

Contamination: The presence of a constituent, impurity, or some other undesirable element in a physical body, or the natural environment resulting in degradation of its quality.

Photocatalysis: A change in the rate of a photoreaction by absorption of light radiations by suitable material.

Water Remediation: Treatment of contaminated water by eliminating the contaminants or turning them into harmless products.

Chapter 12
Antimicrobial Applications of Nanoparticles

Ayesha Kanwal
The Islamia University of Bahawalpur, Pakistan

Zeeshan Ahmad Bhutta
https://orcid.org/0000-0002-0148-425X
Chungbuk National University, South Korea

Ambreen Ashar
University of Agriculture, Faisalabad, Pakistan

Ashar Mahfooz
University of Agriculture, Faisalabad, Pakistan

Rizwan Ahmed
University of Agriculture, Faisalabad, Pakistan

Muhammad Fakhar-e-Alam Kulyar
Huazhong Agricultural University, China

Kun Li
Nanjing Agricultural University, China

ABSTRACT

Human mortality due to drug-resistant infections is becoming more prevalent in our society. Antibiotics are impotent due to abuse and/or misuse, leading to new, more expensive, and more effective medicines and treatments. Therefore, it causes many short-term and long-term side effects in the patient. On the other hand, nanoparticles have exhibited antibacterial activity against various pathogens due to their small size and ability to destroy cells by various mechanisms. Unlike antibiotics for the treatment of patients' diseases and infections, nanomaterials provide an exciting way to limit the growth of microorganisms due to infections in humans. This has led to the development of a number of nanoparticles as active antibacterial agents. Therefore, the authors have carefully reviewed the recent developments in the use of nanomaterials for antibacterial applications and the mechanisms that make them an effective alternate antibacterial agent.

INTRODUCTION

Nanomaterials were being used around 4500 years ago by Egyptians (Heiligtag & Niederberger, 2013) in the form of different nanofibers like PbS for hair dye and Egyptian Blue as a Quartz (Jeevanandam

DOI: 10.4018/978-1-7998-8936-6.ch012

et al., 2018). It has also been used by various countries such as Greece and Mesopotamia for multiple decorative purposes (Johnson-McDaniel et al., 2013). From the studies, it is observed that metals as nanoparticles (NPs) were being used in the past 13th and 14th centuries to make glass, and these could be referred to as the first synthetic metallic nanoparticles (MNPs) (Schaming & Remita, 2015). In Italy, another piece of evidence, red glass, confirms the usage of nanoparticles in 1200 to 1000 BC as it is covered with Cu NPs. Some other examples from history which prove the use of NPs, is connected with the civilization of humans like dichroic glass, Lycurgus cups, and Satsuma statue (Deshmukh, 2019; Jeevanandam et al., 2018). Clay has been reported as a natural nanomaterial for the bleaching of wools and cloths in Cyprus (Babu et al., 2019). After this, in 1857, scientist Michael Faraday was the first to talk about the quantum size, the preparation of gold nanoparticles (AuNP) and reveals that they have different optical properties compared to bulk counterparts. Later on, in 1908, Mie discussed the different colloidal colors (Lee et al., 2020). Multiple products came in the market by different companies with the help of NPs, like Samsung launches Silver Nano™ (containing ionic AgNPs) in 2003 (Putro et al., 2013) while Mercedes company came with a paint based on metallic and non-metallic NPs (providing fine glow and resistance against scratches) (Jeevanandam et al., 2018). In 2005, Abraxane™ (Human Serum Albumin containing NPs) was launched for pharmaceutical purposes, while in 2012, Logitech introduced the first commercial dye-sensitized solar cells (Aglawe et al., 2018). There were about 1814 commercially used products in the 20th century as reported in 2014 (Vance, 2015).

BACKGROUND

NPs have become a unique and novel tool to overcome many bacterial diseases that arise from the misuse of antibiotics. Using nanoparticles could reduce antimicrobial resistance because they can act as antimicrobial agents and react with the cell wall instead of entering the cell. The use of NPs is a targeted, effective, and safe approach in pharmacotherapy (Fernando et al., 2018). There is a higher rate of infections globally and loss of economy due to these diseases, but the reversal is being observed due to development of the advanced medicine in the form of NPs (Shi et al., 2014). The NPs are effective and safer than traditional antibiotics, as well as they do not produce any hazardous byproduct.

In a report, it was mentioned that antimicrobial resistance is a crisis for the health and wealth of nations (Neill, 2014). In 2014, the death rate was 700,000, but it was estimated that the death rate would increase up to 10 million in 2050, and in that case, the cost being used will increase up to 100 trillion US dollars (Huh & Kwon, 2011). In these days, 'superbugs' are being developed, which have formed an excellent resistance against antibiotics and created a lot of trouble for the researchers. Hence, it has become necessary to overcome these resistances and set up some balance between the death rate and the economy being used on healthcare tools (Gardini et al., 2018). NPs could withstand harsh conditions like sterilizing temperature, shelf life on different materials such as fabric, polymer, etc. and are more competent against microbial resistance, toxicity, and much more cost-effective. Moreover, many of the NPs not only kill or inhibit the growth of bacteria but also prove effective against viruses and fungi (Dizaj et al., 2014; Lemire et al., 2013) therefore, can be used as an efficient tool for many purposes like disinfectant for water, surfaces, and medical apparatus. As it clear from the studies that antimicrobial NPs are inert and are not oxidative in water so, they are safe and could not produce harmful byproducts as ClO_3 (Chlorate), BrO_3 (Bromate), and THMs (Trihalomethane i.e., dibromochloromethane $CHBr_2Cl$) (Gardini et al., 2018). In this concern, a brief review on the origin and types of nanoparticles, the mechanism of different antimicrobial nanoparticles, their multiple advantages and disadvantages are given below.

SOURCES OF NANOPARTICLES BASED ON ORIGIN

Natural Nanoparticles

The natural nanoparticles are present in living organisms such as bacteria, viruses, algae, humans, plants, birds, insects, etc. The organisms present at the nanoscale are known as nano-organisms. With the development of the latest techniques, it becomes easier for scientists to search on these particles' morphology, reactivity, toxicity, pH, temperature, etc. The nanostructure-based knowledge is essential in the development of medicines and resistance against many diseases. It could play a crucial role in survival during harsh conditions for insects (Jeevanandam et al., 2018) like butterflies; its beautiful colors complement its wings, and nanostructures give goniochromism, diffraction grating, and interference (Boulenguez et al., 2012). The bacteria present at the nano level are known as nanobacteria; these bacteria combine with heavy toxic metallic NPs and benefit by lowering the toxicity level. *Pseudomonas stutzeri* A259 was the first bacterial strain used to produce NPs; later, many strains like *Thermomonospora sp* and *Rhodococcus sp* have been used to produce multiple NPs such as gold NPs (AuNPs), NP composites, CdS, and ZnS, quantum dots, etc. Some nanostructures are gaining more attention nowadays due to their size range being between 20 to 150 nm and found in Martian meteorite ALH84001. They contain oxygen, DNA, carbon, nitrogen, minerals, and membrane-bounded structures like other microorganisms. Some biocompatible NPs like Fe_3O_4 (magnetite), Fe_2O_3 (maghemite, Iron oxide), etc., are prepared by using the magentotactic bacteria that are helpful in diagnosing the cancer, analyzing the DNA, magnetic resonance imaging (MRI), and gene therapy. Other microorganisms like algae, fungi, yeast, and microbial spores are used to prepare different NPs like AgNPs, Phytochelatin coated CdS, metallic and metallic sulfide NPs, Au and Au-Ag Alloy, and zirconium NPa etc. Plants and humans also have an equal role in the formation of particles. For example, the plants synthesize cellulosic fibrils (size; 100 to 1000 nm) known as nanofibers while human body produce enzymes, antibodies, polypeptides, and many proteins in nanostructure form. The researchers have synthesized nanozyme, which mimic the structure of a natural enzyme, to cure genetic and other cancerous diseases.

Incidental Nanoparticles

Some Nanoparticles are produced incidentally during photochemical reactions, dust storms, volcanic eruptions, forest fires, and ocean water evaporation. It is studied that humans aid to only 10% of the earth's atmosphere, but 90% of it is made by nature, and it produced many nanoparticulate, which may affect air quality in a good or bad way (Jeevanandam et al., 2018). With the help of spectroscopy, it is observed that most of the stardust is made of nanoparticles such as silicate, nitride, carbon, carbide, and some other organic nanoparticles (Barnard & Guo, 2012). Volcanic eruption gives rise to multiple micro and nano-size particles. From studies, it is clear that a single eruption can give approximately 30×10^{30} tons of NPs in an atmosphere (Jeevanandam et al., 2018). $CaCO_3$ NPs form in Lake Michigan under the change of temperature and environmental conditions (Argyilan et al., 2015).

Engineered Nanoparticles

Some nanoparticles are in engineered form, produced during combustion in vehicles, cooking, coal for the requirement of power generation, chemical synthesis, ore refining, anthropogenic activities, and welding (Jeevanandam et al., 2018). Carbon NPs (De Volder et al., 2013) and TiO_2 NPs (Weir et al., 2012) are used in cosmetics, sporting materials, sun blocks, and toothpaste (Jeevanandam et al., 2018). These manufactured NPs have a significant effect on the environment and human health. These manufactured sun blocks contain NPs as an antioxidant (Jeevanandam et al., 2018) and anti-reflectants (Monteiro-Riviere et al., 2011) agent. Some beauty products like wet wipes, shampoos, toothpaste, and food storage containers, and air sanitizer also contain AgNPs (Jeevanandam et al., 2018).

ANTIMICROBIAL MECHANISMS OF NANOPARTICLES

In general terms, the nanomaterial can be defined as a structure with at least one dimension range between 1 to 100 nm. It has different structures, physicochemical and electrical properties, and crystallinity. A detailed description of the mechanism for the antimicrobial action of a nanomaterial is given in Figure 1. This single system can explain different mechanisms of microorganisms and their antimicrobial effectiveness, but it can also negatively impact microorganisms by developing resistance. Moreover, the activity is increased by applying the stress on NPs, which gives a high surface area for the interaction of NPs and microorganisms to provide a productive antimicrobial action. The different mechanisms have been explained below.

Cell Membrane

Nanoparticles interact with the membrane mechanically and cause permanent damage to it. The damage of the cellular wall causes the release of cytoplasm and leads to cell death (Gold et al., 2018).

DNA Damage

Some of the NPs can puncture the cell wall, then invade or alter the shape (folding) of DNA to prevent further cell duplication and lead to cell death.

Release of Heavy Metal Ions

The release of heavy metal ions makes the microorganisms inactive for a long time when the thiol group binds to enzymes and proteins and stops the biological functions. The release of ions could be increased by reducing the size of NPs, and it is the best source for the development of biocidal ions (Mohammadi et al., 2020).

Interrupted Transmembrane Electron/Ion Transport

When NPs interact with the membrane, they release ions that cause resistance in the exchange of electrons and ions through the cellular wall, disruption in cellular respiration, and death (Lemire et al., 2013).

Oxidative Stress

Some NPs can act as a catalyst for the production of reactive oxygen species (ROS) such as OH, $-O^{2-}$, and H_2O_2. They oxidize the cellular membrane, which causes disruption and invades the internal components of the cell. With few exceptions, the process continues until the complete oxidation of microorganisms into water and carbon dioxide (H_2O, CO_2). Oxidative stress can also be generated into the cell instead of ROS with the help of an electron transfer mechanism specific to some NPs like carbon nanostructures (Gardini et al., 2018).

Figure 1. A Detail Description of Antimicrobial Mechanism of Nanoparticles

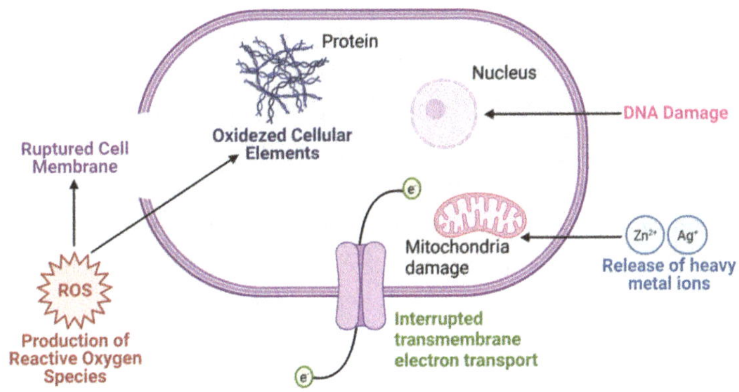

NANOPARTICLES BEING USED AS ANTIMICROBIAL AGENTS

Metallic Nanoparticles

Some essential metals like sodium (Na), calcium (Ca), iron (Fe), and potassium (K) are vital for the survival of life, take part in the formation of many enzymes, metabolic pathways, to stable cellular structures, and in many other reactions act as a catalyst (Yun et al., 2018). Although all metals, either they are essential or non-essential, can be very dangerous if they cross their optimum level concentration. Some heavy metals like Arsenic (As), Chromium (Cr), Mercury (Hg), Cadmium (Cd), and plumbum or lead (Pb) are the most toxic (Tchounwou et al., 2012). Despite that, these toxic metals could be used as an antimicrobial agent against some living organisms to benefit human life. In this way, these NPs have the advantage of a large surface area and change their chemical or physical properties during the synthesis phase. These NPs can attach or be immobilized on multiple types of surfaces, as they have importance in numerous fields such as medical, water, and food manufacturing processes. The essential metals are less in concentration in the body as they are provided from outside the body, but they may cause toxicity if they increase. At the same time, non-essential NPs are low in level, so they do not harm the body (Gardini et al., 2018).

Silver Nanoparticles

Silver (Ag) is very renowned in the medical industry and medicines due to its antibiotic properties (Rajeshkumar et al., 2019; Tang & Zheng, 2018). The toxicity of silver is still in the discussion so, its toxicity can be used as a benefit against many species. The Ag ions react with -SH (sulfhydryl) groups in protein which contains an essential element of respiratory enzyme NDH dehydrogenase II. In addition, silver ions inhibit DNA replication, casue trouble for penetration and integrity of the cellular membrane (Gardini et al., 2018). Similarly, Ag ions are efficient in their antioxidant property and have an intense antibacterial activity based on their different mechanisms. It disturbs the normal functioning of transmembrane pathways of bacteria and inhibits the electron transport mechanisms; after this, the accumulation of electrons is occurred due to AgNP, which is transferred for the synthesis of Oxygen and Hydrogen peroxide (O_2, H_2O_2), and at the end, this leads to oxidative damage and leakage of the cellular membrane. This is confirmed by the observation that there is less concentration of Ag ions in solution while it has a high concentration of ROS. In addition, when AgNP interacts with the cellular membrane, it starts degradation of lipopolysaccharide molecules to form "pits", accumulation inside the cellular membrane, and then disturb the whole membrane permeability. In the last step, the nanoscale Ag is partially dissolved then release as a cytotoxic Ag ion. Multiple organisms have been tested against AgNPs, including *E. coli, Staphylococcus aureus, Staphylococcus epidermis, Staphylococcus aureus* [Methicillin Resistant Staphylococcus aureus (MRSA)] *Bacillus subtilis, Enterococcus faecalis, Enterococcus faecium* [Vancomycin resistant enterococcus (VRE)], *Klebsiella pneumoniae* [Extended Spectrum β-lactose (ESBL)] and *Staphylococcus aeruginosa*. AgNPs are being used for coatings of implants (Bosetti et al., 2002), water purification (Hossain et al., 2014), antimicrobial urinary catheters (Saint et al., 1998), and for the bone cement against bacterial activities (Alt et al., 2004).

Gold Nanoparticles

The potential of gold nanoparticles to act as an antimicrobial is less than the other metallic NPs; in that case, Au (I) and Au (III) is used as initial raw material for the preparation of low bactericidal active NPs for the Gram positive and negative bacteria. On the other hand, AuNPs having less than 2 nm size show strong antimicrobial activity (Zheng et al., 2017). It is supposed that the bactericidal activity of AuNPs is not entirely removed during synthesis because of chemicals present in co-existence. Moreover, a new technique, photothermal therapy (PTT) is being used to improve the bactericidal activity of AuNPs (Riley & Day, 2017). These AuNPs are irradiated by the laser energy to release heat and then excited to start oscillation. After this, they are used as an anticancer and antibacterial material. It is observed that when AuNPs combine with *S. aureus*, it increases its effects when it reacts with laser energy and is mainly being used for the drugs and gene delivery in cancer (Fernando et al., 2018).

Copper Nanoparticles

The generation of copper nanoparticles (CuNPs) is very sensitive in air. There is the formation of an oxide layer, which reduces the effectiveness of the antimicrobial activity. Many efforts have been made to form a composite to make it more stable and increase its antimicrobial activity. The chelated NPs are more durable as organic compounds (chelating agents) around the desired metal ions (Chatterjee et al., 2014). They are ten times more robust and stable than the simple one CuNPs. Cu-EDTA chelated molecule has been proved as an efficient and effective antimicrobial against the *E. faecalis* than that of the *E. coli and S. aureus* (DeAlba-Montero et al., 2017). Its activity has been tested against many microbes such as methicillin resistant *S. aureus* (MRSA), *S. cholerasuis, Bacillus species, Candida albicans, Candida parapsilosis* (Yeast), and *P. aeroguinosa* (Wang et al., 2017). The bactericidal activity depends only on the agglomeration of CuNPs; if there is less agglomeration, then it produces small size CuNPs which provide a large surface area for the solubilization of Cu ions and interactions with the cellular membrane, which leads to toxicity. These metallic and ionic types of Cu synthesized hydroxyl (-OH) radicals disturb or alter the essential protein and DNA structures (Chatterjee et al., 2014).

It is getting attention nowadays because it is much cheaper than silver for antimicrobial activities. Like silver, it can also be inserted into the different materials to prepare a composite nanomaterial (Manikandan & Sathiyabama, 2015; Palza et al., 2015) for multiple applications. The mechanism involved behind the bactericidal activity is not fully understood, but it is believed that the mechanism is as similar as described for AgNPs; it attaches and damages the cellular membrane of bacteria, after which it alters the shape of DNA and other biomolecules. From the current studies (Chatterjee et al., 2014), it is clear that either due to *in vitro* or *in vivo* causes, the bacterial filamentation occurs by the depolarization of the cellular membrane of bacteria, and generation of ROS leads to death (Gardini et al., 2018). It is being used for the coatings of medical tools to make them resistant against resistant microbes. These CuNPs are used as coatings against the fungus (Kanhed et al., 2014; Vanti et al., 2020) and for water purification (Yang et al., 2019).

Iron Nanoparticles

The iron nanoparticles (FeNPs) work in two steps, i.e., cellular membrane disruption and then production of oxidative stress ROS. FeNPs are gaining more attention in the form of zero-valent iron nanoparticles (ZVINPs). They can be differentiated from iron oxide and iron-dropped nanostructures. Its bactericidal characteristics against *E. coli* have been observed and this nanoparticle is less effective in oxidative stress. When there is the oxidation of iron in the presence of oxygen, it forms an iron oxide layer, called magnetite, on the cellular membrane. In addition, the magnetite itself and all possible forms of iron released from the ZVINPs layer show no or less biocidal activity against *E. coli*. Its activity relationship between the bactericidal and ZVINPs oxidation is further ratified by (Hsueh et al., 2017; Singh & Singh, 2016). This could be explained by an equation given below:

It causes oxidative stress by releasing hydrogen peroxide and Ferric ion (Fe^{+2}) (Marsalek et al., 2012)

$$Fe^0 + O_2 + 2H^+ \rightarrow Fe^{+2} + H_2O_2$$

At the same time, it reacts with internal hydrogen peroxide and produces oxygen free radicals, and the equation is given below:

$$Fe^0 + H_2O_2 \rightarrow Fe^{+2} + 2OH^-$$

$$Fe^{+2} + H_2O_2 \rightarrow Fc^{+3} + OH + OH^-$$

$$Fe^{+2} + O_2 \rightarrow Fe^{+3} + O_2^-$$

It is confirmed that there is inactivation of MS2 coliphages under the ZVINPs in the presence or absence of oxidizing or non-oxidizing atmosphere (Kim et al., 2011). Similarly, it is found that the virus is also inactivated in a higher oxygen level as it happens for bacteria. This is observed for viruses to produce ROS by damaging the membrane with the help of ZVINPs and Fe^{+2} in the presence of oxygen (Gardini et al., 2018). Many of the microorganisms are being tested for this purpose, including *Escherichia coli, Bacillus subtilis, Microcystis aeruginosa, MS-2 phage Virus,* and *Pseudomonas fluorescens.*

Metallic Oxides Nanostructures

TiO_2 Nanostructures

Titanium dioxide (TiO_2) shows a greater range of antimicrobial and antibiofilm activities against many pathogens like bacteria, viruses, and fungi. It has multiple applications in the food industry, drugs, cosmetics, and it is approved by the Food and Drug Administration (FDA). Like other NPs, it also has some unique qualities such as photocatalytic activity and quantum size that make it more efficient against the microbes for multiple purposes like air and water purification and coatings for biomedical devices (Seil & Webster, 2012). It is very famous for its antimicrobial activities against Gram positive and negative bacteria (Senarathna et al., 2017). It kills microorganisms when beam light falls on them because of its photocatalytic activity. The synthesis of ROS by TiO_2 disturbs the cellular membrane and leads to cell death. The studies also observed that TiO_2 performs actions in the absence of light, direct contact, and adsorption of TiO_2NPs in cellular membrane leads to cellular damage and death (Seil & Webster, 2012). The microorganisms that have been tested include *E. coli, Staphylococcus aureus, P. aeruginosa, Micrococcus luteus, Enterococcus faecium,* and the applications are self-disinfecting coatings, water purification, and antibacterial textiles (Kang et al., 2019).

ZnO Nanostructures

Zinc Oxide (ZnO) is generally recognized as a safe compound by the US FDA department and has minimum toxicity against humans. ZnONPs have greater photocatalytic activity than of TiO_2NPs. Like TiO_2NPs, it is also used in the food industry to preserve food colors and prevent deterioration. Its advantage is that it has a high surface-to-volume ratio and coarse-grained nanostructures and shows bacteriostatic activity against *S. aureus, Enterococcus,* and *Streptococcus pyogenes* (Sirelkhatim et al., 2015). *Campylobacter jejuni, Salmonella spp., E. coli (strain O157:H7)* and *Listeria monocytogenes* are the common food infectious agents that could be halted by the ZnONPs (Seil & Webster, 2012). In addition, it has many other unique applications in cosmetics (creams, deodorants, ointments, and lotions), ceramic and glass cleaning solutions (McNair et al., 2014). These properties of ZnONPs are due to the accumulation of NPs inside the membrane, which leads to damage of cellular wall by the production of ROS in a high amount and disturbs the DNA replication. At the same time, it enhances the membrane permeability, making a path for NPs to move inside in the absence of proton motive force (PMF) (Sirelkhatim et al., 2015).

MgO Nanostructures

Magnesium oxide (MgO) are important NPs with multiple uses like industrial (catalysis, semiconductors, batteries) pharmaceutics, waste remediation, antireflection coating, and inactivation of microorganisms by the generation of superoxide radicals O^{2-} (Rashid et al., 2018). It produces ROS condition by non-photocatalytic process, this shows that it has antimicrobial activity even in the absence of external optical beam light. When these NPs comes in contact with humidity, they produce $Mg(OH)_2$ on the surface, and lead to a reduction reaction with oxygen to produce superoxide radicals (Gardini et al., 2018). The experiments further confirm on *E. coli* bacteria that these radicals generate ROS condition, then cause damage to the cellular membrane, and then change the shape of bacteria (Jin & He, 2011). MgO nanoparticles oxidize essential internal components and increase the local pH because of hydration in poor water environment, which helps in the antibacterial activity (Monzavi et al., 2015). The microorganisms used for this purpose are *E. coli, C. albicans, S. aureus, B. subtilis var. niger,* and *E. faecalis.*

Carbonaceous Nanomaterials

Carbon-based nanoparticles are also known as carbon nanostructures (CNSs) including fullerene, graphene, carbon nanotubes (CNTs), diamond-like carbon (DLC), which have an extensive range of antibacterial activities. The process involved in the antibacterial activity of CNSs or bacterial lysis is complex. Fullerene causes cellular membrane DNA damage, disturbs the energy (ATP) metabolic pathways to inactive microorganisms. It also has photolysis activity, which helps to develop ROS by the exposure of beam light to enhance the antibacterial activity. From the current studies, it is clear that fullerene is less efficient than carbon nanotubes, made of single and multi-layered carbon walls. Chemical bonding develops between the NPs and bacterial surface to generate ROS species. Due to these reasons, carbon has a great range of antibacterial properties.

Graphene Nanostructures

There are a lot of forms of graphene, such as graphene oxide (GO), Graphite, Graphene nanosheets, and reduced graphene oxide (rGO) (Al-Jumaili et al., 2017). It has a single layer of carbon joined by sp^2 hybridization, and it is known as one atomic dimension of graphite, and the single sheet was first observed in 2003 (Maas, 2016). It is well known for its high electrical conductance and tensile strength. The graphene is converted into graphene oxide after partial oxidation. Its dispersion increases in a liquid medium; it alters its surface structure, breaks hexagonal bonding, and then carboxylic groups containing oxygen are introduced at the sheet's corners. The metal-reducing bacteria and hydrazine chemical convert back graphite into graphene (De Silva et al., 2017). After reduction, it is known as reduced graphene oxide (rGO), but its hexagonal shape does not recover. Both of them have well-known antibacterial activities. There is a loss of release of cytoplasm and cellular membrane integrity with 2-D nanosheets (Liu et al., 2020). As physical contact can produce membrane stress, graphene shapes can also become the reason for cytotoxicity with the help of the oxidative process (Chowdhury et al., 2013). Usually, it is known that oxidative stress is the reason for ROS generation. However, a study (Liu et al., 2011) showed that graphene derivatives take part in the inactivation of microorganisms through a ROS-independent oxidative pathway and through a process that transfers membrane electrons to the carbon shapes from the microbial membrane (Li et al., 2014). These structures are good semi-metallic in nature so, and they have application as an oxidative stress indicator and oxidation of glutathione (Gardini et al., 2018). These graphene sheets can separate the microorganism (bacteria) from the microenvironment, which is harmful to them. There are a lot of factors that affect its activity like nanopores, functional groups (-R), layer number, shapes, and sheet size (Al-Jumaili et al., 2017). The mainly used bacteria are *E. coli, PC12* (neuronal cell line) and *S. aureus.*

Carbon Nanotubes

The carbon nanotubes (CNTs) are hollow and made of multiple graphene sheets in the cylindrical allotropes. CNTs activity depends on the length, catalyst, functional group (-R), diameter, electronic shapes, coatings, surface chemical nature, etc. If the length of CNTs is approximately 50 μm, then it wraps around the bacterial cellular membrane for antibacterial activity (Chen et al., 2013). These long nanotubes accumulate in liquid medium are considered more active and have more efficient antibiofilm activity, mainly in the initial phase of biofilm formation (Dong & Yang, 2014). Firstly, they develop a single sheet of graphene known as single-walled nanotubes (SWNTs). In contrast, multiple sheets are known as multi-walled nanotubes (MWNTs) and detailed studies reveal that they show strong antibacterial activity against *E. coli* only after the incubation period of 60 min. The literature showed that SWNTs (thin and rigid) are more efficient than the MWNTs against the bacteria as they have a smaller diameter and large surface area for penetration into the membrane (Gardini et al., 2018). They prove these nanotubes effective against the five strains of human gut bacteria (Chen et al., 2013). It is revealed that the nanotube's electronic property strongly affects the cytotoxicity and provides evidence that the metallic CNTs action rate against the bacterial activity is higher than that of the semiconducting CNTs. The reason is that the metallic CNTs have high independent oxidative stress generating capacity like the graphene. This process for ROS generation is further increased by photocatalytic activity (Chae et al., 2011). It is used as an antibacterial coating for implants (Haung et al., 2013). For this purpose, multiple microorganisms have been tested including *E. coli, Bifidobacterium adolescentis, S. aureus, Lactobacillus acidophilus,* and *E. faecalis.*

Nanodiamonds

Nanodiamonds (NDs) can be produced either by the bottom-up (explosive) or bottom down (milling) techniques, and their arrangement of carbon is in sp^2 hybridization. The NDs with a diameter of 2 to 10 nm are smaller and produced by doing explosions under the low oxygen level atmosphere, which has sp^3 hybridized carbon atoms. At the same time, semi-graphitic shells are oxidized partially with other species of oxygen. Based on their synthesis conditions, NDs are produced with different properties, i.e., crystalline purity, surface chemicals significantly impact physical activity, and chemical properties. NDs are observed nontoxic for multicellular organisms (Paget et al., 2014). From the studies, it is clear that the usage of explosive NDs for the inhibition activity of bacteria (*E. coli* and *B. subtilis*) is lower (Beranová et al., 2012), and also give information that smaller size (approximately 200 nm) NDs are much effective than the larger one that are produced by milling (Beranová et al., 2014). The carboxylated NDs are considered inert and have less antibacterial activity than partially oxidized NDs. NDs having acid anhydride groups are considered more efficient for antibacterial activity. The -R (functional) groups also play an essential role in reactions. They attract nucleophilic additions and lead to various reactions like bonding between protein and intercellular molecules or walls. This mechanism is known as clustering and leads to almost complete separation of microorganisms from the nutrient environment medium and causes hindrance for the biological actions (Rekha Deka et al., 2015). The model microorganisms being used for this purpose are *E. coli* and *B. subtilis*.

Buckminster-fullerenes

It directly induces dependent and independent ROS under oxidative stress. The microorganisms used are *E. coli, B. subtilis,* and *MS2 phage*.

Chitosan

It first causes cellular disruption, then does chelation of microelements, and in the last step stops the synthesis of protein and RNA. The microorganisms used for this purpose are *Drechstera Sorokiana, Salmonella typhimurium, Listeria monocytogenes, Pseudomonas fluorescens, Vibrio parahaemolyticus, Bacillus megaterium, Lactobacillus plantarum, Fusarium oxysporum, Lactobacillus brevis, Botrytis cinerea, L. aureus* and *Bacillus cereus*. It is being used in food preservatives (Gomes et al., 2017), agriculture industry (fungicidal and insecticidal) (Oh et al., 2019), and pharmecutical sector with its antibacterial activity (Ma et al., 2017).

Other Nanoparticles

Silicon is found as a nontoxic and promising nanocomposite. Silica nanowires or silicon composites can act as biocidal for bacteria by disturbing the cellular functions i.e., adhesion (cause hindrance of bacteria to dental biofilms), growth of bacteria, cellular differentiation, antimicrobial activity, and act as a carrier for biomedical applications (Shevchenko et al., 2017). CaO has potent activity against bacteria for the generation of superoxides on the surface of particles by increasing pH because of the hydration of CaO with water. Bacterial cellular membrane penetration and attachment could be increased by using Al_2O_3 NPs (Beyth et al., 2015).

ENGINEERING THE NANOMATERIALS FOR AN ENHANCED ANTIMICROBIAL ACTION

The basic antimicrobial activity actions are based on the physical and chemical properties, i.e., crystallinity, structure, and composition. The formation of free radicals increases the antimicrobial activity of NPs; for this objective, the computer tools give functional and density theories to predict the electronic characteristics of particles at the nanoscale. This importance is increasing day by day, specifically in heterogeneous catalytic research. The emergence of new advanced techniques in nanotechnology makes it easier for researchers to synthesize and characterize NPs on their desires with enhanced antimicrobial activities. For this purpose, some properties like size, shape, and crystallinity are necessary to understand (Figure 2) (Gardini et al., 2018).

There is a large surface for the interaction of NPs as compared to micro and macro-structures. It could be explained that the 1 μm of silver equals $10^6 \times 10$ nm particles, which means there is an increase in surface area of 100 times. As it has a large surface area, so it exposes many of its functional groups, ions, and reactions to eliminate and inhibit the desired microorganisms; its dissolution potential is also increased simultaneously (Zhang et al., 2009). This action increases the activity of AgNPs (Gliga et al., 2014) to penetrate the cellular membrane and release Ag ions. Another example is ZnONPs which are more effective in small size (12 nm) than large size (45 nm), making it a more efficient disinfectant and quickly penetrating the cellular membrane. They have a large surface area, and they have more significant photocatalytic activity interaction with the reducing and oxidizing species, and there would be a faster generation of ROS. It is similar in TiO_2, MgO and ZnO NPs (Hossain et al., 2014).

Figure 2. A brief description of different properties of Nanoparticles

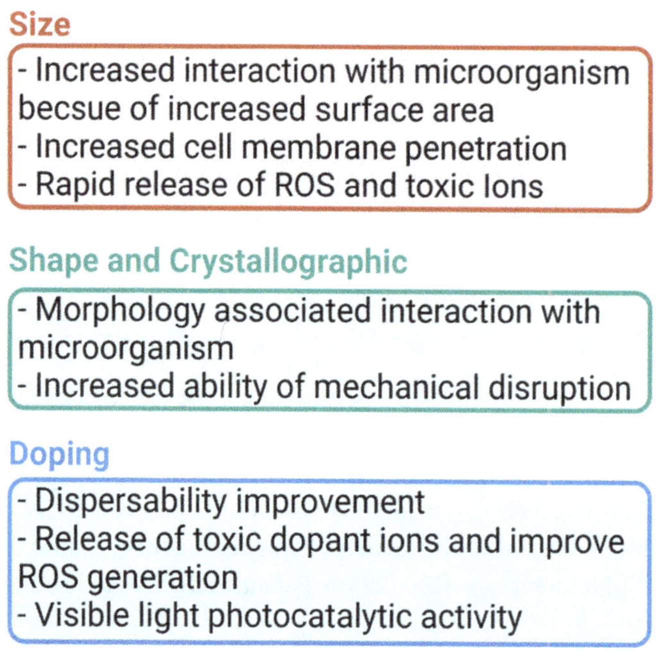

DISADVANTAGES OF NANOPARTICLES

As it has many advantages in medical and other industries, it also has some disadvantages. NPs have well-known applications in the medical field, i.e., preventing and treating contagious diseases. The release of these NPs in water, soil, and atmosphere is sometimes considered beneficial for antibacterial activities. These bacteria are essential to maintain the life cycle of life as they are involved in bioremediation, nitrogen fixation, and cycling. AgNPs are considered toxic for nitrogen fixing bacteria (Ammonia and Nitrite) (Gardini et al., 2018). Still, it may face difficulties during the treatment because of interaction with humans' tissues, cells, and other components. There is also a problem with the release of ions and electrons. Another disadvantage is that there is an aggregation of NPs in the human body. This exposure can be through the three routes, i.e., oral, intravenous, and inhalation (a most dangerous form of exposure) (Djurišić et al., 2015). Some smaller NPs could penetrate the cellular wall and pass from the endothelial (vascular and lymphatic), and epithelial cells (present at both internal and external surfaces, e.g., skin, hair, blood vessels, urinary tract, etc.). The intravenously applied injections containing NPs start accumulation in the lungs, liver, bone marrow, spleen, colon, and epithelial lymphatic system while inhaled NPs start accumulation at the lungs, brain, heart, etc.

NPs toxicity is a hot topic for research in recent years. Their mechanisms are in discussion for the days, and their antimicrobial activity is dependent on it. The generation of ROS by oxidative stress in different organs of humans, i.e., liver and lung cells, may cause hepatic and pulmonary toxicity (Huh & Kwon, 2011). In addition, it becomes clear that these NPs induced metabolic pathway changes like ketogenesis, glycolysis, nephrotoxicity, mitochondrial, and fatty acid oxidation. Some NPs may pose as the most dangerous because they change heartbeat rate and affect the reproductive system. Although, its toxicity is still in discussion and its different parameters are under in detailed discussion.

FUTURE RESEARCH DIRECTIONS

NPs have many benefits for humankind, but on the other, they have some toxic issues for the environment, so their use becomes limited in the current situation. The process of toxicity of NPs in the human body is still very complex, but it could be understood very well by the researchers and soon solve this difficult task. The safety protocols (desired cells and administration route etc.,) should be made and followed to prepare NPs on a large scale by controlling the different factors such as structure, size, surface area, etc. Some methods should be developed to lower the toxicity level when NPs released into the environment. This could be improved by improving the surface area, i.e., chemical functionalization (-R) and nanocoating.

CONCLUSION

The new changes are being developed day by day in microorganisms, and they are more resistant to disinfectants and antibiotics than previously. In this way, antimicrobial NPs are considered essential for fighting against these multi drug resistant microorganisms and using them for multiple applications. These NPs should be used used in the medical sector to sterilize different tools, as it is the main point of spreading these most resistant microbes in the environment. Other applications of NPs include water remediation, the food industry, and all those industries that have hygiene as a significant problem.

REFERENCES

Aglawe, S., Gayke, A., Kadam, P., Kherud, R., Kanawade, S., Garud, Y., & Surase, R. (2018). *Review on nanoparticles*. Academic Press.

Al-Jumaili, A., Alancherry, S., Bazaka, K., & Jacob, M. (2017). *Review on the antimicrobial properties of carbon nanostructures*. Academic Press.

Alt, V., Bechert, T., Steinrücke, P., Wagener, M., Seidel, P., Dingeldein, E., Domann, E., & Schnettler, R. (2004). *An in vitro assessment of the antibacterial properties and cytotoxicity of nanoparticulate silver bone cement*. Academic Press.

Argyilan, E. P., Avis, P. G., Krekeler, M. P., & Morris, C. C. (2015). The origin of collapse features appearing in a migrating parabolic dune along the southern coast of Lake Michigan. *Aeolian Research, 19*, 137–149. doi:10.1016/j.aeolia.2015.09.008

Babu, S. S., Radhakrishnan, E., Thomas, S., & Kalarikkal, N. (2019). Nanotechnology in Ancient Indian Holistic Healthcare. In Holistic Healthcare. Volume 2: Possibilities and Challenges (pp. 65-76). Apple Academic Press. doi:10.1201/9780429487804-5

Barnard, A. S., & Guo, H. (2012). *Nature's Nanostructures*. CRC Press. doi:10.1201/b11618

Beranová, J., Seydlová, G., Kozak, H., Benada, O., Fišer, R., Artemenko, A., Konopásek, I., & Kromka, A. (2014). *Sensitivity of bacteria to diamond nanoparticles of various size differs in gram-positive and gram-negative cells*. Academic Press.

Beranová, J., Seydlová, G., Kozak, H., Potocký, Š., Konopásek, I., & Kromka, A. (2012). *Antibacterial behavior of diamond nanoparticles against Escherichia coli*. Academic Press.

Beyth, N., Houri-Haddad, Y., Domb, A., Khan, W., & Hazan, R. (2015). Alternative Antimicrobial Approach: Nano-Antimicrobial Materials. *Evidence-Based Complementary and Alternative Medicine, 2015*, 1–16. doi:10.1155/2015/246012 PMID:25861355

Bosetti, M., Massè, A., Tobin, E., & Cannas, M. (2002). *Silver coated materials for external fixation devices: in vitro biocompatibility and genotoxicity*. Academic Press.

Boulenguez, J., Berthier, S., & Leroy, F. (2012). *Multiple scaled disorder in the photonic structure of Morpho rhetenor butterfly*. Academic Press.

Chae, S.-R., Watanabe, Y., & Wiesner, M. R. (2011). *Comparative photochemical reactivity of spherical and tubular fullerene nanoparticles in water under ultraviolet (UV) irradiation*. Academic Press.

Chatterjee, A. K., Chakraborty, R., & Basu, T. (2014). *Mechanism of antibacterial activity of copper nanoparticles*. Academic Press.

Chen, H., Wang, B., Gao, D., Guan, M., Zheng, L., Ouyang, H., Chai, Z., Zhao, Y., & Feng, W. (2013). *Broad-spectrum antibacterial activity of carbon nanotubes to human gut bacteria*. Academic Press.

Chowdhury, S. M., Lalwani, G., Zhang, K., Yang, J. Y., Neville, K., & Sitharaman, B. (2013). Cell specific cytotoxicity and uptake of graphene nanoribbons. *Biomaterials, 34*(1), 283–293. doi:10.1016/j.biomaterials.2012.09.057 PMID:23072942

De Silva, K., Huang, H.-H., Joshi, R., & Yoshimura, M. (2017). Chemical reduction of graphene oxide using green reductants. *Carbon, 119*, 190–199. doi:10.1016/j.carbon.2017.04.025

De Volder, M. F., Tawfick, S. H., Baughman, R. H., & Hart, A. (2013). *Carbon nanotubes: present and future commercial applications*. Academic Press.

DeAlba-Montero, I., Guajardo-Pacheco, J., Morales-Sánchez, E., Araujo-Martínez, R., Loredo-Becerra, G., Martínez-Castañón, G.-A., Ruiz, F., & Compeán Jasso, M. (2017). *Antimicrobial properties of copper nanoparticles and amino acid chelated copper nanoparticles produced by using a soya extract*. Academic Press.

Deshmukh, K. (2019). Nanotechnology in Ancient Era. In *Biotechnology products in everyday life* (pp. 3–14). Springer. doi:10.1007/978-3-319-92399-4_1

Dizaj, S. M., Lotfipour, F., Barzegar-Jalali, M., Zarrintan, M. H., & Adibkia, K. (2014). *Antimicrobial activity of the metals and metal oxide nanoparticles*. Academic Press.

Djurišić, A. B., Leung, Y. H., Ng, A. M., Xu, X. Y., Lee, P. K., Degger, N., & Wu, R. (2015). *Toxicity of metal oxide nanoparticles: mechanisms, characterization, and avoiding experimental artefacts*. Academic Press.

Dong, X., & Yang, L. (2014). *Inhibitory effects of single-walled carbon nanotubes on biofilm formation from Bacillus anthracis spores*. Academic Press.

Fernando, S., Gunasekara, T., & Holton, J. (2018). *Antimicrobial Nanoparticles: Applications and mechanisms of action*. Academic Press.

Gardini, D., Lüscher, C. J., Struve, C., & Krogfelt, K. A. (2018). Tailored nanomaterials for antimicrobial applications. In *Fundamentals of Nanoparticles* (pp. 71–104). Elsevier. doi:10.1016/B978-0-323-51255-8.00004-5

Gliga, A. R., Skoglund, S., Wallinder, I. O., Fadeel, B., & Karlsson, H. L. (2014). *Size-dependent cytotoxicity of silver nanoparticles in human lung cells: the role of cellular uptake, agglomeration and Ag release*. Academic Press.

Gold, K., Slay, B., Knackstedt, M., & Gaharwar, A. K. (2018). Antimicrobial activity of metal and metal-oxide based nanoparticles. *Advanced Therapeutics*, *1*(3), 1700033. doi:10.1002/adtp.201700033

Gomes, L., Paschoalin, V., & Del Aguila, E. (2017). Chitosan nanoparticles: Production, physicochemical characteristics and nutraceutical applications. *Rev. Virtual Quim*, *9*(1), 387–409. doi:10.21577/1984-6835.20170022

Haung, C.-F., Chan, Y.-H., Chen, L.-K., Liu, C.-M., Huang, W.-C., Ou, S.-F., Ou, K.-L., & Wang, D.-Y. (2013). *Preparation, characterization, and properties of anticoagulation and antibacterial films of carbon-based nanowires fabricated on surfaces Of Ti implants*. Academic Press.

Heiligtag, F. J., & Niederberger, M. (2013). *The fascinating world of nanoparticle research*. Academic Press.

Heron, M. (2018). Deaths: Leading Causes for 2016. *National Vital Statistics Reports*, *67*(6), 1–77. PMID:30248017

Hossain, F., Perales-Perez, O. J., Hwang, S., & Román, F. (2014). *Antimicrobial nanomaterials as water disinfectant: applications, limitations and future perspectives*. Academic Press.

Hsueh, Y.-H., Tsai, P.-H., Lin, K.-S., Ke, W.-J., & Chiang, C.-L. (2017). Antimicrobial effects of zero-valent iron nanoparticles on gram-positive Bacillus strains and gram-negative Escherichia coli strains. *Journal of Nanobiotechnology*, *15*(1), 1–12. doi:10.118612951-017-0314-1 PMID:29100510

Huh, A. J., & Kwon, Y. (2011). *"Nanoantibiotics": a new paradigm for treating infectious diseases using nanomaterials in the antibiotics resistant era*. Academic Press.

Jeevanandam, J., Barhoum, A., Chan, Y. S., Dufresne, A., & Danquah, M. K. (2018). Review on nanoparticles and nanostructured materials: History, sources, toxicity and regulations. *Beilstein Journal of Nanotechnology*, *9*(1), 1050–1074. doi:10.3762/bjnano.9.98 PMID:29719757

Jin, T., & He, Y. (2011). *Antibacterial activities of magnesium oxide (MgO) nanoparticles against food-borne pathogens*. Academic Press.

Johnson-McDaniel, D., Barrett, C. A., Sharafi, A., & Salguero, T. T. (2013). *Nanoscience of an ancient pigment*. Academic Press.

Kang, X., Liu, S., Dai, Z., He, Y., Song, X., & Tan, Z. (2019). Titanium dioxide: From engineering to applications. *Catalysts*, *9*(2), 191. doi:10.3390/catal9020191

Kanhed, P., Birla, S., Gaikwad, S., Gade, A., Seabra, A. B., Rubilar, O., Duran, N., & Rai, M. (2014). In vitro antifungal efficacy of copper nanoparticles against selected crop pathogenic fungi. *Materials Letters*, *115*, 13–17. doi:10.1016/j.matlet.2013.10.011

Kievit, F. M., & Zhang, M. (2011). Surface engineering of iron oxide nanoparticles for targeted cancer therapy. *Accounts of Chemical Research*, *44*(10), 853–862. doi:10.1021/ar2000277 PMID:21528865

Kim, J. Y., Lee, C., Love, D. C., Sedlak, D. L., Yoon, J., & Nelson, K. L. (2011). *Inactivation of MS2 coliphage by ferrous ion and zero-valent iron nanoparticles*. Academic Press.

Lee, S.-J., Kumar, S., Choi, J. W., & Lee, J.-S. (2020). Coloration of colloidal polymer particles through selective extraction of Mie backscattering for cation-responsible colorimetric sensors. *Journal of Colloid and Interface Science*, *560*, 894–901. doi:10.1016/j.jcis.2019.10.073 PMID:31718790

Lemire, J. A., Harrison, J. J., & Turner, R. (2013). *Antimicrobial activity of metals: mechanisms, molecular targets and applications*. Academic Press.

Li, J., Wang, G., Zhu, H., Zhang, M., Zheng, X., Di, Z., Liu, X., & Wang, X. (2014). *Antibacterial activity of large-area monolayer graphene film manipulated by charge transfer*. Academic Press.

Liu, L., Zhang, M., Zhang, Q., & Jiang, W. (2020). Graphene nanosheets damage the lysosomal and mitochondrial membranes and induce the apoptosis of RBL-2H3 cells. *The Science of the Total Environment*, *734*, 139229. doi:10.1016/j.scitotenv.2020.139229 PMID:32450398

Liu, S., Zeng, T. H., Hofmann, M., Burcombe, E., Wei, J., Jiang, R., Kong, J., & Chen, Y. (2011). *Antibacterial activity of graphite, graphite oxide, graphene oxide, and reduced graphene oxide: membrane and oxidative stress*. Academic Press.

Ma, Z., Garrido-Maestu, A., & Jeong, K. C. (2017). Application, mode of action, and in vivo activity of chitosan and its micro-and nanoparticles as antimicrobial agents: A review. *Carbohydrate Polymers*, *176*, 257–265. doi:10.1016/j.carbpol.2017.08.082 PMID:28927606

Maas, M. (2016). *Carbon nanomaterials as antibacterial colloids*. Academic Press.

Manikandan, A., & Sathiyabama, M. (2015). Green synthesis of copper-chitosan nanoparticles and study of its antibacterial activity. *Journal of Nanomedicine & Nanotechnology*, *6*(1), 1.

Marsalek, B., Jancula, D., Marsalkova, E., Mashlan, M., Safarova, K., Tucek, J., & Zboril, R. (2012). *Multimodal action and selective toxicity of zerovalent iron nanoparticles against cyanobacteria*. Academic Press.

McNair, O. D., Brent, D. P., Sparks, B. J., Patton, D. L., & Savin, D. A. (2014). *Sequential thiol click reactions: formation of ternary thiourethane/thiol–ene networks with enhanced thermal and mechanical properties*. Academic Press.

Mohammadi, A., Doctorsafaei, A. H., Burujeny, S. B., Rudbari, H. A., Kordestani, N., & Najafabadi, S. A. A. (2020). Silver (I) complex with a Schiff base ligand extended waterborne polyurethane: A developed strategy to obtain a highly stable antibacterial dispersion impregnated with in situ formed silver nanoparticles. *Chemical Engineering Journal, 381*, 122776. doi:10.1016/j.cej.2019.122776

Monteiro-Riviere, N. A., Wiench, K., Landsiedel, R., Schulte, S., Inman, A. O., & Riviere, J. E. (2011). *Safety evaluation of sunscreen formulations containing titanium dioxide and zinc oxide nanoparticles in UVB sunburned skin: an in vitro and in vivo study*. Academic Press.

Monzavi, A., Eshraghi, S., Hashemian, R., & Momen-Heravi, F. (2015). *In vitro and ex vivo antimicrobial efficacy of nano-MgO in the elimination of endodontic pathogens*. Academic Press.

O'Neill, J. (2014, December). Antimicrobial Resistance: Tackling a crisis for the health and wealth of nations. http://www.jpiamr.eu/wp-content/uploads/2014/12/AMR-Review-Paper-Tackling-a-crisis-for-the-health-and-wealth-of-nations_1-2.pdf

Oh, J.-W., Chun, S. C., & Chandrasekaran, M. (2019). Preparation and in vitro characterization of chitosan nanoparticles and their broad-spectrum antifungal action compared to antibacterial activities against phytopathogens of tomato. *Agronomy (Basel), 9*(1), 21. doi:10.3390/agronomy9010021

Paget, V., Sergent, J., Grall, R., Altmeyer-Morel, S., Girard, H., Petit, T., Gesset, C., Mermoux, M., Bergonzo, P., & Arnault, J.-C. (2014). *Carboxylated nanodiamonds are neither cytotoxic nor genotoxic on liver, kidney, intestine and lung human cell lines*. Academic Press.

Palza, H., Quijada, R., & Delgado, K. (2015). Antimicrobial polymer composites with copper micro-and nanoparticles: Effect of particle size and polymer matrix. *Journal of Bioactive and Compatible Polymers, 30*(4), 366–380. doi:10.1177/0883911515578870

Putro, D. Y., Saputra, R., & Albab, M. F. (2013). *Silver Nano Perfume Ejector to Destroy Bacteria for Clothes.* ASEAN/Asian Academic Society International Conference Proceeding Series.

Rajeshkumar, S., Bharath, L., & Geetha, R. (2019). Broad spectrum antibacterial silver nanoparticle green synthesis: Characterization, and mechanism of action. In *Green synthesis, characterization and applications of nanoparticles* (pp. 429–444). Elsevier.

Rashid, F. L., Hadi, A., Al-Garah, N. H., & Hashim, A. (2018). Novel phase change materials, MgO nanoparticles, and water based nanofluids for thermal energy storage and biomedical applications. *Int Pharmaceu Phytopharmacol Res, 8*, 46–56.

Rekha Deka, S., Kumar Sharma, A., & Kumar, P. (2015). *Cationic polymers and their self-assembly for antibacterial applications*. Academic Press.

Riley, R., & Day, E. (2017). *Wiley interdiscip*. Academic Press.

Saint, S., Elmore, J. G., Sullivan, S. D., Emerson, S. S., & Koepsell, T. D. (1998). *The efficacy of silver alloy-coated urinary catheters in preventing urinary tract infection: a meta-analysis*. Academic Press.

Schaming, D., & Remita, H. (2015). *Nanotechnology: from the ancient time to nowadays*. Academic Press.

Seil, J. T., & Webster, T. (2012). *Antimicrobial applications of nanotechnology: methods and literature*. Academic Press.

Senarathna, U., Fernando, S., Gunasekara, T., Weerasekera, M., Hewageegana, H., Arachchi, N., Siriwardena, H., & Jayaweera, P. (2017). *Enhanced antibacterial activity of TiO 2 nanoparticle surface modified with Garcinia zeylanica extract*. Academic Press.

Shevchenko, S. N., Burkhardt, M., Sheval, E. V., Natashina, U. A., Grosse, C., Nikolaev, A. L., Gopin, A. V., Neugebauer, U., Kudryavtsev, A. A., & Sivakov, V. (2017). *Antimicrobial effect of biocompatible silicon nanoparticles activated using therapeutic ultrasound*. Academic Press.

Shi, L.-E., Li, Z.-H., Zheng, W., Zhao, Y.-F., Jin, Y.-F., & Tang, Z.-X. (2014). Synthesis, antibacterial activity, antibacterial mechanism and food applications of ZnO nanoparticles: A review. *Food Additives & Contaminants: Part A*, *31*(2), 173–186. doi:10.1080/19440049.2013.865147 PMID:24219062

Singh, A. K., & Singh, K. P. (2016). Evaluation of phosphate removal capacity of Fe 3 O 4–ZVINPs from aqueous solution: Optimization using response surface analysis. *Research on Chemical Intermediates*, *42*(10), 7397–7415. doi:10.100711164-016-2543-6

Sirelkhatim, A., Mahmud, S., Seeni, A., Kaus, N. H. M., Ann, L. C., Bakhori, S. K. M., Hasan, H., & Mohamad, D. (2015). *Review on zinc oxide nanoparticles: antibacterial activity and toxicity mechanism*. Academic Press.

Tang, S., & Zheng, J. (2018). Antibacterial activity of silver nanoparticles: Structural effects. *Advanced Healthcare Materials*, *7*(13), 1701503. doi:10.1002/adhm.201701503 PMID:29808627

Tchounwou, P. B., Yedjou, C. G., Patlolla, A. K., & Sutton, D. (2012). *Heavy metal toxicity and the environment*. Academic Press.

Vance, M. J. N., Kuiken, T., Vejerano, E. P., McGinnis, S. P., Hochella, M. F. Jr, Rejeski, D., & Hull, M. S. (2015). Nanotechnology in the real world: Redeveloping the nanomaterial consumer products inventory. *Beilstein J.*, *6*, 1769–1178. doi:10.3762/bjnano.6.181

Vanti, G. L., Masaphy, S., Kurjogi, M., Chakrasali, S., & Nargund, V. B. (2020). Synthesis and application of chitosan-copper nanoparticles on damping off causing plant pathogenic fungi. *International Journal of Biological Macromolecules*, *156*, 1387–1395. doi:10.1016/j.ijbiomac.2019.11.179 PMID:31760011

Wang, L., Hu, C., & Shao, L. (2017). *The antimicrobial activity of nanoparticles: present situation and prospects for the future*. Academic Press.

Weir, A., Westerhoff, P., Fabricius, L., Hristovski, K., & Von Goetz, N. (2012). *Titanium dioxide nanoparticles in food and personal care products*. Academic Press.

Xing, Y., Xiong, W., Zhu, L., Osawa, E., Hussin, S., & Dai, L. (2011). *DNA damage in embryonic stem cells caused by nanodiamonds*. Academic Press.

Yang, Z., Huang, X., Ma, X., Zhou, Z., Guo, H., Yao, Z., Feng, S.-P., & Tang, C. Y. (2019). Fabrication of a novel and green thin-film composite membrane containing nanovoids for water purification. *Journal of Membrane Science*, *570*, 314–321. doi:10.1016/j.memsci.2018.10.057

Yun, G., Pan, S., Wang, T. Y., Guo, J., Richardson, J. J., & Caruso, F. (2018). Synthesis of metal nanoparticles in metal-phenolic networks: Catalytic and antimicrobial applications of coated textiles. *Advanced Healthcare Materials*, *7*(5), 1700934. doi:10.1002/adhm.201700934 PMID:29024556

Zhang, H., & Chen, G. (2009). *Potent antibacterial activities of Ag/TiO2 nanocomposite powders synthesized by a one-pot sol- gel method*. Academic Press.

Zheng, K., Setyawati, M. I., Leong, D. T., & Xie, J. (2017). *Antimicrobial gold nanoclusters*. Academic Press.

ADDITIONAL READING

Bai, X., Wang, Y., Song, Z., Feng, Y., Chen, Y., Zhang, D., & Feng, L. (2020). The Basic Properties of Gold Nanoparticles and their Applications in Tumor Diagnosis and Treatment. *International Journal of Molecular Sciences*, *21*(7), 2480. doi:10.3390/ijms21072480 PMID:32260051

Deshmukh, S. P., Patil, S., Mullani, S., & Delekar, S. (2019). Silver nanoparticles as an effective disinfectant: A review. *Materials Science and Engineering C*, *97*, 954–965. doi:10.1016/j.msec.2018.12.102 PMID:30678983

Flieger, J., Flieger, W., Baj, J., & Maciejewski, R. (2021). Antioxidants: Classification, Natural Sources, Activity/Capacity Measurements, and Usefulness for the Synthesis of Nanoparticles. *Materials (Basel)*, *14*(15), 4135. doi:10.3390/ma14154135 PMID:34361329

Gacem, M. A., Gacem, H., & Ould-El-Hadj-Khelil, A. (2020). 22 - Nanocarbons: Antibacterial, antifungal, and antiviral activity and the underlying mechanism. In K. A. Abd-Elsalam (Ed.), *Carbon Nanomaterials for Agri-Food and Environmental Applications* (pp. 505–533). Elsevier. doi:10.1016/B978-0-12-819786-8.00022-0

Ijaz, M., Zafar, M., Islam, A., Afsheen, S., & Iqbal, T. (2020). A review on antibacterial properties of biologically synthesized zinc oxide nanostructures. *Journal of Inorganic and Organometallic Polymers and Materials*, *30*(8), 2815–2826. doi:10.100710904-020-01603-9

Jamkhande, P. G., Ghule, N. W., Bamer, A. H., & Kalaskar, M. G. (2019). Metal nanoparticles synthesis: An overview on methods of preparation, advantages and disadvantages, and applications. *Journal of Drug Delivery Science and Technology*, *53*, 101174. doi:10.1016/j.jddst.2019.101174

Noman, M. T., Amor, N., & Petru, M. (2021). Synthesis and applications of ZnO nanostructures (ZONSs): A review. *Critical Reviews in Solid State and Material Sciences*, 1–43. doi:10.1080/10408436.2021.1886041

Patil, M. P., & Kim, G.-D. (2017). Eco-friendly approach for nanoparticles synthesis and mechanism behind antibacterial activity of silver and anticancer activity of gold nanoparticles. *Applied Microbiology and Biotechnology*, *101*(1), 79–92. doi:10.100700253-016-8012-8 PMID:27915376

Reshma, V. G., Syama, S., Sruthi, S., Reshma, S. C., Remya, N. S., & Mohanan, P. V. (2017). Engineered nanoparticles with antimicrobial property. *Current Drug Metabolism*, *18*(11), 1040–1054. doi:10.2174/1389200218666170925122201 PMID:28952436

Tang, S., & Zheng, J. (2018). Antibacterial activity of silver nanoparticles: Structural effects. *Advanced Healthcare Materials*, *7*(13), 1701503. doi:10.1002/adhm.201701503 PMID:29808627

Vines, J. B., Yoon, J.-H., Ryu, N.-E., Lim, D.-J., & Park, H. (2019). Gold Nanoparticles for Photothermal Cancer Therapy [Review]. *Frontiers in Chemistry*, *7*(167), 167. Advance online publication. doi:10.3389/fchem.2019.00167 PMID:31024882

Xu, L., Wang, Y.-Y., Huang, J., Chen, C.-Y., Wang, Z.-X., & Xie, H. (2020). Silver nanoparticles: Synthesis, medical applications and biosafety. *Theranostics*, *10*(20), 8996–9031. doi:10.7150/thno.45413 PMID:32802176

KEY TERMS AND DEFINITIONS

Chitosan: Chitosan is a decomposable, biocompatible polymer is considered as nontoxic, and safe for human digestion and medical use. It is also used as a carrier for drug delivery via various route of administration.

Engineered Nanoparticles: Engineered nanomaterials are made by the researcher for special size, shape and functionality; these are not made by nature or accindently, they are larger than 1 nm and smaller than 100 nm.

Goniochromism: It is also known as Iridescence, a process of surface that changes colors whenever it alters its angle of view. It produces multiple of colors while pearlescence produce white color.

Graphene Nanostructures: It is a two-dimensional (2D) element composed of carbon atoms packed in a honeycomb lattice. It has excellent mechanical, electronic and thermal properties. For example, nanoribbons, etc.

Nanoparticles: The nanoparticle (NP) is a small particle ranging in size from 1 to 100 nm, and invisible to the human naked eye. Sometimes, it is just made up of one of the few hundred atoms, and it possess different physical and chemical properties as compare to the bulky one.

Phytochelatins: Enzyme phytochelatin synthase produces polymer of phytochelatins, found in plants, nematodes, algae, cyanobacteria and fungi. It is very useful for the detoxification of heavy metals.

Reactive Oxygen Species (ROS): An oxygen containing unstable molecule which react with other molecules in a cell that causes damage to the DNA, proteins and leads to cell death. They are present in free radical form like peroxides, superoxide, singlet oxygen, etc.

Chapter 13
Biomedical Applications of Nanoparticles

Raghavv Raghavender Suresh
SASTRA University (Deemed), India

Suwetha Bharathirengan
SASTRA University (Deemed), India

Shruthee Sankarlinkam
SASTRA University (Deemed), India

Dinesh Kumar Subbiah
SASTRA University (Deemed), India

Sai Rakshana Karuppusami
SASTRA University (Deemed), India

Soorya Srinivasan
SASTRA University (Deemed), India

Niraimathi Pandiyan
SASTRA University (Deemed), India

Arockia Jayalatha Kulandaisamy
SASTRA University (Deemed), India

Noel Nesakumar
SASTRA University (Deemed), India

ABSTRACT

In recent years, there has been significant growth and burgeoning interest in utilizing nanoparticles for various biomedical applications, including medical diagnostics, targeted drug delivery, tissue engineering, regenerative medicine, and biomedical textiles. In particular, nanoparticles functionalized with biological molecules have unique properties and are very effective in medical diagnostics. Besides that, nanoparticles have a wide range of therapeutic applications, including the development of nanodrug delivery systems, the design of novel drugs, as well as their contribution to the design of therapeutic materials. This chapter provides an overview of recent advancements in the biomedical applications of nanoparticles. Finally, this chapter discusses the challenges of the toxicological evaluation of engineered nanoparticles and the importance of conducting detailed studies on the synthesis of future nanomaterials to develop cutting-edge technologies for addressing a wide range of biomedical issues.

DOI: 10.4018/978-1-7998-8936-6.ch013

INTRODUCTION

Nanotechnology-derived products are becoming invaluable in biomedicine, resulting in a hybrid science known as nanobiotechnology (Rahman et al., 2021). Nanomaterials are used extensively in nanobiotechnology for implants, prostheses, drug delivery systems, and diagnosis. Since most biological systems are often only a few nanometers in size, nanomaterials can be easily incorporated into biomedical devices. Conventionally, liposomes, carbon nanotubes (CNTs), and metal Nps are the common materials employed in developing these nanotechnology products. By utilizing specific biological reactions, including DNA-DNA hybridization, ligand-receptor interaction, and antigen-antibody interaction, as well as physical or chemical methods, it is plausible to combine nanoparticles (Nps) and biomolecules.

In recent years, a multitude of research has been carried out on the applications of nanobiotechnology in biomedical research, covering the functionalization, visualization, and development of nanomaterials at the scale of less than 100 nm. Nanomaterials also have a significant impact on human health, and their increasing use in therapeutic applications can revolutionize and personalize medicine. In general, 2D materials, such as nanosheets, nanofilms, nanodisks, and nanobelts, 1D materials, such as nanowires and nanorods, and 0D materials, also known as quantum dots, which vary in diameter and shape, are the three types of nanomaterials that are of particular interest. Based on their source of origin, nano-scale materials can be classified into three categories: biodegradable, non-metallic, and metallic. These nano-scale materials are made from micro-sized materials that have been functionalized with different functional groups, allowing them to bind to drugs, ligands, and antibodies of interest, opening the door to a diverse variety of biomedical research applications.

Figure 1. Biomedical applications of nanoparticles.

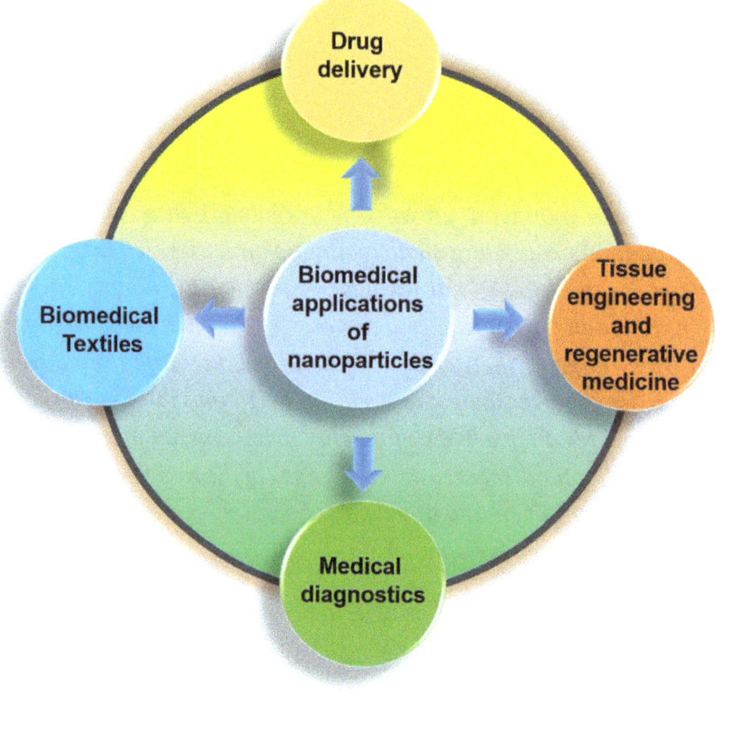

In this chapter, nanomaterials employed extensively in a variety of biomedical applications have been focused on. This book chapter also paid attention to the technological innovations that are being successfully implemented for biomedical applications for therapeutic and diagnostic purposes.

BACKGROUND

Nanomaterials of various sorts have numerous applications in the biomedical domain (Figure 1), particularly in biomaterials as implants and prostheses, tissue engineering, biomedical textiles, regenerative medicine, drug delivery, imaging, and diagnostics, due to their nanometric size and ease of integration into the biological system (Jiang et al., 2021).

Nanomaterials can be obtained from various sources, including plants, microbes, chemicals, nonmetals, and metals. Even though non-metallic and metallic Nps have been more widely employed in various nanotechnology-based therapeutics, Nps consisting of proteins and lipids are becoming more popular as nanocarriers for drug delivery. Currently, a wide range of nanomedicine products is being developed and manufactured, with the majority of them being used to treat cancer and microbial infections.

DRUG DELIVERY

An emerging area in which nanomaterials have started to play an important role is "drug delivery" at the cellular level (He et al., 2021). This approach of administering the active substance enhances the drug's pharmacokinetic and pharmacodynamic properties, resulting in rapid absorption and dissolution kinetics, allowing the target drug to be transported to the target tissue. A targeted drug delivery system can efficiently deliver drugs to target cells at a sustained rate over time, improving patient adherence and prolonging drug life cycles. On the other hand, nano-delivery systems have many pharmacokinetic benefits, including increased membrane permeability, bioavailability, and stability. The modified physicochemical properties of nanodrugs, including solubility, size, and surface charge, enable them to enter specific sites via endocytotic processes.

Furthermore, the surface ligand functionalization in Nps improves the transport of encapsulated drugs into the cell interior. It has been reported that functionalizing biodegradable nanospheres with carbohydrate-binding ligands increases binding to target cells, and binding of specific proteins like antibodies to Nps allows for more specific immunologically guided targeting of particles (Bartneck, 2021). The goal of drug entrapment in nanospheres is to improve cell-specific drug delivery while reducing the toxic effects of the free drug on nontarget sites, resulting in less toxicity to other organ systems and a higher therapeutic index. Thereby, drugs can be loaded through entrapping, surface attachment, or encapsulation. The attachment technique is determined by the architecture of the Nps, the material, the type of drug, and the target area. Typically, the drugs that are encapsulated can be shielded from degradation. The drug could exist in the crystalline or amorphous form in Nps with coatings as thin as a few nanometers. Drugs are typically implanted or ingested and are engineered to provide a controlled release of drugs that can last for weeks to months and can be initiated at various sites throughout the body. Nanopores can function as nanoscopic gates for drug release. Despite the amount of drug remaining within a capsule, the diffusion rate of molecules can be controlled by tailoring the dimensions of nanopores that are marginally wider than drug molecules. Drugs in nanocrystalline form can be delivered in small amounts because they can be transported to the cell/tissue and in regulated doses.

Even though Nps are being used to deliver drugs for a variety of illnesses, the primary emphasis has been on improving cancer treatment. Nevertheless, notwithstanding concerted efforts, the outcomes have indeed been underwhelming. The effectiveness of the onco-drug is ascertained by blood circulation, accretion, and permeation into tumour cells and cellular endocytosis followed by the release of the drug into the cells, known as the CAPIR cascade. An ideal nanomaterial used to transport cancer drugs must be stealthy and stable in circulation, as well as the nanocarriers should carry the drug firmly throughout transport to the tumour tissues, where the nanocarriers should accrue in the tumour tissue and release its cargo at a target site, which is then effectively absorbed by the tumour cells. Besides that, the nanocarrier should be capable of reaching all tumour cells, even those located far from blood vessels (Cheng et al., 2022). Although cancer has been the primary focus, a large number of nanocarriers have been evaluated for the treatment of infectious diseases, neurodegenerative disorders, and cardiovascular disease. Besides cancer drugs, which are tiny and easily stuffed within Nps, the macro-sized molecules that are efficacious against many other human illnesses are bulky, swiftly metabolized, and thus cannot enter diverse biological barriers, including cellular membranes and tissues.

Properties of Nps

Properties of Nps that are useful for the drug delivery system have been discussed in detail ahead (Figure 2).

Particle Size

Particle size is one of the most significant characteristics of Nps because it influences targeting ability, in vivo distribution, toxicity, and biological fate (Yadav & Dalal, 2021). In addition, Nps influence the release, stability, and loading of the drug. Unlike micro-sized particles, Nps can reach intracellular and cellular targets due to their rapid mobility and tiny size. Nps also have the ability to cross the blood brain barrier, providing localized and prolonged delivery of drugs for dreadful brain diseases. The size of the particles is critical for drug release. The rapid release rate of the drug from the nanocarrier is only possible if the nanocarrier has a high surface-area-to-volume-ratio.

Figure 2. Criteria for the selection of nanomaterial for drug delivery system.

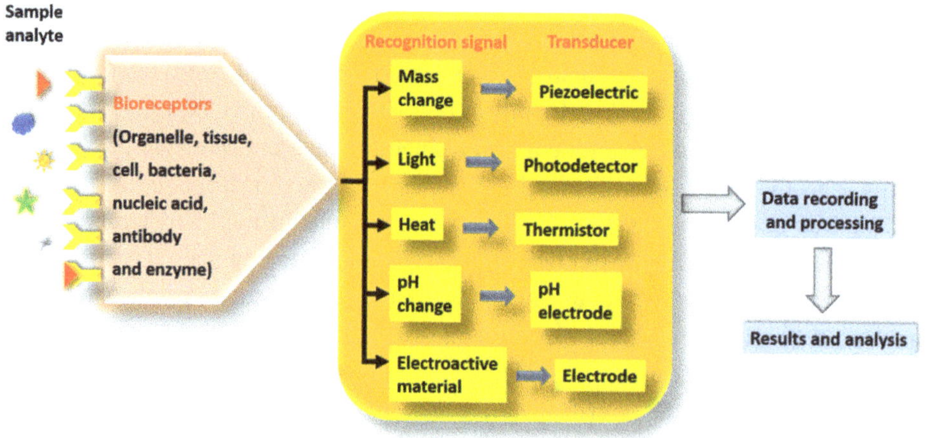

Surface Properties

Hydrophobicity and surface charge are the two main surface properties of Nps. They influence blood flow, biodistribution, opsonization, phagocytosis. Surface properties also influence the ability to interact with cells as well as the stability of Nps.

Surface Charge

The surface charge of Nps has a strong influence on plasma protein adsorption, which impacts biodistribution (Xie & Liu, 2020). Nps with a positive charge induce a dramatic increase in phagocytosis, resulting in better interaction between the Nps and the anionic cell membrane. On the contrary, anionic Nps cause a significant decrease in phagocytic uptake, which improves blood circulation time. Thus, it is concluded that Nps with a negative charge are effective in decreasing the unwanted clearance of the reticuloendothelial system in the body, enhancing blood compatibility, and ensuring adequate delivery of the drug to the area of interest. On the other hand, positively charged Nps reduce the acceptance of particles in the reticuloendothelial system. As a result, cationic particles are less rapidly removed from the blood, guaranteeing their biocompatibility and effective functionality with a long-term drug delivery system.

Particle Shape

The shape and size of Nps play an important role in delivering the drugs efficiently to the target site (Mei et al., 2022). These encompass intracellular transport, endocytosis, phagocytosis, targeting efficiency, circulation half-life, and vasculature transport, all of which directly impact the targeting capability of Nps (Sharifianjazi et al., 2021). The shape of Nps has received a lot of attention in drug delivery research. For instance, platinum Nps in various forms, including monodisperse cubic dendrites and a quintuple star form, have been reported to be unprecedented.

Surface Hydrophobicity

The surface hydrophobicity is essential in opsonization. In our human body, water-insoluble particles with extremely low solubility are bound by plasma proteins and immunoglobulin complement proteins before being eliminated through the reticuloendothelial system (Maji et al., 2021). This promotes the adsorption of proteins to the surface, increasing the hydrophobicity of the surface of the Nps. Typically, polyethylene glycol functionalized Nps are employed to increase the accumulation of Nps in tumour blood vessels and to prolong the circulating half-life of Nps.

Drug Loading

An ideal nano-delivery system should have a wider range of drug loading capacity (Long et al., 2021). There are two basic approaches to load a drug into the core of the nanoparticle. The first is the incorporation method in which the drug is introduced during the nanoparticle formulation process. The second is the absorption/adsorption method in which the drug is absorbed upon nanoparticle formation, which is accomplished by incubating the nanocarrier with a high concentration of drug solution. The solubility of the drug in the matrix carrier determines the efficiency of drug loading. A suitable polymer, like polyethylene glycol, should be chosen to have minimal or no drug loading effect without contaminating the environment. Polymers that exhibit zero drug loading effect have the highest loading efficiency when drugs are loaded into the core of Nps near or at their isoelectric point. Besides polyethylene glycol, several other polymers are extensively utilized to synthesize Nps such as chitosan for illnesses, including allergic rhinitis and cancer.

Drug Release

When designing a targeted drug delivery system, it is necessary to consider the degradation of the polymer and the release of the drug. The drug release rate is typically determined by the rate of desorption of the adsorbed drug, the solubility of the drug, the erosion, degradation, and diffusion properties of nanoparticle matrices (Moradi Kashkooli et al., 2020). Especially in nanospheres, the drug is evenly distributed throughout the volume of nanospheres and from which it is released through matrix erosion or diffusion.

Dialysis bag diffusion, agitation accompanied by centrifugation/ultracentrifugation, reverse dialysis bag diffusion, side-by-side diffusion cells with biological or synthetic membranes and ultra-filtration are some of the techniques extensively employed to study the release of drugs from Nps (Yu et al., 2019).

TARGETED DELIVERY OF NANOPARTICLES

For an efficient drug delivery system, Nps must be able to reach the target site (organ/tissue/cell) with a negligible decrease in activity or volume. Moreover, once inside the target organ/tissue/cell, they must be able to perform the planned therapeutic action. Nps must accurately reach the anticipated site of action, after which the binding of the drug to the target tissue occurs and effectively delivers the therapeutic drug to the target tissues while reducing the toxic side effects caused by the drugs on healthy tissues. The most widely accepted approach is to keep coating-specific ligands on the surface of Nps because ligands can bind to specific receptors found at the target site. This ligand can take the form of nucleic acid aptamers, protein biomolecules, antibodies, tiny molecules, etc.

Passive Targeting by Nanoparticles

Nps with the desired surface, size, and shape characteristics can remain in the bloodstream for a prolonged time, increasing the likelihood that they will reach the targeted cells or tumour sites (Zou et al., 2021). The unique pathologic characteristics of tumour vessels enable the selective accretion of Nps in tumour tissues. Concerning the microenvironment of tumour, cancer cells have a high rate of metabolism and thus utilize glycolysis to obtain additional energy and oxygen, resulting in an acidic environment. Liposomes, which are pH-sensitive, remain unaffected at physiological pH but degrade to deliver the encapsulated drug to the target cell/tissue where the pH is not as close to 7.4.

Active Targeting by Nanoparticles

Passive targeting processes are constrained by their specificity. To resolve this problem, a targeting ligand or a polymer-drug conjugated antibody can be employed (Ghalehkhondabi et al., 2021). Liposomes and polymers, which were recently introduced and developed as drug delivery carriers, increase the number of drugs bound to Nps without affecting their targeting affinity. To do the same, specific conditions must be met, such as receptors and cell surface antigens being uniformly expressed only in tumour cells but not in healthy cells. Additionally, once the Nps bind to surface receptors on the target cells, the targeted conjugates must be introduced via receptor-mediated endocytosis, which also selects the appropriate targeted ligands.

TYPES OF NANODRUG DELIVERY SYSTEM

The various kinds of nanodrug delivery systems include micelles, carbon-based nanocarriers, dendrimers, polymeric Nps and liposomes (Figure 3).

Figure 3. Nanoparticle-based drug delivery system for delivering drugs to the target site.

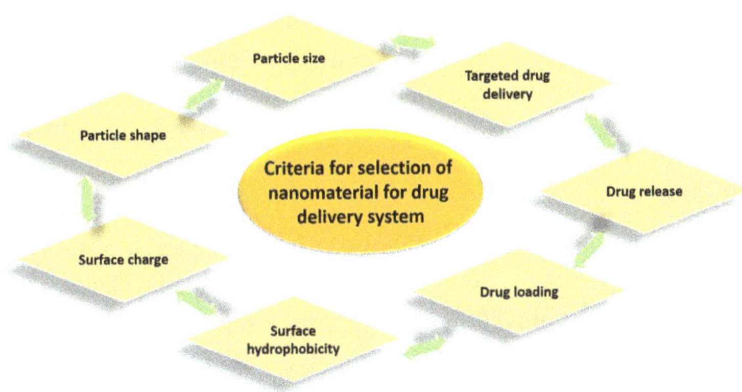

Liposomes

Liposomes, which have already been portrayed as artificial lipid vesicles by Bangham in 1961, have been widely used as drug delivery carriers because they emulate the properties of biological cells. Liposomes are artificial spherical amphiphilic vesicles made of two layers of lipid on the outside and an aqueous compartment on the inside (Guimarães et al., 2021). The aqueous compartment of liposomes can entrap water-soluble therapeutic drugs, while the lipid bilayer can uptake hydrophobic drugs. Liposomes with varying characteristics can be developed by altering their size, surface charge and composition. By selecting specific ligands, the fluidity and rigidity can also be engineered. These artificial spherical amphiphilic vesicles can be categorized based on their diameter. The size of large unilamellar vesicles varies between 200 and 1000 nm, while the size of small unilamellar vesicles varies between 20 and 100 nm. Lipoplexes are DNA-liposome complexes that have been engineered for gene therapy. They are commonly used as drug delivery vehicles to deliver DNA to target cells.

Based on their composition and intracellular administration, liposomes are classified into different types: cationic liposomes, immunoliposomes, long circulating, pH sensitive and conventional (Kooijmans et al., 2021). The modification of the surface of nanoliposomes with site-directing ligands has been constructed to improve the specificity of liposomal drug delivery to target sites. Liposomes typically do not elicit an immune response in humans and are eventually nontoxic and biodegradable in biological environments, allowing them to deliver bioactive compounds to the target site at a controlled degradation rate without affecting normal cells.

In addition to being drug delivery materials, these liposomes can also be utilized as biomimetic models to investigate how proteins and hydrophobic drugs interact with membranes. In general, liposomes with proteins incorporated into the lipid bilayer, known as proteoliposomes, can be used to study the activity of photosensitive dyes used in photodynamic therapy. Furthermore, proteoliposomes can also be employed to investigate how the interactions between lipid and protein significantly impact the surface of the membrane model.

Polymeric Nanoparticles

Polymeric Nps obtained from natural or artificial polymers offer a wide variety of applications in drug delivery to the anticipated site of action because they are highly biodegradable and biocompatible by nature (İçhedef et al., 2021). Although Nps made of any polymer can be referred to as polymer Nps, the term is most commonly associated with nanocapsules and nanospheres. The potential of polymer Nps for delivering drugs at a controlled rate to the site of action necessitates the integration of imaging and therapy to provide drug protection and targeted drug delivery, allowing for an increase in therapeutic index. The Food and Drug Administration has also recommended non-biodegradable polymers for delivering drugs to the target site, where non-biodegradable polymers release drugs in a controlled manner via diffusion control rather than burst release, as biodegradable polymers do. The most common polymeric Nps used for drug delivery are ethyl vinyl acetate, polyvinyl acetate, and silicone. They were developed to treat diabetic retinopathy, uveitis, and retinitis. Polymers are not commonly available in bulk, but they can be used as coating materials on implantable implants or carriers containing drugs of interest, particularly in nanodrug delivery systems. The thin-layer coatings made from polymeric materials can be layer-by-layer deposited films, cross-linked, or homogeneous coatings. The polymeric Nps can be formed into nanocapsules, nanogels, micelles and dendrimers. It has been found that the surface properties of polymer Nps influence their subcellular and cellular fate, biodistribution, biocompatibility, biodegradability, and stability (Urimi et al., 2021). The introduction of functional groups with variable and selective reactivity on the surfaces of polymeric Nps enables chemical modification, which in turn modifies the reaction of polymeric Nps in biological microenvironments with varying redox conditions, enzymatic, pH, or in response to extrinsic stimuli including magnetic field activation, radiation, and temperature variations. Polymeric Nps derived from natural compounds outperform conventional drug delivery approaches in terms of targeted drug delivery and efficacy. Numerous natural polymers, including alginate, chitosan, albumin, gelatin, and dextran, have been extensively utilized to synthesise Nps.

Dendrimers

Dendrimers are three-dimensional artificial nano-sized repetitively branched molecules inspired by nature with a homogeneous and well-defined structure formulated in 1978 for drug delivery. Dendrimers are appealing and potentially valuable nanocarriers, which have a molecular weight in the range of 5000 – 50000 $gmol^{-1}$ and are distinguished by layers referred to as "generations" (Bartusik-Aebisher et al., 2021). They are composed of an inner core surrounded by branches or tree-like arms linked to specific functional groups. Controlled delivery of therapeutic drugs to the biological site of action is essential for lowering drug toxicity, which is usually accomplished by modifying the surface of dendrimers with specific functional groups like peptides, antibodies, and folic acid. Furthermore, loading the therapeutic drug into dendrimers enhances solubilization as well as in vitro drug release. Polyamidoamines are the most commonly utilized dendrimers in drug delivery because they are hydrophilic, nonimmunogenic, and biocompatible. The inner core of these nano-sized dendrimers is composed of hydrophobic molecules with tree-like arms formed of methyl acrylate and ethylenediamine that are terminated with amine and carboxyl groups. Polyamidoamines dendrimers are being employed as diagnostic agents, transfection agents, and drug carriers. Polyamidoamine dendrimers contain tertial amines that can form noncovalent interactions and hydrogen bonds with neighbouring molecules, enabling them to be better suited to soluble drugs that are hydrophobic in nature. The properties of dendrimers, such as good biological compatibility, varying chemical composition, and multiple drug loading ability, make them promising carriers in drug delivery applications.

Carbon-based Nanocarriers

CNTs were developed in 1976 using a chemical vapour deposition method and later evolved into multiwalled carbon nanotubes, which offer numerous advantages in diagnostics, imaging, and targeted drug delivery (Chadar et al., 2021). CNTs are extensively employed in biomedicine and science due to their unique physical, chemical and electrochemical properties. These distinctive characteristics make CNTs useful in a wide range of fields, including biomedical, optical, electronic, thermal, structural, and mechanical. The surface of CNTs that have been functionalized with special biomolecules is utilized as targeted drug delivery carriers because they have displayed great promise in targeting cancer/tumour cells while causing minimal harm to healthy cells in the body. Modifying the surface of CNTs with targeting ligands allows CNT-DNA complex formation, making them ideal carriers for gene delivery. CNTs have also been widely utilized to target breast cancer cells by delivering tiny proteins like recombined ricin A chain protein toxin to the anticipated site of action. Buckyball, also widely recognized as buckminsterfullerene (C60), is a cage-like ring fused structure explored in 1985; it is particularly fascinating due to its diagnostic applications, gene delivery, carrier for drug delivery, and anti-human immunodeficiency virus activity. A wide range of carbon-based nanorods and nanocapsules have also been used as drug carriers. Although CNTs have been demonstrated to be effective nanocarriers for delivering therapeutic drugs, their reliability in delivering drug to the target site remains a challenge, limiting their commercial viability as drug delivery vehicles.

Polymeric Micelles

Polymeric micelles are nanocarriers that are supramolecular, colloidal, and spherical in nature and are made from nano-sized amphiphilic triblock or diblock copolymers for the delivery of genes and drugs (Rezaeisadat et al., 2021). The hydrophilic polymer forms the shell of the micelles, whereas the hydrophobic drugs are entrapped in the core of the block copolymer micelles. They display enhanced stability at low critical micelle concentrations, allow a longer time for circulation of the drug in the blood, and prevent reticuloendothelial absorption due to their tinier size and inability to be degraded by blood components. Polymeric micelles bearing targeting ligands are the effective nanocarriers for drug delivery because they remain typically stable and have a low critical micelle concentration, resulting in increased drug levels at the anticipated site of action. Micelles are available in a multitude of forms, which also affects the efficacy of the formulation. The introduction of hydrophobic drugs into micellar structures not only enhances the therapeutic efficiency of chemo-drugs, but also significantly increases the bioavailability of the drug. Micelles have been developed for biomedical applications using two approaches: passive or active targeting for enhanced targeted drug delivery. Peptide-based ligands, proteins, aptamers, antibodies, folate molecules, and carbohydrates are examples of active targeting. The latest developments include multifunctional micelles with 2-3 targeted ligands, which amalgamate many features within a single carrier, allowing each element to function effectively in precise conjunction with the other elements.

DIAGNOSIS

In recent decades, contrast agents have been used to visualize anatomical structures to recognize diseased tissue and assist doctors in determining the best course of treatment (Jiang et al., 2021). The materials, which are less than 10 nm in size, can be tailored to construct contrast agents with enhanced sensitivity and specificity while reducing toxicity in various imaging modalities. Although X-rays have made significant contributions to medical imaging over the past decades, computed tomography (CT) is the most popularly used imaging modality due to its rapid response and better signal-to-noise ratio. Nevertheless, the CT's reduced sensitivity is mainly owing to the inefficient interaction between incident x-rays and soft tissues. The use of Nps with a high electron density and atomic number, including gold, as promising contrast agents that attach or accrue in specific cells has been shown to enhance the sensitivity of CT. The characteristics of these Nps can be altered by varying their shape and size. Besides that, high specificity for tissue or cell types can be achieved by labelling nanoparticle-based contrast agents with specific targeting biomolecules, including folic acid, making them suitable for imaging and targeting. With the advancement of technology, CT has greatly evolved. Phase-contrast CT is an advanced CT that uses scattering or phase shift rather than X-ray attenuation to reduce X-ray dosage, making phase-contrast CT a safe diagnostic tool for patients (Seitz et al., 2021). Magnetic resonance imaging (MRI), on the other hand, does not rely on the use of ionizing radiation; rather, it depends on variations in the time it takes for field-aligned protons to come back to equilibrium after being exposed to pulses of radio waves. MRI can provide both functional and anatomical information and has higher spatial resolution than CT. It varies depending on the tissue, enabling higher depth penetration, specificity, greater soft-tissue contrast and superior spatial resolution. To enhance MRI resolution, superparamagnetic iron oxide Nps are used to cause negative contrast, while paramagnetic lanthanide metal chelates like gadolinium are used to provide positive contrast in the image. However, as these Nps are highly toxic to humans, their application in MRI is restricted. In the last decade, a plethora of research findings on MRI studies using dual-mode contrast agents with paramagnetic material as a shell and superparamagnetic material as a core have been published, but signal interference remains a major issue. It is possible to resolve this difficulty by incorporating a segregating layer between the shell and the core. Conversely, a bridge can be employed in a dumbbell layout to isolate the two contrast agents, thereby reducing magnetic interference and enabling their surfaces to be engineered for further targeting. Furthermore, smart nanoparticle-based MRI probes can detect minute changes in temperature and the cell's redox status, allowing the MRI to distinguish between tumours and tissues.

A variety of radiolabelled Nps have been investigated for use in single-photon emission computed tomography/positron emission tomography (SPECT/PET) imaging. The majority of these radiolabelled Nps have a core-shell structure. The core is a radiolabel, and the shell being an organic material surrounded by numerous ligands for specific tissue/cell targeting. Typically, the surfaces of the Nps are labelled with various radiolabels of different energies to assess the transfer of Nps from one location to another for *in vivo* imaging to identify targeted tissues/cells and investigate the removal of the Nps. This kind of surface labelling will be extremely helpful in the development of effective nanoparticle-based targeted drug delivery systems. Conventional fluorophores employed in fluorescence imaging suffer from proximity-induced quenching and aggregation-induced quenching, and are therefore swiftly photobleached. Inorganic fluorescent quantum dots possess a narrow emission bandwidth and are resilient to photobleaching, enabling individual biological entities to be visualized at the same time via multiplexing. The utilization of quantum dots is restricted due to their potential toxicity, which can be mitigated to a certain extent by using quantum dots with diameters less than 10 nm. On the other hand, the characteristics of recently developed aggregation-induced emission fluorochromes, such as intense fluorescence emission after accumulation and high photobleaching resistance, allow for real-time imaging. The functionality and targeting of aggregation-induced emission fluorochromes can be increased by introducing various receptors, ligands and functional groups to the surface, allowing them to be used for assessing drug release profiling and targeted drug delivery.

In addition, a wide variety of nanomaterials have been tested and used for fluorescence imaging. Among these nanomaterials, nanodiamonds and carbon dots are very popular owing to their nontoxicity, environmental friendliness, variable surface chemistry, excellent photostability and particle size at the nano-scale (Seitz et al., 2021). Smart fluorophores that alter their emission properties based on the microenvironment, like pH and temperature, are also being investigated for their ability to detect changes in the cell's internal environment for diagnostic purposes.

In addition to detecting disease and localizing the desired biomolecules into the target cell/tissue, Nps are being employed in biological fluids, including sweat, saliva, urine and blood, to diagnose disease and localize specific molecule at a particular site. Typically, ligand-modified Nps are used to detect target biomolecules such as DNA moieties and proteins, which has led to their widespread application due to substantial improvements in specificity and detection sensitivity. Fluorescent quantum dots have been utilized to build sensing devices capable of detecting multiple molecules simultaneously with rapidity, simplicity and high throughput due to their precise and unique emission characteristics and enhanced stability. Another method for detecting multiple analytes rapidly requires many "noses", each of which is modified with a recognition element specific to the target analyte (Figure 4). This method has also been used to detect cancer cells in human blood serum and bacteria in human internal organs. Furthermore, many microfluidic-based efficient platforms for sensitive and specific detection of multiple biomolecules such as nucleic acids and proteins have been developed. Many efforts are being taken to build microfluidic chips capable of simultaneously detecting ultra-low concentrations of several biomolecules while consuming low sample volumes as low as a microliter to provide a highly reliable point-of-care diagnosis (Kaushal et al., 2022).

Although fluorescent Nps are very effective for passive sensing, the use of functional Nps for active sensing that responds to changes in intracellular biomolecule levels has proven to be more difficult. For instance, metal ion-based, carbon-based and ZnS/CdSe quantum dots have been extensively used to detect changes in intracellular pH. Nevertheless, research findings on these quantum dots have revealed that the quantum dots mentioned above are limited in detecting intracellular pH changes due to their poor ability to enter cells and intrinsic toxicity. In addition to quantum dots, organic and inorganic Nps have been studied for their ability to detect pH changes within cells. Due to the availability of non-toxic precursor materials and the ability to tailor the surface to be hydrophilic or hydrophobic, organic polymer Nps are more versatile for sensing intracellular pH changes than inorganic Nps.

In addition to fluorescent detection, plasmonic Nps are most often used to identify biological targets. In this instance, the change in surface plasmon resonance of the plasmon Nps after coupling to the biomolecule of interest is recognized (Kiba et al., 2021). This approach has been employed to identify changes in enzyme concentrations in cells and specific biomarkers associated with cancer. Surface-enhanced Raman scattering (SERS) is another effective method for detecting biomolecules that makes extensive use of Nps (Kiba et al., 2021). Although their ability to recognize specific biomolecules in complex biological samples is limited, this has been partially addressed by using chemoreceptors as mediators for the target molecules. Highly sensitive biosensors are built using the SERS technique to detect intracellular changes in pH, toxic ions, metabolites associated with disease conditions, microbial pathogens and drugs.

Figure 4. Schematic representation of principle of biosensor operation.

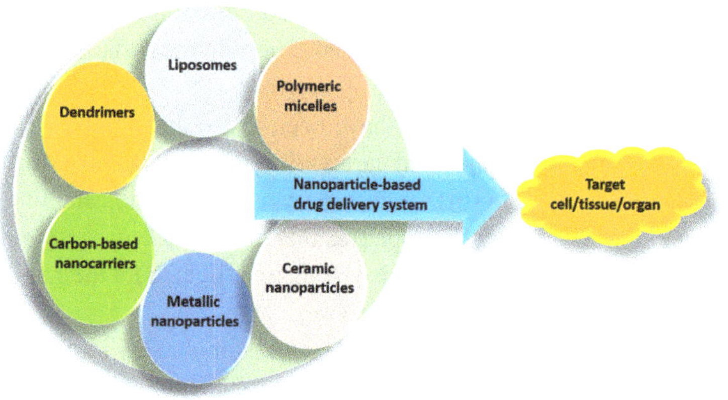

NANOPARTICLES FOR TISSUE ENGINEERING APPLICATIONS

Tissue engineering is a sub-discipline of biomedical engineering that integrates engineering and biology principles to aid in the restoration of organ function or the development of biological substitutes for damaged tissues (Godoy-Gallardo et al., 2021). TE is based on isolating cells derived from the tissues of living animals or people, seeding them onto a biocompatible synthetic scaffold that serves as an adhesion surface for the cells, and regenerating new tissue that resembles the structure and function of the patient's own tissue. In tissue engineering, nanotechnology is used to regenerate hepatic cells, heart valves, corneas, vascular cells, cartilage, bone tissue, and neural cells (Figure 5). Nanomaterials offer adequate mechanical strength and act as a scaffold for cells, allowing them to multiply and differentiate to repair damaged tissue. These biomaterials, which act as a substitute for the complex cell microenvironment (extracellular matrix), also regulate the differentiation of the seeded cells. Tissue-engineered scaffolds mimic the extracellular matrix by providing biological and physicochemical support in the proper regeneration of injured tissue in vivo rather than simply providing temporary mechanical support. Once damaged tissue has been repaired, these biocompatible scaffolds should degrade into harmless products at a rate that coincides with the regeneration of new tissue. The use of nanomaterials in tissue engineering, like nanopatterns and nanofibers, has paved the way for developing more biomimetic scaffolds (Jian et al., 2021). Since many biomolecules are found at the submicron and nanoscale, the biomimicking scaffolds provide the ideal internal microenvironment for cell proliferation. These scaffolds are fabricated with fibers to generate an interconnected porosity architecture using techniques like self-assembly and electrospinning, and the surface is tailored with peptides and other biomolecules that drive cell differentiation and proliferation. Nanomaterials derived from a natural source are now recommended because they are non-toxic and closely match the structure of extracellular matrices, where TE is performed by promoting cellular attachment and proliferation for tissue regeneration. Among the different types of nanomaterials utilized as scaffolds, natural polymers are the most commonly recognized due to their excellent characteristics, including biodegradability, biocompatibility, and high surface area-to-volume ratio. Furthermore, natural polymeric Nps are one of the best scaffold materials to overcome barriers in TERM due to their reduced cytotoxic effects, increased permeability, improved targeting ability, controlled release of drug substances, and sustained bioactivity of therapeutic drug concentrations in target cells. The key properties of polymeric Nps are their rate of degradation, solubility, mechanical strength, porosity, surface charge, shape, size, and formulation, which can be tailored for a variety of applications in TERM. Until now, polymeric Nps have been engineered to be susceptible to specific chemical and physical stimuli, including oxidizing agents, light, pH, enzymes, temperature, magnetic field, and reducing agents, aiding in the controlled drug release and targeted drug delivery.

Figure 5. Nanomaterials commonly employed for multifaceted tissue engineering applications.

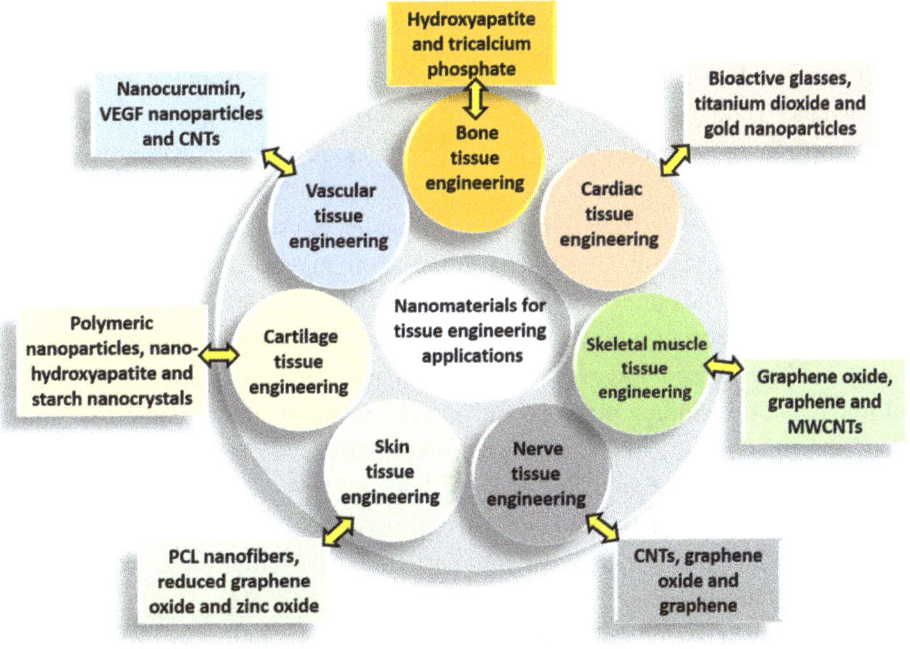

Nanomaterials have several benefits, including a large surface area for adsorption of proteins, as well as remarkable optical, magnetic, and mechanical characteristics that regulate cell adhesion and enable signals to integral membrane proteins for cell differentiation, proliferation, and tissue regeneration. Due to their intrinsic benefits, natural polymers have been widely used in tissue regeneration and regenerative medicine (Sunasee, 2021). Natural materials being investigated for TE include polysaccharides like chitosan, alginate, and hyaluronan, and proteins like fibronectin, gelatin, and collagen. On the other hand, natural polymers possess inferior mechanical characteristics by nature. Due to their origin, there are still difficulties with consistency in terms of physicochemical characteristics. In certain instances, biomaterials derived from nature may elicit an immune response in humans after implantation. Researchers are investigating whether the biomimetics of these naturally derived biomaterials can be employed as a substitute for TE to resolve the intrinsic disadvantages of natural polymers. Another appealing alternative for TE is the utilization of artificial polymers because their biodegradability, tensile strength, shape, size, and other characteristics can be engineered and modified to meet specific needs. The most commonly employed artificial polymers for TE are poly (ε-caprolactone), polyethylene glycol, poly (lactic-co-glycolic acid), polyglycolic acid, polylactic acid, and poly (a-hydroxy acids).

On the other hand, wound healing is a fairly complex process that involves numerous different components of the extracellular and cellular matrix over time and the involvement of distinct signalling pathways (Maroufi et al., 2020). The natural process of wound healing introduces new barriers to tissue engineering, although it cannot imitate the natural healing process, which requires extensive cellular interactions, as well as the ability of distinct cells to interact with the extracellular matrix. Blended nanocomposites integrating artificially and naturally derived polymers are being evaluated to resolve these concerns. Another major issue is the structural complexity caused by the complex three-dimensional architecture of regenerated tissues, which is being resolved using 3D printing techniques, enabling accurate and specific arrangement of various cell types within a delineated three-dimensional extracellular environment. Nevertheless, the strategies adopted in the regeneration process in TE are time-intensive to be useful for therapeutic interventions. Besides that, the approaches, including the utilization of fiber-assisted moulding and polymer brushes, are being investigated. Aside from the extracellular matrix and cell types, the wound healing process necessitates the formation and secretion of biomolecules like cytokines and growth factors, posing a new set of challenges. A wide range of advanced drug delivery systems is being used to resolve this issue. While the issues need to be addressed, they are being resolved as we gain a better quantitative understanding of the complex physiological process of wound healing.

Although different types of Nps are utilized in tissue engineering, metallic, ceramic and polymer Nps have been widely used and applied in human clinical trials. Metallic Nps like gold and silver can be produced and tailored by incorporating various moieties that allow for the conjugation of drugs, ligands, and antibodies. Gold Nps are a colloid of gold particles of nanometric size. Because of their high affinity for gold, multiple ligands, including proteins, antibodies, and polypeptide sequences can be conjugated on the gold surface with a variety of functional groups like thiols, amines, and phosphines (H. Li et al., 2020). A promising application of gold Nps in TE is that gold Nps enhance stem cell differentiation for bone tissue engineering. Furthermore, gold Nps can improve the cell-adhesive and mechanical characteristics of TE scaffolds. The nanometric size of gold particles also promotes the use of three-dimensional scaffolds to control cell behaviour.

Silver Nps are nanometer-sized colloidal particles with sizes ranging from 1 to 100 nm. They are among the most extensively utilized metallic Nps in TERM due to their antimicrobial activity. Since bacterial infection hampers clinical TE applications, the utilization of silver Nps as a protective measure against bacterial infections could be a viable solution. Silver Nps are typically formed using either chemical or physical approaches. Ultrasonic irradiation, gamma irradiation, laser ablation, and evaporation condensation process are the physical methods commonly used to produce silver Nps. On the other hand, chemical approaches employed to synthesize silver Nps involve the use of reducing agents like polyol and sodium borohydride to reduce silver ions into metallic silver. Over the last few decades, silver Nps have been widely employed for their antimicrobial activities against a wide variety of microbes. The antimicrobial properties of silver Nps demonstrate their ability to inhibit the respiratory chain system of microorganisms. The attributes of silver Nps have also been studied in TERM, primarily for wound healing applications. In particular, burn wounds and chronic wounds are more susceptible to bacterial infection. Furthermore, the increasing likelihood of microbes resistant to multiple drugs is a serious health concern, necessitating the inclusion of Nps with antimicrobial properties in the design of TE scaffolds for wound dressing applications. Silver Nps of controlled size have also been integrated into TE scaffolds made of various materials, including cellulose acetate, alginate-chitosan, gelatin, polyvinyl chloride, and polyvinyl alcohol. It has been shown that all of the scaffolds produced using the aforementioned materials have a high level of antimicrobial activity. It has been proved that all of the scaffolds produced using the materials mentioned above have a high level of antimicrobial activity. It is important to note that silver Nps can also have cytotoxic effects on cancer cells. These research findings illustrate the ability of silver Nps for TERM.

Ceramic Nps, which are primarily composed of inorganic solids such as phosphates, carbonates, carbides, and oxides other than metal sulfides, metal oxides, and metals, can be utilized to develop nanomaterials with varying porosity, size, and shape. Generally, ceramic Nps are categorized as magnetic Nps, resorbable or bioactive ceramics, and inert ceramics, depending on the tissue response. Calcium, oxygen, phosphorus, magnesium, potassium, sodium, and silicone are chemical elements that cells can absorb and can be used to develop bioactive glass ceramic Nps. Conventionally, gas-phase synthesis, sol-gel, laser spinning, and microemulsion are the most commonly used methods for constructing bioactive glass-ceramic Nps. Bioactive glass ceramic Nps have a higher surface area per volume than bulk bioactive glass ceramics, which results in better protein adsorption and bioactivity. Furthermore, glass-ceramic Nps with bimodal or unimodal pore size distributions have a higher bioactivity than bulk glass ceramics. Glass ceramic Nps have been recognized as promising candidates for dentin regeneration because of their exceptional bioactivity, angiogenic and antibacterial characteristics. Recent research results also show promising results for the applicability of bioactive glass-ceramic Nps in bone TE applications.

BIOMEDICAL TEXTILES

Biomedical nanotechnology has advanced considerably in recent years. A wide range of advanced techniques are being developed to improve health outcomes. Biomedical textiles are one of the promising candidates in the biomedical industry that several scientists are focussing on these days, and they are commonly employed to replace damaged organs or tissues using highly developed biomedical technology (Liu et al., 2021). Biomedical textiles are nanoengineered with advanced biocompatible materials to enhance the quality and rate of the natural healing process in the human body. Biomedical textiles include various subtypes, such as sutures & ligatures, vascular implants, hard tissue implants, tissue-engineered soft tissue implants, hernia repair, nerve guidance channels, artificial skin, heart patches, biomaterials in ophthalmology & dental, and smart biomedical fabrics (Figure 6). In particular, incorporating nanomaterials into textiles to develop biomedical implants and devices has attracted researchers and scientists due to their desired properties, such as sustainability, biocompatibility, high surface-area-to-volume-ratio, good stiffness, easy production, low cost, simplicity, and high accessibility to large molecules. Nanotextured fabrics can perform various functions, including tissue engineering, drug delivery, pressure and thermal therapy, and artificial implants. These functionalities can be achieved through the use of surface modification or functionalization processing techniques. Besides their functional properties, fiber selection is critical in the fabrication of biomedical textiles, and it is classified into two categories: commodity fiber and speciality fiber. Commodity fibers are high-performance fibers produced from materials like nylon, polyester or cotton. On the other hand, speciality fibers are more distinct and explicitly engineered for a wide variety of biomedical applications than commodity fibers. These fibers employed in the design of biomedical textiles must have non-carcinogenic, non-allergic, fire-resistant and antimicrobial properties. Conventionally, knitting, non-weaving and weaving techniques have been utilized to fabricate biomedical textiles for the development of extracorporeal devices, as well as implantable and non-implantable biomedical applications. Bandages like cohesive, adhesive and gauze are examples of non-implantable applications. On the other hand, implantable biomedical textiles produced from biodegradable fibers are commonly utilized in complex biological environments where implantable biomedical textiles are stable and biocompatible with biological fluids and cells. Surgical sutures are manufactured from non-biodegradable and biodegradable fibers and are employed in surgery to close wounds. Furthermore, biomedical textiles are being used as mechanical organs in extracorporeal devices to purify blood.

Figure 6. Transdermal drug delivery system based on smart textiles using hydrogel containing drug loaded temperature/pH sensitive polymers.

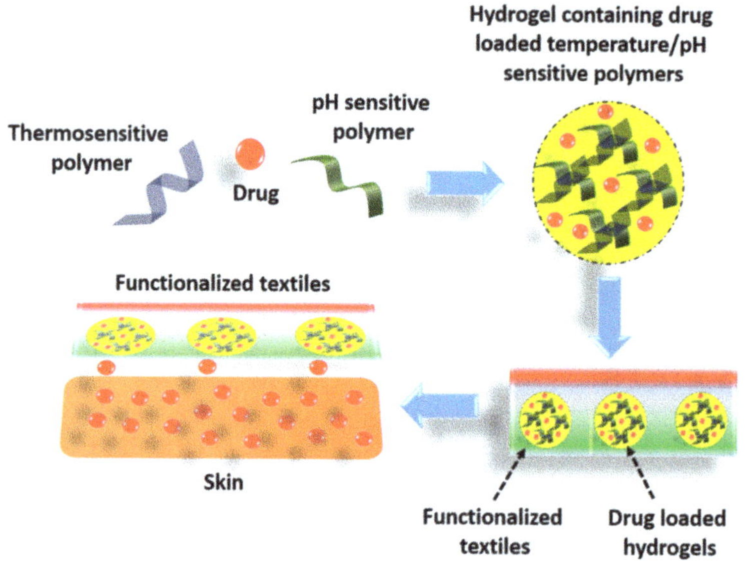

The high surface area and capability of functionalized textiles to preserve moisture allow microorganisms to grow, leading to a variety of adverse effects on the functionalized textile and the consumer. In recent years, extensive investigation into the toxic effects of pathogens on human health has been conducted to reduce microbial population growth on functionalized textiles due to public health awareness. As a result, various techniques have been explored to bring antimicrobial activity to functionalized textiles, which are primarily based on the incorporation of antimicrobial agents into functionalized woven fabrics. Antimicrobial agents of various kinds have been employed, including natural polymers, polybiguanides, metal salts, triclosan, and quaternary ammonium compounds (Table 1). Depending on the method employed, the woven fabric carrying antimicrobial agents can operate in two distinct ways: diffusion and/or contact. In the case of contact, the antimicrobial agent is applied to the woven fabric and does not disperse, so it will only respond if the pathogenic microorganism comes into contact with the surface of the woven fabric. On the other hand, in the case of diffusion, the microbial agent is applied on the surface of the woven fabric. As a result, the microbial agent will move from the woven fabric to the external environment to fight against pathogenic microorganisms. Apart from being effective against microorganisms, any antimicrobial treatment applied to a woven fabric must meet a number of requirements, the most important of which is the necessity for non-toxicity to the user, specifically sensitization, irritation or allergy, and cytotoxicity.

Table 1. List of recently reported antimicrobial coated textiles

Fabric Type	Antimicrobial Agent	Reference
Cotton	prodigiosin	(Metwally et al., 2021)
Cotton/polyester	CuO and ZnO nanoparticles	(Bengalli et al., 2021)
Cotton	Zein and ellagic acid	(Gonçalves et al., 2020)
Eri/eri, eri/ mulberry and eri/ cotton	Methanolic extracts of *Mikania micrantha* and *Drymaria cordata*	(Neog & Kola, 2021)
Cotton	Quaternary ammonium salt	(Z. Li et al., 2020)
Cotton	MgO nanoparticles	(Fouda et al., 2021)
Cotton	$Mg_{0.8-x}Zn_xFe_2O_4$	(Bessy et al., 2022)
Cotton, glace cotton, silk and rayon	bacterial red pigments of *R. psychrophilium* GL8	(Bisht et al., 2020)
Cotton	Carbon quantum dots	(Ahmed et al., 2021)
Cotton	Graphene	(Kumar et al., 2020)
Cotton	Extracts of *Acalypha indica, Vitex negundo* and *Datura metel*	(Jesica Roshima & Jayalakshmi, 2020)
Knitted cotton textile	PCN-244 and Ag	(Nie et al., 2021)
Cotton	P(DAM-co-DA)	(Xu et al., 2021)
Polyester	P(DMAMS-co-PFDA) and PDP	(Song et al., 2021)

NANOPARTICLE TOXICITY

Although Nps have shown promising results in biomedical applications, the understanding of their potential toxicity is limited. The toxicity of Nps is determined by various factors such as shape, crystal structure, roughness, surface charge, size, composition, and dose (He et al., 2021). On the other hand, the physical and chemical characteristics of nanomaterials are determined by their dimensions. Accordingly, the characteristic toxicity of the macro-sized material elicits a non-specific toxic response. The word "nanotoxicology" has been coined to broaden our understanding of the physical and chemical characteristics of Nps, as well as their potential toxicity. The majority of nanotoxicological analyses are focused on cell culture studies. Many studies have used biomimetic models of cell membranes to engineer specific properties like antifouling, controlled selectivity, and hierarchical organizations (He et al., 2021). Nevertheless, since *in vitro* results do not always accurately reflect *in vivo* effects, *in vivo* testing is required to assess how biological systems interact with nanomaterials. Despite various tests, the lack of consistency in methodologies has resulted in erroneous toxicological interpretations. For instance, the level of dose and the time of exposure are investigated in classic toxicology, however, the toxicological effects of nanomaterials do not support this hypothesis. Therefore, the use of appropriate dose metrics should be thoroughly examined. Several researchers have proposed that the total surface area needs to be considered as the dose to assess the influence of the size of the nanoparticle on its toxicity because the main phenomena take place on the exterior of the Nps rather than in the macro-sized particles, as in the case of routine toxicological assays (He et al., 2021). In this context, an extensive characterization of nanomaterials is essential to correlate toxic effects with the appropriate variables and attributes of nanomaterials and to enable comparison between previous results found in the literature.

Another notion that should be included in toxicity testing is the modification of nanomaterials due to interaction with the biological microenvironment. These modifications are critical to comprehending their biological behaviour. Before interacting with cells in biological media, nanomaterials come into contact with body fluids containing biomolecules and proteins. The nanomaterial is surrounded by a protein in biological media which tends to form a corona, thus causing a significant change in the surface characteristics of the Nps. The morphology, shape, size, surface chemistry, surface area, and surface charge of the Nps, which modulate surface properties and protein binding capabilities, determine the corona's composition. According to research findings, cells recognize this corona as the first layer. Hence, the physical and chemical properties of the protein corona associated with nanomaterials must be considered while prognosticating their biological fate (He et al., 2021). It is difficult to ascertain the true impact of nanomaterials in biological systems. Regardless of these obstacles, a comprehensive assessment of the toxic effects of nanomaterials on healthy cells is needed before exploring the potential applications of nanomaterials in TE and drug delivery. Nevertheless, the correlation between physio-chemical properties and toxicity must always be carefully evaluated to reduce negative results.

FUTURE RESEARCH DIRECTIONS

Since engineered Nps are being developed almost exclusively to address specific human health issues, nanomedicine has a promising future. Even though nanomedicine holds promise for improving human health, extensive *in vivo* and *in vitro* toxicological studies are needed to ensure its safety and effectiveness. The primary challenge in diagnostics is to establish reliable and cost-effective point-of-care diagnostic systems capable of detecting human diseases with enhanced selectivity and sensitivity. Until now, the majority of attention has been focused on cancer treatment and diagnosis; nevertheless, the biggest challenge is gaining in-depth information and developing cutting-edge technologies to resolve the microbial infectious diseases that are the leading causes of death. Ideally, diagnostic devices for infectious diseases that can recognize numerous species of pathogens must be developed. In addition, point-of-care diagnostic systems must be able to recognize various pathogenic microorganisms inflicting multiple kinds of communicable diseases that are highly prevalent within a specific group of people or in a specific geographical location. These point-of-care diagnostic systems must be inexpensive and portable with enhanced sensitivity for early diagnosis. In particular, these diagnostic platforms must be simple to use, and the output readings must be swift and easy to interpret for a class of moderate or low-skilled healthcare workers. Such early detection of human illnesses through the use of point-of-care diagnostic systems will enable their inclusion in healthcare programs to support a diverse group of people, particularly those who are economically disadvantaged, live-in deplorable conditions, and are most vulnerable to numerous pathogens. Following that, inexpensive, non-toxic harmless drugs and drug delivery systems must be developed. Even though the use of individual drugs is currently promoted, this may not be advantageous to a significant proportion of the target population. For instance, tumours are intrinsically mobile and heterogeneous in nature; therefore, an individual drug is sometimes unable to target certain tumour cells, allowing unaffected tumour cells to multiply and cause recurrence. In addition, long-term benefits can be achieved by using a variety of drugs that target tumour-causing agents, such as proteins or extracellular matrix components, or by deliberately targeting specific receptors on the surface of cancer cells. Similarly, most patients are infected with numerous pathogenic infections at the same time, making it difficult to obtain effective treatment. As far as TE is concerned, the major challenges in TE encompass (i) the delivery of a diverse range of drugs at various levels and in varying proportions to target cells, (ii) cell differentiation and growth in three-dimensional scaffolds to mimic the hierarchical cellular organization of native cells that is distinctive for specific organs and tissues, and (iii) the affordability of reliable biomaterials that can be employed as TE scaffolds.

CONCLUSION

In summary, nanobiotechnology has immense potential to offer a wide range of therapeutic applications and point-of-care diagnostic platforms capable of detecting multiple analytes, thus dramatically transforming the public health scenario. Significant efforts are being made to translate these technological innovations into clinical practice, which should lead to a new paradigm in biomedicine.

ACKNOWLEDGMENT

Authors are grateful to SASTRA Deemed University, India for providing infrastructural support.

REFERENCES

Ahmed, H. B., Abualnaja, K. M., Ghareeb, R. Y., Ibrahim, A. A., Abdelsalam, N. R., & Emam, H. E. (2021). Technical textiles modified with immobilized carbon dots synthesized with infrared assistance. *Journal of Colloid and Interface Science*, *604*, 15–29. doi:10.1016/j.jcis.2021.07.014 PMID:34261016

Bartneck, M. (2021). Lipid nanoparticle formulations for targeting leukocytes with therapeutic RNA in liver fibrosis. *Advanced Drug Delivery Reviews*, *173*, 70–88. doi:10.1016/j.addr.2021.03.009 PMID:33774114

Bartusik-Aebisher, D., Chrzanowski, G., Bober, Z., & Aebisher, D. (2021). An analytical study of Trastuzumab-dendrimer-fluorine drug delivery system in breast cancer therapy in vitro. *Biomedicine and Pharmacotherapy*, *133*, 111053. doi:10.1016/j.biopha.2020.111053 PMID:33378959

Bengalli, R., Colantuoni, A., Perelshtein, I., Gedanken, A., Collini, M., Mantecca, P., & Fiandra, L. (2021). In vitro skin toxicity of CuO and ZnO nanoparticles: Application in the safety assessment of antimicrobial coated textiles. *NanoImpact*, *21*, 100282. doi:10.1016/j.impact.2020.100282

Bessy, T. C., Bindhu, M. R., Johnson, J., Chen, S.-M., Chen, T.-W., & Almaary, K. S. (2022). UV light assisted photocatalytic degradation of textile waste water by Mg0.8-xZnxFe2O4 synthesized by combustion method and in-vitro antimicrobial activities. *Environmental Research, 204*, 111917. doi:10.1016/j.envres.2021.111917 PMID:34453899

Bisht, G., Srivastava, S., Kulshreshtha, R., Sourirajan, A., Baumler, D. J., & Dev, K. (2020). Applications of red pigments from psychrophilic Rhodonellum psychrophilum GL8 in health, food and antimicrobial finishes on textiles. *Process Biochemistry, 94*, 15–29. doi:10.1016/j.procbio.2020.03.021

Chadar, R., Afzal, O., Alqahtani, S. M., & Kesharwani, P. (2021). Carbon nanotubes as an emerging nanocarrier for the delivery of doxorubicin for improved chemotherapy. *Colloids and Surfaces. B, Biointerfaces, 208*, 112044. doi:10.1016/j.colsurfb.2021.112044 PMID:34419810

Cheng, H., Jiang, Z., Sun, C., Wang, Z., Han, G., Chen, X., Li, T., Fan, Z., Zhang, F., Yang, X., Lv, L., Zhang, H., Zhou, J., & Ding, Y. (2022). Protein stabilized polymeric nanoparticles inspired relay drug delivery for tackling post-chemotherapeutic metastasis. *Chemical Engineering Journal, 427*, 131672. doi:10.1016/j.cej.2021.131672

Fouda, A., Hassan, S. E.-D., Saied, E., & Hamza, M. F. (2021). Photocatalytic degradation of real textile and tannery effluent using biosynthesized magnesium oxide nanoparticles (MgO-NPs), heavy metal adsorption, phytotoxicity, and antimicrobial activity. *Journal of Environmental Chemical Engineering, 9*(4), 105346. doi:10.1016/j.jece.2021.105346

Ghalehkhondabi, V., Soleymani, M., & Fazlali, A. (2021). Folate-targeted nanomicelles containing silibinin as an active drug delivery system for liver cancer therapy. *Journal of Drug Delivery Science and Technology, 61*, 102157. doi:10.1016/j.jddst.2020.102157

Godoy-Gallardo, M., Eckhard, U., Delgado, L. M., de Roo Puente, Y. J. D., Hoyos-Nogués, M., Gil, F. J., & Perez, R. A. (2021). Antibacterial approaches in tissue engineering using metal ions and nanoparticles: From mechanisms to applications. *Bioactive Materials, 6*(12), 4470–4490. doi:10.1016/j.bioactmat.2021.04.033 PMID:34027235

Gonçalves, J., Torres, N., Silva, S., Gonçalves, F., Noro, J., Cavaco-Paulo, A., Ribeiro, A., & Silva, C. (2020). Zein impart hydrophobic and antimicrobial properties to cotton textiles. *Reactive & Functional Polymers, 154*, 104664. doi:10.1016/j.reactfunctpolym.2020.104664

Guimarães, D., Cavaco-Paulo, A., & Nogueira, E. (2021). Design of liposomes as drug delivery system for therapeutic applications. *International Journal of Pharmaceutics, 601*, 120571. doi:10.1016/j.ijpharm.2021.120571 PMID:33812967

He, S., Leanse, L. G., & Feng, Y. (2021). Artificial intelligence and machine learning assisted drug delivery for effective treatment of infectious diseases. *Advanced Drug Delivery Reviews, 113922*, 113922. doi:10.1016/j.addr.2021.113922 PMID:34461198

İçhedef, Ç., Teksöz, S., Çetin, O., Aydın, B., Sarıkavak, İ., Parlak, Y., & Bilgin, B. E. (2021). Design of 99mTc radiolabeled gemcitabine polymeric nanoparticles as drug delivery system and in vivo evaluation. *Materials Chemistry and Physics, 263*, 124380. doi:10.1016/j.matchemphys.2021.124380

Jesica Roshima, A., & Jayalakshmi, I. (2020). Effect of herbal microencapsulated antimicrobial and antilarvicidal finishes on selected natural fabrics. *Materials Today: Proceedings.*

Jian, Z., Zhuang, T., Qinyu, T., Liqing, P., Kun, L., Xujiang, L., Diaodiao, W., Zhen, Y., Shuangpeng, J., Xiang, S., Jingxiang, H., Shuyun, L., Libo, H., Peifu, T., Qi, Y., & Quanyi, G. (2021). 3D bioprinting of a biomimetic meniscal scaffold for application in tissue engineering. *Bioactive Materials, 6*(6), 1711–1726. doi:10.1016/j.bioactmat.2020.11.027 PMID:33313450

Jiang, K., Zhang, L., & Bao, G. (2021). Magnetic Iron Oxide Nanoparticles for Biomedical Applications. *Current Opinion in Biomedical Engineering, 100330.* doi:10.1016/j.cobme.2021.100330

Kaushal, A., Seth, A., Kala, D., Gupta, S., Krishnia, L., & Verma, V. (2022). *Miniaturized devices for point-of-care testing/miniaturization and integration with microfluidic systems.* Academic Press.

Kiba, T., Masui, K., Inomata, Y., Furumoto, A., Kawamura, M., Abe, Y., & Kim, K. H. (2021). Control of localized surface plasmon resonance of Ag nanoparticles by changing its size and morphology. *Vacuum, 192*, 110432. doi:10.1016/j.vacuum.2021.110432

Kooijmans, S. A. A., de Jong, O. G., & Schiffelers, R. M. (2021). Exploring interactions between extracellular vesicles and cells for innovative drug delivery system design. *Advanced Drug Delivery Reviews, 173*, 252–278. doi:10.1016/j.addr.2021.03.017 PMID:33798644

Kumar, A., Sharma, K., & Dixit, A. R. (2020). Role of graphene in biosensor and protective textile against viruses. *Medical Hypotheses, 144*, 110253. doi:10.1016/j.mehy.2020.110253 PMID:33254558

Li, H., Pan, S., Xia, P., Chang, Y., Fu, C., Kong, W., Yu, Z., Wang, K., Yang, X., & Qi, Z. (2020). Advances in the application of gold nanoparticles in bone tissue engineering. *Journal of Biological Engineering, 14*(1), 1–15. doi:10.118613036-020-00236-3 PMID:32391080

Li, Z., Cheng, J., Yang, X., Liu, H., Xu, X., Ma, L., Shang, S., & Song, Z. (2020). Construction of antimicrobial and biocompatible cotton textile based on quaternary ammonium salt from rosin acid. *International Journal of Biological Macromolecules, 150*, 1–8. doi:10.1016/j.ijbiomac.2020.01.259 PMID:32007552

Liu, J., Zhai, H., Sun, Y., Wu, S., & Chen, S. (2021). Developing high strength poly(L-lactic acid) nanofiber yarns for biomedical textile materials: A comparative study of novel nanofiber yarns and traditional microfiber yarns. *Materials Letters, 300*, 130229. doi:10.1016/j.matlet.2021.130229

Long, W., Ouyang, H., Zhou, C., Wan, W., Yu, S., Qian, K., Liu, M., Zhang, X., Feng, Y., & Wei, Y. (2021). Simultaneous surface functionalization and drug loading: A novel method for fabrication of cellulose nanocrystals-based pH responsive drug delivery system. *International Journal of Biological Macromolecules, 182*, 2066–2075. doi:10.1016/j.ijbiomac.2021.05.193 PMID:34087297

Maji, I., Mahajan, S., Sriram, A., Medtiya, P., Vasave, R., Khatri, D. K., Kumar, R., Singh, S. B., Madan, J., & Singh, P. K. (2021). Solid self emulsifying drug delivery system: Superior mode for oral delivery of hydrophobic cargos. *Journal of Controlled Release, 337*, 646–660. doi:10.1016/j.jconrel.2021.08.013 PMID:34384795

Maroufi, N. F., Ashouri, N., Mortezania, Z., Ashoori, Z., Vahedian, V., Amirzadeh-Iranaq, M. T., Fattahi, A., Kazemzadeh, H., Bizzarri, M., Akbarzadeh, M., Nejabati, H. R., Faridvand, Y., Rashidi, M.-R., & Nouri, M. (2020). The potential therapeutic effects of melatonin on breast cancer: An invasion and metastasis inhibitor. *Pathology, Research and Practice*, *216*(10), 153226. doi:10.1016/j.prp.2020.153226 PMID:32987338

Mei, H., Cai, S., Huang, D., Gao, H., Cao, J., & He, B. (2022). Carrier-free nanodrugs with efficient drug delivery and release for cancer therapy: From intrinsic physicochemical properties to external modification. *Bioactive Materials*, *8*, 220–240. doi:10.1016/j.bioactmat.2021.06.035 PMID:34541398

Metwally, R. A., El Sikaily, A., El-Sersy, N. A., Ghozlan, H. A., & Sabry, S. A. (2021). Antimicrobial activity of textile fabrics dyed with prodigiosin pigment extracted from marine Serratia rubidaea RAM_Alex bacteria. *The Egyptian Journal of Aquatic Research*.

Moradi Kashkooli, F., Soltani, M., & Souri, M. (2020). Controlled anti-cancer drug release through advanced nano-drug delivery systems: Static and dynamic targeting strategies. *Journal of Controlled Release*, *327*, 316–349. doi:10.1016/j.jconrel.2020.08.012 PMID:32800878

Neog, R., & Kola, N. (2021). In vitro antimicrobial screening of methanolic extract of Mikania micrantha and Drymaria cordata against S. Aureus and its textile application. *Materials Today: Proceedings*, *42*, 916–920. doi:10.1016/j.matpr.2020.11.835

Nie, X., Wu, S., Liao, S., Chen, J., Huang, F., Li, W., Wang, Q., & Wei, Q. (2021). Light-driven self-disinfecting textiles functionalized by PCN-224 and Ag nanoparticles. *Journal of Hazardous Materials*, *416*, 125786. doi:10.1016/j.jhazmat.2021.125786 PMID:33873032

Rahman, A., Kang, S., Wang, W., Garg, A., Maile-Moskowitz, A., & Vikesland, P. J. (2021). Nanobiotechnology enabled approaches for wastewater based epidemiology. *Trends in Analytical Chemistry*, *143*, 116400. doi:10.1016/j.trac.2021.116400 PMID:34334850

Rezaeisadat, M., Bordbar, A.-K., & Omidyan, R. (2021). Molecular dynamics simulation study of curcumin interaction with nano-micelle of PNIPAAm-b-PEG co-polymer as a smart efficient drug delivery system. *Journal of Molecular Liquids*, *332*, 115862. doi:10.1016/j.molliq.2021.115862

Seitz, P., Krieghoff, C., Gottschling, S., Lücke, C., Abdel-Wahab, M., Holzhey, D., Gohmann, R. F., & Gutberlet, M. (2021). Pre-procedural high-pitch coronary CT angiography assessment of patients undergoing transcatheter aortic valve implantation (TAVI) without patient-specific adjustment: analysis of diagnostic performance. *Clinical Radiology*.

Sharifianjazi, F., Irani, M., Esmaeilkhanian, A., Bazli, L., Asl, M. S., Jang, H. W., Kim, S. Y., Ramakrishna, S., Shokouhimehr, M., & Varma, R. S. (2021). Polymer incorporated magnetic nanoparticles: Applications for magnetoresponsive targeted drug delivery. *Materials Science and Engineering B*, *272*, 115358. doi:10.1016/j.mseb.2021.115358

Song, Q., Zhao, R., Liu, T., Gao, L., Su, C., Ye, Y., Chan, S. Y., Liu, X., Wang, K., Li, P., & Huang, W. (2021). One-step vapor deposition of fluorinated polycationic coating to fabricate antifouling and anti-infective textile against drug-resistant bacteria and viruses. *Chemical Engineering Journal*, *418*, 129368. doi:10.1016/j.cej.2021.129368 PMID:33746567

Sunasee, R. (2021). *4.20 - Nanocellulose: Preparation, Functionalization and Applications*. Elsevier.

Urimi, D., Widenbring, R., Pérez García, R. O., Gedda, L., Edwards, K., Loftsson, T., & Schipper, N. (2021). Formulation development and upscaling of lipid nanocapsules as a drug delivery system for a novel cyclic GMP analogue intended for retinal drug delivery. *International Journal of Pharmaceutics*, *602*, 120640. doi:10.1016/j.ijpharm.2021.120640 PMID:33901599

Xie, P., & Liu, P. (2020). pH-responsive surface charge reversal carboxymethyl chitosan-based drug delivery system for pH and reduction dual-responsive triggered DOX release. *Carbohydrate Polymers*, *236*, 116093. doi:10.1016/j.carbpol.2020.116093 PMID:32172895

Xu, J., Xie, Z., Du, F., & Wang, X. (2021). One-step anti-superbug finishing of cotton textiles with dopamine-menthol. *Journal of Materials Science and Technology*, *69*, 79–88. doi:10.1016/j.jmst.2020.08.007

Yadav, K. S., & Dalal, D. C. (2021). The heterogeneous multiscale method to study particle size and partitioning effects in drug delivery. *Computers & Mathematics with Applications (Oxford, England)*, *92*, 134–148. doi:10.1016/j.camwa.2021.03.025

Yu, M., Yuan, W., Li, D., Schwendeman, A., & Schwendeman, S. P. (2019). Predicting drug release kinetics from nanocarriers inside dialysis bags. *Journal of Controlled Release*, *315*, 23–30. doi:10.1016/j.jconrel.2019.09.016 PMID:31629038

Zou, T., Lu, W., Mezhuev, Y., Lan, M., Li, L., Liu, F., Cai, T., Wu, X., & Cai, Y. (2021). A review of nanoparticle drug delivery systems responsive to endogenous breast cancer microenvironment. *European Journal of Pharmaceutics and Biopharmaceutics*, *166*, 30–43. doi:10.1016/j.ejpb.2021.05.029 PMID:34098073

KEY TERMS AND DEFINITIONS

Biosensor: A biosensor is an analytical instrument that integrates a biological component with a physicochemical detector to detect a chemical molecule.

Dendrimers: It is a synthetic polymer with a chain structure that branches frequently, producing spherical macromolecules.

Drug Delivery: The technique or process by which a pharmaceutical substance is administered to produce a therapeutic effect in people or animals is referred to as drug delivery.

Nanocomposites: A nanocomposite is a multiphase solid substance in which at least one phase has 1D, 2D or 3D of less than 100 nm, or structures with nanoscale repeat intervals between the material's constituent phases.

Nanoparticles: A nanoparticle or ultrafine particle is often described as a particle with a diameter of between one and one hundred nanometres.

Textiles: A textile is a flexible material created by interlacing a network of yarns or threads spun from raw fibres into long and twisted lengths.

Tissue Engineering: Tissue engineering is a branch of bioengineering that combines cells, engineering, materials science, and appropriate biochemical and physicochemical aspects in order to repair, maintain, enhance, or replace many kinds of biological tissues.

Chapter 14
Applications of Metal and Metal Oxide-Based Nanomaterials in Medical and Biological Activities

Ashajyothi C.
https://orcid.org/0000-0002-6927-681X
Vijayanagara Sri Krishnadevaraya University, India

Harish K. Handral
*Bioprocessing Technology Institute, A*STAR, Singapore*

Prabhurajeshwar C.
Davangere University, India

ABSTRACT

Nanomaterials have been attracting the attention of many researchers because of their size, high stability, affinity, and selectivity nature. Over the past decades, considerable intensive studies on many metal and metal-oxide nanomaterials have drawn consideration through their significant properties like size, shape, surface mass proportion, and their reactivity; all these properties are fundamental cornerstones for the turn of events and use of nanomaterials and nanoscale gadgets in biomedical areas. There is also a vast scope for a broad range of biofunctional applications such as antibacterial, antiviral, antifungal, antitumor, bioimaging, tissue engineering, biosensors, gene, and drug delivery. The authors review the nature, forms, and synthesis of nanomaterials here, with a thorough biological synthesis assessment. They also address the development of nanoparticles by microorganisms in depth, and this chapter also includes updates on different biological and biomedical applications of these bionanomaterials.

INTRODUCTION

Nanomaterials have been receiving more attention in biological applications due to their nano size (1–100 nm) and unique physicochemical properties. The properties of nanomaterials are size-dependent, and

DOI: 10.4018/978-1-7998-8936-6.ch014

they have a high surface area with enhanced electrical, mechanical, optical, magnetic properties and thermal stability. These upgraded properties have empowered nanomaterials to be utilized in various fields including electrical, attractive, optical, and electronic gadgets (Khoshnevisan et al., 2019). Ongoing advancements have been brought about by the improvement in the demonstrating and planning of various clinical and organic devices and applications (Rathore et al., 2019). Besides, some metal and metal-oxide nanomaterials are designed and utilized for insightful purposes in drugs, demonstrative packs, imaging, attractive reverberation imaging (MRI), drug conveyance, and in a lot more clinical gear and methodology (da Rocha et al., 2019; do Amaral et al., 2019; Ferreira et al., 2019). The most commonly used nanomaterials are classified based on their size as coarse, fine, and nanomaterials. Materials ranging from 1 nm to100 nm are categorized as Nanoscale materials, or Nanoparticles. Compare to nanoscale materials, micro/ macro or bulk-sized materials exhibit very peculiar characteristics. Based on chemical nature is another type of classification and are classified into metal and metal-oxide based nanomaterials. The key group of nanomaterials in their purest form of metal-based nanoparticles, also known as metal nanoparticles (e.g., silver, gold, copper, zinc, iron, titanium, platinum, and magnesium nanoparticles). The other forms of nanomaterials are metal oxide nanoparticles, i.e., silver oxide, copper oxide, zinc oxide, titanium dioxide, and more (Yaqoob et al., 2020; Umar et al., 2013). Majorly, some nanomaterials that have potent applications in various fields of medical sciences are commonly occurring in some basic forms. In contrast to metal and metal oxide nanomaterials, metal-organic framework (MOF) and metal sulfide (MS) nanomaterials are also has been creating a center of attention because of such effective features and implementation in the biomedical fields such as antimicrobial and drug delivery.

A few of the nanomaterials will fabricate and characterize to use for various biomedical applications such as gene and drug delivery, as nanosensors, in tissue engineering, bone regeneration, and bio-imaging (Shivaramakrishnan et al., 2017). The application of the different forms of nanomaterials in medicine or as medicine is termed 'nanomedicine'. These nanomedicines emerging as a potential antimicrobial agent to treat microbial diseases with advanced approaches in disease control, prevention, diagnosis, and treatment (Nikalje, 2015, pp. 081-089). Among nanoscale systems, the use of metal and metal oxide nanoparticles has a large influence on disease sensing, imaging, analysis, treatment, and monitoring (Mahmoudi et al., 2011). This chapter describes the all prominent medical and biological applications of various metal and metal oxide bionanomaterials.

BACKGROUND

Bio-imaging, targeted drug deliveries, computed tomography's, and magnetic hyperthermia are important biomedical applications that use to treat cancer. In chemotherapy transport of anticancer drugs to the tumor depends on the circulatory system. Results were not promising from this treatment, for example, non-explicitness and poisonousness of the medication, whereby the medications will likewise affect the healthy cells and organs. Consequently, directed medication conveyance is being created as one safe option in contrast to chemotherapy treatment. The point of focused medication conveyance is to guide the medication to the particular territory where the tumor is found and consequently expand the measure of medication conveyed at the tumor site and decrease the results. Anticancer medications could be delivered to the tumor's precise location, avoiding exposure to healthy cells in the immediate area. Presently different metal and metal oxide nanomaterials are being utilized to convey anticancer medications to target explicit territories (Kango et al., 2013).

In hyperthermia, strategy high temperature is applied to obliterate carcinogenic cells. Whereby temperature of the affected carcinogenic region is increased up to 46 °C to damage the destructive cells without harming the unaffected cells. Carcinogenic cells have a more noteworthy affectability to temperature than solid cells. At the point when the malignant cells are warmed to 41–46 °C the cell apoptosis will happen and this is named hyperthermic impact. Thermoablation induces necrosis as the cells are heated to temperatures above 48°C. Combinations in both radiotherapy and chemotherapy are used to treat hyperthermia conditions. Among various types of hyperthermia treatments, heat is applied to a small region via a local hyperthermia approach, using methods like radio frequency, microwave and ultrasound. Regional hyperthermia is for the most part used to treat enormous tissue regions. Magnetic nanoparticles could be used to target the tumor in local hyperthermia treatment using four different infusion routes: direct, in situ implant, arterial and aggressive targeting. Arterial injection involves administering a solvent containing the enticing nanoparticles into the tumor's bloodstream. The direct injection includes straightforwardly infusing the liquid containing the attractive nanoparticles into the tumor this technique is the most normally utilized. Transplantable sol-gel formulations such as hydrogels and organogels are primarily used for in situ implantation to encapsulate magnetic nanoparticles into tumor MRI and computed tomography use iron oxide nanoparticles to replace gadolinium chelates. (Chen et al., 2016; López-Ortega et al., 2015; Kango et al., 2013; Reddy et al., 2012; Hua et al., 2012). In addition, (McNamara & Tofail, 2017). Furthermore, these iron oxide nanoparticles are being explored for biomedical applications.

Bacteria with antibiotic resistance are one of the key causes of antimicrobial agents' lack of efficacy. Bacterial resistance is induced by shifts in microorganisms' ability to resist antibacterial agents, either by inactivating them or by growing their therapeutic effectiveness. Over time, in microorganisms, these resistances emerge spontaneously due to genetic alteration. Such modifications are considerably favored by the improper usage and misuse of antibiotics (Sánchez-López et al., 2020). This leads to prolonged periods of illness, higher mortality rates, and an additional economic strain on health care systems. Antibiotic-resistant infections have become a significant human health threat nowadays. Among them, various severe diseases and infections are caused by pathogenic microbes such as *Streptococcus, Staphylococcus, Enterococcus,* and *Geobacter* (Olishevska et al., 2019). As per the World Health Organization (WHO), metal-based nanoparticles with a smaller size and specificity are more effective at targeting diseases. (Sánchez-López et al., 2020). Metallic nanoparticles and their derivatives have recently drawn a lot of interest due to their potent antibacterial and antifungal properties. Various metal oxides and metal-doped composites, on the other hand, have inert characteristic features such as spectral, mechanical stability, and excellent antimicrobial profile (Dizaj et al., 2014; Noronha et al., 2017; Zazo et al., 2016).

Participation of pharmaceutical companies involved in the manufacture of non-traditional medicinal products, in particular the nanotechnology industries, which have invested in the production of novel nanomaterials known as potent bacteria resistant inhibitors to traditional antibiotics, has been intensified by several new incentives (Sánchez-López et al., 2020).

METAL-BASED NANOMATERIALS

Metallic nanomaterials have gained the attention of scientists to carry out novel approaches in a long-term environment to synthesize, classify and even figure out their influential applications (Table 1). And such nanomaterials offer a promising solution against resistant pathogens and medical issues (Sánchez-López

Table 1. Metal-based nanomaterials with their potential applications

Sl No.	Nanomaterial	Target Sites (Microbe/Cell/Tissue/Organ)	Applications	Reference
01	AgNPs	Skin	Skin penetration test	(Crosera et al., 2015)
02	AgNPs	Microbe	Antimicrobial and Toxicity assay	(Sabella et al., 2011; Samberg et al., 2012)
03	AgNPs capped with starch	Microbe	Antimicrobial and Toxicity assay	(Wang et al., 2019)
04	AgNPs linked with polyvinylpyrrolidone	Brain	Therapeutic evaluation (Brain cancer)	(Locatelli et al., 2014)
05	AgNPs	Skin	Therapeutic evaluation (Wound healing)	(Austin et al., 2014)
06	Au/AgNPs	Cancer cell	Cancer cell Imaging therapy, photothermal therapy	(Shi et al., 2014)
07	AuNPs	Cancer cell	Cancer cell Radiosensitizer	(Cho et al., 2009; Jain et al., 2012)
08	AuNPs	Cancer cell: glioblastoma-based multiforme	Radiosensitizer applications	(Joh et al., 2013; Sun et al., 2013)
09	Gold-branched shell nanostructures	Breast cancer cell	Breast cancer Imaging therapy, photothermal therapy, chemotherapeutic therapy	(Topete et al., 2014)
10	AuNPs	Intravascular tissue	Therapeutic agents	(Giasuddin et al., 2012)
11	PtNPs	Cancer cell	Toxicity evaluation	(Yamada et al., 2015; Egorova, & Ananikov, 2016)
12	PtNPs	Brain	Toxicity evaluation	(Baptista et al., 2018)
13	PtNPs	Cancer cell	Therapeutic evaluation	(Yang et al., 2020)

et al., 2020). Along with industrial fields, the field of biomedical engineering is currently putting a lot of emphasis on the use of metallic nanomaterials (Edmundson et al., 2014; Wang et al., 2017). Noble metal nanoparticles with biocompatible properties could be combined with biopolymers to target intercellular and cellular signals (Fan et al., 2018).

Silver Nanomaterials (AgNMs)

Silver has been a significant metal used in numerous medicinal procedures since ancient times. One of the most commonly used nanomaterials in the world is these nano sized (AgNMs). Because of their larger surface area, almost all spherical shaped AgNPs are used. AgNPs can coordinate with a wide range of ligands for several fields such as, optical, electrical, and thermal applications (Shivaramakrishnan et al., 2017). Groundbreaking optical properties of AgNPs are currently being integrated into products ranging from photovoltaic to biological sensors (Jamkhande et al., 2019) and they are also working in molecular and photonic diagnostics (Haider & Kang, 2015). AgNPs offer a variety of positive features such as an easy and direct method of synthesis, different morphology, high surface area, inter and intracellular delivery system, a wide plasmon resonance area that recommends them for ideal delivery systems for biosensor, catalyst, or photo-controlled delivery systems. AgNPs are known for potential optimal gene delivery tools in bioengineering applications. AgNPs have been the subject of extensive studies due to their ease of handling and storage in biological uses (Natsuki et al., 2015). Prominent medicinal uses of AgNPs are identified as antibacterial activity (Yadav et al., 2017; Lalitha et al., 2013), anticancer therapy (Jeyaraj et al., 2013; Ong et al., 2015), anti-microbial activity (Prabhu & Poulose, 2012; Durán et al., 2016; Mittal et al., 2013; Rudramurthy et al., 2016; Martins et al., 2014; Hajipour et al., 2012; Patil et al., 2015; Le et al., 2019), anti-fungal therapy (Taware et al., 2020; Tran & Le, 2013), antiviral activities (Kango et al., 2013), wound healing (Galdiero et al., 2014), wound dressing (Chernousova & Epple, 2013), implanted material (Wilkinson et al., 2011), tissue engineering, medical devices, diagnostic applications in biosensing (Rafique et al., 2017), anti-permeability test, and dental preparations (Li et al., 2014).

Gold Nanomaterials (AuNMs)

Gold nanomaterials (AuNMs) are colloidal or clustered particles composed of a gold core are known as gold nanoparticles (AuNPs), these are inert and biocompatible nanomaterials (Sánchez-López et al., 2020). AuNPs are brown in color and spherical in shape with high surface area and possess distinctive optical and physical properties. AuNPs are commonly synthesized and stored in colloidal suspension form. In water gold colloidal suspension shows intense red color. The localized plasmonic resonance on their surface is responsible for this color. At its surface plasmon resonance frequency, AuNP's most dominant property is intense absorption potential and scattering of incident radiation. Due to surface chemistry and vibrant physicochemical properties, these particles can be used in diagnostic tools and functionalization of ligands in order to bind particular biomarkers as well as for anti-microbial screening assays (Shivaramakrishnan et al., 2017) (Jamdagni et al., 2018; Syafiuddin et al., 2017; Lima et al., 2013), biocatalysts and biomarkers (Netto et al., 2013), targeted drug deliver (Elahi et al., 2018; Dreaden et al., 2012; Shivaramakrishnan et al., 2017), anticancer therapy (Cheng et al., 2014; Dykman & Khlebtsov, 2012; Goel et al., 2017), contrast agent in medical imaging (Chen et al., 2016; Cheng et al., 2014), antiviral treatments (Mishra et al., 2014; Vijayakumar & Ganesan, 2012), and in hyperthermic effects to treat tumors (Hainfeld et al., 2014; Verma et al., 2014).

Copper Nanomaterials (CuNMs)

Copper, a promising metal for the synthesis of metallic nanomaterials. CuNMs show electrically conductive, higher melting point, and heat resistant profiles additionally production cost is affordable (Durán et al., 2016; Shivaramakrishnan et al., 2017). CuNPs, nanodots, and nanopowder are the most common forms of CuNMs. They are black-brown with spherical shape and have a higher surface area. CuNPs are commercially available in sizes from 10 to 100 nm, with surface areas ranging from 5 to 70 m^2/g. Other uncommon forms include Copper nanorods, nanowhiskers, nanohorns, nanopyramids, and other nanocomposites. In the biological and medicinal applications of CuNPs, antibacterial and anti-fungal functions are primarily found (Y. Abboud et al., 2014; Shivaramakrishnan et al., 2017), medical diagnosis (Goel et al., 2014), molecular imaging (Subhankari & Nayak 2013), imaging & therapy of cancer cells

(Ku et al., 2012), photothermal ablation of tumor cells (Zhou et al., 2016), as catalysts and theranostic applications (Zhang et al., 2015).

Gadolinium Nanoparticles (GdNPs)

Gadolinium, present in minerals such as monazite and bastnasite, is a Block F, Period 6 element. It is a rare-earth metal, malleable and ductile. These particles have a smooth, glossy, spherical morphology, and they look like black powder. Gadolinium nanoparticles are categorized as poisonous and emit flammable gases which can combust spontaneously when in contact with water. The most popular biomedical applications of GdNPs are in Magnetic Resonance Imaging (MRI) to target tumors in the hopes of detecting early-stage cancer, especially breast cancer. They have recently been found to be effective radiosensitizers under many forms of irradiations, such as radiotherapy and neutron therapy, in addition to their imaging properties. (Tamanoi et al., 2020).

Platinum Nanoparticles (PtNPs)

Platinum nanoparticles (PtNPs) are biocompatible, electrically conductive, radiopacity, and durable (Gurunathan et al., 2019) gained scientific interest in various biomedical uses. Because PtNPs have distinct optical features that can be improved by varying their sizes and shapes and also these PtNPs play a vital role in drug delivery applications, particularly in cancer research (Johnstone et al., 2016). Today, PtNPs are also recommended as sensors for the identification of biologically significant molecules. PtNPs provide perspective for the application towards addressing oxidative stress-related diseases (Jeyaraj et al., 2019). Another study also reveals that immobilized form of platinum films can be used for potential catalytic activity (Inbasekaran et al., 2014). PtNPs have a variety of medicinal applications, including anti-tumor therapy and contrast agents in medical imaging (Johnstone et al., 2016), antimicrobial activity (Jeyaraj et al., 2019), anti-bacterial activities (Deyhimihaghighi et al., 2014), biosensors and intracellular analysis (Yusof & Ismail 2015) and as biomarkers (Raghuwanshi et al., 2019; Manikandan et al., 2013).

Palladium Nanomaterials (PdNMs)

Palladium (Pd) has a face-centered cubic crystalline form and is a lustrous silver-white metal. Palladium is known to have low toxicity and slow adsorption when swallowed. It contains a number of compounds and other complex salts. Palladium nanostructures exhibit outstanding catalytic and physicochemical characteristics. Palladium nanorods are elongated materials with size ranging from a diameter of 10 to 120 nm. Other form of palladium is spherical shaped nanomaterials known as Palladium nanoparticles (PdNPs), and they have a high value as nanocatalysts, as well as for hydrogen storage and sensing. Major biological and medical applications of PdNPs include anti-cancer therapy (Alam & Huq, 2016), anti-microbial activities (Egorova & Ananikov, 2016), biosensors (Aioub et al., 2017) & biocatalysts (Maduraiveeran et al., 2018).

Metal Oxide Nanomaterials

Major role of various metal oxide nanomaterials in the several biological and medical applications has been studied with exceptional results (Discussed in Table 2).

Table 2. Various metal oxide nanomaterials with their potential applications

Sl No.	Nanomaterial	Target Sites (Microbe/Cell/Tissue/Organ...)	Applications	Reference
01	ZnONPs	Gram-negative bacteria's	Antimicrobial agent	(Slavin et al., 2017; Nyangiwe & Ouma, 2019; Sankar et al., 2014)
02	ZnONPs	MCF-7 cancer cells	Anticancer agent (through extrinsic and intrinsic apoptotic pathways)	(Boroumand et al., 2017)
03	TiO_2NPs	*Candida albicans* biofilms Methicillin-resistant *S. aureus.*	Antifungal agent Antibacterial agent	(Goyal et al., 2016)
04	TiO_2NPs	Cancer cells	Active inhibition of tumor	(Miller et al., 2017)
05	IONPs	Breast cancer cells	*In vivo* bio-imaging	(Gao et al., 2018)
06	IONPs	Osteoblast cells	In treating Low bone mineral density condition through drug delivery	(Shi et al., 2012)
07	IONPs	*E.coli, S. aureus*	Higher bactericidal activity	(Kanagasubbulakshmi & Kadirvelu, 2017)

Zinc Oxide Nanomaterials (ZnONMs)

Zinc oxide (ZnO), are the most important metal oxide nanomaterials, which can exhibit nanostructures, having special chemical, semiconductor, and optical features for different applications (Matinise et al., 2017). Low toxicity, biocompatibility, and biodegradability properties of the substance are important for biomedical and pro-ecological systems (Ludi & Niederberger, 2013). And these low toxicity and biodegradability are the most significant properties of ZnONMs. ZnO outer segment will be rich in -OH groups that can be quickly functionalized by different molecules and also decorate the surface. ZnO nanoparticles (ZnONPs) are now evolving novel structures among ZnNMs and are a promising medium for use in biomedical science, particularly with regard to their activities in the fields of anticancer (Sirelkhatim et al., 2015; Fan et al., 2017), antimicrobial, antioxidant, antidiabetic (Kitture et al., 2015), and anti-inflammatory (Thatoi et al., 2016). These activities are associated with the ability of ZnONPs to produce Reactive Oxygen Species (ROS) and induce apoptosis. ZnONPs also have acquired tremendous interest in drug delivery (Li et al., 2018; Wang et al., 2017) and bioimaging applications.

Titanium-dioxide Nanomaterials (TiO_2NMs)

Titanium-dioxide nanoparticle (TiO_2NPs) is commonly used in cosmetics, sunscreen products, sanitary cleaning applications, antimicrobial packaging, and even in cement products (Sycheva et al., 2011) and well-studied photocatalyst (Wang et al., 2015). Another characteristic of the TiO_2NPs is when affected by UV light, they will turn highly reactive and absorb pollutants, thus it fits in among environmentally friendly materials (Ola & Maroto-Valer, 2015). TiO_2NP as a photocatalyst is widely used as an alternative to conventional disinfection methods for removing micropollutants from liquids, and contaminated air samples. Molecules of airborne pollutants adsorb and react to TiO_2NP surfaces with radicals, decomposing them chemically and forming carbon (CO_2) and water (H_2O). TiO_2NP induced photocatalysis is one more effective process in comparison to photolysis for the degradation of several drug micropollutants in water. TiO_2NPs are known widely for antimicrobial activity. Based on the study, a variety of experiments on photocatalytic disinfection have been performed on various pathogenic gram-positive and gram-negative bacteria, viruses, fungi, and algae (Foster et al., 2011; Byrne et al., 2015, Krishna, 2021, pp. 3707-3717).

Iron-oxide/(Magnetic) Nanoparticles (IONPs)

IONPs belong to the magnetic family and are used in genetics, medicine, and drug delivery notions (Mohammed et al., 2017). Different forms of iron oxide or magnetic-based nanoparticles are categorized as magnetite (Fe_3O_4), maghemite (y-Fe_2O_3), and mixed ferrites ($CoFe_2O_4$, $MnFe_2O_4$, Ni/$ZnFe_2O_4$) materials (Hedayatnasab et al., 2017). Magnetic Particle Imaging (MPI) (Ferguson et al., 2014), Magnetic Resonance Imaging (MRI) (Gilro et al., 2016), proteins (Chen et al., 2016), antibodies (Schleich et al., 2015), targeted delivery of drugs (Gai et al., 2014), and nucleic acids (Xu et al., 2012), hyperthermia (Périgo et al., 2015), biosensing (Wu et al., 2015), tissue repair (Siddiqi & Husen, 2016) and analysis of biomolecules (Dadfar et al., 2019) are the widespread list of applications of surface-modified superparamagnetic IONPs. IONPs are exquisitely ideal for theranostic applications due to their nano range size and inherent imaging competencies. Several clinical proof-of-concept trials have been already completed in this respect, (Dadfar et al., 2019) several studies on IONPs for theranostic focused towards validation and drug delivery applications have been conducted. Particularly, for cancer (Miller et al., 2017), and sometimes combined with targeting of magnetic drugs or hyperthermia of magnetic fluids.

Aluminum Oxide Nanoparticles (Al_2O_3NPs)

Aluminum oxide nanoparticles (Al_2O_3NPs) are a class of porous nanomaterials and are commonly known as alumina. Al_2O_3NPs are configured like corundum, where six oxygen atoms encircle one aluminum atom (Al-Tayyar et al., 2020). Alumina oxide nanoparticles are thermodynamically stable and are incredibly high chemical durability and could sustain harsh environments, including abrasive surroundings. The researchers have observed that these particles' behavior was completely influenced by their size, morphology, arrangements, and charge on the surface. In hard water and seawater, these nanoparticles tend to aggregate due to particle interaction with organic matter present in water (Lucky et al., 2015). Also, these nanomaterials exhibit higher surface area and greater mechanical strength with low cost in availability; further, they also possess a low electrical conductivity (Sirelkhatim et al., 2015). Recent studies show that Al_2O_3NPs are strategic nanomaterials in various aspects of life and are growing rapidly in the biomedical markets (Fouly & Alkalla, 2020; Hassanpour et al., 2018), and also they have been found in various applications, particularly in biomedicine, wherein drug delivery (Hassanpour et al., 2018; Tarlani et al., 2017), biosensing (Aramesh et al., 2017; Ito et al., 2017; Guzik et al., 2014), cancer therapy (Hassanpour et al., 2018; Kepp et al., 2014), as an antimicrobial agent, immunotherapy (Beyzayet al., 2017), and biomolecular stabilization (Volodina et al., 2017).

APPLICATIONS

Antimicrobial Agent

World Health Organization (WHO, 2020) reports on obstruction towards antimicrobial medications jeopardizes the viability of medicines for an expanding arrangement of diseases by pathogenic microorganisms. Principally bacterial obstruction is brought about by adjustments in the capacity of microorganisms to oppose against antibacterial specialists either by inactivating them or by causing abatement in their remedial viability. After some time, these protections show up unexpectedly in microorganisms because of hereditary changes (Sánchez-López et al., 2020). Then again, the spread of multidrug-resistant infections will build the requirement for research center tests and the utilization of more costly medications, hence expanding the expense of medical services (Ifeanyichukwu et al., 2020). Metallic nanoparticles are the most mainstream inorganic nanoparticles and address a promising arrangement against microbial protection from conventional antimicrobials or to uncouth antimicrobial specialists. They show the expected antimicrobial movement against pathogens that have just evolved opposition; however, they additionally focus on different biomolecules bargaining the advancement of safe strains (Sánchez et al., 2020; Slavin et al., 2017). Since from old time, metal-based nanomaterials, for example, silver and copper, has been utilized as antimicrobial specialists, and explicitly these metal nanoparticles give solid and broadened antimicrobial movement at more modest doses against a wide scope of microorganisms because of their measurements and shapes (Amini, 2019, pp. 109809). Among a couple of metal nanomaterials, silver nanoparticles have been considered the most fascinating antibacterial nano drug because of their high proficiency against microbes, growths, and infections. As indicated by a recent report reports AgNPs and CuNPs demonstrated expected antibacterial features inhibiting the human pathogenic bacteria, for example, *Escherichia coli* (Konvičková et al., 2019), *S. aureus* (Bankier et al., 2019), *P. aeruginosa* (Nanda et al., 2018) and *C. albicans* (Rao et al., 2019). Metal-oxide nanoparticles are often presented in relation to some pathogenic microorganisms with relevant antimicrobial properties. ZnONPs, TiO_2NPs, CuO, and NiO_2 are the classical nanoparticles with effective antimicrobial and antifungal properties (Garcia et al., 2018; Slavin et al., 2017) and explored the possible antibacterial activities of size based CuO, Fe_2O_3, and ZnO nanoparticles on Gram-negative bacteria, such as *P. aeruginosa, E. coli,* etc. including Gram-positive bacteria such as *B. subtilis* and *S. aureus.*

Drug Delivery

A recent development in nanotechnology has shown that many nanomaterials have the potential to act as good drug carriers and have overcome these disadvantages by reducing side effects, increasing the solubility, *in vivo* stability, pharmacokinetics, biodistribution, and enhancing the efficiency of free drugs in cancer therapy. Reactivity, surface area stability, and capability to trigger and move inside the living system to consider as approved material for targeted drug delivery. Drug therapeutic concentration, safety over time, and drug release in an effective way are all essential features of an optimal drug delivery mechanism (Godwin et al., 2015). Most of the studies depicted that metal and metal-oxide nanomaterials can be used actively for successful drug delivery (Sun et al., 2014; Shrivastava et al., 2019; Syed & Sailaja, 2019). Commonly studied nanoparticles for drug delivery are silver, gold, titanium oxide, zinc oxide, and magnetic nanoparticles. These nanoparticles are embedded with biopolymers such as gold with polymers that contain the drug to be transported to enable either the surface of the anticancer

Figure 1. Surface attached and encapsulated drugs for targeted drug delivery system

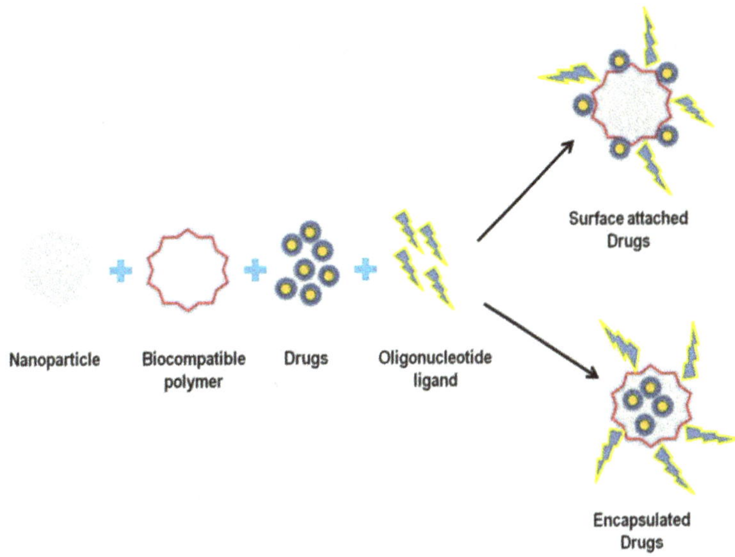

drug, or the embedded nanoparticles (Figure 1). If the drug-nanoparticle conjugation complex has been applied, an external magnetic/optical/thermal/chemical field could be used to direct the conjugate to a particular tumor site. Eventually, the drug can be eluted by the activity of the enzyme or by changes in external variables/parameters i.e. pH, temperature, or osmolality (McNamara & Tofail, 2017; Syed & Sailaja, 2019).

Bio-imaging

To locate and diagnose the tissue or organ of our interest, different types of bioimaging methods are using such as ultrasound, Magnetic resonance imaging, computed tomography, and positron emission tomography. Distinguishing the stable tissue from diseased tissues can also be possible in these bioimaging techniques. Both magnetic and fluorescent nanoparticles have greatly contributed to the development of bioimaging techniques among these imaging instruments (Li et al., 2011). Magnetic nanoparticles are used for MRI and fluorescent nanoparticles such as AuNPs are commonly used for optical imaging (OI). Iron oxide nanoparticles are also identified in addition to metal oxide for imaging purposes (Parodi et al., 2013). MRI, most effective and recommended non-invasive imaging method for the development of high-resolution photographs of soft tissues or internal organs. In MRI, different nanoparticles and comparing chemicals, based on metal and oxide, would be used for enhanced contrast prior to screening. Here magnetite or iron oxide nanoparticles may form the contrast agents (Karlapudi et al., 2018). Few studies have reported on the use of magnetic nanoparticles to boost MRI signals, and iron oxide has also been embedded in AuNPs by researchers. The iron oxide produces magnetism in this case, while the Au ion integrates the optical features of the AuNPs (Chen et al., 2016; Sobhani et al., 2013).

Therapeutic Applications

Nanomaterials of metal and metal oxide have many essential and useful characteristics for therapeutic applications such as vaccine production, diabetic treatment, antibacterial agent, antiviral agent, medication/gene delivery, drug screening, cancer therapy, and much more. Hepatitis and Herpes Simplex viruses have been primarily utilized in combating human immunodeficiency virus (HIV), viral nanoparticles, and drug models (HSV-1&2). The metal nanoparticles have recently been researched and developed as a preventive-therapeutic human immunodeficiency virus (HIV) (Liu & Chen, 2016; Saravanan et al., 2018). And it has also been seen that by glycoprotein interconnection, AgNPs can directly act on viral infection. The multiple ways of viral attachment inhibit this sort of interconnection, which essentially decreases HIV infection (Wang et al., 2020). A few other experiments have also been undertaken in the fields of human rotavirus, papilloma virus, and *Japanese encephalitis* virus diagnosis and treatment. Several experiments were reported on using metal and metal-oxide nanoparticles (Abdul Ghaffar et al., 2014; Unzueta et al., 2014). In general, Parkinson's and Alzheimer's like neurodegenerative disorders, have a long pathological injury process, causing invaluable damage to patients and their families (Galluzzi et al., 2015). Thanks to the distinctive nature of nanomaterials, many researchers' hope of discovering an appropriate cure for these neurodegenerative disorders has steadily entered the field of nanotechnology. The research path of nanomaterials in neurodegenerative diseases is currently predominantly focused on drug distribution, protection of drug concentration, and early detailed detection and treatment of diseases dependent on nanomaterial characteristics (Mavlonov et al., 2020). A medical condition of joints that occurs when the immune system of the body does not function effectively and it strikes the bones-connecting joints this condition is known as rheumatoid arthritis. The AuNPs were used by University of Wollongong researchers to treat arthritic disease (Ensafi et al., 2016; Soni & Yadav, 2015) but it does have some slight pitfalls. Further study has also been documented that AuNPs can target macrophages and stop arthritic disease inflammation (Byeon et al., 2016).

FUTURE RESEARCH DIRECTIONS

Tissue engineering protocols are rapidly in demand due to certain limitations of tissue and organ transplantation. Researchers created an artificial structure that mimics the real tissue/organ through tissue engineering protocols for the development of the implantable device. Many tissue engineering protocols face multiple limitations; the greatest blockade is challenging to produce engineered products which could replicate the tissues like anatomy (Hasan et al., 2015). Scaffolds that could allow tuning the mechanical stability, regulating cellular behavior, and dispersion of bioactive agents in target sites can be controlled by using suitable nanoparticles (Mi et al., 2016). In specific, for bone and cardiac tissue regeneration, AuNPs and TiO_2NPs are commonly used nanoparticles. AuNPs have demonstrated superior biocompatibility, increases the cell proliferation rates and surface adjustment capabilities (Dykman & Khlebtsov, 2012). AuNPs helps adoptively in osteogenic differentiation of the osteoblast cell line precursor in bone tissue engineering (Chen et al., 2015). In another study, TiO_2NPs applied scaffold showed increased proliferation of cardiomyocytes (Tallawi et al., 2015). Therefore, the integration into scaffolds with some forms of nanoparticles will substantially modify the biological properties of scaffolds and therefore assist in tissue engineering.

CONCLUSION

Due to rigorous medical and biological uses, many nanomaterials have gained great interest. Mechanical, chemical stability, reactivity, sensitivity, large surface area, non-toxicity, biocompatibility, and catalytic activities are key features of nanomaterials and are vital and beneficial for the production of effective and high-performance technologies in the medical and biological sectors. Nanomaterials such as AgNPs, CuNPs, AuNPs, ZnONPs, TiO_2NPs, and a few more are widely used for medical-biological applications. As an effective replacement for antimicrobial experiments against bacterial, fungal, and virus infections, metal and metal oxide or pure metal nanoparticles such as AgNPs, AuNPs, ZnONPs, CuNPs are possible medicinal agents. It is highly feasible and concluded that simple low-cost antimicrobial agents can be produced with nano-sized structures of metal and metal oxide.

REFERENCES

Abboud, Y., Saffaj, T., Chagraoui, A., El Bouari, A., Brouzi, K., Tanane, O., & Ihssane, B. (2014). Biosynthesis, characterization and antimicrobial activity of copper oxide nanoparticles (CONPs) produced using brown alga extract (Bifurcaria bifurcata). *Applied Nanoscience*, *4*(5), 571–576. doi:10.100713204-013-0233-x

Abdul Ghaffar, K., Kumar Giddam, A., Zaman, M., Skwarczynski, M., & Toth, I. (2014). Liposomes as nanovaccine delivery systems. *Current Topics in Medicinal Chemistry*, *14*(9), 1194–1208. doi:10.2174/1568026614666140329232757 PMID:24678703

Aioub, M., Panikkanvalappil, S. R., & El-Sayed, M. A. (2017). Platinum-coated gold nanorods: Efficient reactive oxygen scavengers that prevent oxidative damage toward healthy, untreated cells during plasmonic photothermal therapy. *ACS Nano*, *11*(1), 579–586. doi:10.1021/acsnano.6b06651 PMID:28029783

Al-Tayyar, N. A., Youssef, A. M., & Al-Hindi, R. (2020). Antimicrobial food packaging based on sustainable Bio-based materials for reducing foodborne Pathogens: A review. *Food Chemistry*, *310*, 125915. doi:10.1016/j.foodchem.2019.125915 PMID:31841936

Alam, M. N., & Huq, F. (2016). Comprehensive review on tumour active palladium compounds and structure–activity relationships. *Coordination Chemistry Reviews*, *316*, 36–67. doi:10.1016/j.ccr.2016.02.001

Amini, S. M. (2019). Preparation of antimicrobial metallic nanoparticles with bioactive compounds. *Materials Science and Engineering C*, *103*, 109809. doi:10.1016/j.msec.2019.109809 PMID:31349497

Aramesh, M., Tran, P. A., Ostrikov, K. K., & Prawer, S. (2017). Conformal nanocarbon coating of alumina nanocrystals for biosensing and bioimaging. *Carbon*, *122*, 422–427. doi:10.1016/j.carbon.2017.06.101

Austin, L. A., Mackey, M. A., Dreaden, E. C., & El-Sayed, M. A. (2014). The optical, photothermal, and facile surface chemical properties of gold and silver nanoparticles in biodiagnostics, therapy, and drug delivery. *Archives of Toxicology*, *88*(7), 1391–1417. doi:10.100700204-014-1245-3 PMID:24894431

Bankier, C., Matharu, R. K., Cheong, Y. K., Ren, G. G., Cloutman-Green, E., & Ciric, L. (2019). Synergistic antibacterial effects of metallic nanoparticle combinations. *Scientific Reports*, *9*(1), 1–8. doi:10.103841598-019-52473-2 PMID:31690845

Baptista, P. V., McCusker, M. P., Carvalho, A., Ferreira, D. A., Mohan, N. M., Martins, M., & Fernandes, A. R. (2018). Nano-strategies to fight multidrug resistant bacteria—"A Battle of the Titans. *Frontiers in Microbiology*, *9*, 1441. doi:10.3389/fmicb.2018.01441 PMID:30013539

Beyzay, F., Hosseini, A. Z., & Soudi, S. (2017). Alpha alumina nanoparticle conjugation to cysteine peptidase A and B: An efficient method for autophagy induction. *Avicenna Journal of Medical Biotechnology*, *9*(2), 71. PMID:28496946

Boroumand Moghaddam, A., Moniri, M., Azizi, S., Abdul Rahim, R., Bin Ariff, A., Navaderi, M., & Mohamad, R. (2017). Eco-friendly formulated zinc oxide nanoparticles: Induction of cell cycle arrest and apoptosis in the MCF-7 cancer cell line. *Genes*, *8*(10), 281. doi:10.3390/genes8100281 PMID:29053567

Byeon, H. J., Lee, C., Lee, S., Lee, E. S., Choi, H. G., Park, E. S., & Youn, Y. S. (2016). Pharmaceutical potential of tacrolimus-loaded albumin nanoparticles having targetability to rheumatoid arthritis tissues. *International Journal of Pharmaceutics*, *497*(1-2), 268–276. doi:10.1016/j.ijpharm.2015.12.004 PMID:26657273

Byrne, J. A., Dunlop, P. S. M., Hamilton, J. W. J., Fernández-Ibáñez, P., Polo-López, I., Sharma, P. K., & Vennard, A. S. M. (2015). A review of heterogeneous photocatalysis for water and surface disinfection. *Molecules (Basel, Switzerland)*, *20*(4), 5574–5615. doi:10.3390/molecules20045574 PMID:25830789

Chen, G., Roy, I., Yang, C., & Prasad, P. N. (2016). Nanochemistry and nanomedicine for nanoparticle-based diagnostics and therapy. *Chemical Reviews*, *116*(5), 2826–2885. doi:10.1021/acs.chemrev.5b00148 PMID:26799741

Chen, Y., Li, N., Yang, Y., & Liu, Y. (2015). A dual targeting cyclodextrin/gold nanoparticle conjugate as a scaffold for solubilization and delivery of paclitaxel. *RSC Advances*, *5*(12), 8938–8941. doi:10.1039/C4RA13135E

Cheng, L., Wang, C., Feng, L., Yang, K., & Liu, Z. (2014). Functional nanomaterials for phototherapies of cancer. *Chemical Reviews*, *114*(21), 10869–10939. doi:10.1021/cr400532z PMID:25260098

Chernousova, S., & Epple, M. (2013). Silver as antibacterial agent: Ion, nanoparticle, and metal. *Angewandte Chemie International Edition*, *52*(6), 1636–1653. doi:10.1002/anie.201205923 PMID:23255416

Cho, S. H., Jones, B. L., & Krishnan, S. (2009). The dosimetric feasibility of gold nanoparticle-aided radiation therapy (GNRT) via brachytherapy using low-energy gamma-/x-ray sources. *Physics in Medicine and Biology*, *54*(16), 4889–4905. doi:10.1088/0031-9155/54/16/004 PMID:19636084

Crosera, M., Prodi, A., Mauro, M., Pelin, M., Florio, C., Bellomo, F., Adami, G., Apostoli, P., De Palma, G., Bovenzi, M., Campanini, M., & Filon, F. L. (2015). Titanium dioxide nanoparticle penetration into the skin and effects on HaCaT cells. *International Journal of Environmental Research and Public Health*, *12*(8), 9282–9297. doi:10.3390/ijerph120809282 PMID:26262634

da Rocha, S. R., Heyder, R. S., Bielski, E. R., Guo, A., Steinmaurer, M., & Reineke, J. J. (2019). Inhalation drug products containing nanomaterials. In *Pharmaceutical Inhalation Aerosol Technology* (pp. 403–423). CRC Press. doi:10.1201/9780429055201-17

Dadfar, S. M., Roemhild, K., Drude, N. I., von Stillfried, S., Knüchel, R., Kiessling, F., & Lammers, T. (2019). Iron oxide nanoparticles: Diagnostic, therapeutic and theranostic applications. *Advanced Drug Delivery Reviews*, *138*, 302–325. doi:10.1016/j.addr.2019.01.005 PMID:30639256

Deyhimihaghighi, N., Noor, N. M., Soltani, N., Jorfi, R., Haghir, M. E., Adenan, M. Z., Saion, E., & Khandaker, M. U. (2014, November). Contrast enhancement of magnetic resonance imaging (MRI) of polymer gel dosimeter by adding Platinum nano-particles. *Journal of Physics: Conference Series*, *546*(1), 012013. doi:10.1088/1742-6596/546/1/012013

Dizaj, S. M., Lotfipour, F., Barzegar-Jalali, M., Zarrintan, M. H., & Adibkia, K. (2014). Antimicrobial activity of the metals and metal oxide nanoparticles. *Materials Science and Engineering C*, *44*, 278–284. doi:10.1016/j.msec.2014.08.031 PMID:25280707

do Amaral, D. F., Guerra, V., Motta, A. G. C., Silva, D. D. M., & Rocha, T. L. (2019). Ecotoxicity of nanomaterials in amphibians: A critical review. *The Science of the Total Environment*, *686*, 332–344. doi:10.1016/j.scitotenv.2019.05.487 PMID:31181520

Dreaden, E. C., Austin, L. A., Mackey, M. A., & El-Sayed, M. A. (2012). Size matters: Gold nanoparticles in targeted cancer drug delivery. *Therapeutic Delivery*, *3*(4), 457–478. doi:10.4155/tde.12.21 PMID:22834077

Durán, N., Durán, M., De Jesus, M. B., Seabra, A. B., Fávaro, W. J., & Nakazato, G. (2016). Silver nanoparticles: A new view on mechanistic aspects on antimicrobial activity. *Nanomedicine; Nanotechnology, Biology, and Medicine*, *12*(3), 789–799. doi:10.1016/j.nano.2015.11.016 PMID:26724539

Dykman, L., & Khlebtsov, N. (2012). Gold nanoparticles in biomedical applications: Recent advances and perspectives. *Chemical Society Reviews*, *41*(6), 2256–2282. doi:10.1039/C1CS15166E PMID:22130549

Edmundson, M. C., Capeness, M., & Horsfall, L. (2014). Exploring the potential of metallic nanoparticles within synthetic biology. *New Biotechnology*, *31*(6), 572–578. doi:10.1016/j.nbt.2014.03.004 PMID:24681407

Egorova, K. S., & Ananikov, V. P. (2016). Which metals are green for catalysis? Comparison of the toxicities of Ni, Cu, Fe, Pd, Pt, Rh, and Au salts. *Angewandte Chemie International Edition*, *55*(40), 12150–12162. doi:10.1002/anie.201603777 PMID:27532248

Elahi, N., Kamali, M., & Baghersad, M. H. (2018). Recent biomedical applications of gold nanoparticles: A review. *Talanta*, *184*, 537–556. doi:10.1016/j.talanta.2018.02.088 PMID:29674080

Ensafi, A. A., Amini, M., Rezaei, B., & Talebi, M. (2016). A novel diagnostic biosensor for distinguishing immunoglobulin mutated and unmutated types of chronic lymphocytic leukemia. *Biosensors & Bioelectronics*, *77*, 409–415. doi:10.1016/j.bios.2015.09.063 PMID:26436328

Fan, G., Dundas, C. M., Zhang, C., Lynd, N. A., & Keitz, B. K. (2018). Sequence-dependent peptide surface functionalization of metal–organic frameworks. *ACS Applied Materials & Interfaces*, *10*(22), 18601–18609. doi:10.1021/acsami.8b05148 PMID:29762004

Fan, W., Yung, B., Huang, P., & Chen, X. (2017). Nanotechnology for multimodal synergistic cancer therapy. *Chemical Reviews*, *117*(22), 13566–13638. doi:10.1021/acs.chemrev.7b00258 PMID:29048884

Ferguson, R. M., Khandhar, A. P., Kemp, S. J., Arami, H., Saritas, E. U., Croft, L. R., Konkle, J., Goodwill, P. W., Halkola, A., Rahmer, J., Borgert, J., Conolly, S. M., & Krishnan, K. M. (2014). Magnetic particle imaging with tailored iron oxide nanoparticle tracers. *IEEE Transactions on Medical Imaging*, *34*(5), 1077–1084. doi:10.1109/TMI.2014.2375065 PMID:25438306

Ferreira, C. A., Ni, D., Rosenkrans, Z. T., & Cai, W. (2019). Radionuclide-activated nanomaterials and their biomedical applications. *Angewandte Chemie International Edition*, *58*(38), 13232–13252. doi:10.1002/anie.201900594 PMID:30779286

Foster, H. A., Ditta, I. B., Varghese, S., & Steele, A. (2011). Photocatalytic disinfection using titanium dioxide: Spectrum and mechanism of antimicrobial activity. *Applied Microbiology and Biotechnology*, *90*(6), 1847–1868. doi:10.100700253-011-3213-7 PMID:21523480

Fouly, A., & Alkalla, M. G. (2020). Effect of low nanosized alumina loading fraction on the physicomechanical and tribological behavior of epoxy. *Tribology International*, *152*, 106550. doi:10.1016/j.triboint.2020.106550

Gai, S., Li, C., Yang, P., & Lin, J. (2014). Recent progress in rare earth micro/nanocrystals: Soft chemical synthesis, luminescent properties, and biomedical applications. *Chemical Reviews*, *114*(4), 2343–2389. doi:10.1021/cr4001594 PMID:24344724

Galdiero, S., Falanga, A., Cantisani, M., Ingle, A., Galdiero, M., & Rai, M. (2014). Silver nanoparticles as novel antibacterial and antiviral agents. In Handbook of Nanobiomedical Research: Fundamentals, Applications and Recent Developments: Volume 1. Materials for Nanomedicine (pp. 565-594). doi:10.1142/9789814520652_0015

Galluzzi, L., Bravo-San Pedro, J. M., Vitale, I., Aaronson, S. A., Abrams, J. M., Adam, D., Alnemri, E. S., Altucci, L., Andrews, D., Annicchiarico-Petruzzelli, M., Baehrecke, E. H., Bazan, N. G., Bertrand, M. J., Bianchi, K., Blagosklonny, M. V., Blomgren, K., Borner, C., Bredesen, D. E., Brenner, C., ... Kroemer, G. (2015). Essential versus accessory aspects of cell death: Recommendations of the NCCD 2015. *Cell Death and Differentiation*, *22*(1), 58–73. doi:10.1038/cdd.2014.137 PMID:25236395

Gao, Q., Xie, W., Wang, Y., Wang, D., Guo, Z., Gao, F., Zhao, L., & Cai, Q. (2018). A theranostic nanocomposite system based on radial mesoporous silica hybridized with Fe 3 O 4 nanoparticles for targeted magnetic field responsive chemotherapy of breast cancer. *RSC Advances*, *8*(8), 4321–4328. doi:10.1039/C7RA12446E

Garcia, C. V., Shin, G. H., & Kim, J. T. (2018). Metal oxide-based nanocomposites in food packaging: Applications, migration, and regulations. *Trends in Food Science & Technology*, *82*, 21–31. doi:10.1016/j.tifs.2018.09.021

Giasuddin, A. S. M., Jhuma, K. A., & Haq, A. M. (2012). Use of gold nanoparticles in diagnostics, surgery and medicine: A review. *Bangladesh Journal of Medical Biochemistry*, *5*(2), 56–60. doi:10.3329/bjmb.v5i2.13346

Gilroy, K. D., Ruditskiy, A., Peng, H. C., Qin, D., & Xia, Y. (2016). Bimetallic nanocrystals: Syntheses, properties, and applications. *Chemical Reviews*, *116*(18), 10414–10472. doi:10.1021/acs.chemrev.6b00211 PMID:27367000

Godwin, M. A., Shri, K. M., & Balaji, M. (2015). Engineering and Bioscience. *Int. J. Res. Eng. Biosci.*, *3*(5), 11–29.

Goel, S., Chen, F., & Cai, W. (2014). Synthesis and biomedical applications of copper sulfide nanoparticles: From sensors to theranostics. *Small*, *10*(4), 631–645. doi:10.1002mll.201301174 PMID:24106015

Goel, S., England, C. G., Chen, F., & Cai, W. (2017). Positron emission tomography and nanotechnology: A dynamic duo for cancer theranostics. *Advanced Drug Delivery Reviews*, *113*, 157–176. doi:10.1016/j.addr.2016.08.001 PMID:27521055

Goyal, R., Macri, L. K., Kaplan, H. M., & Kohn, J. (2016). Nanoparticles and nanofibers for topical drug delivery. *Journal of Controlled Release*, *240*, 77–92. doi:10.1016/j.jconrel.2015.10.049 PMID:26518723

Gurunathan, S., Jeyaraj, M., Kang, M. H., & Kim, J. H. (2019). Tangeretin-assisted platinum nanoparticles enhance the apoptotic properties of doxorubicin: Combination therapy for osteosarcoma treatment. *Nanomaterials (Basel, Switzerland)*, *9*(8), 1089. doi:10.3390/nano9081089 PMID:31362420

Guzik, U., Hupert-Kocurek, K., & Wojcieszyńska, D. (2014). Immobilization as a strategy for improving enzyme properties-application to oxidoreductases. *Molecules (Basel, Switzerland)*, *19*(7), 8995–9018. doi:10.3390/molecules19078995 PMID:24979403

Haider, A., & Kang, I. K. (2015). Preparation of silver nanoparticles and their industrial and biomedical applications: A comprehensive review. *Advances in Materials Science and Engineering*, *2015*, 2015. doi:10.1155/2015/165257

Hainfeld, J. F., Lin, L., Slatkin, D. N., Dilmanian, F. A., Vadas, T. M., & Smilowitz, H. M. (2014). Gold nanoparticle hyperthermia reduces radiotherapy dose. *Nanomedicine; Nanotechnology, Biology, and Medicine*, *10*(8), 1609–1617. doi:10.1016/j.nano.2014.05.006 PMID:24990355

Hajipour, M. J., Fromm, K. M., Ashkarran, A. A., de Aberasturi, D. J., de Larramendi, I. R., Rojo, T., ... Mahmoudi, M. (2012). Antibacterial properties of nanoparticles. *Trends in Biotechnology*, *30*(10), 499–511. doi:10.1016/j.tibtech.2012.06.004 PMID:22884769

Hasan, A., Paul, A., Memic, A., & Khademhosseini, A. (2015). A multilayered microfluidic blood vessel-like structure. *Biomedical Microdevices*, *17*(5), 1–13. doi:10.100710544-015-9993-2 PMID:26256481

Hassanpour, P., Panahi, Y., Ebrahimi-Kalan, A., Akbarzadeh, A., Davaran, S., Nasibova, A. N., Khalilov, R., & Kavetskyy, T. (2018). Biomedical applications of aluminium oxide nanoparticles. *Micro & Nano Letters*, *13*(9), 1227–1231. doi:10.1049/mnl.2018.5070

Hedayatnasab, Z., Abnisa, F., & Daud, W. M. A. W. (2017). Review on magnetic nanoparticles for magnetic nanofluid hyperthermia application. *Materials & Design*, *123*, 174–196. doi:10.1016/j.matdes.2017.03.036

Hua, M., Zhang, S., Pan, B., Zhang, W., Lv, L., & Zhang, Q. (2012). Heavy metal removal from water/wastewater by nanosized metal oxides: A review. *Journal of Hazardous Materials*, *211*, 317–331. doi:10.1016/j.jhazmat.2011.10.016 PMID:22018872

Ifeanyichukwu, U. L., Fayemi, O. E., & Ateba, C. N. (2020). Green synthesis of zinc oxide nanoparticles from pomegranate (Punica granatum) extracts and characterization of their antibacterial activity. *Molecules (Basel, Switzerland), 25*(19), 4521. doi:10.3390/molecules25194521 PMID:33023149

Inbasekaran, S., Senthil, R., Ramamurthy, G., & Sastry, T. P. (2014). Biosensor using zinc oxide nanoparticles. *International Journal of Innovative Research in Science, Engineering and Technology, 3*(1), 8601–8606.

Ito, T., Matsuda, Y., Jinba, T., Asai, N., Shimizu, T., & Shingubara, S. (2017). Fabrication and characterization of nano porous lattice biosensor using anodic aluminum oxide substrate. *Japanese Journal of Applied Physics, 56*(6S1), 06GG02.

Jahromi, E. Z., Divsalar, A., Saboury, A. A., Khaleghizadeh, S., Mansouri-Torshizi, H., & Kostova, I. (2016). Palladium complexes: New candidates for anti-cancer drugs. *Journal of the Indian Chemical Society, 13*(5), 967–989.

Jain, S., Hirst, D. G., & O'sullivan, J. M. (2012). Gold nanoparticles as novel agents for cancer therapy. *The British Journal of Radiology, 85*(1010), 101–113. doi:10.1259/bjr/59448833 PMID:22010024

Jamdagni, P., Khatri, P., & Rana, J. S. (2018). Green synthesis of zinc oxide nanoparticles using flower extract of Nyctanthes arbor-tristis and their antifungal activity. *Journal of King Saud University. Science, 30*(2), 168–175. doi:10.1016/j.jksus.2016.10.002

Jamkhande, P. G., Ghule, N. W., Bamer, A. H., & Kalaskar, M. G. (2019). Metal nanoparticles synthesis: An overview on methods of preparation, advantages and disadvantages, and applications. *Journal of Drug Delivery Science and Technology, 53*, 101174. doi:10.1016/j.jddst.2019.101174

Jeyaraj, M., Gurunathan, S., Qasim, M., Kang, M. H., & Kim, J. H. (2019). A comprehensive review on the synthesis, characterization, and biomedical application of platinum nanoparticles. *Nanomaterials (Basel, Switzerland), 9*(12), 1719. doi:10.3390/nano9121719 PMID:31810256

Joh, D. Y., Sun, L., Stangl, M., Al Zaki, A., Murty, S., Santoiemma, P. P., Davis, J. J., Baumann, B. C., Alonso-Basanta, M., Bhang, D., Kao, G. D., Tsourkas, A., & Dorsey, J. F. (2013). Selective targeting of brain tumors with gold nanoparticle-induced radiosensitization. *PLoS One, 8*(4), e62425. doi:10.1371/journal.pone.0062425 PMID:23638079

Johnstone, T. C., Suntharalingam, K., & Lippard, S. J. (2016). The next generation of platinum drugs: Targeted Pt (II) agents, nanoparticle delivery, and Pt (IV) prodrugs. *Chemical Reviews, 116*(5), 3436–3486. doi:10.1021/acs.chemrev.5b00597 PMID:26865551

Kanagasubbulakshmi, S., & Kadirvelu, K. (2017). Green synthesis of iron oxide nanoparticles using Lagenaria siceraria and evaluation of its antimicrobial activity. *Defence Life Science Journal, 2*(4), 422–427. doi:10.14429/dlsj.2.12277

Kango, S., Kalia, S., Celli, A., Njuguna, J., Habibi, Y., & Kumar, R. (2013). Surface modification of inorganic nanoparticles for development of organic–inorganic nanocomposites—A review. *Progress in Polymer Science, 38*(8), 1232–1261. doi:10.1016/j.progpolymsci.2013.02.003

Karlapudi, A. P., Venkateswarulu, T. C., Tammineedi, J., Kanumuri, L., Ravuru, B. K., Dirisala, V., & Kodali, V. P. (2018). Role of biosurfactants in bioremediation of oil pollution-a review. *Petroleum, 4*(3), 241–249. doi:10.1016/j.petlm.2018.03.007

Kepp, O., Senovilla, L., Vitale, I., Vacchelli, E., Adjemian, S., Agostinis, P., Apetoh, L., Aranda, F., Barnaba, V., Bloy, N., Bracci, L., Breckpot, K., Brough, D., Buqué, A., Castro, M. G., Cirone, M., Colombo, M. I., Cremer, I., Demaria, S., ... Galluzzi, L. (2014). Consensus guidelines for the detection of immunogenic cell death. *OncoImmunology, 3*(9), e955691. doi:10.4161/21624011.2014.955691 PMID:25941621

Khoshnevisan, K., Maleki, H., Honarvarfard, E., Baharifar, H., Gholami, M., Faridbod, F., Larijani, B., Faridi Majidi, R., & Khorramizadeh, M. R. (2019). Nanomaterial based electrochemical sensing of the biomarker serotonin: A comprehensive review. *Mikrochimica Acta, 186*(1), 49. doi:10.100700604-018-3069-y PMID:30610391

Kitture, R., Chordiya, K., Gaware, S., Ghosh, S., More, P. A., Kulkarni, P., Chopade, B. A., & Kale, S. N. (2015). ZnO nanoparticles-red sandalwood conjugate: A promising anti-diabetic agent. *Journal of Nanoscience and Nanotechnology, 15*(6), 4046–4051. doi:10.1166/jnn.2015.10323 PMID:26369011

Konvičková, Z., Barabaszová, K. Č., Holišová, V., Kratošová, G., & Seidlerová, J. (2019). Phytosynthesis of Ag, ZnO and ZrO2 Nanoparticles Using Linden: Changes in Their Physical-Chemical Nature Over Time. *Journal of Nanoscience and Nanotechnology, 19*(12), 7926–7933. doi:10.1166/jnn.2019.15854 PMID:31196311

Krishna, G. (2021). Tio2 Nanocomposites And Its Applications; A Brief Review. [TURCOMAT]. *Turkish Journal of Computer and Mathematics Education, 12*(10), 3707–3717.

Ku, G., Zhou, M., Song, S., Huang, Q., Hazle, J., & Li, C. (2012). Copper sulfide nanoparticles as a new class of photoacoustic contrast agent for deep tissue imaging at 1064 nm. *ACS Nano, 6*(8), 7489–7496. doi:10.1021/nn302782y PMID:22812694

Lalitha, A., Subbaiya, R., & Ponmurugan, P. (2013). Green synthesis of silver nanoparticles from leaf extract Azhadirachta indica and to study its anti-bacterial and antioxidant property. *International Journal of Current Microbiology and Applied Sciences, 2*(6), 228–235.

Le, V. T., Bach, L. G., Pham, T. T., Le, N. T. T., Ngoc, U. T. P., Tran, D. H. N., & Nguyen, D. H. (2019). Synthesis and antifungal activity of chitosan-silver nanocomposite synergize fungicide against Phytophthora capsici. Journal of Macromolecular Science. *Part A, 56*(6), 522–528.

Li, H., He, X., Liu, Y., Huang, H., Lian, S., Lee, S. T., & Kang, Z. (2011). One-step ultrasonic synthesis of water-soluble carbon nanoparticles with excellent photoluminescent properties. *Carbon, 49*(2), 605–609. doi:10.1016/j.carbon.2010.10.004

Li, L., Wu, G., Hong, T., Yin, Z., Sun, D., Abdel-Halim, E. S., & Zhu, J. J. (2014). Graphene quantum dots as fluorescence probes for turn-off sensing of melamine in the presence of Hg2+. *ACS Applied Materials & Interfaces, 6*(4), 2858–2864. doi:10.1021/am405305r PMID:24460139

Li, Y., Zhang, C., Liu, L., Gong, Y., Xie, Y., & Cao, Y. (2018). The effects of baicalein or baicalin on the colloidal stability of ZnO nanoparticles (NPs) and toxicity of NPs to Caco-2 cells. *Toxicology Mechanisms and Methods*, *28*(3), 167–176. doi:10.1080/15376516.2017.1376023 PMID:28868948

Lima, E., Guerra, R., Lara, V., & Guzmán, A. (2013). Gold nanoparticles as efficient antimicrobial agents for Escherichia coli and Salmonella typhi. *Chemistry Central Journal*, *7*(1), 1–7. doi:10.1186/1752-153X-7-11 PMID:23331621

Liu, Y., & Chen, C. (2016). Role of nanotechnology in HIV/AIDS vaccine development. *Advanced Drug Delivery Reviews*, *103*, 76–89. doi:10.1016/j.addr.2016.02.010 PMID:26952542

Locatelli, E., Naddaka, M., Uboldi, C., Loudos, G., Fragogeorgi, E., Molinari, V., Pucci, A., Tsotakos, T., Psimadas, D., Ponti, J., & Franchini, M. C. (2014). Targeted delivery of silver nanoparticles and alisertib: In vitro and in vivo synergistic effect against glioblastoma. *Nanomedicine (London)*, *9*(6), 839–849. doi:10.2217/nnm.14.1 PMID:24433240

López-Ortega, A., Estrader, M., Salazar-Alvarez, G., Roca, A. G., & Nogués, J. (2015). Applications of exchange coupled bi-magnetic hard/soft and soft/hard magnetic core/shell nanoparticles. *Physics Reports*, *553*, 1–32. doi:10.1016/j.physrep.2014.09.007

Lucky, S. S., Soo, K. C., & Zhang, Y. (2015). Nanoparticles in photodynamic therapy. *Chemical Reviews*, *115*(4), 1990–2042. doi:10.1021/cr5004198 PMID:25602130

Ludi, B., & Niederberger, M. (2013). Zinc oxide nanoparticles: Chemical mechanisms and classical and non-classical crystallization. *Dalton Transactions (Cambridge, England)*, *42*(35), 12554–12568. doi:10.1039/c3dt50610j PMID:23652237

Maduraiveeran, G., Sasidharan, M., & Ganesan, V. (2018). Electrochemical sensor and biosensor platforms based on advanced nanomaterials for biological and biomedical applications. *Biosensors & Bioelectronics*, *103*, 113–129. doi:10.1016/j.bios.2017.12.031 PMID:29289816

Mahmoudi, M., Hosseinkhani, H., Hosseinkhani, M., Boutry, S., Simchi, A., Journeay, W. S., Subramani, K., & Laurent, S. (2011). Magnetic resonance imaging tracking of stem cells in vivo using iron oxide nanoparticles as a tool for the advancement of clinical regenerative medicine. *Chemical Reviews*, *111*(2), 253–280. doi:10.1021/cr1001832 PMID:21077606

Manikandan, M., Hasan, N., & Wu, H. F. (2013). Platinum nanoparticles for the photothermal treatment of Neuro 2A cancer cells. *Biomaterials*, *34*(23), 5833–5842. doi:10.1016/j.biomaterials.2013.03.077 PMID:23642996

Martins, P., Lopes, A. C., & Lanceros-Mendez, S. (2014). Electroactive phases of poly (vinylidene fluoride) Determination, processing and applications. *Progress in Polymer Science*, *39*(4), 683–706. doi:10.1016/j.progpolymsci.2013.07.006

Matinise, N., Fuku, X. G., Kaviyarasu, K., Mayedwa, N., & Maaza, M. J. A. S. S. (2017). ZnO nanoparticles via Moringa oleifera green synthesis: Physical properties & mechanism of formation. *Applied Surface Science*, *406*, 339–347. doi:10.1016/j.apsusc.2017.01.219

Mavlonov, A., Razykov, T., Raziq, F., Gan, J., Chantana, J., Kawano, Y., Nishimura, T., Wei, H., Zakutayev, A., Minemoto, T., Zu, X., Li, S., & Qiao, L. (2020). A review of Sb2Se3 photovoltaic absorber materials and thin-film solar cells. *Solar Energy*, *201*, 227–246. doi:10.1016/j.solener.2020.03.009

McNamara, K., & Tofail, S. A. (2017). Nanoparticles in biomedical applications. *Advances in Physics: X*, *2*(1), 54–88. doi:10.1080/23746149.2016.1254570

Mi, P., Kokuryo, D., Cabral, H., Wu, H., Terada, Y., Saga, T., Aoki, I., Nishiyama, N., & Kataoka, K. (2016). A pH-activatable nanoparticle with signal-amplification capabilities for non-invasive imaging of tumour malignancy. *Nature Nanotechnology*, *11*(8), 724–730. doi:10.1038/nnano.2016.72 PMID:27183055

Miller, M. A., Arlauckas, S., & Weissleder, R. (2017). Prediction of anti-cancer nanotherapy efficacy by imaging. *Nanotheranostics*, *1*(3), 296–312. doi:10.7150/ntno.20564 PMID:29071194

Miller-Kleinhenz, J., Guo, X., Qian, W., Zhou, H., Bozeman, E. N., Zhu, L., Ji, X., Wang, Y. A., Styblo, T., O'Regan, R., Mao, H., & Yang, L. (2018). Dual-targeting Wnt and uPA receptors using peptide conjugated ultra-small nanoparticle drug carriers inhibited cancer stem-cell phenotype in chemo-resistant breast cancer. *Biomaterials*, *152*, 47–62. doi:10.1016/j.biomaterials.2017.10.035 PMID:29107218

Mishra, A., Kumari, M., Pandey, S., Chaudhry, V., Gupta, K. C., & Nautiyal, C. S. (2014). Biocatalytic and antimicrobial activities of gold nanoparticles synthesized by Trichoderma sp. *Bioresource Technology*, *166*, 235–242. doi:10.1016/j.biortech.2014.04.085 PMID:24914997

Mittal, A. K., Chisti, Y., & Banerjee, U. C. (2013). Synthesis of metallic nanoparticles using plant extracts. *Biotechnology Advances*, *31*(2), 346–356. doi:10.1016/j.biotechadv.2013.01.003 PMID:23318667

Mohammed, L., Gomaa, H. G., Ragab, D., & Zhu, J. (2017). Magnetic nanoparticles for environmental and biomedical applications: A review. *Particuology*, *30*, 1–14. doi:10.1016/j.partic.2016.06.001

Nanda, A., Nayak, B. K., & Krishnamoorthy, M. (2018). Antimicrobial properties of biogenic silver nanoparticles synthesized from phylloplane fungus, Aspergillus tamarii. *Biocatalysis and Agricultural Biotechnology*, *16*, 225–228. doi:10.1016/j.bcab.2018.08.002

Natsuki, J., Natsuki, T., & Hashimoto, Y. (2015). A review of silver nanoparticles: Synthesis methods, properties and applications. *International Journal of Materials Science and Applications*, *4*(5), 325–332. doi:10.11648/j.ijmsa.20150405.17

Netto, C. G., Toma, H. E., & Andrade, L. H. (2013). Superparamagnetic nanoparticles as versatile carriers and supporting materials for enzymes. *Journal of Molecular Catalysis. B, Enzymatic*, *85*, 71–92. doi:10.1016/j.molcatb.2012.08.010

Nikalje, A. P. (2015). Nanotechnology and its applications in medicine. *Med chem, 5*(2), 81-89.

Noronha, V. T., Paula, A. J., Duran, G., Galembeck, A., Cogo-Mueller, K., Franz-Montan, M., & Duran, N. (2017). Silver nanoparticles in dentistry. *Dental Materials*, *33*(10), 1110–1126. doi:10.1016/j.dental.2017.07.002 PMID:28779891

Nyangiwe, N. N., & Ouma, C. N. M. (2019). Modelling the adsorption of natural organic matter on Ag (111) surface: Insights from dispersion corrected density functional theory calculations. *Journal of Molecular Graphics & Modelling*, *92*, 313–319. doi:10.1016/j.jmgm.2019.08.013 PMID:31442937

Ola, O., & Maroto-Valer, M. M. (2015). Review of material design and reactor engineering on TiO2 photocatalysis for CO2 reduction. *Journal of Photochemistry and Photobiology C, Photochemistry Reviews*, *24*, 16–42. doi:10.1016/j.jphotochemrev.2015.06.001

Olishevska, S., Nickzad, A., & Déziel, E. (2019). Bacillus and Paenibacillus secreted polyketides and peptides involved in controlling human and plant pathogens. *Applied Microbiology and Biotechnology*, *103*(3), 1189–1215. doi:10.100700253-018-9541-0 PMID:30603850

Ong, C., Yung, L. Y. L., Cai, Y., Bay, B. H., & Baeg, G. H. (2015). Drosophila melanogaster as a model organism to study nanotoxicity. *Nanotoxicology*, *9*(3), 396–403. doi:10.3109/17435390.2014.940405 PMID:25051331

Parodi, A., Quattrocchi, N., Van De Ven, A. L., Chiappini, C., Evangelopoulos, M., Martinez, J. O., Brown, B. S., Khaled, S. Z., Yazdi, I. K., Enzo, M. V., Isenhart, L., Ferrari, M., & Tasciotti, E. (2013). Synthetic nanoparticles functionalized with biomimetic leukocyte membranes possess cell-like functions. *Nature Nanotechnology*, *8*(1), 61–68. doi:10.1038/nnano.2012.212 PMID:23241654

Patil, S., Sivaraj, R., Rajiv, P., Venckatesh, R., & Seenivasan, R. (2015). Green synthesis of silver nanoparticle from leaf extract of Aegle marmelos and evaluation of its antibacterial activity. *International Journal of Pharmacy and Pharmaceutical Sciences*, *7*(6), 169–173.

Périgo, E. A., Hemery, G., Sandre, O., Ortega, D., Garaio, E., Plazaola, F., & Teran, F. J. (2015). Fundamentals and advances in magnetic hyperthermia. *Applied Physics Reviews*, *2*(4), 041302. doi:10.1063/1.4935688

Prabhu, S., & Poulose, E. K. (2012). Silver nanoparticles: Mechanism of antimicrobial action, synthesis, medical applications, and toxicity effects. *International Nano Letters*, *2*(1), 1–10. doi:10.1186/2228-5326-2-32

Rafique, M., Sadaf, I., Rafique, M. S., & Tahir, M. B. (2017). A review on green synthesis of silver nanoparticles and their applications. *Artificial Cells, Nanomedicine, and Biotechnology*, *45*(7), 1272–1291. doi:10.1080/21691401.2016.1241792 PMID:27825269

Raghuwanshi, N., Srivastava, A. K., Yadav, T. C., Gupta, S., Khatri, K., Pruthi, V., & Pruthi, V. (2019). Biogenic nanoparticles as theranostic agents: prospects and challenges. *Integrating Green Chemistry and Sustainable Engineering*, 647-684.

Rao, T. N., Babji, P., Ahmad, N., Khan, R. A., Hassan, I., Shahzad, S. A., & Husain, F. M. (2019). Green synthesis and structural classification of Acacia nilotica mediated-silver doped titanium oxide (Ag/TiO2) spherical nanoparticles: Assessment of its antimicrobial and anticancer activity. *Saudi Journal of Biological Sciences*, *26*(7), 1385–1391. doi:10.1016/j.sjbs.2019.09.005 PMID:31866742

Rathore, B., Sunwoo, K., Jangili, P., Kim, J., Kim, J. H., Huang, M., Xiong, J., Sharma, A., Yang, Z., Qu, J., & Kim, J. S. (2019). Nanomaterial designing strategies related to cell lysosome and their biomedical applications: A review. *Biomaterials*, *211*, 25–47. doi:10.1016/j.biomaterials.2019.05.002 PMID:31078050

Reddy, L. H., Arias, J. L., Nicolas, J., & Couvreur, P. (2012). Magnetic nanoparticles: Design and characterization, toxicity and biocompatibility, pharmaceutical and biomedical applications. *Chemical Reviews*, *112*(11), 5818–5878. doi:10.1021/cr300068p PMID:23043508

Rudramurthy, G. R., Swamy, M. K., Sinniah, U. R., & Ghasemzadeh, A. (2016). Nanoparticles: Alternatives against drug-resistant pathogenic microbes. *Molecules (Basel, Switzerland)*, *21*(7), 836. doi:10.3390/molecules21070836 PMID:27355939

Sabella, S., Galeone, A., Vecchio, G., Cingolani, R., & Pompa, P. (2011). AuNPs are toxic in vitro and in vivo: A review. *Journal of Nanoscience Letters*, *1*(3).

Samberg, M. E., Loboa, E. G., Oldenburg, S. J., & Monteiro-Riviere, N. A. (2012). Silver nanoparticles do not influence stem cell differentiation but cause minimal toxicity. *Nanomedicine (London)*, *7*(8), 1197–1209. doi:10.2217/nnm.12.18 PMID:22583572

Sánchez-López, E., Gomes, D., Esteruelas, G., Bonilla, L., Lopez-Machado, A. L., Galindo, R., Cano, A., Espina, M., Ettcheto, M., Camins, A., Silva, A. M., Durazzo, A., Santini, A., Garcia, M. L., & Souto, E. B. (2020). Metal-based nanoparticles as antimicrobial agents: An overview. *Nanomaterials (Basel, Switzerland)*, *10*(2), 292. doi:10.3390/nano10020292 PMID:32050443

Sankar, R., Maheswari, R., Karthik, S., Shivashangari, K. S., & Ravikumar, V. (2014). Anticancer activity of Ficus religiosa engineered copper oxide nanoparticles. *Materials Science and Engineering C*, *44*, 234–239. doi:10.1016/j.msec.2014.08.030 PMID:25280701

Saravanan, M., Asmalash, T., Gebrekidan, A., Gebreegziabiher, D., Araya, T., Hilekiros, H., Barabadi, H., & Ramanathan, K. (2018). Nano-medicine as a newly emerging approach to combat human immunodeficiency virus (HIV). *Pharmaceutical Nanotechnology*, *6*(1), 17–27. doi:10.2174/2211738506666180209095710 PMID:29424324

Schleich, N., Danhier, F., & Préat, V. (2015). Iron oxide-loaded nanotheranostics: Major obstacles to in vivo studies and clinical translation. *Journal of Controlled Release*, *198*, 35–54. doi:10.1016/j.jconrel.2014.11.024 PMID:25481448

Shi, H., Ye, X., He, X., Wang, K., Cui, W., He, D., Li, D., & Jia, X. (2014). Au@ Ag/Au nanoparticles assembled with activatable aptamer probes as smart "nano-doctors" for image-guided cancer thermotherapy. *Nanoscale*, *6*(15), 8754–8761. doi:10.1039/C4NR01927J PMID:24953128

Shi, S. F., Jia, J. F., Guo, X. K., Zhao, Y. P., Chen, D. S., Guo, Y. Y., ... Zhang, X. L. (2012). Biocompatibility of chitosan-coated iron oxide nanoparticles with osteoblast cells. *International Journal of Nanomedicine*, *7*, 5593. PMID:23118539

Shivaramakrishnan, B., Gurumurthy, B., & Balasubramanian, A. (2017). Potential biomedical applications of metallic nanobiomaterials: A review. *International Journal of Pharmaceutical Sciences and Research*, *8*(3), 985.

Shrivastava, N., Shrivastava, V., Jyoti, A., & Tomar, R. S. (2019). Promises and cons of nanobiotechnology: A critical. *Plant Archives*, *19*(1), 1–11.

Siddiqi, K. S., & Husen, A. (2016). Green synthesis, characterization and uses of palladium/platinum nanoparticles. *Nanoscale Research Letters*, *11*(1), 1–13. doi:10.118611671-016-1695-z PMID:27807824

Sirelkhatim, A., Mahmud, S., Seeni, A., Kaus, N. H. M., Ann, L. C., Bakhori, S. K. M., Hasan, H., & Mohamad, D. (2015). Review on zinc oxide nanoparticles: Antibacterial activity and toxicity mechanism. *Nano-Micro Letters*, *7*(3), 219–242. doi:10.100740820-015-0040-x PMID:30464967

Slavin, Y. N., Asnis, J., Häfeli, U. O., & Bach, H. (2017). Metal nanoparticles: Understanding the mechanisms behind antibacterial activity. *Journal of Nanobiotechnology*, *15*(1), 1–20. doi:10.118612951-017-0308-z PMID:28974225

Sobhani, A., Knight, M. W., Wang, Y., Zheng, B., King, N. S., Brown, L. V., Fang, Z., Nordlander, P., & Halas, N. J. (2013). Narrowband photodetection in the near-infrared with a plasmon-induced hot electron device. *Nature Communications*, *4*(1), 1–6. doi:10.1038/ncomms2642 PMID:23535664

Soni, G., & Yadav, K. S. (2015). Applications of nanoparticles in treatment and diagnosis of leukemia. *Materials Science and Engineering C*, *47*, 156–164. doi:10.1016/j.msec.2014.10.043 PMID:25492184

Subhankari, I., & Nayak, P. L. (2013). Antimicrobial activity of copper nanoparticles synthesized by ginger (Zingiber officinale) extract. *World Journal of Nano Science & Technology*, *2*(1), 10–13.

Sun, L., Joh, D., Stangl, M., Al-Zaki, A., Murty, S., Alonso-Basanta, M., Kao, G., Dorsey, J., & Tsourkas, A. (2013). Selective targeting of brain tumors with nanoparticle-induced radiosensitization and contrast enhancement. *Practical Radiation Oncology*, *3*(2), S19–S20. doi:10.1016/j.prro.2013.01.067 PMID:24674504

Sun, T., Zhang, Y. S., Pang, B., Hyun, D. C., Yang, M., & Xia, Y. (2014). Engineered nanoparticles for drug delivery in cancer therapy. *Angewandte Chemie International Edition*, *53*(46), 12320–12364. doi:10.1002/anie.201403036 PMID:25294565

Syafiuddin, A., Salim, M. R., Beng Hong Kueh, A., Hadibarata, T., & Nur, H. (2017). A review of silver nanoparticles: Research trends, global consumption, synthesis, properties, and future challenges. *Journal of the Chinese Chemical Society (Taipei)*, *64*(7), 732–756. doi:10.1002/jccs.201700067

Sycheva, L. P., Zhurkov, V. S., Iurchenko, V. V., Daugel-Dauge, N. O., Kovalenko, M. A., Krivtsova, E. K., & Durnev, A. D. (2011). Investigation of genotoxic and cytotoxic effects of micro-and nanosized titanium dioxide in six organs of mice in vivo. *Mutation Research/Genetic Toxicology and Environmental Mutagenesis*, *726*(1), 8–14. doi:10.1016/j.mrgentox.2011.07.010 PMID:21871579

Syed, H., & Sailaja, A. K. (2019). Preparation and Characterization of Nanoparticulate Drug Delivery System for Naproxen Sodium Using Various Desolvating Agents. *Nano Biomedicine and Engineering*, *11*(3). Advance online publication. doi:10.5101/nbe.v11i3.p254-263

Tallawi, M., Rosellini, E., Barbani, N., Cascone, M. G., Rai, R., Saint-Pierre, G., & Boccaccini, A. R. (2015). Strategies for the chemical and biological functionalization of scaffolds for cardiac tissue engineering: A review. *Journal of the Royal Society, Interface*, *12*(108), 20150254. doi:10.1098/rsif.2015.0254 PMID:26109634

Tamanoi, F., Matsumoto, K., Doan, T. L. H., Shiro, A., & Saitoh, H. (2020). Studies on the Exposure of Gadolinium Containing Nanoparticles with Monochromatic X-rays Drive Advances in Radiation Therapy. *Nanomaterials (Basel, Switzerland)*, *10*(7), 1341. doi:10.3390/nano10071341 PMID:32660093

Tarlani, A., Isari, M., Khazraei, A., & Eslami Moghadam, M. (2017). New sol-gel derived aluminum oxide-ibuprofen nanocomposite as a controlled releasing medication. *Nanomedicine Research Journal, 2*(1), 28-35.

Taware, P. V., Chautmal, P. G., Waghmare, A. D., Kate, S. A., Kharat, O. R., & Potdukhe, D. A. (2020). Mycosynthesis of silver nanoparticles from mycelium extract of penicillium species and its antimicrobial activity against clinical isolates. *Journal of Global Biosciences, 9*(8), 7808–7825.

Thatoi, P., Kerry, R. G., Gouda, S., Das, G., Pramanik, K., Thatoi, H., & Patra, J. K. (2016). Photo-mediated green synthesis of silver and zinc oxide nanoparticles using aqueous extracts of two mangrove plant species, Heritiera fomes and Sonneratia apetala and investigation of their biomedical applications. *Journal of Photochemistry and Photobiology. B, Biology, 163*, 311–318. doi:10.1016/j.jphotobiol.2016.07.029 PMID:27611454

Topete, A., Alatorre-Meda, M., Iglesias, P., Villar-Alvarez, E. M., Barbosa, S., Costoya, J. A., Taboada, P., & Mosquera, V. (2014). Fluorescent drug-loaded, polymeric-based, branched gold nanoshells for localized multimodal therapy and imaging of tumoral cells. *ACS Nano, 8*(3), 2725–2738. doi:10.1021/nn406425h PMID:24571629

Tran, Q. H., & Le, A. T. (2013). Silver nanoparticles: Synthesis, properties, toxicology, applications and perspectives. *Advances in Natural Sciences: Nanoscience and Nanotechnology, 4*(3), 033001. doi:10.1088/2043-6262/4/3/033001

Umar, K., Haque, M. M., Mir, N. A., Muneer, M., & Farooqi, I. H. (2013). Titanium dioxide-mediated photocatalysed mineralization of two selected organic pollutants in aqueous suspensions. *Journal of Advanced Oxidation Technologies, 16*(2), 252–260. doi:10.1515/jaots-2013-0205

Unzueta, U., Saccardo, P., Domingo-Espín, J., Cedano, J., Conchillo-Solé, O., García-Fruitós, E., Céspedes, M. V., Corchero, J. L., Daura, X., Mangues, R., Ferrer-Miralles, N., Villaverde, A., & Vázquez, E. (2014). Sheltering DNA in self-organizing, protein-only nano-shells as artificial viruses for gene delivery. *Nanomedicine; Nanotechnology, Biology, and Medicine, 10*(3), 535–541. doi:10.1016/j.nano.2013.11.006 PMID:24269989

Verma, J., Lal, S., & Van Noorden, C. J. (2014). Nanoparticles for hyperthermic therapy: Synthesis strategies and applications in glioblastoma. *International Journal of Nanomedicine, 9*, 2863. PMID:24959075

Vijayakumar, S., & Ganesan, S. (2012). Gold nanoparticles as an HIV entry inhibitor. *Current HIV Research, 10*(8), 643–646. doi:10.2174/157016212803901383 PMID:22954307

Volodina, K. V., Avnir, D., & Vinogradov, V. V. (2017). Alumina nanoparticle-assisted enzyme refolding: A versatile methodology for proteins renaturation. *Scientific Reports, 7*(1), 1–7. doi:10.103841598-017-01436-6 PMID:28469182

Wang, J., Lee, J. S., Kim, D., & Zhu, L. (2017). Exploration of zinc oxide nanoparticles as a multitarget and multifunctional anticancer nanomedicine. *ACS Applied Materials & Interfaces, 9*(46), 39971–39984. doi:10.1021/acsami.7b11219 PMID:29076344

Wang, L., Hu, C., & Shao, L. (2017). The antimicrobial activity of nanoparticles: Present situation and prospects for the future. *International Journal of Nanomedicine*, *12*, 1227–1249. doi:10.2147/IJN.S121956 PMID:28243086

Wang, M., Wang, S., Yang, H., Ku, W., Yang, S., Liu, Z., & Lu, G. (2020). Carbon-based electrocatalysts derived from biomass for oxygen reduction reaction: A minireview. *Frontiers in Chemistry*, *8*, 116. doi:10.3389/fchem.2020.00116 PMID:32185161

Wang, S., Li, J., Zhou, Z., Zhou, S., & Hu, Z. (2019). Micro-/nano-scales direct cell behavior on biomaterial surfaces. *Molecules (Basel, Switzerland)*, *24*(1), 75. doi:10.3390/molecules24010075 PMID:30587800

Wang, S., Liu, K., Yao, X., & Jiang, L. (2015). Bioinspired surfaces with superwettability: New insight on theory, design, and applications. *Chemical Reviews*, *115*(16), 8230–8293. doi:10.1021/cr400083y PMID:26244444

Wilkinson, L. J., White, R. J., & Chipman, J. K. (2011). Silver and nanoparticles of silver in wound dressings: A review of efficacy and safety. *Journal of Wound Care*, *20*(11), 543–549. doi:10.12968/jowc.2011.20.11.543 PMID:22240850

World Health Organization. (2020). *Antimicrobial resistance*. https://www.who.int/news/item/20-11-2020-world-leaders-join-forces-to-fight-the-accelerating-crisis-of-antimicrobial-resistance

Wu, W., Wu, Z., Yu, T., Jiang, C., & Kim, W. S. (2015). Recent progress on magnetic iron oxide nanoparticles: Synthesis, surface functional strategies and biomedical applications. *Science and Technology of Advanced Materials*, *16*(2), 023501. doi:10.1088/1468-6996/16/2/023501 PMID:27877761

Xu, P., Zeng, G. M., Huang, D. L., Feng, C. L., Hu, S., Zhao, M. H., Lai, C., Wei, Z., Huang, C., Xie, G. X., & Liu, Z. F. (2012). Use of iron oxide nanomaterials in wastewater treatment: A review. *The Science of the Total Environment*, *424*, 1–10. doi:10.1016/j.scitotenv.2012.02.023 PMID:22391097

Yadav, K. K., Singh, J. K., Gupta, N., & Kumar, V. (2017). A review of nanobioremediation technologies for environmental cleanup: A novel biological approach. *J Mater Environ Sci*, *8*(2), 740–757.

Yamada, M., Foote, M., & Prow, T. W. (2015). Therapeutic gold, silver, and platinum nanoparticles. *Wiley Interdisciplinary Reviews. Nanomedicine and Nanobiotechnology*, *7*(3), 428–445. doi:10.1002/wnan.1322 PMID:25521618

Yang, X., Salado-Leza, D., Porcel, E., González-Vargas, C. R., Savina, F., Dragoe, D., Remita, H., & Lacombe, S. (2020). A facile one-pot synthesis of versatile PEGylated platinum nanoflowers and their application in radiation therapy. *International Journal of Molecular Sciences*, *21*(5), 1619. doi:10.3390/ijms21051619 PMID:32120829

Yaqoob, A. A., Parveen, T., Umar, K., & Mohamad Ibrahim, M. N. (2020). Role of nanomaterials in the treatment of wastewater: A review. *Water (Basel)*, *12*(2), 495. doi:10.3390/w12020495

Yusof, F., & Ismail, N. A. S. (2015). Antioxidants effects of platinum nanoparticles: A potential alternative treatment to lung diseases. *Journal of Applied Pharmaceutical Science*, *5*(7), 140–145. doi:10.7324/JAPS.2015.50722

Zazo, H., Colino, C. I., & Lanao, J. M. (2016). Current applications of nanoparticles in infectious diseases. *Journal of Controlled Release*, *224*, 86–102. doi:10.1016/j.jconrel.2016.01.008 PMID:26772877

Zhang, R., Pan, D., Cai, X., Yang, X., Senpan, A., Allen, J. S., Lanza, G. M., & Wang, L. V. (2015). ανβ3-targeted copper nanoparticles incorporating an Sn 2 lipase-labile fumagillin prodrug for photoacoustic neovascular imaging and treatment. *Theranostics*, *5*(2), 124–133. doi:10.7150/thno.10014 PMID:25553103

Zhou, M., Tian, M., & Li, C. (2016). Copper-based nanomaterials for cancer imaging and therapy. *Bioconjugate Chemistry*, *27*(5), 1188–1199. doi:10.1021/acs.bioconjchem.6b00156 PMID:27094828

ADDITIONAL READING

Makvandi, P., Wang, C. Y., Zare, E. N., Borzacchiello, A., Niu, L. N., & Tay, F. R. (2020). Metal-based nanomaterials in biomedical applications: Antimicrobial activity and cytotoxicity aspects. *Advanced Functional Materials*, *30*(22), 1910021. doi:10.1002/adfm.201910021

Singh, K. R., Nayak, V., Singh, J., Singh, A. K., & Singh, R. P. (2021). Potentialities of bioinspired metal and metal oxide nanoparticles in biomedical sciences. *RSC Advances*, *11*(40), 24722–24746. doi:10.1039/D1RA04273D

KEY TERMS AND DEFINITIONS

Antimicrobial Activity: Inhibition or killing of growth of microorganisms through preventing the formation of colonies.

Biocompatibility: Compatibility of agent (drug) towards living organisms/cells/tissue.

Biodegradability: Eco-friendly degradation of materials by the action of soil microbes.

Biodistribution: Travelling and retaining of the agent (Drug/nanoparticles) inside the living system.

Biosynthesized Nanomaterial: Nano-sized material which is synthesized using biological entities such as microbes and plant extract.

Drug Resistant: Reduction in effectiveness of drug/antibiotic, results in delaying in curing of disease or infection.

Nanomaterials: A metal or metal oxide material, its size ranges from 1-100nm.

Chapter 15
Nanotechnology for Drug Delivery and Cancer Therapy

Israt Jahan
Dicle University, Turkey

ABSTRACT

Nanotechnology facilitates exclusive opportunities for investigating and controlling a wide range of biomedical approaches at nano-scale, which could provide a groundbreaking impact on medicine, biology, and other health-related areas. This latest technology offers certain distinctive benefits in different aspects such as nanocarriers of targeted drug delivery system, diagnosis and treatment of cancer, vaccination and immunotherapy, biosensing and imaging for clinical diagnosis. A wide range of nanomaterials with biomedical prominence have already been exploited for drug delivery and cancer therapy comprising, solid lipid nanoparticles, metallic nanoparticles, polymeric nanoparticles, carbon nanotube, nanogel-based compounds, nanocapsules, magnetic nanoparticles, nanofluids, nanowires, liposomes, etc. Therefore, this chapter reviewed the potential use of different nanomaterials for therapeutic applications, especially for drug delivery and cancer diagnosis and therapy by focusing their suitability and biocompatibility for extensive and safer usages.

INTRODUCTION

With the development of nanotechnology, the formation and production of nanostructured materials with distinctive phenomena, and the utilization of these nanoscale structures in a vast range of fields like biology, chemistry, physics, engineering, computer science, and so on have become more frequent and have been increased day by day. Besides in the last decade, a wide range of profusion investigations has been applied in biomedical science, particularly in drug delivery systems, and diagnostic and therapeutic aspects of human health by using a combination of nanotechnology and biotechnology called nanobiotechnology, which deals with the implication and visualization of materials on the scale of 100 nm down to 1 nm (Faraji & Wipf, 2009). This is due to the fact that compared to the bulk materials, nano-scaled materials possess unique chemical, physical, magnetic, biochemical properties, which facilitate their surface to be modified easily in altered ways for gaining distinctive biological properties

DOI: 10.4018/978-1-7998-8936-6.ch015

and functionalities with enhanced solubility under physiological conditions for accomplishing their specific biomedical applications (A. K. Gupta et al., 2007; Kasemo, 2002). Moreover, nanosized materials can easily be integrated into biomedical devices, and efficiently delivered into most biological systems to interact with different biomolecules both extracellularly and intracellularly. Due to possessing multi-functionality for instance, biocompatible nanostructured materials are considered as the smart biomaterial in tissue engineering (Kim et al., 2013). In such cases, titanium oxide (TiO_2) nanoparticles have been widely utilized for tissue engineering (Fei Yin et al., 2013). The biofluid of bone-substituting materials initiates its first interaction with a thin layer of TiO_2NPs which instinctively occurs on the top surface of metallic titanium (Feng et al., 2003; Hanawa, 2011). On the other hand, metallic nanoparticles can be used for facilitating the stimulation of cellular immune response i.e., T-helper 17 and T-helper

Table 1. Suitable nanosystems for biomedical applications

Nanosystems	Biomedical Applications	References
Carbon based nanostructures	Enhancement of cell specificity	(McDevitt et al., 2007)
Multi-walled and single-walled carbon nanotubes structures	With the purpose of treating breast, ovarian, and bladder cancer, leukemia, lymphoma they are utilized in drug delivery system	(Taghdisi et al., 2011)
Endohedral metallofullerenes	For clinical diagnostic advancement by incorporating them into magnetic resonance imaging and other imaging techniques	(Komatsu, 2005)
Exohedral fullerenes	Suitable photosensitizers can be utilized against various types of malignancies through photodynamic therapy.	(Mroz et al., 2007)
Zinc sulfide quantum dots (ZnS QDs)	Targeted and controlled drugs (5-Fluorouracil) are delivered via ZnS QDs to breast cancer cell	(Iga et al., 2007)
Cadmium telluride quantum dots (CdTe QDs)	Capable of enhancing drug uptake for inhabiting leukemia cancer cells	(Xu et al., 2013)
Near-infrared quantum dots (NIR QDs)	Utilized in visible spectral imaging for lymphoma, pancreatic cancer, liver cancer, and colorectal cancer diagnosis.	(Dubertret, 2002; Parungo et al., 2004)
Dendrimers	Suitable agents for image contrasting. Restructured dendrimers can be utilized as nano-drugs for tumor cell suppression, antiviral and antibacterial activities.	(Svenson & Tomalia, 2005)
Nanoshells	In UV-infrared emission arrays for imaging tumor structures like tumors responsible for breast cancer.	(Nunes et al., 2019) (Nunes et al., 2019)
Nanocrystals	Used to increase the bioavailability of poorly soluble drugs by expanding the construction. Also utilized for overcoming instability and physicochemical-related difficulties	(Gigliobianco et al., 2018)
Nanowires	In biosensors for detecting gene expression, DNA mutation, and disease-induced protein e.g. protein of prostate cancer detections.	(Doucey & Carrara, 2019)
Metallic nanoparticles	Studies demonstrated their angiogenesis and apoptosis effects. Utilized as permeation enhancers, and as the carrier in target-specific drug delivery, Image-contrasting agent ultrasound and magnetic resonance imaging	(Parvanian et al., 2017)
Titanium oxide (TiO_2) nanoparticles	For adhesion of osteoblast and bone regeneration. Used as biological substitutes for maintaining and restoring the regular functionality of diseased or injured tissues/ organs.	(He et al., 2008; Kubota et al., 2004)
Alginate scaffolds integrated around gold-based nanomaterials i.e., Au-nanowires	Utilized for creating a channel through the electrically resistant pore walls of alginate with the purpose of improving electrical interconnection among adjacent cardiac cells.	(Kim et al., 2013)
Carbon nanotubes (CNTs)	Exploited widely for coating bone and cartilage scaffolds, which support osteoblast proliferation by enhancing their mechanical properties.	(Zhao et al., 2010)
Carbon based nanostructures	Implicated in neuronal tissue regeneration since they are capable of promoting growth, differentiation, attachment, neuron, and long-term survival of brain cells.	(Lovat et al., 2005; Mattson et al., 2000)
Biodegradable polylactide-*co*-glycolide nanosystem	Used for transferring AIDS vaccine (HIV-1P24-Nef peptide along with FLiC i.e., flagellin molecule sequenced from *Pseudomonas aeruginosa*) to immune systems.	(Rostami et al., 2017)
Lipid-structured nanoparticles	Used for delivering mRNA vaccine since they are capable of facilitating endosomal escape, protecting mRNA from degradation	(Reichmuth et al., 2016)
Covalent functionalized carboxylated multi-walled carbon nanotubes (MWCNTs)	As nano-carrier, utilized for delivering recombinant dengue envelope (DENV3E) proteins of dengue vaccine	(Versiani et al., 2017)

1 (Marques Neto et al., 2017). Similarly, through electrostatic interactions, oligonucleotide CpG-gold nanoparticle conjugates (CpG-Au@HBc VLP) and self-assembled engineered virus-like particles (VLP) were encapsulated for developing a characteristic immunostimulatory nanocomposite vaccine, where gold nanoparticles help to produce a cellular and humoral immune response by generating enhanced immunogenicity of virus-like particles and oligonucleotide CpG (Wang et al., 2016). By achieving access to most areas of the body and applying physical and chemical approaches for specific biological interaction of nanomaterials, particles at nano-level possess the potential to screen cells within the living system, identify and diagnose illness, and deliver target-specific therapeutics into the disease sites for treatment, which were previously considered as unachievable ways (Patra et al., 2008). Such as a tumor vascular targeting system is developed using a 200 nm-sized perfluorocarbon-based nanostructure, instantaneously capable of delivering drug molecules along with magnetic resonance imaging (MRI) and ultrasound distinction agents (Sumer & Gao, 2008). Moreover, another study also confirmed that polymeric micelles nanostructure of less than 100 nm-sized has possible multifunctional capacity in cancer therapy by detecting tumor cells through ultrasensitive MRI and affecting cancer cells through the release of pH-sensitive medications to target specific sites (Nasongkla et al., 2006). Additionally, nanomaterials are also found to retain various therapeutic applications i.e., for the treatment of tumor and cancer, photothermal therapy; for treating neurodegeneration, impaired or abnormal function of renal, hepatic and lung, cardiovascular disorders and other infectious diseases (J. Li et al., 2016). The nanomaterials commonly used to develop nanotechnology products for biomedical purposes are inorganic and metallic nanoparticles, carbon nanotubes, quantum dots (QDs), dendrimers, fullerenes, polymeric nanoparticles, liposomes, nanowires, nanoshells, nanofibers, nanocrystals, nanosuspension metallic surfaces etc. (Baetke et al., 2015; Canfarotta & Piletsky, 2014; Lalani et al., 2015; Rostami et al., 2014; Liu et al., 2016). A brief description of the biomedical applications of different nanoparticles and nanostructured materials is given by Table 1.

BACKGROUND

Lately, the utilization of nanomaterials for drug delivery and cancer therapy has gained huge attention since they comprise enormous potential for offering a revolutionary and progressive approach to resolve weaknesses and problems of existing treatments (Gupta et al., 2021). This is due to the fact that therapeutic efficacy can be improved by loading chemotherapy agents into the nanostructured carrier, which also decreases non-specific toxicity, and hence, confirms safe delivery of drugs with more effective cancer treatment (Li et al., 2021). Consequently, remarkable energies have been implied in recent times for developing multi-functional nanosystems ensuring target-based drug delivery for cancer therapy, which can also observer and follow-up therapeutic response of the drugs to the tumor by the visual screening of cancer cells (Nayak et al., 2021). Hence, this chapter will review the potential use of different nanomaterials for therapeutic applications, especially for drug delivery and cancer diagnosis and therapy. Moreover, this discussion will focus on the eminent compensations of nano-based drug carriers and the molecular mechanism of nanomedicine for targeting cancer cells and also present the future directions and perspectives of nanotechnology in these aspects.

NANOMATERIALS IN DRUG DELIVERY SYSTEM

Because of retaining smaller dimensions and structural similarities, nanomaterials demonstrate excellent intracellular uptake like many other biomolecules, which offers them to become suitable candidates in target-specific drug delivery for the treatment of cancer and other critical diseases (Mitchell et al., 2021; Zhang et al., 2020). Besides, materials at nano-scale retain a wider surface area ratio to their total volume that helps them to load a large number of targeted agents compared to other carriers; for instance, a polymer-drug conjugate can contain 9 drug molecules per molecule while a polymeric nanoparticle with 70 nm diameter surface area is capable of carrying nearly 2,000 drug molecules (Raj et al., 2021). Considering these extraordinary capacities, nanomaterials therefore can be utilized as nanovectors/nanocarriers in the form of solid lipid nanoparticles, nanostructured lipid carriers, nanoshells, liposomes with the aim of delivering target-specific drug to treat cancer other serious health-related disorders (Shaw et al., 2014). Nano-drug preparation for drug delivery requires intensive processes of NPs including nanoformulations i.e., controlled fabrication of nanoparticles, optimization of particle size, shape, flexibility, core and surface structure, stabilization of particle charge with the intention of determining their targeting-ability and toxicity as well as their *in vivo* dispersal capacity (Sumer & Gao, 2008). Moreover, nanoparticle-drug complexes must accomplish specific conditions for their successful and target-specific action, 1) nanoparticles must have binding capacity to targeted drugs; 2) for allowing extensive and pervasive delivery of the nanoparticle-drug complex, it should stay constant and unchangeable in the body serum; 3) by enhancing permeability and increasing receptor-mediated interactions or retention (EPR) effect, a successful transportation of the complex to tumor cells is needed to be done in order to decrease the risk of any undesirable difficulties due to non-specific transfer; 4) once the complex is delivered to the targeted cells, the nanoparticle of the complex must have the ability to discharge the drug for achieving its maximum action; 5) for allowing safe degradation, the carrier residues (nanoparticles) must be biocompatible and must have limited lifespan with proper channel of disposal from biological system. In case of non-biodegradable nanoparticles moreover, dose-dependent nontoxic level must be optimized or proven to be safe.

The Nanoparticle-Drug Complex

For using nanomaterials as carriers in nanoparticle-drug complex, the targeted drug can be encapsulated inside the NPs or tagged on their surface for avoiding denaturation or degradation of the drug molecules. Liposomal nanoparticles, inorganic nanoparticles, mesoporous silica, polymeric nanomaterials, solid lipid are vastly employed as nano-carriers for drug delivery systems.

A liposome is a spherical shaped self-assembling concentric lipid i.e., phospholipids bilayer structure that is mainly fabricated from natural non-toxic phospholipids and cholesterol. This nano-carrier retains an interior aqueous space, which can be utilized for encoding hydrophilic drugs in a nanocapsulation system with the purpose of releasing drugs in the cells after the attachment with the cell membrane and the process of endocytosis (Akbarzadeh et al., 2013). The liposome is one of the most biocompatible carriers suitable for drug or nutrient delivery e.g., lipid nanoparticles in DNA vaccines and mRNA vaccines (Cevc, 2012). The best example of liposome-based drug delivery is target-specific delivery of amphotericin B for treating leishmaniasis and severe fungal infections with the aim of reducing its neurotoxicity (Akbarzadeh et al., 2013).

Solid lipid nanoparticles (SLNs) are developed as colloidal nano-carrier, alternative to previously available carrier system i.e., polymeric nanoparticles, liposomes, and emulsions. SLNs consist of liquid lipid (oil) containing rigid core surrounded by a phospholipid solid monolayer, which is comparatively cheaper than polymeric/surfactant based carriers. This structure provides many distinctive features like having improved pharmaceutical stability, target-specific and well-organized drug release, high drug loading capacity, large surface area, smaller size, which especially liposomes (Cavalli et al., 1993; Mukherjee et al., 2009). SLNs are utilized for topical applications with the purpose of delivering numerous medications e.g. glucocorticoids (Cervantes et al., 2019), flurbiprofen (Jain et al., 2005), DNA (Choi et al., 2006), ketoconazole (Santos Maia et al., 2002), isotretinoin (J. Liu et al., 2007), vitamin A (Jenning et al., 2000), imidazole antifungals (Souto & Müller, 2006), tropolide (Mei et al., 2005). Moreover, SLNs are also used for target specific anticancer drug delivery to solid tumor, lymph node metastases and breast cancer as well as for antitubercular chemotherapy (Wong et al., 2006; Pandey et al., 2005).

Along with nanodiamonds (NDs), graphene, fullerenols, nanohorns, and nanotubes, various carbon nanostructures are extensively utilized as a carrier for target specific delivery of a wide spectrum of therapeutic agents because of possessing versatile surface properties and high loading capacity of these carbon-based nanostructures (Ajima et al., 2008; Chaudhuri et al., 2009; Chow et al., 2011; Liu et al., 2008; Podesta et al., 2009). Moreover, the distinctive surface property of these nano-carbons provide the capacity of absorbing light from the near-infrared (NIR) region, which allows them heating up easily due to their electrical and thermal conductivity, and consequently, capable of targeting tumor cells (Brennan et al., 2003). For instance, polyhydroxylated fullerenes (fullerenols) have been utilized for loading a cancer chemotherapeutic agent (doxorubicin) and successfully delivered to tumor cells with the purpose of extending these chemotherapeutic agents and reducing toxicity to healthy cells (Chaudhuri et al., 2009).

Polymeric nanoparticles (PNPs) are communally referred to as polymer micelles, nanocapsules, nanospheres, or polymer nanoparticle, which were the first reported nanomaterials used as the carrier for drug delivery, especially for hydrophobic drugs (Jin et al., 2020). Having the capacity of interacting through a hydrophobic core or forming covalent bond, they are able to incarcerate hydrophobic drugs. In an aqueous solution, polymeric nanoparticles can create a hydrophobic and hydrophilic block by assembling from amphiphilic polymers, which helps them performing a quick self-assembly through hydrophobic interfaces (Tian et al., 2018). These blocks can be switched which helps PNPs to carry hydrophilic charged molecules like nucleic acids, peptides, proteins by allowing interactions in the core and neutralizing the charge (Duncan, 2003). Moreover, modifications of nanomaterial-based drug delivery system of polymeric nanoparticles are performed by incorporating a number of target-based ligands and stimuli responsive groups into polymeric chains to increase their loading capacity Poly(alkylcyanoac rylate), polyglycolic acid, poly(alkylcyanoac rylate), poly(lacti de-coglycolide)(PLGA), poly(e-cap-rolactone), polylactic acid (PLA) are the most important synthetic biodegradable polymeric nanoparticles that have been intensively studied in recent times in drug delivery systems (Faraji & Wipf, 2009). For example, polylactic acid (PLA) and poly(lacti de-coglycolide)(PLGA) has been approved as biodegradable polymeric nanoparticles by U.S food and drug administration (FDA) for human use (Faraji & Wipf, 2009). Besides, FDA also approved a natural polymeric nanoparticle-bound paclitaxel (derived from albumin) for treating metastatic breast cancer patients (Schoppa et al., 2021; Stylianopoulos et al., 2012).

Because of retaining unique physical encasement for sheltering and guarding a drug from denaturation or degeneration, mesoporous silica nanoparticles have become quite popular for their extensive studies in drug delivery system. This is due to the fact that the surface pores of these nanomaterials can serve as the central reservoir for loading the drugs or the whole nanoparticle itself is capable of forming worm-like network complex of channels, which in a controlled manner permit delivering comparatively larger doses of a drug (Botterhuis et al., 2006). Moreover, the release kinetics of the drug payload is related to the distribution of the pore size, and studies have been done to measure capping capacity of the surface pores of this nanosystem as well as to determine the "zero premature release" functionality i.e., determining the physically blocking capability of mesoporous silica nanoparticle by holding the dugs until it reached its targeted place (Botterhuis et al., 2006). Mesoporous silica nanoparticles like carrier systems are capable of carrying membrane-impermeable drugs to the targeted biological system; hence as universal transmembrane carriers, they can be utilized not only for the purpose of intracellular drug delivery but also for imaging procedures (Radu et al., 2004; Zhao et al., 2010).

NANOMATERIALS IN CANCER THERAPY

Cancer is the second most lethal disease that causes approximately 15.7% of annual deaths worldwide (Wang et al., 2016). The most frequent cancer types in females are cervical cancer, breast cancer, lung cancer, and colorectal cancer while in the male are stomach cancer, colon cancer, liver cancer, colorectal cancer, prostate cancer, and lung cancer (Jayasekara et al., 2016). Existing drugs for chemotherapy, like cisplatin, bleomycin, daunorubicin, doxorubicin, etc. retain limitations i.e., having non-specific action, cellular resistance and poor penetration capacity of drugs into tumors, poor water solubility, toxicity to healthy cells, costly maintenances, severe side effects, and by developing resistance to these chemotherapeutic drugs (Boca et al., 2011). To conquer this situation, therefore, discovering and developing alternative treatments, kits, implements, and drugs are as important as progressing early detection of cancer. Applications of nanotechnology for developing nanoformulations and nano-drug particles can open a new promising opportunity. In recent times, the utilization of nanomaterials for detecting, screening, and treating cancer biology purposes has attracted huge attention from many researchers (Huang et al., 2011). Regarding tumor treatments, on the other hand, nanomaterial-based drugs retain versatile capabilities since they are capable of reaching targeted sites by invading lots of obstacles through blood flow, releasing drugs in a controlled manner with minimum loss of drug, which thereby, help to exert effective anti-cancer effects (Cheng et al., 2021). Besides, biocompatible nanomaterials at particular intracellular spaces are capable of interacting with biocompounds, which allows their interference in cancer development and proliferation pathways (Bae et al., 2011).

For all kinds of cancer therapy, medical imaging is the prerequisite step for their clinical diagnosis, or treating diseases. Small molecules, particularly about 1 nm and ≥2000 daltons are utilized in traditional imaging modalities as imaging agents for clinical diagnosis. Nevertheless, nonspecific interaction capacity, poor stability, the low signal intensity of these molecules, and their speedy clearance from the circulation system create a visual problem for imaging. Moreover, nanomaterials as molecular imaging agents are capable of overcoming these restrictions and can provide the lucidity of a large number of cells to be imaged at greater tissue depths by increasing signal intensity, and delivering stable imaging signals over longer periods (Shilo et al., 2012; Thakor et al., 2011). Nanomaterials are capable of playing a vivacious role in cancer diagnosis by incorporating them in the imaging process for the early detection of tumors cells. Nanomaterials have long been employed as contrasting agents, image-guided intercessions, labeling organs, grafts, transplants, and implants, tracking and labeling cancer cells for disease scanning and follow-up (Kiessling et al., 2014). Different experiments have indicated that imaging contrast can be upgraded by acquainting the nano-sized particles with targeted tissues, which could

Figure 1. Diagrammatic illustration showing how nanoparticles can enhance imaging process (Yu et al., 2021)

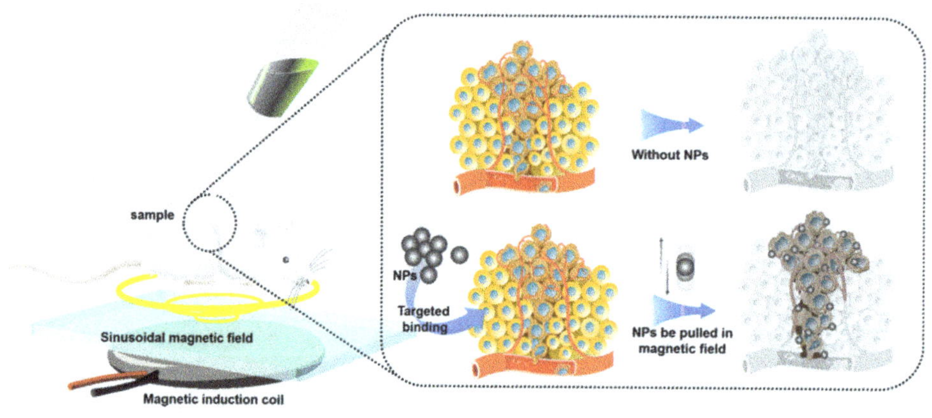

Table 2. Nanosystem used for cancer imaging

Imaging Modality for Cancer Detection	Nanosystem	Cancer Type Detected by Nano-mediated Imaging	References
Magnetic resonance imaging (MRI)	Immune super-paramagnetic iron oxide nanoparticles	Breast cancer, Lung cancer, Liver tumors (ie, hepatocellular carcinoma, liver metastases)	(Jafari et al., 2015; Wan et al., 2016)
Fluorescence/T1-MRI	N-(trimethoxysilylpropyl) ethylenediamine triacetic acid modified manganese(II) oxide nanoparticles (MnO-TETT)	Tumor originating cells in the brain (glioma cells)	(Lai et al., 2018)
MRI/fluorescence imaging	Polyethylene glycol (PEG)-coated and Gadolinium (Gd)-loaded fluorescent nano-silica	Prostate cancer cells	(Jiang et al., 2019)
Computed Tomography (CT)	Tantalum, lanthanide, gold nanoparticles	Solid tumors in different organs	(Shilo et al., 2012)
MRI-guided chemotherapy	Mesoporous silica nanoparticles capped by ultra-small manganese oxide-(USMO@MSNs)	Cervical cancer cells	(Wang et al., 2018)
Ultrasound (US) imaging system	Prodrug-loaded phase-transitional nanoparticles (Glutathione (GSH)-sensitive platinum (IV))	Human ovarian cancer cell	(Huang et al., 2019)
Optoacoustic imaging	Single-walled carbon nanotubes	Solid tumors in different organs	(de la Zerda et al., 2012)
Optical imaging	silica-gold nanoparticles capable of enhancing surface of Raman scattering imaging	Colorectal cancer	(Thakor et al., 2011)

Table 3. Nanostructured materials used for delivering anticancer drugs

Drugs	Nanomaterials as Carrier	Targeted Cancer or Tumor Cells	References
1:5 daunorubicin and cytarabine	Liposome	Acute myeloid leukemia	(Tardi et al., 2009)
Doxorubicin	Solid lipid nanoparticles (SLNs)	Solid tumor	(Wong et al., 2006)
Daunorubicin and doxorubicin (chemotherapeutic drugs)	Liposome	Solid tumor	(Mamot et al., 2012)
1:1 floxuridine and irinotecan	Liposome	Colorectal cancer	(Harasym et al., 2006)
Doxorubicin and gemcitabine	Polymeric nanoparticle	Prostate cancer and several other tumor cells	(Lammers et al., 2009)
Doxorubicin, BCL2-targeted siRNA, and MRP1-targeted siRNA	Liposome	Lung cancer	(Saad et al., 2008)
1:7 cisplatin and irinotecan	Liposome	Small cell lung carcinoma	(Tardi et al., 2009)
siRNA and Doxorubicin	Dendrimers	Malignant brain cancer (glioblastoma)	(Kaneshiro & Lu, 2009)
Verapamil and vincristine	Polymeric nanoparticle	Breast cancer	(Song et al., 2009)
Doxorubicin and combretastatin	Polymeric nanoparticle	Malignant melanoma, lung carcinoma, and other cancer types	(Sengupta et al., 2005)
Cisplatin	Single-wall carbon nanohorns (conical shaped nanocarbon)	Human lung cancer	(Ajima et al., 2008)
BCL2-targeted siRNA	Gold-based spherical nucleic acid nanoparticles	Human brain tumor (glioma)	(Jensen et al., 2013)
ID4 siRNA	Peptide nanocomplexes	Human ovarian cancer oncogene	(Ren et al., 2012)
Epirubicin	Nanodiamond-lipid hybrid particles (NDLPs)	Breast cancers	(Moore et al., 2013)
siRNA (Proprietary)	Multiwalled carbon nanotubes (MWNTs)	Human lung cancer	(Podesta et al., 2009)
Paclitaxel (PTX)	Single-walled carbon nanotubes (SWNTs)	Breast cancers	(Liu et al., 2008)
Methotrexate and all-trans retinoic acid	Polyamidoamine (PAMAM) dendrimers	Leukemia	(Tekade et al., 2009)
Bcl-2-targeted siRNA and Paclitaxel	Polymeric nanoparticles (cationic core–shell nanoparticles)	Breast cancer	(Wang et al., 2006) `
Daunorubicin and 6-mercaptopurine (dual incorporation)	Liposome	Acute lymphocytic leukemia	(Agrawal et al., 2005)
Doxorubicin	Polyhydroxylated fullerenes (fullerenols)	Non-specific tumor cells	(Chaudhuri et al., 2009)
Doxorubicin	Nanodiamond	Liver tumor and mammary carcinoma	(Chow et al., 2011)

offer advanced image guidance for disease detection, even for critical surgery (Smith & Gambhir, 2017). In cryosurgery for improving the therapeutic effect, the imaging eminence of ice ball edges and tumor can be enhanced by using nanoparticles in the imaging system, and thereby, can be covered the ice balls precisely to overcome inadequate freezing to the targeted cancer cells without causing any inevitable injury to healthy tissue (Hou et al., 2018).

Metallic nanoparticles are mainly utilized for cancer imaging, such as iron oxide nanoparticles exhibit distinctive mechanical, acoustic, magnetic, and optical characteristics, capable of enhancing images, which is crucial for any kind of critical tumor cell diagnosis (Fig. 1). In magnetic resonance imaging (MRI) system, immune super-paramagnetic iron oxide nanoparticles can be costumed for detecting cancer cell lines, which is suitable for detecting metastases in lung cancer without any known side effects (Wan et al., 2016). Moreover, this type of super-paramagnetic nanosystem is suitable for MRI imaging, and can also be used as building blocks for aerosols in breast cancer diognosis (Jafari et al., 2015). Moreover, Table 2 is providing a brief list of some latest examples of nanomaterials-mediated imaging modality used for cancer detection.

With the help of recent technology, on the other hand, non-cytotoxic pro-drugs can be attached with nanoparticles and delivered to the targeted tumor cells. Pro-drugs are usually chemotherapeutic agents that are inserted in the biological system in their inactive form and can be activated after reaching the target cancer cells. For example, nano-platinum (PtNPs) are used as nano-carried to attached anticancer drugs, and inside the targeted tumor cells, photo-reduction of Pt[IV] pro-drug to the active Pt [II] pro-drug is completed using visible light to avoid unwanted toxicity to healthy cells (Blanco et al., 2009). Since biodegradable polyester nanomaterials possess excellent biocompatibility, hemocompatibility, they are widely are utilized as vehicles for delivering chemotherapeutic agents in most cancers treatment (Begines et al., 2020).

Solid lipid nanoparticles (SLNs) as a carrier was utilized for loading mitoxantrone, and formulated for local injections with the purpose of stopping the growth and spreading of cancer cells with cells, and thereby, reduced toxicity to healthy cells by improving the safe and biocompatible release of drugs (Lu et al., 2006). Similarly, SLNs have been reported to enhance the proficiency of doxorubicin (Dox), a chemotherapy medicine for cancer treatment (Wong et al., 2006). Moreover, the progressing clinical trials are designed for treating hematological malignancies and solid tumors patients by exploring the effects of liposomes loaded chemotherapeutic drugs i.e., daunorubicin and doxorubicin (Mamot et al., 2012). Moreover, the application of nanosystems in drug delivery and cancer therapy are given in Table 3.

Cytotoxic Mechanism of Inorganic Metallic Nanoparticles Against Cancer Cells

Nanoparticles are capable of showing more toxicity on cancerous cells than healthy cells; for instance, biologically synthesized silver nanoparticles from *Beta vulgaris* showed higher toxicity to human larynx carcinoma cancer (Hep-2), lung cancer (A549), and breast cancer (MCF-7) cell lines compared to healthy cells (human breast cell (Hbl100) and African green monkey kidney (Vero) cell lines) (Venugopal et al., 2017).

At the cellular level, nanoparticles are able to combine with cellular macro structures, and DNA and protein like intercellular macromolecules, which increases electron receptivity of cellular oxygen, and escalates the accumulation of highly reactive chemical molecules like ROS (Reactive oxygen species) as a consequence of stimulation of ROS bursts in cells (Akter et al., 2018; Lee et al., 2018). Along with the peroxisomes and endoplasmic reticulum (ER), mitochondria are the predominant organelle in the cell that primarily generate ROS (Murphy, 2009). Ions released by nanoparticles, particularly inorganic

NPs promote the ROS accumulation by increasing the NADP+/NADPH ratio (Lee et al., 2018), reducing the levels of redox potential, interrupting electron-shuttling process (Holt & Bard, 2005), diminishing mitochondrial functions and membrane potentials (MMP) (Kang et al., 2012).

It is remarkable that certain sizes and shapes of nanoparticles have higher capacity of accelerating ROS production in cells (Huang et al., 2019); for instance, anticancer activities of three different sized biogenic silver nanoparticles were evaluated on the human prostate (PC3) and liver (HepG2) cancer cell line (Prasannaraj et al., 2016). Apoptotic cells observation at IC_{50} concentration after 48 hours of incubation, the higher inhibitions of all AgNPs were found against liver cancer cell lines; and the maximum cytotoxicity (IC_{50} values of 10.04 and 42.77 μg/ml for HepG2 and PC3 cell lines, respectively)

Figure 2. Nanoparticle-induced cellular events. ① Lipid peroxidation and cell membrane disruption by the cellular release of nanoparticles; ② NPs induces releasing lysosomal contains and enzymes by destroying their membrane; ③ & ④ mitochondrial matrixes are also discharged because of NPs-induced mitochondrial membrane damage, which decreases ATP generation, ROS production is enlarged by NPs as a result of RNA mistranslation; ⑤ attachment of transfer ribonucleic acid (tRNA) to the ribosome is also prohibited by NPs; ⑥ & ⑦ polymerization of proteins and DNA is induced by increased ROS, which causes DNA mutations; ⑧ by destroying nuclear membrane, nanoparticles provoke the disruption of nuclear structure and release its contents
(Yu et al., 2020)

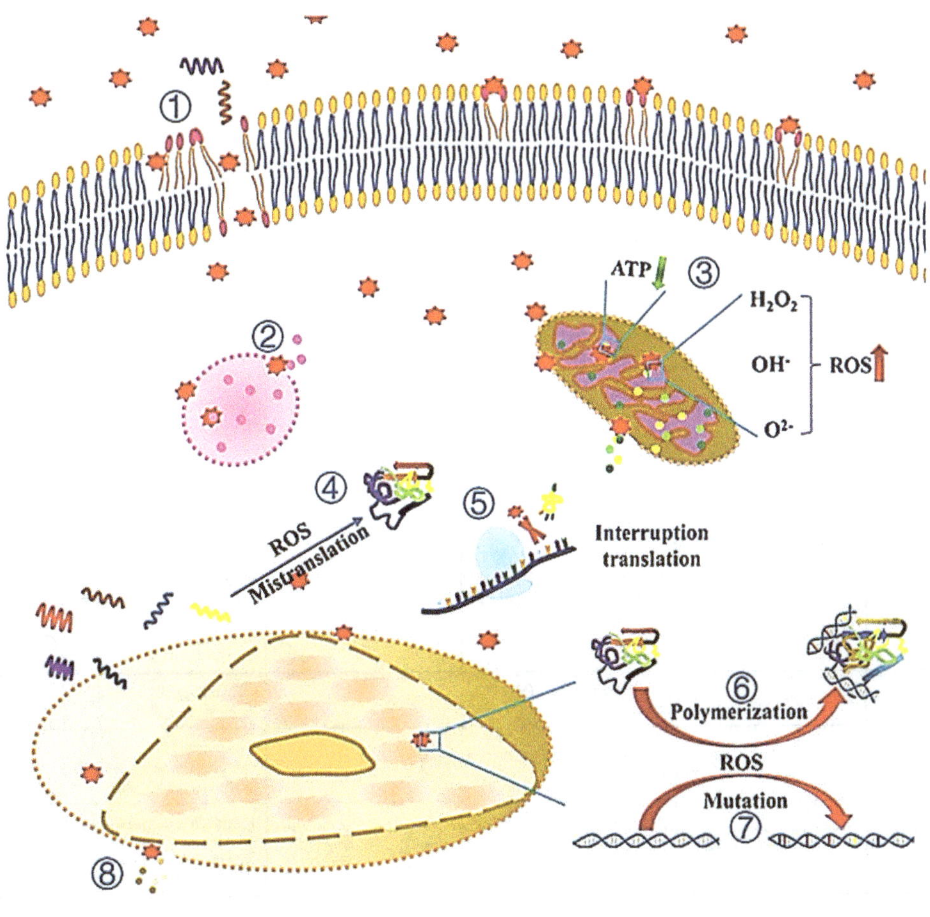

was observed by 60-95 nm-sized silver nanoparticles, which was synthesized from nut extracts of *Semecarpus anacardium* (Prasannaraj et al., 2016). Similarly, intracellular ROS generation can be induced by TiO_2NPs, which harms cell metabolism by damaging protein and nucleic acid (Pelgrift & Friedman, 2013). Genotoxicity of different sized TiO_2NPs was evaluated in study, and the result indicated that 10 nm particle-sized nanoparticles exhibited the maximum toxicity among all tested sizes by inducing higher ROS generation (Liao et al., 2019).

Nanoparticles also provoke oxidative stress by expressing oxidative stress-related genes, i.e., ahpC, oxyR, soxR, and soxS, reduced glutathione level (GSH), increase lipid peroxidation, affects the normal functions of cells, eventually damages membrane structure and important cellular components, increase oxidative damage to biomacromolecules, which leads to cellular apoptosis (Pan et al., 2009, Yu et al., 2020). Pro-apoptosis gene expressions in the cells can be influenced by the dose-dependent applica-

Table 4. In vitro cytotoxic effects of metallic nanoparticles on different cancer cell lines

Nanoparticles	Shape and Size (nm) of NPs	Cancer Cell Lines	Cell Viability Assay for Measuring Cytotoxicity	Effective Cytotoxic Concentration of NPs	References
AgNPs	Cuboid; 60-95 nm	human prostate (PC3) and liver (HepG2) cancer cell line	MTT assay	IC_{50} values of 10.04 and 42.77 µg/ml for HepG2 and PC3 cell lines, respectively	(Prasannaraj et al., 2016)
AgNPs	Spherical; 22 nm	Breast cancer (MCF-7 breast adenocarcinoma cell line)	Sulforhodamine B (SRB) assays	13% toxicity at 6.25 µM	(Loutfy et al., 2015)
AuNPs	Spherical; 36 nm	Breast cancer (MCF-7 breast adenocarcinoma cell line)	Sulforhodamine B (SRB) assays	15% toxicity at 100 µM	(Loutfy et al., 2015)
AgNPs	irregular shaped; 20–50 nm	Breast cancer (MCF-7 cell line)	MTT assay	50% cell viability at 51.3 µg/g concentration	(Devi & Sathishkumar, 2017)
AuNPs	sphere-shaped, circular and triangular; 20–50 nm	Breast cancer (MCF-7 cell line)	MTT assay	IC_{50} value at 44.8 µg/ml concentration	(Devi & Sathishkumar, 2017)
AgNPs	spherical, circular and triangular; ~20 nm	Human Larynx carcinoma cancer (Hep-2), breast cancer (MCF-7), and lung cancer (A549) cells	MTT assay	IC_{50} values were observed at 47.1, 48.2, and 47.6 µg/ml for Hep2, A549, and MCF-7 cell lines	(Venugopal et al., 2017)
AgNPs	5–40 nm; polydispersed	Breast cancer cells (MCF-7)	MTT assay	IC_{50} values from 94.91 µg/ml to 34.51 µg/ml	(Kulandaivelu & Gothandam, 2016)
AgNPs	5 -50 nm; polydispersed	breast cancer cells (MCF-7) and cervical cancer cells (HeLa)	MTT assay	At 100 µg/ml NPs concentration, 58 and 47% mortalities of breast cancer and cervical cancer cell lines, correspondingly	(Kelkawi et al., 2017)
AgNPs	5–25 nm; spherical	colorectal cancer (HT29) cells	MTT assay	The significant IC_{50} values of the HT29 cells at 38.55 µg/ml concentration of AgNPs.	(Ghanbar et al., 2017)
AgNPs	5-50 nm; spherical	human gastric carcinoma (AGS) cells	MTT assay	The IC_{50} value of AgNPs was 21.05 µg/ml on gastric cancer AGS cell line.	(Ardestani et al., 2016)
ZnO	-	lung cancer; SK-MES-1, NCI-H460 and A549 cell lines; cervical cancer cells (HeLa)	cell colorimetric assays	For all the studied cell lines ZnO was highly toxic; lethal dose of ≥100 µg/ml with at least 50% cell mortality.	(Lozano et al., 2011)
AgNPs	22 nm; spherical	human breast cancer (MCF-7) cells	MTT assay	50 µg/ml AgNPs induced complete mortality.	(Jeyaraj et al., 2013)
TiO_2		human cervical cancer cells (HeLa)	MTT assay	34.45 µg/ml NPs concentration was identified as the average IC_{50} value	(Narayanan et al., 2021)
Cd NPs	11.2–18.6 nm; hexagonal	glioblastoma (U87), colon cancer (HT29), breast cancer (MCF-7), lung cancer (A549), cell lines	MTT assay	Maximum toxicity on glioblastoma (U87) cells with the IC_{50} value of 62.9 ± 3.3 µg/ml	(Shakibaie et al., 2021)

tion of nanoparticles. In a study, silver nanoparticles were applied on human stomach cancer (gastric carcinoma (AGS)) cells line and measured the expression levels of pro-apoptotic Bax and anti-apoptotic Bcl-2genes at the mRNA level using real-time polymerase chain reaction (Ardestani et al., 2016). Silver nanoparticles expose to cells considerably reduced the anti-apoptotic Bcl-2 gene expression while up-regulated the expression of pro-apoptotic Bax gene; and thereby, the ratio of Bax/Bcl-2 was increased (Ardestani et al., 2016). Moreover, the possible mechanism of cytotoxicity induction at the cellular level by NPs is presented in Figure 2.

Several studies have already confirmed the significant anticancer potential of nanoparticles although most NPs showed their dose-dependent effects; such as chemically synthesized silver (22 nm) and gold (36 nm) nanoparticles were applied against breast cancer (MCF-7 breast adenocarcinoma cell line), where both the NPs showed their maximum cytotoxicity after 24 h of treatment at the concentration of 6.25 μM and 100 μM NPs, respectively (Loutfy et al., 2015). In another similar study, AuNPs displayed stronger cytotoxic effect then AgNPs at different experimental concentrations (1, 10, 25, 50 and 100 μg/ml of 1 mM NPs) (Devi & Sathishkumar, 2017). After 24 hours of treatment with both the 20–50 nm size ranged nanoparticles, MCF7 breast cancer cells showed a 50% reduction in viability (IC_{50}) at the concentration of 51.3 and 44.8 μg/ml for silver and gold nanoparticles, respectively (Devi & Sathishkumar, 2017). A brief list of *in vitro* cytotoxic effects of inorganic metallic nanoparticles on different cancer cell lines is given in Table 4.

FUTURE RESEARCH DIRECTIONS

Nanomaterials possess diverse composition and morphology, immunogenicity, hydrophobicity, magnetism, and other unique properties, which could be utilized vastly in drug delivery, cancer therapies and other biomedical applications by enhancing their utility through more expended and intensive research. For instance, surface modifications of different nano-structured materials can be accomplished to enhance the loading capacity of nanocarriers with the purpose of delivering conventional chemotherapy drugs. Compared to traditional immunotherapy moreover, nanoformulations have established their improved efficiency; however, *in vivo* tolerant testing in models and side effects of this nanomaterial based therapeutics through clinical trials remains insufficient, and hence, further investigations are required to confirm the broad-spectrum utility as well as to evaluate the safely of these new approaches. Achieving these however, it is necessary to understand and evaluate the potential toxicity of nanomaterials since their interactive effect in biological systems is crucial for their potential applications in biomedical science. Therefore, association of physicochemical properties of nanomaterials and their toxic effects should continuously be monitored and measured for minimizing false results as well as confirming their progressive utility for advanced clinical therapy.

CONCLUSION

Nano-based technology has become progressively advantageous in healthcare sectors, which led to their multipurpose utilization in different biomedical innovations such as for employing implantable nano-devices in the biological system, in drug delivery systems including nanocarriers and nano-delivery of medicine, nucleic acid delivery, for marker identification, in biosensing for locating the site of nanomedicine and recognizing their spreading nature, in imaging for target and site-specific binding identification of particles, and diagnosis of diseases. This chapter presents an overall scenario of the biomedical applications of nanostructured materials regarding the types of biomedical nanosystems and their utilizations in imaging, drug delivery, cancer therapy, etc. Previously published literature, nanotechnology reports, and other pharmacological evidence suggested that medical science might be upgraded dramatically by exploiting nanotechnology for the progression of clinical monitoring and diagnosis as well as for developing advanced medicine with minor hazards.

REFERENCES

Agrawal, V., Paul, M. K., & Mukhopadhyay, A. K. (2005). 6-mercaptopurine and daunorubicin double drug liposomes—Preparation, drug-drug interaction and characterization. *Journal of Liposome Research*, *15*(3–4), 141–155. doi:10.1080/08982100500364081 PMID:16393906

Ajima, K., Murakami, T., Mizoguchi, Y., Tsuchida, K., Ichihashi, T., Iijima, S., & Yudasaka, M. (2008). Enhancement of in vivo anticancer effects of cisplatin by incorporation inside single-wall carbon nanohorns. *ACS Nano*, *2*(10), 2057–2064. doi:10.1021/nn800395t PMID:19206452

Akbarzadeh, A., Rezaei-Sadabady, R., Davaran, S., Joo, S. W., Zarghami, N., Hanifehpour, Y., Samiei, M., Kouhi, M., & Nejati-Koshki, K. (2013). Liposome: Classification, preparation, and applications. *Nanoscale Research Letters*, *8*(1), 102. doi:10.1186/1556-276X-8-102 PMID:23432972

Akter, M., Sikder, M., Rahman, M. M., Ullah, A. K. M. A., Hossain, K. F. B., Banik, S., Hosokawa, T., Saito, T., & Kurasaki, M. (2018). A systematic review on silver nanoparticles-induced cytotoxicity: Physicochemical properties and perspectives. *Journal of Advanced Research*, *9*, 1–16. doi:10.1016/j.jare.2017.10.008 PMID:30046482

Ardestani, M. S., Sadat Shandiz, S. A., Salehi, S., Ghanbar, F., Darvish, M. R., Mirzaie, A., & Jafari, M. (2016). Phytosynthesis of silver nanoparticles using Artemisia marschalliana Sprengel aerial part extract and assessment of their antioxidant, anticancer, and antibacterial properties. *International Journal of Nanomedicine*, *1835*, 1835. Advance online publication. doi:10.2147/IJN.S99882

Bae, K. H., Chung, H. J., & Park, T. G. (2011). Nanomaterials for cancer therapy and imaging. *Molecules and Cells*, *31*(4), 295–302. doi:10.100710059-011-0051-5 PMID:21360197

Baetke, S. C., Lammers, T., & Kiessling, F. (2015). Applications of nanoparticles for diagnosis and therapy of cancer. *The British Journal of Radiology*, *88*(1054), 20150207. doi:10.1259/bjr.20150207 PMID:25969868

Begines, B., Ortiz, T., Pérez-Aranda, M., Martínez, G., Merinero, M., Argüelles-Arias, F., & Alcudia, A. (2020). Polymeric nanoparticles for drug delivery: Recent developments and future prospects. *Nanomaterials (Basel, Switzerland)*, *10*(7), 1403. doi:10.3390/nano10071403 PMID:32707641

Blanco, N. G., Maldonado, C. R., & Mareque-Rivas, J. C. (2009). Effective photoreduction of a Pt(iv) complex with quantum dots: A feasible new light-induced method of releasing anticancer Pt(ii) drugs. *Chemical Communications*, *35*(35), 5257. doi:10.1039/b910000h PMID:19707638

Boca, S. C., Potara, M., Gabudean, A.-M., Juhem, A., Baldeck, P. L., & Astilean, S. (2011). Chitosan-coated triangular silver nanoparticles as a novel class of biocompatible, highly effective photothermal transducers for in vitro cancer cell therapy. *Cancer Letters, 311*(2), 131–140. doi:10.1016/j.canlet.2011.06.022 PMID:21840122

Botterhuis, N. E., Sun, Q., Magusin, P. C. M. M., van Santen, R. A., & Sommerdijk, N. A. J. M. (2006). Hollow silica spheres with an ordered pore structure and their application in controlled release studies. *Chemistry (Weinheim an der Bergstrasse, Germany), 12*(5), 1448–1456. doi:10.1002/chem.200500588 PMID:16308879

Brennan, M. E., Coleman, J. N., Drury, A., Lahr, B., Kobayashi, T., & Blau, W. J. (2003). Nonlinear photoluminescence from van Hove singularities in multiwalled carbon nanotubes. *Optics Letters, 28*(4), 266. doi:10.1364/OL.28.000266 PMID:12653367

Canfarotta, F., & Piletsky, S. A. (2014). Engineered magnetic nanoparticles for biomedical applications. *Advanced Healthcare Materials, 3*(2), 160–175. doi:10.1002/adhm.201300141 PMID:24497448

Cavalli, R., Caputo, O., & Gasco, M. R. (1993). Solid lipospheres of doxorubicin and idarubicin. *International Journal of Pharmaceutics, 89*(1), R9–R12. doi:10.1016/0378-5173(93)90313-5

Cervantes, B., Arana, L., Murillo-Cuesta, S., Bruno, M., Alkorta, I., & Varela-Nieto, I. (2019). Solid lipid nanoparticles loaded with glucocorticoids protect auditory cells from cisplatin-induced ototoxicity. *Journal of Clinical Medicine, 8*(9), 1464. doi:10.3390/jcm8091464 PMID:31540035

Cevc, G. (2012). Rational design of new product candidates: The next generation of highly deformable bilayer vesicles for noninvasive, targeted therapy. *Journal of Controlled Release, 160*(2), 135–146. doi:10.1016/j.jconrel.2012.01.005 PMID:22266051

Chaudhuri, P., Paraskar, A., Soni, S., Mashelkar, R. A., & Sengupta, S. (2009). Fullerenol−cytotoxic conjugates for cancer chemotherapy. *ACS Nano, 3*(9), 2505–2514. doi:10.1021/nn900318y PMID:19681636

Cheng, Z., Li, M., Dey, R., & Chen, Y. (2021). Nanomaterials for cancer therapy: Current progress and perspectives. *Journal of Hematology & Oncology, 14*(1), 85. doi:10.118613045-021-01096-0 PMID:34059100

Choi, M., Kim, J., & Maibach, H. (2006). Topical DNA vaccination with DNA/Lipid based complex. *Current Drug Delivery, 3*(1), 37–45. doi:10.2174/156720106775197484 PMID:16472092

Chow, E. K., Zhang, X.-Q., Chen, M., Lam, R., Robinson, E., Huang, H., Schaffer, D., Osawa, E., Goga, A., & Ho, D. (2011). Nanodiamond therapeutic delivery agents mediate enhanced chemoresistant tumor treatment. *Science Translational Medicine, 3*(73), 73ra21-73ra21. doi:10.1126/scitranslmed.3001713

de la Zerda, A., Bodapati, S., Teed, R., May, S. Y., Tabakman, S. M., Liu, Z., Khuri-Yakub, B. T., Chen, X., Dai, H., & Gambhir, S. S. (2012). Family of enhanced photoacoustic imaging agents for high-sensitivity and multiplexing studies in living mice. *ACS Nano, 6*(6), 4694–4701. doi:10.1021/nn204352r PMID:22607191

Devi, G. K., & Sathishkumar, K. (2017). Synthesis of gold and silver nanoparticles using Mukia maderaspatna plant extract and its anticancer activity. *IET Nanobiotechnology / IET*, *11*(2), 143–151. doi:10.1049/iet-nbt.2015.0054 PMID:28476996

Doucey, M.-A., & Carrara, S. (2019). Nanowire sensors in cancer. *Trends in Biotechnology*, *37*(1), 86–99. doi:10.1016/j.tibtech.2018.07.014 PMID:30126620

Duncan, R. (2003). The dawning era of polymer therapeutics. *Nature Reviews. Drug Discovery*, *2*(5), 347–360. doi:10.1038/nrd1088 PMID:12750738

Faraji, A. H., & Wipf, P. (2009). Nanoparticles in cellular drug delivery. *Bioorganic & Medicinal Chemistry*, *17*(8), 2950–2962. doi:10.1016/j.bmc.2009.02.043 PMID:19299149

Fei Yin, Z., Wu, L., Gui Yang, H., & Hua Su, Y. (2013). Recent progress in biomedical applications of titanium dioxide. *Physical Chemistry Chemical Physics*, *15*(14), 4844. doi:10.1039/c3cp43938k PMID:23450160

Feng, B., Weng, J., Yang, B. C., Qu, S. X., & Zhang, X. D. (2003). Characterization of surface oxide films on titanium and adhesion of osteoblast. *Biomaterials*, *24*(25), 4663–4670. doi:10.1016/S0142-9612(03)00366-1 PMID:12951009

Ghanbar, F., Mirzaie, A., Ashrafi, F., Noorbazargan, H., Dalirsaber Jalali, M., Salehi, S., & Sadat Shandiz, S. A. (2017). Antioxidant, antibacterial and anticancer properties of phytosynthesised Artemisia quttensis Podlech extract mediated AgNPs. *IET Nanobiotechnology / IET*, *11*(4), 485–492. doi:10.1049/iet-nbt.2016.0101 PMID:28530200

Gigliobianco, M., Casadidio, C., Censi, R., & Di Martino, P. (2018). Nanocrystals of poorly soluble drugs: Drug bioavailability and physicochemical stability. *Pharmaceutics*, *10*(3), 134. doi:10.3390/pharmaceutics10030134 PMID:30134537

Gupta, A. K., Naregalkar, R. R., Vaidya, V. D., & Gupta, M. (2007). Recent advances on surface engineering of magnetic iron oxide nanoparticles and their biomedical applications. *Nanomedicine (London)*, *2*(1), 23–39. doi:10.2217/17435889.2.1.23 PMID:17716188

Gupta, P. K., Gahtori, R., Govarthanan, K., Sharma, V., Pappuru, S., Pandit, S., Mathuriya, A. S., Dholpuria, S., & Bishi, D. K. (2021). Recent trends in biodegradable polyester nanomaterials for cancer therapy. *Materials Science and Engineering C*, *127*, 112198. doi:10.1016/j.msec.2021.112198 PMID:34225851

Hanawa, T. (2011). A comprehensive review of techniques for biofunctionalization of titanium. *Journal of Periodontal & Implant Science*, *41*(6), 263. doi:10.5051/jpis.2011.41.6.263 PMID:22324003

Harasym, T. O., Tardi, P. G., Harasym, N. L., Harvie, P., Johnstone, S. A., & Mayer, L. D. (2006). Increased preclinical efficacy of irinotecan and floxuridine coencapsulated inside liposomes is associated with tumor delivery of synergistic drug ratios. *Oncology Research Featuring Preclinical and Clinical Cancer Therapeutics*, *16*(8), 361–374. doi:10.3727/000000006783980937 PMID:17913044

He, J., Zhou, W., Zhou, X., Zhong, X., Zhang, X., Wan, P., Zhu, B., & Chen, W. (2008). The anatase phase of nanotopography titania plays an important role on osteoblast cell morphology and proliferation. *Journal of Materials Science. Materials in Medicine*, *19*(11), 3465–3472. doi:10.100710856-008-3505-3 PMID:18592349

Holt, K. B., & Bard, A. J. (2005). Interaction of silver(I) ions with the respiratory chain of Escherichia coli: An electrochemical and scanning electrochemical microscopy study of the antimicrobial mechanism of micromolar Ag+. *Biochemistry*, *44*(39), 13214–13223. doi:10.1021/bi0508542 PMID:16185089

Hou, Y., Sun, Z., Rao, W., & Liu, J. (2018). Nanoparticle-mediated cryosurgery for tumor therapy. *Nanomedicine; Nanotechnology, Biology, and Medicine*, *14*(2), 493–506. doi:10.1016/j.nano.2017.11.018 PMID:29197593

Huang, H., Dong, Y., Zhang, Y., Ru, D., Wu, Z., Zhang, J., Shen, M., Duan, Y., & Sun, Y. (2019). GSH-sensitive Pt(IV) prodrug-loaded phase-transitional nanoparticles with a hybrid lipid-polymer shell for precise theranostics against ovarian cancer. *Theranostics*, *9*(4), 1047–1065. doi:10.7150/thno.29820 PMID:30867815

Huang, H.-C., Barua, S., Sharma, G., Dey, S. K., & Rege, K. (2011). Inorganic nanoparticles for cancer imaging and therapy. *Journal of Controlled Release*, *155*(3), 344–357. doi:10.1016/j.jconrel.2011.06.004 PMID:21723891

Huang, T., Holden, J. A., Heath, D. E., O'Brien-Simpson, N. M., & O'Connor, A. J. (2019). Engineering highly effective antimicrobial selenium nanoparticles through control of particle size. *Nanoscale*, *11*(31), 14937–14951. doi:10.1039/C9NR04424H PMID:31363721

Iga, A. M., Robertson, J. H. P., Winslet, M. C., & Seifalian, A. M. (2007). Clinical potential of Quantum Dots. *Journal of Biomedicine & Biotechnology*, *2007*, 1–10. doi:10.1155/2007/76087 PMID:18317518

Jafari, A., Salouti, M., Shayesteh, S. F., Heidari, Z., Rajabi, A. B., Boustani, K., & Nahardani, A. (2015). Synthesis and characterization of Bombesin-superparamagnetic iron oxide nanoparticles as a targeted contrast agent for imaging of breast cancer using MRI. *Nanotechnology*, *26*(7), 075101. doi:10.1088/0957-4484/26/7/075101 PMID:25642737

Jain, S. K., Chourasia, M. K., Masuriha, R., Soni, V., Jain, A., Jain, N. K., & Gupta, Y. (2005). Solid lipid nanoparticles bearing flurbiprofen for transdermal delivery. *Drug Delivery*, *12*(4), 207–215. doi:10.1080/10717540590952591 PMID:16036715

Jayasekara, H., MacInnis, R. J., Room, R., & English, D. R. (2016). Long-term alcohol consumption and breast, upper aero-digestive tract and colorectal cancer risk: A systematic review and meta-analysis. *Alcohol and Alcoholism (Oxford, Oxfordshire)*, *51*(3), 315–330. doi:10.1093/alcalc/agv110 PMID:26400678

Jenning, V., Schäfer-Korting, M., & Gohla, S. (2000). Vitamin A-loaded solid lipid nanoparticles for topical use: Drug release properties. *Journal of Controlled Release*, *66*(2–3), 115–126. doi:10.1016/S0168-3659(99)00223-0 PMID:10742573

Jensen, S. A., Day, E. S., Ko, C. H., Hurley, L. A., Luciano, J. P., Kouri, F. M., Merkel, T. J., Luthi, A. J., Patel, P. C., Cutler, J. I., Daniel, W. L., Scott, A. W., Rotz, M. W., Meade, T. J., Giljohann, D. A., Mirkin, C. A., & Stegh, A. H. (2013). Spherical nucleic acid nanoparticle conjugates as an RNAi-based therapy for glioblastoma. *Science Translational Medicine, 5*(209), 209ra152-209ra152. doi:10.1126/scitranslmed.3006839

Jeyaraj, M., Sathishkumar, G., Sivanandhan, G., MubarakAli, D., Rajesh, M., Arun, R., Kapildev, G., Manickavasagam, M., Thajuddin, N., Premkumar, K., & Ganapathi, A. (2013). Biogenic silver nanoparticles for cancer treatment: An experimental report. *Colloids and Surfaces. B, Biointerfaces, 106*, 86–92. doi:10.1016/j.colsurfb.2013.01.027 PMID:23434696

Jiang, W., Fang, H., Liu, F., Zhou, X., Zhao, H., He, X., & Guo, D. (2019). PEG-coated and Gd-loaded fluorescent silica nanoparticles for targeted prostate cancer magnetic resonance imaging and fluorescence imaging. *International Journal of Nanomedicine, 14*, 5611–5622. doi:10.2147/IJN.S207098 PMID:31413566

Jin, C., Wang, K., Oppong-Gyebi, A., & Hu, J. (2020). Application of nanotechnology in cancer diagnosis and therapy—A mini-review. *International Journal of Medical Sciences, 17*(18), 2964–2973. doi:10.7150/ijms.49801 PMID:33173417

Kaneshiro, T. L., & Lu, Z.-R. (2009). Targeted intracellular codelivery of chemotherapeutics and nucleic acid with a well-defined dendrimer-based nanoglobular carrier. *Biomaterials, 30*(29), 5660–5666. doi:10.1016/j.biomaterials.2009.06.026 PMID:19595449

Kang, S. J., Lee, Y. J., Lee, E.-K., & Kwak, M.-K. (2012). Silver nanoparticles-mediated G2/M cycle arrest of renal epithelial cells is associated with NRF2-GSH signaling. *Toxicology Letters, 211*(3), 334–341. doi:10.1016/j.toxlet.2012.04.016 PMID:22546375

Kasemo, B. (2002). Biological surface science. *Surface Science, 500*(1–3), 656–677. doi:10.1016/S0039-6028(01)01809-X

Kelkawi, A. H. A., Abbasi Kajani, A., & Bordbar, A. (2017). Green synthesis of silver nanoparticles using Mentha pulegium and investigation of their antibacterial, antifungal and anticancer activity. *IET Nanobiotechnology / IET, 11*(4), 370–376. doi:10.1049/iet-nbt.2016.0103 PMID:28530184

Kiessling, F., Mertens, M. E., Grimm, J., & Lammers, T. (2014). Nanoparticles for imaging: Top or flop? *Radiology, 273*(1), 10–28. doi:10.1148/radiol.14131520 PMID:25247562

Kim, N. J., Lee, S. J., & Atala, A. (2013). Biomedical nanomaterials in tissue engineering. In *Nanomaterials in Tissue Engineering* (pp. 1–25e). Elsevier. doi:10.1533/9780857097231.1

Komatsu, K., Murata, M., & Murata, Y. (2005). Encapsulation of molecular hydrogen in fullerene C60 by organic synthesis. *Science, 307*(5707), 238–240. doi:10.1126cience.1106185 PMID:15653499

Kubota, S., Johkura, K., Asanuma, K., Okouchi, Y., Ogiwara, N., Sasaki, K., & Kasuga, T. (2004). Titanium oxide nanotubes for bone regeneration. *Journal of Materials Science. Materials in Medicine, 15*(9), 1031–1035. doi:10.1023/B:JMSM.0000042689.78768.77 PMID:15448411

Kulandaivelu, B., & Gothandam, K. M. (2016). Cytotoxic effect on cancerous cell lines by biologically synthesized silver nanoparticles. *Brazilian Archives of Biology and Technology*, *59*(0). Advance online publication. doi:10.1590/1678-4324-2016150529

Lai, J., Wang, T., Wang, H., Shi, F., Gu, W., & Ye, L. (2018). MnO nanoparticles with unique excitation-dependent fluorescence for multicolor cellular imaging and MR imaging of brain glioma. *Mikrochimica Acta*, *185*(4), 244. doi:10.100700604-018-2779-5 PMID:29610993

Lalani, J., Patil, S., Kolate, A., Lalani, R., & Misra, A. (2015). Protein-functionalized PLGA nanoparticles of lamotrigine for neuropathic pain management. *AAPS PharmSciTech*, *16*(2), 413–427. doi:10.120812249-014-0235-3 PMID:25354788

Lammers, T., Subr, V., Ulbrich, K., Peschke, P., Huber, P. E., Hennink, W. E., & Storm, G. (2009). Simultaneous delivery of doxorubicin and gemcitabine to tumors in vivo using prototypic polymeric drug carriers. *Biomaterials*, *30*(20), 3466–3475. doi:10.1016/j.biomaterials.2009.02.040 PMID:19304320

Lee, A.-R., Lee, S.-J., Lee, M., Nam, M., Lee, S., Choi, J., Lee, H.-J., Kim, D.-U., & Hoe, K.-L. (2018). Editor's highlight: A Genome-wide screening of target genes against silver nanoparticles in fission yeast. *Toxicological Sciences*, *161*(1), 171–185. doi:10.1093/toxsci/kfx208 PMID:29294138

Li, J., Wang, S., Wu, C., Dai, Y., Hou, P., Han, C., & Xu, K. (2016). Activatable molecular MRI nanoprobe for tumor cell imaging based on gadolinium oxide and iron oxide nanoparticle. *Biosensors & Bioelectronics*, *86*, 1047–1053. doi:10.1016/j.bios.2016.07.044 PMID:27501342

Li, Y., Zheng, X., & Chu, Q. (2021). Bio-based nanomaterials for cancer therapy. *Nano Today*, *38*, 101134. doi:10.1016/j.nantod.2021.101134

Liu, D., Yang, F., Xiong, F., & Gu, N. (2016). The smart drug delivery system and its clinical potential. *Theranostics*, *6*(9), 1306–1323. doi:10.7150/thno.14858 PMID:27375781

Liu, J., Hu, W., Chen, H., Ni, Q., Xu, H., & Yang, X. (2007). Isotretinoin-loaded solid lipid nanoparticles with skin targeting for topical delivery. *International Journal of Pharmaceutics*, *328*(2), 191–195. doi:10.1016/j.ijpharm.2006.08.007 PMID:16978810

Liu, Z., Chen, K., Davis, C., Sherlock, S., Cao, Q., Chen, X., & Dai, H. (2008). Drug delivery with carbon nanotubes for in vivo cancer treatment. *Cancer Research*, *68*(16), 6652–6660. doi:10.1158/0008-5472.CAN-08-1468 PMID:18701489

Loutfy, S. A., Al-Ansary, N. A., Abdel-Ghani, N. T., Hamed, A. R., Mohamed, M. B., Craik, J. D., Eldin, T. A. S., Abdellah, A. M., Hussein, Y., Hasanin, M., & Elbehairi, S. E. I. (2015). Anti-proliferative activities of metallic nanoparticles in an in vitro breast cancer model. *Asian Pacific Journal of Cancer Prevention*, *16*(14), 6039–6046. doi:10.7314/APJCP.2015.16.14.6039 PMID:26320493

Lovat, V., Pantarotto, D., Lagostena, L., Cacciari, B., Grandolfo, M., Righi, M., Spalluto, G., Prato, M., & Ballerini, L. (2005). Carbon nanotube substrates boost neuronal electrical signaling. *Nano Letters*, *5*(6), 1107–1110. doi:10.1021/nl050637m PMID:15943451

Lozano, T., Rey, M., Rojas, E., Moya, S., Fleddermann, J., Estrela-Lopis, I., Donath, E., Wang, B., Mao, Z., Gao, C., & González-Fernández, Á. (2011). Cytotoxicity effects of metal oxide nanoparticles in human tumor cell lines. *Journal of Physics: Conference Series*, *304*, 012046. doi:10.1088/1742-6596/304/1/012046

Lu, B., Xiong, S.-B., Yang, H., Yin, X.-D., & Chao, R.-B. (2006). Solid lipid nanoparticles of mitoxantrone for local injection against breast cancer and its lymph node metastases. *European Journal of Pharmaceutical Sciences*, *28*(1–2), 86–95. doi:10.1016/j.ejps.2006.01.001 PMID:16472996

Mamot, C., Ritschard, R., Wicki, A., Stehle, G., Dieterle, T., Bubendorf, L., Hilker, C., Deuster, S., Herrmann, R., & Rochlitz, C. (2012). Tolerability, safety, pharmacokinetics, and efficacy of doxorubicin-loaded anti-EGFR immunoliposomes in advanced solid tumours: A phase 1 dose-escalation study. *The Lancet. Oncology*, *13*(12), 1234–1241. doi:10.1016/S1470-2045(12)70476-X PMID:23153506

Marques Neto, L. M., Kipnis, A., & Junqueira-Kipnis, A. P. (2017). Role of metallic nanoparticles in vaccinology: Implications for infectious disease vaccine development. *Frontiers in Immunology*, *8*. Advance online publication. doi:10.3389/fimmu.2017.00239 PMID:28337198

Mattson, M. P., Haddon, R. C., & Rao, A. M. (2000). Molecular functionalization of carbon nanotubes and use as substrates for neuronal growth. *Journal of Molecular Neuroscience*, *14*(3), 175–182. doi:10.1385/JMN:14:3:175 PMID:10984193

McDevitt, M. R., Chattopadhyay, D., Kappel, B. J., Jaggi, J. S., Schiffman, S. R., Antczak, C., Njardarson, J. T., Brentjens, R., & Scheinberg, D. A. (2007). Tumor targeting with antibody-functionalized, radiolabeled carbon nanotubes. *Journal of Nuclear Medicine*, *48*(7), 1180–1189. doi:10.2967/jnumed.106.039131 PMID:17607040

Mei, Z., Wu, Q., Hu, S., Lib, X., & Yang, X. (2005). Triptolide loaded solid lipid nanoparticle hydrogel for topical application. *Drug Development and Industrial Pharmacy*, *31*(2), 161–168. doi:10.1081/DDC-200047791 PMID:15773283

Mitchell, M. J., Billingsley, M. M., Haley, R. M., Wechsler, M. E., Peppas, N. A., & Langer, R. (2021). Engineering precision nanoparticles for drug delivery. *Nature Reviews. Drug Discovery*, *20*(2), 101–124. doi:10.103841573-020-0090-8 PMID:33277608

Moore, L., Chow, E. K.-H., Osawa, E., Bishop, J. M., & Ho, D. (2013). Diamond-lipid hybrids enhance chemotherapeutic tolerance and mediate tumor regression. *Advanced Materials*, *25*(26), 3532–3541. doi:10.1002/adma.201300343 PMID:23584895

Mroz, P., Pawlak, A., Satti, M., Lee, H., Wharton, T., Gali, H., Sarna, T., & Hamblin, M. R. (2007). Functionalized fullerenes mediate photodynamic killing of cancer cells: Type I versus Type II photochemical mechanism. *Free Radical Biology & Medicine*, *43*(5), 711–719. doi:10.1016/j.freeradbiomed.2007.05.005 PMID:17664135

Mukherjee, S., Ray, S., & Thakur, R. (2009). Solid lipid nanoparticles: A modern formulation approach in drug delivery system. *Indian Journal of Pharmaceutical Sciences*, *71*(4), 349. doi:10.4103/0250-474X.57282 PMID:20502539

Murphy, M. P. (2009). How mitochondria produce reactive oxygen species. *The Biochemical Journal, 417*(1), 1–13. doi:10.1042/BJ20081386 PMID:19061483

Narayanan, M., Vigneshwari, P., Natarajan, D., Kandasamy, S., Alsehli, M., Elfasakhany, A., & Pugazhendhi, A. (2021). Synthesis and characterization of TiO2 NPs by aqueous leaf extract of Coleus aromaticus and assess their antibacterial, larvicidal, and anticancer potential. *Environmental Research, 200*, 111335. doi:10.1016/j.envres.2021.111335 PMID:34051200

Nasongkla, N., Bey, E., Ren, J., Ai, H., Khemtong, C., Guthi, J. S., Chin, S.-F., Sherry, A. D., Boothman, D. A., & Gao, J. (2006). Multifunctional polymeric micelles as cancer-targeted, MRI-ultrasensitive drug delivery systems. *Nano Letters, 6*(11), 2427–2430. doi:10.1021/nl061412u PMID:17090068

Nayak, P. P., S, N., Narayanan, A., Badekila, A. K., & Kini, S. (2021). Nanomedicine in cancer clinics: Are we there yet? *Current Pathobiology Reports, 9*(2), 43–55. doi:10.100740139-021-00220-6

Nunes, T., Pons, T., Hou, X., Van Do, K., Caron, B., Rigal, M., Di Benedetto, M., Palpant, B., Leboeuf, C., Janin, A., & Bousquet, G. (2019). Pulsed-laser irradiation of multifunctional gold nanoshells to overcome trastuzumab resistance in HER2-overexpressing breast cancer. *Journal of Experimental & Clinical Cancer Research, 38*(1), 306. doi:10.118613046-019-1305-x PMID:31299997

Pan, Y., Leifert, A., Ruau, D., Neuss, S., Bornemann, J., Schmid, G., Brandau, W., Simon, U., & Jahnen-Dechent, W. (2009). Gold nanoparticles of diameter 1.4 nm trigger necrosis by oxidative stress and mitochondrial damage. *Small, 5*(18), 2067–2076. doi:10.1002mll.200900466 PMID:19642089

Pandey, R., Sharma, S., & Khuller, G. K. (2005). Oral solid lipid nanoparticle-based antitubercular chemotherapy. *Tuberculosis (Edinburgh, Scotland), 85*(5–6), 415–420. doi:10.1016/j.tube.2005.08.009 PMID:16256437

Parungo, C. P., Ohnishi, S., De Grand, A. M., Laurence, R. G., Soltesz, E. G., Colson, Y. L., Kang, P. M., Mihaljevic, T., Cohn, L. H., & Frangioni, J. V. (2004). In vivo optical imaging of pleural space drainage to lymph nodes of prognostic significance. *Annals of Surgical Oncology, 11*(12), 1085–1092. doi:10.1245/ASO.2004.03.054 PMID:15545502

Parvanian, S., Mostafavi, S. M., & Aghashiri, M. (2017). Multifunctional nanoparticle developments in cancer diagnosis and treatment. *Sensing and Bio-Sensing Research, 13*, 81–87. doi:10.1016/j.sbsr.2016.08.002

Patra, C. R., Cao, S., Safgren, S., Bhattacharya, R., Ames, M. M., Shah, V., Reid, J. M., & Mukherjee, P. (2008). Intracellular fate of a targeted delivery system. *Journal of Biomedical Nanotechnology, 4*(4), 508–514. doi:10.1166/jbn.2008.016

Pelgrift, R. Y., & Friedman, A. J. (2013). Nanotechnology as a therapeutic tool to combat microbial resistance. *Advanced Drug Delivery Reviews, 65*(13–14), 1803–1815. doi:10.1016/j.addr.2013.07.011 PMID:23892192

Podesta, J. E., Al-Jamal, K. T., Herrero, M. A., Tian, B., Ali-Boucetta, H., Hegde, V., Bianco, A., Prato, M., & Kostarelos, K. (2009). *Antitumor activity and prolonged survival by carbon-nanotube-mediated therapeutic siRNA silencing in a human lung xenograft model. Small.* doi:10.1002mll.200801572

Prasannaraj, G., Sahi, S. V., Ravikumar, S., & Venkatachalam, P. (2016). Enhanced cytotoxicity of biomolecules loaded metallic silver nanoparticles against human liver (HepG2) and prostate (PC3) cancer cell lines. *Journal of Nanoscience and Nanotechnology*, *16*(5), 4948–4959. doi:10.1166/jnn.2016.12336 PMID:27483851

Radu, D. R., Lai, C.-Y., Jeftinija, K., Rowe, E. W., Jeftinija, S., & Lin, V. S.-Y. (2004). A polyamidoamine dendrimer-capped mesoporous silica nanosphere-based gene transfection reagent. *Journal of the American Chemical Society*, *126*(41), 13216–13217. doi:10.1021/ja046275m PMID:15479063

Raj, S., Khurana, S., Choudhari, R., Kesari, K. K., Kamal, M. A., Garg, N., Ruokolainen, J., Das, B. C., & Kumar, D. (2021). Specific targeting cancer cells with nanoparticles and drug delivery in cancer therapy. *Seminars in Cancer Biology*, *69*, 166–177. doi:10.1016/j.semcancer.2019.11.002 PMID:31715247

Reichmuth, A. M., Oberli, M. A., Jaklenec, A., Langer, R., & Blankschtein, D. (2016). MRNA vaccine delivery using lipid nanoparticles. *Therapeutic Delivery*, *7*(5), 319–334. doi:10.4155/tde-2016-0006 PMID:27075952

Ren, Y., Cheung, H. W., von Maltzhan, G., Agrawal, A., Cowley, G. S., Weir, B. A., Boehm, J. S., Tamayo, P., Karst, A. M., Liu, J. F., Hirsch, M. S., Mesirov, J. P., Drapkin, R., Root, D. E., Lo, J., Fogal, V., Ruoslahti, E., Hahn, W. C., & Bhatia, S. N. (2012). Targeted tumor-penetrating siRNA nanocomplexes for credentialing the ovarian cancer oncogene ID4. *Science Translational Medicine, 4*(147), 147ra112-147ra112. doi:10.1126/scitranslmed.3003778

Rostami, E., Kashanian, S., Azandaryani, A. H., Faramarzi, H., Dolatabadi, J. E. N., & Omidfar, K. (2014). Drug targeting using solid lipid nanoparticles. *Chemistry and Physics of Lipids*, *181*, 56–61. doi:10.1016/j.chemphyslip.2014.03.006 PMID:24717692

Rostami, H., Ebtekar, M., Ardestani, M. S., Yazdi, M. H., & Mahdavi, M. (2017). Co-utilization of a TLR5 agonist and nano-formulation of HIV-1 vaccine candidate leads to increased vaccine immunogenicity and decreased immunogenic dose: A preliminary study. *Immunology Letters*, *187*, 19–26. doi:10.1016/j.imlet.2017.05.002 PMID:28479111

Saad, M., Garbuzenko, O. B., & Minko, T. (2008). Co-delivery of siRNA and an anticancer drug for treatment of multidrug-resistant cancer. *Nanomedicine (London)*, *3*(6), 761–776. doi:10.2217/17435889.3.6.761 PMID:19025451

Saji, V. S., Choe, H. C., & Yeung, K. W. K. (2010). Nanotechnology in biomedical applications: A review. *International Journal of Nano and Biomaterials*, *3*(2), 119. doi:10.1504/IJNBM.2010.037801

Santos Maia, C., Mehnert, W., Schaller, M., Korting, H. C., Gysler, A., Haberland, A., & Schäfer-korting, M. (2002). Drug targeting by solid lipid nanoparticles for dermal use. *Journal of Drug Targeting*, *10*(6), 489–495. doi:10.1080/1061186021000038364 PMID:12575739

Schoppa, T., Jung, D., Rust, T., Mulac, D., Kuckling, D., & Langer, K. (2021). Light-responsive polymeric nanoparticles based on a novel nitropiperonal based polyester as drug delivery systems for photosensitizers in PDT. *International Journal of Pharmaceutics*, *597*, 120326. doi:10.1016/j.ijpharm.2021.120326 PMID:33540003

Sengupta, S., Eavarone, D., Capila, I., Zhao, G., Watson, N., Kiziltepe, T., & Sasisekharan, R. (2005). Temporal targeting of tumour cells and neovasculature with a nanoscale delivery system. *Nature*, *436*(7050), 568–572. doi:10.1038/nature03794 PMID:16049491

Shakibaie, M., Riahi-Madvar, S., Ameri, A., Amiri-Moghadam, P., Adeli-Sardou, M., & Forootanfar, H. (2021). Microwave assisted biosynthesis of cadmium nanoparticles: Characterization, antioxidant and cytotoxicity studies. *Journal of Cluster Science*. Advance online publication. doi:10.100710876-021-02107-3

Shaw, J., Raja, S. O., & Dasgupta, A. K. (2014). Modulation of cytotoxic and genotoxic effects of nanoparticles in cancer cells by external magnetic field. *Cancer Nanotechnology*, *5*(1), 2. doi:10.118612645-014-0002-x PMID:26561510

Shilo, M., Reuveni, T., Motiei, M., & Popovtzer, R. (2012). Nanoparticles as computed tomography contrast agents: Current status and future perspectives. *Nanomedicine (London)*, *7*(2), 257–269. doi:10.2217/nnm.11.190 PMID:22339135

Smith, B. R., & Gambhir, S. S. (2017). Nanomaterials for in vivo imaging. *Chemical Reviews*, *117*(3), 901–986. doi:10.1021/acs.chemrev.6b00073 PMID:28045253

Song, X. R., Cai, Z., Zheng, Y., He, G., Cui, F. Y., Gong, D. Q., Hou, S. X., Xiong, S. J., Lei, X. J., & Wei, Y. Q. (2009). Reversion of multidrug resistance by co-encapsulation of vincristine and verapamil in PLGA nanoparticles. *European Journal of Pharmaceutical Sciences*, *37*(3–4), 300–305. doi:10.1016/j.ejps.2009.02.018 PMID:19491019

Souto, E. B., & Müller, R. H. (2006). The use of SLN and NLC as topical particulate carriers for imidazole antifungal agents. *Die Pharmazie - An International. Journal of Pharmaceutical Sciences*, *61*(5), 431–437. PMID:16724541

Stylianopoulos, T., Wong, C., Bawendi, M. G., Jain, R. K., & Fukumura, D. (2012). Multistage nanoparticles for improved delivery into tumor tissue. In *Methods in Enzymology* (Vol. 508, pp. 109–130). Elsevier. doi:10.1016/B978-0-12-391860-4.00006-9

Sumer, B., & Gao, J. (2008). Theranostic nanomedicine for cancer. *Nanomedicine (London)*, *3*(2), 137–140. doi:10.2217/17435889.3.2.137 PMID:18373419

Svenson, S., & Tomalia, D. (2005). Dendrimers in biomedical applications—Reflections on the field. *Advanced Drug Delivery Reviews*, *57*(15), 2106–2129. doi:10.1016/j.addr.2005.09.018 PMID:16305813

Taghdisi, S. M., Lavaee, P., Ramezani, M., & Abnous, K. (2011). Reversible targeting and controlled release delivery of daunorubicin to cancer cells by aptamer-wrapped carbon nanotubes. *European Journal of Pharmaceutics and Biopharmaceutics*, *77*(2), 200–206. doi:10.1016/j.ejpb.2010.12.005 PMID:21168488

Tardi, P., Johnstone, S., Harasym, N., Xie, S., Harasym, T., Zisman, N., Harvie, P., Bermudes, D., & Mayer, L. (2009). In vivo maintenance of synergistic cytarabine:daunorubicin ratios greatly enhances therapeutic efficacy. *Leukemia Research*, *33*(1), 129–139. doi:10.1016/j.leukres.2008.06.028 PMID:18676016

Tardi, P. G., Dos Santos, N., Harasym, T. O., Johnstone, S. A., Zisman, N., Tsang, A. W., Bermudes, D. G., & Mayer, L. D. (2009). Drug ratio–dependent antitumor activity of irinotecan and cisplatin combinations in vitro and in vivo. *Molecular Cancer Therapeutics*, *8*(8), 2266–2275. doi:10.1158/1535-7163.MCT-09-0243 PMID:19671743

Tekade, R. K., Dutta, T., Gajbhiye, V., & Jain, N. K. (2009). Exploring dendrimer towards dual drug delivery: PH responsive simultaneous drug-release kinetics. *Journal of Microencapsulation*, *26*(4), 287–296. doi:10.1080/02652040802312572 PMID:18791906

Thakor, A. S., Luong, R., Paulmurugan, R., Lin, F. I., Kempen, P., Zavaleta, C., Chu, P., Massoud, T. F., Sinclair, R., & Gambhir, S. S. (2011). The fate and toxicity of Raman-active silica-gold nanoparticles in mice. *Science Translational Medicine*, *3*(79), 79ra33-79ra33. doi:10.1126/scitranslmed.3001963

Tian, C., Feng, J., Cho, H. J., Datta, S. S., & Prud'homme, R. K. (2018). Adsorption and denaturation of structured polymeric nanoparticles at an interface. *Nano Letters*, *18*(8), 4854–4860. doi:10.1021/acs.nanolett.8b01434 PMID:29975545

Venugopal, K., Ahmad, H., Manikandan, E., Thanigai Arul, K., Kavitha, K., Moodley, M. K., Rajagopal, K., Balabhaskar, R., & Bhaskar, M. (2017). The impact of anticancer activity upon Beta vulgaris extract mediated biosynthesized silver nanoparticles (ag-NPs) against human breast (MCF-7), lung (A549) and pharynx (Hep-2) cancer cell lines. *Journal of Photochemistry and Photobiology. B, Biology*, *173*, 99–107. doi:10.1016/j.jphotobiol.2017.05.031 PMID:28570910

Versiani, A. F., Astigarraga, R. G., Rocha, E. S. O., Barboza, A. P. M., Kroon, E. G., Rachid, M. A., Souza, D. G., Ladeira, L. O., Barbosa-Stancioli, E. F., Jorio, A., & Da Fonseca, F. G. (2017). Multi-walled carbon nanotubes functionalized with recombinant Dengue virus 3 envelope proteins induce significant and specific immune responses in mice. *Journal of Nanobiotechnology*, *15*(1), 26. doi:10.118612951-017-0259-4 PMID:28376812

Wan, X., Song, Y., Song, N., Li, J., Yang, L., Li, Y., & Tan, H. (2016). The preliminary study of immune superparamagnetic iron oxide nanoparticles for the detection of lung cancer in magnetic resonance imaging. *Carbohydrate Research*, *419*, 33–40. doi:10.1016/j.carres.2015.11.003 PMID:26649917

Wang, D., Lin, H., Zhang, G., Si, Y., Yang, H., Bai, G., Yang, C., Zhong, K., Cai, D., Wu, Z., Wang, R., & Zou, D. (2018). Effective pH-activated theranostic platform for synchronous magnetic resonance imaging diagnosis and chemotherapy. *ACS Applied Materials & Interfaces*, *10*(37), 31114–31123. doi:10.1021/acsami.8b11408 PMID:30141893

Wang, H., Naghavi, M., Allen, C., Barber, R. M., Bhutta, Z. A., Carter, A., Casey, D. C., Charlson, F. J., Chen, A. Z., Coates, M. M., Coggeshall, M., Dandona, L., Dicker, D. J., Erskine, H. E., Ferrari, A. J., Fitzmaurice, C., Foreman, K., Forouzanfar, M. H., Fraser, M. S., ... Murray, C. J. L. (2016). Global, regional, and national life expectancy, all-cause mortality, and cause-specific mortality for 249 causes of death, 1980–2015: A systematic analysis for the Global Burden of Disease Study 2015. *Lancet*, *388*(10053), 1459–1544. doi:10.1016/S0140-6736(16)31012-1 PMID:27733281

Wang, W., Li, J., Liu, R., Zhang, A., & Yuan, Z. (2016). Size effect of Au/PAMAM contrast agent on CT imaging of reticuloendothelial system and tumor tissue. *Nanoscale Research Letters*, *11*(1), 429. doi:10.118611671-016-1650-z PMID:27671016

Wang, Y., Gao, S., Ye, W.-H., Yoon, H. S., & Yang, Y.-Y. (2006). Co-delivery of drugs and DNA from cationic core–shell nanoparticles self-assembled from a biodegradable copolymer. *Nature Materials*, *5*(10), 791–796. doi:10.1038/nmat1737 PMID:16998471

Wong, H. L., Rauth, A. M., Bendayan, R., Manias, J. L., Ramaswamy, M., Liu, Z., Erhan, S. Z., & Wu, X. Y. (2006). A new polymer–lipid hybrid nanoparticle system increases cytotoxicity of doxorubicin against multidrug-resistant human breast cancer cells. *Pharmaceutical Research*, *23*(7), 1574–1585. doi:10.100711095-006-0282-x PMID:16786442

Xu, P., Chen, B., Li, J., & Wang, X. (2013). Gambogic acid combined with CdTe QDs for leukemia cancer cells inhibition and their bio-safety for rat brain. *Blood*, *122*(21), 5028–5028. doi:10.1182/blood.V122.21.5028.5028

Yu, Z., Gao, L., Chen, K., Zhang, W., Zhang, Q., Li, Q., & Hu, K. (2021). Nanoparticles: A new approach to upgrade cancer diagnosis and treatment. *Nanoscale Research Letters*, *16*(1), 88. doi:10.118611671-021-03489-z PMID:34014432

Yu, Z., Li, Q., Wang, J., Yu, Y., Wang, Y., Zhou, Q., & Li, P. (2020). Reactive oxygen species-related nanoparticle toxicity in the biomedical field. *Nanoscale Research Letters*, *15*(1), 115. doi:10.118611671-020-03344-7 PMID:32436107

Zhang, Y., Fang, F., Li, L., & Zhang, J. (2020). Self-Assembled Organic Nanomaterials for Drug Delivery, Bioimaging, and Cancer Therapy. *ACS Biomaterials Science & Engineering*, *6*(9), 4816–4833. doi:10.1021/acsbiomaterials.0c00883 PMID:33455214

Zhao, H. Y., Xu, X. X., Zhang, J. X., Zheng, W., & Zheng, Y. F. (2010). Carbon nanotube–hydroxyapatite–hemoglobin nanocomposites with high bioelectrocatalytic activity. *Bioelectrochemistry (Amsterdam, Netherlands)*, *78*(2), 124–129. doi:10.1016/j.bioelechem.2009.08.009 PMID:19762287

Zhao, Y., Vivero-Escoto, J. L., Slowing, I. I., Trewyn, B. G., & Lin, V. S.-Y. (2010). Capped mesoporous silica nanoparticles as stimuli-responsive controlled release systems for intracellular drug/gene delivery. *Expert Opinion on Drug Delivery*, *7*(9), 1013–1029. doi:10.1517/17425247.2010.498816 PMID:20716017

ADDITIONAL READING

Kim, N. J., Lee, S. J., & Atala, A. (2013). Biomedical nanomaterials in tissue engineering. In *Nanomaterials in Tissue Engineering* (pp. 1–25e). Elsevier. doi:10.1533/9780857097231.1

Mozafari, M. R. (Ed.). (2007). *Nanomaterials and Nanosystems for Biomedical Applications*. Springer Netherlands. doi:10.1007/978-1-4020-6289-6

Rajendran, S., Naushad, Mu., Durgalakshmi, D., & Lichtfouse, E. (Eds.). (2021). *Metal, Metal Oxides and Metal Sulphides for Biomedical Applications* (Vol. 58). Springer International Publishing. doi:10.1007/978-3-030-56413-1

Santra, T. S., & Mohan, L. (Eds.). (2021). *Nanomaterials and Their Biomedical Applications* (Vol. 16). Springer Singapore. doi:10.1007/978-981-33-6252-9

Srivastava, R., Thakur, M., Kumawat, M. K., & Bahadur, R. (2021). *Next Generation Graphene Nanomaterials for Cancer Theranostic Applications*. Springer Singapore. doi:10.1007/978-981-33-6303-8

Stoika, R. S. (Ed.). (2022). *Biomedical Nanomaterials: From design and synthesis to imaging, application and environmental impact*. Springer International Publishing. doi:10.1007/978-3-030-76235-3

Stylianopoulos, T., Wong, C., Bawendi, M. G., Jain, R. K., & Fukumura, D. (2012). Multistage nanoparticles for improved delivery into tumor tissue. In *Methods in Enzymology* (Vol. 508, pp. 109–130). Elsevier. doi:10.1016/B978-0-12-391860-4.00006-9

Tekinay, A. B. (Ed.). (2019). *Nanomaterials for Regenerative Medicine*. Springer International Publishing. doi:10.1007/978-3-030-31202-2

Vo-Dinh, T. (Ed.). (2021). *Nanoparticle-Mediated Immunotherapy* (Vol. 12). Springer International Publishing. doi:10.1007/978-3-030-78338-9

KEY TERMS AND DEFINITIONS

Apoptosis: A form of cell death in which a programmed sequence of events leads to the elimination of cells without releasing harmful substances into the surrounding area.

Chemotherapeutic Agent: Drugs or effects that are used directly or indirectly for inhibiting the uncontrolled growth and proliferation of cancer cells.

Drug Delivery System (DDS): A formulation or a device that enables a therapeutic substance to selectively reach its site of action without reaching the non-target cells, organs, or tissues.

Fluorescence Imaging: Is a type of non-invasive imaging technique that can help visualize biological processes taking place in a living organism.

Magnetic Resonance Imaging (MRI): Is a non-invasive imaging technology that produces three dimensional detailed anatomical images.

Nanobiotechnology: Is a discipline in which tools from nanotechnology are developed and applied to study biological phenomena.

Nanoformulations: The formulation of a nanomedicine.

Nanovectors/Nanocarriers: Nanoparticles with a diameter loosely between 1 and 100 nm in at least one dimension capable of carrying and delivering one or more bioactive molecules, including therapeutic agents and imaging contrast enhancer.

Photothermal Therapy (PTT): Use of electromagnetic radiation for the treatment of various medical conditions, including cancer.

Super-Paramagnetic Nanosystem: A type of metallic nanoparticles that show magnetic properties in the presence of an external magnetic field.

Chapter 16
Role of Nanoparticles in Cancer Therapy

Zeeshan Ahmad Bhutta
https://orcid.org/0000-0002-0148-425X
Chungbuk National University, South Korea

Ayesha Kanwal
The Islamia University of Bahawalpur, Pakistan

Ambreen Ashar
University of Agriculture, Faisalabad, Pakistan

Moazam Ali
University of Agriculture, Faisalabad, Pakistan

Ashar Mahfooz
University of Agriculture, Faisalabad, Pakistan

Muhammad Fakhar-e-Alam Kulyar
Huazhong Agricultural University, China

Muhammad Shoaib
Lanzhou Institute of Husbandry and Pharmaceutical Sciences, Graduate School of the Chinese Academy of Agricultural Sciences, China

Rizwan Ahmed
University of Agriculture, Faisalabad, Pakistan

Kun Li
Nanjing Agricultural University, China

ABSTRACT

The rapid growth of nanotechnology towards the development of nanomedicines has improved cancer treatment. Nanomedicine provides the opportunity to implement complex and targeted multifunctional strategies. Today, nanoparticles (NPs) have many uses in a number of scientific fields. In recent years, it has been repeatedly reported that NPs hold a significant place in the regulation of modern medicine by implementing a varying number of clinical approaches like drug carrying substances, genetic material delivery to tumors, as well as in radiography as a contrast media agent. Various nanomaterials based on organic, inorganic, lipid or glycan compounds, and synthetic polymers have been used to develop and improve new cancer treatments. In this chapter, the authors discussed the role of NPs in cancer treatment among various anticancer drug delivery methods.

DOI: 10.4018/978-1-7998-8936-6.ch016

INTRODUCTION

In the United States, cancer is now becoming the second most dangerous disease that causes approximately 591,699 deaths in 2017 (Vines et al., 2019). The basic understanding of cancer biology has improved significantly, which has led to better diagnosis and treatment methods. One of such advances includes nanotechnology which put forwards a universal platform for cancer treatment and imaging. Cancer nanomedicine has significant benefits in overcoming systemic and tumor obstructions by employing targeted delivery of therapeutic drugs to cancer cells with minimal impact on healthy cells (Prabhu et al., 2015). Nano formulations having a specific targeted delivery to a particular cell receptor permit an increased dose of the drug at the target cells (Ambade et al., 2005). This chapter examines how NPs contribute to cancer treatment by utilizing various cancer drug delivery strategies.

Cancer is one of the most serious diseases in the world and among the leading cause of death. Every year millions of individuals die because of this disease. According to data from the National Cancer Institute, from 2012 to 2030, the number of cancer cases are expected to rise from 14 million to 21 million, an increase of 50%, and the ratio of death rate is going to be raised expectedly by 60% i.e., 8 to 13 million (Siegel et al., 2016). Cancers are developed due to the accretion of various genetic and somatic mutations in genes (tumor suppressor genes and protooncogenes) that control cell growth and apoptosis (tumor suppression genes and offspring). Due to the heterogeneity of tumors, tumor cells show genetic diversity and drug resistance and require very complicated treatment regimes. As a result, different ways of treating cancer have been used, and all known therapeutic strategies have major limitations and side effects (Holohan et al., 2013).

Recently used therapies for cancer are in combination form either it is surgery plus radiation or surgery plus chemotherapy. The use of surgeries for tumor tissues is limited because there is not enough information to clearly distinguish between the cancerous and normal cells. Other limitations of available treatments include insufficient specificity due to inappropriate exposure to normal cells, leading to increased systemic toxicity. Chemotherapeutic agents inhibit DNA replication or induce cell apoptosis, disrupt normal cellular functions, and induce rapidly proliferating normal cells to produce the most harmful and lethal cytotoxic effects (Gupta et al., 2016).

According to the International Union of Pure and Applied Chemistry (IUPAC), nanoparticles are microscopic substances varying in size from 1 to 100 nm (Shi et al., 2017) and possesses unique physical characters like stability, optical and electricity conduction, making them an ultimate candidate for biological and material sciences. Nanomaterials have an extensive array of applications in the fields of optics, biology, and electronics and have become a viable choice in cell engineering, tissue engineering, protein diagnosis, biomolecules, pharmacy, medicine, electronics, energy, environment, and forward. They are divided into several groups according to their characteristics, size, and shape. These groups include polymeric nanoparticles, ceramic nanoparticles metal nanoparticles, and fullerenes.

Among traditional cancer treatment methods, targeted therapy is the best as compared to other therapies (conventional chemotherapy). It is more effective with weaker viability, requires lower doses, has fewer side effects, lower therapeutic index, and minimal problem of multidrug resistance (Senapati et al., 2018). In addition to medications, contrast agents (with diagnostic capabilities), and genes, recent studies have shown that NPs have many benefits in diagnosing and treating tumors (Dang & Guan, 2020). Over the last few decades, the nature of NPs has become the basis for many NPs employing therapies in clinical trials (Bi et al., 2016).

BACKGROUND

In recent times, it is being considered that NPs are the best carrier for drug delivery. Nanocarriers can alter the pharmacokinetics of drugs, thereby increasing their efficacy, reducing their negative effects. Metallic particles, lipids, and polymers are substances that can be used as NPs for drug delivery. Different forms and sizes can be developed in accordance with the structure of NPs. Nanocarriers-based Drug delivery systems (DDS) have already entered the global pharmaceutical marketplace. DDS applications are rapidly growing, and future research must emphasize several NPs utilities, i.e., target-oriented drug delivery and concomitant imaging techniques.

The main purpose to develop the DDS is to enhance the efficacy and therapeutic chattels of medicines by frequently storing the medicine inside as a reservoir (Karra & Benita, 2012). Such a system can release medication in a timely manner towards appropriate places by affecting the pharmacokinetics and drug distribution in the body (Tiwari et al., 2012). Over the past half-century, various advances in polymers, chemistry, biology, mechanical engineering, and physics have influenced the variations of nanocarriers. As a result, several vectors with exceptional belongings have been launched into the field of medicine (Khan et al., 2019).

CLASSIFICATION OF DRUGS NANO-CARRIERS

Organic and Inorganic carriers are the two generalized categories for drug delivery carriers as shown in Figure 1. Organic particles serve as the fundamental structure for a drug carrier, are classified in the 1st group (polymeric nanoparticles), and a drug carrier comprising mineral particles as a core is classified in the 2nd group (lipid nanoparticles). Quantum dots, metallic NPs, ceramic NPs, carbon nanotubes and fullerenes are considered as inorganic NPs.

ORGANIC NANOCARRIERS

Lipid Nanoparticles

Liposomes

From various lipid-based nano arsenal compounds available nowadays, liposomes are widely known for their unique properties and are most versatile. Alec D. Bangham first discovered the liposomes in 1961 and the first prelude of liposome formulations was in 1995. These are PEGylated liposomes that contain doxorubicin. The half-life of doxorubicin flow is prolonged in the presence of polyethylene glycol (Bulbake et al., 2017). The presence of an aqueous core in liposomes besieged by one or more coatings of cholesterol and phospholipids forming a lipid bilayer. This distinctive structure permits liposomes to upload and retain the hydrophilic nature agent in the aqueous space and the hydrophobic nature agent in the triglyceride compartment (Bozzuto & Molinari, 2015).

Other benefits of liposomes as nanocarriers in drug delivery applications include the delayed release of the drug, drug protection from degradation, and prevention of drug exposure to the environment. Certain lipids, like cholesterol and stiff saturated lipids, have ability to stabilize the lipid bilayer ar-

Figure 1. An overview of the Nanoparticle-based drug carrier system

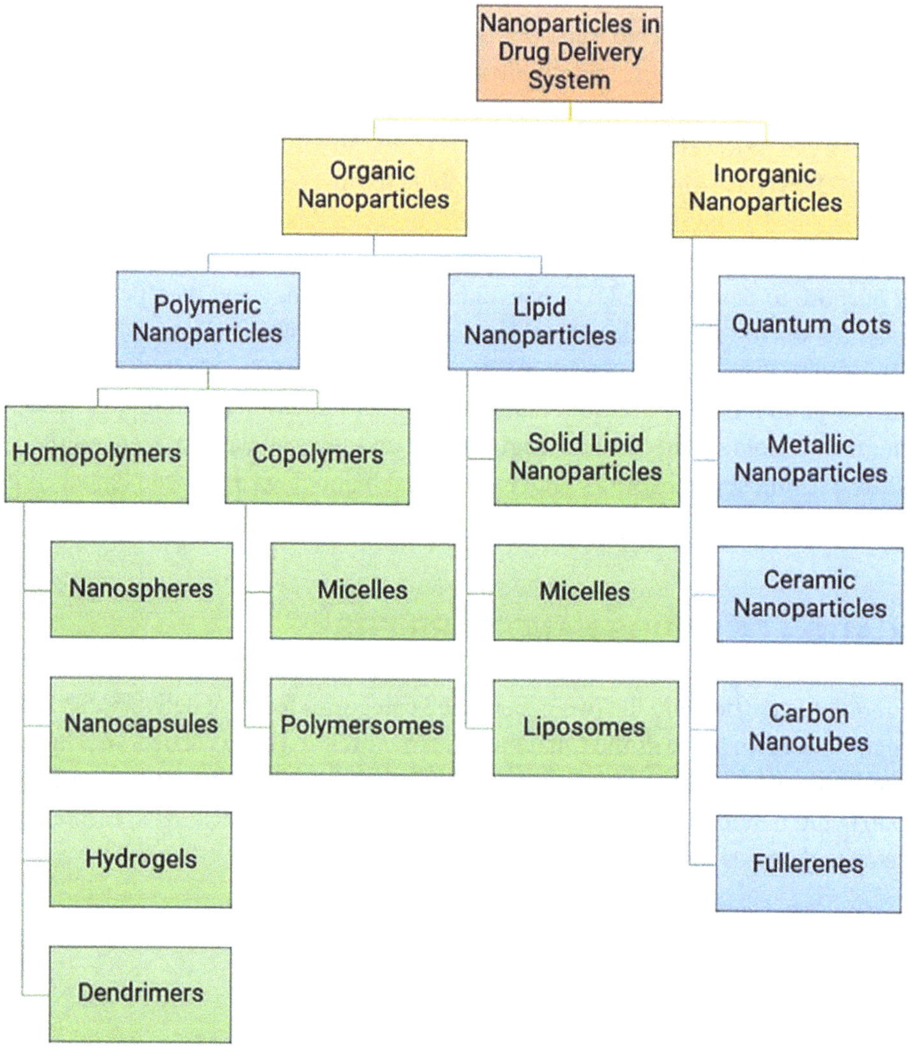

rangement and resist the invasion of plasma proteins by reducing the leakage of the drug (Daraee et al., 2016). The release mechanism of liposomal drugs has been under consideration for the last few years. For example, liposomes can be triggered to release drugs by enzymes, ultrasound, light, high temperature, or magnetism. These liposomes with drug-releasing capability can also be bound to ligand-mediated targeted nucleic acid delivery (Barba et al., 2019). In addition, the multicomponent and multifunctional formulations (Sun et al., 2019) configured improved localizing selectivity by addressing unique tissue types. Chen et al. (2015) used glycyrrhetinic acid (GA) modified liposomes, laden with oxaliplatin (OX) for liver biodistribution trials and concluded that the area under the curve (AUC) ratio was 3.84 between GA-OX liposome and AUC of OX liposome. Thus, such eminent results indicated that liposomes display outstanding targeting of organs and tissue.

Apart from enhancing the drug absorption, liposomes can also be used to transform cancer therapy drugs, antibiotics as well as DNA too. Liu et al. (2014) used the AAN-TAT (AAN: alanine–alanine–as-

Table 1. Different liposome formulations available for the treatment of cancer [Data collected from European Medicines Agency (https://www.ema.europa.eu/) and U.S. Food and Drug Administration (https://www.fda.gov/)]

Brand Name	Composition of Liposome	Drug Encapsulated	Cancer	Year of Approval
Doxil® Caelyx®	DSPE-PEG$_{2000}$/ Chol/HSPC	Doxorubicin	Ovarian cancer, multiple myeloma	USA 1995, Europe 1996
Lipo-dox®	DSPE-PEG$_{2000}$/ Chol/DSPC	Doxorubicin	Breast cancer, Ovarian cancer	USA 2013
Marqibo®	Chol/SM	Vincristine	Acute lymphoblastic leukaemia	USA 2012
Mepact®	DOPS/POPC	Mifamurtide	Osteosarcoma	Europe 2009
Myocet®	Chol/EPC	Doxorubicin	Breast cancer	Europe 2000
Onivyde®	DSPE-PEG$_{2000}$/ Chol/DSPC	Irinotecan	Pancreatic adenocarcinoma	USA 2015, Europe 2011
Visudyne®	EPG/DMPC	Verteporfin	Subfoveal choroidal neovascularisation	USA 1996
Vyxeos	Chol/ DSPG/DSPC	Daunorubicin, cytarabine	Acute myeloid leukaemia	USA 2017

Chol: cholesterol; DSPC: 1,2-distearoyl-sn-glycero-3-phosphocholine; DSPE-PEG$_{2000}$: 1,2-distearoyl-sn-glycero-3-phosphoethanolamine-N-[carboxy(polyethylene glycol)-2000]; SM: sphingomyelin; POPC: 1-palmitoyl-2-oleoyl-sn-glycero-3-phosphocholine; DOPS: 1,2-dioleoyl-sn-glycero-3-phospho-L-serine; EPC: egg phosphatidylcholine; DSPG: 1,2-distearoyl-sn-glycero-3-phospho-(1′-rac-glycerol)

paragine, TAT: trans-activating factor)" after AAN-TAT as "AAN-TAT (AAN: alanine–alanine–asparagine, TAT: trans-activating factor) liposome platform for the development of a doxorubicin transporter that can boost the tumoricidal influence of the drug by reducing the systemic side effects. The design of modern cell prototypes aims to combine different types of changes to accomplish a full nano-drug delivery system. For example, chemotherapeutic agents, micro RNAs, and nanoparticles can be combined with or encapsulated with nanoporous silica core to mimic chemotherapy through drug delivery on specific sites. On the basis of the kind of cancerous or vascular structure, a supportive lipid bilayer arrangement can also be modified with target molecules located on the surface such as fusion peptides and polyethylene glycol. For the sake of gene transfer, liposomes can also be used as non-viral vectors, while liposome / DNA complexes are one of the most capable gears in cancer gene therapy (Cong et al., 2020). Table 1 enlists different products for cancer drug delivery using liposomes.

Solid Lipid NPs

It is widely recognized that the toxicity and biocompatibility of lipid-based nanoparticles are lower than polymeric or inorganic nanoparticles. For instance, solid lipid nanoparticles (SLNPs) have become a promising and effective alternative. They are submicron colloidal particles with a diameter of 50-1000 nm. They are composed of lipid matrix solids, surfactants, and in some cases in combination with co-surfactants at physiological temperature (Figure 2) (Geszke-Moritz & Moritz, 2016).

Lipid nanoparticles are of particular interest because of their biocompatibility. In this sense, although liposomes have low physical and chemical stability, they are still one of the most widely used lipid DDS. However, SLNPs are expected to be an effective nanocarrier to control drug delivery, because they can

Figure 2. Proposed schematic illustration of the structure of SLNPs.

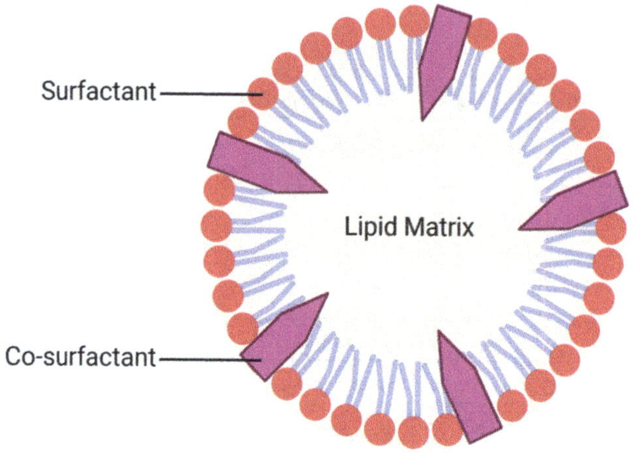

be an alternative to liposomes, which are composed of biocompatible and biodegradable lipids having the ability to solidify at body temperature (Shah et al., 2014).

In addition, they can control the release of integrated drugs and provide chemical protection to ensure the high stability of integrated compounds (Scioli Montoto et al., 2020). Liquid crystal nanoparticles (LCNPs) also have these useful properties and they are deliberated as very favorable nanocarriers (Angelova et al., 2017). Due to the high cost of the enormous energy consumption required in the production process, LCNPs have not yet been popularized (Lee et al., 2016). In this way, SLNPs provides simple and inexpensive mass production, by employing widely used methods i.e., high-pressure homogenization and microemulsion (Geszke-Moritz & Moritz, 2016). Another advantage of SLNPs is that there are many effective routes of administration of these NPs, including oral, parenteral, nasal, ocular, rectal, etc. (Ramteke et al., 2012).

Furthermore, SLN can contain both hydrophilic as well as lipophilic drugs. Contingent upon the arrangement and manufacturing technique of the nanoparticles, the drug can be integrated in three means (Figure 3): (i) evenly distributed in lipid matrix, (ii) penetration in the adjacent matrix layer, or (iii) spread over in the external shell (ud Din et al., 2017).

Figure 3. Distribution of the drug in SLNPs. Drug (represented in white dot) can be incorporated in SLNPs as (a) in the outer (b) in the shell matrix and (c) dispersion in lipid matrix

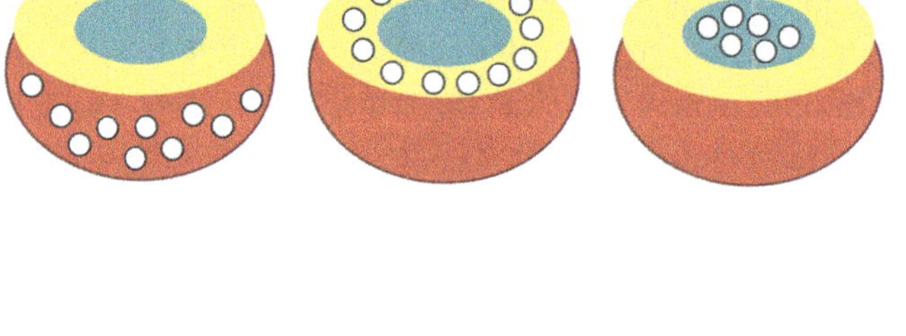

Table 2. Use of SLNPs coupled with different anticancer drugs against various types of cancerous cells

SLNPs Composition With Incorporated Drugs	Type of Cancer	Reference
Paclitaxel-NLC and miRN-200c incorporated in Poloxamer 188, monestearin, and DOTAP (1,2-dioleoyl-3-trimethylammonium-propane)	Breast	(Liu et al., 2017)
Erlotinib incorporated in Poloxamer 188, Glycerol monostearate, and transcutol	Lung	(Naseri et al., 2017)
Linalool embedded in Pluronic F68, and cetyl palmitate/ myristyl myristate/cetyl esters/	Liver	(Rodenak-Kladniew et al., 2017)
Oxaliplatin incorporated in Lipoid S75, DSPE (1,2-distearoyl-sn-glycero-3-phosphoethanolamine), Tween 80, Folic acid, and Tristearin	Colon	(Rajpoot & Jain, 2018)
Indirubin embedded in polysorbate 80 and cetyl palmitate	Brain	(Rahiminejad et al., 2019)
α-Linolenic acid in α-Tocopheryl linolenate, Tween 20, sodium taurocholate, and butanol	Melanoma	(Cassano et al., 2017)

Docetaxel is a natural lipophilic drug with mitotic inhibitory function and less water solubility. Therefore, preparations containing SLNPs may enhance its administration. This has been shown in pharmacokinetic studies that calculated plasma concentrations in rats and found that SLNPs may provide increased absorption and decreased excretion. In addition, these systems may enhance the cytotoxicity-inducing effects of the commercially available form (micelle) and free-form drugs in MCF-7 breast tumor cells (Qureshi et al., 2017).

Doxorubicin is another natural drug, closely linked to anti-cancer compounds. However, its use has some related drawbacks including the advancement of resistance in cancer cells and heart problems. Studies showed that doxorubicin can be incorporated into SLNPs, and when used in amalgamation with α-tocopherol succinate (which also has anti-cancer role), have higher acceptance aptitude and toxicity in MCF-7-resistant human breast and NCI ovarian cancerous cells (Oliveira et al., 2016).

The incorporation of hormonal compounds into SLNPs has also been reported. For example, the tamoxifen targeting estrogen receptors loaded SLNPs verified in contrast to tamoxifen tolerant MCF-7 breast cancerous cells. Table 2 summarizes the examples of the use of SLNPs being utilized for the therapy of various types of tumors.

Polymer NPs (PNPs)

A polymer is a macromolecular structure composed of a large number of monomer units, and the composition, structure, properties, and function of these monomer units differ from each other. Polymeric nanoparticles are used more than only for drug delivery, but also as a medical diagnostic tool for biodetection and bioimaging. Polymer nanoparticles have recently attained much attention due to their special properties, efficiency, and non-toxicity. The resulting nanoparticle drug compound may have a capsule (polymeric nanoparticle), dendrimer (hyperbranched polymer), or polymeric micellar structure (Moreno-Vega et al., 2012). The ideal components required for producing polymer-carriers for NPs are simple synthesis, excellent biocompatibility, flexibility, biodegradability, solubility, and non-toxicity. Polymer nanoparticles have many benefits as compared to delivery system, such as;

Figure 4. Classification of polymeric based nanoparticles (nanospheres and nanocapsule)

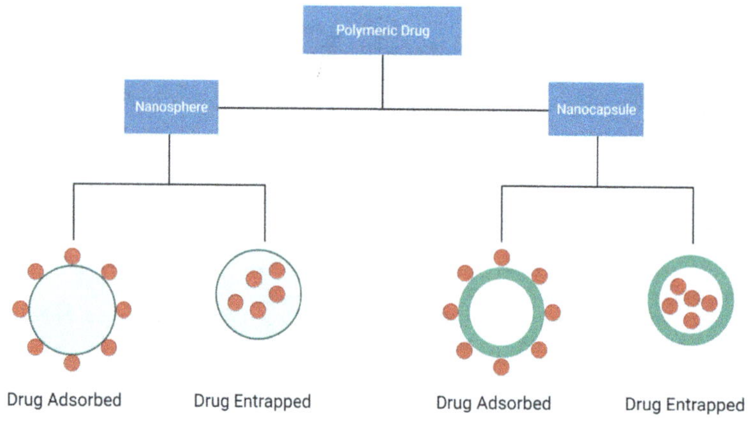

- Therapeutic substances with high concentrations can be given directly to the target site, thereby minimizing toxicity.
- Polymer nanoparticles can protect bioactive molecules and drugs from enzymatic and hydrolytic breakdown.
- The nanoparticles also have the capacity to cross the blood-brain barrier (BBB).

Polymer nanomaterials can be divided into bio-degradable PNP and non-biodegradable PNP. Polymer nanoparticles, on the other hand, can be categorized into two types, depending on their manufacturing methods. (Figure 4).

Nano-Capsule

An ultra-thin drug carrier system comprising of an aqueous or oil core enclosed by a thin polymer film has certain characteristics. Synthetic or natural polymers can be used to create the membrane. The active substance is absorbed by the surface, after its dissolution in the inner core.

Nanospheres

Nanospheres contain a matrix structure of polymer chains, which can disperse, trap, dissolve in or adsorb drugs onto the nanoparticles.

Abraxane was the first polymer nanomedicine to enter the drug market and was released in 2005. It was comprised of NPs of the paclitaxel drug bound to albumin. The formula does not contain the substance chromophore electroluminescent (C-EL). Although C-EL surges the solubility aptitude of paclitaxel in some patients, it can cause severe allergic reactions and fatal symptoms (Weissig et al., 2014). Table 3 summarizes the drug-loaded polymeric nanoparticles that have been employed to treat various malignancies.

Table 3. Drug-loaded polymer nanoparticles used for the treatment of cancers

SLN Composition With Incorporated Drugs	Type of Cancer	Reference
Rapamycin and piperine embedded in Poly(d,l-lactide-co-glycolide)	Breast	(Katiyar et al., 2016)
doxorubicin and triptolide embedded in poly(D,L-lactide-co-glycolide), soybean lecithin, and monomethoxy-poly(ethylene glycol)-S-S-hexadecyl (mPEG-S-S-C16),	squamous cell carcinoma cells	(Wu et al., 2017)
Cisplatin incorporated poly(D,L-lactide-co-glycolide)	human head and neck squamous carcinoma cell line (UM-SCC-47)	(Reardon et al., 2017)
Gambogic acid incorporated poly(D,L-lactide-co-glycolide) and red blood-cell membrane	colorectal cancer	(Zhang et al., 2017)
Triptolide and Paclitaxel coloaded lipid polymer hybrid nanoparticles	Lung Cancer	(Liu et al., 2018)
Paclitaxel embedded in albumin nanoparticles (Abraxane® - FDA Approved)	pancreatic cancer, breast cancer	(Miele et al., 2009)

Polymer Micelles

Polymer micelles are amphiphilic copolymer-based nanoscale (5-100 nm) colloidal particles that develop spontaneously in aqueous environments (Figure 5). Above the critical micelle concentration (CMC), polymeric chains are combined to form a core-shell micelle structure, thereby minimizing the contact between the hydrophobic moiety and the aqueous medium (Yokoyama, 2011).

Figure 5. Schematic diagram of (a) polymeric micelle structure (b) a monomer of polymeric micelle

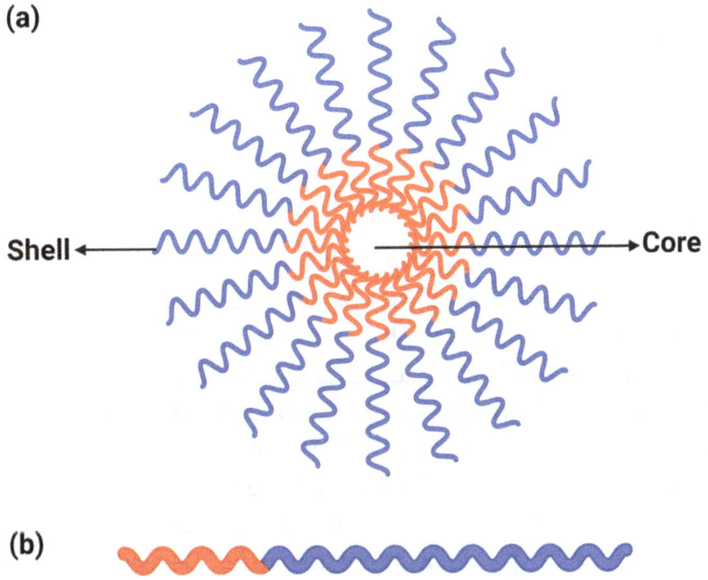

Each component of the polymer micelle contributes to drug delivery. The outer shell helps in the interaction with different biological components in blood circulation (e.g. proteins) and the cell membrane. This layer sets out the stealth properties that reduce phagocytosis of micelles of the reticuloendothelial system (RES). This prolongs blood circulation, increases the possibility of drug delivery via EPR, and improves drug efficacy (Yokoyama, 2011). These properties of the outer shell determine the stability of polymer micelles, biocompatibility, pharmacokinetics, longevity in blood stream, and biodistribution behavior of integrated drugs. That's why, the action of a drug in the body can be controlled by the outer layer and has nothing to do with the inner core.

Blanco et al. (2009) developed an anticancer drug using PEG-PLA micelles that contain the readily soluble compound β-lapacon (β-lap). This combination of PLA and β-lap micellar core is very stable in the micelle during the blood circulation hence resulting in an increased amount of the drug at the tumor site. PEG-PLA-β-lap has also been shown to be very effective in the therapy of subcutaneous lung cancer (NSCLC) A549 and orthotopic Lewis lung cancer (CCL) in vitro. Furthermore, Mu et al. (2010) load the docetaxel in PEG-PLA (MPP) for therapy of multi-drug resistant (MDR) oral epidermoid carcinoma and noted that micelles increase the circulation period of docetaxel in the blood of mice (clinical trial) and inhibit the tumor growth as compared to commercially available docetaxel (Taxotere®).

Dendrimers

Dendrimers are branched polymers of uniform size and shape. The basic structure consists of 3 major elements: a central core, repeated branches, and a set of terminals that give changeable surface features. The formation of dendrimers is done by the rapidly increasing branching units, further, it participates in the creation of circular structures. The strong control on architecture makes it a more effective choice for drug delivery (Kalomiraki et al., 2016) and gene delivery systems (Yang et al., 2015). Oligonucleotides and medicine can be enclosed in internal cavities, or they can be attached to the surface through electrostatic or hydrophobic interactions.

Since the 1980s, various dendrimers have been developed and used, but polyamidoamine (PAMAM) derivatives are the most popular. They are hydrophilic, non-immunogenic and biocompatible systems that can be used for medicine delivery. The key of PAMAM is usually ethylenediamine, but many hydrophobic molecules such as diaminobutane, diaminohexane, and diaminododecane can be utilized (Chang et al., 2014). Their branch units depend on ethylenediamine, methyl acrylate, as well as, they have terminating groups of amines (in half generations) and carboxyl (in full generations) (Gillani et al., 2020).

Poly-L-lysine (PLL) dendrimers are a kind of peptide dendrimers, which are mainly utilized as gene transporters because of their good condensation with oligonucleotides (Wu et al., 2013). PLL dendrimers are almost asymmetric, which is a variant of the usual concept of PAMAM and PPI dendrimers. However, it is a thin molecule, which branches off into the nucleus and controls the amount of lysine with terminal amine residues. PLL's terminal group, lysine, comprises two main amines that are frequently altered to improve biological characteristics (Kaminskas et al., 2011).

Dendrimers can use a variety of strategies to transport drugs, and these strategies can be roughly classified into chemical and physical interactions (Choudhary et al., 2017). Many effective anticancer agents, including camptothecin, paclitaxel, 5-fluorouracil, methotrexate, and doxorubicin are known to be highly hydrophobic, so it is difficult to find an appropriate mean for their administration. In addition, these substances are not tumor-specific and therefore toxic. Dendrimers have been shown to be a promising platform for these molecules (Zhong et al., 2016). Table 4 summarizes some of the studies to treat cancer with dendrimer-based drug delivery systems.

Table 4. Different drug carrier systems developed based on dendrimers

Polymer	Payload	Application	Modification	Reference
PAMAM	Doxorubicin	Lung metastasis	-	(Zhong et al., 2016)
	Docetaxel and Paclitaxel	Breast cancer	Trastuzumab	(Marcinkowska et al., 2019)
PPI	Paclitaxel	Ovarian cancer	mAbK1	(Jain et al., 2015)
	Melphalan	Breast cancer	Folic acid	(Kesharwani et al., 2015)
PLL	Doxorubicin	Rectum cancer	Pentapeptide PEG	(Niidome et al., 2014)
	Doxorubicin	Melanoma	-	(Al-Jamal et al., 2013)

Polymersomes

The therapeutic drug delivery strategy based on polymersomes has shown great potential in cancer treatment because of its physio-chemical stability, higher drug piling capacity, higher biocompatibility, increased colloidal constancy, and aptitude to encapsulate both hydrophobic and hydrophilic molecules (Mohammadi et al., 2017). Polymersomes are circular nanostructures composed of amphiphilic block copolymers, where a hydrophobic bilayer surrounds an aqueous core (Hu & Qiu, 2019). Modification of the surface of the polymersomes with target particular ligands that attach to receptors that are up-regulated in tumor cells can enhance the absorption of cancer molecules by the cells (Alibolandi et al., 2019). Compared with healthier cells, polymersomes are gradually absorbed by cancerous cells and are even less noxious at elevated levels (Pegoraro et al., 2013). Following systemic administration, due to the absorbency and withholding effect, the polymersomes passively gather at the tumor place and secrete the substance at the target site (Shahriari et al., 2019). The inclusion of stimuli sensitive agents to polymersomes can also control the secretion of drugs from them (Wang et al., 2020). Due to these advantages, polymer bodies have been thoroughly studied in multiple biomedical functions, such as gene plus drug delivery, and diagnostics.

The mixture forms a fine dispersal of polymer in the liquid phase, and amphiphiles arrange themselves to produce polymersome (Balasubramanian et al., 2016). Nanoprecipitation, film rehydration, electro-formation, and emulsification are some of the methods currently being employed for the preparation of polymersomes (Matoori & Leroux, 2020).

Release of drug from polymersome is usually caused by passive diffusion, driven through concentration difference. The stimuli sensitive polymersome responds to chemical (mainly related to pH and molecular redox chain arrangement) and physical (light, heat, electrical, magnetic, and ultrasonic) triggers at the target position have been designed to release therapeutic agents (Sharma et al., 2020). There are usually stimuli characterized by irregular heat levels, pH, oxidation stress, and irregular glucose Sintensities in the affected cells, thereby promoting intracellular lysis of the polymersome and releasing encapsulated drug molecules.

NPs Assorted Hydrogels

Hydrogel NPs are 3-D polymer structures utilized for encapsulation and delivery of drugs. These edifices swell in aquatic or organic settings, as it absorbs more fluid (Salatin et al., 2016). In addition to hydrogel ensnarement, the construction of a three-dimensional hydrogel network (where the nanoparticles are used directly as a crosslinking agent) also provides another way of assembling nanoparticles to achieve hydrogel-like properties (e.g. nanoparticle colloidal hydrogel) (Li et al., 2011). In this process, NPs are bounded by powerful hydrophobic connections or by combining nanoparticles with inverse surface charges.

Responsive polymers discharge preparations in the feedback to variations in pH and temperature. Such schemes are used for the delivery of protein, DNA, biosensors, wound healing, and tissue engineering (Buwalda et al., 2017). Gold nanoparticles enclosed into a heat-sensitive hydrogel prepared with glycidylmethacrylated chitooligosaccharide, methoxylpoly(ethylene glycol)-poly(ε-caprolactone)-acryloyl chloride, acrylamide, and N-isopropylacrylamide were capable to initiate the contraction of this thermo sensitive hydrogels and triggering the secretion of doxorubicin to stop breast cancer through NIR radiation (Qu et al., 2015).

INORGANIC NANO-CARRIERS

Quantum Dots (QD)

When exposed to ultraviolet radiation or a light source, a quantum dot (semiconductor) emits fluorescence. They have powerful optical properties such as altered wavelength, maximum luminescence, and resistive photo leaching. The nucleus of a single QD contains 100 to 100,000 atoms from elements of the periodic table, Groups II and VI (zinc, technetium, cadmium, and selenium), Group III (tantalum), and Group V (indium) in their central nucleus with a diameter of 2 to 10 nm (Zhu et al., 2013). The core can be replaced with nano-medicine or scanning probe. Anticancer drugs can diffuse into the cell nucleus through biological and chemical binding reactions. Other cancer-specific biomarker ligands (such as biomolecules) that bind to the target surface can be linked to QDs through chemical bonds, such as, DNA conjugation with QDs drug complex through amine linkage (Radenkovic et al., 2016) injected in body blood stream by injection or directly at the cancer site. Zhou et al. created the CdTe QD (MPA-CdTe QD) coated with 3-mercaptopropionic acid (MPA) that promoted the interaction between the cancer drug daunorubicin (DNR) and leukemia cells and kills the drug-tolerant K562 / A02 leukemia cell line (Zhou et al., 2010). QDs have also been used as a drug carrier for the doxorubicin for leukemia and breast cancer treatment (Yang et al., 2016).

Metallic Nanoparticles

Nano-scale analogs of metals have aroused great interest in biomedical applications because of their special thermal, electrical, optical, magnetic, and catalytic characteristics. The experimental results show that the absorption of radiation by ZnO NPs can significantly improve the photo-oxidation ability (Caputo et al., 2014). The physicochemical features of nanoparticles contribute to their anticancer potential due to their extrinsic or intrinsic features. Intrinsic or endogenous anti cancerous effects act as antioxidant function, which may slow the growth rate of tumors (Porcel et al., 2010). Extrinsic functions include the

Table 5. Summary of the outcomes of different metallic nanoparticles in cancer therapy

Nanoparticle Type	Cancer	Description	Reference
Silica gold nanoshells having superparamagnetic iron oxide NPs	Cell lines of head and neck cancer	Photothermal therapy	(Melancon et al., 2011)
Gold nanoparticles conjugated with glucose	breast MCF-7 adenocarcinoma	Radiosensitiser	(Roa et al., 2012)
Silver nanoshell coupled with a carbon core	Prostate adenocarcinoma cell line	Photothermal ablation	(Kleinauskas et al., 2013)
Platinum NPs	Colon cancer cell line	Toxicity evaluation	(Gehrke et al., 2011)
Titanium dioxide (TiO_2) NPs	LL2 and CT26 (cell line)	Increase oxidative stress and UV radiations	(Fujiwara et al., 2015)
Copper NPs	MCF-7 (Breast cancer cell line)	Cytotoxic effect	(Kamble et al., 2016)
Cerium oxide NPs	A2780 (Ovarian cancer cell line)	Apoptosis, ROS production	(Giri et al., 2013)
ZnO NPs	Mice cell lines	Cytotoxic effects	(Namvar et al., 2015)

use of external stimuli (e.g., hyperthermia), where metal NPs being used as co-adjuvants, and external radiations such as IR and X-rays as excipients to generate free radicals to kill cancer cells as well as increases the cytotoxic effect of ionizing radiations (Danhier et al., 2010). Table 5 summarizes the use of different nanoparticles in cancer treatment.

Fullerenes

Fullerenes (C_{60}) can be made by evaporating solid graphite rods at high temperature or heating resistively in the presence of several torr of inert gases. According to the evaporation conditions, the evaporated smoke contains different contents of the fullerene (Chen et al., 2012).

The laboratory therapeutic effect of C_{60} contained in paclitaxel on inhibiting the growth of MCF-7 breast cancer cells was compared with Abraxane (a commercially available nanoparticle albumin binder for paclitaxel). The outcome showed that paclitaxel containing buckysomes in cell viability studies is as effective as Abraxane, indicating that empty buckysomes is not cytotoxic. A conjugate with C_{60} and DOX is prepared to make the resulting conjugate hydrophilic. Due to the combination of amide bonds, the antitumor activity of DOX and the unique properties of fullerene cells are retained. Fullerenol-DOX has shown in vitro antitumor effects similar to free drugs without systemic toxicity of free DOX. In another study, water-soluble lysine-based C_{60}-fullerene nanoconjugates (CF-LYS-TEG-MMF) were developed to deliver monomethyl fumarate (MMF) by using Prato reaction to treat the neuroblastoma cells (Kumar et al., 2018).

Carbon Nanotubes

Carbon nanotubes (CNT) are categorized into two groups on the basis of structure and diameter: single-walled CNTs (SWCNTs) and multi-walled CNT (MWCNTs). SWCNTs are composed of a single layer of cylindrical graphene, while MWCNTs consist of many concentric graphene sheets (Kesharwani &

Iyer, 2015). As the CNTs absorb near-infrared (NIR) light and heat the nanotubes, so the thermal effect is a feature that is employed to destroy cancerous cells. Overexpression of folate (FA) receptors occurs in cancer cells, and several research groups have produced nanocarriers made from biomaterials related to FA derivatives. Compared to spherical nanocarriers, CNTs have been shown to stay longer in lymph nodes (Zandi et al., 2021). The researcher in another study, loaded gemcitabine (an anti-cancer molecule) into magnetic MWCNTs and introduced it subcutaneously into mice, reporting that it has high activity in lymph nodes (Elhissi et al., 2012). The introduction of CNT-based drugs improves the biodistribution and blood circulation of therapeutic drugs, reduces their dosage, and increases their effectiveness.

Considering the longer circulation time of CNT, Liu and his colleagues created branched PEG-functionalized SWNTs with DOX and injected the SWNT-DOX complex into the cancerous site of mice. They found that DOX can infect tumors, and SWNTs can be removed from the systemic circulation by renal excretion. In addition, researchers modified SWNTs to epidermal growth factor (EGF) mediated SWNTs carriers to improve the delivery efficiency of cancer drugs (Guo et al., 2017). MWNTs is a carbon-based nanomaterial that can kill cancer cells when heated and can be used for thermal ablation. Drug delivery of CNTs has many benefits, including minimizing side effects and reducing cytotoxic drugs (Li et al., 2017).

Ceramic Nanoparticles

Ceramic nanoparticles are inorganic systems with porous properties. Considerable research has been conducted to investigate typical biocompatible ceramic nanoparticles, such as titania, silica, and alumina (Mondal et al., 2018). The benefits of ceramic nanoparticles include easy access to the desired shape, size, and porosity without affecting swelling or porosity due to changes in pH (Singh et al., 2016). Nano-sized ceramics such as calcium carbonate ($CaCO_3$), hydroxyapatite (HA), silica (SiO_2), zirconia (ZrO_2), alumina (Al_2O_3), and titanium dioxide (TiO_2) have been developed using new methodologies to improve the physical and chemical properties of the material hence reducing their cytotoxicity in biological systems. Table 6 summarizes the current research advances in the field of ceramic nanoparticles for cancer therapy.

TUMOR TARGETING BY NPs

In general, NPs mediate drug delivery through active and passive mechanisms (Figure 6). Through passive targeting, the systems can be delivered to the target site according to the body's physio-anatomical circumstances. This current information system can effectively treat patients with liver and spleen diseases. First, the substance enters macrophages, accumulates, and exerts the effect. Therefore, these macrophages can serve as a shield for liver and splenic ailments (Khodabandehloo et al., 2016). Another example is carcinoid (sarcoma) with defective vascular permeability in the lymphatic and vascular systems. Due to the defects in the lymphatic system of these diseases, the drugs that leave the circulation and enter the tumor-affected area will accumulate over a long period of time, resulting in a therapeutic effect. Liposomes, polymer particles, and micelles use this technique for drug delivery to tissues. In addition, this technology alters the environment in which carcinoid tissue thrives. In carcinoid tissues, the degree of temperature increased (above 40 °C), and the pH value is subordinate that in neighboring tissues (approx. 5.4). These properties can be used in pH and temperature-sensitive inactive drug delivery methods.

Table 6. Summary of the researches using ceramic nanoparticles for the treatment of cancer

Type of Nanoparticle	Type of Cancer	Reference
calcium phosphate nanoparticles coupled with doxorubicin in conjunction with diethylenetriaminepentaacetic acid gadolinium (A54-CaP/Gd-DTPA/DOX)	Hepatic Cancer	(Zhang et al., 2018)
Epirubicin and melittin embedded in calcium carbonate nanoparticles	MCF-7 (breast cancer) and C26 (colon cancer) cell lines	(Yazdian-Robati et al., 2019)
Silica nanoparticles containing curcumin	HepG2 (liver cancer) and HeLa (cervical cancer) cell lines	(Elbialy et al., 2020)
amorph, brookite, anatase, and rutile forms of TiO_2 nanoparticles coated with polyethylene glycol (PEG) and bovine serum albumin (BSA)	MKN-45 (gastric cancer cell line)	(Nasr et al., 2018)
Modified core-shell magnetic mesoporous zirconia nanoparticles loaded with doxorubicin	HepG2 (liver cancer) cell line	(Chen et al., 2019)

Compared to passive drug delivery, active drug delivery has the potential to target more specific cells and tissues (after identifying specific receptors and transports) for drug delivery.

Figure 6. Targeted drug distribution categories. Active and Passive Aiming leads to the distribution of carriers in tumor tissue. (a) In passive targeting, nanoparticles are created to be transported through permeable vessels and through the inherent intra organ tumor pressure. (b) In active targeting of (1) cancer cells or (2) tumor endothelium, NPs are formulated to cling to particular biological structures in tumors by using molecular recognition of surface bounded ligands. (c) Triggered release allows NPs to muster if revealed to any external stimuli like light or a magnetic field.

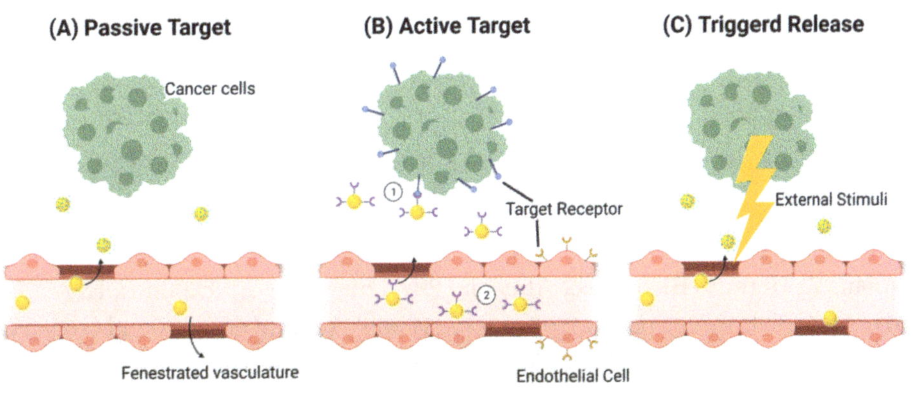

FUTURE RESEARCH DIRECTIONS

Aforementioned cited data indicates that the future research should not only focus on the production of different NPs, but extensive rapid research should be conducted both in vitro and in vivo to assess the NPs efficacy. Precise targeting of NPs along with drug release strategies, drug holding capacity, and drug bioavailability should also be a focus of research in future. Moreover, scientists should also focus on monitoring the toxicological effects associated with the use of NPs. Overall, with the rapid development in the field of nanotechnology and interdisciplinary research, we are hopeful that in near future there will be a remarkable nanotechnology-based drug to treat cancer.

CONCLUSION

The combination of advances in nanotechnology and other sciences has led to the emergence of different types of nanoparticles with different structures. Each has its advantages and disadvantages. NPs are performing the most competent role in DDS, as carriers for lipids, polymers, ceramics, metals, etc. to treat diseases, especially those that are difficult to cure like cancer. Disease diagnosis, treatment, drug delivery, and biomedical imaging can be done by using the NPs. Advances in nano-technology have increased the ability to target multiple molecules in a tumor sample simultaneously and adopt appropriate treatment strategies. Soon, nanotechnology will change all fields of medicine, not just oncology.

REFERENCES

Al-Jamal, K. T., Al-Jamal, W. T., Wang, J. T., Rubio, N., Buddle, J., Gathercole, D., Zloh, M., & Kostarelos, K. (2013). Cationic poly-L-lysine dendrimer complexes doxorubicin and delays tumor growth in vitro and in vivo. *ACS Nano*, *7*(3), 1905–1917. doi:10.1021/nn305860k PMID:23527750

Alibolandi, M., Shahriari, M., & Ramezani, M. (2019). Smart Polymersomes as Intelligent Nanomedicines in Cancer Treatment. *Polymeric Nanoparticles as a Promising Tool for Anti-Cancer Therapeutics*, 343-371.

Ambade, A. V., Savariar, E. N., & Thayumanavan, S. (2005). Dendrimeric micelles for controlled drug release and targeted delivery. *Molecular Pharmaceutics*, *2*(4), 264–272. doi:10.1021/mp050020d PMID:16053329

Angelova, A., Garamus, V. M., Angelov, B., Tian, Z., Li, Y., & Zou, A. (2017). Advances in structural design of lipid-based nanoparticle carriers for delivery of macromolecular drugs, phytochemicals and anti-tumor agents. *Advances in Colloid and Interface Science*, *249*, 331–345. doi:10.1016/j.cis.2017.04.006 PMID:28477868

Balasubramanian, V., Herranz-Blanco, B., Almeida, P. V., Hirvonen, J., & Santos, H. A. (2016). Multifaceted polymersome platforms: Spanning from self-assembly to drug delivery and protocells. *Progress in Polymer Science*, *60*, 51–85. doi:10.1016/j.progpolymsci.2016.04.004

Barba, A. A., Bochicchio, S., Dalmoro, A., & Lamberti, G. (2019). Lipid Delivery Systems for Nucleic-Acid-Based-Drugs: From Production to Clinical Applications. *Pharmaceutics*, *11*(8), 360. Advance online publication. doi:10.3390/pharmaceutics11080360 PMID:31344836

Bi, Y., Hao, F., Yan, G., Teng, L. J., Lee, R., & Xie, J. (2016). Actively targeted nanoparticles for drug delivery to tumor. *Current Drug Metabolism*, *17*(8), 763–782. doi:10.2174/13892002176661606191 91853 PMID:27335116

Blanco, E., Kessinger, C. W., Sumer, B. D., & Gao, J. (2009). Multifunctional micellar nanomedicine for cancer therapy. *Experimental Biology and Medicine, 234*(2), 123–131. doi:10.3181/0808-MR-250 PMID:19064945

Bozzuto, G., & Molinari, A. (2015). Liposomes as nanomedical devices. *International Journal of Nanomedicine, 10*, 975–999. doi:10.2147/IJN.S68861 PMID:25678787

Bulbake, U., Doppalapudi, S., Kommineni, N., & Khan, W. (2017). Liposomal Formulations in Clinical Use: An Updated Review. *Pharmaceutics, 9*(2), 12. doi:10.3390/pharmaceutics9020012 PMID:28346375

Buwalda, S. J., Vermonden, T., & Hennink, W. E. (2017). Hydrogels for therapeutic delivery: Current developments and future directions. *Biomacromolecules, 18*(2), 316–330. doi:10.1021/acs.biomac.6b01604 PMID:28027640

Caputo, F., De Nicola, M., & Ghibelli, L. (2014). Pharmacological potential of bioactive engineered nanomaterials. *Biochemical Pharmacology, 92*(1), 112–130. doi:10.1016/j.bcp.2014.08.015 PMID:25175739

Cassano, R., Mellace, S., Marrelli, M., Conforti, F., & Trombino, S. (2017). α-Tocopheryl linolenate solid lipid nanoparticles for the encapsulation, protection, and release of the omega-3 polyunsaturated fatty acid: In vitro anti-melanoma activity evaluation. *Colloids and Surfaces. B, Biointerfaces, 151*, 128–133. doi:10.1016/j.colsurfb.2016.11.043 PMID:27988473

Chang, H., Wang, H., Shao, N., Wang, M., Wang, X., & Cheng, Y. (2014). Surface-engineered dendrimers with a diaminododecane core achieve efficient gene transfection and low cytotoxicity. *Bioconjugate Chemistry, 25*(2), 342–350. doi:10.1021/bc400496u PMID:24410081

Chen, J., Jiang, H., Wu, Y., Li, Y., & Gao, Y. (2015). A novel glycyrrhetinic acid-modified oxaliplatin liposome for liver-targeting and in vitro/vivo evaluation. *Drug Design, Development and Therapy, 9*, 2265. PMID:25945038

Chen, L., Zhong, H., Qi, X., Shao, H., & Xu, K. (2019). Modified core–shell magnetic mesoporous zirconia nanoparticles formed through a facile "outside-to-inside" way for CT/MRI dual-modal imaging and magnetic targeting cancer chemotherapy. *RSC Advances, 9*(23), 13220–13233. doi:10.1039/C9RA01063G

Chen, Z., Mao, R., & Liu, Y. (2012). Fullerenes for cancer diagnosis and therapy: Preparation, biological and clinical perspectives. *Current Drug Metabolism, 13*(8), 1035–1045. doi:10.2174/138920012802850128 PMID:22380017

Choudhary, S., Gupta, L., Rani, S., Dave, K., & Gupta, U. (2017). Impact of Dendrimers on Solubility of Hydrophobic Drug Molecules. *Frontiers in Pharmacology, 8*, 261. doi:10.3389/fphar.2017.00261 PMID:28559844

Cong, X., Tian, H., Liu, S., Mao, K., Chen, H., Xin, Y., Liu, F., Wang, X., Meng, X., Zhu, G., Wang, J., Gao, X., Tan, H., Yang, Y.-G., & Sun, T. (2020). Cationic Liposome/DNA complexes mediate antitumor immunotherapy by promoting immunogenic tumor cell death and dendritic cell activation. *ACS Applied Materials & Interfaces, 12*(25), 28047–28056. doi:10.1021/acsami.0c08112 PMID:32478501

Dang, Y., & Guan, J. (2020). Nanoparticle-based drug delivery systems for cancer therapy. *Smart Materials in Medicine, 1*, 10–19. doi:10.1016/j.smaim.2020.04.001 PMID:34553138

Danhier, F., Feron, O., & Préat, V. (2010). To exploit the tumor microenvironment: Passive and active tumor targeting of nanocarriers for anti-cancer drug delivery. *Journal of Controlled Release, 148*(2), 135–146. doi:10.1016/j.jconrel.2010.08.027 PMID:20797419

Daraee, H., Etemadi, A., Kouhi, M., Alimirzalu, S., & Akbarzadeh, A. (2016). Application of liposomes in medicine and drug delivery. *Artificial Cells, Nanomedicine, and Biotechnology, 44*(1), 381–391. doi:10.3109/21691401.2014.953633 PMID:25222036

Elbialy, N. S., Aboushoushah, S. F., Sofi, B. F., & Noorwali, A. (2020). Multifunctional curcumin-loaded mesoporous silica nanoparticles for cancer chemoprevention and therapy. *Microporous and Mesoporous Materials, 291*, 109540. doi:10.1016/j.micromeso.2019.06.002

Elhissi, A., Ahmed, W., Hassan, I. U., Dhanak, V., & D'Emanuele, A. (2012). Carbon nanotubes in cancer therapy and drug delivery. *Journal of Drug Delivery*. PMID:22028974

Fujiwara, R., Luo, Y., Sasaki, T., Fujii, K., Ohmori, H., & Kuniyasu, H. (2015). Cancer therapeutic effects of titanium dioxide nanoparticles are associated with oxidative stress and cytokine induction. *Pathobiology, 82*(6), 243–251. doi:10.1159/000439404 PMID:26485713

Gehrke, H., Pelka, J., Hartinger, C. G., Blank, H., Bleimund, F., Schneider, R., Gerthsen, D., Bräse, S., Crone, M., Türk, M., & Marko, D. (2011). Platinum nanoparticles and their cellular uptake and DNA platination at non-cytotoxic concentrations. *Archives of Toxicology, 85*(7), 799–812. doi:10.100700204-010-0636-3 PMID:21229235

Geszke-Moritz, M., & Moritz, M. (2016). Solid lipid nanoparticles as attractive drug vehicles: Composition, properties and therapeutic strategies. *Materials Science and Engineering C, 68*, 982–994. doi:10.1016/j.msec.2016.05.119 PMID:27524099

Gillani, S. S., Munawar, M. A., Khan, K. M., & Chaudhary, J. A. (2020). Synthesis, characterization and applications of poly-aliphatic amine dendrimers and dendrons. *Journal of the Indian Chemical Society, 17*(11), 2717–2736. doi:10.100713738-020-01973-4

Giri, S., Karakoti, A., Graham, R. P., Maguire, J. L., Reilly, C. M., Seal, S., Rattan, R., & Shridhar, V. (2013). Nanoceria: A rare-earth nanoparticle as a novel anti-angiogenic therapeutic agent in ovarian cancer. *PLoS One, 8*(1), e54578. doi:10.1371/journal.pone.0054578 PMID:23382918

Guo, Q., Shen, X., Li, Y.-y., & Xu, S. (2017). Carbon nanotubes-based drug delivery to cancer and brain. *Current Medical Science, 37*(5), 635–641. doi:10.100711596-017-1783-z PMID:29058274

Gupta, A., Ahmad, A., Dar, A., Bhatt, A., & Khan, R. (2016). Selective targeting of cancer cells using personalized nanomedicine. *Toxicol Forensic Med Open J, 1*(1), e12–e13. doi:10.17140/TFMOJ-1-e005

Holohan, C., Van Schaeybroeck, S., Longley, D. B., & Johnston, P. G. (2013). Cancer drug resistance: An evolving paradigm. *Nature Reviews. Cancer, 13*(10), 714–726. doi:10.1038/nrc3599 PMID:24060863

Hu, Y., & Qiu, L. (2019). Polymersomes: Preparation and Characterization. In Pharmaceutical Nanotechnology (pp. 247-265). Springer.

Jain, N. K., Tare, M. S., Mishra, V., & Tripathi, P. K. (2015). The development, characterization and in vivo anti-ovarian cancer activity of poly(propylene imine) (PPI)-antibody conjugates containing encapsulated paclitaxel. *Nanomedicine (London), 11*(1), 207–218. doi:10.1016/j.nano.2014.09.006 PMID:25262579

Kalomiraki, M., Thermos, K., & Chaniotakis, N. A. (2016). Dendrimers as tunable vectors of drug delivery systems and biomedical and ocular applications. *International Journal of Nanomedicine, 11*, 1. PMID:26730187

Kamble, S., Utage, B., Mogle, P., Kamble, R., Hese, S., Dawane, B., & Gacche, R. (2016). Evaluation of curcumin capped copper nanoparticles as possible inhibitors of human breast cancer cells and angiogenesis: A comparative study with native curcumin. *AAPS PharmSciTech, 17*(5), 1030–1041. doi:10.120812249-015-0435-5 PMID:26729534

Kaminskas, L. M., Kelly, B. D., McLeod, V. M., Sberna, G., Owen, D. J., Boyd, B. J., & Porter, C. J. (2011). Characterisation and tumour targeting of PEGylated polylysine dendrimers bearing doxorubicin via a pH labile linker. *Journal of Controlled Release, 152*(2), 241–248. doi:10.1016/j.jconrel.2011.02.005 PMID:21315119

Karra, N., & Benita, S. (2012). The ligand nanoparticle conjugation approach for targeted cancer therapy. *Current Drug Metabolism, 13*(1), 22–41. doi:10.2174/138920012798356899 PMID:21892918

Katiyar, S. S., Muntimadugu, E., Rafeeqi, T. A., Domb, A. J., & Khan, W. (2016). Co-delivery of rapamycin- and piperine-loaded polymeric nanoparticles for breast cancer treatment. *Drug Delivery, 23*(7), 2608–2616. doi:10.3109/10717544.2015.1039667 PMID:26036652

Kesharwani, P., & Iyer, A. K. (2015). Recent advances in dendrimer-based nanovectors for tumor-targeted drug and gene delivery. *Drug Discovery Today, 20*(5), 536–547. doi:10.1016/j.drudis.2014.12.012 PMID:25555748

Kesharwani, P., Tekade, R. K., & Jain, N. K. (2015). Generation dependent safety and efficacy of folic acid conjugated dendrimer based anticancer drug formulations. *Pharmaceutical Research, 32*(4), 1438–1450. doi:10.100711095-014-1549-2 PMID:25330744

Khan, I., Saeed, K., & Khan, I. (2019). Nanoparticles: Properties, applications and toxicities. *Arabian Journal of Chemistry, 12*(7), 908–931. doi:10.1016/j.arabjc.2017.05.011

Khodabandehloo, H., Zahednasab, H., & Hafez, A. A. (2016). Nanocarriers usage for drug delivery in cancer therapy. *Iranian Journal of Cancer Prevention, 9*(2). Advance online publication. doi:10.17795/ijcp-3966 PMID:27482328

Kleinauskas, A., Rocha, S., Sahu, S., Sun, Y.-P., & Juzenas, P. (2013). Carbon-core silver-shell nanodots as sensitizers for phototherapy and radiotherapy. *Nanotechnology, 24*(32), 325103. doi:10.1088/0957-4484/24/32/325103 PMID:23868054

Kumar, M., Sharma, G., Kumar, R., Singh, B., Katare, O. P., & Raza, K. (2018). Lysine-based C60-fullerene nanoconjugates for monomethyl fumarate delivery: A novel nanomedicine for brain cancer cells. *ACS Biomaterials Science & Engineering, 4*(6), 2134–2142. doi:10.1021/acsbiomaterials.7b01031 PMID:33435037

Lee, D. R., Park, J. S., Bae, I. H., Lee, Y., & Kim, B. M. (2016). Liquid crystal nanoparticle formulation as an oral drug delivery system for liver-specific distribution. *International Journal of Nanomedicine, 11*, 853. PMID:27042053

Li, F., Josephson, D. P., & Stein, A. (2011). Colloidal assembly: The road from particles to colloidal molecules and crystals. *Angewandte Chemie International Edition, 50*(2), 360–388. doi:10.1002/anie.201001451 PMID:21038335

Li, Z., de Barros, A. L. B., Soares, D. C. F., Moss, S. N., & Alisaraie, L. (2017). Functionalized single-walled carbon nanotubes: Cellular uptake, biodistribution and applications in drug delivery. *International Journal of Pharmaceutics, 524*(1-2), 41–54. doi:10.1016/j.ijpharm.2017.03.017 PMID:28300630

Liu, B., Han, L., Liu, J., Han, S., Chen, Z., & Jiang, L. (2017). Co-delivery of paclitaxel and TOS-cisplatin via TAT-targeted solid lipid nanoparticles with synergistic antitumor activity against cervical cancer. *International Journal of Nanomedicine, 12*, 955–968. doi:10.2147/IJN.S115136 PMID:28203075

Liu, J., Cheng, H., Han, L., Qiang, Z., Zhang, X., Gao, W., Zhao, K., & Song, Y. (2018). Synergistic combination therapy of lung cancer using paclitaxel- and triptolide-coloaded lipid-polymer hybrid nanoparticles. *Drug Design, Development and Therapy, 12*, 3199–3209. doi:10.2147/DDDT.S172199 PMID:30288024

Liu, Z., Xiong, M., Gong, J., Zhang, Y., Bai, N., Luo, Y., Li, L., Wei, Y., Liu, Y., Tan, X., & Xiang, R. (2014). Legumain protease-activated TAT-liposome cargo for targeting tumours and their microenvironment. *Nature Communications, 5*(1), 1–11. doi:10.1038/ncomms5280 PMID:24969588

Marcinkowska, M., Stanczyk, M., Janaszewska, A., Sobierajska, E., Chworos, A., & Klajnert-Maculewicz, B. (2019). Multicomponent Conjugates of Anticancer Drugs and Monoclonal Antibody with PAMAM Dendrimers to Increase Efficacy of HER-2 Positive Breast Cancer Therapy. *Pharmaceutical Research, 36*(11), 154. doi:10.100711095-019-2683-7 PMID:31482205

Matoori, S., & Leroux, J.-C. (2020). Twenty-five years of polymersomes: Lost in translation? *Materials Horizons, 7*(5), 1297–1309. doi:10.1039/C9MH01669D

Melancon, M. P., Lu, W., Zhong, M., Zhou, M., Liang, G., Elliott, A. M., Hazle, J. D., Myers, J. N., Li, C., & Stafford, R. J. (2011). Targeted multifunctional gold-based nanoshells for magnetic resonance-guided laser ablation of head and neck cancer. *Biomaterials, 32*(30), 7600–7608. doi:10.1016/j.biomaterials.2011.06.039 PMID:21745689

Miele, E., Spinelli, G. P., Miele, E., Tomao, F., & Tomao, S. (2009). Albumin-bound formulation of paclitaxel (Abraxane® ABI-007) in the treatment of breast cancer. *International Journal of Nanomedicine, 4*, 99. PMID:19516888

Mohammadi, M., Ramezani, M., Abnous, K., & Alibolandi, M. (2017). Biocompatible polymersomes-based cancer theranostics: Towards multifunctional nanomedicine. *International Journal of Pharmaceutics, 519*(1-2), 287–303. doi:10.1016/j.ijpharm.2017.01.037 PMID:28115259

Mondal, S., Dorozhkin, S. V., & Pal, U. (2018). Recent progress on fabrication and drug delivery applications of nanostructured hydroxyapatite. *WIREs Nanomedicine and Nanobiotechnology, 10*(4), e1504.

Moreno-Vega, A.-I., Gomez-Quintero, T., Nunez-Anita, R.-E., Acosta-Torres, L.-S., & Castaño, V. (2012). Polymeric and ceramic nanoparticles in biomedical applications. *Journal of Nanotechnology, 2012*, 2012. doi:10.1155/2012/936041

Mu, C.-F., Balakrishnan, P., Cui, F.-D., Yin, Y.-M., Lee, Y.-B., Choi, H.-G., Yong, C. S., Chung, S.-J., Shim, C.-K., & Kim, D.-D. (2010). The effects of mixed MPEG–PLA/Pluronic® copolymer micelles on the bioavailability and multidrug resistance of docetaxel. *Biomaterials, 31*(8), 2371–2379. doi:10.1016/j.biomaterials.2009.11.102 PMID:20031202

Namvar, F., Rahman, H. S., Mohamad, R., Azizi, S., Tahir, P. M., Chartrand, M. S., & Yeap, S. K. (2015). Cytotoxic effects of biosynthesized zinc oxide nanoparticles on murine cell lines. *Evidence-Based Complementary and Alternative Medicine, 2015*, 2015. doi:10.1155/2015/593014 PMID:25784947

Naseri, N., Zakeri-Milani, P., Hamishehkar, H., Pilehvar-Soltanahmadi, Y., & Valizadeh, H. (2017). Development, In Vitro Characterization, Antitumor and Aerosol Performance Evaluation of Respirable Prepared by Self-nanoemulsification Method. *Drug Research (Stuttgart), 67*(06), 343–348. doi:10.1055-0043-102404 PMID:28288490

Nasr, R., Hasanzadeh, H., Khaleghian, A., Moshtaghian, A., Emadi, A., & Moshfegh, S. (2018). Induction of Apoptosis and Inhibition of Invasion in Gastric Cancer Cells by Titanium Dioxide Nanoparticles. *Oman Medical Journal, 33*(2), 111–117. doi:10.5001/omj.2018.22 PMID:29657679

Niidome, T., Yamauchi, H., Takahashi, K., Naoyama, K., Watanabe, K., Mori, T., & Katayama, Y. (2014). Hydrophobic cavity formed by oligopeptide for doxorubicin delivery based on dendritic poly(L-lysine). *Journal of Biomaterials Science. Polymer Edition, 25*(13), 1362–1373. doi:10.1080/09205063.2014.938979 PMID:25040893

Oliveira, M. S., Aryasomayajula, B., Pattni, B., Mussi, S. V., Ferreira, L. A. M., & Torchilin, V. P. (2016). Solid lipid nanoparticles co-loaded with doxorubicin and α-tocopherol succinate are effective against drug-resistant cancer cells in monolayer and 3-D spheroid cancer cell models. *International Journal of Pharmaceutics, 512*(1), 292–300. doi:10.1016/j.ijpharm.2016.08.049 PMID:27568499

Pegoraro, C., Cecchin, D., Gracia, L. S., Warren, N., Madsen, J., Armes, S. P., Lewis, A., MacNeil, S., & Battaglia, G. (2013). Enhanced drug delivery to melanoma cells using PMPC-PDPA polymersomes. *Cancer Letters, 334*(2), 328–337. doi:10.1016/j.canlet.2013.02.007 PMID:23402813

Porcel, E., Liehn, S., Remita, H., Usami, N., Kobayashi, K., Furusawa, Y., Le Sech, C., & Lacombe, S. (2010). Platinum nanoparticles: A promising material for future cancer therapy? *Nanotechnology, 21*(8), 85103. doi:10.1088/0957-4484/21/8/085103 PMID:20101074

Prabhu, R. H., Patravale, V. B., & Joshi, M. D. (2015). Polymeric nanoparticles for targeted treatment in oncology: Current insights. *International Journal of Nanomedicine, 10*, 1001–1018. doi:10.2147/IJN.S56932 PMID:25678788

Qu, Y., Chu, B. Y., Peng, J. R., Liao, J. F., Qi, T. T., Shi, K., Zhang, X. N., Wei, Y. Q., & Qian, Z. Y. (2015). A biodegradable thermo-responsive hybrid hydrogel: Therapeutic applications in preventing the post-operative rccurrence of breast cancer. *NPG Asia Materials, 7*(8), e207–e207. doi:10.1038/am.2015.83

Qureshi, O. S., Kim, H.-S., Zeb, A., Choi, J.-S., Kim, H.-S., Kwon, J.-E., Kim, M.-S., Kang, J.-H., Ryou, C., Park, J.-S., & Kim, J.-K. (2017). Sustained release docetaxel-incorporated lipid nanoparticles with improved pharmacokinetics for oral and parenteral administration. *Journal of Microencapsulation, 34*(3), 250–261. doi:10.1080/02652048.2017.1337247 PMID:28557649

Radenkovic, D., Kobayashi, H., Remsey-Semmelweis, E., & Seifalian, A. M. (2016). Quantum dot nanoparticle for optimization of breast cancer diagnostics and therapy in a clinical setting. *Nanomedicine; Nanotechnology, Biology, and Medicine, 12*(6), 1581–1592. doi:10.1016/j.nano.2016.02.014 PMID:27013132

Rahiminejad, A., Dinarvand, R., Johari, B., Nodooshan, S. J., Rashti, A., Rismani, E., Mahdaviani, P., Saltanatpour, Z., Rahiminejad, S., Raigani, M., & Khosravani, M. (2019). Preparation and investigation of indirubin-loaded SLN nanoparticles and their anti-cancer effects on human glioblastoma U87MG cells. *Cell Biology International, 43*(1), 2–11. doi:10.1002/cbin.11037 PMID:30080277

Rajpoot, K., & Jain, S. K. (2018). Colorectal cancer-targeted delivery of oxaliplatin via folic acid-grafted solid lipid nanoparticles: Preparation, optimization, and in vitro evaluation. *Artificial Cells, Nanomedicine, and Biotechnology, 46*(6), 1236–1247. doi:10.1080/21691401.2017.1366338 PMID:28849671

Ramteke, K., Joshi, S., & Dhole, S. (2012). Solid lipid nanoparticle: A review. *IOSR Journal of Pharmacy, 2*(6), 34–44. doi:10.9790/3013-26103444

Reardon, P. J., Parhizkar, M., Harker, A. H., Browning, R. J., Vassileva, V., Stride, E., Pedley, R. B., Edirisinghe, M., & Knowles, J. C. (2017). Electrohydrodynamic fabrication of core-shell PLGA nanoparticles with controlled release of cisplatin for enhanced cancer treatment. *International Journal of Nanomedicine, 12*, 3913–3926. doi:10.2147/IJN.S134833 PMID:28579777

Roa, W., Xiong, Y., Chen, J., Yang, X., Song, K., Yang, X., Kong, B., Wilson, J., & Xing, J. Z. (2012). Pharmacokinetic and toxicological evaluation of multi-functional thiol-6-fluoro-6-deoxy-D-glucose gold nanoparticles in vivo. *Nanotechnology, 23*(37), 375101. doi:10.1088/0957-4484/23/37/375101 PMID:22922305

Rodenak-Kladniew, B., Islan, G. A., de Bravo, M. G., Durán, N., & Castro, G. R. (2017). Design, characterization and in vitro evaluation of linalool-loaded solid lipid nanoparticles as potent tool in cancer therapy. *Colloids and Surfaces. B, Biointerfaces, 154*, 123–132. doi:10.1016/j.colsurfb.2017.03.021 PMID:28334689

Salatin, S., Barar, J., Barzegar-Jalali, M., Adibkia, K., Milani, M. A., & Jelvehgari, M. (2016). Hydrogel nanoparticles and nanocomposites for nasal drug/vaccine delivery. *Archives of Pharmacal Research, 39*(9), 1181–1192. doi:10.100712272-016-0782-0 PMID:27352214

Scioli Montoto, S., Muraca, G., & Ruiz, M. E. (2020). Solid Lipid Nanoparticles for Drug Delivery: Pharmacological and Biopharmaceutical Aspects [Review]. *Frontiers in Molecular Biosciences, 7*(319), 587997. Advance online publication. doi:10.3389/fmolb.2020.587997 PMID:33195435

Senapati, S., Mahanta, A. K., Kumar, S., & Maiti, P. (2018). Controlled drug delivery vehicles for cancer treatment and their performance. *Signal Transduction and Targeted Therapy, 3*(1), 7. doi:10.103841392-017-0004-3 PMID:29560283

Shah, R., Eldridge, D., Palombo, E., & Harding, I. (2014). Optimisation and stability assessment of solid lipid nanoparticles using particle size and zeta potential. *The Journal of Physiological Sciences; JPS, 25*(1).

Shahriari, M., Taghdisi, S. M., Abnous, K., Ramezani, M., & Alibolandi, M. (2019). Synthesis of hyaluronic acid-based polymersomes for doxorubicin delivery to metastatic breast cancer. *International Journal of Pharmaceutics, 572*, 118835. doi:10.1016/j.ijpharm.2019.118835 PMID:31726198

Sharma, A. K., Prasher, P., Aljabali, A. A., Mishra, V., Gandhi, H., Kumar, S., Mutalik, S., Chellappan, D. K., Tambuwala, M. M., Dua, K., & Kapoor, D. N. (2020). Emerging era of "somes": Polymersomes as versatile drug delivery carrier for cancer diagnostics and therapy. *Drug Delivery and Translational Research, 10*(5), 1171–1190. doi:10.100713346-020-00789-2 PMID:32504410

Shi, J., Kantoff, P. W., Wooster, R., & Farokhzad, O. C. (2017). Cancer nanomedicine: Progress, challenges and opportunities. *Nature Reviews. Cancer, 17*(1), 20–37. doi:10.1038/nrc.2016.108 PMID:27834398

Siegel, R. L., Miller, K. D., & Jemal, A. (2016). Cancer statistics, 2016. *CA: a Cancer Journal for Clinicians, 66*(1), 7–30. doi:10.3322/caac.21332 PMID:26742998

Singh, D., Singh, S., Sahu, J., Srivastava, S., & Singh, M. R. (2016). Ceramic nanoparticles: Recompense, cellular uptake and toxicity concerns. *Artificial Cells, Nanomedicine, and Biotechnology, 44*(1), 401–409. doi:10.3109/21691401.2014.955106 PMID:25229834

Sun, H., Guo, X., Zeng, S., Wang, Y., Hou, J., Yang, D., & Zhou, S. (2019). A multifunctional liposomal nanoplatform co-delivering hydrophobic and hydrophilic doxorubicin for complete eradication of xenografted tumors. *Nanoscale, 11*(38), 17759-17772. doi:10.1039/C9NR04669K

Tiwari, G., Tiwari, R., Sriwastawa, B., Bhati, L., Pandey, S., Pandey, P., & Bannerjee, S. K. (2012). Drug delivery systems: An updated review. *International Journal of Pharmaceutical Investigation, 2*(1), 2. doi:10.4103/2230-973X.96920 PMID:23071954

ud Din, F., Aman, W., Ullah, I., Qureshi, O. S., Mustapha, O., Shafique, S., & Zeb, A. (2017). Effective use of nanocarriers as drug delivery systems for the treatment of selected tumors. *International Journal of Nanomedicine, 12*, 7291.

Vines, J. B., Yoon, J.-H., Ryu, N.-E., Lim, D.-J., & Park, H. (2019). Gold Nanoparticles for Photothermal Cancer Therapy [Review]. *Frontiers in Chemistry, 7*(167). https://doi.org/10.3389/fchem.2019.00167

Wang, X., Yao, C., Zhang, G., & Liu, S. (2020). Regulating vesicle bilayer permeability and selectivity via stimuli-triggered polymersome-to-PICsome transition. *Nature Communications, 11*(1), 1–13.

Weissig, V., Pettinger, T. K., & Murdock, N. (2014). Nanopharmaceuticals (part 1): Products on the market. *International Journal of Nanomedicine, 9*, 4357.

Wu, B., Lu, S. T., Zhang, L. J., Zhuo, R. X., Xu, H. B., & Huang, S. W. (2017). Codelivery of doxorubicin and triptolide with reduction-sensitive lipid-polymer hybrid nanoparticles for in vitro and in vivo synergistic cancer treatment. *International Journal of Nanomedicine, 12*, 1853–1862. https://doi.org/10.2147/ijn.S131235

Wu, J., Huang, W., & He, Z. (2013). Dendrimers as carriers for siRNA delivery and gene silencing: a review. *Scientific World Journal, 2013*, 630654. doi:10.1155/2013/630654

Yang, J., Zhang, Q., Chang, H., & Cheng, Y. (2015). Surface-engineered dendrimers in gene delivery. *Chemical Reviews*, *115*(11), 5274–5300. https://doi.org/10.1021/cr500542t

Yang, K., Feng, L., & Liu, Z. (2016). Stimuli responsive drug delivery systems based on nano-graphene for cancer therapy. *Advanced Drug Delivery Reviews*, *105*, 228–241.

Yazdian-Robati, R., Arab, A., Ramezani, M., Rafatpanah, H., Bahreyni, A., Nabavinia, M. S., Abnous, K., & Taghdisi, S. M. (2019). Smart aptamer-modified calcium carbonate nanoparticles for controlled release and targeted delivery of epirubicin and melittin into cancer cells in vitro and in vivo. *Drug Development and Industrial Pharmacy*, *45*(4), 603–610. https://doi.org/10.1080/03639045.2019.1569029

Yokoyama, M. (2011). Clinical applications of polymeric micelle carrier systems in chemotherapy and image diagnosis of solid tumors. *Journal of Experimental and Clinical Medicine*, *3*(4), 151–158.

Zandi, A., Sh, Z. D., Shojaeian, F., Mousavi-Kiasary, S. S., Abbasvandi, F., Zandi, A., Gilani, A., Saghafi, Z., Kordehlachin, Y., & Mamdouh, A. (2021). The design and fabrication of nanoengineered platinum needles with laser welded carbon nanotubes (CNTs) for the electrochemical biosensing of cancer lymph nodes. *Biomaterials Science*.

Zhang, N.-n., Yu, R.-s., Xu, M., Cheng, X.-y., Chen, C.-m., Xu, X.-l., Lu, C.-y., Lu, K.-j., Chen, M.-j., & Zhu, M.-l. (2018). Visual targeted therapy of hepatic cancer using homing peptide modified calcium phosphate nanoparticles loading doxorubicin guided by T1 weighted MRI. *Nanomedicine; Nanotechnology, Biology, and Medicine*, *14*(7), 2167–2178.

Zhang, Y., Cao, J., & Yuan, Z. (2020). Strategies and challenges to improve the performance of tumor-associated active targeting. *Journal of Materials Chemistry. B, Materials for Biology and Medicine*, *8*(18), 3959–3971.

Zhang, Z., Qian, H., Yang, M., Li, R., Hu, J., Li, L., Yu, L., Liu, B., & Qian, X. (2017). Gambogic acid-loaded biomimetic nanoparticles in colorectal cancer treatment. *International Journal of Nanomedicine*, *12*, 1593–1605. https://doi.org/10.2147/ijn.S127256

Zhong, Q., Bielski, E. R., Rodrigues, L. S., Brown, M. R., Reineke, J. J., & da Rocha, S. R. (2016). Conjugation to Poly(amidoamine) Dendrimers and Pulmonary Delivery Reduce Cardiac Accumulation and Enhance Antitumor Activity of Doxorubicin in Lung Metastasis. *Molecular Pharmaceutics*, *13*(7), 2363–2375. https://doi.org/10.1021/acs.molpharmaceut.6b00126

Zhou, Y., Shi, L., Li, Q., Jiang, H., Lv, G., Zhao, J., Wu, C., Selke, M., & Wang, X. (2010). Imaging and inhibition of multi-drug resistance in cancer cells via specific association with negatively charged CdTe quantum dots. *Biomaterials*, *31*(18), 4958–4963.

Zhu, J.-J., Li, J.-J., Huang, H.-P., & Cheng, F.-F. (2013). *Quantum dots for DNA biosensing*. Springer.

ADDITIONAL READING

Aghebati-Maleki, A., Dolati, S., Ahmadi, M., Baghbanzhadeh, A., Asadi, M., Fotouhi, A., Yousefi, M., & Aghebati-Maleki, L. (2020). Nanoparticles and cancer therapy: Perspectives for application of nanoparticles in the treatment of cancers. *Journal of Cellular Physiology*, *235*(3), 1962–1972. doi:10.1002/jcp.29126 PMID:31441032

Aljabali, A. A., & Obeid, M. A. (2020). Inorganic-organic Nanomaterials for Therapeutics and Molecular Imaging Applications. *Nanoscience & Nanotechnology-Asia*, *10*(6), 748–765. doi:10.2174/22106812 09666190807145229

Babu, L. T., & Paira, P. (2017). Current application of quantum dots (QD) in cancer therapy: A review. *Mini-Reviews in Medicinal Chemistry*, *17*(14), 1406–1415. doi:10.2174/13895575176661703151255 04 PMID:28302038

Dang, Y., & Guan, J. (2020). Nanoparticle-based drug delivery systems for cancer therapy. *Smart Materials in Medicine*, *1*, 10–19. doi:10.1016/j.smaim.2020.04.001 PMID:34553138

Hossen, S., Hossain, M. K., Basher, M. K., Mia, M. N. H., Rahman, M. T., & Uddin, M. J. (2019). Smart nanocarrier-based drug delivery systems for cancer therapy and toxicity studies: A review. *Journal of Advanced Research*, *15*, 1–18. doi:10.1016/j.jare.2018.06.005 PMID:30581608

Jamkhande, P. G., Ghule, N. W., Bamer, A. H., & Kalaskar, M. G. (2019). Metal nanoparticles synthesis: An overview on methods of preparation, advantages and disadvantages, and applications. *Journal of Drug Delivery Science and Technology*, *53*, 101174. doi:10.1016/j.jddst.2019.101174

Kulkarni, N. S., Guererro, Y., Gupta, N., Muth, A., & Gupta, V. (2019). Exploring potential of quantum dots as dual modality for cancer therapy and diagnosis. *Journal of Drug Delivery Science and Technology*, *49*, 352–364. doi:10.1016/j.jddst.2018.12.010

Liu, Y., Bravo, K. M. C., & Liu, J. (2021). Targeted liposomal drug delivery: A nanoscience and biophysical perspective. *Nanoscale Horizons*, *6*(2), 78–94. doi:10.1039/D0NH00605J PMID:33400747

Wong, K. H., Lu, A., Chen, X., & Yang, Z. (2020). Natural Ingredient-Based Polymeric Nanoparticles for Cancer Treatment. *Molecules (Basel, Switzerland)*, *25*(16), 3620. doi:10.3390/molecules25163620 PMID:32784890

Yu, L., Xu, M., Xu, W., Xiao, W., Jiang, X., Wang, L., & Gao, H. (2020). Enhanced cancer-targeted drug delivery using precoated nanoparticles. *Nano Letters*, *20*(12), 8903–8911. doi:10.1021/acs.nanolett.0c03982 PMID:33232167

Zhang, Y., Cao, J., & Yuan, Z. (2020). Strategies and challenges to improve the performance of tumor-associated active targeting. *Journal of Materials Chemistry. B, Materials for Biology and Medicine*, *8*(18), 3959–3971. doi:10.1039/D0TB00289E PMID:32222756

KEY TERMS AND DEFINITIONS

Active and Passive Targeting: Active and passive are the two means of targeting nanoparticles. In active targeting, the nanoparticles are actively uptake by the tumor cells while in passive targeting, a location is specified for the nanoparticles in a small place of tumor.

Inorganic Nanoparticles: In organic nanoparticles are more vital in modern techniques as it can be synthesized easily and cheaply. Therefore, it can be used in multiple of applications.

Nanocapsules: It is a unique circular shaped nanocarriers, size ranges between 5 to 1000nm, and it is commonly found in the range of 100 to 500 nm. It is usually used for targeted and controlled drug delivery and protect other materials like enzyme, protein, and foreign cells.

Nanocarriers: These are the nanomaterials used as a carrier to transport some material like drug, as it become unstable when it reaches in gastrointestinal tract (GIT), and also useful for site specific targeting.

Nanoparticles Assorted Hydrogel: Hydrogels are compatible to living organs and also has the ability to store proteins in their original state, so they considered as a good source to deliver the drugs into human body. It is more interesting because it is mainly made up of water and that's why they become shrink in dry state so, they can bypass the small intestine. In that case, drugs are loaded to them and then released in the body in a controlled manner.

Organic Nanoparticles: These compounds are one or two dimensional and formed from proteins, lipids, and carbohydrates, and size of the radius is smaller than the 100 nm. It has more importance in life sciences as nanobeads.

Quantum Dots: It is a semiconductor particle with core shell shape of nanometer size, and it has very specific optical properties. When light passed from the quantum dots then they emit light of different wavelengths.

Chapter 17
Recent Advances on Nanomaterials for Diagnostic, Treatment, and Prevention of COVID-19

Rafael Vargas-Bernal
https://orcid.org/0000-0003-4865-4575
Instituto Tecnológico Superior de Irapuato, Mexico

ABSTRACT

Since the end of 2019, the SARS-CoV-2 virus, commonly known as COVID-19, has become an infectious disease that has produced a severe global public health problem that requires solutions from different scientific initiatives. Nanomaterials are used to diagnose, treat, and prevent the SARS-CoV-2 virus. Its potential is highlighted through specific applications and developments carried out by researchers around the world. In addition to highlighting the characteristics and properties of the COVID-19 virus, the purpose of the chapter is focused on describing the role that nanomaterials are playing and that must be overcome due to the pandemic. Future research directions are described to guide readers on the innovations that are required related to the diagnosis, treatment, and prevention of SARS-CoV-2 and its variants. The impact that nanomaterials will have on COVID-19 will reduce the time for humans to return to the rhythm of life before the pandemic.

INTRODUCTION

The emergence of the SARS-CoV-2 (severe acute respiratory syndrome coronavirus 2) virus, also known as COVID-19 (Coronavirus disease 2019), toward the end of 2019, was one of the events that redefined our medical safety globally (Khan, 2021). It is an infectious illness (Choi, 2020). Since its discovery, researchers have undertaken countless activities dedicated to the prevention, treatment, and diagnosis of this virus. Several scientific publishing houses joined forces to disseminate the latest developments related to COVID-19. Immediately, pharmaceutical laboratories and research centers looked for solutions

DOI: 10.4018/978-1-7998-8936-6.ch017

to offer alternatives that would allow the world population to return to normality before the appearance of the virus. One of the most important tasks was the development of vaccines that would allow the virus's transmission to be halted, at least for the known variants (Chauhan, 2020; Chung, 2020; Shin, 2020). Fortunately, several pharmaceutical companies are offering vaccine alternatives that seek to immunize the world's population. Unfortunately, the security problem has not been completely solved, since not all countries have the same economic and health possibilities. In addition, multiple variants of the virus have emerged in different countries and it is not yet possible to establish whether the vaccines will eliminate the current risk. Some experts comment on the possibility that this event is the beginning of a novel health strategy that will force a permanent change in the way things have been done so far (Kostarelos, 2020; Pathak, 2021).

Coronaviruses are a type of virus that causes illness in the upper respiratory system and can cause more serious problems than a regular cold (Antiochia, 2021). Hundreds of coronaviruses have been found so far, with camels, pigs, cats, dogs, and bats being the most common carriers. However, through a mechanism known as spillover, these coronaviruses can sometimes reach humans (HCoV) and cause sickness. A total of seven HCoVs have been discovered and classified into two groups. The common flu is caused by the first group of viruses, which includes HCoV-HKU1, HCoV-OC43, HCoV-NL63, and HCoV-229E, which is considered a mild illness. The MERS coronavirus (MERS-CoV), the SARS coronavirus (SARS-CoV), and the new SARS coronavirus (SARS-CoV-2) are all members of the second category, and they all have catastrophic consequences that can even lead to death.

As a complement to the pharmaceutical strategy of vaccines, researchers are using nanotechnology and specifically nanomaterials to develop solutions that allow the prevention, treatment, and diagnosis of COVID-19 (Rai, M. 2020; Rai, P.K. 2020; Tang, 2021). Until now, the first test for COVID-19 has been based on a real-time reverse-transcription polymerase chain reaction (RT-PCR). This detection takes about 3 to 8 hours, and its reagent kits as well as the infrastructure to install and measure many samples, and its low effectiveness in detecting asymptomatic patients make it impractical (Adhikari, 2020). Other quick tests based on the detection of human stool, blood serum, antibodies, or nasal swabs to identify COVID-19 in about 30 minutes have been proposed, however, they rely on certain test patterns and stages of infection to make a sure diagnosis. COVID-19 is diagnosed at the point of care using protein and nucleic acid-based test kits. As of April 2020, 29 therapeutic drugs had been proposed to treat COVID-19. Only Favipiravir, Chloroquine, Hydroxychloroquine, and Remdersivir have been examined for antiviral cases, and the results were promising. The use of these drugs depends on person to person, due to the pre-existing medical conditions of the patient.

The SARS-CoV-2 virus possesses a compact core-shell nanostructure of 60 to 140 nm (Adhikari, 2020). Nanomaterials can passivate, encapsulate, and bind the COVID-19 virus. In this way, the use of nanomaterials will make it possible to study, design, or manipulate the virus for its prevention, treatment, and diagnosis of COVID-19 (Carvalho, 2021; Chaudhary, 2021; Jastrzębska, 2021). Among the physical properties that can be used are variable bandgap, large surface-to-volume ratio, high luminescence, and ease of integrating into nano-dimensional materials (Rangayasami, 2021). Nanomaterials can be used to make detection kits (Weiss, 2020; Tavakol, 2021), nanobiosensors (Ali, 2020; Bhalla, 2020), nanocoatings (Rai, M. 2020), nanocomposites (Palestino, 2020; Palmieri, 2020), nanomedicines (Sivansankarapillai, 2020), and specific viruses' vaccines (Ruiz-Hitzky, 2020; Vazquez-Munoz, 2020). Nanomaterials can be conjugated into antibodies and can be used to rapidly detect targeted virus proteins. These materials can be used as disinfectants through the inactivation of the virus by heat or light (Palmieri, 2020). Although the surfaces must be cleaned regularly, it is nearly impossible to apply the sanitizer and ensure that it does not become infected again. The researchers intend through nanomaterials to develop chemically active coatings to inactivate viral glycoproteins and nucleotides. Given the versatility that nanomaterials (nanocomposites and two-dimensional materials) must contribute to the development of emerging science and technology, our research group is studying the possibility of expanding the range of application of two-dimensional materials in nanomedicine towards the diagnosis and prevention of COVID-19. This chapter provides an overview of the current state of nanomaterial's application in the prevention, treatment, and diagnosis of COVID-19. This document compiles some of the main contributions of scientists related to nanomaterials developed around the world so that their understanding is accessible to the entire population that can read this chapter.

BACKGROUND

The severe acute respiratory syndrome coronavirus (SARS-CoV-2) or COVID-19, as it is generally called, was discovered in the Hubei province of China, and spread like a pandemic in December 2019, and spread like a pandemic since then (Udugama, 2020). Shortness of breath, cough, fever and other indicia were evident in the initial patients. They were scanned with computed tomography (CT), and physicians found opacities (dense, confluent, and profuse) that do not occur in healthy lungs, suggesting a diagnosis of pneumonia. To round up the research, the pathogen's nucleic acid (DNA) was analyzed using a real-time polymerase chain reaction, which yielded negative results for previously recognized infections, which suggested that the source was unknown. The samples' analysis necessitates direct PCR, whole-genome sequencing, and culture of the patients' bronchoalveolar lavage fluid (BAL) (Zhu, 2020). The pathogen had a genetic sequence that was 50%, 80%, and 96% like the genomes of the severe Middle Eastern respiratory syndrome virus (MERS-CoV), acute respiratory syndrome virus (SARS-CoV), and bat coronavirus RaTG13, according to the study. As of December 2020, the infection had spread to at least 202 nations, with an estimated 82 million individuals affected and 1.8 million deaths worldwide. These numbers are underestimated because many asymptomatic or mild patients go undetected. Furthermore, these individuals are so infectious that they can spread the disease.

The origin of the first transmission is not yet determined, although it is assumed to come from animal origin. When the virus infects humans, it spreads from person to person. Each person can infect up to 3.28 other persons, according to estimates (Liu, 2020). Symptoms vary from patient to patient and might be mistaken for influenza or the common flu in some cases. Direct touch and diffusion by bodily fluid droplets are the most likely modes of transmission (Li, 2020). It's also been suggested that the virus may be discovered in aerosols (smaller than 5 micrometers in diameter) for up to 3 hours and that it spends more time active on stainless steel and plastic than on cardboard and copper (Van Doremalen, 2020). The following subsections describe the biological properties of COVID-19, symptoms for COVID-19, and common COVID-10 diagnostic techniques.

Biological Properties of COVID-19

Epithelial cells from the human respiratory tract were extracted from COVID-19 infected patients to be cultured and studied through the analysis of their negative strain concerning healthy cells through transmission electron microscopy (TEM) (Zhu, 2020). The COVID-19 virus has a diameter of 60 to 140 nm, an envelope with protein spikes, and genetic material inside, according to images of human cells hosting the virus. Furthermore, the virus's spikes range in length from 9 to 12 nanometers, and the virions resemble a solar corona. Phylogenetic study on human cells revealed that 2019-*n*-COV is a member of the beta-coronavirus genus, which contains coronaviruses (bat SARS-like CoV, SARS-CoV, and other variants) that were discovered in humans before 2019. In this study, unbiased high-throughput DNA sequencing was used to understand and study the occurrence and transmission of this infectious disease. Several precise and sensitive assays targeting the E, N, and ORF1ab segments of the genome were also designed to identify viral RNA in clinical specimens, which might be used as detection standards in future cases. The significance of doing this epidemiological investigation was to determine the mechanisms of transmission, reproduction interval, and clinical spectrum to inform and develop ways to prevent, control, and stop COVID-19 from spreading.

The SARS-CoV-2 virus has a 30,000-nucleotide genome with positive-sense single-stranded RNA (Udugama, 2020). There are encodes 27 proteins encoded by the virus genome, as well as four structural proteins, and an RNA-dependent RNA polymerase (*RdRP*). To maintain genomic fidelity, this polymerase collaborates with non-structural proteins. After further investigation, it was discovered that a portion of the *RdRP* gene in SARS-CoV-2 is 96 percent identical to the bat coronavirus RaTG13. An envelope protein (E), a matrix protein (M), a nucleocapsid protein (N), and a spike glycoprotein (S) are the structural proteins of the coronavirus. A schematic representation of the basic structure of COVID-19 is outlined in Figure 1. The S gene serves as the receptor link that allows the virus to infect cells and gives it its transmission capabilities. The other three structural proteins (E, M, and N) develop functions such as pathogenesis, envelope formation, protein assembly, budding, and RNA encasing. To enter cells, SARS-CoV-2 must be angiotensin-converting in enzyme 2 (ACE2) in a receptor. Because ACE2 *m*RNA is distributed in numerous human organs, an infection can occur from several different directions. Brain, kidney parietal epithelial cells, liver bile ducts, lymph nodes, skin, colon, enterocytes of the small intestine, stomach, alveolar epithelial cells of the lungs, arterial and venous endothelial cells, and arterial soft muscle cells in the lungs can all contain the ACE2 enzyme. Furthermore, it is important to note that upper respiratory tract tissues, including the oral and nasal mucosa as well as the nasopharynx, do not present the ACE2 enzyme, and therefore, become unlikely sites of infection by SARS-COV-2. Oral swabs, stools, and bronchoalveolar-lavage fluid can all be used to isolate COVID-19. In both asymptomatic and symptomatic patients, however, larger virus loads are stored in the nose through the throat. Researchers can create strategies for diagnosis, treatment, and prevention by better understanding the biological features of SARS-CoV-2.

Symptoms for COVID-19

COVID-19 causes symptoms that are not specific, so they cannot be utilized to accurately diagnose infectious infections (Udugama, 2020). Only 44% of 1,099 Chinese patients with COVID-19 had a fever when they were admitted to the hospital, but 89 percent acquired one while hospitalized, according to one study (Guan, 2020). Furthermore, the patients presented other symptoms such as difficulty breathing (19%), sputum production (34%), fatigue (38%), and cough (68%). However, these symptoms can also be presented in patients with other types of respiratory infections, which does not necessarily imply being infected with COVID-19.

Figure 1. Basic biological structure of COVID-19
(adapted from Udugama, 2020)

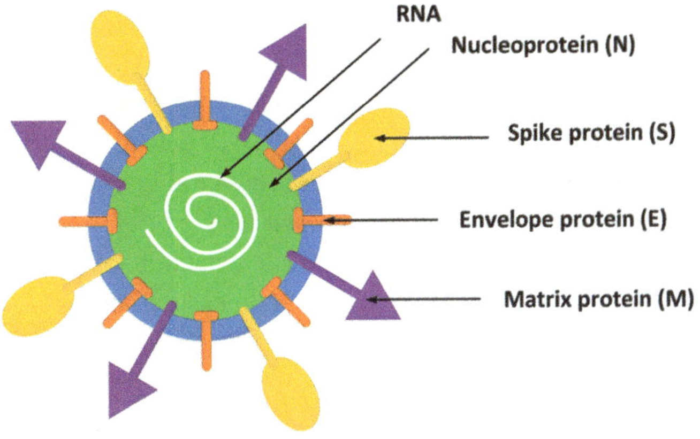

Common COVID-19 Diagnostic Techniques

The traditional techniques applied so far for the detection and screening of COVID-19 are DNA tests and computed tomography (CT) (Udugama, 2020). Molecular techniques such as DNA and/or RNA testing are more convenient than syndromic tests, where the physician tests a patient for multiple pathogens with overlapping symptoms, that is, the test is not specific. Computed tomography allows the diagnosis required to identify and target specific pathogens. Conventional COVID-19 diagnosis systems are costly and only available in hospitals or specialist laboratories. The amount of time it takes to obtain the results of a COVID-19 test depends on the type of test and the place where it is performed, it can take minutes, hours, or even days (Carter, 2020). When using molecular approaches, it is important to understand the pathogen's proteome and genomic makeup on the one hand, and the host's induction of changes in protein and gene expression during and after infection on the other. The GenBank sequence repository and the Global Initiative on Sharing All Influenza Data (GISAID) had previously released and publicly publicized advances linked to SARS-CoV-2 genomic chains for researchers all over the world (Zhu, 2020; Wu, 2020; Rockett, 2020; Wang, 2020). To construct primers and probe sequences for the COVID-19 genome sequencing, both PCR and other nucleic acid assays (DNA and RNA) are required.

The most common method for diagnosing COVID-19 is a nucleic acid test (Udugama, 2020). Different reverse transcription-polymerase chain reactions (RT-PCR) have been designed to genetically detect SARS-CoV-2 in the nation and the laboratory. The genes targeted are N, ORF1ab, RdRP, E, and/or ORF1b-bsp14. RT-PCR transcription of the complementary DNA strands (*c*DNA) of SARS-CoV-2 RNA is required, followed by amplification of regions of the *c*DNA. This process is divided into two parts: (1) sequence alignment and primer design, and (2) assay tuning and testing. Three common sections have been identified within the sequences: (1) the RdRP gene in the ORF1ab region's open reading frame, (2) the N gene, and (3) the E gene. The RdRP and E genes have higher sensitivity than the N gene in the analytical detection of COVID by PCR. This assay necessitates the employment of a two-target method to detect the presence of a coronavirus in general, as well as the identification of SARS-CoV-2 in a second stage.

Once the primers and probe sequences have been designed it is necessary to optimize the assay conditions including temperatures, incubation times, and reagent conditions before the PCR test is completed

(Udugama, 2020). A one-step or a two-step experiment is implied by RT-PCR. In the case of the one-step reaction, both transcription and PCR amplification are performed in a single step, which produces a high throughput analysis in a fast and reproducible way, but unfortunately, it leads to a lower target amplification. The two-step reaction, on the other hand, necessitates two processes in distinct sections, which increases sensitivity, takes longer to adjust performance parameters, and requires more rigorous control to ensure the assay's dependability. Negative COVID-19 detections can occur due to events such as viral genome mutations, poor sampling techniques, or low viral levels in the studied tissue.

Chest computed tomography was the initial approach utilized before the widespread adoption of RT-PCR for COVID-19 detection and the use of large-scale detection kits (Udugama, 2020). The scans generated with this method are non-invasive and use X-ray measurements at various angles to produce images of the cross-section of the chest. Radiologists examine these images for aberrant abnormalities that can help them make a more accurate diagnosis. The images obtained present different characteristics depending on the symptoms and level of infection reached. The presence of COVID-19 can appear in the images as peripherical and bilateral blurs that appear as cloudy areas or illuminated areas illustrating fluid or solid materials of the lung tissue. Computer tomography has higher sensitivities than RT-PCR and similar false negative ratios and has low specificity for diagnosing SARS-CoV-2.

MAIN FOCUS OF THE CHAPTER

The diagnosis of COVID-19 allows its containment by implementing control measures that limit its dissemination through the identification of cases, isolation, and the tracing of human contacts (Udugama, 2020). To prevent and diagnose SARS-CoV-2, nanomaterials such as composite materials, quantum dots, and low-dimensional semiconductor structures have been proposed (Adhikari, 2020). Through a core-shell structure with a range of 60 to 140 nm, it is possible to bind, encapsulate and passivate the COVID-19. Among the useful properties of nanomaterials are a high luminescence, a variable bandgap, a high surface-to-volume ratio, and simplicity of integration that allows the design of detection kits, nanomedicines, and vaccines for coronavirus.

Related Problems With Diagnostic Methods Against COVID-19

Once an infectious disease appears in society, it is necessary to control its spread to achieve a timely and accurate diagnosis (Deng, 2021). Devices used in the detection of viruses that exploit nanomaterials and biosensing mechanisms are nanobiosensors. They develop a selective, sensitive, and short response time diagnosis by exploiting the physical, biological, chemical, and medical properties involved. The most common infectious pathogens are parasites, fungi, bacteria, and viruses. The traditional detection of the virus implies that the pathogens are cultured and identified by microscopy, but this is slow and limited to the resolution of the equipment. Later, with the discovery of biomarkers, these were used to diagnose infections through the antibodies, antigens, proteins, and genome of the pathogens. The polymerase chain reaction (PCR) and the enzyme-linked immunosorbent assay (ELISA) are two biomarker-based approaches for this identification. The PCR technique detects nucleic acids, while the ELISA technique detects antigens or antibodies (Carter, 2020). The use of these techniques involves expensive equipment and highly skilled technicians to develop the diagnosis. To combat these infectious diseases, the optimal approach involves fast, simple, low-resource, and low-cost kits to be available to people all over the world. Due to the previous requirements, the detection of pathogens remains a challenge until now.

The versatile analytical device for diagnosing infectious diseases is the biosensor. They are sensitive to detect biochemical phenomena and convert these into thermal, magnetic, optical, electrical, or electrochemical signals (Liu, 2019). These are composed of two parts: (1) biological or bioreceptor recognition, and (2) physicochemical transduction. It is necessary to recognize or bind a specific analyte through biomimetic or biologically derived materials to develop biological recognition. Biological recognition involves the recognition and interaction of viruses with target analytes. This is done through a nucleic acid, cell, enzyme, or antibody. Target analytes can be antigens, complementary DNA, or enzyme substrates. The sensitivity of the bioreceptor is strategic to avoid interference from external agents or systems from the sample matrix. The recognition event is performed through the transduction part so that a physicochemical signal can be measured. In this way, biosensors offer a cost-effective solution, quick response, and simple operation.

Related Problems With Prevention Methods Against COVID-19

The quintessential reliable preventive measure against COVID-19 and other viral infections is the surgical face mask. The key materials to make these masks are non-woven materials, which act as barriers against viruses by filtering them. However, its extensive use can lead to cross-infection due to the lack of inherent virucidal properties which makes it a source of infection for its user (Karmacharya, 2021). Disposable surgical masks have the role of preventing respiratory droplets from entering the lungs, thereby reducing the risk of being infected, along with hand cleaning methods provide proper hygiene against the possible spread of COVID-19 (Zhong, 2020). The use of these masks became mandatory for doctors, patients, and even the public anywhere given the high levels of the spread of SARS-CoV-2. Reuse or recycling ordinary masks is not feasible because they cannot be self-sterilized. This has resulted in high environmental and economic costs not only involved in its production but also its disposal. Some elements of standard surgical masks limit their usage from a technical standpoint: (1) Virus-containing water droplets remain on their surfaces despite their hydrophobic surfaces; (2) they have a melting point below 130 °C, making reuse difficult even with steam sterilization methods because they can degrade; and (3) their recycling because they can contain viruses on their surfaces. Masks are produced at a pace of 40 million per day over the world, resulting in a waste stream of more than 15,000 tons per day. The method of disposing of them that is used in most countries is incineration. The poisonous chemicals and massive carbon emissions cause significant environmental damage. Nanomaterials can produce superhydrophobic surfaces with self-cleaning characteristics. Materials such as fluorinated polymers, metal nanowires, and graphene have been proposed.

Nanomaterials can tune their surface electrical charge to interact with the S-gene and thus block a possible viral infection caused by a coronavirus (Carvalho, 2021). While the most effective commercially available surgical masks filter up to 95% of airborne particles, multi-layer wearable masks can block up to 50-70% of the particles and fine droplets that humans release into the environment and limit the spreading droplets and aerosolized particles that they do not capture. The benefits of the double mask or the addition of a wearable mask to the surgical mask are twofold. First, it improves the fit of a mask, preventing respiratory droplets containing the virus from escaping or entering the wearer of the mask. Second, it increases the filtering capacity of the mask, making it difficult for droplets containing the virus to travel through the mouth and nose. Correct and highly compliant use of surgical masks ensures their effectiveness in reducing the number of COVID-19 reproduction. To reduce the reproduction number from 3 to below 1, minimum compliance of 60 to 70 percent is required even to reduce the effect of the most infectious variants of SARS-CoV-2.

Related Problems With Treatment Methods Against COVID-19

Nanomaterials are used in antiviral drugs delivery due to their tunable surface charge (different types of drugs can be encapsulated), nanometric size (which makes them bioavailable and decreases release time), and the huge surface-area-to-volume ratio (which improves the solubility) (Carvalho, 2021). In addition to direct use, nanomaterial-based platforms including carbon nanotubes, nano-emulsions, cobalt, and gold nanoparticles are used against different types of viruses acting from inside or outside the host cells. So far it has been shown that nanomaterials can control viral human pathogens such as Zika virus, influenza, hepatitis, herpes simplex, and human immunodeficiency virus type 1 (HIV-1). Nanospheres, nanocapsules, nanocrystals, nanogels, nano-sheets, nanoclusters, nanostars, nanoparticles, nanowires, nanobars, and carriers of nanostructured lipids (NLCs), as well as nanotubes, have all been utilized to treat the different types of coronaviruses.

SOLUTIONS AND RECOMMENDATIONS

This section is focused mostly on describing how nanomaterials are currently being used or explored to detect the presence of SARS-CoV-2 as well as treat and preventing its infections. Nanomaterials offer improved physicochemical properties and feature versatile chemical functionalization which is valuable in combating SARS-CoV-2 with a size in the nanometer range. Nanoparticles (NPs), nanocomposites, carbon nanotubes (CNTs), quantum dots (QDs), and graphene or graphene-based nanomaterials are among the nanomaterials employed in COVID-19 biosensing. Figure 2 is a schematic illustration of the nanomaterials that have been proposed to combat SARS-CoV-2.

Related Solutions with Diagnostic Methods Against COVID-19

The state-of-the-art of the wide range of nanobiosensors for the diagnosis of COVID-19 is discussed in this subsection. The use of nanomaterials increases the analytical properties of biosensors like selectivity, sensitivity, and response time to detect analytes of interest in the size of 1 to 100 nm. Nanobiosensors' sensitivity and selectivity will be improved thanks to their huge surface-to-volume ratio and the ability to incorporate vast amounts of functional surface groups that act as biological recognition components. Additionally, some unique properties of nanomaterials can enhance their analytical performance. Gold nanoparticles (AuNPs), for example, have larger attenuation coefficients (more light absorption at a given wavelength) than conventional dyes, which is useful for COVID-19 colorimetric detection (Moitra, 2020).

Colorimetry (Moitra, 2020) and electrical conductivity (field effect transistor, FET) have been utilized to identify the presence of COVID-19 up to this point (Seo, 2020). Gold nanoparticles (AuNPs) and graphene are the active nanomaterials employed to detect COVID-19 (Deng, 2021). The recognition elements of COVID-19 are oligonucleotides and antibodies. Table 1 summarizes some of the proposals based on nanomaterials for implementing nanobiosensors for COVID-19.

Figure 2. Proposed nanomaterials against COVID-19
(adapted from Carvalho, 2021)

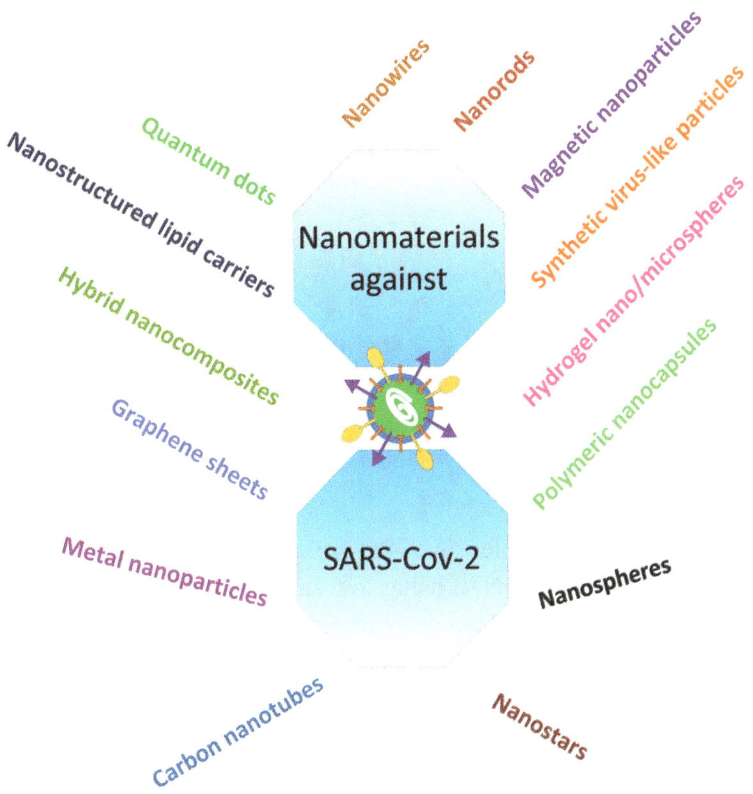

Table 1. Different approaches for implementing nanobiosensors to realize COVID-19 diagnosis

Detection Approach	Nanomaterial	Target	Recognition Element	Reference
Colorimetry	AuNPs	N-gene	Oligonucleotides	Moitra, 2020
Electrical conductivity	Graphene	S-gene	Antibody	Seo, 2020
Electrochemical biosensor	Reduced graphene oxide	S-gene	Antibody	Ali, 2020
Refraction index	AuNPs	G-gene	Antibody	Abid, 2021

Bioassay-based biosensors using nanomaterials with colorimetric principles are reliable and convenient due to their reduced instrumentation complexity, visual output, and simplicity (Moitra, 2020). Gold nanoparticles have unique optical properties such as high attenuation coefficient, inherent photostability, and localized surface plasmon resonance (SPR). The biological and chemical biosensing targets are nucleic acids, metal ions, proteins, and small molecules. Through these methods, the nanoparticles change their color in response to their reactivity to external environmental conditions. These changes can be visualized through human eyes or a portable optical detector is required to perform a quantitative measurement. However, using biosensing necessitates the creation of ssDNA (single-stranded DNA) probes as well as the fulfillment of numerous intermediate stages, such as time-consuming denaturation

Figure 3. Colorimetric detection of COVID-19 by gold nanoparticle (AuNPs) assay (adapted from Moitra, 2020)

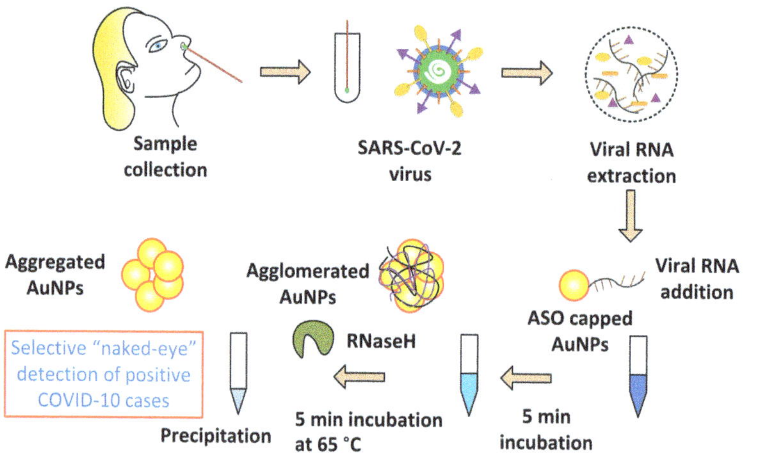

and post-PCR annealing. Au nanoparticles (AuNPs) are also utilized to stabilize single-stranded and double-stranded DNA (ssDNA and dsDNA) using low salt concentrations, as well as cationic agents for probe detection. For the SARS-CoV-2 gene-N to diagnose COVID-19 positive cases in minutes, the AuNPs must be coated with appropriate thiol-modified antisense oligonucleotides (ASOs). This technique uses an all-inclusive directed approach based on four ASO trajectories to cover two areas of the N-gene sequence at the same time. The surface plasmon resonance (SPR) of AuNPs coated with thiolated ASO changes, which is enhanced with RNaseH to create visually observable precipitate, and these are agglomerated due to the target SARS-CoV-2 RNA sequence and AuNPs. The material was examined using enhanced dark-field hyperspectral microscopy (EDFHM), transmission electron microscopy (TEM), and ultraviolet-visible absorbance spectroscopy (UV-Vis) to confirm the findings. As a result, the assay takes advantage of AuNPs' plasmonic capabilities and ASOs' targeted ability to develop the naked eye and selective detection of COVID-19. The inherent advantage of the assay is that its reliability and reproducibility ensure that two of the SAR-CoV-2 N-gene regions are simultaneously tested to ensure that if a virus mutation occurs, the gene can be determined in at least one of the two regions. Figure 3 shows a schematic illustration of the assay based on the colorimetric detection of COVID-19 using gold nanoparticles.

The AuNPs-based assay described above required 10 minutes to deliver a positive result. The absorbance spectra of the surface plasmon resonance (SPR) of the virus-containing nanoparticles shifted by 40 nm (Moitra, 2020). The detection limit for SARS-CoV-2 containing RNA was 0.18 ng/μL. The nanoparticles agglomerated and the precipitate was visibly detected. The designed nanobiosensor can be assigned the following qualifications: deliverable to end-users, free of sophisticated equipment, robust, fast, available, user-friendly, specific, and sensitive. The biosensor can be modified to detect the E-gene, the M-gene, and the S-gene, which would reduce false positives to a minimum.

COVID-19 was detected in clinical samples using a biosensor based on a field-effect transistor (FET) (Seo, 2020). The sensor is based on a FET with a gate terminal built of graphene sheets covered with a specific sensitive antibody to the S-gene of SARS-CoV-2. COVID-19 was immobilized for this biosensor using 1-pyrene-butyric acid *N*-hydroxysuccinimide ester (PBASE), which also acted as a probe linker and interface coupling agent. This device was tested using nasopharyngeal swab specimens, cultured virus,

Figure 4. A graphene-based field-effect transistor was used to detect COVID-19 electrically (adapted from Seo, 2020)

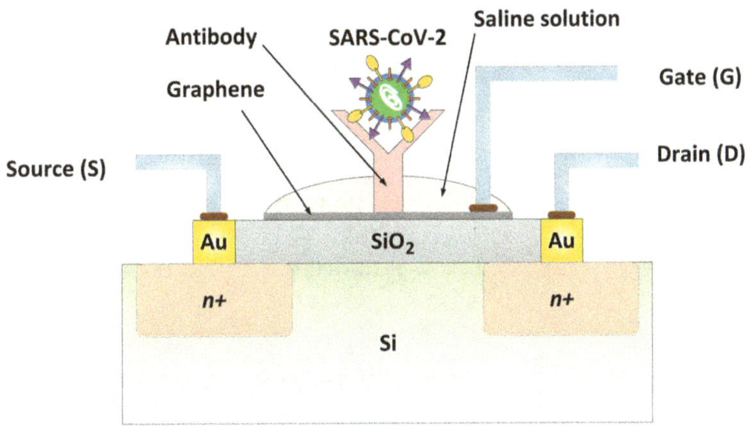

and protein antigens from COVID-19 patients. The presence of the COVID-19 virus in the biosensor was determined by the increase in electrical current (I_{DS}) between the drain (D) and source (S) terminals of the transistor. To maintain an efficient gate effect, the saline solution worked as a buffer for the electrolyte. The SARS-CoV-2 S-gene was detected at quantities of 1 fg/mL in phosphate-buffered saline, and 100 fg/mL in clinical transport medium using the biosensor. Furthermore, this biosensor detected COVID-19 in a culture medium with a limit of detection [LOD] of 1.6×10^1 pfu/mL and clinical samples with a LOD of 2.42×10^2 copies/mL. The key advantages of the engineered device are its high sensitivity and the fact that samples do not need to be pretreated or tagged. Figure 4 shows a schematic illustration of a biosensor based on a field-effect transistor for detecting SARS-CoV-2.

To make the previous biosensor using a field-effect transistor, the following steps were performed. Poly (methyl methacrylate) (PMMA) was covered by spinning graphene on a copper film. First, the polymer was independently chemically etched and then placed on glass surfaces. To place the polymer, the glass slides must be washed with deionized water (DI) and then deposited on the SiO_2/Si substrate, and dried overnight in ambient conditions (Seo, 2020). After removing the PMMA layer in an acetone bath and leaving the transferred graphene on the substrate, then the graphene was washed in isopropyl alcohol and dried in an N_2 atmosphere. After placing these materials, graphene is deposited using photolithography and reactive ion chemical etching followed by an Au/Cr electrode layer placed through metallization via thermal vaporization and a lift-off process.

FET-based biosensor represents a highly sensitive portable device capable of quantitatively measuring COVID-19 requiring minimal amounts of analytes (Seo, 2020). On-site detection, point-of-care (POC) testing, and medical diagnosis are all possible with these SARS-CoV-2 sensing devices. The biological sensing material used is graphene, a hexagonally arranged two-dimensional carbon nanosheet whose surface is completely exposed to the biological sample. Furthermore, other properties that make graphene unique are its large surface area, high electrical conductivity, and high carrier mobility. The previously described properties make it an ultrasensitive material to biochemical changes and allow low noise detection, which favors its application in the immunological diagnosis of viruses such as coronavirus.

By 3D nanoprinting of gold electrodes with reduced graphene oxide (rGO) nanosheets and immobilizing certain viral antigens, a novel COVID-19 biosensing platform was designed. The sensing material

Figure 5. COVID-19 has been detected using a variety of biosensors
(adapted from Antiochia, 2021)

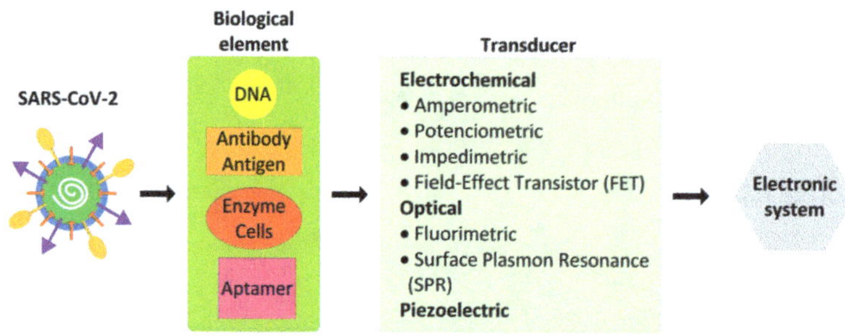

was placed on the active electrode of a microfluidic device that was built to design a typical electrochemical cell. The antigens must be brought into contact with the electrode surface to selectively bind and thus change the electrical impedance of the biosensor is detected by impedance spectroscopy. The antibodies detecting the S-gene reaching a receptor-binding domain (RBD) of 2.8×10^{-15} and a detection limit of 16.9×10^{-15} M, which were read via a smartphone interface. The device's advantage is that it can be regenerated with a low pH chemical that bypasses the antibodies and antigens, allowing for several samples to be sensed with the same sensor (Ali, 2020).

Another category of the COVID-19 sensing device is affinity-based biosensor (ABB), which uses piezoelectric, optical, and electrochemical principles. These biosensors use as biorecognition elements both DNA and antibodies, which have non-covalent interaction in the formation of complexes single-stranded DNA (ssDNA)-oligonucleotide-complementary strand and antigen-antibody pairs for DNA-sensors and immunosensors, respectively (Antiochia, 2021). Because of their higher electrical conductivity and unique photoelectrochemical characteristics, nanomaterials are used as signal amplifiers to boost the sensitivity of biosensors (Mujawar, 2020). For example, the hybrid graphene-gold nanostructures nanoarray was used as a neuro biomarker together with complementary DNA (cDNA) for sensitive and selective gene sensing through surface-enhanced Raman scattering (SERS) phenomenon. To improve the electromagnetic and chemical mechanism for biosensing, a Raman active dye-labeled probe DNA oligonucleotide was attached to the nanoarray. Gold nanostructures produce plasmons through an electromagnetic mechanism to amplify the Raman signal and graphene increases the energy level of the signal through a chemical mechanism utilizing a specific analyte. Figure 5 shows a diagram that outlines the many types of biosensors that have been proposed to detect the presence of COVID-19.

Among the nanomaterials that are being studied to aid in the efficient and reliable sensing of COVID-19 are carbon nanotubes, graphene, quantum dots, and molybdenum disulfide (Sharma, 2021). Two-dimensional materials offer a high surface sensitivity, the presence of multiple reaction sites, optical transparency, improved flexibility, and a large surface-to-volume ratio. Transition metal dichalcogenides (TMDs) are two-dimensional materials with tunable bandgaps that allow them to have stable nature, reduced toxicity, greater electrical conductivity, and photoluminescence. Tungsten disulfide (WS_2), molybdenum disulfide (MoS_2), tungsten diselenide (WSe_2), and molybdenum diselenide ($MoSe_2$) are the most studied dichalcogenides.

Photoluminescence, field-effect transistors (Fathi-Hafshejani, 2021), surface plasmon resonance, and electrochemical devices have all been used to detect COVID-19 using nanocomposites based on transition metal dichalcogenides (TMDs) (Lam, 2021). Through these biosensors, it is possible to sense

Figure 6. Surgical mask reinforced with nanomaterials against COVID-19

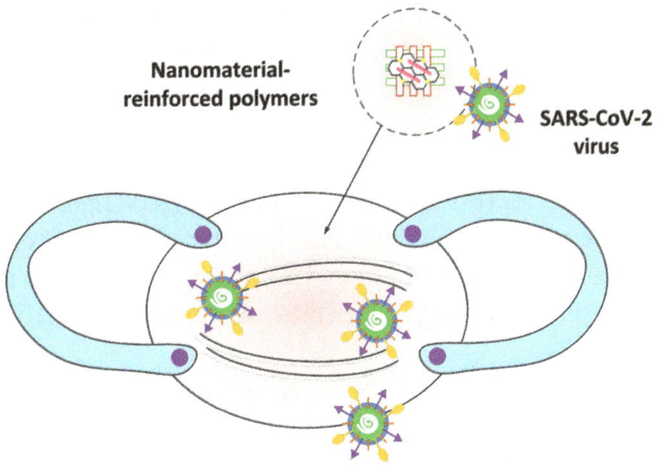

SARS-CoV-2 proteins and nucleic acids. However, its application and commercialization have limited its use at the experimental laboratory level due to its poor uniformity, difficulties for its mass production, and high cost.

In addition to transition metal dichalcogenides (TMDs) or graphene, black phosphorus (BP) or phosphorene is being explored to develop point-of-care (POC) biosensors for medical diagnostics because this material exhibits excellent electrochemical properties, which improve selectivity and sensitivity due to their inherent redox properties (Choi, 2020). Poly-L-lysine (PLL) covered phosphorene can be used to develop aptamer functionalized electrodes using the electron transfer mechanism to make electrochemical biosensors. This type of biosensor detects the S or M gene in patient blood samples, making it sensitive to SARS-CoV-2.

Related Solutions With Prevention Methods Against COVID-19

Surgical face masks exploit filtration and antiviral properties to guarantee reliability and durability, by reducing the possibility of cross-infection (Karmacharya, 2021). To achieve antiviral properties, it is necessary to modify the masks with reinforcements based on nanomaterials. The functions that are intended to improve through the nanomaterials in the masks are to improve the filtration efficiency, obtain a superhydrophobic performance, and develop a photothermal and photocatalytic performance. To achieve these functions, they must present self-cleaning, self-sterilization, electrostatic, mechanical, and antimicrobial effects. A schematic representation of a surgical mask reinforced with nanomaterials against COVID-19 is illustrated in Figure 6.

Current surgical masks can filter particles with a size of 0.3 μm (Karmacharya, 2021). However, COVID-19 has a size of 150 nm, which is smaller. An efficient mask must capture fine particles in the size of nanometers. The polymeric fibers of the masks must be densely packed to allow the virus to be trapped either chemically, electrostatically, or mechanically. To ensure acceptable breathability, these membranes have a porosity of less than 5 nm, have airflow rates higher than 85 L/min, and weight of less than 0.12 g. This version can be improved to exploit the properties of nanomaterials.

The surface of surgical masks can be modified using graphene to give them self-cleaning properties through their superhydrophobic nature by increasing the static contact angle to repel incoming water droplets. COVID-19 can be inactivated at a temperature of 56 °C within 15 minutes by modifying the surfaces of surgical masks using silver nanoparticles (AgNPs) to allow self-sterilization through their photothermal properties. Under sun illumination, graphene can be employed at temperatures higher than 70 °C in less than 40 seconds (Karmacharya, 2021).

Because reactive oxygen species (ROS) are formed following UV light irradiation to eradicate the virus by degrading its nucleic acids, proteins, and lipid membrane, photocatalysis can inactivate SARS-CoV-2 through the presence of oxygen. The use of TiO_2 nanowires increases the recombination ratio between electrons and holes, interfacial charge carrier transfer ratios, and the efficiency of photocatalytic activity (Karmacharya, 2021).

Low-cost graphene precursors such as bakelite, polyimide, and sulfonated polyether ketone ether are synthesized by irradiating a laser beam on its surface to produce superhydrophobic or superhydrophilic materials and controlling the processing environment (Zhong, 2020). Moisture droplets on the human body can be captured through superhydrophilic patterns stamped onto superhydrophobic surfaces. A continuous-wave laser-induced forward transfer (LIFT) technique is used to deposit graphene additively onto the surgical mask material. Since graphene is photothermal, it is possible to apply it in the design of antibacterial coatings. Under sunlight, the surface temperature of the mask can be elevated to around 80 °C, which is required to sterilize the coronavirus. In addition, the mask must provide a better salt rejection, and, in this way, long-term desalination capabilities are achieved.

Related Solutions With Treatment Methods Against COVID-19

Vaccines and antiviral agents are two treatment options based on nanomaterials that have been proposed against SARS-CoV-2. Spike proteins are the priority viral agents thanks to the main binding factor and because they represent the immunodominant antigen in coronaviruses. Nanomedicine to treat COVID-19 is developing strategies such as immunomodulatory therapies, nano vaccines, and antiviral drugs through novel antigens. Metals like copper, zinc, and silver can be used to alter nanomaterials to form specific antiviral bonds against SARS-CoV-2. The key benefit of utilizing nanoparticles for antiviral drug delivery is their high specificity, which means they only function on diseased cells and not on healthy cells. Infection-sensitive drug-delivery through nanoparticles reduces drug loss as well as premature drug release to prevent site-specific infections before, during, and after the process. For the delivery of medications against SARS-CoV-2, nanoformulations based on degradable polymers and biopolymers are being proposed (Carvalho, 2021). Nano-decoys and manganese nanoparticles have been considered as interesting alternatives to develop antiviral agents or vaccine adjuvants due these offer a robust, safe, and simple technology to increase the immune response for in vivo assays.

Nanomaterials, thanks to their physicochemical properties, represent a promising therapeutic approach in nanomedicine against SARS-CoV-2. For the development of medicines and vaccines against COVID-19, nanoparticles (NPs) are impregnated with viral antigens or antibodies via functional groups. Through COVID-19 vaccines, human cells must be exposed either to antigens containing inactivated viruses, live attenuated viruses, or recombinant viral parts based on *m*RNA, DNA, and proteins, or to neutralizing antibodies (Ellah, 2020). In the case of mRNA-based vaccines, it has been suggested that this antigen can be encapsulated in lipid nanoparticles (Chung, 2020). Some of the nanoparticle-based platforms that have been proposed for vaccination against coronavirus are based on nanoparticles with

Figure 7. Nanomaterials are being used in the treatment of COVID-19
(adapted from Ellah, 2020)

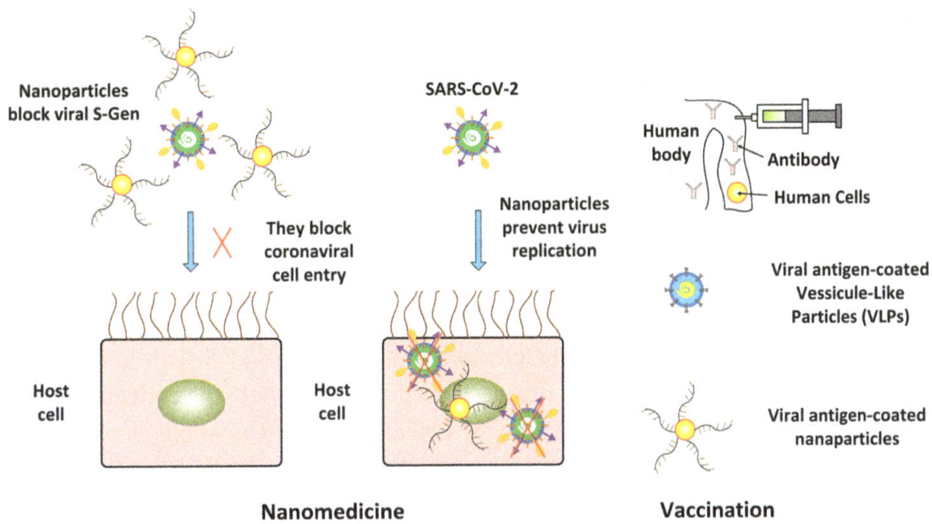

spike protein, virus-like particles displaying spike protein, virus-like particles displaying receptor-binding domains (RBDs), Chaperna-based nanoparticles, polypeptide nanoparticles, gold nanoparticles (AuNPs) with spike protein. So, both organic and inorganic solutions have been proposed (Shin, 2020). Figure 7 depicts a schematic representation of the use of nanomaterials in the treatment of COVID-19.

The COVID-19 pathology resembles typical sepsis, and severe COVID-19 is considered viral sepsis (Chen, 2021). Nanotherapies for sepsis that seeks to regulate signals causing inflammation and control reactive oxygen and nitrogen species (RONS) can be used to treat COVID-19. Nanomaterials with these species have been experimentally tested to produce therapeutic effects unattainable by traditional techniques. Transition metal dichalcogenides (TMDs) including tungsten disulfide (WS_2), molybdenum diselenide ($MoSe_2$) and tungsten diselenide (WSe_2) functionalized with poly (ε-caprolactone), serve as transporters of species that exert effects antioxidants and anti-inflammatories, which relieve multiple damaged tissues and increase the survival rate. These materials offer structural stability, good biocompatibility, low cytotoxicity, and durability of action.

FUTURE RESEARCH DIRECTIONS

This section primarily focuses on discussing the future research activities that must be carried out by the different research groups throughout the world for the treatment, prevention, and diagnosis of COVID-19.

Related Perspectives With Diagnostic Methods Against COVID-19

The challenges and potential that nanobiosensors have for the diagnosis of COVID-19 are discussed in this subsection. To date, COVID-19 detection can be done through point-of-care biosensors, whether based on chips, paper, textiles, and nanomaterials (Rasmi, 2021). The inherent advantages of these devices are their simplicity, user-friendliness, cost-effectiveness, and their ability to replace large and time-consuming technologies as well as their ease of operation without requiring highly trained personnel. Antibody-based tests are good for detecting late-stage or previous infections, whereas nucleic acid-based tests are appropriate for detecting infections in the early stages. The sensitivity and specificity of nucleic acid-based tests are higher, but they need more sophisticated and time-consuming methods. For their part, antibody-based tests have less specificity, require less time, allow faster decision-making, and the management of remote facilities. More research is needed to mass-produce and minimize the cost of biosensors based on two-dimensional materials.

Hitherto developed point-of-care biosensors use G- and/or M-gene test strips to reduce assay time but are not very sensitive and have low specificity (Rasmi, 2021). Multiplexing, sample concentration, fluidic control, and enzyme-based signal augmentation are all microfluidic-based technical solutions that could improve the biosensor's sensitivity. The most difficult task is to develop a single method that combines G and M gene detection with a nucleic acid test to detect SARS-CoV-2 in its early and late stages, allowing for accurate and trustworthy results. This implies that both the collection and preparation of the sample directly from the patient, as well as the detection of the signal are carried out in a biosensor system. To increase the efficiency of the biosensors, microfluidic support processes of the sample must be included, such as washing, incubation, mixing, and breaking of the solution (Bhalla, 2020). Like laboratories on a chip, it is necessary to continue optimizing systems to preserve reagents on the chip, incorporation of portable power supplies, wireless data quantification systems to take advantage of the Internet of Things (IoT) with physicians for monitoring in real-time, and above all looking for its mass production. Furthermore, as diagnostic tools for the COVID-19 pandemic, biosensing devices based on nanomaterials that are supported by artificial intelligence (AI) and the Internet of Things (IoT) are required (Antiochia, 2021).

Related Perspectives With Prevention Methods Against COVID-19

In virus-rich indoor environments, the probability of COVID-19 infection is high, as are highly crowded places and hospitals. High-efficiency surgical masks against SARS-CoV-2 can be developed by manufacturing composite materials whose reinforcements based on advanced nanomaterials provide properties that are aimed at reducing the probability of contagion. The protection of the mask is diminished over time, so to prevent the user from being exposed, the integrity of the mask should not be compromised by any condition to which it is subjected. Smart properties such as self-cleaning, self-healing, and self-sanitizing must be guaranteed, therefore, the development of materials that satisfy all these properties should be a top goal in the next years. The future materials that will be used in surgical masks must be reusable, environmentally sustainable, protective, washable, and have smart designs (Karmacharya, 2021). Because the masks will be worn for many hours in the day, no skin discomfort, thermal stress, as well as loss of communication should be presented to be accepted by the users. Prospective materials for the manufacture of face masks based on biodegradable or natural polymers will make it possible to reduce environmental pollution, allow self-sanitization and self-cleaning, which would be invaluable in the short term.

Related Perspectives With Treatment Methods Against COVID-19

One of the difficulties in treating SARS-CoV-2 is many mutations, as well as the possibility that the virus will develop resistance to vaccinations or treatments. The pharmaceutical industry must develop efficient strategies so that solutions using nanomaterials can beat the speed of genetic innovation to SARS-CoV-2. In the field of vaccines and treatments, nanomedicine is providing alternative tactics against SARS-CoV-2 using nanomaterials (Ellah, 2020). However, the safe transfer of the application of nanomaterials from the laboratory to the clinic requires complete mastery of both COVID-19 genomics and proteomics. Safety processes such as pathogenesis and reversion have direct relevance to the potential for S-gene antigen-related liver damage and eosinophilic immunopathology in the lungs. Since viruses have few therapeutic targets, this reduces the effects on host cells. To design specific ligands that must be employed to functionalize the nanoparticles and assault the virus's life cycle, researchers must continue to research SARS-CoV-2-flaws as well as the vulnerabilities of infected cells. The different behavior of the virus from one host to another does not allow a universal strategy for its diagnosis and treatment. As a result, researchers all over the world must continue to look for ways to develop new nanomedicines that are of high quality, safe, and affordable.

CONCLUSION

Nanomaterials offer immense potential to diagnose, treat, and prevent humans from becoming infected with SARS-CoV-2 because its size is on the nanoscale. The application of nanomaterials in medicine continues to open the possibility of promoting the development of new detection methods, protective materials for face masks, and materials for the treatment against COVID-19. This chapter described the role and potential of nanomaterials to address these three main issues in the fight against COVID-19. The advances achieved have provided vaccines with considerable efficiencies, preliminary prototypes of drugs against COVID-19, materials to protect ourselves from the virus, and biosensors for its detection both in its early or consolidated stage within humans. Despite all the research that has been done so far, there is still a lot of work to be done before the COVID-19 pandemic can be properly contained. Emerging nanomaterials like nanocomposites and two-dimensional materials (including graphene, phosphorene, and transition metal dichalcogenides (TMDs)) will bring up high-value technological and scientific prospects for controlling and even eradicating SARS-CoV-2. Several technical fronts must be working to achieve the immunity of the population taking advantage of the physicochemical properties of nanomaterials.

ACKNOWLEDGMENT

The author is grateful for the support of the Instituto Tecnológico Superior de Irapuato (ITESI) and the National Consortium of Scientific and Technological Information Resources (CONRICYT) of the National Council of Science and Technology (CONACYT) for their support for the realization of this publication related to the research topic of Advanced Materials of the Academic Group ITESI-CA-01 "Advanced Materials Applied to Engineering". Also, he appreciates the support of his wife and son to carry out this research on nanomaterials.

REFERENCES

Abid, S. A., Muneer, A. A., Al-Kadmy, I. M. S., Sattar, A. A., Beshbishy, A. M., Batiha, G.-E.-S., & Hetta, H. F. (2021). Biosensors as a Future Diagnostic Approach for COVID-19. *Life Sciences*, *273*, 119117. doi:10.1016/j.lfs.2021.119117 PMID:33508293

Adhikari, S., Adhikari, U., Mishra, A., & Guragain, B. S. (2020). Nanomaterials for Diagnostic, Treatment and Prevention of COVID-19. *Applied Science & Technology Annals*, *1*(1), 155–164. doi:10.3126/asta.v1i1.30295

Ali, M. A., Hu, C., Jahan, S., Yuan, B., Saleh, M. S., Ju, E., Gao, S.-J., & Panat, R. (2020). Sensing of COVID-19 Antibodies in Seconds via Aerosol Jet Nanoprinted Reduced-Graphene-Oxide-coated 3D Electrodes. *Advanced Materials*, *33*(7), 2006647. doi:10.1002/adma.202006647 PMID:33349975

Antiochia, R. (2021). Developments in Biosensors for CoV Detection and Future Trends. *Biosensors & Bioelectronics*, *173*, 112777. doi:10.1016/j.bios.2020.112777 PMID:33189015

Bhalla, N., Pan, Y., Yang, Z., & Payam, F. (2020). Opportunities and Challenges for Biosensors and Nanoscale Analytical Tools for Pandemics: COVID-19. *ACS Nano*, *14*(7), 7783–7807. doi:10.1021/acsnano.0c04421 PMID:32551559

Carter, L. J., Garner, L. V., Smoot, J. W., Li, Y., Zhou, Q., Saveson, C. J., Sasso, J. M., Gregg, A. C., Soares, D. J., Beskid, T. R., Jervey, S. R., & Liu, C. (2020). Assay Techniques and Test Development for COVID-19 Diagnosis. *ACS Central Science*, *6*(5), 591–605. doi:10.1021/acscentsci.0c00501 PMID:32382657

Carvalho, A. P. A., & Conte-Junior, C. A. (2021). Recent Advances on Nanomaterials to COVID-19 Management: A Systematic Review on Antiviral/Virucidal Agents and Mechanisms of SARS-CoV-2 Inhibition/Inactivation. *Global Challenges (Hoboken, NJ)*, *5*(5), 2000115. doi:10.1002/gch2.202000115 PMID:33786199

Chaudhary, V., Royal, A., Chavali, M., & Yadav, S. K. (2021). Advancements in Research and Development to Combat COVID-19 using Nanotechnology. *Nanotechnology for Environmental Engineering*, *6*(1), 8. doi:10.100741204-021-00102-7

Chauhan, G., Madou, M. J., Kalra, S., Chopra, V., Ghosh, D., & Martinez-Chapa, S. O. (2020). Nanotechnology for COVID-19: Therapeutics and Vaccine Research. *ACS Nano*, *14*(7), 7760–7782. doi:10.1021/acsnano.0c04006 PMID:32571007

Chen, L., Huang, Q., Zhao, T., Sui, L., Wang, S., Xiao, Z., Nan, Y., & Ai, K. (2021). Nanotherapies for Sepsis by Regulating Inflammatory Signals and Reactive Oxygen and Nitrogen Species: New Insight for Treating COVID-19. *Redox Biology*, *45*, 102046. doi:10.1016/j.redox.2021.102046 PMID:34174559

Choi, J. R. (2020). Development of Point-of-Care Biosensors for COVID-19. *Frontiers in Chemistry*, *8*, 517. doi:10.3389/fchem.2020.00517 PMID:32574316

Chung, Y. H., Beiss, V., Fiering, S. N., & Steinmetz, N. F. (2020). COVID-19 Vaccine Frontrunners and Their Nanotechnology Design. *ACS Nano*, *14*(10), 12522–12537. doi:10.1021/acsnano.0c07197 PMID:33034449

Deng, J., Zhao, S., Liu, Y., Liu, C., & Sun, J. (2021). Nanosensors for Diagnosis of Infectious Diseases. *ACS Applied Bio Materials*, *4*(5), 3863–3879. doi:10.1021/acsabm.0c01247

Ellah, N. H. A., Gad, S. F., Muhammad, K., Batiha, G. E., & Hetta, H. F. (2020). Nanomedicine as a Promising Approach for Diagnosis, Treatment and Prophylaxis against COVID-19. *Nanomedicine (London)*, *15*(21), 2085–2102. doi:10.2217/nnm-2020-0247 PMID:32723142

Fathi-Hafshejani, P., Azam, N., Wang, L., Kuroda, M. A., Hamilton, M. C., Hasim, S., & Mahjouri-Samani, M. (2021). Two-Dimensional-Material-based Field-Effect Transistor Biosensor for Detecting COVID-19 Virus (SARS-CoV-2). *ACS Nano*, *15*(7), 11461–11469. doi:10.1021/acsnano.1c01188 PMID:34181385

Guan, W.-J., Ni, Z.-Y., Hu, Y., Liang, W.-H., Ou, C.-Q., He, J.-X., Liu, L., Shan, H., Lei, C.-L., Hui, D. S. C., Du, B., Li, L.-J., Zeng, G., Yuen, K.-Y., Chen, R.-C., Tang, C.-L., Wang, T., Chen, P.-Y., Xiang, J., ... Zhong, N.-S. (2020). Clinical Characteristics of Coronavirus Disease 2019 in China. *The New England Journal of Medicine*, *382*(18), 1708–1720. doi:10.1056/NEJMoa2002032 PMID:32109013

Jastrzębska, A. M., & Vasilchenko, A. S. (2021). Smart and Sustainable Nanotechnological Solutions in a Battle against COVID-19 and Beyond: A Critical Review. *ACS Sustainable Chemistry & Engineering*, *9*(2), 601–622. doi:10.1021/acssuschemeng.0c06565 PMID:34192094

Karmacharya, M., Kumar, S., Gulenko, O., & Cho, Y.-K. (2021). Advances in Facemasks during the COVID-19 Pandemic Era. *ACS Applied Bio Materials*, *4*(5), 3891–3908. doi:10.1021/acsabm.0c01329

Khan, A. H., Tirth, V., Fawzy, M., Mahmoud, A. E. D., Khan, N. A., Ahmed, S., Ali, S. S., Akram, M., Hameed, L., Das, G., Roy, S., & Dehghani, M. H. (2021). COVID-19 Transmission, Vulnerability, Persistence and Nanotherapy: A Review. *Environmental Chemistry Letters*, *19*(4), 2773–2787. Advance online publication. doi:10.100710311-021-01229-4 PMID:33846683

Kostarelos, K. (2020). Nanoscale Nights of COVID-19. *Nature Nanotechnology*, *15*(5), 343–344. doi:10.103841565-020-0687-4 PMID:32341506

Lam, C. Y. K., Zhang, Q., Yin, B., Huang, Y., Wang, H., Yang, M., & Wong, S. H. D. (2021). Recent Advances in Two-Dimensional Transition Metal Dichalcogenide Nanocomposites Biosensors for Virus Detection before and during COVID-19 Outbreak. *Journal of Composites Science*, *5*(7), 190. doi:10.3390/jcs5070190

Li, X., Zai, J., Wang, X., & Li, Y. (2020). Potential of Large "First Generation" Human-to-Human Transmission of 2019-nCOV. *Journal of Medical Virology*, *92*(4), 448–454. doi:10.1002/jmv.25693 PMID:31997390

Liu, H., Ge, J., Ma, E., & Yang, L. (2019). Advanced Biomaterials for Biosensor and Theranostics. In Biomaterials in Translational Medicine (pp. 213-255). Elsevier. doi:10.1016/B978-0-12-813477-1.00010-4

Liu, Y., Gayle, A.A., Wilder-Smith, A., & Rocklöv, J. (2020). The Reproductive Number of COVID-19 is Higher Compared to SARS Coronavirus. *Journal of Travel Medicine*, *27*(2). doi:10.1093/jtm/taaa021

Moitra, P., Alafeef, M., Dighe, K., Frieman, M. B., & Pan, D. (2020). Selective Naked-Eye Detection of SARS-CoV-2 Mediated by N Gene Targeted Antisense Oligonucleotide Capped Plasmonic Nanoparticles. *ACS Nano*, *14*(6), 7617–7627. doi:10.1021/acsnano.0c03822 PMID:32437124

Mujawar, M. A., Gohel, H., Bhardwaj, S. K., Srinivasan, S., Hickman, N., & Kaushik, A. (2020). Nano-enabled Biosensing Systems for Intelligent Healthcare: Towards COVID-19 Management. *Materials Today. Chemistry*, *17*, 100306. doi:10.1016/j.mtchem.2020.100306 PMID:32835155

Palestino, G., García-Silva, I., González-Ortega, O., & Rosales-Mendoza, S. (2020). Can Nanotechnology Help in the Fight against COVID-19? *Expert Review of Anti-Infective Therapy*, *18*(9), 849–864. doi:10.1080/14787210.2020.1776115 PMID:32574081

Palmieri, V., & Papi, M. (2020). Can Graphene Take Part in the Fight Against COVID-19? *Nano Today*, *33*, 100883. doi:10.1016/j.nantod.2020.100883 PMID:32382315

Pathak, S., Saha, G. C., Hadi, M. B. A., & Jain, N. K. (2021). Engineered Nanomaterials for Aviation Industry in COVID-19 Context: A Time-Sensitive Review. *Coatings*, *11*(4), 382. doi:10.3390/coatings11040382

Rai, M., Bonde, S., Yadav, A., Plekhanova, Y., Reshetilov, A., Gupta, I., Golińska, P., Pandit, R., & Ingle, A. P. (2020). Nanotechnology-based promising Strategies for the Management of COVID-19: Current Development and Constraints. *Expert Review of Anti-Infective Therapy*, 1–10. Advance online publication. doi:10.1080/14787210.2021.1836961 PMID:33164589

Rai, P. K., Usmani, Z., Thakur, V. K., Gupta, V. K., & Mishra, Y. K. (2020). Tackling COVID-19 Pandemic through Nanocoatings: Confront and Exactitude. *Current Research in Green and Sustainable Chemistry*, *3*, 100011. doi:10.1016/j.crgsc.2020.100011

Rangayasami, A., Kannan, K., Murugesan, S., Radhika, D., Sadasivuni, K. K., Reddy, K. R., & Raghu, A. V. (2021). Influence of Nanotechnology to Combat against COVID-19 for Global Health Emergency: A Review. *Sensors International*, *2*, 100079. doi:10.1016/j.sintl.2020.100079 PMID:34766049

Rasmi, Y., Li, X., Khan, J., Ozer, T., & Choi, J. R. (2021). Emerging Point-of-Care Biosensors for Rapid Diagnosis of COVID-19: Current Progress Challenges, and Future Prospects. *Analytical and Bioanalytical Chemistry*, *413*(16), 4137–4159. doi:10.100700216-021-03377-6 PMID:34008124

Rockett, R. J., Arnott, A., Lam, C., Sadsad, R., Timms, V., Gray, K.-A., Eden, J.-S., Chang, S., Gall, M., Draper, J., Sim, E. M., Bachmann, N. L., Carter, I., Basile, K., Byun, R., O'Sullivan, M. V., Chen, S. C.-A., Maddocks, S., Sorrell, T. C., ... Sintchenko, V. (2020). Revealing COVID-19 Transmission in Australia by SARS-CoV-2 Genome Sequencing and Agent-based Modeling. *Nature Medicine*, *26*(9), 1398–1404. doi:10.103841591-020-1000-7 PMID:32647358

Ruiz-Hitzky, E., Darder, M., Wicklein, B., Ruiz-Garcia, C., Martín-Sampedro, R., del Real, G., & Aranda, P. (2020). Nanotechnology Responses to COVID-19. *Advanced Healthcare Materials*, *9*(19), 2000979. doi:10.1002/adhm.202000979 PMID:32885616

Seo, G., Lee, G., Kim, M. J., Baek, S.-H., Choi, M., Ku, K. B., Lee, C.-S., Jun, S., Park, D., Kim, H. G., Kim, S.-J., Lee, J.-O., Kim, B. T., Park, E. C., & Kim, S. I. (2020). Rapid Detection of COVID-19 Causative Virus (SARS-CoV-2) in Human Nasopharyngeal Swab Specimens using Field-Effect Transistor-based Biosensor. *ACS Nano*, *14*(4), 5135–5142. doi:10.1021/acsnano.0c02823 PMID:32293168

Sharma, S., Saini, S., Khangembam, M., & Singh, V. (2021). Nanomaterials-based Biosensors for COVID-19 Detection – A Review. *IEEE Sensors Journal*, *21*(5), 5598–5611. doi:10.1109/JSEN.2020.3036748

Shin, M. D., Shukla, S., Chung, Y. H., Beiss, V., Chan, S. K., Ortega-Rivera, O. A., Wirth, D. M., Chen, A., Sack, M., Pokorski, J. K., & Steinmetz, N. F. (2020). COVID-19 Vaccine Development and a Potential Nanomaterial Path Forward. *Nature Nanotechnology*, *15*(8), 646–655. doi:10.103841565-020-0737-y PMID:32669664

Sivansankarapillai, V. S., Pillai, A. M., Rahdar, A., Sobha, A. P., Das, S. S., Mitropoulos, A. C., Mokarrar, M. H., & Kyzas, G. Z. (2020). On Facing the SARS-CoV-2 (COVID-19) with Combination of Nanomaterials and Medicine: Possible Strategies and First Challenges. *Nanomaterials (Basel, Switzerland)*, *10*(5), 852. doi:10.3390/nano10050852 PMID:32354113

Tang, Z., Zhang, X., Shu, Y., Guo, M., Zhang, H., & Tao, W. (2021). Insights from Nanotechnology in COVID-19 Treatment. *Nano Today*, *36*, 101019. doi:10.1016/j.nantod.2020.101019 PMID:33178330

Tavakol, S., Zahmatkenshan, M., Mohammadinejad, R., Mehrzadi, S., Joghataei, M. T., Alavijeh, M. S., & Seifalian, A. (2021). The Role of Nanotechnology in Current COVID-19 Outbreak. *Heliyon*, *7*(4), e06841. doi:10.1016/j.heliyon.2021.e06841 PMID:33880422

Udugama, B., Kadhiresan, P., Kozlowski, H. N., Malekjahani, A., Osborne, M., Li, V. Y. C., Chen, H., Mubareka, S., Gubbay, J. B., & Chan, W. C. W. (2020). Diagnosing COVID-19: The Disease and Tools for Detection. *ACS Nano*, *14*(4), 3822–3835. doi:10.1021/acsnano.0c02624 PMID:32223179

Van Doremalen, B. T., Morris, D.H., Holbrook, M.G., Gamble, A. Williamson, B.N., Tamin, A., Harcourt, J.L., Thornburg, N.J., Gerber, S.I., Lloyd-Smith, J.O., De Wit E., & Munster, V.J. (2020). Aerosol and Surface Stability of SARS-Co-V-2 as Compared with SARS-CoV-1. *The New England Journal of Medicine*, *382*(16), 1564-1567. doi:10.1056/NEJMc2004973 PMID:32182409

Vazquez-Munoz, R., & Lopez-Ribot, J. L. (2020). Nanotechnology as an Alternative to Reduce the Spread of COVID-19. *Challenges*, *11*(2), 15. doi:10.3390/challe11020015

Wang, F., Huang, S., Gao, R., Zhou, Y., Lai, C., Li, Z., Xian, W., Qian, X., Li, Z., Huang, Y., Tang, Q., Liu, P., Chen, R., Liu, R., Li, X., Tong, X., Zhou, X., Bai, Y., Duan, G., ... Liu, L. (2020). Initial Whole-Genome Sequencing and Analysis of the Host Genetic Contribution to COVID-19 Severity and Susceptibility. *Cell Discovery*, *6*(1), 83. doi:10.103841421-020-00231-4 PMID:33298875

Weiss, C., Carriere, M., Fusco, L., Capua, I., Regla-Nava, J. A., Pasquali, M., Scott, J. A., Vitale, F., Unal, M. A., Mattevi, C., Bedognetti, D., Merkoçi, A., Tasciotti, E., Yilmazer, A., Gogotsi, Y., Stellacci, F., & Delogu, L. G. (2020). Toward Nanotechnology-enabled Approaches against the COVID-19 Pandemic. *ACS Nano*, *14*(6), 6383–6406. doi:10.1021/acsnano.0c03697 PMID:32519842

Wu, F., Zhao, S., Yu, B., Chen, Y.-M., Wang, W., Song, Z.-G., Hu, Y., Tao, Z.-W., Tian, J.-H., Pei, Y.-Y., Yuan, M.-L., Zhang, Y.-L., Dai, F.-H., Liu, Y., Wang, Q.-M., Zheng, J.-J., Xu, L., Holmes, E. C., & Zhang, Y. Z. (2020). A New Coronavirus Associated with Human Respiratory Disease in China. *Nature*, *579*(7798), 265–269. doi:10.103841586-020-2008-3 PMID:32015508

Zhong, H., Zhu, Z., Lin, J., Cheung, C. F., Lu, V. L., Yan, F., Chan, C.-Y., & Li, G. (2020). Reusable and Recyclable Graphene Masks with Outstanding Superhydrophobic and Photothermal Performances. *ACS Nano*, *14*(5), 6213–6221. doi:10.1021/acsnano.0c02250 PMID:32329600

Zhu, N., Zhang, D., Wang, W., Li, X., Yang, B., Song, J., Zhao, X., Huang, B., Shi, W., Lu, R., Niu, P., Zhan, F., Ma, X., Wang, D., Xu, W., Wu, G., Gao, G. F., & Tan, W. (2020). A Novel Coronavirus from Patients with Pneumonia in China, 2019. *The New England Journal of Medicine*, *382*(8), 727–733. doi:10.1056/NEJMoa2001017 PMID:31978945

ADDITIONAL READING

Ahmad, M. Z., Ahmad, J., Aslam, M., Khan, M. A., Alasmary, M. Y., & Abdel-Wahab, B. A. (2021). Repurposed Drug against COVID-19: Nanomedicine as an Approach for Finding New Hope in Old Medicines. *Nano Express*, *2*(2), 022007. doi:10.1088/2632-959X/abffed

Aydin, A., Demirtas, Z., Ok, M., Erkus, H., Cebi, G., Uysal, E., Gunduz, O., & Ustundag, C. B. (2021). 3D Printing in the Battle against COVID-19. *Emergent Materials*, *4*(1), 363–386. doi:10.100742247-021-00164-y PMID:33585793

Kaushik, A. K., Dhau, J. S., Gohel, H., Mishra, Y. K., Kateb, B., Kim, N.-Y., & Goswami, D. Y. (2020). Electrochemical SARS-CoV-2 Sensing at Point-of-Care and Artificial Intelligence for Intelligent COVID-19 Management. *ACS Applied Bio Materials*, *3*(11), 7306–7325. doi:10.1021/acsabm.0c01004

Poon, C., & Patel, A. A. (2020). Organic and Inorganic Nanoparticle Vaccines for Prevention of Infectious Diseases. *Nano Express*, *1*(1), 012001. doi:10.1088/2632-959X/ab8075

Pullangott, G., Kannan, U., Gayathri, S., Kiran, D. V., & Maliyekkal, S. M. (2021). A Comprehensive Review on Antimicrobial Face Masks: An Emerging Weapon in Fighting Pandemics. *RSC Advances*, *11*(12), 6544–6576. doi:10.1039/D0RA10009A

Shetti, N. P., Mishra, A., Bukkitgar, S. D., Basu, S., Narang, J., Reddy, K. R., & Aminabhavi, T. M. (2021). Conventional and Nanotechnology-based Sensing Methods for SARS Coronavirus (2019-nCoV). *ACS Applied Bio Materials*, *4*(2), 1178–1190. doi:10.1021/acsabm.0c01545 PMID:34192244

Yadav, S., Sadique, M. A., Ranjan, P., Kumar, N., Singhal, A., Srivastava, A. K., & Khan, R. (2021). SERS based Lateral Flow Immunoassay for Point-of-Care Detection of SARS-CoV-2 in Clinical Samples. *ACS Applied Bio Materials*, *4*(4), 2974–2995. doi:10.1021/acsabm.1c00102

Yu, Y., Bu, F., Zhou, H., Wang, Y., Cui, J., Wang, X., Nie, G., & Xiao, H. (2020). Biosafety Materials: An Emerging New Research Direction of Materials Science from the COVID-19 Outbreak. *Materials Chemistry Frontiers*, *4*(7), 1930–1953. doi:10.1039/D0QM00255K

KEY TERMS AND DEFINITIONS

Computed Tomography (CT) Scan: Medical imaging technique used in radiology to obtain detailed images of the body non-invasively for medical diagnostic purposes.

COVID-19: Virus that infects the respiratory system spread through droplets of saliva or discharge from the nose when an infected person sneezes or coughs and can cause death.

Deoxyribonucleic Acid (DNA): A molecule made up of two polynucleotide chains that twist together to create a double helix and carry the genetic instructions for all known organisms and viruses' formation, function, growth, and reproduction.

Infectious Disease: A health disorder caused by organisms such as bacteria, viruses, fungi, or parasites living in or on a human body passed from person to person causing infection, fever, and fatigue that, being serious, may require hospitalization or even cause the death.

Nanomaterial: Material with a size less than 100 nm in at least one of its dimensions.

Nanosensor: A device that can collect data and information about the behavior and characteristics of phenomena at the nanoscale level to make them detectable at the macroscopic level.

Point-of-Care Biosensing: Biosensing at or near the point of care, that is, when and where the patient is being treated.

Polymerase Chain Reaction (PCR): A technique for rapidly making millions to billions of copies of a DNA sample, allowing it to be enlarged for more detailed analysis.

Chapter 18
Use of Nanoparticles to Combat COVID-19

Ashish Kumar
https://orcid.org/0000-0002-2003-6209
Lovely Professional University, India

Abhinay Thakur
Lovely Professional University, India

ABSTRACT

On January 30, 2020, the World Health Organization declared the unexpected coronavirus pneumonia pandemic a public health crisis of global significance. COVID-19 in its most severe form can cause coagulation issues and acute respiratory distress syndrome. Due to viral modifications and the advent of new viral strains, the efficiency of traditional treatments for viral infections is rapidly fading. To overcome the limitations and to improve anti-viral treatments, integrated scientific research toward nanotechnology treatment investigations are anticipated for probable use in the prevention and/or treatment of viral infections. This chapter focuses on nano-based diagnostics, accompanied by a study of nano-based therapeutics and treatments that have been shown to be effective against viruses that are closely related to SARS-CoV-2. Furthermore, the authors anticipate that nanotechnology-based approaches mentioned throughout this chapter will aid researchers in developing new COVID-19 prevention, diagnosis, and treatment methods.

INTRODUCTION

Coronavirus disease 2019 (COVID-19) is a viral disease caused by coronavirus 2 that causes severe acute respiratory syndrome (SARS-CoV-2). The World Health Organization (WHO) proclaimed the COVID-19 outbreak a worldwide public health crisis of global outrage on January 20, 2020. COVID-19 has become much more prevalent, with more than 18 crore cases documented globally and > 4 million fatalities(Allawadhi et al., 2021; Poggio et al., 2020; Prajapati et al., 2020). COVID-19 can cause moderate symptoms like fever, coughs, and asthma to life-threatening conditions including acute respiratory distress

DOI: 10.4018/978-1-7998-8936-6.ch018

syndrome, pneumonia, and sometimes fatality. COVID-19 has a strong tendency to spread, rendering prevention and control complicated. A range of antiviral medicines is already being studied in clinical studies. In India, the anti-malarial medicine 'Hydroxychloroquine' was approved by the government to cure extreme COVID-19 cases, however, several hospitals began treating very ill patients with plasma therapy and vaccinations such as COVISHIELD and COVAXIN. Other medications being tested against COVID-19 include 'Lopinavir' with 'Ritonavir,' which is used to treat HIV; 'Remdesivir,' which binds to viral RNA polymerase, 'Interferon-beta-1a', which combines ritonavir and lopinavir to treat numerous sclerosis. Outside the host cell, the virus is usually inactive, but it replicates rapidly within the host via its replication machinery(Spigaglia, 2020; Zhao et al., 2021). As a result, there is a great need to understand how the virus interacts with the host that can be examined with cutting-edge technology and a reliable system. COVID-19 has three primary challenges: prevention, efficient and early detection, and treatment. Nanotechnology has already been shown to be effective against the human immunodeficiency virus, lyme disease, influenza virus, and numerous respiratory viruses, therefore it could be an effective therapy for COVID-19(Li et al., 2020). Figure 1 shows the basic route of transmission for COVID-19.

BACKGROUND

Numerous researchers have suggested that nanotechnology could be used to detect COVID-19 infection and design successful therapy techniques based on gut microbiome modification. Nanoparticles (NPs) constitute tiny particles with several 100 atoms whose size is measured in nano-units. Such polymeric particles might be naturally occurring or lab-made. NPs are generally made using top-down and bottom-up synthesis techniques like spinning molecular condensation, mechanical milling, biological synthesis, chemical etching puttering, laser pyrolysis, and electro-explosion(El-Megharbel et al., 2021; Kelleni, 2021). NPs are used in a variety of industries, including engineering, healthcare, biotech, and pharmaceuticals. The surface layer of such particles is fabricated with diverse molecules including metal ions, polymers, and surfactants. The shell is the intermediate layer, which is chemically distinct from the other layers, while the core is the innermost layer or central section, which is also known as NP. In the fabrication of the core and shell, a variety of inorganic and organic nanomaterials could be employed, resulting in nano- composites. The core-shell layers contents determine the physiological, photonic, and physicochemical aspects of these nanocomposites. Nanoparticles react rapidly in their natural condition. They absorb a lot of solar energy and show quantum phenomena(Chaudhary et al., 2021; Chintagunta et al., 2021). Metal NPs generally fabricated of metal precursors that have a unique optical property called localized surface plasmon resonance (SPR). These NPs have a wide variety of functions in research: for example, gold NPs are commonly used to enhance the electronic stream in scanning electron microscopy (SEM) materials in order to get a high-quality image. The certain viral particle-like SARS-Cov-2 virus is smaller than antiviral NPs having an avg. size of 120 nm as shown in Figure 2. As a result, NPs may interact directly on the entire viral particles, as well as their surface proteins, triggering viruses inactive.

Furthermore, Nanoparticles are promising representatives for COVID 19 virus treatment because of their small size, which facilitates drug delivery to the targeted location, the charge of surface, which aids drug penetration throughout the plasma membrane, large drug load capacity, enhanced surface area to volume ratio, biomimetic qualities, which mimic biochemical processes, and drug encapsulation. Nanoparticles possess several advantages in the preparation of vaccines: they are comparable in size to several viruses, thereby stimulating immune responses quickly; they can also evolve a preferential

Figure 1. Infected persons generate respiratory fluid drops carrying live viruses, which can infect those nearby by breathing airborne droplets or by direct touch with droplets or an infected hand. In addition, these droplets can infect other surfaces (fomites) and infect individuals through indirect touch
(Figure adapted from the source (Ruiz-Hitzky et al., 2020))

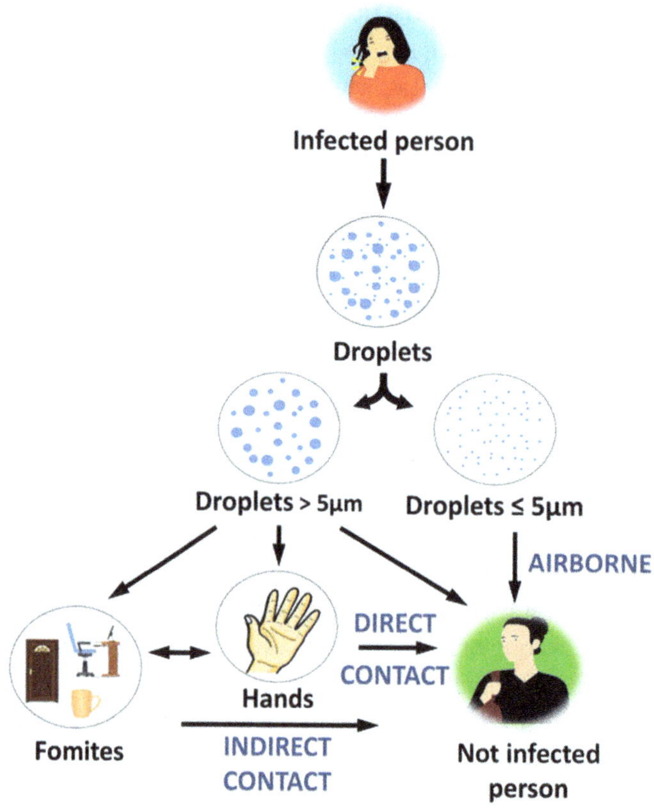

Figure 2. A single silver nanoparticle is compared to the avg. size of the SARS-Cov-2 coronavirus (AgNPs)
(Figure adapted from the source (Vazquez-Munoz & Lopez-Ribot, 2020))

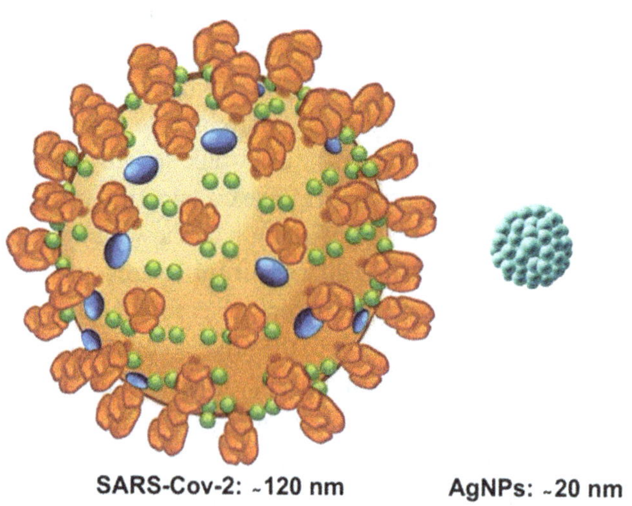

immune response by the ability to bind to receptors of targeted immune cells; and discharge antigen molecules slowly, leading to humoral immune response. Also, various NPs are explored for their ability to suppress encapsulated virus infection and have been created as targeted medications for preventing disease(Kirpotin et al., 2006; Rai et al., 2020). As a result, the utilization of NPs in form of medications or vaccinations may be an efficient way to combat COVID-19 viral infection. Antimicrobial nanoparticle coating in public areas and on personal protective equipment is an efficient prophylactic approach in reducing COVID-19 transmission to vulnerable hosts. Because coronaviruses are nanoscale in size, nanotechnology holds immense promise for better understanding the virus's composition, replication, and lifespan. Nanotechnology has the potential to open up new doors in the production of cost-effective and scalable detection equipment, as well as safe personal protection equipment and modern treatment remedies. The significance and prospects of nanotechnology in addressing the three primary components of the COVID-19 pandemic, detection, diagnosis, and treatment, will be discussed in the beneath sections.

Exploring COVID-19 Through Nanotechnology: Structure and Pathology

The optical microscope and the advancement of cell culture techniques have made it easier to discover and identify novel viruses. The area of diagnostic virology has been transformed by the advent of polymerase chain reaction (PCR) and immunoassays approach. Nanomaterials' distinctive features have been successfully exploited in the efficient development of a number of molecular and genomic detectors for the detection of novel viruses (as stated above). MERS (Middle East respiratory syndrome) and SARS (severe acute respiratory syndrome) viruses, which are linked to SARS-CoV-2 and were concerned for geographic epidemics in the 21st century, have both been detected and understood using nanomaterial-based approaches. They are now being utilized to better understand the configuration and lifespan of SARS-CoV-2 (Cavalcanti & Cajubá de Britto Lira Nogueira, 2020; Kleine-Tebbe et al., 2021). Various studies have shown that SARS-CoV-2 is encapsulated, single-stranded positive-sense RNA virus with pikes over the exterior top layer. The SARS coronavirus, CoV-2, has a genome that is similar to other coronaviruses. The angiotensin-converting enzyme II (ACE2) receptors on the surface of host cells are the virus's primary target. The encapsulating integrated spike (S) protein of SARS-CoV-2 interacts with the ACE2 receptor on the host cell surface. By acting as a protease, transmembrane serine protease 2 aids virus entry into cells. Figure 3 shows the COVID-19 pathology and related nano solutions. After internalization, virus particles penetrate endosomes. The exposed viral particle exposes its genome for protein synthesis because the pH of endosomes is moderate. This led to the production of protein and viral RNA that leads to the development of new infectious particles that are discharged into the host cell (Dube et al., 2021).

CURRENT NANOTECHNOLOGY APPLICATIONS USED TO COMBAT COVID-19

Just like TB, HCV, or even HIV, wherein prompt and concise detection at the early stages of infection could contribute to better management and therapy, the COVID-19 problem arises from the fact that asymptomatic infected people may be a substantial source of transmission. Let's look at AIDS, which is caused by HIV, which is among the century's most serious problems. In 2019, an estimated 690,000 people died as a result of AIDS-related illnesses (Chauhan et al., 2020). In the absence of a potential cure, early detection is essential for improving health. Organizations all over the world utilized quick

Figure 3. COVID-19 pathology and related nano solutions
(Figure adapted from the source (Vahedifard & Chakravarthy, 2021))

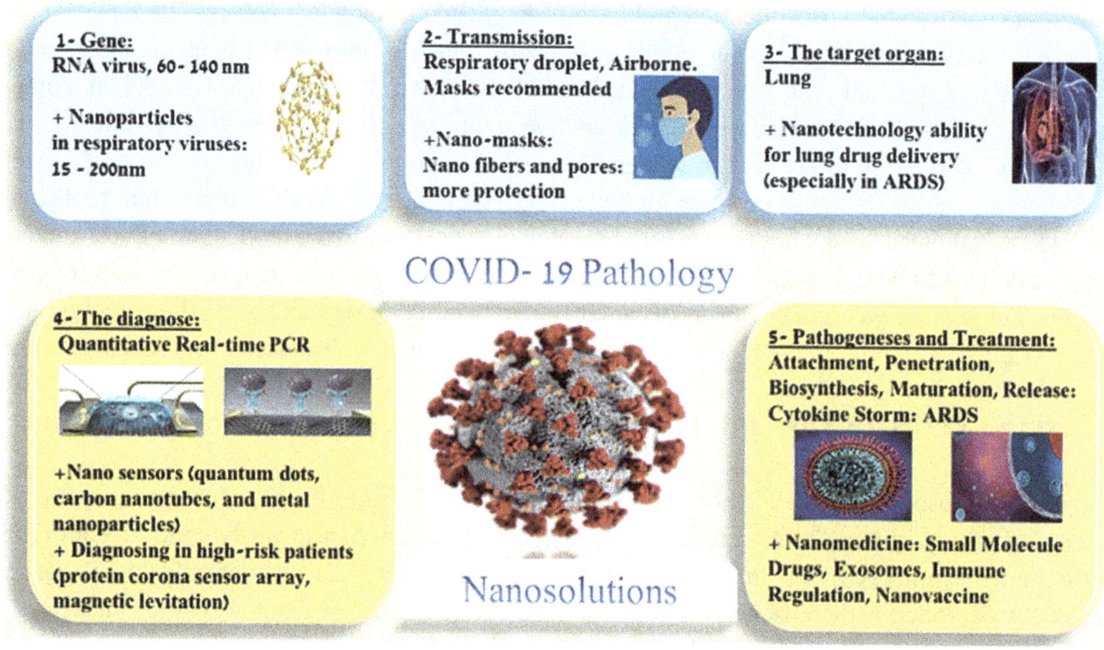

COVID-19 antibody tests to diagnose infections earlier on in the outbreak, but several of them missed the infection stage. Later, more advanced tests based on nanotechnology (colloidal gold method) which can identify very low viral loads with high dedication were developed, assisting in the control of infection and the development of appropriate treatment paths. Nanotechnology has the potential to transform pathogen detection and testing by providing safer, faster, more reliable, cost-effective, and accurate nanodevices and nanoprobes. Nanomaterials offer a variety of physical and chemical features that could aid in the combat over SARS-Cov-2, from minimizing transmission to developing potential treatments. Textiles and polymers absorb nanomaterials for a range of uses, including personal protective equipment (PPE), packaging, medical fabrics, and filter. Nanomaterial dispersions could be utilized for developing numerous coatings, cleaning, medicine, and much more.

Nanotechnology has a lot of promise in terms of diagnosing, treating, and preventing COVID-19. Nanotechnology might aid in the battle against COVID-19 in a variety of ways, including preventing viral contamination and disinfectant by: (a) designing infection-safe personal protective equipment (PPE) for health professionals and developing efficacious antiviral disinfectants and protective coatings that could inhibit the activity of the virus and preclude its dispersion; (b) the development of highly sensitive and specific nano-based sensors to quickly detect an infectious disease or immunological reaction; c) the development of novel drugs with the improved operation, low cytotoxicity, and controlled release, and also tissue-targeting, such as the lung tissue; and d) the development of a nano-based vaccine to strengthen the humoral immune response. Nanomaterials can also guard against virus contamination through the air and contact with infected surfaces, which can be especially important in the disinfection of protective equipment in a medical environment. Nanomaterials contain electronic and optical properties which have been intensively studied in the development of diagnostic methods including point-of-care

Figure 4. The major benefits of nanoparticles as antiviral medication carriers
(Figure adapted from the source (Tortella et al., 2021))

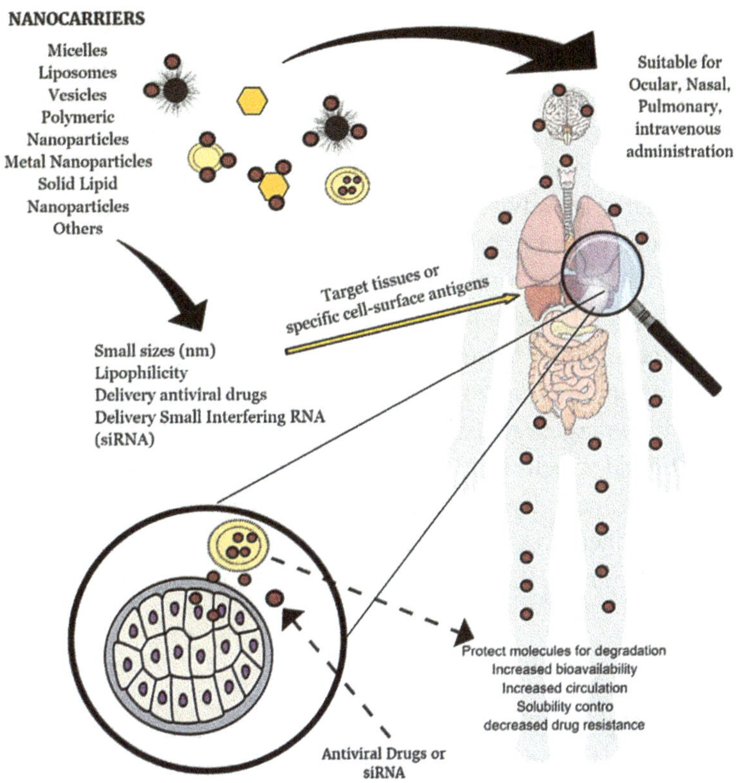

(POC) biosensors. Nanoparticles, which allow us to recognize analyses in little amounts, could improve detection sensitivity. These strategies enable the rapid detection and isolation of SARS-CoV-2 infected individuals, as well as the treatment of those who have been exposed. The interactions of nanomaterials with biological surfaces lay the premise for their potential (bio)medical applications.

Finally, for nanocarriers, diagnostics, and detection, a variety of nanomaterials can be used(Heidari et al., 2020). The production of nanocarriers for antiviral medications has already proven to be effective in preventing or reducing the most important constraints of antiviral drugs, owing to their tailorable size. Nanotechnology brings up new possibilities in combating COVID-19, by offering innovative and improved modes of diagnosis, prevention, and therapy. This chapter examines nanotechnology-related approaches and probable applications that might help control the COVID-19 epidemic based on the characteristics and capabilities of nanostructured materials. Figure 4 shows how the key properties of each nanoparticle may affect the mechanism associated with drug delivery.

Nanoparticles/Nanomaterials for the Detection of COVID-19

Early intervention of a disease is critical for screening abnormalities and triaging patients in order to prevent disease transmission and improve health. Previously, computed CT imaging and electron microscopy were used to detect SARS CoV-2. To establish the presence of SARS-CoV-2 in the body, molecular techniques like reverse transcription-polymerase chain reaction (RT-PCR) tests are performed in the later phases. In 2002, this test has been used to diagnose SARS (Heidari et al., 2020). The RT-PCR-based test is time-consuming and can result in false positives in asymptomatic/recovered individuals. As a result, quick, precise, and dependable diagnostic assays based on RNA/DNA/protein from serum, blood, faeces, and nasopharyngeal swabs are being developed. The rapid diagnostic test (RDT) takes half-hour in an examining and evaluate the specimens. RDT is a point-of-care (POC) test that is commonly utilized in the identification of different pathogens. RDTs have previously been used to identify Rubella, Enterovirus, Salmonella, E. coli, Falciparum, etc. Since the COVID-19 epidemic is rapidly propagating over the world, effective diagnostic procedures are advised to prevent the disease's spread. Nanomaterials are used to create well and potential diagnostic devices against SARS-CoV-2 detection. Several biosensors and rapid test kits for COVID-19 diagnostics use nanoscale materials such as silver, iron oxide, tin oxide (TO). Biosensing applications frequently use gold nanoparticles (due to their biocompatibility). Antibodies and DNA strands that are attached to the surface of gold NPs serve as markers. Such nanoparticles can also be used to provide a potent transduction framework for detecting unique biomolecules. Numerous viruses are being successfully diagnosed using gold nanoparticles. As these allow biological molecules to bind together to form hybrid structures known as Bio- AuNPs, which promote virus target detection while also boosting the test's sensitivity, sensing range, and transmission time. Gold nanoparticles were also employed by UK researchers to construct a quick testing kit for IgM and IgG antibodies, which produced outcomes in just 10 minutes(Chakraborty et al., 2020; Ibrahim Fouad, 2021; Kabir et al., 2021).

Antibody Assay Kit

Antibody assay kits, an alternate to PCR, are used as point-of-care diagnostics for disorders such as Rubella, allergies, autoimmunity, cirrhosis, HIV, and hepatitis. The binding of antigenic epitopes with either IgMor IgG is the primary premise of antibody assay kits. These antibodies are generated mostly during the main immune response, in which IgM provides the first line of defense and IgG provides long-term protection and immunological memory. 7 days following the appearance of IgM, IgG is found. Since 20–80 percent of people are asymptomatic and PCR has a high risk of false positives, antibody testing is crucial in limiting the transmission of infectious disease (Mitrani et al., 2021; Vivekanandhan et al., 2021). The World Nano Foundation has designed a 2nd COVID-19 antibody assay kit featuring distinct IgM and IgG readings that uses gold colloid. The kit's efficacy is increased by the introduction of gold NPs, and it only takes 3 minutes and 15 minutes to get positive and negative findings, correspondingly. The biggest disadvantage of this test is that it does not verify the existence of a specific virus; instead, it merely verifies infection, allowing health practitioners to better triage patients. To resolve this challenge, MIT researchers developed a screening device that can discriminate between two gene sequences that are quite similar, allowing for precise SARS-CoV-2 identification. A DNA probe mounted on Au NPs generates surface plasmon resonance, allowing it to differentiate between particular SARS-CoV-2 RNA sequences.

Lateral Flow Immunoassay(LFIA)

The LFIA is a diagnostic cum detecting equipment that may identify infections in foodstuff, drinks, and biological molecules. Because of its cheap price, the convenience of use, and availability, LFIA is frequently utilized as a point-of-care diagnostic—especially its effective for pregnancy tests. A team of researchers devised a lanthanide-doped polystyrene nanoparticle-based approach for detecting anti-

SARV-CoV-2 IgG in COVID-19 patient's plasma by LFIA concept. To engage with the targeted IgG, the coronavirus synthesized nucleocapsid phosphoprotein was immobilized upon this membrane. The fluorescent sensor acts as a mouse anti-human IgG antibody encapsulated on NPs.

Nanomaterials-Based Biosensors to Detect COVID-19

A biosensor is an experimental tool that detects biomolecules by combining a biomarker, transducers, and a signal receiver. Biosensors are very sensitive, cost-effective, and well-suited to the diagnosis of diseases. Nanomaterial-based biosensors have greater potential for detecting several viral infections, as well as the best optimum alternative to PCR-based diagnostics as they are faster (10–100 minutes).

Aptamer-Based Nano-Biosensor

The aptamer-based nano biosensor technology has emerged as one of the foremost effective monitoring instruments possessing quick identification and detection of a wide range of disease-causing microorganisms with great selectivity and sensitivity. In contrast to traditional diagnostic approaches, it's a very premium, user-friendly, and environmentally friendly method. Any virus's nucleocapsid protein, which has significant specific binding affinities for specific target molecules, can be used to create aptamers. Any virus element, such as amino acids, membrane proteins, cytokines, pesticides, binding agents, or antibodies, could be the target molecule. These manufactured aptamers make use of apta-sensors that may be precisely controlled to any desired design by adding coupling sites and linkers by chemical modification or surface activation. Electrochemical measurements are used to operate this bio-nanogate bifunctional aptamer-based sensor. As a consequence, an aptamer-based nanosensor with great specificity, selectivity, and sensitivity for COVID-19 will have a lot of potential for detecting SARS-CoV-2 even in people who don't have any symptoms. These aptamer-based biosensors for detecting SARS-CoV-2 are said to be in the early stages of development. The SARS-CoV-2 nucleocapsid spike S1 protein, which distinguishes the virus from others, was used to create the aptamer for such biosensors. A viral agent was recovered from the mouth or nasal swab of a person. Electrochemical analysis is used to evaluate the response of such a bio-nanogate bifunctional aptasensor to the spike protein. It claimed to provide the SARS-CoV-2 test results in less than a minute and to be accurate in less than 30 seconds.

Plasmonic Photothermal Dual-Functional Biosensors

The plasmonic photothermal (PPT) action and the localized surface plasmon resonance (LSPR) phenomenon are combined to create dual-functional plasmonic photothermal biosensors. Similar strategies were used by scientists all across the world to create a dual-functional biosensor. They employed DNA receptors as a detector to track the SARS-CoV-2 specific motifs via nucleic acid hybridization. The use of a nanogold-based chip is said to improve sensing performance (AuNIs). AuNIs chips produce thermoplastic heat when illuminated to their plasmonic resonance frequency. This localized PPE heat raises the temperature of in situ hybridization and speeds up the exact distinction of two identical gene sequences. It is said to have a very low detection limit of 0.22 pM, great sensitivity, and high efficiency for the exact identification of a particular gene via a multigene mixture.

Graphene-based Biosensor

The SARS-CoV-2 spike protein is detected by a particular antibody that is coated onto graphene sheets to manufacture these sensors. G. Seo et al. described the use of a graphene FET device to identify SARS-CoV-2 spike protein at a concentration of 0.1 fg/mL in phosphate-buffered saline and 100 fg/mL in a clinical transport medium. This article also claims that the virus has been detected in the air. Groltex and the Medical Discovery Institute of Sanford Burnham Prebys claimed to have developed a graphene-based biosensor system that includes handheld reading units and plastic disposable assessment chips. Layqah et al. described an electrochemical immunosensor for measuring human coronavirus (HCoV) and MERS-CoV proteins in spiked nasal samples, which used altered carbon electrodes with gold NPs. The possibilities for graphene-based sensors to recognize SARS-CoV-2 are great. The basic use of the biosensor for detecting COVID-19 is shown in Figure 5.

Nanomaterials for Diagnosis and Treatment of COVID-19

Nanomaterials are essential in antiviral treatment because they improve drug delivery and usage effectiveness by boosting the distribution of water-insoluble medicines. The US Food and Drug Administration regulates nanomaterials, which are used in drug delivery to improve therapeutic efficacy. Viral infection could be resisted by nanoparticle-based medications that prevent access into the cell, virus attachment as well as viral proliferation, and direct inactivation. Silver, gold, selenium, silicon, zinc oxide, polylactic acid, silver sulfide, and other nanoparticles are commonly utilized to cure COVID-19. Glycyrrhizic acid (GA) nanoparticles (GANPs) premised on GA were produced by Zhaoyan et al.(Zhao et al., 2021) GANPs decrease the growth of the murine coronavirus MHV-A59 and lower the generation of proinflammatory cytokines produced by MHV-A59 or the N protein of SARS-CoV-2 in vitro. GANPs preferentially target sites with severe inflammation, like the lung, in an MHV-A59-induced COVID-19 surrogate mouse model, that seemed to boost GANP formation and therapy efficiency. GANPs additionally exhibit anti-inflammatory and antiviral activities, which reduce organ failure and provide significant protection to sick mice. This research reveals that treating SARS-CoV-2 with antiviral and anti-inflammatory medications at the same time had rather effective and complete effects in reducing the severe inflammation induced by the virus.

Similarly, Alexander et al. (Cohen et al., 2021) developed homotypic nanoparticles that displayed the SARS-CoV-2 receptor-binding domain (RBD) alone or in combination with RBDs from animals beta coronaviruses that pose a hazard to humans (mosaic nanoparticles; 4-8 distinct RBDs). Mice immunized with RBD-nanoparticles but not with soluble antigens elicited cross-reactive binding and neutralization reactions in this study. When comparison to sera from vaccinations with homotypic SARS-CoV-2–RBD-nanoparticles or COVID-19 convalescent human plasmas, mosaic-RBD-nanoparticles generated antibodies with greater cross-reactive detection of heterologous RBDs. Furthermore, sera from mosaic-RBD–immunized animals neutralized heterologous pseudotyped coronaviruses as well as or superior to sera from homotypic SARS-CoV-2–RBD-nanoparticle vaccinations following primed, suggesting that co-display has no effect on immunogenicity against specific RBDs. A single immunization with mosaic-RBD-nanoparticles could be used to defend against SARS-CoV-2 and new zoonotic coronaviruses at the same time. Ozturk Alper et al. (Alper Öztürk et al., 2021) used the nanoprecipitation process to make CEF-loaded Eudragit S100/PLGA NPs. Both Eudragit S100/PLGA polymer mixtures including the utilization of polymers individually would have been suitable nanocarriers for CEF. All developed

Figure 5. COVID-19 point-of-care biosensors
(Figure adapted from the source (Choi, 2020))

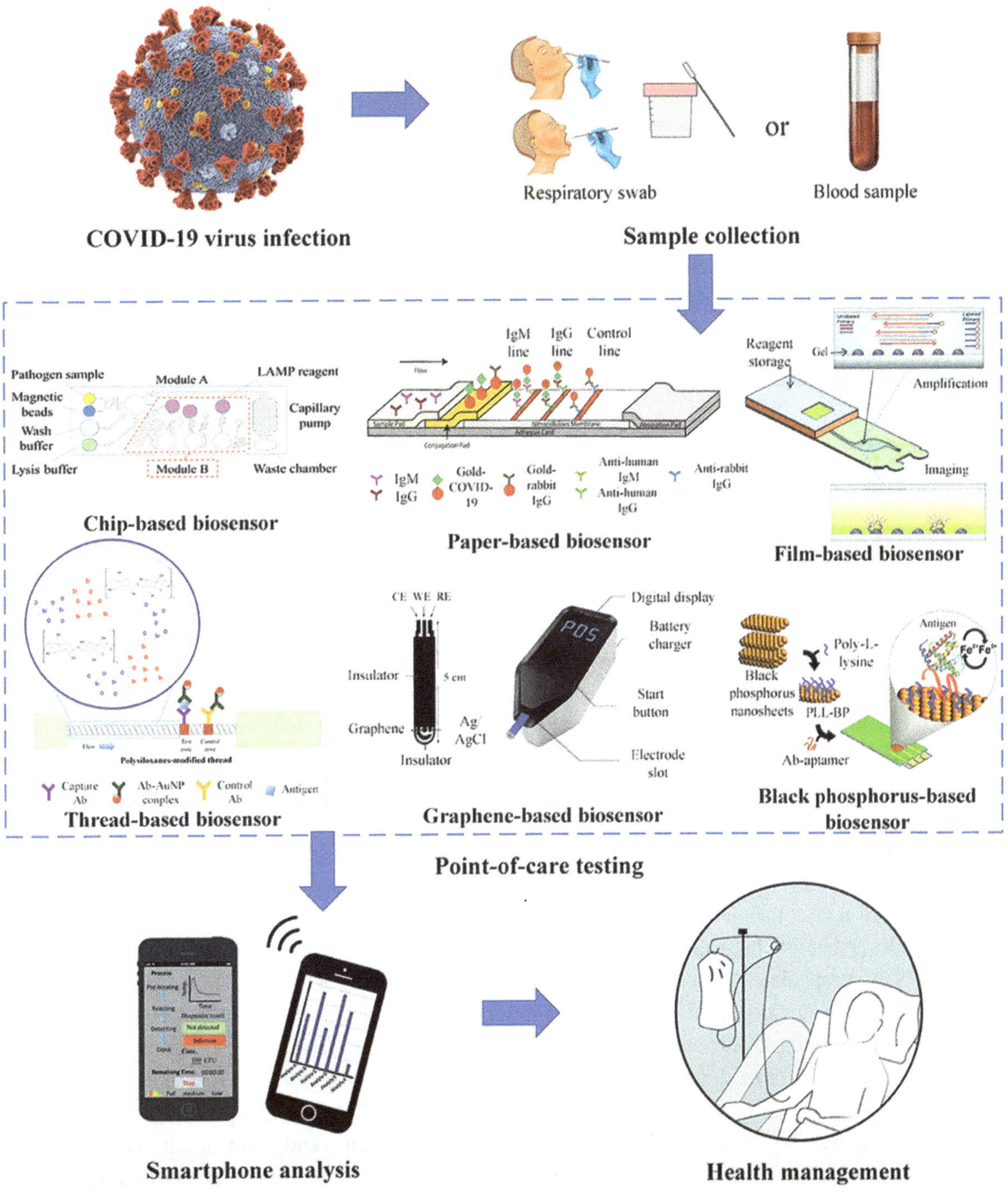

formulations utilizing Eudragit S100 and/or PLGA culminated in nano-range size, with PS of 198.8 nm in this investigation. The size of the NPs was shown to be dependent on the polymer effect, with greater PS achieved in formulations including ES100. All of the findings indicated that the produced

Figure 6. In this SR, the following are the major nano-based techniques towards animal coronaviruses: A) gold nanoparticle-based sVLPs and IBV spike protein incubation; B) silver nanomaterials' inhibitory impacts on TGEV-induced host cell infection and p38-MAPK signaling activation; and C) Ag/GO nanocomposites' interactions with FCoV
(Figure adapted from the source (Carvalho & Conte-Junior, 2021))

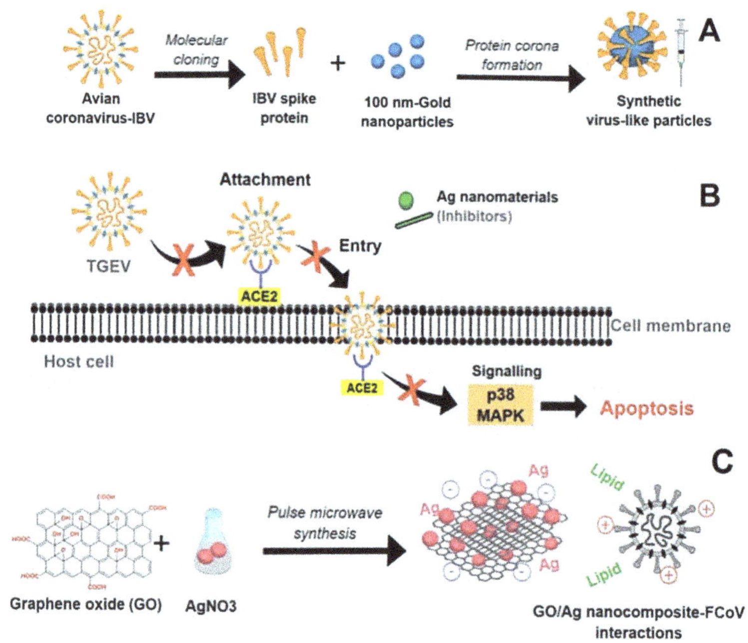

NPs might be utilized orally to treat gram-positive and gram-negative bacteria-induced infections, and also COVID-19-related confections and syndromes. Figure 6 is showing the utilization of some NPs against SARS-CoV-2.

The Role of Gold Nanoparticles During the COVID-19 Crisis

In the fabrication of colorimetric bio-sensors, Au nanoparticles having unique qualities such as optical characteristics, inherent photostability, a high emission spectrum has been commonly utilized. In a recent study, Au NPs encapsulated with thiol customized oligonucleotides were utilized to build a colorimetric bioassay for SARS-nucleocapsid CoV-2's phosphoprotein that was used to quickly detect COVID-19 (10 min). The surface plasmon resonance changes when the virus's target RNA sequence is present, allowing for selective "naked-eye" diagnosis. The report's primary features were that it didn't necessitate any complex experimental treatments and that it produced a low detection limit. Au nanoparticles were used to create a low-cost, simple nanoplasmonic bio-sensor chip and point-of-care equipment for the quick and one-step detection of SARS-CoV-2 infection, that could assist in the containment of the devastating epidemic.

Tedrick et al. (Lew et al., 2021) examined epitope-functionalized gold nanoparticles to identify SARS-CoV2 IgG antibodies quickly and selectively. Short antigenic epitopes attached to Au NPs were used to construct a colorimetric serological test to identify SARS-CoV-2 IgGs in patient plasma (AuNPs). For their IgG binding intensities, 4 immunodominant linear B-cell epitopes were discovered and employed as exceptionally specialized biological patterns on the nanoparticle to detect particular antibodies in SARS-spike CoV-2 (S) and nucleocapsid (N) proteins. A bivalent interaction among epitope-functionalized AuNPs and SARS-CoV-2 antibodies promotes NPs aggregation within 30 minutes of antibody delivery, as evidenced by a significant optical shift in the plasmon properties of the AuNPs. Co-immobilization of 2 epitopes increased analyte sensitivity, with a detection limit of 3.2 nM, which corresponded to IgG concentrations in COVID-19-infected individuals in recovery, compared to single-epitope AuNPs. A passivation technique was investigated further in order to maintain the sensory response in a human plasma medium. The improved nanosensor assay could accurately assess SARS-CoV-2 infection having 100% accuracy and 83 percent sensitivity when evaluated on 35 clinical plasma samples of various sickness grades. The potential of gold nanoparticles (AuNPs) stabilized by various groups, like undecanesulfonic acid (Mus), octanethiol (Ot), 3-mercaptoethylsulfonate (Mes), and a new peptide, to impede SARS-CoV-2 was explored in Aliyeh et al.(Mehranfar & Izadyar, 2020) using molecular dynamics (MD) simulations. The crystalline morphology of angiotensin-converting enzyme 2 (ACE2), which interacts with SARS-CoV-2 receptor-binding domain (RBD), reveals that 15 of ACE2's amino acids interacts substantially with RBD. As a result, a novel peptide was created based on these amino acids to serve as the functional group of AuNP. Functionalized AuNPs have a considerable effect on the RBD and bind extensively with the SARS-CoV-2 protein. The AuNP complexes by novel peptide generate a substantially more persistent complex with RBD than ACE2 that could be the human receptor for SARS-CoV-2, amongst these nanoparticles examined.

The Role of Silver Nanoparticles During the COVID-19 Crisis

Ag has remarkable antibacterial properties towards microbes, viruses, and fungi, and it is now frequently employed to make antibacterial materials for a variety of purposes, including biomedical materials, antimicrobial sprays, and textiles. In essence, this nanoparticle with excellent antimicrobial properties is utilized as a biocide on a variety of surfaces, including medical equipment and antibiotic ointments. Ag nanoparticles have been proven in studies to have antiviral effects against cirrhosis, influenza, and Aids. These could also be useful in the fight against SARS-CoV-2. Ryo and colleagues investigated the effects of Ag NPs on SARS-CoV-2 and revealed the optimal conc. and structure for successful SARS-CoV-2 suppression. At various concentrations and sizes of Ag, particles having10 nm diameter of and conc. of 2-11 ppm were found to be effective against SARS-CoV-2. At doses of 20 ppm and higher, toxicity was detected. Ag has been demonstrated in studies to successfully prevent virus entry into cells. As a result, our investigation verified Ag's antibacterial effects for SARS-CoV-2, however, caution must be exercised when using it owing to its toxicity. Horacio et al.(Almanza-Reyes et al., 2021) investigated AgNPs in vitro and discovered that they inhibited SARS-CoV-2 infection in cultivated microbes. They then revealed the effectiveness through the mouthwash and throat rinse that included ARGOVIT® AgNPs in mitigating SARS-CoV-2 infection in healthcare professionals at high risk of contracting the virus at the General Tijuana Hospital (Mexico), that is devoted to the registration of COVID-19 patients. They presented the findings of a 9-week prospective randomized study with 231 participants. Participants in the "intervention" group were told to use AgNPs mouthwash and nasal rinse, while those in the "control" group were told to use regular mouthwash and nasal rinse. The "experimental" group exhibited an efficiently lower probability of SARS-CoV-2 infection (two individuals out of 114, 1.8 percent) than the "control" group, with an efficiency rate of 84.8 percent (thirty-three out of 117, 28.2 percent).

Sundararaj et al. (Jeremiah et al., 2020) tested a variety of AgNPs of various sizes and concentrations and found that particles with a diameter of roughly 10 nm were efficient in suppressing extracellular SARS-CoV-2 at doses ranging from 1 to 10 ppm, whereas quantities of 20 ppm and above were cytotoxic. AgNPs effectively prevented viral entry by affecting viral integrity, according to a luciferase-based pseudovirus entry experiment. AgNPs appear to be efficient microbicides against SARS-CoV-2. Oron et al.(Zachar, 2020) examined the feasibility of producing an actual efficient minimal inhibitory concentration (MIC) of silver nanoparticles in a variety of respiratory system target locations. Controlling local COVID-19 outbreaks by early-stage home therapy and lowering the incidence of ventilator-associated pneumonia (VAP) in hospital ICUs are two applications. Their main goal is to suggest a first-line intervention that has the ability to prevent viral infection from spreading throughout the respiratory system, providing the immune system more time to respond and reducing the danger of virus exacerbation and transmission. According to the investigation, a colloidal silver with particle sizes ranging from 3 to 7 nanometers could be extremely efficacious in suppressing viral infection pathways. We also find that the MIC concentrations of such colloids are around 10 g/ml, although a target tissue concentration of 25 g/ml is proposed for improved effectiveness.

The Role of Copper and Copper Oxide Nanoparticles During the COVID-19 Crisis

Copper is cheap than noble metals that have been utilized for generations. Copper and copper oxide nanoparticles are employed in a variety of applications due to their outstanding antibacterial and antiviral characteristics. They've also been demonstrated to work well towards the coronavirus. Furthermore, the use of these antimicrobial nanoparticles in the formulation of effective antiviral coatings inhibits the propagation of coronavirus. The SARS-CoV-2 virus may persist for several days on plastic and steel surfaces, but just 4 to 8 hours on copper surfaces. ROS are generated when copper ions are released from such substances, causing pathogen damage and mortality. From a long time ago, due to the antibacterial and antiviral therapies, Cu and CuO have been used to combat SARS-CoV-2. Noemi et al.(Jardón-Maximino et al., 2021) CuNPs conjugated with 4-aminobutyric acid (GABA) and polyethyleneimine (PEI) were utilized for the preparation of composites comprising isotactic polypropylene. Melt mixing was used to make the iPP/CuNPs composites at Cu conc. of 0.25–5.0 wt%, and no trace of CuNP oxidation was seen. Copper ion discharge from iPP/CuNPs composites in aqueous media has also been investigated. In composites with 2.5 and 5.0 weight percent, cupric ion release was increased. These composites had substantial antibacterial activity (AA) against Staphylococcus aureus and Pseudomonas aeruginosa. The addition of CuNP to the iPP polymeric matrix reduced the composite material's thermal stability marginally but enhanced its crystallinity and storage modulus. According to the COVID 19 diagnosis, CuNPs could act as nucleating agents in the iPP crystallization process.

Cu might attack various pathogenic viruses, including bronchitis virus, coronavirus, human immunodeficiency virus type 1 (HIV-1), other enclosed or non-enveloped, single- or double-stranded DNA and RNA viruses, as per Syamal et al.(Raha et al., 2020) thorough overview of copper's capability against COVID-19. Cu also has the ability to kill a variety of viruses when it comes into touch with them, including SARSCoV2. They believe that increasing plasma copper levels will increase people's innate and adaptive immunity, based on current data. Cu may potentially be used as a COVID-19 preventative and treatment regimen due to its powerful antiviral properties.

The Role of Iron-Based Nanoparticles During the COVID-19 Crisis

Magnetic Fe_3O_4 nanoparticles are suitable carriers with acceptable separation performance, with various advantages such as high surface area, remarkable magnetism, and so on. In this context, a new approach for detecting COVID-19-specific IgG and IgM with higher accuracy, precision, and stability was proposed. Furthermore, a fluorescent immunoassay approach based on Fe_3O_4 and quantum dots was developed to assess human IgG in serum samples. Mouse antihuman IgG was added to coated Fe_3O_4 nano-spheres and employed as an immune capture probe to attain this purpose. To improve accuracy, quantum dot nano-beads were coupled with rabbit antihuman IgG as a fluorescence detecting probe. A sandwich-shaped immunological complex is generated when a human IgG antibody is added, and the precipitate provides a spectacular fluorescence signal following magnetic separation. There is no fluorescence response in the lack of a human antibody. This fluorescence-linked immunosorbent assay offered several advantages, including ease of use, a wide linear range, a low detection limit, high specificity and accuracy, and great efficiency when compared to traditional analytical methods.

Yasmin et al.(Abo-zeid et al., 2020) used docking analysis to look into the binding of IONPs (Fe_2O_3 and Fe_3O_4) on the SARS-CoV-2 spike protein receptor-binding domain (S1- RBD), which is essential for virus attachment on the host cell receptors. The glycoproteins E1 and E2 of the hepatitis C virus (HCV) were docked in a similar way. HCV glycoproteins E1 and E2 and SARS-CoV-2 S1-RBD interacted successfully with both Fe_2O_3 and Fe_3O_4. Fe_3O_4 made a stabilized compound with S1-RBD, whereas HCV E1 and E2 were enhanced by Fe_2O_3. These IONP connections are hypothesized to be linked to structural alterations in viral proteins, which lead to viral inactivation. As a result, they recommend that COVID-19 treatment clinical trials be conducted with FDA-approved IONPs.

Other Nanoparticles During the COVID-19 Crisis

Zhenhua et al. (Chen et al., 2020) discuss the fabrication of a quick and accurate lateral flow immunoassay (LFIA) that detects anti-SARV-CoV-2 IgG in human serum using lanthanide-doped polystyrene nanoparticles (LNPs). To collect particular IgG, a recombinant nucleocapsid phosphoprotein of SARS-CoV-2 was distributed over a nitrocellulose membrane. Mouse anti-human IgG antibody was utilized as a fluorescent reporter, and self-assembled LNPs were used to label it. The detection technique took ten minutes and used a 100-L aliquot of serum samples (1:1000 dilution). The validation trial's findings met the clinical diagnostic reagents' standards. By analyzing 51 normal samples, a cutoff value of 0.0666 was obtained. The group of same researchers (Zhao et al., 2021) designed glycyrrhizic acid (GA) nanoparticles (GANPs) built over GA that is extremely biocompatible. GANPs decrease the growth of murine coronavirus MHV-A59, thus lowering the generation of proinflammatory cytokines produced by MHV-A59 and SARS-CoV-2 N protein in vitro. GANPs exhibit anti-inflammatory and antiviral properties, reducing organ failure and giving sick mice a considerable survival benefit.

Lang rao et al. (Rao et al., 2020) used a potent two-stage neutralization technique to create a decoy nanoparticle towards COVID-19: virus neutralization in 1st phase, accompanied by cytokine neutralization in 2nd. The nano-decoy has almost the same antigenic exterior as source cells and is produced by mixing engineered cells that are stable in generating angiotensin-converting enzyme II (ACE2) receptors and cellular membrane nanovesicles derived from human monocytes genetically. These nanodecoys efficiently prevent host cells against infection by pseudoviruses and actual viruses like SARS-CoV-2 by competing for virus contact with host cells. The nanodecoys effectively attach and neutralize inflammatory

Figure 7. Diagrammatic representation of nanodecoys towards COVID-19
(Figure adapted from the source (Rao et al., 2020))

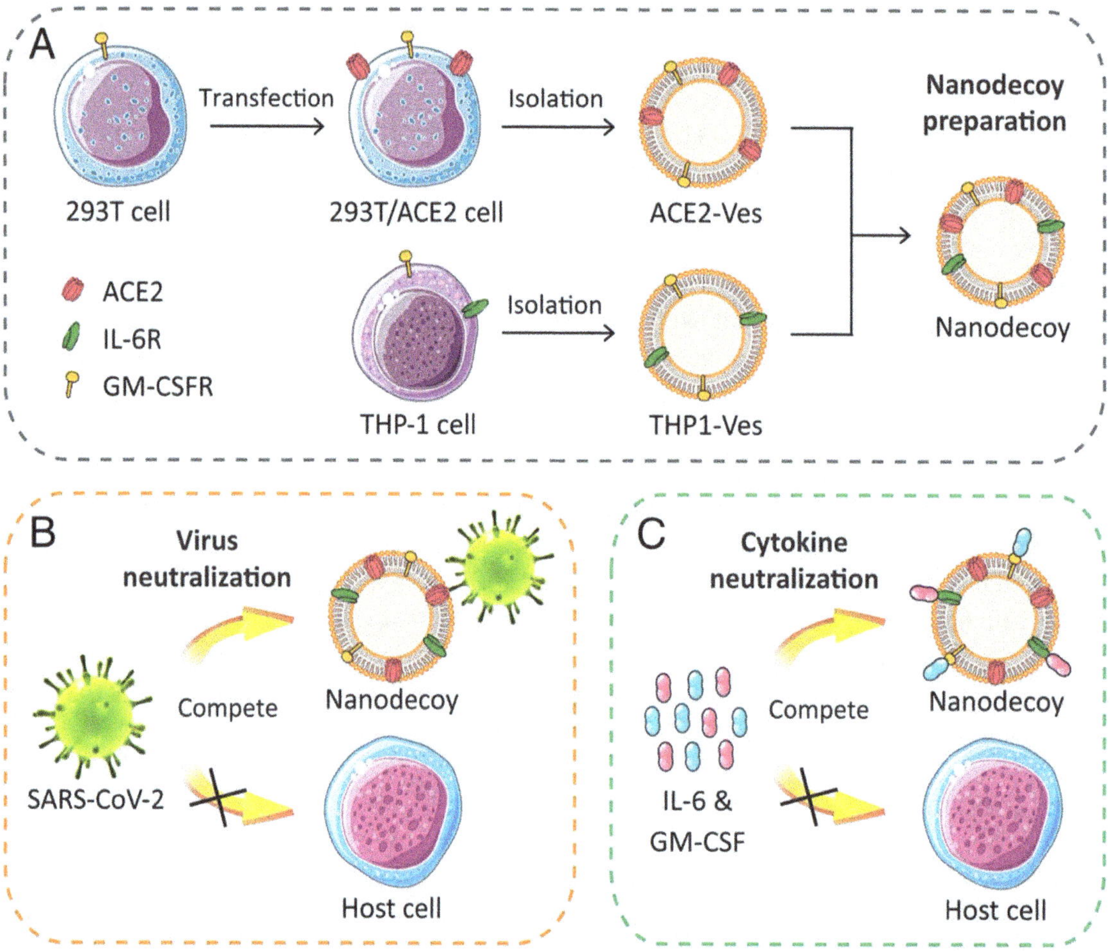

cytokines such as interleukin 6 (IL-6) and granulocyte-macrophage colony-stimulating factor (GM-CSF) due to the abundance of cytokine receptors on the surface, and considerably inhibit immune disability and lung damage in an acute pneumonia mouse model. Figure 7 shows the diagrammatic representation of nanodecoys towards COVID-19.

Titanium dioxide, or TiO_2, is also another metal oxide with unique properties and applications. It has antimicrobial, antiviral, and photocatalytic properties, and it may be useful in the fight against COVID-19. In this context, the photocatalytic activity of nano-TiO_2 was investigated for the deactivation of coronavirus using UV light in a study that might be used in convenience stores and healthcare. On glass coverslips, nano-TiO_2 particles were deposited, and the findings indicated good biocidal effectiveness. In this investigation, the potency of TiO_2-coated surfaces for disinfection of coronaviruses was confirmed. Because of its many advantages, including photocatalytic activity and high surface area, Nano-TiO_2 tends to be a promising metal oxide in order to develop antiviral photocatalysts. In a 0.4 $mWcm^{-2}$ irradiation, Zhong and colleagues investigated the effect of TiO_2 nanoparticles on pathogens like coronavirus. The experiments revealed that photo-activated TiO_2 may effectively inactivate human viral infections like

SARS-CoV-2. The pathogens were rendered inactive by reactive oxygen species (ROS), particularly hydroxyl radicals (-OH). This produced photocatalyst exhibited promise for use in medical devices, air filters, and surface coatings.

Nanomaterials for the Prevention of COVID-19

Vaccines

The production of vaccines for COVI-19 is indeed a challenging perspective that began after the whole sequence of the virus was revealed by a group of researchers working in collaboration with pharmaceutical companies. Several studies have identified the entire S protein or some of its portions as suitable targets for vaccine production although the antigen composition should improve to enhance the immune response up to some extent. Additionally, non-structural proteins and nucleoproteins have been discovered as intriguing candidates for the creation of a SARS-CoV-2 combo vaccine. While vaccine production, the antigenic component of the virus could be embedded in the nanocarrier/linked to the nanoparticle substrate for injection including adjuvan. Antigens are delivered through a variety of delivery mechanisms, including virosomes, polymeric nanoparticles, liposomes, emulsions, immune-stimulating complexes. Altering the shape, size, and charges of the NPs can improve the vaccine's efficacy. On November 18, 2020, BioNTech and Pfizer declared BNT162b2 to be 95 percent efficient. Similarly, Moderna presented their findings and declared that mRNA-1273 had a 94.5 percent efficacy rate. The Fusogenix DNA vaccine is a potential COVID-19 vaccine. Fusogenix is a proteo-lipid vehicle (PLV) that employs a revolutionary fusion technique to transport therapeutic nucleic acid payloads to target cells intact. It was created using Entos proprietary fusion-associated tiny trans-membrane proteins and well-tolerated neutral lipids (FAST proteins). Fusogenix DNA vaccine employs plasmid DNA, that incorporates numerous protein epitopes (antigen) from major immunogenic SARS-CoV-2 proteins, to offer a superior therapeutic dosage with optimum protection. Protein epitopes may contribute in the activation of the body's natural antibody production as well as the creation of a protective immune response against COVID-19. Furthermore, unlike traditional vaccines, DNA-based vaccine is free of infectious agents, has better durability, is easy to mass-produce, and is expected to boost both B and T-cell responses. Sputnik V, a COVID-19 vaccine, has been approved for quick use under pandemic emergency limits by the Russian Ministry of Health on August 11, 2020. It's a vaccine made up of SARS-CoV-2 S protein gene fragments and dual adenovirus vectors (rAd5 and rAd26). At the Gamaleya National Center, the employment of two different types of adenovirus carriers for the first and second immunization dosages is a unique method for ensuring long-lasting protection.

Metal Nanoparticle-Based Disinfecting Agents

Due to their antibacterial and antiviral capabilities, metal nanoparticles (MNPs) like silver, copper, and titanium have a wide range of uses in industries such as medicine, cosmetics, and disinfectant preparation. Viral infection occurs when the virus's nucleic acid is introduced into the host cell, preceded by viral replication. Metal NPs are thought to operate on the virus's surface, physically inhibiting the virus's contact with host cells. MNPs like Ag and Cu NPs, in contrast to alcohol-based disinfectants, are environmental, non-flammable, and manufactured using green sustainable themes. As a result, MNP-based disinfectants and sanitizers have improved formulation, safety, wellness features against COVID-19. As

Figure 8. Inhibitory conc. and cytotoxicity of Zno-NPs against SARS-CoV-2
(Figure adapted from the source(El-Megharbel et al., 2021))

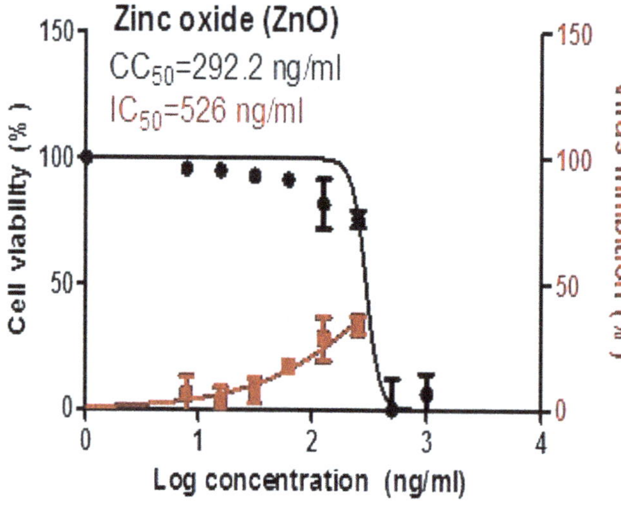

a disinfectant Nano-spray for spreading COVID 19, Samy et al. (El-Megharbel et al., 2021) created zinc (II) oxide nanoparticles (ZnO-NPs) (SARS-CoV-2). Various spectroscopic methodologies (UV, FT-IR) and morphological techniques were used to completely chemically analyze the produced (ZnO) nanomaterial. The antiviral efficacy of the synthesized ZnO-NPs towards (SARS-CoV-2) was determined by utilizing a cytotoxicity assay (CC50) and inhibitory concentrations in 10 percent DMSO and ddH_2O (IC50). The data showed that (ZnO-NPs) possess a considerable anti-SARS-CoV-2 capacity in vitro at cytotoxic quantities having a non-significant selective index (IC50/CC50) and a non-significant selectivity index (CC50/IC50 1). At a level of 526 ng/mL, ZnO-NPs displayed strong antiviral efficacy against SARS-CoV-2. The MTT assay was performed to assess ZnO-NPs' potential cytotoxicity in cells. The lifespan of VERO-E6 cells was dramatically lowered by 292.2 ng/mL ZnO-NPs as shown in figure 8.

SHEPROS SDN BHD has also developed a non-foaming and ecologically friendly nano silver-based multifunctional sanitizer that employs a nano colloidal technique and has antimicrobial, antiviral and antifungal properties. The Italian company NanoTechSurface has developed a disinfection product that uses titanium dioxide and silver ions to treat coronavirus-infected surfaces. By employing biodegradable nanoparticles like lipids or polymers, the negative impacts of several MNPs on human wellness and the ecosystem might be avoided. According to a report posted in Nature India, Weinnovate Biosolutions (Pune) has evolved silver NPs solution-based sanitizers and disinfectants with funding from the Indian government's Department of Science and Technology (DST) and Department of Biotechnology (DBT).

Protective Gear

Nanomaterials having traditionally been known to be the key to producing protective clothing such as lab coats, masks and aprons, as evidenced by huge increasing demand in the COVID-19 pandemic. Due to the incorporation of nanoparticles, the physicochemical qualities of fabrics utilized in the manufacture of PPE have been enhanced. UV protection, antibacterial properties, and heat resistance are just a few of the amplifications available. Nanoengineering has a key role in the fabrication of personal protective equipment (PPE) because it prevents antibacterial activity and offers hydrophobicity. Considering these advantages, the fact that nanoparticles have had no effect on the fabric's breathability is quite remarkable. Because SARS-COV-2 is mostly transmitted via respiratory secretions, hydrophobic nature contributes directly to the perceived enhancement of personal protective equipment (PPE). Surface tension is increased by a cluster of hydrocarbon Nano whiskers that prevents droplets from being absorbed. Microorganisms are prevented from proliferating on the surface of microbial membranes when nanoscale biocides such as peptides or polymers are used to oxidize them. Face masks are an amazing illustration of the life of nanoparticles in PPE. Face masks traditionally have gaps between the fibers that are inadequate for virus prevention. Despite the fact that N95 masks are touted as the most effective to date, they have only efficacy of 85 percent for particles <300 nm. So, Because the COVID-19 virus has a diameter of 60–140 nanometers, further effective masks are required. Electrospinning produces nanofibers, which could be a solution to this problem. Furthermore, gloves imbued with silver nanoparticles have antiviral properties. This field has a huge amount of potential, and there are some exciting advancements in the future.

Masks and PPE Kits

Covering the face with a mask is among the most effective COVID-19 prevention strategies. Face protection is required for both sick and non-infected people. When a mouth is hidden by the mask. A virus-infected person could reduce the risks of the virus being spread through sneezing or coughing, and a virus-uninfected person can protect themselves from the virus in the air as drops or aerosol, surfaces. The use of masks is required for touch-based or air-borne transmission of the virus. N95 masks are regularly used face masks which filter up to 95 percent of fine pollutants with a diameter of at minimum 300 nm. However, any type of face mask can efficiently protect a person from becoming infected, the mask itself gets fomite because the virus has the ability to adhere to the fabric surface. Protracted usage of masks can cause breathing problems and a decrease in oxygen concentration in the body.

Nanofibrous Membrane-Based Masks and PPE Kits

Nanofibrous membranes with a large specific surface area are formed by a dense web-like network of nanofibres. Such nanofibrous membranes are used in breathing masks, allowing for better breathing and purification. The fabrication of a biodegradable mask based on nanocellulose nanofibres that can shield against pollutants as small as 100 nanometers. Nanocellulose nanofibres were derived from discarded plant material and agricultural waste such as sugar cane residue. The Amrita Institute for Nanosciences and Molecular Medicine revealed that they have developed reusable masks that are cost-effective thanks to nanofibrous technology. Two biodegradable and biocompatible textile layers make up these nanofibre-based masks. Face masks fabricated of nanococo-carbon fibers having an ultra-thin membrane < 1 nm could keep up to 99.98 percent of harmful air-borne pollutants out of the body have been developed by Metamasks. Nanopoli, a Korean firm, claims to have produced a nanofibre mask that filters air at a rate of 98.75 percent.

Graphene and Its Derivative-Based Masks and PPE Kits

Graphene is mainly a 2D hexagonal organization of single C atoms, while graphene oxide (GO) is the oxidized form. By withdrawing O_2 groups by using reducing agents, GO can be converted to reduced

GO (rGO). These substances are single-layer materials with a high absorption capacity for incident visible light and a large specific surface area and that have been successfully employed for sterilizing and high-temperature generation. Due to their superhydrophobicity and improved light absorption qualities, graphene and its derivatives can be utilized to make masks, PPE, and other medical devices. Directa Plus PLC claims to have developed masks based on the microbe-static, antimicrobial, and non-toxic characteristics of graphene and its derivatives utilizing nanotechnology. LIGC Applications has also created the "Guardian G-Volt" face mask, which is made of laser-induced microporous graphene. It effectively prevents 99 percent of contaminants with a diameter bigger than 0.3 m by using an electrical current that destroys entrapped germs within the mask's filter.

Nanocomposite-Based Masks and PPE Kits

It's a composite that has a single component dimension on a nanometric scale (less than 100 nm). Phase-separated systems, intercalated systems, and exfoliated systems are three types of nanocomposites which can be categorized depending on their structure. Masks and PPE have also been claimed to be made from a variety of nanocomposite materials. Copper3D claims to have developed the "Nanohack," an antibiotic and antibacterial face mask with 3 filtration and cleaning sheets and 5% copper oxide nanoparticles utilized to make a non-woven polypropylene coating. ZEN Graphene Solutions Ltd has developed a graphene oxide ink with silver nanoparticles. This viricidal graphene-based composite ink could have been utilized in PPE and N95 surgical masks for improved COVID-19 prevention. Respilon Group, a Czech nanofibre technology company, has created masks with a 99.9% filtration effectiveness relying on a three nanofibre membrane. Copper dioxide (CuO) NPs encapsulated in a nanofibre matrix were used as the materials.

FUTURE RESEARCH DIRECTIONS

In assertion, nanoparticles are an effective public management process for COVID-19 transmission. Metallic nanoparticles are a cost-effective and long-term solution for controlling any pandemic, including COVID-19. A multifunctional metallic antibacterial covering can limit the transmission of any devastating disease in society. These broad-spectrum NPs will also prevent the proliferation of microbes that help COVID-19 spread. This procedure will be a novel treatment alternative in the event of future pandemic revival and emergence, which the world might confront ahead. For decades, nanoparticles have been investigated and utilized in the manufacture of effective medications, earlier illness diagnostics, disease prevention, and tailored nano-medicines. As a result, this approach may become the first course of action for obtaining optimal treatments for a variety of ailments in the future. Each technology has its own set of obstacles; every nanotechnology application has its own set of limits. Because nanoparticles decay slowly, there's a chance they'll build up at the injection site, causing aggravation and organ failure. Whenever non-biodegradable nanoparticles are utilized, this issue arises. Biodegradable nanoparticles, which are normally expelled from the body, can be used to solve the problem. Moreover, nanoparticle synthesis is time-consuming and costly. This analysis of newly discovered nanotechnology-based SARS-CoV-2 medications and sensors offers out a roadmap for the production of more efficient sensors and disinfectants for the identification and prevention of this and other coronaviruses. To tackle the problems on the front lines of patient care, more advancement in nano-based techniques is required.

CONCLUSION

The evolution of nanoscale technologies has facilitated growth in various industries since the revelation of nanotechnology. The chapter highlights the substantial array of nanotechnology research and development being done to prevent the COVID-19 outbreak. The globe is rapidly evolving in order to effectively combat the novel virus, but worldwide dedication to research and development in this field is challenging. The COVID-19 pandemic is not the first epidemic and it's not expected to be last. Adequate detailed information on drug nanotechnology significantly facilitated the manufacture of controlled release nanosystems which can effectively target the proper system by contributing to the quick and potential treatment of COVID-19 with the production of PPE and face mask. Prevent SARSCoV2 infection, in this regard, we hope to fight COVID and bring people around the world to recover from daily life. Now a days, by using different spectroscopic tools to carry out the comprehensive chemical characterization of ZnO-NPs, ZnO-NPs show substantial antiviral action against SARS-CoV-2 at low concentrations and can produce a large number of free radicals, causing oxidative stress in SARS-CoV-2 and significant disruption to its cellular membranes, that is an essential unit in the combat over SARS-CoV-2. Recent research has combined specimen procedures into such a unified biosensor that can detect COVID-19 infections by identifying pathogen human antibodies and nucleic acids to the pathogens. Furthermore, recent research has aimed to increase biosensor detection sensitivity, accessibility, and efficiency. Enzyme-based signal amplification, sample concentration, and considerably simpler fluidic-control techniques can all improve assay sensitivity.

REFERENCES

Abo-zeid, Y., Ismail, N. S. M., Mclean, G. R., & Hamdy, N. M. (2020). Since January 2020 Elsevier has created a COVID-19 resource centre with free information in English and Mandarin on the novel coronavirus COVID- 19. The COVID-19 resource centre is hosted on Elsevier Connect, the company's public news and information. *European Journal of Pharmaceutical Sciences*, *153*(1), 105465. doi:10.1016/j.ejps.2020.105465 PMID:32668312

Allawadhi, P., Singh, V., Khurana, A., Khurana, I., Allwadhi, S., Kumar, P., Banothu, A. K., Thalugula, S., Barani, P. J., Naik, R. R., & Bharani, K. K. (2021). Silver nanoparticle based multifunctional approach for combating COVID-19. *Sensors International*, *2*(May), 100101. doi:10.1016/j.sintl.2021.100101 PMID:34766057

Almanza-Reyes, H., Moreno, S., Plascencia-Lopez, I., Alvarado-Vera, M., Patron-Romero, L., Borrego, B., . . . Bogdanchikova, N. (2021). Evaluation of silver nanoparticles for the prevention of SARS-CoV-2 infection in health workers in vitro and in vivo. MedRxiv, 1–25. https://medrxiv.org/cgi/content/short/2021.05.20.21256197

Alper Öztürk, A., Namll, I., & Aygül, A. (2021). Cefaclor Monohydrate-Loaded Colon-Targeted Nanoparticles for Use in COVID-19 Dependent Coinfections and Intestinal Symptoms: Formulation, Characterization, Release Kinetics, and Antimicrobial Activity. *Assay and Drug Development Technologies*, *19*(3), 156–175. doi:10.1089/adt.2020.1014 PMID:33728979

Carvalho, A. P. A., & Conte-Junior, C. A. (2021). Recent Advances on Nanomaterials to COVID-19 Management: A Systematic Review on Antiviral/Virucidal Agents and Mechanisms of SARS-CoV-2 Inhibition/Inactivation. *Global Challenges (Hoboken, NJ)*, *5*(5), 2000115. doi:10.1002/gch2.202000115 PMID:33786199

Cavalcanti, I. D. L., & Cajubá de Britto Lira Nogueira, M. (2020). Pharmaceutical nanotechnology: Which products are been designed against COVID-19? *Journal of Nanoparticle Research*, *22*(9), 276. Advance online publication. doi:10.100711051-020-05010-6 PMID:32922162

Chakraborty, D., Kumar, S., Chandrasekaran, N., & Mukherjee, A. (2020). Viral Diagnostics and Preventive Techniques in the Era of COVID-19: Role of Nanoparticles. *Frontiers in Nanotechnology*, *2*(October), 588795. Advance online publication. doi:10.3389/fnano.2020.588795

Chaudhary, V., Royal, A., Chavali, M., & Yadav, S. K. (2021). Advancements in research and development to combat COVID-19 using nanotechnology. *Nanotechnology for Environmental Engineering*, *6*(1), 1–15. doi:10.100741204-021-00102-7

Chauhan, G., Madou, M. J., Kalra, S., Chopra, V., Ghosh, D., & Martinez-Chapa, S. O. (2020). Nanotechnology for COVID-19: Therapeutics and Vaccine Research. *ACS Nano*, *14*(7), 7760–7782. doi:10.1021/acsnano.0c04006 PMID:32571007

Chen, Z., Zhang, Z., Zhai, X., Li, Y., Lin, L., Zhao, H., Bian, L., Li, P., Yu, L., Wu, Y., & Lin, G. (2020). Rapid and Sensitive Detection of anti-SARS-CoV-2 IgG, Using Lanthanide-Doped Nanoparticles-Based Lateral Flow Immunoassay. *Analytical Chemistry*, *92*(10), 7226–7231. doi:10.1021/acs.analchem.0c00784 PMID:32323974

Chintagunta, A. D., Sai Krishna, M., Nalluru, S., & Sampath Kumar, N. S. (2021). Nanotechnology: An emerging approach to combat COVID-19. *Emergent Materials*, *4*(1), 119–130. doi:10.100742247-021-00178-6 PMID:33615141

Choi, J. R. (2020). Development of Point-of-Care Biosensors for COVID-19. *Frontiers in Chemistry*, *8*(May), 517. Advance online publication. doi:10.3389/fchem.2020.00517 PMID:32574316

Cohen, A. A., Gnanapragasam, P. N. P., Lee, Y. E., Hoffman, P. R., Ou, S., Kakutani, L. M., Keeffe, J. R., Wu, H. J., Howarth, M., West, A. P., Barnes, C. O., Nussenzweig, M. C., & Bjorkman, P. J. (2021). Mosaic nanoparticles elicit cross-reactive immune responses to zoonotic coronaviruses in mice. *Science*, *371*(6530), 735–741. doi:10.1126cience.abf6840 PMID:33436524

Dube, T., Ghosh, A., Mishra, J., Kompella, U. B., & Panda, J. J. (2021). Repurposed Drugs, Molecular Vaccines, Immune-Modulators, and Nanotherapeutics to Treat and Prevent COVID-19 Associated with SARS-CoV-2, a Deadly Nanovector. *Advanced Therapeutics*, *4*(2), 1–30. doi:10.1002/adtp.202000172 PMID:33173808

El-Megharbel, S. M., Alsawat, M., Al-Salmi, F. A., & Hamza, R. Z. (2021). Utilizing of (Zinc oxide nano-spray) for disinfection against "sars-cov-2" and testing its biological effectiveness on some biochemical parameters during (covid-19 pandemic)—"zno nanoparticles have antiviral activity against (sars-cov-2). *Coatings*, *11*(4), 388. Advance online publication. doi:10.3390/coatings11040388

Heidari, A., Caissutti, A., Henderson, M., Schmitt, K., Besana, E., Esposito, J., & Peterson, V. (2020). Recent New Results and Achievements of California South University (CSU) BioSpectroscopy Core Research Laboratory for COVID-19 or 2019-nCoV Treatment: Diagnosis and Treatment Methodologies of "Coronavirus.". *Journal of Current Viruses and Treatment Methodologies*, *1*(1), 3–41. doi:10.14302/issn.2691-8862.jvat-20-3275

Ibrahim Fouad, G. (2021). A proposed insight into the anti-viral potential of metallic nanoparticles against novel coronavirus disease-19 (COVID-19). *Bulletin of the National Research Center*, *45*(1), 36. Advance online publication. doi:10.118642269-021-00487-0 PMID:33564223

Jardón-Maximino, N., Cadenas-Pliego, G., Ávila-Orta, C. A., Comparán-Padilla, V. E., Lugo-Uribe, L. E., Pérez-Alvarez, M., Tavizón, S. F., & Dejesús Sosa Santillán, G. (2021). Antimicrobial property of polypropylene composites and functionalized copper nanoparticles. *Polymers*, *13*(11), 1694. Advance online publication. doi:10.3390/polym13111694 PMID:34067323

Jeremiah, S. S., Miyakawa, K., Morita, T., & Yamaoka, Y. (2020). Since January 2020 Elsevier has created a COVID-19 resource centre with free information in English and Mandarin on the novel coronavirus COVID- 19. The COVID-19 resource centre is hosted on Elsevier Connect, the company ' s public news and information. *Biochemical and Biophysical Research Communications*, *533*(1), 195–200. doi:10.1016/j.bbrc.2020.09.018 PMID:32958250

Kabir, M. A., Ahmed, R., Iqbal, S. M. A., Chowdhury, R., Paulmurugan, R., Demirci, U., & Asghar, W. (2021). Diagnosis for COVID-19: Current status and future prospects. *Expert Review of Molecular Diagnostics*, *21*(3), 269–288. doi:10.1080/14737159.2021.1894930 PMID:33621145

Kelleni, M. T. (2021). Resveratrol-zinc nanoparticles or pterostilbene-zinc: Potential COVID-19 mono and adjuvant therapy. *Biomedicine and Pharmacotherapy*, *139*, 111626. doi:10.1016/j.biopha.2021.111626 PMID:33894625

Kirpotin, D. B., Drummond, D. C., Shao, Y., Shalaby, M. R., Hong, K., Nielsen, U. B., Marks, J. D., Benz, C. C., & Park, J. W. (2006). Antibody targeting of long-circulating lipidic nanoparticles does not increase tumor localization but does increase internalization in animal models. *Cancer Research*, *66*(13), 6732–6740. doi:10.1158/0008-5472.CAN-05-4199 PMID:16818648

Kleine-Tebbe, J., Klimek, L., Hamelmann, E., Pfaar, O., Taube, C., Wagenmann, M., Werfel, T., & Worm, M. (2021). Severe allergic reactions to the COVID-19 vaccine – statement and practical consequences. *Allergologie Select*, *5*(01), 26–28. doi:10.5414/ALX02215E PMID:33426427

Lew, T. T. S., Aung, K. M. M., Ow, S. Y., Amrun, S. N., Sutarlie, L., Ng, L. F. P., & Su, X. (2021). Epitope-Functionalized Gold Nanoparticles for Rapid and Selective Detection of SARS-CoV-2 IgG Antibodies. *ACS Nano*, *15*(7), 12286–12297. Advance online publication. doi:10.1021/acsnano.1c04091 PMID:34133128

Li, S., Jiang, W., Huang, J., Liu, Y., Ren, L., Zhuang, L., Zheng, Q., Wang, M., Yang, R., Zeng, Y., & Wang, Y. (2020). Highly sensitive and specific diagnosis of COVID-19 by reverse transcription multiple cross-displacement amplification-labelled nanoparticles biosensor. *The European Respiratory Journal*, *56*(6), 2002060. Advance online publication. doi:10.1183/13993003.02060-2020 PMID:32859676

Mehranfar, A., & Izadyar, M. (2020). Theoretical Design of Functionalized Gold Nanoparticles as Antiviral Agents against Severe Acute Respiratory Syndrome Coronavirus 2 (SARS-CoV-2). *The Journal of Physical Chemistry Letters*, *11*(24), 10284–10289. doi:10.1021/acs.jpclett.0c02677 PMID:33226815

Mitrani, M. I., Bellio, M. A., Sagel, A., Saylor, M., Kapp, W., VanOsdol, K., Haskell, G., Stewart, D., Abdullah, Z., Santos, I., Milberg, J., Arango, A., Mitrani, A., & Shapiro, G. C. (2021). Case Report: Administration of Amniotic Fluid-Derived Nanoparticles in Three Severely Ill COVID-19 Patients. *Frontiers in Medicine*, *8*(March), 1–8. doi:10.3389/fmed.2021.583842 PMID:33816515

Moitra, P., Alafeef, M., Alafeef, M., Alafeef, M., Dighe, K., Frieman, M. B., Pan, D., Pan, D., & Pan, D. (2020). Selective Naked-Eye Detection of SARS-CoV-2 Mediated by N Gene Targeted Antisense Oligonucleotide Capped Plasmonic Nanoparticles. *ACS Nano*, *14*(6), 7617–7627. doi:10.1021/acsnano.0c03822 PMID:32437124

Poggio, C., Colombo, M., Arciola, C. R., Greggi, T., Scribante, A., & Dagna, A. (2020). Copper-alloy surfaces and cleaning regimens against the spread of SARS-CoV-2 in dentistry and orthopedics. From fomites to anti-infective nanocoatings. *Materials (Basel)*, *13*(15), 1–12. doi:10.3390/ma13153244 PMID:32707757

Prajapati, P. K., Yadav, P., & Gauri, A. (2020). Possible Potential of Tamra Bhasma (Calcined Copper) in COVID-19 Management. *Journal of Research in Ayurvedic Sciences*, *4*(3), 113–120. doi:10.5005/jras-10064-0111

Raha, S., Mallick, R., Basak, S., & Duttaroy, A. K. (2020). Is copper beneficial for COVID-19 patients? *Medical Hypotheses*, *142*(April), 109814. doi:10.1016/j.mehy.2020.109814 PMID:32388476

Rai, P. K., Usmani, Z., Thakur, V. K., Gupta, V. K., & Mishra, Y. K. (2020). Tackling COVID-19 pandemic through nanocoatings: Confront and exactitude. *Current Research in Green and Sustainable Chemistry*, *3*(July), 100011. doi:10.1016/j.crgsc.2020.100011

Rao, L., Xia, S., Xu, W., Tian, R., Yu, G., Gu, C., Pan, P., Meng, Q. F., Cai, X., Qu, D., Lu, L., Xie, Y., Jiang, S., & Chen, X. (2020). Decoy nanoparticles protect against COVID-19 by concurrently adsorbing viruses and inflammatory cytokines. *Proceedings of the National Academy of Sciences of the United States of America*, *117*(44), 27141–27147. doi:10.1073/pnas.2014352117 PMID:33024017

Ruiz-Hitzky, E., Darder, M., Wicklein, B., Ruiz-Garcia, C., Martín-Sampedro, R., del Real, G., & Aranda, P. (2020). Nanotechnology Responses to COVID-19. *Advanced Healthcare Materials*, *9*(19), 1–26. doi:10.1002/adhm.202000979 PMID:32885616

Spigaglia, P. (2020). COVID-19 outbreak: Migration, effects on society, global environment and prevention. *Science of the Total Environment Journal*, *728*(January), 138882. PMID:32335410

Tortella, G. R., Rubilar, O., Diez, M. C., Padrão, J., Zille, A., Pieretti, J. C., & Seabra, A. B. (2021). Advanced Material Against Human (Including Covid-19) and Plant Viruses: Nanoparticles As a Feasible Strategy. *Global Challenges (Hoboken, NJ)*, *5*(3), 2000049. doi:10.1002/gch2.202000049 PMID:33614127

Vahedifard, F., & Chakravarthy, K. (2021). *Nanomedicine for COVID-19: the role of nanotechnology in the treatment and diagnosis of COVID-19*. Emergent Materials. doi:10.100742247-021-00168-8

Vazquez-Munoz, R., & Lopez-Ribot, J. L. (2020). Nanotechnology as an Alternative to Reduce the Spread of COVID-19. *Challenges*, *11*(2), 15. doi:10.3390/challe11020015

Vivekanandhan, K., Shanmugam, P., Barabadi, H., Arumugam, V., Daniel Raj Daniel Paul Raj, D., Sivasubramanian, M., Ramasamy, S., Anand, K., Boomi, P., Chandrasekaran, B., Arokiyaraj, S., & Saravanan, M. (2021). Emerging Therapeutic Approaches to Combat COVID-19: Present Status and Future Perspectives. *Frontiers in Molecular Biosciences*, *8*(March), 1–10. doi:10.3389/fmolb.2021.604447 PMID:33763450

Zachar, O. (2020). Formulations for COVID-19 Early Stage Treatment via Silver Nanoparticles Inhalation Delivery at Home and Hospital. *ScienceOpen* Preprints, 1–14. doi:10.14293/S2199-1006.1.SOR-.PPHBJEO.v1

Zhao, Z., Xiao, Y., Xu, L., Liu, Y., Jiang, G., Wang, W., Li, B., Zhu, T., Tan, Q., Tang, L., Zhou, H., Huang, X., & Shan, H. (2021). Glycyrrhizic Acid Nanoparticles as Antiviral and Anti-inflammatory Agents for COVID-19 Treatment. *ACS Applied Materials & Interfaces*, *13*(18), 20995–21006. doi:10.1021/acsami.1c02755 PMID:33930273

KEY TERMS AND DEFINITIONS

COVAXIN®: COVAXIN®, India's indigenous COVID-19 vaccine by Bharat Biotech is developed in collaboration with the Indian Council of Medical Research (ICMR) - National Institute of Virology (NIV). The indigenous, inactivated vaccine is developed and manufactured in Bharat Biotech's BSL-3 (Bio-Safety Level 3) high containment facility. The vaccine is developed using Whole-Virion Inactivated Vero Cell derived platform technology. Inactivated vaccines do not replicate and are therefore unlikely to revert and cause pathological effects. They contain dead virus, incapable of infecting people but still able to instruct the immune system to mount a defensive reaction against an infection.

COVID-19: Coronavirus disease (COVID-19) is an infectious disease caused by the SARS-CoV-2 virus. Most people infected with the virus will experience mild to moderate respiratory illness and recover without requiring special treatment. However, some will become seriously ill and require medical attention. Older people and those with underlying medical conditions like cardiovascular disease, diabetes, chronic respiratory disease, or cancer are more likely to develop serious illness. Anyone can get sick with COVID-19 and become seriously ill or die at any age.

Covishield: Serum Institute of India's COVID-19 vaccine, called Covishield, is a version of the Oxford-AstraZeneca vaccine that manufacturers in India produce locally. According to AstraZeneca's primary analysis of phase 3 trial data, the vaccine has a 76% efficacy rate after both doses.

DNA: DNA, or deoxyribonucleic acid, is the hereditary material in humans and almost all other organisms. Nearly every cell in a person's body has the same DNA. The information in DNA is stored as a code made up of four chemical bases: adenine (A), guanine (G), cytosine (C), and thymine (T). Human DNA consists of about 3 billion bases, and more than 99 percent of those bases are the same in all people. The order, or sequence, of these bases determines the information available for building and maintaining an organism, similar to the way in which letters of the alphabet appear in a certain order to form words and sentences.

MERS: Middle East respiratory syndrome (MERS) is a viral respiratory disease caused by a novel coronavirus (Middle East respiratory syndrome coronavirus, or MERS-CoV) that was first identified in Saudi Arabia in 2012.

Nanoparticles: A nanoparticle is a small particle that ranges between 1 to 100 nanometers in size. Undetectable by the human eye, nanoparticles can exhibit significantly different physical and chemical properties to their larger material counterparts.

PPE: Personal protective equipment, commonly referred to as "PPE", is equipment worn to minimize exposure to hazards that cause serious workplace injuries and illnesses. These injuries and illnesses may result from contact with chemical, radiological, physical, electrical, mechanical, or other workplace hazards. Personal protective equipment may include items such as gloves, safety glasses and shoes, earplugs or muffs, hard hats, respirators, or coveralls, vests and full body suits.

RNA: Ribonucleic acid (RNA) is a molecule similar to DNA. Unlike DNA, RNA is single-stranded. An RNA strand has a backbone made of alternating sugar (ribose) and phosphate groups. Attached to each sugar is one of four bases--adenine (A), uracil (U), cytosine (C), or guanine (G). Different types of RNA exist in the cell: messenger RNA (mRNA), ribosomal RNA (rRNA), and transfer RNA (tRNA). More recently, some small RNAs have been found to be involved in regulating gene expression.

RT-PCR: Real time RT–PCR is a nuclear-derived method for detecting the presence of specific genetic material in any pathogen, including a virus. Originally, the method used radioactive isotope markers to detect targeted genetic materials, but subsequent refining has led to the replacement of isotopic labelling with special markers, most frequently fluorescent dyes. This technique allows scientists to see the results almost immediately while the process is still ongoing, whereas conventional RT–PCR only provides results at the end of the process.

SARS: Severe acute respiratory syndrome (SARS) is a viral respiratory disease caused by a SARS-associated coronavirus. It was first identified at the end of February 2003 during an outbreak that emerged in China and spread to 4 other countries. SARS is an airborne virus and can spread through small droplets of saliva in a similar way to the cold and influenza. It was the first severe and readily transmissible new disease to emerge in the 21st century and showed a clear capacity to spread along the routes of international air travel.

WHO: The World Health Organization (WHO) is a specialized agency of the United Nations responsible for international public health. The WHO Constitution states its main objective as "the attainment by all peoples of the highest possible level of health". Headquartered in Geneva, Switzerland, it has six regional offices and 150 field offices worldwide.

APPENDIX

Mechanisms of Action of Antiviral NPs

Clinical experiments have revealed that interactions between viruses and nanoparticles result in effective directly or indirectly antiviral effects. NPs showing indirect action don't suppress virus on their own; rather, they promote the activity of antiviral medicines and are employed for a variety of purposes, including transport, stabilization, and increased bioavailability. Furthermore, nanomaterials can elicit an immunological response, resulting in adequate immunity. NPs showing direct action operate as an active element by inactivating viruses on their own, most commonly by modifying the viral structural or genetic code. Inhibitors of viral protein synthesis, viral RNA/DNA synthesis, and permeation into cells are some of the primary mechanisms Figure 9 depicts the pathway by antiviral nanoparticles to devastate viruses.

Blockage of Virus Attachment

When the virus attaches to a receptor on the cell surface, it penetrates the host cell. In SARS-CoV-2 infection, the S protein is required for cell attachment and penetration. There are two subunits in this protein: S1 and S2. S1 aids in adhesion, while S2 aids in membrane fusion and cell penetration. In alveoli, the S1 protein interacts with the human ACE2 receptor. ACE2 is expressed by epithelial cells, esophageal cells, enterocytes, in addition to these cells. Virus attachment can also be inhibited directly by nanoparticles. In particular, Ag NPs prevent respiratory viruses from entering cells. Ag NPs combined with graphene oxide are efficient against the feline coronavirus and certain other contained viruses, whereas graphene prevents virus adhesion in HIV patients. Researchers believe that graphene, in addition to its antiviral capabilities; will be critical in combating COVID-19. Au NPs are preferred over Ag NPs because of their less toxicity. Because of their bioactivity, Si NPs are the most accessible of the synthetic nanoparticles. These biodegradable materials disintegrate to generate non-toxic chemicals. Si NPs stop virus particulates from invading host cells. Mesoporous-SiO_2 nanoparticles having the ability to form hydrophobic or hydrophilic interactions with enveloped viruses, reducing viral entrance into cells. The cationic chitosan, a non-toxic polysaccharide, binds with S protein of SARS-CoV-2 and prevents spike protein from binding to the ACE2 receptor. As a result, nanoparticles can both prevent viral adhesion to the host and serve as medicine delivery vectors. Figure 10 depicts the synthesis of nanoparticles and their modes of action against coronaviruses.

Inhibiting Viral Multiplication

After entering cells, the virus reproduces and multiplies via the host cell's mechanisms, and transmits the infection in the body. The viral concentration might be at low conc. to infect the whole body if viral multiplication is inhibited, and the immune system will be able to combat the virus effectively. Antivirals, such as nanoparticles, are used to prevent viral reproduction. The cell death induced by viral infection is reduced by silver nanoparticles. Ag2S are possible nanoparticles that inhibit coronavirus proliferation by blocking viral particles from emerging from the host cell. Such nanoclusters significantly boost the production of pro-inflammatory cytokines, which aid in viral infection mitigation. The porcine epidemic diarrhea virus (PEDV) was taken as a model because of its structural closeness to the COVID-19 family.

Figure 9. The pathway by an antiviral nanoparticle to devastate viruses.
(Figure adapted from source (Vazquez-Munoz & Lopez-Ribot, 2020))

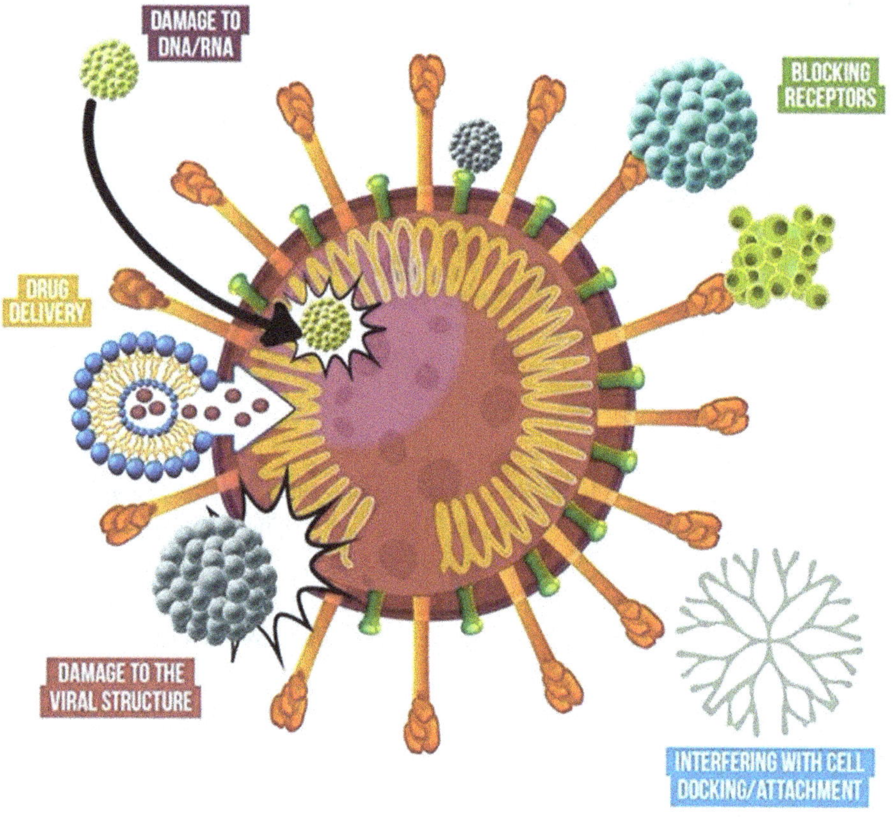

PEDV is suppressed when glutathione-capped Ag2S nanoclusters are used. Inhibition of RNA synthesis is the rationale underpinning this technique. Other viruses which are comparable to SARS-CoV-2 are used as experiments as they are comparable to SARS-CoV-2. Furthermore, ZnO_2 NPs are much effective at preventing H1N1 virus infection. Zinc (Zn) has also been shown to be capable of reducing SARS-CoV-2 replication. Zn aids in the production of antiviral cytokines and boosts the immune system's response, reducing the inflammatory response. The most critical SARS-CoV-2 protein, RNA-dependent RNA polymerase, aids in preserving genome correctness. The viral replication could be inhibited by blocking this protein. Previously, the same method of blocking the HIV reverse transcriptase was utilized to treat the infection. This method consists of at least two distinct aptamer gold NPs, the first one is polymerase and another is HIV reverse transcriptase RNaseH. Ag and Au NPs are more effective and appropriate nanostructures to suppress virus proliferation.

Inactivation of Virus

Another way to prevent a viral infection is to inactivate or destroy the virus. 3-mercaptoethylsulfonate (MES) coated on AuNPs have been proven to effectively suppress the virus at concentrations equal to the EC90. Its viral infectivity was regained after dilution, which is known as reversible viral inhibition.

Figure 10. The synthesis of nanoparticles and their modes of action against coronaviruses
(Figure adapted from the source(Carvalho & Conte-Junior, 2021))

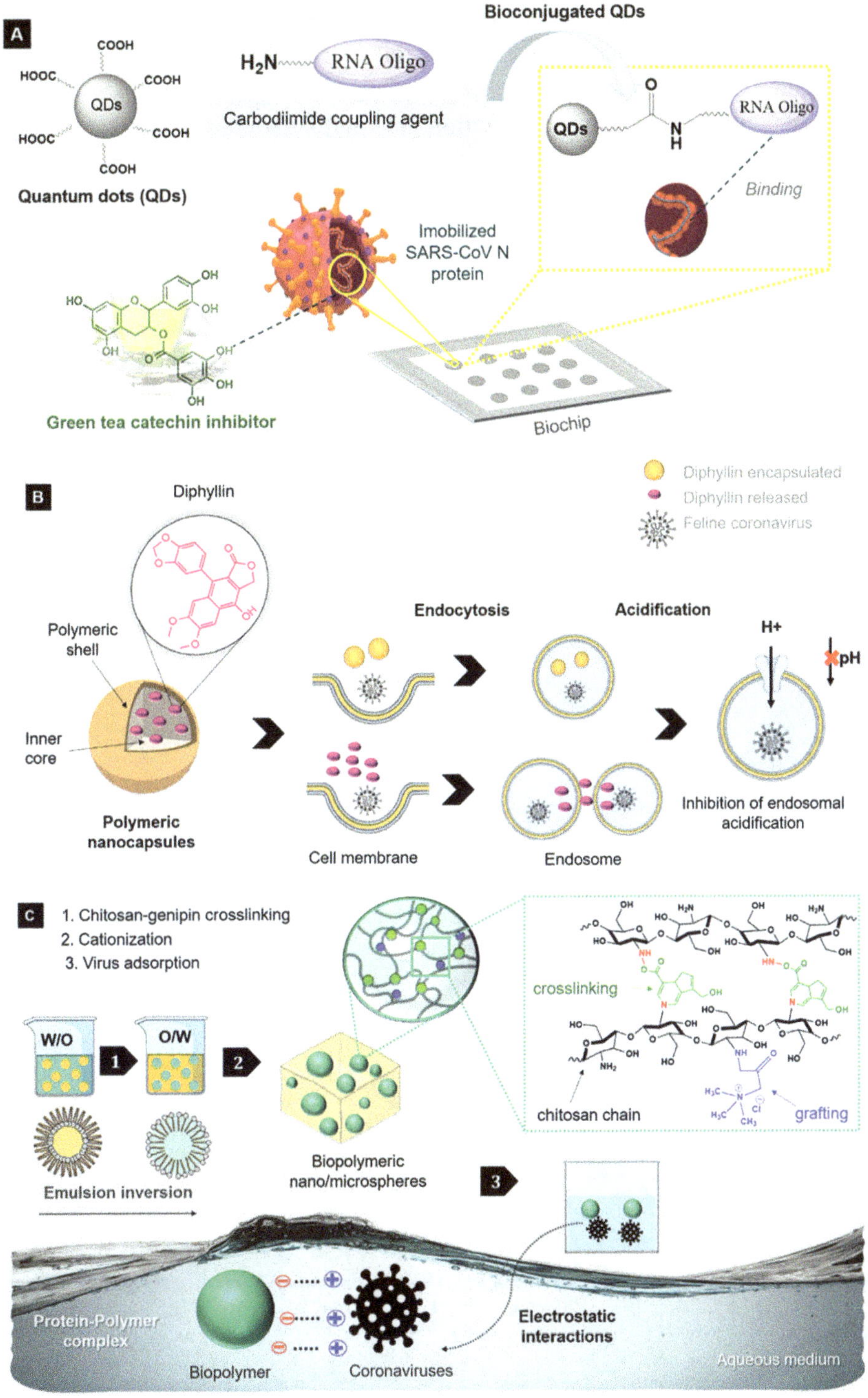

When a 2:1 combination of 1-octanethinol (OT) and undecanesulfonic acid (MUS) is utilized rather than

MES, MUS:OT-AuNPs is utilized for irreversible inactivation of viruses including vesicular stomatitis virus (VSV), herpes simplex virus (HSV) and lentivirus. Electron microscopy as well as other approaches confirmed the strong interaction between virus-MUS:OT-AuNP. Because the viruses examined using these techniques are related to coronaviruses, it's possible that SARS-CoV-2 could benefit from it. Iron oxide nanozymes are iron oxide nanoparticles with a diameter of 200 nanometers (IONzymes). Such IONzymes render the virus ineffective by generating lipid peroxidation that as a result compromises the stability of the virus's envelope proteins. As a result, there's a decent possibility that these nanoparticles can potentially inactivate SARS-CoV-2.

Compilation of References

A, S., Thamer, A., K, R., A, P., V, R., K, M., Murad, A., & P, M. (2020). Facile one pot microwave-assisted green synthesis of Fe_2O_3/Ag nanocomposites by phytoreduction: Potential application as sunlight-driven photocatalyst, antibacterial and anticancer agent. *Journal of Photochemistry and Photobiology B: Biology, 207*(April), 111885. doi:10.1016/j.jphotobiol.2020.111885

Abbas, A. H., Elhag, H. H., Sulaiman, W. R. W., Gbadamosi, A., Pourafshary, P., Ebrahimi, S. S., Alqohaly, O. Y., & Agi, A. (2021). Modelling of continuous surfactant flooding application for marginal oilfields: A case study of Bentiu reservoir. *Journal of Petroleum Exploration and Production Technology*. Advance online publication. doi:10.100713202-020-01077-y

Abboud, Y., Saffaj, T., Chagraoui, A., El Bouari, A., Brouzi, K., Tanane, O., & Ihssane, B. (2014). Biosynthesis, characterization and antimicrobial activity of copper oxide nanoparticles (CONPs) produced using brown alga extract (Bifurcaria bifurcata). *Applied Nanoscience, 4*(5), 571–576. doi:10.100713204-013-0233-x

Abd Ellah, N. H., & Abouelmagd, S. A. (2016). Surface functionalization of polymeric nanoparticles for tumor drug delivery: Approaches and challenges. *Expert Opinion on Drug Delivery*, •••, 1–14. PMID:27426638

Abdallah, Y., Ogunyemi, S. O., Abdelazez, A., Zhang, M., Hong, X., Ibrahim, E., Hossain, A., Fouad, H., Li, B., & Chen, J. (2019). The green synthesis of MgO nano-flowers using Rosmarinus officinalis L. (Rosemary) and the antibacterial activities against Xanthomonas oryzae pv. Oryzae. *BioMed Research International, 2019*, 1–8. doi:10.1155/2019/9395146 PMID:30906776

Abdel Ghani, N. T., El Chaghaby, G. A., & Helal, F. S. (2015). Individual and competitive adsorption of phenol and nickel onto multiwalled carbon nanotubes. *Journal of Advanced Research, 6*(3), 405–415. doi:10.1016/j.jare.2014.06.001 PMID:26257938

Abdel-Halim, K. Y., Osman, S. R., & Abdou, G. Y. (2020). In vivo evaluation of oxidative stress and biochemical alteration as biomarkers in glass clover snail, Monacha cartusiana exposed to zinc oxide nanoparticles. *Environ Pollut, 257*, 113120. doi:10.1016/j.envpol.2019.113120 PMID:31753629

Abd-Rabou, A. A., Abdalla, A. M., Ali, N. A., & Zoheir, K. M. A. (2017). Moringa oleifera root induces cancer apoptosis more effectively than leave nanocomposites and its free counterpart. *Asian Pacific Journal of Cancer Prevention, 18*(8), 2141–2149. doi:10.22034/APJCP.2017.18.8.2141 PMID:28843248

Abdul Ghaffar, K., Kumar Giddam, A., Zaman, M., Skwarczynski, M., & Toth, I. (2014). Liposomes as nanovaccine delivery systems. *Current Topics in Medicinal Chemistry, 14*(9), 1194–1208. doi:10.2174/1568026614666140329232757 PMID:24678703

Abid, S. A., Muneer, A. A., Al-Kadmy, I. M. S., Sattar, A. A., Beshbishy, A. M., Batiha, G.-E.-S., & Hetta, H. F. (2021). Biosensors as a Future Diagnostic Approach for COVID-19. *Life Sciences*, *273*, 119117. doi:10.1016/j.lfs.2021.119117 PMID:33508293

Abo-zeid, Y., Ismail, N. S. M., Mclean, G. R., & Hamdy, N. M. (2020). Since January 2020 Elsevier has created a COVID-19 resource centre with free information in English and Mandarin on the novel coronavirus COVID- 19. The COVID-19 resource centre is hosted on Elsevier Connect, the company's public news and information. *European Journal of Pharmaceutical Sciences*, *153*(1), 105465. doi:10.1016/j.ejps.2020.105465 PMID:32668312

Adedirin, O., Adamu, U., & Nnabuk, E. O. (2011). Removal of Cd(II) from aqueous solution using Bacillus subtilis and Escherichia coli immobilized in agarose gel: Equilibrium, kinetics and thermodynamic study. *Archives of Applied Science Research*, *3*(3), 59–76.

Adelere, I. A., & Lateef, A. (2016). A novel approach to the green synthesis of metallic nanoparticles: The use of agro-wastes, enzymes, and pigments. *Nanotechnology Reviews*, *5*(6). Advance online publication. doi:10.1515/ntrev-2016-0024

Aderibigbe, A., Cheng, K., Heidari, Z., Killough, J., & Fuss-Dezelic, T. (2016). Application of magnetic nanoparticles mixed with propping agents in enhancing near-wellbore fracture detection. *Journal of Petroleum Science Engineering*, *141*, 133–143. Advance online publication. doi:10.1016/j.petrol.2016.01.027

Adesina, A. (2020). Overview of Workability and Mechanical Performance of Cement-Based Composites Incorporating Nanomaterials. Silicon. doi:10.100712633-020-00800-y

Adeyemi, O. S., & Otohinoyi, D. A. (2018). Inorganic nanoparticles restrict viability of metastatic breast cancer cells in vitro. *Comparative Clinical Pathology*, *28*(4), 949–954. doi:10.100700580-018-2841-z

Adhikari, S., Adhikari, U., Mishra, A., & Guragain, B. S. (2020). Nanomaterials for Diagnostic, Treatment and Prevention of COVID-19. *Applied Science & Technology Annals*, *1*(1), 155–164. doi:10.3126/asta.v1i1.30295

Aditha, S. K., Kurdekar, A. D., Chunduri, L. A. A., Patnaik, S., & Kamisetti, V. (2016). Aqueous based reflux method for green synthesis of nanostructures: Application in CZTS synthesis. *MethodsX*, *3*, 35–42. doi:10.1016/j.mex.2015.12.003 PMID:27408826

Adrees, M., Ali, S., Rizwan, M., Ibrahim, M., Abbas, F., Farid, M., Zia-ur-Rehman, M., Irshad, M. K., & Bharwana, S. A. (2015). The effect of excess copper on growth and physiology of important food crops: A review. *Environmental Science and Pollution Research International*, *22*(11), 8148–8162. doi:10.100711356-015-4496-5 PMID:25874438

Aftabtalab, A., Sadabadi, H., Shilpa Chakra, C. H., & Venkateswara Rao, K. (2014). Magnetite nanoparticles (Fe3O4) synthesis for removal of Chromium (VI) from waste water. *International Journal of Scientific & Engineering Research*, *5*(1). https://www.ijser.org/paper/Magnetite-nanoparticles-Fe3O4-synthesis-for-removal.html

Afzal, S., Nikookar, M., Ehsani, M. R., & Roayaei, E. (2014). An Experimental Investigation of the Catalytic Effect of Fe2O3 Nanoparticle on Steam Injection Process of an Iranian Reservoir. *Iranian Journal of Oil & Gas Science and Technology*, *3*(2). Advance online publication. doi:10.22050/ijogst.2014.6033

Agasti, N. (2021). Decontamination of heavy metal ions from water by composites prepared from waste. *Current Research in Green and Sustainable Chemistry*, *4*(October), 100088. doi:10.1016/j.crgsc.2021.100088

Agawane, G. L., Shin, S. W., Vanalakar, S. A., Suryawanshi, M. P., Moholkar, A. V., & Kim, J. H. (2015). Investigations on chemo-mechano stabilities of the molybdenum thin films deposited by DC-sputter technique. *Zeitschrift für Physikalische Chemie*, *229*(3), 377–393. doi:10.1515/zpch-2014-0624

Agista, M. N., Guo, K., & Yu, Z. (2018). A state-of-the-art review of nanoparticles application in petroleum with a focus on enhanced oil recovery. Applied Sciences, 8(6). doi:10.3390/app8060871

Aglawe, S., Gayke, A., Kadam, P., Kherud, R., Kanawade, S., Garud, Y., & Surase, R. (2018). *Review on nanoparticles*. Academic Press.

Agrawal, V., Paul, M. K., & Mukhopadhyay, A. K. (2005). 6-mercaptopurine and daunorubicin double drug liposomes—Preparation, drug-drug interaction and characterization. *Journal of Liposome Research*, *15*(3–4), 141–155. doi:10.1080/08982100500364081 PMID:16393906

Ahamed, M., Javed Akhtar, M., Majeed Khan, M. A., & Alhadlaq, H. A. (2021). Facile green synthesis of ZnO-RGO nanocomposites with enhanced anticancer efficacy. *Methods (San Diego, Calif.)*, (March). Advance online publication. doi:10.1016/j.ymeth.2021.04.020 PMID:33930572

Ahir, N. A., Takaloo, A. V., Nirmal, K. A., Kundale, S. S., Chougale, M. Y., Bae, J., Kim, D., & Dongale, T. D. (2021). Capacitive coupled non-zero I–V and type-II memristive properties of the $NiFe_2O_4$–TiO_2 nanocomposite. *Materials Science in Semiconductor Processing*, *125*, 105646. doi:10.1016/j.mssp.2020.105646

Ahmad, A., Senapati, S., Khan, M. I., Kumar, R., & Sastry, M. (2003). Extracellular biosynthesis of monodisperse gold nanoparticles by a novel extremophilic actinomycete, Thermomonospora sp. *Langmuir*, *19*(8), 3550–3553. doi:10.1021/la0267721

Ahmadi, R., Ghanbarzadeh, B., Ayaseh, A., Kafil, H. S., Özyurt, H., Katourani, A., & Ostadrahimi, A. (2019). The antimicrobial bio-nanocomposite containing non-hydrolyzed cellulose nanofiber (CNF) and Miswak (Salvadora persica L.) extract. *Carbohydrate Polymers*, *214*(March), 15–25. doi:10.1016/j.carbpol.2019.03.010 PMID:30925983

Ahmad, M., Wang, J., Xu, J., Zhang, Q., & Zhang, B. (2020). Magnetic tubular carbon nanofibers as efficient Cu(II) ion adsorbent from wastewater. *Journal of Cleaner Production*, *252*, 119825. doi:10.1016/j.jclepro.2019.119825

Ahmad, N., Sharma, S., Singh, V. N., Shamsi, S. F., Fatma, A., & Mehta, B. R. (2011). Biosynthesis of Silver Nanoparticles from Desmodium triflorum: A Novel Approach Towards Weed Utilization. *Biotechnology Research International*, *454090*, 1–8. Advance online publication. doi:10.4061/2011/454090 PMID:21350660

Ahmed, H. B., Abualnaja, K. M., Ghareeb, R. Y., Ibrahim, A. A., Abdelsalam, N. R., & Emam, H. E. (2021). Technical textiles modified with immobilized carbon dots synthesized with infrared assistance. *Journal of Colloid and Interface Science*, *604*, 15–29. doi:10.1016/j.jcis.2021.07.014 PMID:34261016

Ahmed, S., Ahmad, M., Swami, B. L., & Ikram, S. (2016). A review on plants extract mediated synthesis of silver nanoparticles for antimicrobial applications: A green expertise. *Journal of Advanced Research*, *7*(1), 17–28. doi:10.1016/j.jare.2015.02.007 PMID:26843966

Ahmed, S., Annu, S., & Yudha, S. S. (2016). Biosynthesis of gold nanoparticles: A green approach. *Journal of Photochemical Photobiology B: Biology*, *161*, 141–153. doi:10.1016/j.jphotobiol.2016.04.034 PMID:27236049

Ahuja, T., Chaudhari, K., Paramasivam, G., Ragupathy, G., Mohanty, J. S., & Pradeep, T. (2021). Toward Vibrational Tomography of Citrate on Dynamically Changing Individual Silver Nanoparticles. *The Journal of Physical Chemistry C*, *125*(6), 3553–3566. doi:10.1021/acs.jpcc.0c09981

Ahuja, T., Ghosh, A., Mondal, S., Basuri, P., Jenifer, S. K., Srikrishnarka, P., Mohanty, J. S., Bose, S., & Pradeep, T. (2019). Ambient electrospray deposition Raman spectroscopy (AESD RS) using soft-landed preformed silver nanoparticles for rapid and sensitive analysis. *Analyst (London)*, *144*(24), 7412–7420. doi:10.1039/C9AN01700C PMID:31674602

Aillon, K. L., Xie, Y., El-Gendy, N., Berkland, C. J., & Forrest, M. L. (2009). Effects of nanomaterial physicochemical properties on in-vivo toxicity. *Advanced Drug Delivery*, *61*(6), 457–466. doi:10.1016/j.addr.2009.03.010 PMID:19386275

Aioub, M., Panikkanvalappil, S. R., & El-Sayed, M. A. (2017). Platinum-coated gold nanorods: Efficient reactive oxygen scavengers that prevent oxidative damage toward healthy, untreated cells during plasmonic photothermal therapy. *ACS Nano*, *11*(1), 579–586. doi:10.1021/acsnano.6b06651 PMID:28029783

Aisida, S. O., Madubuonu, N., Alnasir, M. H., Ahmad, I., Botha, S., Maaza, M., & Ezema, F. I. (2019). Biogenic synthesis of iron oxide nanorods using Moringa oleifera leaf extract for antibacterial applications. *Applied Nanoscience*, *10*(1), 305–315. doi:10.100713204-019-01099-x

Ajima, K., Murakami, T., Mizoguchi, Y., Tsuchida, K., Ichihashi, T., Iijima, S., & Yudasaka, M. (2008). Enhancement of in vivo anticancer effects of cisplatin by incorporation inside single-wall carbon nanohorns. *ACS Nano*, *2*(10), 2057–2064. doi:10.1021/nn800395t PMID:19206452

Ajitha, B., Ashok Kumar Reddy, Y., & Sreedhara Reddy, P. (2015). Green synthesis and characterization of silver nanoparticles using Lantana camara leaf extract. *Materials Science and Engineering C*, *49*, 373–381. doi:10.1016/j.msec.2015.01.035 PMID:25686962

Ajmal, A., Majeed, I., Malik, R. N., Idriss, H., & Nadeem, M. A. (2014). Principles and mechanisms of photocatalytic dye degradation on TiO 2 based photocatalysts: A comparative overview. *RSC Advances*, *4*(70), 37003–37026. doi:10.1039/C4RA06658H

Akafu, T., Chimdi, A., & Gomoro, K. (2019). Removal of Fluoride from Drinking Water by Sorption Using Diatomite Modified with Aluminum Hydroxide. *Journal of Analytical Methods in Chemistry*, *2019*, 1–11. Advance online publication. doi:10.1155/2019/4831926 PMID:31886021

Akbarzadeh, A., Rezaei-Sadabady, R., Davaran, S., Joo, S. W., Zarghami, N., Hanifehpour, Y., Samiei, M., Kouhi, M., & Nejati-Koshki, K. (2013). Liposome: Classification, preparation, and applications. *Nanoscale Research Letters*, *8*(1), 102. doi:10.1186/1556-276X-8-102 PMID:23432972

Akhavan, O., Ghaderi, E., Shirazian, S. A., & Rahighi, R. (2016). Rolled graphene oxide foams as three-dimensional scaffolds for growth of neural fibers using electrical stimulation of stem cells. *Carbon*, *97*, 71–77. doi:10.1016/j.carbon.2015.06.079

Akhtar, M. S., Panwar, J., & Yun, Y.-S. (2013). Biogenic synthesis of metallic nanoparticles by plant extracts. *ACS Sustainable Chemistry & Engineering*, *1*(6), 591–602. doi:10.1021c300118u

Akshatha, S., Sreenivasa, S., Parashuram, L., Udaya kumar, V., Alharthi, F. A., Chakrapani Rao, T. M., & Kumar, S. (2020, January). Microwave assisted green synthesis of p-type Co_3O_4@Mesoporous carbon spheres for simultaneous degradation of dyes and photocatalytic hydrogen evolution reaction. *Materials Science in Semiconductor Processing*, *121*, 105432. doi:10.1016/j.mssp.2020.105432

Akter, M., Sikder, M., Rahman, M. M., Ullah, A. K. M. A., Hossain, K. F. B., Banik, S., Hosokawa, T., Saito, T., & Kurasaki, M. (2018). A systematic review on silver nanoparticles-induced cytotoxicity: Physicochemical properties and perspectives. *Journal of Advanced Research*, *9*, 1–16. doi:10.1016/j.jare.2017.10.008 PMID:30046482

Alam, M. N., & Huq, F. (2016). Comprehensive review on tumour active palladium compounds and structure–activity relationships. *Coordination Chemistry Reviews*, *316*, 36–67. doi:10.1016/j.ccr.2016.02.001

Alavi, M., & Karimi, N. (2018). Antiplanktonic, antibiofilm, antiswarming motility and antiquorum sensing activities of green synthesized Ag–TiO_2, TiO_2–Ag, Ag–Cu and Cu–Ag nanocomposites against multi-drug-resistant bacteria. *Artificial Cells, Nanomedicine and Biotechnology*, *46*(sup3), S399–S413. doi:10.1080/21691401.2018.1496923

Alavi, M., & Karimi, N. (2020). Hemoglobin self-assembly and antibacterial activities of bio-modified Ag-MgO nanocomposites by different concentrations of Artemisia haussknechtii and Protoparmeliopsis muralis extracts. *International Journal of Biological Macromolecules*, *152*, 1174–1185. doi:10.1016/j.ijbiomac.2019.10.207 PMID:31760003

Alavi, M., & Rai, M. (2019). Recent advances in antibacterial applications of metal nanoparticles (MNPs) and metal nanocomposites (MNCs) against multidrug-resistant (MDR) bacteria. *Expert Review of Anti-Infective Therapy*, *17*(6), 419–428. doi:10.1080/14787210.2019.1614914 PMID:31046483

Alavi, M., & Varma, R. S. (2021). Phytosynthesis and modification of metal and metal oxide nanoparticles/nanocomposites for antibacterial and anticancer activities: Recent advances. *Sustainable Chemistry and Pharmacy*, *21*(March), 100412. doi:10.1016/j.scp.2021.100412

Albdiry, M. T., Yousif, B. F., Ku, H., & Lau, K. T. (2013). A critical review on the manufacturing processes in relation to the properties of nanoclay/polymer composites. *Journal of Composite Materials*, *47*(9), 1093–1115. doi:10.1177/0021998312445592

Al-Hagri, A., Li, R., Rajput, N. S., Lu, J.-Y., Cong, S., Sloyan, K., Almahri, M. A., Tamalampudi, S. R., Chiesa, M., & Al Ghaferi, A. (2019). Direct growth of single-layer terminated vertical graphene array on germanium by plasma enhanced chemical vapor deposition. *Carbon*, *155*, 320–325. doi:10.1016/j.carbon.2019.08.069

Ali, R., Batool, T., Manzoor, B., Waseem, H., Mehmood, S., Kabeer, A., Ali, Z., Habib, S., Rashid, U., & Iqbal, M. J. (2020). Nanobiotechnology-based drug delivery strategy as a potential weapon against multiple drug-resistant pathogens. In Antibiotics and Antimicrobial Resistance Genes in the Environment (pp. 350-368). doi:10.1016/B978-0-12-818882-8.00022-X

Alibolandi, M., Shahriari, M., & Ramezani, M. (2019). Smart Polymersomes as Intelligent Nanomedicines in Cancer Treatment. *Polymeric Nanoparticles as a Promising Tool for Anti-Cancer Therapeutics*, 343-371.

Ali, H., Khan, E., & Ilahi, I. (2019). Environmental Chemistry and Ecotoxicology of Hazardous Heavy Metals: Environmental Persistence, Toxicity, and Bioaccumulation. *Journal of Chemistry*, *2019*, 1–14. doi:10.1155/2019/6730305

Ali, M. A., Hu, C., Jahan, S., Yuan, B., Saleh, M. S., Ju, E., Gao, S.-J., & Panat, R. (2020). Sensing of COVID-19 Antibodies in Seconds via Aerosol Jet Nanoprinted Reduced-Graphene-Oxide-coated 3D Electrodes. *Advanced Materials*, *33*(7), 2006647. doi:10.1002/adma.202006647 PMID:33349975

Ali, S., Khan, I., Khan, S. A., Sohail, M., Ahmed, R., Rehman, A., Ur Ansari, M. S., & Morsy, M. A. (2017). Electrocatalytic performance of Ni@Pt core–shell nanoparticles supported on carbon nanotubes for methanol oxidation reaction. *Electroanalytical Chemistry*, *795*, 17–25. doi:10.1016/j.jelechem.2017.04.040

Al-Jamal, K. T., Al-Jamal, W. T., Wang, J. T., Rubio, N., Buddle, J., Gathercole, D., Zloh, M., & Kostarelos, K. (2013). Cationic poly-L-lysine dendrimer complexes doxorubicin and delays tumor growth in vitro and in vivo. *ACS Nano*, *7*(3), 1905–1917. doi:10.1021/nn305860k PMID:23527750

Al-Jumaili, A., Alancherry, S., Bazaka, K., & Jacob, M. (2017). *Review on the antimicrobial properties of carbon nanostructures*. Academic Press.

Allard, M. (1992). GEMS/WATER operational guide / prepared by the WHO Collaborating Centre for Surface and Ground Water under the joint sponsorhsip of the United Nations Environment Programme, World Health Organization, United Nations Educational, Scientific and Cultural Organization, World Meteorological Organization (3rd ed.). World Health Organization.

Allawadhi, P., Singh, V., Khurana, A., Khurana, I., Allwadhi, S., Kumar, P., Banothu, A. K., Thalugula, S., Barani, P. J., Naik, R. R., & Bharani, K. K. (2021). Silver nanoparticle based multifunctional approach for combating COVID-19. *Sensors International*, *2*(May), 100101. doi:10.1016/j.sintl.2021.100101 PMID:34766057

Allen, M. J., Tung, V. C., & Kaner, R. B. (2010). Honeycomb Carbon: A Review of Graphene. *Chemical Reviews*, *110*(1), 132–145. doi:10.1021/cr900070d PMID:19610631

Almanza-Reyes, H., Moreno, S., Plascencia-Lopez, I., Alvarado-Vera, M., Patron-Romero, L., Borrego, B., . . . Bogdanchikova, N. (2021). Evaluation of silver nanoparticles for the prevention of SARS-CoV-2 infection in health workers in vitro and in vivo. MedRxiv, 1–25. https://medrxiv.org/cgi/content/short/2021.05.20.21256197

Al-Marri, A. H., Khan, M., Shaik, M. R., Mohri, N., Adil, S. F., Kuniyil, M., Alkhathlan, H. Z., Al-Warthan, A., Tremel, W., Tahir, M. N., Khan, M., & Siddiqui, M. R. H. (2016). Green synthesis of Pd@graphene nanocomposite: Catalyst for the selective oxidation of alcohols. *Arabian Journal of Chemistry*, *9*(6), 835–845. doi:10.1016/j.arabjc.2015.12.007

Alper Öztürk, A., Namll, I., & Aygül, A. (2021). Cefaclor Monohydrate-Loaded Colon-Targeted Nanoparticles for Use in COVID-19 Dependent Coinfections and Intestinal Symptoms: Formulation, Characterization, Release Kinetics, and Antimicrobial Activity. *Assay and Drug Development Technologies*, *19*(3), 156–175. doi:10.1089/adt.2020.1014 PMID:33728979

Al-Qahtani, K. M. (2017). Cadmium removal from aqueous solution by green synthesis zero valent silver nanoparticles with Benjamina leaves extract. *Egyptian Journal of Aquatic Research*, *43*(4), 269–274. doi:10.1016/j.ejar.2017.10.003

Alsaba, M. T., Al Dushaishi, M. F., & Abbas, A. K. (2020). A comprehensive review of nanoparticles applications in the oil and gas industry. Journal of Petroleum Exploration and Production Technology, 10(4). doi:10.100713202-019-00825-z

Alsabagh, A. M., Fathy, M., & Morsi, R.E. (2015). Preparation and characterization of chitosan/silver nanoparticle/copper nanoparticle/carbon nanotube multifunctional nanocomposite for water treatment: Heavy metals removal; kinetics, isotherms, and competitive studies. *Royal Science Chemistry Advance*, *5*, 55774–55783.

Al-Shehri, A. A., Ellis, E. S., Servin, J. M. F., Kosynkin, D. V., Kanj, M. Y., & Schmidt, H. K. (2013). Illuminating the reservoir: Magnetic nanomappers. *SPE Middle East Oil and Gas Show and Conference, MEOS, Proceedings, 3*. 10.2118/164461-MS

Alt, V., Bechert, T., Steinrücke, P., Wagener, M., Seidel, P., Dingeldein, E., Domann, E., & Schnettler, R. (2004). *An in vitro assessment of the antibacterial properties and cytotoxicity of nanoparticulate silver bone cement*. Academic Press.

Al-Tayyar, N. A., Youssef, A. M., & Al-Hindi, R. (2020). Antimicrobial food packaging based on sustainable Bio-based materials for reducing foodborne Pathogens: A review. *Food Chemistry*, *310*, 125915. doi:10.1016/j.foodchem.2019.125915 PMID:31841936

Althomali, R. H., Alamry, K. A., Hussein, M. A., Khan, A., & Alosaimi, A. M. (2021). A green nanocomposite based modified cellulose/TiO_2/Cinnamon bark for the reduction of toxic organic compounds using spectrophotometric technique. *Journal of Materials Research and Technology*, *12*, 947–966. doi:10.1016/j.jmrt.2021.03.002

Amanullah, M., & Al-Tahini, A. M. (2009). Nano-technology- its significance in smart fluid development for oil and gas field application. *Society of Petroleum Engineers - SPE Saudi Arabia Section Technical Symposium 2009*. 10.2118/126102-MS

Ambade, A. V., Savariar, E. N., & Thayumanavan, S. (2005). Dendrimeric micelles for controlled drug release and targeted delivery. *Molecular Pharmaceutics*, *2*(4), 264–272. doi:10.1021/mp050020d PMID:16053329

Amin Alavi, M., & Morsali, A. (2010). Ultrasonic-assisted synthesis of Ca(OH) 2 and CaO nanostructures. *Journal of Experimental Nanoscience*, *5*(2), 93–105. doi:10.1080/17458080903305616

Amin, M.T., Alazba, A.A., & Manzoor, U. (2014). A review of removal of pollutants from water/wastewater using different types of nanomaterials. *Advance Material Science Engineering,* 825-910.

Amini, S. M. (2019). Preparation of antimicrobial metallic nanoparticles with bioactive compounds. *Materials Science and Engineering C, 103*, 109809. doi:10.1016/j.msec.2019.109809 PMID:31349497

Amiri, M., Salavati-Niasari, M., Akbari, A., & Gholami, T. (2017). Removal of malachite green (a toxic dye) from water by cobalt ferrite silica magnetic nanocomposite: Herbal and green sol-gel autocombustion synthesis. *International Journal of Hydrogen Energy, 42*(39), 24846–24860. doi:10.1016/j.ijhydene.2017.08.077

Amooaghaie, R., Saeri, M. R., & Azizi, M. (2015). Synthesis, characterization and biocompatibility of silver nanoparticles synthesized from Nigella sativa leaf extract in comparison with chemical silver nanoparticles. *Ecotoxicology and Environmental Safety, 120*, 400–408. doi:10.1016/j.ecoenv.2015.06.025 PMID:26122733

Anand, K., Gengan, R. M., Phulukdaree, A., & Chuturgoon, A. (2015). Agroforestry waste Moringa oleifera petals mediated green synthesis of gold nanoparticles and their anti-cancer and catalytic activity. *Journal of Industrial and Engineering Chemistry, 21*, 1105–1111. doi:10.1016/j.jiec.2014.05.021

Andeani, K. (2013). Javad, & Mohsenzadeh, S. (2013). Phytosynthesis of cadmium oxide nanoparticles from Achillea wilhelmsii flowers. *Journal of Chemistry, 1–4*, 1–4. Advance online publication. doi:10.1155/2013/147613

Andjelkovic, I., Tran, D. N. H., Kabiri, S., Azari, S., Markovic, M., & Losic, D. (2015). Graphene aerogels decorated with α-FeOOH nanoparticles for efficient adsorption of arsenic from contaminated waters. *ACS Applied Materials & Interfaces, 7*(18), 9758–9766. doi:10.1021/acsami.5b01624 PMID:25871444

Andreescu, D., Wanekaya, A. K., Sadik, O. A., & Wang, J. (2005). Nanostructured polyamic acid membranes as novel electrode materials. *Langmuir, 21*(15), 6891–6899. doi:10.1021/la050141k PMID:16008401

Andualem, W. W., Sabir, F. K., Mohammed, E. T., Belay, H. H., & Gonfa, B. A. (2020). Synthesis of copper oxide nanoparticles using plant leaf extract of Catha edulis and its antibacterial activity. *Journal of Nanotechnology, 2020,* 1–10. doi:10.1155/2020/2932434

Angelova, A., Garamus, V. M., Angelov, B., Tian, Z., Li, Y., & Zou, A. (2017). Advances in structural design of lipid-based nanoparticle carriers for delivery of macromolecular drugs, phytochemicals and anti-tumor agents. *Advances in Colloid and Interface Science, 249*, 331–345. doi:10.1016/j.cis.2017.04.006 PMID:28477868

Angkuratipakorn, T., Chung, C., Koo, C. K. W., Mundo, J. L. M., McClements, D. J., Decker, E. A., & Singkhonrat, J. (2020). Development of food-grade Pickering oil-in-water emulsions: Tailoring functionality using mixtures of cellulose nanocrystals and lauric arginate. *Food Chemistry, 327*, 127039. Advance online publication. doi:10.1016/j.foodchem.2020.127039 PMID:32454273

Ansari, M. A., Khan, H. M., Alzohairy, M. A., Jalal, M., Ali, S. G., Pal, R., & Musarrat, J. (2015). Green synthesis of Al2O3 nanoparticles and their bactericidal potential against clinical isolates of multi-drug resistant Pseudomonas aeruginosa. *World Journal of Microbiology & Biotechnology, 31*(1), 153–164. doi:10.100711274-014-1757-2 PMID:25304025

Antiochia, R. (2021). Developments in Biosensors for CoV Detection and Future Trends. *Biosensors & Bioelectronics, 173*, 112777. doi:10.1016/j.bios.2020.112777 PMID:33189015

Anwar, A., Masri, A., Rao, K., Rajendran, K., Khan, N. A., Shah, M. R., & Siddiqui, R. (2019). Antimicrobial activities of green synthesized gums-stabilized nanoparticles loaded with flavonoids. *Scientific Reports, 9*(1), 3122. doi:10.103841598-019-39528-0 PMID:30816269

Apul, O. G., & Karanfil, T. (2015). Adsorption of synthetic organic contaminants by carbon nanotubes: A critical review. *Water Research*, *68*, 34–55. doi:10.1016/j.watres.2014.09.032 PMID:25462715

Arab, D., & Pourafshary, P. (2013). Nanoparticles-assisted surface charge modification of the porous medium to treat colloidal particles migration induced by low salinity water flooding. *Colloids and Surfaces. A, Physicochemical and Engineering Aspects*, *436*, 803–814. Advance online publication. doi:10.1016/j.colsurfa.2013.08.022

Aramesh, M., Tran, P. A., Ostrikov, K. K., & Prawer, S. (2017). Conformal nanocarbon coating of alumina nanocrystals for biosensing and bioimaging. *Carbon*, *122*, 422–427. doi:10.1016/j.carbon.2017.06.101

Arancibia-Miranda, N., Baltazar, S. E., García, A., Muñoz-Lira, D., Sepúlveda, P., Rubio, M. A., & Altbir, D. (2016). Nanoscale zero valent supported by Zeolite and Montmorillonite: Template effect of the removal of lead ion from an aqueous solution. *Journal of Hazardous Materials*, *301*, 371–380. doi:10.1016/j.jhazmat.2015.09.007 PMID:26384998

Araújo, I. M. S., Silva, R. R., Pacheco, G., Lustri, W. R., Tercjak, A., Gutierrez, J., Júnior, J. R. S., Azevedo, F. H. C., Figuêredo, G. S., Vega, M. L., Ribeiro, S. J. L., & Barud, H. S. (2018). Hydrothermal synthesis of bacterial cellulose–copper oxide nanocomposites and evaluation of their antimicrobial activity. *Carbohydrate Polymers*, *179*(September), 341–349. doi:10.1016/j.carbpol.2017.09.081

Araujo, J. F. D. F., Tahir, Arsalani, S., Freire, F. L. Jr, Mariotto, G., Cremona, M., Mendoza, L. A. F., Luz-Lima, C., Zaman, Q., Del Rosso, T., Baffa, O., & Bruno, A. C. (2020). Novel scanning magnetic microscopy method for the characterization of magnetic nanoparticles. *Journal of Magnetism and Magnetic Materials*, *499*, 166300. Advance online publication. doi:10.1016/j.jmmm.2019.166300

Archana, B., Manjunath, K., Nagaraju, G., Chandra Sekhar, K. B., & Kottam, N. (2017). Enhanced photocatalytic hydrogen generation and photostability of ZnO nanoparticles obtained via green synthesis. *International Journal of Hydrogen Energy*, *42*(8), 5125–5131. doi:10.1016/j.ijhydene.2016.11.099

Archana, S. (2021). A Comparative Study of Iron Oxide Nanoparticles Surface Modified Using Carboxylic Acids. *International Journal for Research in Applied Sciences and Biotechnology*, *8*(1), 116–125. Advance online publication. doi:10.31033/ijrasb.8.1.13

Ardestani, M. S., Sadat Shandiz, S. A., Salehi, S., Ghanbar, F., Darvish, M. R., Mirzaie, A., & Jafari, M. (2016). Phytosynthesis of silver nanoparticles using Artemisia marschalliana Sprengel aerial part extract and assessment of their antioxidant, anticancer, and antibacterial properties. *International Journal of Nanomedicine*, *1835*, 1835. Advance online publication. doi:10.2147/IJN.S99882

Argyilan, E. P., Avis, P. G., Krekeler, M. P., & Morris, C. C. (2015). The origin of collapse features appearing in a migrating parabolic dune along the southern coast of Lake Michigan. *Aeolian Research*, *19*, 137–149. doi:10.1016/j.aeolia.2015.09.008

Armendariz, V., Herrera, I., peralta-videa, J. R., Jose-yacaman, M., Troiani, H., Santiago, P., & Gardea-Torresdey, J. L. (2004). Size controlled gold nanoparticle formation by Avena sativa biomass: Use of plants in nanobiotechnology. *Journal of Nanoparticle Research*, *6*(4), 377–382. doi:10.100711051-004-0741-4

Aromal, S. A., Vidhu, V. K., & Philip, D. (2012). Green synthesis of well-dispersed gold nanoparticles using Macrotyloma uniflorum. *Spectrochimica Acta. Part A: Molecular and Biomolecular Spectroscopy*, *85*(1), 99–104. doi:10.1016/j.saa.2011.09.035 PMID:22018585

Arshad, R., Barani, M., Rahdar, A., Sargazi, S., Cucchiarini, M., Pandey, S., & Kang, M. (2021). Multi-Functionalized Nanomaterials and Nanoparticles for Diagnosis and Treatment of Retinoblastoma. *Biosensors (Basel)*, *11*(4), 97. Advance online publication. doi:10.3390/bios11040097 PMID:33810621

Asadi, F., Forootanfar, H., & Ranjbar, M. (2020). A facile one-step preparation of $Ca_{10}(PO_4)_6(OH)_2$/Li-BioMOFs resin nanocomposites with Glycyrrhiza glabra (licorice) root juice as green capping agent and mechanical properties study. *Artificial Cells, Nanomedicine, and Biotechnology, 48*(1), 1331–1339. doi:10.1080/21691401.2020.1842748 PMID:33170039

Ashby, N. P., & Binks, B. P. (2000). Pickering emulsions stabilised by Laponite clay particles. *Physical Chemistry Chemical Physics, 2*(24), 5640–5646. Advance online publication. doi:10.1039/b007098j

Ashour, M., Wink, M., & Gershenzon, J. (2010). Biochemistry of terpenoids: Monoterpenes, sesquiterpenes and diterpenes. In M. Wink (Ed.), *Biochemistry of Plant Secondary Metabolism* (pp. 258–303). Wiley-Blackwell. doi:10.1002/9781444320503.ch5

Ashraf, S., Abbasi, A. Z., Pfeiffer, C., Hussain, S. Z., Khalid, Z. M., Gil, P. R., Parak, W. J., & Hussain, I. (2013). Protein-mediated synthesis, pH-induced reversible agglomeration, toxicity and cellular interaction of silver nanoparticles. *Colloids and Surfaces. B, Biointerfaces, 102*, 511–518. doi:10.1016/j.colsurfb.2012.09.032 PMID:23107938

Auclair, J., Turcotte, P., Gagnon, C., Peyrot, C., Wilkinson, K. J., & Gagne, F. (2019). The influence of surface coatings on the toxicity of silver nanoparticle in rainbow trout. *Comparative Biochemistry and Physiology. Toxicology & Pharmacology: CBP, 226*, 108623. doi:10.1016/j.cbpc.2019.108623 PMID:31505268

Auffan, M., Bottero, J. Y., Chaneac, C., & Rose, J. (2010). Inorganic manufactured nanoparticles: How their physicochemical properties influence their biological effects in aqueous environments. *Nanomedicine (London), 5*(6), 999–1007. doi:10.2217/nnm.10.61 PMID:20735233

Auffan, M., Rose, J., Bottero, J. Y., Lowry, G. V., Jolivet, J. P., & Wiesner, M. R. (2009). Towards a definition of inorganic nanoparticles from an environmental, health and safety perspective. *Nature Nanotechnology, 4*(10), 634–641. doi:10.1038/nnano.2009.242 PMID:19809453

Austin, L. A., Mackey, M. A., Dreaden, E. C., & El-Sayed, M. A. (2014). The optical, photothermal, and facile surface chemical properties of gold and silver nanoparticles in biodiagnostics, therapy, and drug delivery. *Archives of Toxicology, 88*(7), 1391–1417. doi:10.100700204-014-1245-3 PMID:24894431

Awad, M. A. A., Hendi, A., Ortashi, K. M., Alzahrani, B., Soliman, D., Alanazi, A., Alenazi, W., Taha, R. M., Ramadan, R., El-Tohamy, M., AlMasoud, N., & Alomar, T. S. (2021). Biogenic synthesis of silver nanoparticles using Trigonella foenum-graecum seed extract: Characterization, photocatalytic and antibacterial activities. *Sensors and Actuators. A, Physical, 323*. Advance online publication. doi:10.1016/j.sna.2021.112670

Ayyoob, M., Lee, D. H., Kim, J. H., Nam, S. W., & Kim, Y. J. (2017). Synthesis of poly(glycolic acids) via solution polycondensation and investigation of their thermal degradation behaviors. *Fibers and Polymers, 18*(3), 407–415. doi:10.100712221-017-6889-1

Azeez, L., Lateef, A., Adebisi, S. A., & Oyedeji, A. O. (2018). Novel biosynthesized silver nanoparticles from cobweb as adsorbent for Rhodamine B: Equilibrium isotherm, kinetic and thermodynamic studies. *Applied Water Science, 8*(1), 32. doi:10.100713201-018-0676-z

Babichev, A. V, Rykov, S. A., Tchernycheva, M., Smirnov, A. N., Davydov, V. Y., Kumzerov, Y. A., & Butko, V. Y. (2015). *Influence of Substrate Microstructure on the Transport Properties of CVD-Graphene*. doi:10.1021/acsami.5b08479

Babu, S. S., Radhakrishnan, E., Thomas, S., & Kalarikkal, N. (2019). Nanotechnology in Ancient Indian Holistic Healthcare. In Holistic Healthcare. Volume 2: Possibilities and Challenges (pp. 65-76). Apple Academic Press. doi:10.1201/9780429487804-5

Bae, K. H., Chung, H. J., & Park, T. G. (2011). Nanomaterials for cancer therapy and imaging. *Molecules and Cells, 31*(4), 295–302. doi:10.100710059-011-0051-5 PMID:21360197

Baetke, S. C., Lammers, T., & Kiessling, F. (2015). Applications of nanoparticles for diagnosis and therapy of cancer. *The British Journal of Radiology*, *88*(1054), 20150207. doi:10.1259/bjr.20150207 PMID:25969868

Bai, R. G., Sabouni, R., & Husseini, G. (2018). Green Nanotechnology—A Road Map to Safer Nanomaterials. In Applications of Nanomaterials (pp. 133-159). doi:10.1016/B978-0-08-101971-9.00006-5

Baig, N., Kammakakam, I., Falath, W., & Kammakakam, I. (2021). Nanomaterials: A review of synthesis methods, properties, recent progress, and challenges. Materials Advances, 2(6). doi:10.1039/D0MA00807A

Bai, H.-J., & Zhang, Z.-M. (2009). Microbial synthesis of semiconductor lead sulfide nanoparticles using immobilized Rhodobacter sphaeroides. *Materials Letters*, *63*(9), 764–766. doi:10.1016/j.matlet.2008.12.050

Baiocco, D., Lavecchia, R., Natali, S., & Zuorro, A. (2016). Production of metal nanoparticles by agro-industrial wastes: A green opportunity for nanotechnology. *Chemical Engineering Transactions*, *47*, 67–72. doi:10.3303/CET1647012

Bajpai, V. K., Kamle, M., Shukla, S., Mahato, D. K., Chandra, P., Hwang, S. K., Kumar, P., Huh, Y. S., & Han, Y.-K. (2018). Prospects of using nanotechnology for food preservation, safety, and security. *Yao Wu Shi Pin Fen Xi*, *26*(4), 1201–1214. doi:10.1016/j.jfda.2018.06.011 PMID:30249319

Baker, B., Aldridge, C., & Omer, A. M. H. (2019, November). *Water: Availability and Use*. Mississippi State University Extension Service. http://extension.msstate.edu/publications/water-availability-and-use

Baker, S., Rakshith, D., Shanthamma Kavitha, K., Santosh, P., Umapathy Kavitha, Y., Rao, Y., & Satish, S. (2013). Plants: Emerging as nanofactories towards facile route in synthesis of nanoparticles. *BioImpacts*. Advance online publication. doi:10.5681/BI.2013.012 PMID:24163802

Bakshi, M., & Abhilash, P. C. (2020). Nanotechnology for soil remediation: Revitalizing the tarnished resource. In *Nano-Materials as Photocatalysts for Degradation of Environmental Pollutants* (pp. 345–370). Elsevier Inc. doi:10.1016/B978-0-12-818598-8.00017-1

Baladi, M., Hajizadeh-Oghaz, M., Amiri, O., Valian, M., & Salavati-Niasari, M. (2021). Enhanced photocatalytic activity of $Sr_7Mn_7O_{19.62}$-Dy_2O_3 nanocomposite synthesized via a green method. *International Journal of Hydrogen Energy*, *46*(5), 3763–3779. doi:10.1016/j.ijhydene.2020.10.167

Balali-Mood, M., Naseri, K., Tahergorabi, Z., Khazdair, M. R., & Sadeghi, M. (2021). Toxic Mechanisms of Five Heavy Metals: Mercury, Lead, Chromium, Cadmium, and Arsenic. *Frontiers in Pharmacology*, *12*, 643972. doi:10.3389/fphar.2021.643972 PMID:33927623

Balasooriya, E. R., Jayasinghe, C. D., Jayawardena, U. A., Ruwanthika, R. W. D., Mendis de Silva, R., & Udagama, P. V. (2017). Honey Mediated Green Synthesis of Nanoparticles: New Era of Safe Nanotechnology. *Journal of Nanomaterials*, *2017*, 1–10. doi:10.1155/2017/5919836

Balasubramanian, V., Herranz-Blanco, B., Almeida, P. V., Hirvonen, J., & Santos, H. A. (2016). Multifaceted polymersome platforms: Spanning from self-assembly to drug delivery and protocells. *Progress in Polymer Science*, *60*, 51–85. doi:10.1016/j.progpolymsci.2016.04.004

Bankar, A., Joshi, B., Kumar, A. R., & Zinjarde, S. (2010). Banana peel extract mediated novel route for the synthesis of palladium nanoparticles. *Materials Letters*, *64*(18), 1951–1953. doi:10.1016/j.matlet.2010.06.021

Bankier, C., Matharu, R. K., Cheong, Y. K., Ren, G. G., Cloutman-Green, E., & Ciric, L. (2019). Synergistic antibacterial effects of metallic nanoparticle combinations. *Scientific Reports*, *9*(1), 1–8. doi:10.103841598-019-52473-2 PMID:31690845

Bano, S., & Pillai, S. (2020). Green synthesis of calcium oxide nanoparticles at different calcination temperatures. *World Journal of Science, Technology and Sustainable Development, 17*(3), 283–295. doi:10.1108/WJSTSD-12-2019-0087

Bansal, M., Singh, D., & Garg, V. K. (2009). A comparative study for the removal of hexavalent chromium from aqueous solution by agriculture wastes' carbons. *Journal of Hazardous Materials, 171*(1–3), 83–92. doi:10.1016/j.jhazmat.2009.05.124 PMID:19553015

Bansal, M., Singh, D., & Garg, V. K. (2012). Use of chemically treated sawdust for chromium removal from aqueous solutions: Effect of process parameters. *International Journal of Environment and Waste Management, 10*(2–3), 190–200. doi:10.1504/IJEWM.2012.048370

Bansal, V., Ahmad, A., & Sastry, M. (2006). Fungus-mediated biotransformation of amorphous silica in rice husk to nanocrystalline silica. *Journal of the American Chemical Society, 128*(43), 14059–14066. doi:10.1021/ja062113+ PMID:17061888

Bao, Y., Zhang, Y., Liu, P., Ma, J., Zhang, W., Liu, C., & Simion, D. (2019). Novel fabrication of stable Pickering emulsion and latex by hollow silica nanoparticles. *Journal of Colloid and Interface Science, 553*, 83–90. Advance online publication. doi:10.1016/j.jcis.2019.06.008 PMID:31195217

Baptista, P. V., McCusker, M. P., Carvalho, A., Ferreira, D. A., Mohan, N. M., Martins, M., & Fernandes, A. R. (2018). Nano-strategies to fight multidrug resistant bacteria—"A Battle of the Titans. *Frontiers in Microbiology, 9*, 1441. doi:10.3389/fmicb.2018.01441 PMID:30013539

Barba, A. A., Bochicchio, S., Dalmoro, A., & Lamberti, G. (2019). Lipid Delivery Systems for Nucleic-Acid-Based-Drugs: From Production to Clinical Applications. *Pharmaceutics, 11*(8), 360. Advance online publication. doi:10.3390/pharmaceutics11080360 PMID:31344836

Barnard, A. S., & Guo, H. (2012). *Nature's Nanostructures*. CRC Press. doi:10.1201/b11618

Bartneck, M. (2021). Lipid nanoparticle formulations for targeting leukocytes with therapeutic RNA in liver fibrosis. *Advanced Drug Delivery Reviews, 173*, 70–88. doi:10.1016/j.addr.2021.03.009 PMID:33774114

Bartusik-Aebisher, D., Chrzanowski, G., Bober, Z., & Aebisher, D. (2021). An analytical study of Trastuzumab-dendrimer-fluorine drug delivery system in breast cancer therapy in vitro. *Biomedicine and Pharmacotherapy, 133*, 111053. doi:10.1016/j.biopha.2020.111053 PMID:33378959

Baruah, D., Goswami, M., Yadav, R. N. S., Yadav, A., & Das, A. M. (2018). Biogenic synthesis of gold nanoparticles and their application in photocatalytic degradation of toxic dyes. *Journal of Photochemistry and Photobiology. B, Biology, 186*, 51–58. doi:10.1016/j.jphotobiol.2018.07.002 PMID:30015060

Baruwati, B., & Varma, R. S. (2009). High value products from waste: Grape pomace extract—A three-in-One package for the synthesis of metal nanoparticles. *ChemSusChem, 2*(11), 1041–1044. doi:10.1002/cssc.200900220 PMID:19842157

Basavegowda, N., & Rok Lee, Y. (2013). Synthesis of silver nanoparticles using Satsuma mandarin (Citrus unshiu) peel extract: A novel approach towards waste utilization. *Materials Letters, 109*, 31–33. doi:10.1016/j.matlet.2013.07.039

Bashambu, L., Singh, R., & Verma, J. (2021). Metal/metal oxide nanocomposite membranes for water purification. *Materials Today: Proceedings, 44*(January), 538–545. doi:10.1016/j.matpr.2020.10.213

Basumatary, K., Daimary, P., Das, S. K., Thapa, M., Singh, M., Mukherjee, A., & Kumar, S. (2018). Lagerstroemia speciosa fruit-mediated synthesis of silver nanoparticles and its application as filler in agar-based nanocomposite films for antimicrobial food packaging. *Food Packaging and Shelf Life, 17*(June), 99–106. doi:10.1016/j.fpsl.2018.06.003

Begines, B., Ortiz, T., Pérez-Aranda, M., Martínez, G., Merinero, M., Argüelles-Arias, F., & Alcudia, A. (2020). Polymeric nanoparticles for drug delivery: Recent developments and future prospects. *Nanomaterials (Basel, Switzerland), 10*(7), 1403. doi:10.3390/nano10071403 PMID:32707641

Behravan, M., Hossein Panahi, A., Naghizadeh, A., Ziaee, M., Mahdavi, R., & Mirzapour, A. (2019). Facile green synthesis of silver nanoparticles using Berberis vulgaris leaf and root aqueous extract and its antibacterial activity. *International Journal of Biological Macromolecules, 124*, 148–154. doi:10.1016/j.ijbiomac.2018.11.101 PMID:30447360

Bengalli, R., Colantuoni, A., Perelshtein, I., Gedanken, A., Collini, M., Mantecca, P., & Fiandra, L. (2021). In vitro skin toxicity of CuO and ZnO nanoparticles: Application in the safety assessment of antimicrobial coated textiles. *NanoImpact, 21*, 100282. doi:10.1016/j.impact.2020.100282

Benn, T. M., & Westerhoff, P. (2008). Nanoparticle silver released into water from commercially available sock fabrics. *Environmental Science & Technology, 42*(11), 4133–4139. doi:10.1021/es7032718 PMID:18589977

Beranová, J., Seydlová, G., Kozak, H., Benada, O., Fišer, R., Artemenko, A., Konopásek, I., & Kromka, A. (2014). *Sensitivity of bacteria to diamond nanoparticles of various size differs in gram-positive and gram-negative cells.* Academic Press.

Beranová, J., Seydlová, G., Kozak, H., Potocký, Š., Konopásek, I., & Kromka, A. (2012). *Antibacterial behavior of diamond nanoparticles against Escherichia coli.* Academic Press.

Berezovska, I., Kapilov, K., Dhavalikar, P., Cosgriff-Hernandez, E., & Silverstein, M. S. (2021). Reactive Surfactants for Achieving Open-Cell PolyHIPE Foams from Pickering Emulsions. *Macromolecular Materials and Engineering, 306*(6), 2000825. Advance online publication. doi:10.1002/mame.202000825

Berry, V. (2013). Impermeability of graphene and its applications. Carbon. doi:10.1016/j.carbon.2013.05.052

Besenhard, M. O., Panariello, L., Kiefer, C., Lagrow, A. P., Storozhuk, L., Perton, F., Begin, S., Mertz, D., Thanh, N. T. K., & Gavriilidis, A. (2021). Small iron oxide nanoparticles as MRI T 1contrast agent: Scalable inexpensive water-based synthesis using a flow reactor. *Nanoscale, 13*(19), 8795–8805. Advance online publication. doi:10.1039/D1NR00877C PMID:34014243

Bessy, T. C., Bindhu, M. R., Johnson, J., Chen, S.-M., Chen, T.-W., & Almaary, K. S. (2022). UV light assisted photocatalytic degradation of textile waste water by Mg0.8-xZnxFe2O4 synthesized by combustion method and in-vitro antimicrobial activities. *Environmental Research, 204*, 111917. doi:10.1016/j.envres.2021.111917 PMID:34453899

Beyth, N., Houri-Haddad, Y., Domb, A., Khan, W., & Hazan, R. (2015). Alternative Antimicrobial Approach: Nano-Antimicrobial Materials. *Evidence-Based Complementary and Alternative Medicine, 2015*, 1–16. doi:10.1155/2015/246012 PMID:25861355

Beyzay, F., Hosseini, A. Z., & Soudi, S. (2017). Alpha alumina nanoparticle conjugation to cysteine peptidase A and B: An efficient method for autophagy induction. *Avicenna Journal of Medical Biotechnology, 9*(2), 71. PMID:28496946

Bhadane, M. S., Gavhane, K. H., Kulkarni, P. P., Dahiwale, S. S., Bhoraskar, V. N., More, M. A., Patil, P. S., & Dhole, S. D. (2020). Particle size dependent TL response of CaF_2: Dy phosphor for gamma dosimetry. *Journal of Luminescence, 223*, 117168. doi:10.1016/j.jlumin.2020.117168

Bhalla, N., Pan, Y., Yang, Z., & Payam, F. (2020). Opportunities and Challenges for Biosensors and Nanoscale Analytical Tools for Pandemics: COVID-19. *ACS Nano, 14*(7), 7783–7807. doi:10.1021/acsnano.0c04421 PMID:32551559

Bharagava, R. N., Purchase, D., Saxena, G., & Mulla, S. I. (2019). Applications of Metagenomics in Microbial Bioremediation of Pollutants. In *Microbial Diversity in the Genomic Era* (pp. 459–477). Elsevier. doi:10.1016/B978-0-12-814849-5.00026-5

Bhatt, D., Gupta, E., Kaushik, S., Srivastava, V. K., Saxena, J., & Jyoti, A. (2018). Bio-fabrication of silver nanoparticles by Pseudomonas aeruginosa: Optimisation and antibacterial activity against selected waterborne human pathogens. *IET Nanobiotechnology / IET, 12*(7), 981–986. doi:10.1049/iet-nbt.2018.0051 PMID:30247141

Bhavya, C., Yogendra, K., Mahadevan, K. M., & Madhusudhana, N. (2016). Synthesis of Calcium Oxide Nanoparticles and Its Mortality Study on Fresh Water Fish Cyprinus Carpio. *IOSR Journal of Environmental Science, Toxicology and Food Technology, 10*(12), 55–60. doi:10.9790/2402-1012015560

Bilal, M., Zhao, Y., Rasheed, T., Ahmed, I., Hassan, S. T., Nawaz, M. Z., & Iqbal, H. (2019). Biogenic nanoparticle–chitosan conjugates with antimicrobial, antibiofilm, and anticancer potentialities: Development and characterization. *International Journal of Environmental Research and Public Health, 16*(4), 598. doi:10.3390/ijerph16040598 PMID:30791374

Binks, B. P. (2002). Particles as surfactants - Similarities and differences. *Current Opinion in Colloid & Interface Science, 7*(1–2), 21–41. Advance online publication. doi:10.1016/S1359-0294(02)00008-0

Bisht, G., Srivastava, S., Kulshreshtha, R., Sourirajan, A., Baumler, D. J., & Dev, K. (2020). Applications of red pigments from psychrophilic Rhodonellum psychrophilum GL8 in health, food and antimicrobial finishes on textiles. *Process Biochemistry, 94*, 15–29. doi:10.1016/j.procbio.2020.03.021

Bi, Y., Hao, F., Yan, G., Teng, L. J., Lee, R., & Xie, J. (2016). Actively targeted nanoparticles for drug delivery to tumor. *Current Drug Metabolism, 17*(8), 763–782. doi:10.2174/1389200217666160619191853 PMID:27335116

Blackie, E. J., Le Ru, E. C., & Etchegoin, P. G. (2009). Single-Molecule Surface-Enhanced Raman Spectroscopy of Nonresonant Molecules. *JACS, 131*(40), 14466–14472. doi:10.1021/ja905319w PMID:19807188

Blake, R., Gun, Y. K., Coleman, J., Cadek, M., Fonseca, A., Nagy, J. B., & Blau, W. J. (2004). A Generic Organometallic Approach toward Ultra-Strong Carbon Nanotube Polymer Composites. *JACS, 126*(33), 10226–10227. doi:10.1021/ja0474805 PMID:15315418

Blanco, E., Kessinger, C. W., Sumer, B. D., & Gao, J. (2009). Multifunctional micellar nanomedicine for cancer therapy. *Experimental Biology and Medicine, 234*(2), 123–131. doi:10.3181/0808-MR-250 PMID:19064945

Blanco, N. G., Maldonado, C. R., & Mareque-Rivas, J. C. (2009). Effective photoreduction of a Pt(iv) complex with quantum dots: A feasible new light-induced method of releasing anticancer Pt(ii) drugs. *Chemical Communications, 35*(35), 5257. doi:10.1039/b910000h PMID:19707638

Boca, S. C., Potara, M., Gabudean, A.-M., Juhem, A., Baldeck, P. L., & Astilean, S. (2011). Chitosan-coated triangular silver nanoparticles as a novel class of biocompatible, highly effective photothermal transducers for in vitro cancer cell therapy. *Cancer Letters, 311*(2), 131–140. doi:10.1016/j.canlet.2011.06.022 PMID:21840122

Bolat, G., Mareci, D., Chelariu, R., Izquierdo, J., González, S., & Souto, R. M. (2013). Investigation of the electrochemical behaviour of TiMo alloys in simulated physiological solutions. *Electrochimica Acta, 113*, 470–480. doi:10.1016/j.electacta.2013.09.116

Boretti, A., & Rosa, L. (2019). Reassessing the projections of the World Water Development Report. *Npj Clean Water, 2*(1), 15. doi:10.103841545-019-0039-9

Borgohain, X., Boruah, A., Sarma, G. K., & Rashid, M. H. (2020). Rapid and extremely high adsorption performance of porous MgO nanostructures for fluoride removal from water. *Journal of Molecular Liquids, 305*, 112799. doi:10.1016/j.molliq.2020.112799

Boroumand Moghaddam, A., Moniri, M., Azizi, S., Abdul Rahim, R., Bin Ariff, A., Navaderi, M., & Mohamad, R. (2017). Eco-friendly formulated zinc oxide nanoparticles: Induction of cell cycle arrest and apoptosis in the MCF-7 cancer cell line. *Genes*, *8*(10), 281. doi:10.3390/genes8100281 PMID:29053567

Bosetti, M., Massè, A., Tobin, E., & Cannas, M. (2002). *Silver coated materials for external fixation devices: in vitro biocompatibility and genotoxicity.* Academic Press.

Botterhuis, N. E., Sun, Q., Magusin, P. C. M. M., van Santen, R. A., & Sommerdijk, N. A. J. M. (2006). Hollow silica spheres with an ordered pore structure and their application in controlled release studies. *Chemistry (Weinheim an der Bergstrasse, Germany)*, *12*(5), 1448–1456. doi:10.1002/chem.200500588 PMID:16308879

Bottero, J.Y. & M.R. (2010). Wiesner Considerations in evaluating the physicochemical properties and transformations of inorganic nanoparticles in water. *Nanomedicine, 5*(6).

Bou Hamdan, K. F., Harkouss, R., & Chakra, H. A. (2015, November 16). An overview of Extended Reach Drilling: Focus on design considerations and drag analysis. *Mediterranean Gas and Oil International Conference, MedGO 2015 - Conference Proceedings*. 10.1109/MedGO.2015.7330328

Bou-Hamdan, K. (2020). *Key Design Considerations for Maximizing the Recovery Rate of Unconventional Reservoirs. The Way Ahead.* SPE.

Boulenguez, J., Berthier, S., & Leroy, F. (2012). *Multiple scaled disorder in the photonic structure of Morpho rhetenor butterfly.* Academic Press.

Boxall, A., Rudd, M. A., Brooks, B. W., Caldwell, D. J., Choi, K., Hickmann, S., Innes, E., Ostapyk, K., Staveley, J. P., Verslycke, T., Ankley, G. T., Beazley, K. F., Belanger, S. E., Berninger, J. P., Carriquiriborde, P., Coors, A., DeLeo, P. C., Dyer, S. D., Ericson, J. F., ... Van Der Kraak, G. (2012). Pharmaceuticals and Personal Care Products in the Environment: What Are the Big Questions? *Environmental Health Perspectives*, *120*(9), 1221–1229. doi:10.1289/ehp.1104477 PMID:22647657

Bozzuto, G., & Molinari, A. (2015). Liposomes as nanomedical devices. *International Journal of Nanomedicine*, *10*, 975–999. doi:10.2147/IJN.S68861 PMID:25678787

Braithwaite, E. R. (1994). Molybdenum: an outline of its chemistry and users. In Occurrence, Extraction, Production and Uses of Molybdenum. Elsevier Science.

Brar, S. K., Verma, M., Tyagi, R. D., & Surampalli, R. Y. (2010). Engineered nanoparticles in wastewater and wastewater sludge - Evidence and impacts. *Waste Management (New York, N.Y.)*, *30*(3), 504–520. doi:10.1016/j.wasman.2009.10.012 PMID:19926463

Brennan, M. E., Coleman, J. N., Drury, A., Lahr, B., Kobayashi, T., & Blau, W. J. (2003). Nonlinear photoluminescence from van Hove singularities in multiwalled carbon nanotubes. *Optics Letters*, *28*(4), 266. doi:10.1364/OL.28.000266 PMID:12653367

Briggs, N., Raman, A. K. Y., Barrett, L., Brown, C., Li, B., Leavitt, D., Aichele, C. P., & Crossley, S. (2018). Stable pickering emulsions using multi-walled carbon nanotubes of varying wettability. *Colloids and Surfaces. A, Physicochemical and Engineering Aspects*, *537*, 227–235. Advance online publication. doi:10.1016/j.colsurfa.2017.10.010

Brodie, B. C. (1860). Researches on the atomic weight of graphite. *Quarterly Journal of the Chemical Society of London*, *12*(1), 261–268. Advance online publication. doi:10.1039/QJ8601200261

Brown, P. O., Botstein, D., Lapidot, M., Pilpel, Y., Mazo, A., Hodgson, J. W., Petruk, S., Sedkov, Y., Brock, H. W., Timmons, J. A., Good, L., Kapranov, P., Willingham, A. T., Gingeras, T. R., Yazgan, O., Krebs, J. E., Fahey, M. E., Moore, T. F., Higgins, D. G., ... Schadt, E. E. (2008). Label-Free Biomedical Imaging with. *Science*, *322*(5909), 1857–1862. doi:10.1126cience.1165758 PMID:19095943

Brusotti, G., Calleri, E., Colombo, R., Massolini, G., Rinaldi, F., & Temporini, C. (2018). Advances on Size Exclusion Chromatography and Applications on the Analysis of Protein Biopharmaceuticals and Protein Aggregates : A Mini Review. *Chromatographia*, *81*(1), 3–23. doi:10.100710337-017-3380-5

Buazar, F., Baghlani-Nejazd, M. H., Badri, M., Kashisaz, M., Khaledi-Nasab, A., & Kroushawi, F. (2016). Facile one-pot phytosynthesis of magnetic nanoparticles using potato extract and their catalytic activity. *Stärke*, *68*(7-8), 796–804. doi:10.1002tar.201500347

Bulbake, U., Doppalapudi, S., Kommineni, N., & Khan, W. (2017). Liposomal Formulations in Clinical Use: An Updated Review. *Pharmaceutics*, *9*(2), 12. doi:10.3390/pharmaceutics9020012 PMID:28346375

Burdus, A.-C., Gherasim, O., Grumezescu, A.M., Mogoantă, L., Ficai, A., & Andronescu, E. (2018). Biomedical applications of silver nanoparticles: An up-to-date overview. *Nanomaterials, 8*, 681.

Buwalda, S. J., Vermonden, T., & Hennink, W. E. (2017). Hydrogels for therapeutic delivery: Current developments and future directions. *Biomacromolecules*, *18*(2), 316–330. doi:10.1021/acs.biomac.6b01604 PMID:28027640

Byeon, H. J., Lee, C., Lee, S., Lee, E. S., Choi, H. G., Park, E. S., & Youn, Y. S. (2016). Pharmaceutical potential of tacrolimus-loaded albumin nanoparticles having targetability to rheumatoid arthritis tissues. *International Journal of Pharmaceutics*, *497*(1-2), 268–276. doi:10.1016/j.ijpharm.2015.12.004 PMID:26657273

Byrne, J. A., Dunlop, P. S. M., Hamilton, J. W. J., Fernández-Ibáñez, P., Polo-López, I., Sharma, P. K., & Vennard, A. S. M. (2015). A review of heterogeneous photocatalysis for water and surface disinfection. *Molecules (Basel, Switzerland)*, *20*(4), 5574–5615. doi:10.3390/molecules20045574 PMID:25830789

Camilli, L., Yu, F., Cassidy, A., Hornekær, L., & Bøggild, P. (2019). Challenges for continuous graphene as a corrosion barrier. In 2D Materials (Vol. 6, Issue 2, p. 022002). IOP Publishing Ltd. doi:10.1088/2053-1583/ab04d4

Canfarotta, F., & Piletsky, S. A. (2014). Engineered magnetic nanoparticles for biomedical applications. *Advanced Healthcare Materials*, *3*(2), 160–175. doi:10.1002/adhm.201300141 PMID:24497448

Cao, J., Elliott, D., & Zhang, W. (2005). Perchlorate reduction by nanoscale iron particles. *Journal of Nanoparticle Research*, *7*(4-5), 499–506. doi:10.100711051-005-4412-x

Caputo, F., De Nicola, M., & Ghibelli, L. (2014). Pharmacological potential of bioactive engineered nanomaterials. *Biochemical Pharmacology*, *92*(1), 112–130. doi:10.1016/j.bcp.2014.08.015 PMID:25175739

Cardinal, M. F., Vander Ende, E., Hackler, R. A., McAnally, M. O., Stair, P. C., Schatz, G. C., & Van Duyne, R. P. (2017). Expanding applications of SERS through versatile nanomaterials engineering. *Chemical Society Reviews*, *46*(13), 3886–3903. doi:10.1039/C7CS00207F PMID:28640313

Carolin, C. F., Kumar, P. S., Saravanan, A., Joshiba, G. J., & Naushad, M. (2017). Efficient techniques for the removal of toxic heavy metals from aquatic environment: A review. *Journal of Environmental Chemical Engineering*, *5*(3), 2782–2799. doi:10.1016/j.jece.2017.05.029

Carpenter, C. (2017). New Magnetite Nanoparticles Allow Smart Drilling Fluids with Superior Properties. *Journal of Petroleum Technology*, *69*(11), 41–42. Advance online publication. doi:10.2118/1117-0041-JPT

Carter, L. J., Garner, L. V., Smoot, J. W., Li, Y., Zhou, Q., Saveson, C. J., Sasso, J. M., Gregg, A. C., Soares, D. J., Beskid, T. R., Jervey, S. R., & Liu, C. (2020). Assay Techniques and Test Development for COVID-19 Diagnosis. *ACS Central Science*, *6*(5), 591–605. doi:10.1021/acscentsci.0c00501 PMID:32382657

Carvalho, A. P. A., & Conte-Junior, C. A. (2021). Recent Advances on Nanomaterials to COVID-19 Management: A Systematic Review on Antiviral/Virucidal Agents and Mechanisms of SARS-CoV-2 Inhibition/Inactivation. *Global Challenges (Hoboken, NJ)*, *5*(5), 2000115. doi:10.1002/gch2.202000115 PMID:33786199

Cassano, R., Mellace, S., Marrelli, M., Conforti, F., & Trombino, S. (2017). α-Tocopheryl linolenate solid lipid nanoparticles for the encapsulation, protection, and release of the omega-3 polyunsaturated fatty acid: In vitro anti-melanoma activity evaluation. *Colloids and Surfaces. B, Biointerfaces*, *151*, 128–133. doi:10.1016/j.colsurfb.2016.11.043 PMID:27988473

Cavalcanti, I. D. L., & Cajubá de Britto Lira Nogueira, M. (2020). Pharmaceutical nanotechnology: Which products are been designed against COVID-19? *Journal of Nanoparticle Research*, *22*(9), 276. Advance online publication. doi:10.100711051-020-05010-6 PMID:32922162

Cavalli, R., Caputo, O., & Gasco, M. R. (1993). Solid lipospheres of doxorubicin and idarubicin. *International Journal of Pharmaceutics*, *89*(1), R9–R12. doi:10.1016/0378-5173(93)90313-5

Cervantes, B., Arana, L., Murillo-Cuesta, S., Bruno, M., Alkorta, I., & Varela-Nieto, I. (2019). Solid lipid nanoparticles loaded with glucocorticoids protect auditory cells from cisplatin-induced ototoxicity. *Journal of Clinical Medicine*, *8*(9), 1464. doi:10.3390/jcm8091464 PMID:31540035

Cevc, G. (2012). Rational design of new product candidates: The next generation of highly deformable bilayer vesicles for noninvasive, targeted therapy. *Journal of Controlled Release*, *160*(2), 135–146. doi:10.1016/j.jconrel.2012.01.005 PMID:22266051

Chadar, R., Afzal, O., Alqahtani, S. M., & Kesharwani, P. (2021). Carbon nanotubes as an emerging nanocarrier for the delivery of doxorubicin for improved chemotherapy. *Colloids and Surfaces. B, Biointerfaces*, *208*, 112044. doi:10.1016/j.colsurfb.2021.112044 PMID:34419810

Chae, S.-R., Watanabe, Y., & Wiesner, M. R. (2011). *Comparative photochemical reactivity of spherical and tubular fullerene nanoparticles in water under ultraviolet (UV) irradiation*. Academic Press.

Chakraborty, D., Kumar, S., Chandrasekaran, N., & Mukherjee, A. (2020). Viral Diagnostics and Preventive Techniques in the Era of COVID-19: Role of Nanoparticles. *Frontiers in Nanotechnology*, *2*(October), 588795. Advance online publication. doi:10.3389/fnano.2020.588795

Chandra, H., Kumari, P., Bontempi, E., & Yadav, S. (2020). Medicinal plants: Treasure trove for green synthesis of metallic nanoparticles and their biomedical applications. *Biocatalysis and Agricultural Biotechnology*, *24*, 101518. doi:10.1016/j.bcab.2020.101518

Chandran, S. P., Chaudhary, M., Pasricha, R., Ahmad, A., & Sastry, M. (2006). Synthesis of gold nanotriangles and silver nanoparticles using Aloe vera plant extract. *Biotechnology Progress*, *22*(2), 577–583. doi:10.1021/bp0501423 PMID:16599579

Chang Chien, S. W., Chang, C. H., Chen, S. H., Wang, M. C., Madhava Rao, M., & Satya Veni, S. (2011). Effect of sunlight irradiation on photocatalytic pyrene degradation in contaminated soils by micro-nano size TiO2. *The Science of the Total Environment*, *409*(19), 4101–4108. doi:10.1016/j.scitotenv.2011.06.050 PMID:21762957

Chang, H., Wang, H., Shao, N., Wang, M., Wang, X., & Cheng, Y. (2014). Surface-engineered dendrimers with a diaminododecane core achieve efficient gene transfection and low cytotoxicity. *Bioconjugate Chemistry*, *25*(2), 342–350. doi:10.1021/bc400496u PMID:24410081

Chang, X., Wang, T., & Gong, J. (2016). CO2 photo-reduction: Insights into CO2 activation and reaction on surfaces of photocatalysts. *Energy & Environmental Science*, *9*(7), 2177–2196. doi:10.1039/C6EE00383D

Chatterjee, A. K., Chakraborty, R., & Basu, T. (2014). *Mechanism of antibacterial activity of copper nanoparticles.* Academic Press.

Chaturvedi, S., & Dave, P. N. (2020). Application of nanotechnology in foods and beverages. In Nanoengineering in the Beverage Industry (pp. 137-162). doi:10.1016/B978-0-12-816677-2.00005-3

Chaudhary, S., Sharma, P., Chauhan, P., Kumar, R., & Umar, A. (2019). Functionalized nanomaterials: A new avenue for mitigating environmental problems. *International Journal of Environmental Science and Technology*, *16*(9), 5331–5358. doi:10.100713762-019-02253-2

Chaudhary, V., Royal, A., Chavali, M., & Yadav, S. K. (2021). Advancements in Research and Development to Combat COVID-19 using Nanotechnology. *Nanotechnology for Environmental Engineering*, *6*(1), 8. doi:10.100741204-021-00102-7

Chaudhuri, P., Paraskar, A., Soni, S., Mashelkar, R. A., & Sengupta, S. (2009). Fullerenol–cytotoxic conjugates for cancer chemotherapy. *ACS Nano*, *3*(9), 2505–2514. doi:10.1021/nn900318y PMID:19681636

Chauhan, G., Madou, M. J., Kalra, S., Chopra, V., Ghosh, D., & Martinez-Chapa, S. O. (2020). Nanotechnology for COVID-19: Therapeutics and Vaccine Research. *ACS Nano*, *14*(7), 7760–7782. doi:10.1021/acsnano.0c04006 PMID:32571007

Chen, H., Wang, B., Gao, D., Guan, M., Zheng, L., Ouyang, H., Chai, Z., Zhao, Y., & Feng, W. (2013). *Broad-spectrum antibacterial activity of carbon nanotubes to human gut bacteria.* Academic Press.

Chen, K., Li, C., Chen, Z., Shi, L., Reddy, S., Meng, H., Ji, Q., Zhang, Y., & Liu, Z. (n.d.). *Bioinspired synthesis of CVD graphene flakes and graphene-supported molybdenum sulfide catalysts for hydrogen evolution reaction.* doi:10.1007/s12274-016-1013-1

Chen, L., Han, Q., Li, W., Zhou, Z., Fang, Z., Xu, Z., Wang, Z., & Qian, X. (2018). Three-dimensional graphene-based adsorbents in sewage disposal: a review. In Environmental Science and Pollution Research (Vol. 25, Issue 26, pp. 25840–25861). Springer Verlag. doi:10.100711356-018-2767-7

Chen, G., Roy, I., Yang, C., & Prasad, P. N. (2016). Nanochemistry and nanomedicine for nanoparticle-based diagnostics and therapy. *Chemical Reviews*, *116*(5), 2826–2885. doi:10.1021/acs.chemrev.5b00148 PMID:26799741

Cheng, H., Jiang, Z., Sun, C., Wang, Z., Han, G., Chen, X., Li, T., Fan, Z., Zhang, F., Yang, X., Lv, L., Zhang, H., Zhou, J., & Ding, Y. (2022). Protein stabilized polymeric nanoparticles inspired relay drug delivery for tackling post-chemotherapeutic metastasis. *Chemical Engineering Journal*, *427*, 131672. doi:10.1016/j.cej.2021.131672

Cheng, K. K., Chan, P. S., Fan, S., Kwan, S. M., Yeung, K. L., Wáng, Y. X. J., Chow, A. H. L., Wu, E. X., & Baum, L. (2015). Curcumin-conjugated magnetic nanoparticles for detecting amyloid plaques in Alzheimer's disease mice using magnetic resonance imaging (MRI). *Biomaterials*, *44*, 155–172. Advance online publication. doi:10.1016/j.biomaterials.2014.12.005 PMID:25617135

Cheng, L., Wang, C., Feng, L., Yang, K., & Liu, Z. (2014). Functional nanomaterials for phototherapies of cancer. *Chemical Reviews*, *114*(21), 10869–10939. doi:10.1021/cr400532z PMID:25260098

Cheng, P., Zhang, S., Wang, Q., Feng, X., Zhang, S., Sun, Y., & Wang, F. (2021). Contribution of Nano-Zero-Valent Iron and Arbuscular Mycorrhizal Fungi to Phytoremediation of Heavy Metal-Contaminated Soil. *Nanomaterials (Basel, Switzerland)*, *11*(5), 1264. Advance online publication. doi:10.3390/nano11051264 PMID:34065026

Cheng, Z., Li, M., Dey, R., & Chen, Y. (2021). Nanomaterials for cancer therapy: Current progress and perspectives. *Journal of Hematology & Oncology*, *14*(1), 85. doi:10.118613045-021-01096-0 PMID:34059100

Chen, H., Kou, X., Yang, Z., Ni, W., & Wang, J. (2008). Shape- and size dependent refractive index sensitivity of gold nanoparticles. *Langmuir*, *24*(10), 5233–5237. doi:10.1021/la800305j PMID:18435552

Chen, J., Jiang, H., Wu, Y., Li, Y., & Gao, Y. (2015). A novel glycyrrhetinic acid-modified oxaliplatin liposome for liver-targeting and in vitro/vivo evaluation. *Drug Design, Development and Therapy*, *9*, 2265. PMID:25945038

Chen, L., Huang, Q., Zhao, T., Sui, L., Wang, S., Xiao, Z., Nan, Y., & Ai, K. (2021). Nanotherapies for Sepsis by Regulating Inflammatory Signals and Reactive Oxygen and Nitrogen Species: New Insight for Treating COVID-19. *Redox Biology*, *45*, 102046. doi:10.1016/j.redox.2021.102046 PMID:34174559

Chen, L., Zhong, H., Qi, X., Shao, H., & Xu, K. (2019). Modified core–shell magnetic mesoporous zirconia nanoparticles formed through a facile "outside-to-inside" way for CT/MRI dual-modal imaging and magnetic targeting cancer chemotherapy. *RSC Advances*, *9*(23), 13220–13233. doi:10.1039/C9RA01063G

Chen, R., Zhang, Q., Gu, Y., Tang, L., Li, C., & Zhang, Z. (2015). One-pot green synthesis of Prussian blue nanocubes decorated reduced graphene oxide using mushroom extract for efficient 4-nitrophenol reduction. *Analytica Chimica Acta*, *853*(1), 579–587. doi:10.1016/j.aca.2014.10.049 PMID:25467506

Chen, S., Brown, L., Levendorf, M., Cai, W., Ju, S. Y., Edgeworth, J., Li, X., Magnuson, C. W., Velamakanni, A., Piner, R. D., Kang, J., Park, J., & Ruoff, R. S. (2011). Oxidation resistance of graphene-coated Cu and Cu/Ni alloy. *ACS Nano*, *5*(2), 1321–1327. doi:10.1021/nn103028d PMID:21275384

Chen, S.-H., Jiang, Y.-S., & Lin, H. (2020). Easy Synthesis of BiVO4 for Photocatalytic Overall Water Splitting. *ACS Omega*, *5*(15), 8927–8933. doi:10.1021/acsomega.0c00699 PMID:32337456

Chen, W., Weimin, H., Li, D., Chen, S., & Dai, Z. (2018). A critical review on the development and performance of polymer/graphene nanocomposites. *Science and Engineering of Composite Materials*, *25*(6), 1059–1073. doi:10.1515ecm-2017-0199

Chen, W., & Yan, L. (2011). In situ self-assembly of mild chemical reduction graphene for three-dimensional architectures. *Nanoscale*, *3*(8), 3132–3137. doi:10.1039/c1nr10355e PMID:21698339

Chen, X., Li, X., Xu, D., Yang, W., & Bai, S. (2020). Application of nanoscale zero-valent iron in hexavalent chromium-contaminated soil: A review. *Nanotechnology Reviews*, *9*(1), 736–750. doi:10.1515/ntrev-2020-0059

Chen, Y., Li, N., Yang, Y., & Liu, Y. (2015). A dual targeting cyclodextrin/gold nanoparticle conjugate as a scaffold for solubilization and delivery of paclitaxel. *RSC Advances*, *5*(12), 8938–8941. doi:10.1039/C4RA13135E

Chen, Z., Mao, R., & Liu, Y. (2012). Fullerenes for cancer diagnosis and therapy: Preparation, biological and clinical perspectives. *Current Drug Metabolism*, *13*(8), 1035–1045. doi:10.2174/138920012802850128 PMID:22380017

Chen, Z., Zhang, Z., Zhai, X., Li, Y., Lin, L., Zhao, H., Bian, L., Li, P., Yu, L., Wu, Y., & Lin, G. (2020). Rapid and Sensitive Detection of anti-SARS-CoV-2 IgG, Using Lanthanide-Doped Nanoparticles-Based Lateral Flow Immunoassay. *Analytical Chemistry*, *92*(10), 7226–7231. doi:10.1021/acs.analchem.0c00784 PMID:32323974

Cheraghian, G., Rostami, S., & Afrand, M. (2020). Nanotechnology in enhanced oil recovery. Processes, 8(9). doi:10.3390/pr8091073

Cheriyamundath, S., & Vavilala, S. L. (2021). Nanotechnology-based wastewater treatment. *Water and Environment Journal: the Journal / the Chartered Institution of Water and Environmental Management*, *35*(1), 123–132. doi:10.1111/wej.12610

Chernousova, S., & Epple, M. (2013). Silver as antibacterial agent: Ion, nanoparticle, and metal. *Angewandte Chemie International Edition*, *52*(6), 1636–1653. doi:10.1002/anie.201205923 PMID:23255416

Chevalier, Y., & Bolzinger, M. A. (2013). Emulsions stabilized with solid nanoparticles: Pickering emulsions. *Colloids and Surfaces. A, Physicochemical and Engineering Aspects*, *439*, 23–34. Advance online publication. doi:10.1016/j.colsurfa.2013.02.054

Chintagunta, A. D., Sai Krishna, M., Nalluru, S., & Sampath Kumar, N. S. (2021). Nanotechnology: An emerging approach to combat COVID-19. *Emergent Materials*, *4*(1), 119–130. doi:10.100742247-021-00178-6 PMID:33615141

Choi, J. R. (2020). Development of Point-of-Care Biosensors for COVID-19. *Frontiers in Chemistry*, *8*, 517. doi:10.3389/fchem.2020.00517 PMID:32574316

Choi, M., Kim, J., & Maibach, H. (2006). Topical DNA vaccination with DNA/Lipid based complex. *Current Drug Delivery*, *3*(1), 37–45. doi:10.2174/156720106775197484 PMID:16472092

Chong, M. N., Jin, B., Chow, C. W. K., & Saint, C. (2010). Recent developments in photocatalytic water treatment technology: A review. *Water Research*, *44*(10), 2997–3027. doi:10.1016/j.watres.2010.02.039 PMID:20378145

Cho, S. H., Jones, B. L., & Krishnan, S. (2009). The dosimetric feasibility of gold nanoparticle-aided radiation therapy (GNRT) via brachytherapy using low-energy gamma-/x-ray sources. *Physics in Medicine and Biology*, *54*(16), 4889–4905. doi:10.1088/0031-9155/54/16/004 PMID:19636084

Choudhary, S., Gupta, L., Rani, S., Dave, K., & Gupta, U. (2017). Impact of Dendrimers on Solubility of Hydrophobic Drug Molecules. *Frontiers in Pharmacology*, *8*, 261. doi:10.3389/fphar.2017.00261 PMID:28559844

Choudhury, P. R., Bhattacharya, P., Ghosh, S., Majumdar, S., Saha, S., & Sahoo, G. C. (2017). Removal of Cr (VI) by synthesized titania embedded dead yeast nanocomposite: Optimization and modeling by response surface methodology. *Journal of Environmental Chemical Engineering*, *5*(1), 214–221. doi:10.1016/j.jece.2016.11.041

Chow, E. K., Zhang, X.-Q., Chen, M., Lam, R., Robinson, E., Huang, H., Schaffer, D., Osawa, E., Goga, A., & Ho, D. (2011). Nanodiamond therapeutic delivery agents mediate enhanced chemoresistant tumor treatment. *Science Translational Medicine, 3*(73), 73ra21-73ra21. doi:10.1126/scitranslmed.3001713

Chowdhury, S. M., Lalwani, G., Zhang, K., Yang, J. Y., Neville, K., & Sitharaman, B. (2013). Cell specific cytotoxicity and uptake of graphene nanoribbons. *Biomaterials*, *34*(1), 283–293. doi:10.1016/j.biomaterials.2012.09.057 PMID:23072942

Chuah, T. G., Jumasiah, A., Azni, I., Katayon, S., & Thomas Choong, S. Y. (2005). Rice husk as a potentially low-cost biosorbent for heavy metal and dye removal: An overview. *Desalination*, *175*(3), 305–316. doi:10.1016/j.desal.2004.10.014

Chugh, H., Sood, D., Chandra, I., Tomar, V., Dhawan, G., & Chandra, R. (2018). Role of gold and silver nanoparticles in cancer nanomedicine. *Artificial Cells, Nanomedicine, and Biotechnology*, *46*(sup1), 1210–1220. doi:10.1080/21691401.2018.1449118 PMID:29533101

Chung, Y. H., Beiss, V., Fiering, S. N., & Steinmetz, N. F. (2020). COVID-19 Vaccine Frontrunners and Their Nanotechnology Design. *ACS Nano*, *14*(10), 12522–12537. doi:10.1021/acsnano.0c07197 PMID:33034449

Çıplak, Z., Getiren, B., Gökalp, C., Yıldız, A., & Yıldız, N. (2020). Green synthesis of reduced graphene oxide-AgAu bimetallic nanocomposite: Catalytic performance. *Chemical Engineering Communications*, *207*(4), 559–573. doi:10.1080/00986445.2019.1613227

Çıplak, Z., Yıldız, A., & Yıldız, N. (2020). Green preparation of ternary reduced graphene oxide-au@polyaniline nanocomposite for supercapacitor application. *Journal of Energy Storage*, *32*(August), 101846. doi:10.1016/j.est.2020.101846

Cohen, A. A., Gnanapragasam, P. N. P., Lee, Y. E., Hoffman, P. R., Ou, S., Kakutani, L. M., Keeffe, J. R., Wu, H. J., Howarth, M., West, A. P., Barnes, C. O., Nussenzweig, M. C., & Bjorkman, P. J. (2021). Mosaic nanoparticles elicit cross-reactive immune responses to zoonotic coronaviruses in mice. *Science, 371*(6530), 735–741. doi:10.1126cience.abf6840 PMID:33436524

Compton, O. C., & Nguyen, S. T. (2010). Graphene oxide, highly reduced graphene oxide, and graphene: Versatile building blocks for carbon-based materials. Small. doi:10.1002mll.200901934

Cong, X., Tian, H., Liu, S., Mao, K., Chen, H., Xin, Y., Liu, F., Wang, X., Meng, X., Zhu, G., Wang, J., Gao, X., Tan, H., Yang, Y.-G., & Sun, T. (2020). Cationic Liposome/DNA complexes mediate antitumor immunotherapy by promoting immunogenic tumor cell death and dendritic cell activation. *ACS Applied Materials & Interfaces, 12*(25), 28047–28056. doi:10.1021/acsami.0c08112 PMID:32478501

Contreras, O., Hareland, G., Husein, M., Nygaard, R., & Alsaba, M. (2014). Application of in-house prepared nanoparticles as filtration control additive to reduce formation damage. *SPE - European Formation Damage Conference, Proceedings, EFDC*, 1. 10.2118/168116-MS

Corres, M. Á., Mayor, Á., Sangroniz, A., del Río, J., Iriarte, M., & Etxeberria, A. (2020). Blends based on biodegradable poly(caprolactone) with outstanding barrier properties for packaging applications: The role of free volume and interactions. *European Polymer Journal, 135*(July), 109869. doi:10.1016/j.eurpolymj.2020.109869

Courty, A., Mermet, A., Albouy, P. A., Duval, E., & Pileni, M. P. (2005). Vibrational coherence of self-organized silver nanocrystals in f.c.c. supra-crystals. *Nature Materials, 4*(5), 395–398. doi:10.1038/nmat1366 PMID:15834414

Crosera, M., Prodi, A., Mauro, M., Pelin, M., Florio, C., Bellomo, F., Adami, G., Apostoli, P., De Palma, G., Bovenzi, M., Campanini, M., & Filon, F. L. (2015). Titanium dioxide nanoparticle penetration into the skin and effects on HaCaT cells. *International Journal of Environmental Research and Public Health, 12*(8), 9282–9297. doi:10.3390/ijerph120809282 PMID:26262634

da Rocha, S. R., Heyder, R. S., Bielski, E. R., Guo, A., Steinmaurer, M., & Reineke, J. J. (2019). Inhalation drug products containing nanomaterials. In *Pharmaceutical Inhalation Aerosol Technology* (pp. 403–423). CRC Press. doi:10.1201/9780429055201-17

da Rosa, M. B., Oreste, E. Q., Bönemann, D. H., & Rodrigues, A. A., Vendruscolo, C. T., Moreira, A. S., Ribeiro, A. S., & Nunes, A. M. (2016). Evaluation of the use of a reflux system for sample preparation of xanthan gum and subsequent determination of Ca, Cu, K, Mg, Na and Zn by atomic spectrometry techniques. *Journal of the Brazilian Chemical Society, 27*(5), 919–924. doi:10.1016/j.foodchem.2017.07.112

Dadfar, S. M., Roemhild, K., Drude, N. I., von Stillfried, S., Knüchel, R., Kiessling, F., & Lammers, T. (2019). Iron oxide nanoparticles: Diagnostic, therapeutic and theranostic applications. *Advanced Drug Delivery Reviews, 138*, 302–325. doi:10.1016/j.addr.2019.01.005 PMID:30639256

Dallinger, D., & Kappe, C. O. (2007). Microwave-Assisted Synthesis in Water as Solvent. *Chemical Reviews, 107*(6), 2563–2591. doi:10.1021/cr0509410 PMID:17451275

Danesh, N., Ghorbani, M., & Marjani, A. (2021). Separation of copper ions by nanocomposites using adsorption process. *Scientific Reports, 11*(1), 1–23. doi:10.103841598-020-80914-w PMID:33462314

Dang, Y., & Guan, J. (2020). Nanoparticle-based drug delivery systems for cancer therapy. *Smart Materials in Medicine, 1*, 10–19. doi:10.1016/j.smaim.2020.04.001 PMID:34553138

Danhier, F., Feron, O., & Préat, V. (2010). To exploit the tumor microenvironment: Passive and active tumor targeting of nanocarriers for anti-cancer drug delivery. *Journal of Controlled Release, 148*(2), 135–146. doi:10.1016/j.jconrel.2010.08.027 PMID:20797419

Daraee, H., Etemadi, A., Kouhi, M., Alimirzalu, S., & Akbarzadeh, A. (2016). Application of liposomes in medicine and drug delivery. *Artificial Cells, Nanomedicine, and Biotechnology, 44*(1), 381–391. doi:10.3109/21691401.2014.953633 PMID:25222036

Darroudi, M. (2011). Time-dependent effect in green synthesis of silver nanoparticles. *International Journal of Nanomedicine, 677*, 677. Advance online publication. doi:10.2147/IJN.S17669 PMID:21556342

Das, O. P., & Al-Enezi, K. (2014). Successful application of innovative reversible invert emulsion reservoir drilling fluid and improved filter-cake cleanup agent increased sustained well productivity in smart multilateral wells with inflow mart completion in high mobility reservoir, west Kuwait. *Society of Petroleum Engineers - 30th Abu Dhabi International Petroleum Exhibition and Conference, ADIPEC 2014: Challenges and Opportunities for the Next 30 Years, 5.* 10.2118/172030-MS

Das, A., Kamle, M., Bharti, A., & Kumar, P. (2019). Nanotechnology and it's applications in environmental remediation: An overview. *Vegetos, 32*(3), 1–11. doi:10.100742535-019-00040-5

Das, T. R., & Sharma, P. K. (2020). Hydrothermal-assisted green synthesis of Ni/Ag@rGO nanocomposite using Punica granatum juice and electrochemical detection of ascorbic acid. *Microchemical Journal, 156*(March), 104850. doi:10.1016/j.microc.2020.104850

Dauthal, P., & Mukhopadhyay, M. (2015). Biofabrication, characterization, and possible bio-reduction mechanism of platinum nanoparticles mediated by agro-industrial waste and their catalytic activity. *Journal of Industrial and Engineering Chemistry, 22*, 185–191. doi:10.1016/j.jiec.2014.07.009

Dave, P. N., & Chopda, L. V. (2014). Application of iron oxide nanomaterials for the removal of heavy metals. *Journal of Nanotechnology, 2014*, 1–14. Advance online publication. doi:10.1155/2014/398569

David, L., & Moldovan, B. (2020). Green Synthesis of Biogenic Silver Nanoparticles for Efficient Catalytic Removal of Harmful Organic Dyes. *Nanomaterials (Basel, Switzerland), 10*(2), 202. Advance online publication. doi:10.3390/nano10020202 PMID:31991548

Day, J., Senthilarasu, S., & Mallick, T. K. (2019). Improving spectral modification for applications in solar cells: A review. *Renewable Energy, 132*, 186–205. doi:10.1016/j.renene.2018.07.101

De Gusseme, B., Sintubin, L., Baert, L., Thibo, E., Hennebel, T., Vermeulen, G., Uyttendaele, M., Verstraete, W., & Boon, N. (2010). Biogenic silver for disinfection of water contaminated with viruses. *Applied and Environmental Microbiology, 76*(4), 1082–1087. doi:10.1128/AEM.02433-09 PMID:20038697

De Heer, W. A., Berger, C., Wu, X., First, P. N., Conrad, E. H., Li, X., Li, T., Sprinkle, M., Hass, J., Sadowski, M. L., Potemski, M., Martinez, G., & Macdonald, H. (2007). Epitaxial graphene. *Solid State Communications, 143*(1-2), 92–100. doi:10.1016/j.ssc.2007.04.023

de la Zerda, A., Bodapati, S., Teed, R., May, S. Y., Tabakman, S. M., Liu, Z., Khuri-Yakub, B. T., Chen, X., Dai, H., & Gambhir, S. S. (2012). Family of enhanced photoacoustic imaging agents for high-sensitivity and multiplexing studies in living mice. *ACS Nano, 6*(6), 4694–4701. doi:10.1021/nn204352r PMID:22607191

de Medeiros, T. V., Manioudakis, J., Noun, F., Macairan, J.-R., Victoria, F., & Naccache, R. (2019). Microwave-assisted synthesis of carbon dots and their applications. *Journal of Materials Chemistry. C, Materials for Optical and Electronic Devices, 7*(24), 7175–7195. doi:10.1039/C9TC01640F

De Silva, K. K. H., Huang, H. H., Joshi, R. K., & Yoshimura, M. (2017). Chemical reduction of graphene oxide using green reductants. In *Carbon* (Vol. 119, pp. 190–199). Elsevier Ltd. doi:10.1016/j.carbon.2017.04.025

de Souza Farias, S. A., da Costa, K. S., & Martins, J. B. L. (2021). Analysis of conformational, structural, magnetic, and electronic properties related to antioxidant activity: Revisiting flavan, anthocyanidin, flavanone, flavonol,isoflavone, flavone, and flavan-3-ol. *ACS Omega*, *6*(13), 8908–8918. doi:10.1021/acsomega.0c06156 PMID:33842761

De Volder, M. F., Tawfick, S. H., Baughman, R. H., & Hart, A. (2013). *Carbon nanotubes: present and future commercial applications*. Academic Press.

DeAlba-Montero, I., Guajardo-Pacheco, J., Morales-Sánchez, E., Araujo-Martínez, R., Loredo-Becerra, G., Martínez-Castañón, G.-A., Ruiz, F., & Compeán Jasso, M. (2017). *Antimicrobial properties of copper nanoparticles and amino acid chelated copper nanoparticles produced by using a soya extract*. Academic Press.

DeLassus, P. (2002). Barrier polymers. In *Encyclopedia of Chemical Technology* (pp. 375–407). John Wiley & Sons, Inc.

Demissie, M. G., Sabir, F. K., Edossa, G. D., & Gonfa, B. A. (2020). Synthesis of zinc oxide nanoparticles using leaf extract of Lippia adoensis (Koseret) and evaluation of its antibacterial activity. *Journal of Chemistry*, *2020*, 1–9. doi:10.1155/2020/7459042

Deng, J., Zhao, S., Liu, Y., Liu, C., & Sun, J. (2021). Nanosensors for Diagnosis of Infectious Diseases. *ACS Applied Bio Materials*, *4*(5), 3863–3879. doi:10.1021/acsabm.0c01247

Deng, N. N., Wang, W., Ju, X. J., Xie, R., Weitz, D. A., & Chu, L. Y. (2013). Wetting-induced formation of controllable monodisperse multiple emulsions in microfluidics. *Lab on a Chip*, *13*(20), 4047–4052. Advance online publication. doi:10.1039/C3LC50638J PMID:23948718

Deng, Z., Wang, Y., Yu, C., Li, G., Lang, X., Wang, S., & Fan, S. (2021). Promoting methane hydrate formation with expanded graphite additives: Application to solidified natural gas storage. *Fuel*, *299*, 120867. Advance online publication. doi:10.1016/j.fuel.2021.120867

Deshmukh, K. (2019). Nanotechnology in Ancient Era. In *Biotechnology products in everyday life* (pp. 3–14). Springer. doi:10.1007/978-3-319-92399-4_1

Devience, S. J., Pham, L. M., Lovchinsky, I., Sushkov, A. O., Bar-gill, N., Belthangady, C., Casola, F., Corbett, M., Zhang, H., Lukin, M., Park, H., Yacoby, A., & Walsworth, R. L. (2015). Nanoscale NMR spectroscopy and imaging of multiple nuclear species. *Nature Nanotechnology*, *10*(2), 129–134. doi:10.1038/nnano.2014.313 PMID:25559712

Devi, G. K., & Sathishkumar, K. (2017). Synthesis of gold and silver nanoparticles using Mukia maderaspatna plant extract and its anticancer activity. *IET Nanobiotechnology / IET*, *11*(2), 143–151. doi:10.1049/iet-nbt.2015.0054 PMID:28476996

Deyhimihaghighi, N., Noor, N. M., Soltani, N., Jorfi, R., Haghir, M. E., Adenan, M. Z., Saion, E., & Khandaker, M. U. (2014, November). Contrast enhancement of magnetic resonance imaging (MRI) of polymer gel dosimeter by adding Platinum nano-particles. *Journal of Physics: Conference Series*, *546*(1), 012013. doi:10.1088/1742-6596/546/1/012013

Dhandapani, P., Prakash, A. A., AlSalhi, M. S., Maruthamuthu, S., Devanesan, S., & Rajasekar, A. (2020). Ureolytic bacteria mediated synthesis of hairy ZnO nanostructure as photocatalyst for decolorization of dyes. *Materials Chemistry and Physics*, *243*, 122619. doi:10.1016/j.matchemphys.2020.122619

Dhand, C., Dwivedi, N., Loh, X. J., Jie Ying, A. N., Verma, N. K., Beuerman, R. W., Lakshminarayanan, R., & Ramakrishna, S. (2015). Methods and strategies for the synthesis of diverse nanoparticles and their applications: A comprehensive overview. *RSC Advances*, *5*(127), 105003–105037. doi:10.1039/C5RA19388E

Dharmalingam, K., & Anandalakshmi, R. (2020). Functionalization of cellulose-based nanocomposite hydrogel films with zinc oxide complex and grapefruit seed extract for potential applications in treating chronic wounds. *Polymer, 202*(April), 122620. doi:10.1016/j.polymer.2020.122620

Dhas, S. D., Maldar, P. S., Patil, M. D., Hubali, K. M., Shembade, U. V., Abitkar, S. B., Waikar, M. R., Sonkawade, R. G., Agawane, G. L., & Moholkar, A. V. (2021). Hydrothermal synthesis of mesoporous $NiMnO_3$ nanostructures for supercapacitor application: Effect of electrolyte. *Journal of Energy Storage, 35*, 102277.

Dhoot, S. N., Freeman, B. D., & Stewart, M. E. (2002). Barrier Polymers. In *Encyclopedia of Polymer Science and Technology*. John Wiley & Sons, Inc. doi:10.1002/0471440264.pst025

DHV Consultants & DELFT Hydraulics. (2003). *Operational Manual-water quality analysis (SW)*. Author.

Dieng, S. M., Anton, N., Bouriat, P., Thioune, O., Sy, P. M., Massaddeq, N., Enharrar, S., Diarra, M., & Vandamme, T. (2019). Pickering nano-emulsions stabilized by solid lipid nanoparticles as a temperature sensitive drug delivery system. *Soft Matter, 15*(40), 8164–8174. Advance online publication. doi:10.1039/C9SM01283D PMID:31593197

Digraskar, R., Sapner, V. S., Ghule, A. V., & Sathe, B. R. (2021). CZTS/MoS_2-rGO Heterostructures: An efficient and highly stable electrocatalyst for enhanced hydrogen generation reactions. *Journal of Electroanalytical Chemistry (Lausanne, Switzerland), 882*, 114983. doi:10.1016/j.jelechem.2021.114983

Ding, Y., Cheng, H., Zhou, C., Fan, Y., Zhu, J., Shao, H., & Qu, L. (2012). Functional microspheres of graphene quantum dots. *Nanotechnology, 23*(25), 255605. doi:10.1088/0957-4484/23/25/255605 PMID:22653222

Ding, Y., Zhu, J., Wang, C., Dai, B., Li, Y., Qin, Y., Xu, F., Peng, Q., Yang, Z., Bai, J., Cao, W., Yuan, Y., & Li, Y. (2016). Multifunctional three-dimensional graphene nanoribbons composite sponge. *Carbon, 104*, 133–140. doi:10.1016/j.carbon.2016.03.058

Diniz-Ferreira, E. L., & Torres-Verdín, C. (2012). Improved estimation of pore connectivity and permeability in deepwater carbonates with the construction of multi-layer static and dynamic petrophysical models. *SPWLA 53rd Annual Logging Symposium 2012*.

Dizaj, S. M., Lotfipour, F., Barzegar-Jalali, M., Zarrintan, M. H., & Adibkia, K. (2014). *Antimicrobial activity of the metals and metal oxide nanoparticles*. Academic Press.

Dizaj, S. M., Lotfipour, F., Barzegar-Jalali, M., Zarrintan, M. H., & Adibkia, K. (2014). Antimicrobial activity of the metals and metal oxide nanoparticles. *Materials Science and Engineering C, 44*, 278–284. doi:10.1016/j.msec.2014.08.031 PMID:25280707

Djurišić, A. B., Leung, Y. H., Ng, A. M., Xu, X. Y., Lee, P. K., Degger, N., & Wu, R. (2015). *Toxicity of metal oxide nanoparticles: mechanisms, characterization, and avoiding experimental artefacts*. Academic Press.

do Amaral, D. F., Guerra, V., Motta, A. G. C., Silva, D. D. M., & Rocha, T. L. (2019). Ecotoxicity of nanomaterials in amphibians: A critical review. *The Science of the Total Environment, 686*, 332–344. doi:10.1016/j.scitotenv.2019.05.487 PMID:31181520

do Nascimento, J. M., Cruz, N. D., de Oliveira, G. R., Sá, W. S., de Oliveira, J. D., Ribeiro, P. R. S., & Leite, S. G. F. (2021). Evaluation of the kinetics of gold biosorption processes and consequent biogenic synthesis of AuNPs mediated by the fungus Trichoderma harzianum. *Environmental Technology & Innovation, 21*, 101238. Advance online publication. doi:10.1016/j.eti.2020.101238

Dong, X., & Yang, L. (2014). *Inhibitory effects of single-walled carbon nanotubes on biofilm formation from Bacillus anthracis spores*. Academic Press.

Doucey, M.-A., & Carrara, S. (2019). Nanowire sensors in cancer. *Trends in Biotechnology*, *37*(1), 86–99. doi:10.1016/j.tibtech.2018.07.014 PMID:30126620

Dou, S., Tao, L., Huo, J., Wang, S., & Dai, L. (2016). Etched and doped Co9S8/graphene hybrid for oxygen electrocatalysis. *Energy & Environmental Science*, *9*(4), 1320–1326. doi:10.1039/C6EE00054A

Dreaden, E. C., Austin, L. A., Mackey, M. A., & El-Sayed, M. A. (2012). Size matters: Gold nanoparticles in targeted cancer drug delivery. *Therapeutic Delivery*, *3*(4), 457–478. doi:10.4155/tde.12.21 PMID:22834077

Duan, H., Wang, D., & Li, Y. (2015). Green chemistry for nanoparticle synthesis. *Chemical Society Reviews*, *44*(16), 5778–5792. doi:10.1039/C4CS00363B PMID:25615873

Duan, Z., & Thomas, N. L. (2014). Water vapour permeability of poly(lactic acid): Crystallinity and the tortuous path model. *Journal of Applied Physics*, *115*(6), 064903. Advance online publication. doi:10.1063/1.4865168

Dube, T., Ghosh, A., Mishra, J., Kompella, U. B., & Panda, J. J. (2021). Repurposed Drugs, Molecular Vaccines, Immune-Modulators, and Nanotherapeutics to Treat and Prevent COVID-19 Associated with SARS-CoV-2, a Deadly Nanovector. *Advanced Therapeutics*, *4*(2), 1–30. doi:10.1002/adtp.202000172 PMID:33173808

Dubey, S. P., Lahtinen, M., & Sillanpää, M. (2010). Tansy fruit mediated greener synthesis of silver and gold nanoparticles. *Process Biochemistry*, *45*(7), 1065–1071. doi:10.1016/j.procbio.2010.03.024

Dubochet, J., Adrian, M., Chang, J., Homo, J., Lepault, J., Mcdowall, A. W., & Schultz, P. (1988). Cryo-electron microscopy of vitrified specimens. *Quarterly Reviews of Biophysics*, *2*(2), 129–228. doi:10.1017/S0033583500004297 PMID:3043536

Du, J.-J., Chen, C., Gan, Y.-L., Zhang, R.-H., Yang, C.-Y., & Zhou, X.-W. (2014). Facile one-pot hydrothermal synthesis of Pt nanoparticles and their electrocatalytic performance. *International Journal of Hydrogen Energy*, *39*(31), 17634–17637. doi:10.1016/j.ijhydene.2014.08.044

Du, L., Jiang, H., Liu, X., & Wang, E. (2007). Biosynthesis of gold nanoparticles assisted by Escherichia coli DH5α and its application on direct electrochemistry of hemoglobin. *Electrochemistry Communications*, *9*(5), 1165–1170. doi:10.1016/j.elecom.2007.01.007

Duman, O., Tun, S., Polat, T. G., & Bozoglan, B. K. I. (2016). Synthesis of magnetic oxidized multiwalled carbon nanotube- k-carrageenan-Fe_3O_4 nanocomposite adsorbent and its application in cationic Methylene Blue dye adsorption. *Carbohydrate Polymers*, *147*, 79–88. doi:10.1016/j.carbpol.2016.03.099 PMID:27178911

Duncan, R. (2003). The dawning era of polymer therapeutics. *Nature Reviews. Drug Discovery*, *2*(5), 347–360. doi:10.1038/nrd1088 PMID:12750738

Duque, J. G., Parra-Vasquez, A. N. G., Behabtu, N., Green, M. J., Higginbotham, A. L., Price, B. K., Leonard, A. D., Schmidt, H. K., Lounis, B., Tour, J. M., Doorn, S. K., Cognet, L., & Pasquali, M. (2010). Diameter-dependent solubility of single-walled carbon nanotubes. *ACS Nano*, *4*(6), 3063–3072. Advance online publication. doi:10.1021/nn100170f PMID:20521799

Durán, N., Durán, M., De Jesus, M. B., Seabra, A. B., Fávaro, W. J., & Nakazato, G. (2016). Silver nanoparticles: A new view on mechanistic aspects on antimicrobial activity. *Nanomedicine; Nanotechnology, Biology, and Medicine*, *12*(3), 789–799. doi:10.1016/j.nano.2015.11.016 PMID:26724539

Duran, V., Grabski, E., Hozsa, C., Becker, J., Yasar, H., Monteiro, J. T., Costa, B., Koller, N., Lueder, Y., Wiegmann, B., Brandes, G., Kaever, V., Lehr, C. M., Lepenies, B., Tampe, R., Forster, R., Bosnjak, B., Furch, M., Graalmann, T., & Kalinke, U. (2021). Fucosylated lipid nanocarriers loaded with antibiotics efficiently inhibit mycobacterial propagation in human myeloid cells. *Journal of Controlled Release*, *334*, 201–212. doi:10.1016/j.jconrel.2021.04.012 PMID:33865899

Dursun, S., Akay, R. G., & Yazici, M. S. (2020). CVD graphene supported cobalt (II) phthalocyanine as cathode electrocatalyst for PEM fuel cells. *International Journal of Hydrogen Energy*, *45*(60), 34837–34844. doi:10.1016/j.ijhydene.2020.04.168

Du, X., Liu, H. Y., & Mai, Y. W. (2016). Ultrafast synthesis of multifunctional n-doped graphene foam in an ethanol flame. *ACS Nano*, *10*(1), 453–462. doi:10.1021/acsnano.5b05373 PMID:26635121

Dykman, L., & Khlebtsov, N. (2012). Gold nanoparticles in biomedical applications: Recent advances and perspectives. *Chemical Society Reviews*, *41*(6), 2256–2282. doi:10.1039/C1CS15166E PMID:22130549

Eceolaza, S., Sangroniz, A., del Río, J., Iriarte, M., & Etxeberria, A. (2019). Improving the barrier character of poly(caprolactone): Transport properties and free volume of immiscible blends. *Journal of Applied Polymer Science*, *136*(40), 1–11. doi:10.1002/app.48018

Eddy, N. O., & Ekop, A. S. (2007). Assessment of the quality of water treated and distributed by the Akwa Ibom State Water Company. *E-Journal of Chemistry*, *4*(2), 180–186. https://doi.org/10.1155/2007/343042

Edison, T. J. I., & Sethuraman, M. G. (2012). Instant green synthesis of silver nanoparticles using Terminalia chebula fruit extract and evaluation of their catalytic activity on reduction of methylene blue. *Process Biochemistry*, *47*(9), 1351–1357. doi:10.1016/j.procbio.2012.04.025

Edling, H. E., Sun, H., Paschke, E., Schiraldi, D. A., Tanko, J. M., Paradzinsky, M., & Turner, S. R. (2020). High barrier biosourced polyester from dimethyl [2,2′-bifuran]-5,5′-dicarboxylate. *Polymer*, *191*(December). doi:10.1016/j.polymer.2020.122258

Edmundson, M. C., Capeness, M., & Horsfall, L. (2014). Exploring the potential of metallic nanoparticles within synthetic biology. *New Biotechnology*, *31*(6), 572–578. doi:10.1016/j.nbt.2014.03.004 PMID:24681407

Edris, A. E. (2007). Pharmaceutical and therapeutic Potentials of essential oils and their individual volatile constituents: A review. *Phytotherapy Research*, *21*(4), 308–323. doi:10.1002/ptr.2072 PMID:17199238

Egorova, K. S., & Ananikov, V. P. (2016). Which metals are green for catalysis? Comparison of the toxicities of Ni, Cu, Fe, Pd, Pt, Rh, and Au salts. *Angewandte Chemie International Edition*, *55*(40), 12150–12162. doi:10.1002/anie.201603777 PMID:27532248

Ehrampoush, M. H., Miria, M., Salmani, M. H., & Mahvi, A. H. (2015). Cadmium removal from aqueous solution by green synthesis iron oxide nanoparticles with tangerine peel extract. *Journal of Environmental Health Science & Engineering*, *13*(1), 84. doi:10.118640201-015-0237-4 PMID:26682059

Eigler, S., Grimm, S., Hof, F., & Hirsch, A. (2013). Graphene oxide: A stable carbon framework for functionalization. *Journal of Materials Chemistry. A, Materials for Energy and Sustainability*, *1*(38), 11559–11562. doi:10.1039/c3ta12975f

Eisenbarth, E., Velten, D., Müller, M., Thull, R., & Breme, J. (2004). Biocompatibility of β-stabilizing elements of titanium alloys. *Biomaterials*, *25*(26), 5705–5713. doi:10.1016/j.biomaterials.2004.01.021 PMID:15147816

El Sherbeny, W., Al-Baddaly, H., Rahal, A., Said, M., Hardman, D., & Henry, T. (2014). Nanotechnology applications to minimize geo-mechanical related challenges while drilling intercalated sediments, Western Desert, Egypt. *Society of Petroleum Engineers - SPE Saudi Arabia Section Technical Symposium and Exhibition*. 10.2118/172242-MS

Elahi, N., Kamali, M., & Baghersad, M. H. (2018). Recent biomedical applications of gold nanoparticles: A review. *Talanta*, *184*, 537–556. doi:10.1016/j.talanta.2018.02.088 PMID:29674080

Elbialy, N. S., Aboushoushah, S. F., Sofi, B. F., & Noorwali, A. (2020). Multifunctional curcumin-loaded mesoporous silica nanoparticles for cancer chemoprevention and therapy. *Microporous and Mesoporous Materials*, *291*, 109540. doi:10.1016/j.micromeso.2019.06.002

Elhissi, A., Ahmed, W., Hassan, I. U., Dhanak, V., & D'Emanuele, A. (2012). Carbon nanotubes in cancer therapy and drug delivery. *Journal of Drug Delivery*. PMID:22028974

El-hoshoudy, A. N., Gomaa, S., & Selim, O. A. (2019). Application of acrylamide polymer grafted with SiO2 nanoparticles in enhanced oil recovery-design project. *Petroleum and Coal*, *61*(6).

Elias, D. C., Nair, R. R., Mohiuddin, T. M. G., Morozov, S. V., Blake, P., Halsall, M. P., Ferrari, A. C., Boukhvalov, D. W., Katsnelson, M. I., Geim, A. K., & Novoselov, K. S. (2009). Control of graphene's properties by reversible hydrogenation: Evidence for graphane. *Science*, *323*(5914), 610–613. Advance online publication. doi:10.1126cience.1167130 PMID:19179524

Eliseev, A. A., Lukashin, A. V., Vertegel, A. A., Heifets, L. I., Zhirov, A. I., & Tretyakov, Y. D. (2000). Complexes of Cu(II) with polyvinyl alcohol as precursors for the preparation of CuO/SiO 2 nanocomposites. *Materials Research Innovations*, *3*(5), 308–312. https://doi.org/10.1007/PL00010877

Ellah, N. H. A., Gad, S. F., Muhammad, K., Batiha, G. E., & Hetta, H. F. (2020). Nanomedicine as a Promising Approach for Diagnosis, Treatment and Prophylaxis against COVID-19. *Nanomedicine (London)*, *15*(21), 2085–2102. doi:10.2217/nnm-2020-0247 PMID:32723142

El-Megharbel, S. M., Alsawat, M., Al-Salmi, F. A., & Hamza, R. Z. (2021). Utilizing of (Zinc oxide nano-spray) for disinfection against "sars-cov-2" and testing its biological effectiveness on some biochemical parameters during (covid-19 pandemic)—"zno nanoparticles have antiviral activity against (sars-cov-2). *Coatings*, *11*(4), 388. Advance online publication. doi:10.3390/coatings11040388

El-Rafie, H. M., El-Rafie, M. H., & Zahran, M. K. (2013). Green synthesis of silver nanoparticles using polysaccharides extracted from marine macro algae. *Carbohydrate Polymers*, *96*(2), 403–410. doi:10.1016/j.carbpol.2013.03.071 PMID:23768580

El-Sayed, E. I., Al-Ghamdi, A. A., Al-Heniti, S., Al-Marzouki, F., & El-Tantawy, F. (2011). Synthesis of ultrafine β-Ga2O3 nanopowder via hydrothermal approach: A strong UV "excimer-like" emission. *Materials Letters*, *65*(2), 317–321. doi:10.1016/j.matlet.2010.10.007

El-Temsah, Y. S., & Joner, E. J. (2013). Effects of nano-sized zero-valent iron (nZVI) on DDT degradation in soil and its toxicity to collembola and ostracods. *Chemosphere*, *92*(1), 131–137. doi:10.1016/j.chemosphere.2013.02.039 PMID:23522781

Emima Jeronsia, J., Ragu, R., Sowmya, R., Mary, A. J., & Jerome Das, S. (2020). Comparative investigation on Camellia Sinensis mediated green synthesis of Ag and Ag/GO nanocomposites for its anticancer and antibacterial efficacy. *Surfaces and Interfaces*, *21*(November), 100787. doi:10.1016/j.surfin.2020.100787

Enescu, D., Cerqueira, M. A., Fucinos, P., & Pastrana, L. M. (2019). Recent advances and challenges on applications of nanotechnology in food packaging. A literature review. *Food and Chemical Toxicology*, *134*, 110814. doi:10.1016/j.fct.2019.110814 PMID:31520669

Ensafi, A. A., Amini, M., Rezaei, B., & Talebi, M. (2016). A novel diagnostic biosensor for distinguishing immunoglobulin mutated and unmutated types of chronic lymphocytic leukemia. *Biosensors & Bioelectronics*, *77*, 409–415. doi:10.1016/j.bios.2015.09.063 PMID:26436328

EPA. (2017). *For Underground Storage Tank Sites A Guide For Corrective Action Plan Reviewers: Chapter V Landfarming*. EPA 510-B-17-003, October. https://www.epa.gov/sites/production/files/2014-03/documents/tum_ch9.pdf

Ergen, O., Gilbert, S. M., Pham, T., Turner, S. J., Tan, M. T. Z., Worsley, M. A., & Zettl, A. (2017). Graded bandgap perovskite solar cells. In Nature Materials (Vol. 16, Issue 5, pp. 522–525). Nature Publishing Group. doi:10.1038/nmat4795

Erickson, C. S., Bradshaw, L. R., McDowall, S., Gilbertson, J. D., Gamelin, D. R., & Patrick, D. L. (2014). Zero-reabsorption doped-nanocrystal luminescent solar concentrators. *ACS Nano*, *8*(4), 3461–3467. doi:10.1021/nn406360w PMID:24621014

Ersan, G., Apul, O. G., Perreault, F., & Karanfil, T. (2017). Adsorption of organic contaminants by graphene nanosheets: A review. *Water Research*, *126*, 385–398. doi:10.1016/j.watres.2017.08.010 PMID:28987890

Ersan, G., & Ersan, M. S. (2021). Are carbon-based nanomaterials for the adsorption of organic contaminants perform better than nanoplastics (NPs) and microplastics (MPs)? *Journal of International Environmental Engineering and Science*, *16*(2), 72–81.

Ersan, G., Kaya, Y., Apul, O. G., & Karanfil, T. (2016). Adsorption of organic contaminants by graphene nanosheets, carbon nanotubes and granular activated carbons under different natural organic matter preloading conditions. *The Science of the Total Environment*, *565*, 811–817. doi:10.1016/j.scitotenv.2016.03.224 PMID:27107611

Ersan, G., Kaya, Y., Ersan, M. S., Apul, O. G., & Karanfil, T. (2019). Adsorption kinetics and aggregation for three classes of carbonaceous adsorbents in the presence of natural organic matter. *Chemosphere*, *229*, 515–524. doi:10.1016/j.chemosphere.2019.05.014 PMID:31100622

Esan, V., Mahboob, S., Al-Ghanim, K. A., Elanchezhiyan, C., Al-Misned, F., Ahmed, Z., & Govindarajan, M. (2020). Novel Biogenic Synthesis of Silver Nanoparticles Using Alstonia venenata Leaf Extract: An Enhanced Mosquito Larvicidal Agent with Negligible Impact on Important Eco-biological Fish and Insects. *Journal of Cluster Science*, *32*(2), 489–497. doi:10.100710876-020-01808-5

Essien, E. R., Atasie, V. N., Okeafor, A. O., & Nwude, D. O. (2020). Biogenic synthesis of magnesium oxide nanoparticles using Manihot esculenta (Crantz) leaf extract. *International Nano Letters*, *10*(1), 43–48. doi:10.100740089-019-00290-w

Fabrega, J., Luoma, S. N., Tyler, C. R., Galloway, T. S., & Lead, J. R. (2011). Silver nanoparticles: Behavior and effects in the aquatic environment. *Environment International*, *37*(2), 517–531. doi:10.1016/j.envint.2010.10.012 PMID:21159383

Fadlallah, S., Gad El-Rab, S. M. F., & Halwani, E. M. (2020). Innovative Nanoporous Titania Surface with Stabilized Antimicrobial Ag-Nanoparticles via Salvadora persica L. Roots (Miswak) Extract for Dental Applications. *BioNanoScience*, *10*(4), 998–1009. doi:10.100712668-020-00765-7

Fahandezhsaadi, M., Amooie, M. A., Hemmati-Sarapardeh, A., Ayatollahi, S., Schaffie, M., & Ranjbar, M. (2019). Laboratory evaluation of nitrogen injection for enhanced oil recovery: Effects of pressure and induced fractures. *Fuel*, *253*, 607–614. Advance online publication. doi:10.1016/j.fuel.2019.05.039

Fan, G., Dundas, C. M., Zhang, C., Lynd, N. A., & Keitz, B. K. (2018). Sequence-dependent peptide surface functionalization of metal–organic frameworks. *ACS Applied Materials & Interfaces*, *10*(22), 18601–18609. doi:10.1021/acsami.8b05148 PMID:29762004

Fang, W., Zhang, N., Fan, L., & Sun, K. (2016). Bi2O3 nanoparticles encapsulated by three-dimensional porous nitrogen-doped graphene for high-rate lithium ion batteries. *Journal of Power Sources*, *333*, 30–36. doi:10.1016/j.jpowsour.2016.09.155

Fan, P., Liu, L., Guo, Q., Wang, J., Yang, J., Guan, X., Chen, S., & Hou, H. (2017). Three-dimensional N-doped carbon nanotube@carbon foam hybrid: An effective carrier of enzymes for glucose biosensors. *RSC Advances*, *7*(43), 26574–26582. doi:10.1039/C7RA02592K

Fan, W., Yung, B., Huang, P., & Chen, X. (2017). Nanotechnology for multimodal synergistic cancer therapy. *Chemical Reviews*, *117*(22), 13566–13638. doi:10.1021/acs.chemrev.7b00258 PMID:29048884

Faraji, A. H., & Wipf, P. (2009). Nanoparticles in cellular drug delivery. *Bioorganic & Medicinal Chemistry*, *17*(8), 2950–2962. doi:10.1016/j.bmc.2009.02.043 PMID:19299149

Faried, M., Shameli, K., Miyake, M., Hajalilou, A., Zamanian, A., Zakaria, Z., Abouzari-lotf, E., Hara, H., Ahmad Khairudin, N. B. B., & Binti Mad Nordin, M. F. (2016). A green approach for the synthesis of silver nanoparticles using ultrasonic radiation's times in sodium alginate media: Characterization and antibacterial evaluation. *Journal of Nanomaterials*, *2016*, 1–11. doi:10.1155/2016/4941231

Farooqi, Z. H., Akram, M. W., Begum, R., Wu, W., & Irfan, A. (2021). Inorganic nanoparticles for reduction of hexavalent chromium: Physicochemical aspects. *Journal of Hazardous Materials*, *402*, 123535. doi:10.1016/j.jhazmat.2020.123535 PMID:33254738

Fathi-Hafshejani, P., Azam, N., Wang, L., Kuroda, M. A., Hamilton, M. C., Hasim, S., & Mahjouri-Samani, M. (2021). Two-Dimensional-Material-based Field-Effect Transistor Biosensor for Detecting COVID-19 Virus (SARS-CoV-2). *ACS Nano*, *15*(7), 11461–11469. doi:10.1021/acsnano.1c01188 PMID:34181385

Fatima, R., Priya, M., Indurthi, L., Radhakrishnan, V., & Sudhakaran, R. (2020). Biosynthesis of silver nanoparticles using red algae Portieria hornemannii and its antibacterial activity against fish pathogens. *Microbial Pathogenesis*, *138*, 103780. doi:10.1016/j.micpath.2019.103780 PMID:31622663

Fato, F. P., Li, D.-W., Zhao, L.-J., Qiu, K., & Long, Y.-T. (2019). Simultaneous Removal of Multiple Heavy Metal Ions from River Water Using Ultrafine Mesoporous Magnetite Nanoparticles. *ACS Omega*, *4*(4), 7543–7549. doi:10.1021/acsomega.9b00731 PMID:31459847

Fedele, L., Colla, L., Bobbo, S., Barison, S., & Agresti, F. (2011). Experimental stability analysis of different water-based nanofluids. *Nanoscale Research Letters*, *6*(1), 300. Advance online publication. doi:10.1186/1556-276X-6-300 PMID:21711817

Fei Yin, Z., Wu, L., Gui Yang, H., & Hua Su, Y. (2013). Recent progress in biomedical applications of titanium dioxide. *Physical Chemistry Chemical Physics*, *15*(14), 4844. doi:10.1039/c3cp43938k PMID:23450160

Feng, B., Weng, J., Yang, B. C., Qu, S. X., & Zhang, X. D. (2003). Characterization of surface oxide films on titanium and adhesion of osteoblast. *Biomaterials*, *24*(25), 4663–4670. doi:10.1016/S0142-9612(03)00366-1 PMID:12951009

Ferguson, R. M., Khandhar, A. P., Kemp, S. J., Arami, H., Saritas, E. U., Croft, L. R., Konkle, J., Goodwill, P. W., Halkola, A., Rahmer, J., Borgert, J., Conolly, S. M., & Krishnan, K. M. (2014). Magnetic particle imaging with tailored iron oxide nanoparticle tracers. *IEEE Transactions on Medical Imaging*, *34*(5), 1077–1084. doi:10.1109/TMI.2014.2375065 PMID:25438306

Fernando, S., Gunasekara, T., & Holton, J. (2018). *Antimicrobial Nanoparticles: Applications and mechanisms of action*. Academic Press.

Fernes, T. D., Hunter, D. L., & Paul, D. R. (2004). Nylon-6 Nanocomposites from Alkylammonium-Modified Clay: The Role of Alky1 Tails on Exfoliation. *Macromolecules, 37*(5), 1793–1798. doi:10.1021/ma0305481

Ferreira, A., Kunh, S. S., Fagnani, K. C., De Souza, T. A., Tonezer, C., Dos Santos, G. R., & Coimbra-Araújo, C. H. (2018). Economic overview of the use and production of photovoltaic solar energy in brazil. *Renewable & Sustainable Energy Reviews, 81*, 181–191. doi:10.1016/j.rser.2017.06.102

Ferreira, C. A., Ni, D., Rosenkrans, Z. T., & Cai, W. (2019). Radionuclide-activated nanomaterials and their biomedical applications. *Angewandte Chemie International Edition, 58*(38), 13232–13252. doi:10.1002/anie.201900594 PMID:30779286

Fleischmann, M., Hendra, P. J., & McQuillan, A. J. (1974). Raman spectra of pyridine adsorbed at a silver electrode. *Chemical Physics Letters, 26*(2), 2–5. doi:10.1016/0009-2614(74)85388-1

Fleming, N. (2014). Technology Focus: Nanotechnology (February 2014). *Journal of Petroleum Technology, 66*(02), 116. Advance online publication. doi:10.2118/0214-0116-JPT

Fogarassy, Z., Rümmeli, M. H., Gorantla, S., Bachmatiuk, A., Dobrik, G., Kamarás, K., Biró, L. P., Havancsák, K., & Lábár, J. L. (2014). Dominantly epitaxial growth of graphene on Ni (1 1 1) substrate. *Applied Surface Science, 314*, 490–499. doi:10.1016/j.apsusc.2014.06.197

Foster, C. W., Down, M. P., Zhang, Y., Ji, X., Rowley-Neale, S. J., Smith, G. C., Kelly, P. J., & Banks, C. E. (2017). 3D Printed Graphene Based Energy Storage Devices. *Scientific Reports, 7*(1), 1–11. doi:10.1038rep42233 PMID:28256602

Foster, H. A., Ditta, I. B., Varghese, S., & Steele, A. (2011). Photocatalytic disinfection using titanium dioxide: Spectrum and mechanism of antimicrobial activity. *Applied Microbiology and Biotechnology, 90*(6), 1847–1868. doi:10.100700253-011-3213-7 PMID:21523480

Fouda, A., Hassan, S. E.-D., Saied, E., & Hamza, M. F. (2021). Photocatalytic degradation of real textile and tannery effluent using biosynthesized magnesium oxide nanoparticles (MgO-NPs), heavy metal adsorption, phytotoxicity, and antimicrobial activity. *Journal of Environmental Chemical Engineering, 9*(4), 105346. doi:10.1016/j.jece.2021.105346

Fouly, A., & Alkalla, M. G. (2020). Effect of low nanosized alumina loading fraction on the physicomechanical and tribological behavior of epoxy. *Tribology International, 152*, 106550. doi:10.1016/j.triboint.2020.106550

Frangioni, J. V. (2008). New technologies for human cancer imaging. *Journal of Clinical Oncology, 26*(24), 4012–4021. doi:10.1200/JCO.2007.14.3065 PMID:18711192

Fujiwara, R., Luo, Y., Sasaki, T., Fujii, K., Ohmori, H., & Kuniyasu, H. (2015). Cancer therapeutic effects of titanium dioxide nanoparticles are associated with oxidative stress and cytokine induction. *Pathobiology, 82*(6), 243–251. doi:10.1159/000439404 PMID:26485713

Fu, X., Choi, J. Y., Zamani, P., Jiang, G., Hoque, M. A., Hassan, F. M., & Chen, Z. (2016). Co-N Decorated Hierarchically Porous Graphene Aerogel for Efficient Oxygen Reduction Reaction in Acid. *ACS Applied Materials & Interfaces, 8*(10), 6488–6495. doi:10.1021/acsami.5b12746 PMID:26937737

Gabriel, S. B., Dille, J., Rezende, M. C., Mei, P., de Almeida, L. H., Baldan, R., & Nunes, C. A. (2015). Mechanical characterization of Ti-12Mo-13Nb alloy for biomedical application hot swaged and aged. *Materials Research, 18*(Suppl 2), 8–12. doi:10.1590/1516-1439.329614

Gai, S., Li, C., Yang, P., & Lin, J. (2014). Recent progress in rare earth micro/nanocrystals: Soft chemical synthesis, luminescent properties, and biomedical applications. *Chemical Reviews, 114*(4), 2343–2389. doi:10.1021/cr4001594 PMID:24344724

Galdiero, S., Falanga, A., Cantisani, M., Ingle, A., Galdiero, M., & Rai, M. (2014). Silver nanoparticles as novel antibacterial and antiviral agents. In Handbook of Nanobiomedical Research: Fundamentals, Applications and Recent Developments: Volume 1. Materials for Nanomedicine (pp. 565-594). doi:10.1142/9789814520652_0015

Galluzzi, L., Bravo-San Pedro, J. M., Vitale, I., Aaronson, S. A., Abrams, J. M., Adam, D., Alnemri, E. S., Altucci, L., Andrews, D., Annicchiarico-Petruzzelli, M., Baehrecke, E. H., Bazan, N. G., Bertrand, M. J., Bianchi, K., Blagosklonny, M. V., Blomgren, K., Borner, C., Bredesen, D. E., Brenner, C., ... Kroemer, G. (2015). Essential versus accessory aspects of cell death: Recommendations of the NCCD 2015. *Cell Death and Differentiation*, *22*(1), 58–73. doi:10.1038/cdd.2014.137 PMID:25236395

Gálvez-Vergara, A., Fresco-Cala, B., & Cárdenas, S. (2020). Switchable Pickering emulsions stabilized by polystyrene-modified magnetic nanoparticles. *Colloids and Surfaces. A, Physicochemical and Engineering Aspects*, *606*, 125462. Advance online publication. doi:10.1016/j.colsurfa.2020.125462

Gamal Snousy, M., & Mf, Z. (2017). Nanoparticles Restrictions in Environmental Cleanup. *Nano Research & Applications*, *03*(01). Advance online publication. doi:10.21767/2471-9838.100018

Ganaie, S. U., Abbasi, T., & Abbasi, S. A. (2016). Rapid and green synthesis of bimetallic Au–Ag nanoparticles using an otherwise worthless weed Antigonon leptopus. *Journal of Experimental Nanoscience*, *11*(6), 395–417. doi:10.1080/17458080.2015.1070311

Gandhi, N., Shruthi, Y., Sirisha, G., & Anusha, C. R. (2021). Facile and Eco-Friendly Method for Synthesis of Calcium Oxide (CaO) Nanoparticles and its Potential Application in Agriculture. *Haya: The Saudi Journal of Life Sciences, 6*(5), 89–103. doi:10.36348/sjls.2021.v06i05.003

Gangadhar, L., Reddy, K. B., Garg, A. P., & Sana, S. S. (2020). Green Synthesis of Bio-polymer Composites of Iron for Pharmaceutical Applications. *Journal of Nanomedicine & Nanotechnology*, *11*, 551. doi:10.35248/2157-7439.20.11.551

Gan, P. P., Ng, S. H., Huang, Y., & Li, S. F. Y. (2012). Green synthesis of gold nanoparticles using palm oil mill effluent (POME): A low-cost and eco-friendly viable approach. *Bioresource Technology*, *113*, 132–135. doi:10.1016/j.biortech.2012.01.015 PMID:22297042

Gao, W. (2012). *Graphite Oxide: Structure, Reduction and Applications*. https://scholarship.rice.edu/handle/1911/64614

Gao, C., Meng, Q., Zhao, K., Yin, H., Wang, D., Guo, J., Zhao, S., Chang, L., He, M., Li, Q., Zhao, H., Huang, X., Gao, Y., & Tang, Z. (2016). Co3O4 Hexagonal Platelets with Controllable Facets Enabling Highly Efficient Visible-Light Photocatalytic Reduction of CO2. *Advanced Materials*, *28*(30), 6485–6490. doi:10.1002/adma.201601387 PMID:27171564

Gao, H. L., Zhu, Y. B., Mao, L. B., Wang, F. C., Luo, X. S., Liu, Y. Y., Lu, Y., Pan, Z., Ge, J., Shen, W., Zheng, Y. R., Xu, L., Wang, L. J., Xu, W. H., Wu, H. A., & Yu, S. H. (2016). Super-elastic and fatigue resistant carbon material with lamellar multi-arch microstructure. *Nature Communications*, *7*(1), 1–8. doi:10.1038/ncomms12920 PMID:27676215

Gao, Q., Xie, W., Wang, Y., Wang, D., Guo, Z., Gao, F., Zhao, L., & Cai, Q. (2018). A theranostic nanocomposite system based on radial mesoporous silica hybridized with Fe 3 O 4 nanoparticles for targeted magnetic field responsive chemotherapy of breast cancer. *RSC Advances*, *8*(8), 4321–4328. doi:10.1039/C7RA12446E

Garcia, C. V., Shin, G. H., & Kim, J. T. (2018). Metal oxide-based nanocomposites in food packaging: Applications, migration, and regulations. *Trends in Food Science & Technology*, *82*, 21–31. doi:10.1016/j.tifs.2018.09.021

García-Serrano, J., Pal, U., Herrera, A. M., Salas, P., & Ángeles-Chávez, C. (2008). One-step "Green" synthesis and stabilization of Au and Ag nanoparticles using ionic polymers. *Chemistry of Materials*, *20*(16), 5146–5153. doi:10.1021/cm703201d

Gardini, D., Lüscher, C. J., Struve, C., & Krogfelt, K. A. (2018). Tailored nanomaterials for antimicrobial applications. In *Fundamentals of Nanoparticles* (pp. 71–104). Elsevier. doi:10.1016/B978-0-323-51255-8.00004-5

Garg, R., Kumari, M., Kumar, M., Dhiman, S., & Garg, R. (2020). Green synthesis of calcium carbonate nanoparticles using waste fruit peel extract. *Materials Today: Proceedings, 46*, 6665–6668. doi:10.1016/j.matpr.2021.04.124

Gavrilescu, M., Demnerová, K., Aamand, J., Agathos, S., & Fava, F. (2015). Emerging pollutants in the environment: Present and future challenges in biomonitoring, ecological risks and bioremediation. *New Biotechnology, 32*(1), 147–156. https://doi.org/10.1016/j.nbt.2014.01.001

Gebreeyessus, G. D., & Zewge, F. (2019). A review on environmental selenium issues. *SN Applied Sciences, 1*(1), 55. doi:10.100742452-018-0032-9

Gehrke, H., Pelka, J., Hartinger, C. G., Blank, H., Bleimund, F., Schneider, R., Gerthsen, D., Bräse, S., Crone, M., Türk, M., & Marko, D. (2011). Platinum nanoparticles and their cellular uptake and DNA platination at non-cytotoxic concentrations. *Archives of Toxicology, 85*(7), 799–812. doi:10.100700204-010-0636-3 PMID:21229235

Gehrke, I., Geiser, A., & Somborn-Schulz, A. (2015). Innovations in nanotechnology for water treatment. *Nanotechnology, Science and Applications, 8*, 1–17. doi:10.2147/NSA.S43773 PMID:25609931

Geim, A. K., & Novoselov, K. S. (2007). The rise of graphene. *Nature Materials, 6*(3), 183–191. doi:10.1038/nmat1849 PMID:17330084

Geoff, W. (n.d.). *Richard Fenyman*. Academic Press.

George, M. J., Sichilongo, K. F., Ramabulana, T., Madala, N. E., & Dubery, I. A. (2019). Comparison of Soxhlet and reflux techniques for extraction and characterisation of potential endocrine-disrupting compounds from solid waste dumpsite soil. *Environmental Monitoring and Assessment, 191*(3), 1–11. doi:10.100710661-019-7294-6 PMID:30739205

Gericke, M., & Pinches, A. (2006). Biological synthesis of metal nanoparticles. *Hydrometallurgy, 83*(1–4), 132–140. doi:10.1016/j.hydromet.2006.03.019

Geszke-Moritz, M., & Moritz, M. (2016). Solid lipid nanoparticles as attractive drug vehicles: Composition, properties and therapeutic strategies. *Materials Science and Engineering C, 68*, 982–994. doi:10.1016/j.msec.2016.05.119 PMID:27524099

Ghaedi, M., Ghayedi, M., Kokhdan, S. N., Sahraei, R., & Daneshfar, A. (2013). Palladium, silver, and zinc oxide nanoparticles loaded on activated carbon as adsorbent for removal of bromophenol red from aqueous solution. *Journal of Industrial and Engineering Chemistry, 19*(4), 1209–1217. https://doi.org/10.1016/j.jiec.2012.12.020

Ghalehkhondabi, V., Soleymani, M., & Fazlali, A. (2021). Folate-targeted nanomicelles containing silibinin as an active drug delivery system for liver cancer therapy. *Journal of Drug Delivery Science and Technology, 61*, 102157. doi:10.1016/j.jddst.2020.102157

Ghanbar, F., Mirzaie, A., Ashrafi, F., Noorbazargan, H., Dalirsaber Jalali, M., Salehi, S., & Sadat Shandiz, S. A. (2017). Antioxidant, antibacterial and anticancer properties of phytosynthesised Artemisia quttensis Podlech extract mediated AgNPs. *IET Nanobiotechnology / IET, 11*(4), 485–492. doi:10.1049/iet-nbt.2016.0101 PMID:28530200

Ghanshyam, D. S. (2015). Advantages of green technology. *Social Issues and Environmental Problems, 3*(9), 2394–3629.

Ghaseminezhad, H., & Sahraei, E. (2020). Experimental investigation of foam assisted water alternate gas injection in relation to enhanced oil recovery applications. *Petroleum Science and Technology, 38*(17), 859–869. Advance online publication. doi:10.1080/10916466.2020.1790599

Ghoreishi, S. M., Behpour, M., & Khayatkashani, M. (2011). Green synthesis of silver and gold nanoparticles using Rosa damascena and its primary application in electrochemistry. *Physica E, Low-Dimensional Systems and Nanostructures*, *44*(1), 97–104. doi:10.1016/j.physe.2011.07.008

Ghosh, A., Ahuja, T., Chaudhari, K., & Pradeep, T. (2020). Probing Subtle Changes in Molecular Orientations Using Ambient Electrospray Deposition Raman Spectroscopy (AESD RS). *The Journal of Physical Chemistry C*, *124*(30), 16644–16651. doi:10.1021/acs.jpcc.0c04116

Giasuddin, A. S. M., Jhuma, K. A., & Haq, A. M. (2012). Use of gold nanoparticles in diagnostics, surgery and medicine: A review. *Bangladesh Journal of Medical Biochemistry*, *5*(2), 56–60. doi:10.3329/bjmb.v5i2.13346

Gigliobianco, M., Casadidio, C., Censi, R., & Di Martino, P. (2018). Nanocrystals of poorly soluble drugs: Drug bioavailability and physicochemical stability. *Pharmaceutics*, *10*(3), 134. doi:10.3390/pharmaceutics10030134 PMID:30134537

Gil-Díaz, M., Alonso, J., Rodríguez-Valdés, E., Gallego, J. R., & Lobo, M. C. (2017). Comparing different commercial zero valent iron nanoparticles to immobilize As and Hg in brownfield soil. *The Science of the Total Environment*, *584–585*, 1324–1332. doi:10.1016/j.scitotenv.2017.02.011 PMID:28190571

Gillani, S. S., Munawar, M. A., Khan, K. M., & Chaudhary, J. A. (2020). Synthesis, characterization and applications of poly-aliphatic amine dendrimers and dendrons. *Journal of the Indian Chemical Society*, *17*(11), 2717–2736. doi:10.100713738-020-01973-4

Gilroy, K. D., Ruditskiy, A., Peng, H. C., Qin, D., & Xia, Y. (2016). Bimetallic nanocrystals: Syntheses, properties, and applications. *Chemical Reviews*, *116*(18), 10414–10472. doi:10.1021/acs.chemrev.6b00211 PMID:27367000

Girija, K., Thirumalairajan, S., Avadhani, G. S., Mangalaraj, D., Ponpandian, N., & Viswanathan, C. (2013). Synthesis, morphology, optical and photocatalytic performance of nanostructured β-Ga2O3. *Materials Research Bulletin*, *48*(6), 2296–2303. doi:10.1016/j.materresbull.2013.02.047

Giri, S., Karakoti, A., Graham, R. P., Maguire, J. L., Reilly, C. M., Seal, S., Rattan, R., & Shridhar, V. (2013). Nanoceria: A rare-earth nanoparticle as a novel anti-angiogenic therapeutic agent in ovarian cancer. *PLoS One*, *8*(1), e54578. doi:10.1371/journal.pone.0054578 PMID:23382918

Gkika, D. A., Vordos, N., Nolan, J. W., Mitropoulos, A. C., Vansant, E. F., Cool, P., & Braet, J. (2017). Price tag in nanomaterials? *Journal of Nanoparticle Research*, *19*(5), 177. doi:10.100711051-017-3875-x

Gliga, A. R., Skoglund, S., Wallinder, I. O., Fadeel, B., & Karlsson, H. L. (2014). *Size-dependent cytotoxicity of silver nanoparticles in human lung cells: the role of cellular uptake, agglomeration and Ag release*. Academic Press.

Gnanasekaran, L., Rajendran, S., Priya, A. K., Durgalakshmi, D., Vo, D. V. N., Cornejo-Ponce, L., Gracia, F., & Soto-Moscoso, M. (2021). Photocatalytic degradation of 2,4-dichlorophenol using bio-green assisted TiO_2–CeO_2 nanocomposite system. *Environmental Research*, *195*(February), 110852. Advance online publication. doi:10.1016/j.envres.2021.110852 PMID:33556356

Godoy-Gallardo, M., Eckhard, U., Delgado, L. M., de Roo Puente, Y. J. D., Hoyos-Nogués, M., Gil, F. J., & Perez, R. A. (2021). Antibacterial approaches in tissue engineering using metal ions and nanoparticles: From mechanisms to applications. *Bioactive Materials*, *6*(12), 4470–4490. doi:10.1016/j.bioactmat.2021.04.033 PMID:34027235

Godwin, M. A., Shri, K. M., & Balaji, M. (2015). Engineering and Bioscience. *Int. J. Res. Eng. Biosci.*, *3*(5), 11–29.

Goel, S., Chen, F., & Cai, W. (2014). Synthesis and biomedical applications of copper sulfide nanoparticles: From sensors to theranostics. *Small*, *10*(4), 631–645. doi:10.1002mll.201301174 PMID:24106015

Goel, S., England, C. G., Chen, F., & Cai, W. (2017). Positron emission tomography and nanotechnology: A dynamic duo for cancer theranostics. *Advanced Drug Delivery Reviews*, *113*, 157–176. doi:10.1016/j.addr.2016.08.001 PMID:27521055

Gold, K., Slay, B., Knackstedt, M., & Gaharwar, A. K. (2018). Antimicrobial activity of metal and metal-oxide based nanoparticles. *Advanced Therapeutics*, *1*(3), 1700033. doi:10.1002/adtp.201700033

Goldstein, A., Soroka, Y., Frušić-Zlotkin, M., Popov, I., & Kohen, R. (2014). High resolution SEM imaging of gold nanoparticles in cells and tissues. *Journal of Microscopy*, *256*(3), 237–247.

Gomes, L., Paschoalin, V., & Del Aguila, E. (2017). Chitosan nanoparticles: Production, physicochemical characteristics and nutraceutical applications. *Rev. Virtual Quim*, *9*(1), 387–409. doi:10.21577/1984-6835.20170022

Gonçalves, J., Torres, N., Silva, S., Gonçalves, F., Noro, J., Cavaco-Paulo, A., Ribeiro, A., & Silva, C. (2020). Zein impart hydrophobic and antimicrobial properties to cotton textiles. *Reactive & Functional Polymers*, *154*, 104664. doi:10.1016/j.reactfunctpolym.2020.104664

Gonçalves, R. A., Toledo, R. P., Joshi, N., & Berengue, O. M. (2021). Green Synthesis and Applications of ZnO and TiO2 Nanostructures. *Molecules (Basel, Switzerland)*, *26*(8), 2236. Advance online publication. doi:10.3390/molecules26082236 PMID:33924397

Gong, N., Shao, K., Feng, W., Lin, Z., Liang, C., & Sun, Y. (2011). Biotoxicity of nickel oxide nanoparticles and bio-remediation by microalgae Chlorella vulgaris. *Chemosphere*, *83*(4), 510–516. doi:10.1016/j.chemosphere.2010.12.059 PMID:21216429

González, J. E. G., & Mirza-Rosca, J. C. (1999). Study of the corrosion behavior of titanium and some of its alloys for biomedical and dental implant applications. *Journal of Electroanalytical Chemistry (Lausanne, Switzerland)*, *471*(2), 109–115. doi:10.1016/S0022-0728(99)00260-0

Goodwin, H., Jellicoe, T. C., Davis, N. J. L. K., & Böhm, M. L. (2018). Multiple exciton generation in quantum dot-based solar cells. *Nanophotonics*, *7*(1), 111–126. doi:10.1515/nanoph-2017-0034

Gopinath, K., Gowri, S., Karthika, V., & Arumugam, A. (2014). Green synthesis of gold nanoparticles from fruit extract of Terminalia arjuna, for the enhanced seed germination activity of Gloriosa superba. *Journal of Nanostructure in Chemistry*, *4*(3), 115. Advance online publication. doi:10.100740097-014-0115-0

Gottschalk, F., & Nowack, B. (2011). The release of engineered nanomaterials to the environment. *Journal of Environmental Monitoring*, *13*(5), 1145. doi:10.1039/c0em00547a PMID:21387066

Gottschalk, F., Sonderer, T., Scholz, R. W., & Nowack, B. (2009). Modeled environmental concentrations of engineered nanomaterials (TiO2, ZnO, Ag, CNT, Fullerenes) for different regions. *Environmental Science & Technology*, *43*(24), 9216–9222. doi:10.1021/es9015553 PMID:20000512

Gour, A., & Jain, N. K. (2019). Advances in green synthesis of nanoparticles. *Artificial Cells, Nanomedicine, and Biotechnology*, *47*(1), 844–851. doi:10.1080/21691401.2019.1577878 PMID:30879351

Goutam, S. P., Saxena, G., Singh, V., Yadav, A. K., Bharagava, R. N., & Thapa, K. B. (2018). Green synthesis of TiO2 nanoparticles using leaf extract of Jatropha curcas L. for photocatalytic degradation of tannery wastewater. *Chemical Engineering Journal*, *336*, 386–396. doi:10.1016/j.cej.2017.12.029

Goyal, R., Macri, L. K., Kaplan, H. M., & Kohn, J. (2016). Nanoparticles and nanofibers for topical drug delivery. *Journal of Controlled Release*, *240*, 77–92. doi:10.1016/j.jconrel.2015.10.049 PMID:26518723

Greer, H. F., Yu, F., & Zhou, W. (2011). Early stages of non-classic crystal growth. *Science China. Chemistry*, *54*(12), 1867–1876. doi:10.100711426-011-4441-5

Guan, W.-J., Ni, Z.-Y., Hu, Y., Liang, W.-H., Ou, C.-Q., He, J.-X., Liu, L., Shan, H., Lei, C.-L., Hui, D. S. C., Du, B., Li, L.-J., Zeng, G., Yuen, K.-Y., Chen, R.-C., Tang, C.-L., Wang, T., Chen, P.-Y., Xiang, J., ... Zhong, N.-S. (2020). Clinical Characteristics of Coronavirus Disease 2019 in China. *The New England Journal of Medicine*, *382*(18), 1708–1720. doi:10.1056/NEJMoa2002032 PMID:32109013

Guerra, F. D., Attia, M. F., Whitehead, D. C., & Alexis, F. (2018). Nanotechnology for Environmental Remediation: Materials and Applications. *Molecules (Basel, Switzerland)*, *23*(7), 1760. Advance online publication. doi:10.3390/molecules23071760 PMID:30021974

Guimarães, D., Cavaco-Paulo, A., & Nogueira, E. (2021). Design of liposomes as drug delivery system for therapeutic applications. *International Journal of Pharmaceutics*, *601*, 120571. doi:10.1016/j.ijpharm.2021.120571 PMID:33812967

Gul, M., Kotak, Y., & Muneer, T. (2016). Review on recent trend of solar photovoltaic technology. *Energy Exploration and Exploitation*, *34*(4), 485–526. doi:10.1177/0144598716650552

Guo, J., & Kim, J. (2017). Modifications of polyethersulfone membrane by doping sulfated-TiO 2 nanoparticles for improving anti-fouling property in wastewater treatment. *RSC Advances*, *7*(54), 33822–33828. doi:10.1039/C7RA06406C

Guo, Q., Shen, X., Li, Y.-y., & Xu, S. (2017). Carbon nanotubes-based drug delivery to cancer and brain. *Current Medical Science*, *37*(5), 635–641. doi:10.100711596-017-1783-z PMID:29058274

Guo, T. H., Liu, Y., Zhang, Y. C., & Zhang, M. (2011). Green hydrothermal synthesis and optical absorption properties of ZnO2 nanocrystals and ZnO nanorods. *Materials Letters*, *65*(4), 639–641. doi:10.1016/j.matlet.2010.11.032

Gupta, A. K., Naregalkar, R. R., Vaidya, V. D., & Gupta, M. (2007). Recent advances on surface engineering of magnetic iron oxide nanoparticles and their biomedical applications. *Nanomedicine (London)*, *2*(1), 23–39. doi:10.2217/17435889.2.1.23 PMID:17716188

Gupta, A., Ahmad, A., Dar, A., Bhatt, A., & Khan, R. (2016). Selective targeting of cancer cells using personalized nanomedicine. *Toxicol Forensic Med Open J*, *1*(1), e12–e13. doi:10.17140/TFMOJ-1-e005

Gupta, A., Tandon, M., & Kaur, A. (2020). Role of metallic nanoparticles in water remediation with special emphasis on sustainable synthesis: A review. *Nanotechnology for Environmental Engineering*, *5*(3), 27. doi:10.100741204-020-00092-y

Gupta, O., & Roy, S. (2019). Recent progress in the development of nanocomposite membranes. In *Nanocomposite Membranes for Water and Gas Separation* (pp. 29–67). Elsevier. doi:10.1016/B978-0-12-816710-6.00002-X

Gupta, P. K., Gahtori, R., Govarthanan, K., Sharma, V., Pappuru, S., Pandit, S., Mathuriya, A. S., Dholpuria, S., & Bishi, D. K. (2021). Recent trends in biodegradable polyester nanomaterials for cancer therapy. *Materials Science and Engineering C*, *127*, 112198. doi:10.1016/j.msec.2021.112198 PMID:34225851

Gupta, V. K., Moradi, O., Tyagi, I., Agarwal, S., Sadegh, H., Shahryari-Ghoshekandi, R., Makhlouf, A. S. H., Goodarzi, M., & Garshasbi, A. (2016). Study on the removal of heavy metal ions from industry waste by carbon nanotubes: Effect of the surface modification: A review. *Critical Reviews in Environmental Science and Technology*, *46*(2), 93–118. doi:10.1080/10643389.2015.1061874

Gupta, V. K., Tyagi, I., Sadegh, H., Shahryari-Ghoshekand, R., Makhlouf, A. S. H., & Maazinejad, B. (2015). Nanoparticles as adsorbent; a positive approach for removal of noxious metal ions, a review. *Science. Technology and Development*, *34*(3), 195–214. doi:10.3923td.2015.195.214

Gurluk, M. R., Wang, G., Nasr-Ei-din, H. A., & Crews, J. (2013). The effect of different brine solutions on the viscosity of VES micelles. *SPE - European Formation Damage Conference, Proceedings, EFDC*, 2. 10.2118/165164-MS

Gurunathan, S., Jeyaraj, M., Kang, M. H., & Kim, J. H. (2019). Tangeretin-assisted platinum nanoparticles enhance the apoptotic properties of doxorubicin: Combination therapy for osteosarcoma treatment. *Nanomaterials (Basel, Switzerland)*, *9*(8), 1089. doi:10.3390/nano9081089 PMID:31362420

Gurunathan, S., Park, J. H., Han, J. W., & Kim, J. H. (2015). Comparative assessment of the apoptotic potential of silver nanoparticles synthesized by Bacillus tequilensis and Calocybe indica in MDA-MB-231 human breast cancer cells: Targeting p53 for anticancer therapy. *International Journal of Nanomedicine*, *10*, 4203–4222. doi:10.2147/IJN.S83953 PMID:26170659

Guzik, U., Hupert-Kocurek, K., & Wojcieszyńska, D. (2014). Immobilization as a strategy for improving enzyme properties-application to oxidoreductases. *Molecules (Basel, Switzerland)*, *19*(7), 8995–9018. doi:10.3390/molecules19078995 PMID:24979403

H, S. (2015). Synthesis of Carbon Nanotubes for Oil-water Interfacial Tension Reduction. *Oil & Gas Research*, *1*(1). doi:10.4172/2472-0518.1000104

Habibi, S., Jafari, A., & Fakhroueian, Z. (2020). Application of novel functionalized Al2O3/silica by organosiloxane and amine reagents for enhanced oil recovery. *Applied Nanoscience (Switzerland)*, *10*(7), 2085–2100. Advance online publication. doi:10.100713204-020-01337-7

Habte, L., Shiferaw, N., Mulatu, D., Thenepalli, T., Chilakala, R., & Ahn, J. (2019). Synthesis of Nano-Calcium Oxide from Waste Eggshell by Sol-Gel Method. *Sustainability*, *11*(11), 3196. https://doi.org/10.3390/su11113196

Haghighi Poudeh, L., Saner Okan, B., Seyyed Monfared Zanjani, J., Yildiz, M., & Menceloglu, Y. (2015). Design and fabrication of hollow and filled graphene-based polymeric spheres via core-shell electrospraying. *RSC Advances*, *5*(111), 91147–91157. doi:10.1039/C5RA19581K

Haider, A., & Kang, I. K. (2015). Preparation of silver nanoparticles and their industrial and biomedical applications: A comprehensive review. *Advances in Materials Science and Engineering*, *2015*, 2015. doi:10.1155/2015/165257

Haider, F. U., Liqun, C., Coulter, J. A., Cheema, S. A., Wu, J., Zhang, R., Wenjun, M., & Farooq, M. (2021). Cadmium toxicity in plants: Impacts and remediation strategies. *Ecotoxicology and Environmental Safety*, *211*, 111887. doi:10.1016/j.ecoenv.2020.111887 PMID:33450535

Hainfeld, J. F., Lin, L., Slatkin, D. N., Dilmanian, F. A., Vadas, T. M., & Smilowitz, H. M. (2014). Gold nanoparticle hyperthermia reduces radiotherapy dose. *Nanomedicine; Nanotechnology, Biology, and Medicine*, *10*(8), 1609–1617. doi:10.1016/j.nano.2014.05.006 PMID:24990355

Haiss, W., Thanh, N. T. K., Aveyard, J., & Fernig, D. G. (2007). Determination of size and concentration of gold nanoparticles from UV-Vis spectra. *Analytical Chemistry*, *79*(11), 4215–4221. doi:10.1021/ac0702084 PMID:17458937

Hajipour, M. J., Fromm, K. M., Ashkarran, A. A., de Aberasturi, D. J., de Larramendi, I. R., Rojo, T., ... Mahmoudi, M. (2012). Antibacterial properties of nanoparticles. *Trends in Biotechnology*, *30*(10), 499–511. doi:10.1016/j.tibtech.2012.06.004 PMID:22884769

Hale, W., Lee, J., Keskkula, H., & Paul, D. R. (1999). Effect of PBT melt viscosity on the morphology and mechanical properties of compatibilized and uncompatibilized blends with ABS. *Polymer*, *40*(13), 3621–3629. doi:10.1016/S0032-3861(98)00583-7

Hamad, K., Kaseem, M., Ayyoob, M., Joo, J., & Deri, F. (2018). Polylactic acid blends : The future of green, light and tough. *Progress in Polymer Science*, *85*, 83–127. doi:10.1016/j.progpolymsci.2018.07.001

Hamouda, A. A., & Chughtai, S. (2018). Miscible CO2 flooding for EOR in the presence of natural gas components in displacing and displaced fluids. *Energies*, *11*(2), 391. Advance online publication. doi:10.3390/en11020391

Hanano, A., Almousally, I., Shaban, M., & Murphy, D. J. (2018). Arabidopsis plants exposed to dioxin result in a WRINKLED seed phenotype due to 20S proteasomal degradation of WRI1. *Journal of Experimental Botany*, *69*(7), 1781–1794. doi:10.1093/jxb/ery027 PMID:29394403

Hanawa, T. (2011). A comprehensive review of techniques for biofunctionalization of titanium. *Journal of Periodontal & Implant Science*, *41*(6), 263. doi:10.5051/jpis.2011.41.6.263 PMID:22324003

Han, R., Zou, W., Wang, Y., & Zhu, L. (2007). Removal of uranium(VI) from aqueous solutions by manganese oxide coated zeolite: Discussion of adsorption isotherms and pH effect. *Journal of Environmental Radioactivity*, *93*(3), 127–143. doi:10.1016/j.jenvrad.2006.12.003 PMID:17258360

Harasym, T. O., Tardi, P. G., Harasym, N. L., Harvie, P., Johnstone, S. A., & Mayer, L. D. (2006). Increased pre-clinical efficacy of irinotecan and floxuridine coencapsulated inside liposomes is associated with tumor delivery of synergistic drug ratios. *Oncology Research Featuring Preclinical and Clinical Cancer Therapeutics*, *16*(8), 361–374. doi:10.3727/000000006783980937 PMID:17913044

Harish, B. S., Uppuluri, K. B., & Anbazhagan, V. (2015). Synthesis of fibrinolytic active silver nanoparticle using wheat bran xylan as a reducing and stabilizing agent. *Carbohydrate Polymers*, *132*, 104–110. doi:10.1016/j.carbpol.2015.06.069 PMID:26256330

Hasan, A., Paul, A., Memic, A., & Khademhosseini, A. (2015). A multilayered microfluidic blood vessel-like structure. *Biomedical Microdevices*, *17*(5), 1–13. doi:10.100710544-015-9993-2 PMID:26256481

Hasanuzzaman, M., Nahar, K., Hakeem, K. R., Öztürk, M., & Fujita, M. (2015). Arsenic Toxicity in Plants and Possible Remediation. In *Soil Remediation and Plants* (pp. 433–501). Elsevier. doi:10.1016/B978-0-12-799937-1.00016-4

Hasnain, M. S., & Nayak, A. K. (2019). Applications of carbon nanotubes. Springer Briefs Application Science Technology, 33–36.

Hassani, S. S., Daraee, M., & Sobat, Z. (2020). Advanced development in upstream of petroleum industry using nanotechnology. Chinese Journal of Chemical Engineering, 28(6). doi:10.1016/j.cjche.2020.02.030

Hassanin, H. A., Taha, A., & Afkar, E. (2021). Novel bio-mediated Ag/Co_3O_4 nanocomposites of different weight ratios using aqueous neem leaf extract: Catalytic and microbial behaviour. *Ceramics International*, *47*(3), 3099–3107. doi:10.1016/j.ceramint.2020.09.146

Hassanpour, P., Panahi, Y., Ebrahimi-Kalan, A., Akbarzadeh, A., Davaran, S., Nasibova, A. N., Khalilov, R., & Kavetskyy, T. (2018). Biomedical applications of aluminium oxide nanoparticles. *Micro & Nano Letters*, *13*(9), 1227–1231. doi:10.1049/mnl.2018.5070

Ha, T., Kim, S. K., Choi, J. W., Chang, H., & Jang, H. D. (2019). pH controlled synthesis of porous graphene sphere and application to supercapacitors. *Advanced Powder Technology*, *30*(1), 18–22. doi:10.1016/j.apt.2018.10.002

Haung, C.-F., Chan, Y.-H., Chen, L.-K., Liu, C.-M., Huang, W.-C., Ou, S.-F., Ou, K.-L., & Wang, D.-Y. (2013). *Preparation, characterization, and properties of anticoagulation and antibacterial films of carbon-based nanowires fabricated on surfaces Of Ti implants*. Academic Press.

He, F., Zhang, M. J., Wang, W., Cai, Q. W., Su, Y. Y., Liu, Z., Faraj, Y., Ju, X. J., Xie, R., & Chu, L. Y. (2019). Designable Polymeric Microparticles from Droplet Microfluidics for Controlled Drug Release. Advanced Materials Technologies, 4(6). doi:10.1002/admt.201800687

Hedayatnasab, Z., Abnisa, F., & Daud, W. M. A. W. (2017). Review on magnetic nanoparticles for magnetic nanofluid hyperthermia application. *Materials & Design, 123*, 174–196. doi:10.1016/j.matdes.2017.03.036

Heidari, A., Caissutti, A., Henderson, M., Schmitt, K., Besana, E., Esposito, J., & Peterson, V. (2020). Recent New Results and Achievements of California South University (CSU) BioSpectroscopy Core Research Laboratory for COVID-19 or 2019-nCoV Treatment: Diagnosis and Treatment Methodologies of "Coronavirus.". *Journal of Current Viruses and Treatment Methodologies, 1*(1), 3–41. doi:10.14302/issn.2691-8862.jvat-20-3275

Heiligtag, F. J., & Niederberger, M. (2013). *The fascinating world of nanoparticle research.* Academic Press.

Heinrich. (2011). Factors affecting the geometry of silver nanoparticles synthesis in Chrysosporium tropicum and Fusarium oxysporum. *American Journal of Nanotechnology, 2*(1), 112–121. doi:10.3844/ajnsp.2011.112.121

He, J., Chen, Y., Lv, W., Wen, K., Li, P., Wang, Z., Zhang, W., Qin, W., & He, W. (2016). Three-Dimensional Hierarchical Graphene-CNT@Se: A Highly Efficient Freestanding Cathode for Li-Se Batteries. *ACS Energy Letters, 1*(1), 16–20. doi:10.1021/acsenergylett.6b00015

He, J., Zhou, W., Zhou, X., Zhong, X., Zhang, X., Wan, P., Zhu, B., & Chen, W. (2008). The anatase phase of nanotopography titania plays an important role on osteoblast cell morphology and proliferation. *Journal of Materials Science. Materials in Medicine, 19*(11), 3465–3472. doi:10.100710856-008-3505-3 PMID:18592349

Hering, J. G., Waite, T. D., Luthy, R. G., Drewes, J. E., & Sedlak, D. L. (2013). A changing framework for urban water systems. *Environmental Science & Technology, 47*(19), 10721–10726. doi:10.1021/es4007096 PMID:23650975

Heron, M. (2018). Deaths: Leading Causes for 2016. *National Vital Statistics Reports, 67*(6), 1–77. PMID:30248017

Herrera-Becerra, R., Zorrilla, C., Rius, J. L., & Ascencio, J. A. (2008). Electron microscopy characterization of biosynthesized iron oxide nanoparticles. *Applied Physics. A, Materials Science & Processing, 91*(2), 241–246. doi:10.100700339-008-4420-7

He, S., Leanse, L. G., & Feng, Y. (2021). Artificial intelligence and machine learning assisted drug delivery for effective treatment of infectious diseases. *Advanced Drug Delivery Reviews, 113922*, 113922. doi:10.1016/j.addr.2021.113922 PMID:34461198

Hessien, M., Da'na, E., & Taha, A. (2021). Phytoextract assisted hydrothermal synthesis of ZnO–NiO nanocomposites using neem leaves extract. *Ceramics International, 47*(1), 811–816. doi:10.1016/j.ceramint.2020.08.192

Heydari, R., & Rashidipour, M. (2015). Green synthesis of silver nanoparticles using extract of oak fruit hull (Jaft): Synthesis and in vitro cytotoxic effect on MCF-7 cells. *International Journal of Breast Cancer, 2015*, 1–6. doi:10.1155/2015/846743 PMID:25685560

Hoelscher, K. P., De Stefano, G., Riley, M., & Young, S. (2012). Application of nanotechnology in drilling fluids. *Society of Petroleum Engineers - SPE International Oilfield Nanotechnology Conference 2012.* 10.2118/157031-MS

Holohan, C., Van Schaeybroeck, S., Longley, D. B., & Johnston, P. G. (2013). Cancer drug resistance: An evolving paradigm. *Nature Reviews. Cancer, 13*(10), 714–726. doi:10.1038/nrc3599 PMID:24060863

Holt, K. B., & Bard, A. J. (2005). Interaction of silver(I) ions with the respiratory chain of Escherichia coli: An electrochemical and scanning electrochemical microscopy study of the antimicrobial mechanism of micromolar Ag+. *Biochemistry, 44*(39), 13214–13223. doi:10.1021/bi0508542 PMID:16185089

Honarmand, M., Golmohammadi, M., & Naeimi, A. (2020). Green synthesis of SnO_2-bentonite nanocomposites for the efficient photodegradation of methylene blue and eriochrome black-T. *Materials Chemistry and Physics, 241*(May), 122416. doi:10.1016/j.matchemphys.2019.122416

Hossain, F., Perales-Perez, O. J., Hwang, S., & Román, F. (2014). *Antimicrobial nanomaterials as water disinfectant: applications, limitations and future perspectives*. Academic Press.

Hou, L., Wang, L., Royer, S., & Zhang, H. (2016). Ultrasound-assisted heterogeneous Fenton-like degradation of tetracycline over a magnetite catalyst. *Journal of Hazardous Materials*, *302*, 458–467. doi:10.1016/j.jhazmat.2015.09.033 PMID:26521091

Hou, Y., Sun, Z., Rao, W., & Liu, J. (2018). Nanoparticle-mediated cryosurgery for tumor therapy. *Nanomedicine; Nanotechnology, Biology, and Medicine*, *14*(2), 493–506. doi:10.1016/j.nano.2017.11.018 PMID:29197593

Hsueh, Y.-H., Tsai, P.-H., Lin, K.-S., Ke, W.-J., & Chiang, C.-L. (2017). Antimicrobial effects of zero-valent iron nanoparticles on gram-positive Bacillus strains and gram-negative Escherichia coli strains. *Journal of Nanobiotechnology*, *15*(1), 1–12. doi:10.118612951-017-0314-1 PMID:29100510

Hu, Y. B., Zhang, M., & Li, X. Y. (2019). Improved longevity of nanoscale zero-valent iron with a magnesium hydroxide coating shell for the removal of Cr(VI) in sand columns. *Environ Int, 133*(Pt B), 105249. doi:10.1016/j.envint.2019.105249

Hu, Y., & Qiu, L. (2019). Polymersomes: Preparation and Characterization. In Pharmaceutical Nanotechnology (pp. 247-265). Springer.

Hua, M., Zhang, S., Pan, B., Zhang, W., Lv, L., & Zhang, Q. (2012). Heavy metal removal from water/wastewater by nanosized metal oxides: A review. *Journal of Hazardous Materials*, *211*, 317–331. doi:10.1016/j.jhazmat.2011.10.016 PMID:22018872

Huang, T., Evans, B. A., Crews, J. B., & Belcher, C. K. (2010). Field case study on formation fines control with nanoparticles in offshore wells. *Proceedings - SPE Annual Technical Conference and Exhibition, 5*. 10.2118/135088-MS

Huang, W. (2018). Clay Nanopapers. In Nanopapers: From Nanochemistry and Nanomanufacturing to Advanced Applications (pp. 59–86). Elsevier Inc. doi:10.1016/B978-0-323-48019-2.00003-7

Huang, B., Li, C., Zhang, Y., Ding, W., Yang, M., Yang, Y., Zhai, H., Xu, X., Wang, D., Debnath, S., Jamil, M., Li, H. N., Ali, H. M., Gupta, M. K., & Said, Z. (2021). Advances in fabrication of ceramic corundum abrasives based on sol–gel process. *Chinese Journal of Aeronautics*, *34*(6), 1–17. Advance online publication. doi:10.1016/j.cja.2020.07.004

Huang, F., Liang, Y., & He, Y. (2019). On the Pickering emulsions stabilized by calcium carbonate particles with various morphologies. *Colloids and Surfaces. A, Physicochemical and Engineering Aspects*, *580*, 123722. Advance online publication. doi:10.1016/j.colsurfa.2019.123722

Huang, H. D., Ren, P. G., Chen, J., Zhang, W. Q., Ji, X., & Li, Z. M. (2012). High barrier graphene oxide nanosheet/poly(vinyl alcohol) nanocomposite films. *Journal of Membrane Science*, *409-410*, 156–163. Advance online publication. doi:10.1016/j.memsci.2012.03.051

Huang, H.-C., Barua, S., Sharma, G., Dey, S. K., & Rege, K. (2011). Inorganic nanoparticles for cancer imaging and therapy. *Journal of Controlled Release*, *155*(3), 344–357. doi:10.1016/j.jconrel.2011.06.004 PMID:21723891

Huang, H., Dong, Y., Zhang, Y., Ru, D., Wu, Z., Zhang, J., Shen, M., Duan, Y., & Sun, Y. (2019). GSH-sensitive Pt(IV) prodrug-loaded phase-transitional nanoparticles with a hybrid lipid-polymer shell for precise theranostics against ovarian cancer. *Theranostics*, *9*(4), 1047–1065. doi:10.7150/thno.29820 PMID:30867815

Huang, J., Li, Q., Sun, D., Lu, Y., Su, Y., Yang, X., Wang, H., Wang, Y., Shao, W., He, N., Hong, J., & Chen, C. (2007). Biosynthesis of silver and gold nanoparticles by novel sundried Cinnamomum camphora leaf. *Nanotechnology*, *18*(10), 105104. doi:10.1088/0957-4484/18/10/105104

Huang, T., Holden, J. A., Heath, D. E., O'Brien-Simpson, N. M., & O'Connor, A. J. (2019). Engineering highly effective antimicrobial selenium nanoparticles through control of particle size. *Nanoscale*, *11*(31), 14937–14951. doi:10.1039/C9NR04424H PMID:31363721

Huang, X., Jain, P. K., El-Sayed, I. H., & El-Sayed, M. A. (2007). Gold nanoparticles: Interesting optical properties and recent applications in cancer diagnostics and therapy. *Nanomedicine (London)*, *2*(5), 681–693. doi:10.2217/17435889.2.5.681 PMID:17976030

Huang, Y., Li, C., & Lin, Z. (2014). EDTA-induced self-assembly of 3D graphene and its superior adsorption ability for paraquat using a teabag. *ACS Applied Materials & Interfaces*, *6*(22), 19766–19773. doi:10.1021/am504922v PMID:25359004

Huang, Z., Wang, X., & Yang, D. (2015). Adsorption of Cr(VI) in wastewater using magnetic multi-wall carbon nanotubes. *Water Science and Engineering*, *8*(3), 226–232. doi:10.1016/j.wse.2015.01.009

Hua, S., de Matos, M. B. C., Metselaar, J. M., & Storm, G. (2018). Current Trends and Challenges in the Clinical Translation of Nanoparticulate Nanomedicines: Pathways for Translational Development and Commercialization. *Frontiers in Pharmacology*, *9*, 790. Advance online publication. doi:10.3389/fphar.2018.00790 PMID:30065653

Hu, G., Xu, C., Sun, Z., Wang, S., Cheng, H. M., Li, F., & Ren, W. (2016). 3D Graphene-Foam-Reduced-Graphene-Oxide Hybrid Nested Hierarchical Networks for High-Performance Li-S Batteries. *Advanced Materials*, *28*(8), 1603–1609. doi:10.1002/adma.201504765 PMID:26677000

Huh, A. J., & Kwon, Y. (2011). *"Nanoantibiotics": a new paradigm for treating infectious diseases using nanomaterials in the antibiotics resistant era*. Academic Press.

Hu, H., Zhang, R., Wang, J., Ying, W., Shi, L., Yao, C., Kong, Z., Wang, K., & Zhu, J. (2019). A mild method to prepare high molecular weight poly(butylene furandicarboxylate-co-glycolate) copolyesters: Effects of the glycolate content on thermal, mechanical, and barrier properties and biodegradability. *Green Chemistry*, *21*(11), 3013–3022. doi:10.1039/C9GC00668K

Hulla, J. E., Sahu, S. C., & Hayes, A. W. (2015). Nanotechnology: History and future. *Human and Experimental Toxicology*, *34*(12), 1318–1321. doi:10.1177/0960327115603588 PMID:26614822

Hummers, W. S. Jr, & Offeman, R. E. (1958). Preparation of Graphitic Oxide. *Journal of the American Chemical Society*, *80*(6), 1339. Advance online publication. doi:10.1021/ja01539a017

Hussain, A., Abidi, I. H., Tso, C. Y., Chan, K. C., Luo, Z., & Chao, C. Y. H. (2018). Thermal management of lithium ion batteries using graphene coated nickel foam saturated with phase change materials. *International Journal of Thermal Sciences*, *124*, 23–35. doi:10.1016/j.ijthermalsci.2017.09.019

Hussain, A., Oves, M., Alajmi, M. F., Hussain, I., Amir, S., Ahmed, J., Rehman, M. T., El-Seedi, H. R., & Ali, I. (2019). Biogenesis of ZnO nanoparticles using Pandanus odorifer leaf extract: Anticancer and antimicrobial activities. *RSC Advances*, *9*(27), 15357–15369. doi:10.1039/C9RA01659G

Hussain, I., Singh, N. B., Singh, A., Singh, H., & Singh, S. C. (2016). Green synthesis of nanoparticles and its potential application. *Biotechnology Letters*, *38*(4), 545–560. doi:10.100710529-015-2026-7 PMID:26721237

Ibrahim Fouad, G. (2021). A proposed insight into the anti-viral potential of metallic nanoparticles against novel coronavirus disease-19 (COVID-19). *Bulletin of the National Research Center*, *45*(1), 36. Advance online publication. doi:10.118642269-021-00487-0 PMID:33564223

İçhedef, Ç., Teksöz, S., Çetin, O., Aydın, B., Sarıkavak, İ., Parlak, Y., & Bilgin, B. E. (2021). Design of 99mTc radiolabeled gemcitabine polymeric nanoparticles as drug delivery system and in vivo evaluation. *Materials Chemistry and Physics*, *263*, 124380. doi:10.1016/j.matchemphys.2021.124380

Ifeanyichukwu, U. L., Fayemi, O. E., & Ateba, C. N. (2020). Green synthesis of zinc oxide nanoparticles from pomegranate (Punica granatum) extracts and characterization of their antibacterial activity. *Molecules (Basel, Switzerland)*, *25*(19), 4521. doi:10.3390/molecules25194521 PMID:33023149

Iga, A. M., Robertson, J. H. P., Winslet, M. C., & Seifalian, A. M. (2007). Clinical potential of Quantum Dots. *Journal of Biomedicine & Biotechnology*, *2007*, 1–10. doi:10.1155/2007/76087 PMID:18317518

Ijaz, U., Bhatti, I. A., Mirza, S., & Ashar, A. (2017). Characterization and evaluation of antibacterial activity of plant mediated calcium oxide (CaO) nanoparticles by employing Mentha pipertia extract. *Materials Research Express*, *4*(10), 105402. https://doi.org/10.1088/2053-1591/aa8603

Imtiaz, A., Farrukh, M. A., Khaleeq-ur-rahman, M., & Adnan, R. (2013). Micelle-Assisted Synthesis of Al 2 O 3 ·CaO Nanocatalyst: Optical Properties and Their Applications in Photodegradation of 2,4,6-Trinitrophenol. *TheScientificWorldJournal*, *2013*, 1–11. https://doi.org/10.1155/2013/641420

Inbasekaran, S., Senthil, R., Ramamurthy, G., & Sastry, T. P. (2014). Biosensor using zinc oxide nanoparticles. *International Journal of Innovative Research in Science, Engineering and Technology*, *3*(1), 8601–8606.

Indira, J., & Sreeja, V. (2020). Synthesis of silver/hydroxyapatite/tryptophan nanocomposite particles by biological method. *Materials Today: Proceedings*. Advance online publication. doi:10.1016/j.matpr.2020.09.579

Iosob, G.-A., Prisecaru, M., Stoica, I., & Călin, M. T. O. C. (2016). Biological Remediation of Soil Polluted With Oil Products : an Overview of Available Technologies. *Biological Remediation of Soil Polluted With Oil Products : An Overview of Available Technologie*, 98. https://www.researchgate.net/publication/309901995_BIOLOGICAL_REMEDIATION_OF_SOIL_POLLUTED_WITH_OIL_PRODUCTS_AN_OVERVIEW_OF_AVAILABLE_TECHNOLOGIES

Iravani, S. (2011). Green synthesis of metal nanoparticles using plants. *Green Chemistry*, *13*(10), 2638. doi:10.1039/c1gc15386b

Iravani, S. (2014). Bacteria in Nanoparticle Synthesis: Current Status and Future Prospects. *International Scholarly Research Notices*, *2014*, 1–18. doi:10.1155/2014/359316 PMID:27355054

Ishak, N. A. I., Kamarudin, S. K., & Timmiati, S. N. (2019). Green synthesis of metal and metal oxide nanoparticles via plant extracts: An overview. *Materials Research Express*, *6*(11), 112004. doi:10.1088/2053-1591/ab4458

Ito, T., Matsuda, Y., Jinba, T., Asai, N., Shimizu, T., & Shingubara, S. (2017). Fabrication and characterization of nano porous lattice biosensor using anodic aluminum oxide substrate. *Japanese Journal of Applied Physics*, *56*(6S1), 06GG02.

Itoh, N., Kimoto, A., Yamamoto, E., Higashi, T., Santa, T., Funatsu, T., & Kato, M. (2017). High performance liquid chromatography analysis of 100-nm liposomal nanoparticles using polymer-coated, silica monolithic columns with aqueous mobile phase. *Journal of Chromatography. A*, *1484*, 34–40. doi:10.1016/j.chroma.2016.12.080 PMID:28089273

Jadoun, S., Arif, R., Jangid, N. K., & Meena, R. K. (2021). Green synthesis of nanoparticles using plant extracts: A review. *Environmental Chemistry Letters*, *19*(1), 355–374. doi:10.100710311-020-01074-x

Jafari, A., Salouti, M., Shayesteh, S. F., Heidari, Z., Rajabi, A. B., Boustani, K., & Nahardani, A. (2015). Synthesis and characterization of Bombesin-superparamagnetic iron oxide nanoparticles as a targeted contrast agent for imaging of breast cancer using MRI. *Nanotechnology*, *26*(7), 075101. doi:10.1088/0957-4484/26/7/075101 PMID:25642737

Jagwani, D., & Hari Krishna, P. (2021). Nature's nano-assets: Green synthesis, characterization techniques and applications – A graphical review. *Materials Today: Proceedings*, 2307–2317. Advance online publication. doi:10.1016/j.matpr.2021.04.185

Jahan, I., Erci, F., Cakir-Koc, R., & Isildak, I. (2021). Microwave-irradiated green synthesis of metallic silver and copper nanoparticles using fresh ginger (Zingiber officinale) rhizome extract and evaluation of their antibacterial potentials and cytotoxicity. *Inorganic and Nano-Metal Chemistry*, *51*(5), 722–732. doi:10.1080/24701556.2020.1808017

Jahan, I., Erci, F., & Isildak, I. (2019). Microwave-assisted green synthesis of non-cytotoxic silver nanoparticles using the aqueous extract of Rosa santana (rose) petals and their antimicrobial activity. *Analytical Letters*, *52*(12), 1860–1873. doi:10.1080/00032719.2019.1572179

Jahan, I., Erci, F., & Isildak, I. (2021a). Rapid green synthesis of non-cytotoxic silver nanoparticles using aqueous extracts of 'Golden Delicious' apple pulp and cumin seeds with antibacterial and antioxidant activity. *SN Applied Sciences*, *3*(1), 94. doi:10.100742452-020-04046-6

Jahan, I., Erci, F., & Isildak, I. (2021b). Facile microwave-mediated green synthesis of non-toxic copper nanoparticles using Citrus sinensis aqueous fruit extract and their antibacterial potentials. *Journal of Drug Delivery Science and Technology*, *61*, 102172. doi:10.1016/j.jddst.2020.102172

Jahromi, E. Z., Divsalar, A., Saboury, A. A., Khaleghizadeh, S., Mansouri-Torshizi, H., & Kostova, I. (2016). Palladium complexes: New candidates for anti-cancer drugs. *Journal of the Indian Chemical Society*, *13*(5), 967–989.

Jain, N. K., Tare, M. S., Mishra, V., & Tripathi, P. K. (2015). The development, characterization and in vivo anti-ovarian cancer activity of poly(propylene imine) (PPI)-antibody conjugates containing encapsulated paclitaxel. *Nanomedicine (London)*, *11*(1), 207–218. doi:10.1016/j.nano.2014.09.006 PMID:25262579

Jain, P. K., Lee, K. S., El-Sayed, I. H., & El-Sayed, M. A. (2006). Calculated absorption and scattering properties of gold nanoparticles of different size, shape, and composition: Applications in biological imaging and biomedicine. *The Journal of Physical Chemistry B*, *110*(14), 7238–7248. doi:10.1021/jp057170o PMID:16599493

Jain, S. K., Chourasia, M. K., Masuriha, R., Soni, V., Jain, A., Jain, N. K., & Gupta, Y. (2005). Solid lipid nanoparticles bearing flurbiprofen for transdermal delivery. *Drug Delivery*, *12*(4), 207–215. doi:10.1080/10717540590952591 PMID:16036715

Jain, S., Hirst, D. G., & O'sullivan, J. M. (2012). Gold nanoparticles as novel agents for cancer therapy. *The British Journal of Radiology*, *85*(1010), 101–113. doi:10.1259/bjr/59448833 PMID:22010024

Jain, T. K., Roy, I., De, T. K., & Maitra, A. N. J. (1998). Nanometer Silica Particles Encapsulating Active Compounds: A Novel Ceramic Drug Carrier. *Journal of the American Chemical Society*, *120*(43), 11092–11095. doi:10.1021/ja973849x

Jaiswal, K. K., Dutta, S., Pohrmen, C. B., Verma, R., Kumar, A., & Ramaswamy, A. P. (2021). Bio-waste chicken eggshell-derived calcium oxide for photocatalytic application in methylene blue dye degradation under natural sunlight irradiation. *Inorganic and Nano-Metal Chemistry*, *51*(7), 995–1004. doi:10.1080/24701556.2020.1813769

Jaiswal, P. V., Ijeri, V. S., & Srivastava, A. K. (2002). *Voltammetric behavior of certain vitamins and their determination in surfactant media*. The Japan Society for Analytical Chemistry. doi:10.14891/analscisp.17icas.0.i741.0

Jalu, R. G., Chamada, T. A., & Kasirajan, D. R. (2021). Calcium oxide nanoparticles synthesis from hen eggshells for removal of lead (Pb(II)) from aqueous solution. *Environmental Challenges*, *4*(June), 100193. doi:10.1016/j.envc.2021.100193

Jamdagni, P., Khatri, P., & Rana, J. S. (2018). Green synthesis of zinc oxide nanoparticles using flower extract of Nyctanthes arbor-tristis and their antifungal activity. *Journal of King Saud University. Science*, *30*(2), 168–175. doi:10.1016/j.jksus.2016.10.002

Jamil, B., Habib, H., Abbasi, S., Nasir, H., Rahman, A., Rehman, A., Bokhari, H., & Imran, M. (2016). Cefazolin loaded chitosan nanoparticles to cure multi drug resistant Gram-negative pathogens. *Carbohydrate Polymers*, *136*, 682–691. doi:10.1016/j.carbpol.2015.09.078 PMID:26572401

Jamkhande, P. G., Ghule, N. W., Bamer, A. H., & Kalaskar, M. G. (2019). Metal nanoparticles synthesis: An overview on methods of preparation, advantages and disadvantages, and applications. Journal of Drug Delivery Science and Technology, 53. doi:10.1016/j.jddst.2019.101174

Jardón-Maximino, N., Cadenas-Pliego, G., Ávila-Orta, C. A., Comparán-Padilla, V. E., Lugo-Uribe, L. E., Pérez-Alvarez, M., Tavizón, S. F., & Dejesús Sosa Santillán, G. (2021). Antimicrobial property of polypropylene composites and functionalized copper nanoparticles. *Polymers*, *13*(11), 1694. Advance online publication. doi:10.3390/polym13111694 PMID:34067323

Jastrzębska, A. M., & Vasilchenko, A. S. (2021). Smart and Sustainable Nanotechnological Solutions in a Battle against COVID-19 and Beyond: A Critical Review. *ACS Sustainable Chemistry & Engineering*, *9*(2), 601–622. doi:10.1021/acssuschemeng.0c06565 PMID:34192094

Jatykov, T., Sagindykov, B., & Bimuratkyzy, K. (2021). Hydraulic fracturing design improvements by proppant flowback mitigation method application during hydrocarbon production. *Rudarsko Geolosko Naftni Zbornik*, *36*(1), 99–109. Advance online publication. doi:10.17794/rgn.2021.1.8

Jayasekara, H., MacInnis, R. J., Room, R., & English, D. R. (2016). Long-term alcohol consumption and breast, upper aero-digestive tract and colorectal cancer risk: A systematic review and meta-analysis. *Alcohol and Alcoholism (Oxford, Oxfordshire)*, *51*(3), 315–330. doi:10.1093/alcalc/agv110 PMID:26400678

Jebril, S., Khanfir Ben Jenana, R., & Dridi, C. (2020). Green synthesis of silver nanoparticles using Melia azedarach leaf extract and their antifungal activities: In vitro and in vivo. *Materials Chemistry and Physics*, *248*, 122898. doi:10.1016/j.matchemphys.2020.122898

Jeevanandam, J., Barhoum, A., Chan, Y. S., Dufresne, A., & Danquah, M. K. (2018). Review on nanoparticles and nanostructured materials: History, sources, toxicity and regulations. *Beilstein Journal of Nanotechnology*, *9*(1), 1050–1074. doi:10.3762/bjnano.9.98 PMID:29719757

Jenning, V., Schäfer-Korting, M., & Gohla, S. (2000). Vitamin A-loaded solid lipid nanoparticles for topical use: Drug release properties. *Journal of Controlled Release*, *66*(2–3), 115–126. doi:10.1016/S0168-3659(99)00223-0 PMID:10742573

Jensen, S. A., Day, E. S., Ko, C. H., Hurley, L. A., Luciano, J. P., Kouri, F. M., Merkel, T. J., Luthi, A. J., Patel, P. C., Cutler, J. I., Daniel, W. L., Scott, A. W., Rotz, M. W., Meade, T. J., Giljohann, D. A., Mirkin, C. A., & Stegh, A. H. (2013). Spherical nucleic acid nanoparticle conjugates as an RNAi-based therapy for glioblastoma. *Science Translational Medicine, 5*(209), 209ra152-209ra152. doi:10.1126/scitranslmed.3006839

Jeremiah, S. S., Miyakawa, K., Morita, T., & Yamaoka, Y. (2020). Since January 2020 Elsevier has created a COVID-19 resource centre with free information in English and Mandarin on the novel coronavirus COVID- 19. The COVID-19 resource centre is hosted on Elsevier Connect, the company ' s public news and information. *Biochemical and Biophysical Research Communications*, *533*(1), 195–200. doi:10.1016/j.bbrc.2020.09.018 PMID:32958250

Jesica Roshima, A., & Jayalakshmi, I. (2020). Effect of herbal microencapsulated antimicrobial and antilarvicidal finishes on selected natural fabrics. *Materials Today: Proceedings*.

Jeyaraj, M., Gurunathan, S., Qasim, M., Kang, M. H., & Kim, J. H. (2019). A comprehensive review on the synthesis, characterization, and biomedical application of platinum nanoparticles. *Nanomaterials (Basel, Switzerland)*, *9*(12), 1719. doi:10.3390/nano9121719 PMID:31810256

Jeyaraj, M., Sathishkumar, G., Sivanandhan, G., MubarakAli, D., Rajesh, M., Arun, R., Kapildev, G., Manickavasagam, M., Thajuddin, N., Premkumar, K., & Ganapathi, A. (2013). Biogenic silver nanoparticles for cancer treatment: An experimental report. *Colloids and Surfaces. B, Biointerfaces*, *106*, 86–92. doi:10.1016/j.colsurfb.2013.01.027 PMID:23434696

Jha, A. K., Prasad, K., & Kulkarni, A. R. (2009). Synthesis of TiO2 nanoparticles using microorganisms. *Colloids and Surfaces. B, Biointerfaces*, *71*(2), 226–229. doi:10.1016/j.colsurfb.2009.02.007 PMID:19285838

Jia, L., Zhang, Q., Li, Q., & Song, H. (2009). The biosynthesis of palladium nanoparticles by antioxidants in Gardenia jasminoides Ellis: Long lifetime nanocatalysts for p-nitrotoluene hydrogenation. *Nanotechnology*, *20*(38), 385601. doi:10.1088/0957-4484/20/38/385601 PMID:19713585

Jiang, K., Zhang, L., & Bao, G. (2021). Magnetic Iron Oxide Nanoparticles for Biomedical Applications. *Current Opinion in Biomedical Engineering*, *100330*. doi:10.1016/j.cobme.2021.100330

Jiang, L., Liu, Y., Liu, S., Zeng, G., Hu, X., Hu, X., Guo, Z., Tan, X., Wang, L., & Wu, Z. (2017). Adsorption of Estrogen Contaminants by Graphene Nanomaterials under Natural Organic Matter Preloading: Comparison to Carbon Nanotube, Biochar, and Activated Carbon. *Environmental Science & Technology*, *51*(11), 6352–6359. doi:10.1021/acs.est.7b00073 PMID:28494154

Jiang, W., Fang, H., Liu, F., Zhou, X., Zhao, H., He, X., & Guo, D. (2019). PEG-coated and Gd-loaded fluorescent silica nanoparticles for targeted prostate cancer magnetic resonance imaging and fluorescence imaging. *International Journal of Nanomedicine*, *14*, 5611–5622. doi:10.2147/IJN.S207098 PMID:31413566

Jian, Z., Zhuang, T., Qinyu, T., Liqing, P., Kun, L., Xujiang, L., Diaodiao, W., Zhen, Y., Shuangpeng, J., Xiang, S., Jingxiang, H., Shuyun, L., Libo, H., Peifu, T., Qi, Y., & Quanyi, G. (2021). 3D bioprinting of a biomimetic meniscal scaffold for application in tissue engineering. *Bioactive Materials*, *6*(6), 1711–1726. doi:10.1016/j.bioactmat.2020.11.027 PMID:33313450

Jia, X., & Wei, F. (2017). Advances in Production and Applications of Carbon Nanotubes. *Topics in Current Chemistry (Z)*, *375*(1), 18. doi:10.100741061-017-0102-2 PMID:28138948

Ji, L., Chen, W., Xu, Z., Zheng, S., & Zhu, D. (2013). Graphene nanosheets and graphite oxide as promising adsorbents for removal of organic contaminants from aqueous solution. *Journal of Environmental Quality*, *42*(1), 191–198. doi:10.2134/jeq2012.0172 PMID:23673754

Jin, T., & He, Y. (2011). *Antibacterial activities of magnesium oxide (MgO) nanoparticles against foodborne pathogens*. Academic Press.

Jin, C., Wang, K., Oppong-Gyebi, A., & Hu, J. (2020). Application of nanotechnology in cancer diagnosis and therapy—A mini-review. *International Journal of Medical Sciences*, *17*(18), 2964–2973. doi:10.7150/ijms.49801 PMID:33173417

Ji, Y., Liu, X., Li, J., Xie, X., Huang, M., Jiang, J., Liao, Y. P., Donahue, T., & Meng, H. (2020). Use of ratiometrically designed nanocarrier targeting CDK4/6 and autophagy pathways for effective pancreatic cancer treatment. *Nature Communications*, *11*(1), 4249. doi:10.103841467-020-17996-7 PMID:32843618

Joh, D. Y., Sun, L., Stangl, M., Al Zaki, A., Murty, S., Santoiemma, P. P., Davis, J. J., Baumann, B. C., Alonso-Basanta, M., Bhang, D., Kao, G. D., Tsourkas, A., & Dorsey, J. F. (2013). Selective targeting of brain tumors with gold nanoparticle-induced radiosensitization. *PLoS One*, *8*(4), e62425. doi:10.1371/journal.pone.0062425 PMID:23638079

Johnson-McDaniel, D., Barrett, C. A., Sharafi, A., & Salguero, T. T. (2013). *Nanoscience of an ancient pigment*. Academic Press.

Johnstone, T. C., Suntharalingam, K., & Lippard, S. J. (2016). The next generation of platinum drugs: Targeted Pt (II) agents, nanoparticle delivery, and Pt (IV) prodrugs. *Chemical Reviews*, *116*(5), 3436–3486. doi:10.1021/acs.chemrev.5b00597 PMID:26865551

Jones, R. R., Hooper, D. C., Zhang, L., Wolverson, D., & Valev, V. K. (2019). Raman Techniques : Fundamentals and Frontiers. *Nanoscale Research Letters*, *14*(231). PMID:31300945

Jose, A. J., Jacob, A. M., Manjush, K. C., & Kappen, J. (2018). Chitosan in water purification technology. In S. Ahmad & C. M. Hussain (Eds.), *Green and sustainable advanced materials: Applications*. doi:10.1002/9781119528463.ch5

Joshi, M., Banerjee, K., & Thakare, V. (2006). Polymer/clay nanocomposite based coatings for enhanced gas barrier property. Indian Journal of Fibre & Textile Research, 31.

Joulaei, M., Hedayati, K., & Ghanbari, D. (2019). Investigation of magnetic, mechanical and flame retardant properties of polymeric nanocomposites: Green synthesis of $MgFe_2O_4$ by lime and orange extracts. *Composites. Part B, Engineering*, *176*(July), 107345. doi:10.1016/j.compositesb.2019.107345

Kabcum, S., Channei, D., Tuantranont, A., Wisitsoraat, A., Liewhiran, C., & Phanichphant, S. (2016). Ultra-responsive hydrogen gas sensors based on PdO nanoparticle-decorated WO3 nanorods synthesized by precipitation and impregnation methods. *Sensors and Actuators. B, Chemical*, *226*, 76–89. doi:10.1016/j.snb.2015.11.120

Kabir, M. A., Ahmed, R., Iqbal, S. M. A., Chowdhury, R., Paulmurugan, R., Demirci, U., & Asghar, W. (2021). Diagnosis for COVID-19: Current status and future prospects. *Expert Review of Molecular Diagnostics*, *21*(3), 269–288. doi:10.1080/14737159.2021.1894930 PMID:33621145

Kadiyala, N. K., Mandal, B. K., Ranjan, S., & Dasgupta, N. (2018). Bioinspired gold nanoparticles decorated reduced graphene oxide nanocomposite using Syzygium cumini seed extract: Evaluation of its biological applications. *Materials Science and Engineering C*, *93*(October), 191–205. doi:10.1016/j.msec.2018.07.075

Kaegi, R., Voagelin, A., Ort, C., Sinnet, B., Thalmann, B., Krismer, J., Hagendorfer, H., Elumelu, M., & Mueller, E. (2013). Fate and Transformation of Silver Nanoparticles in Urban Wastewater Systems. *Water Research*, *47*(12), 3866–3877. doi:10.1016/j.watres.2012.11.060 PMID:23571111

Kalita, E., & Baruah, J. (2020). Environmental remediation. In *Colloidal Metal Oxide Nanoparticles*. Elsevier Inc., doi:10.1016/B978-0-12-813357-6.00014-0

Kalomiraki, M., Thermos, K., & Chaniotakis, N. A. (2016). Dendrimers as tunable vectors of drug delivery systems and biomedical and ocular applications. *International Journal of Nanomedicine*, *11*, 1. PMID:26730187

Kalpana, V. N., & Devi Rajeswari, V. (2018). A Review on Green Synthesis, Biomedical Applications, and Toxicity Studies of ZnO NPs. *Bioinorganic Chemistry and Applications*, *3569758*, 1–12. Advance online publication. doi:10.1155/2018/3569758 PMID:30154832

Kamat, P. V. (2002). Photophysical, photochemical and photocatalytic aspects of metal nanoparticles. *The Journal of Physical Chemistry B*, *106*(32), 7729–7744. doi:10.1021/jp0209289

Kamble, G., Kashale, A., Rasal, A., Dengale, S., Kolekar, S., Chang, J. Y., Han, S.-H., & Ghule, A. (2021). Investigating the Influence of Reflux Condensation Reaction Temperature on the Growth of FeCo2O4 Thin Film for Flexible Supercapacitor. *ChemistrySelect*, *6*(8), 1838–1844. doi:10.1002lct.202004544

Kamble, S. P., Jagtap, S., Labhsetwar, N. K., Thakare, D., Godfrey, S., Devotta, S., & Rayalu, S. S. (2007). Defluoridation of drinking water using chitin, chitosan and lanthanum-modified chitosan. *Chemical Engineering Journal, 129*(1–3), 173–180. https://doi.org/10.1016/j.cej.2006.10.032

Kamble, S., Utage, B., Mogle, P., Kamble, R., Hese, S., Dawane, B., & Gacche, R. (2016). Evaluation of curcumin capped copper nanoparticles as possible inhibitors of human breast cancer cells and angiogenesis: A comparative study with native curcumin. *AAPS PharmSciTech, 17*(5), 1030–1041. doi:10.120812249-015-0435-5 PMID:26729534

Kamboj, A., Amjad, M., Ahmad, W., & Singh, A. (2020). A general survey on Green synthesis and application of calcium oxide nanoparticles. *International Journal of Health and Clinical Research, 3*(2), 41–48. https://ijhcr.com/index.php/ijhcr/article/view/63

Kaminskas, L. M., Kelly, B. D., McLeod, V. M., Sberna, G., Owen, D. J., Boyd, B. J., & Porter, C. J. (2011). Characterisation and tumour targeting of PEGylated polylysine dendrimers bearing doxorubicin via a pH labile linker. *Journal of Controlled Release, 152*(2), 241–248. doi:10.1016/j.jconrel.2011.02.005 PMID:21315119

Kanagasubbulakshmi, S., & Kadirvelu, K. (2017). Green synthesis of iron oxide nanoparticles using Lagenaria siceraria and evaluation of its antimicrobial activity. *Defence Life Science Journal, 2*(4), 422–427. doi:10.14429/dlsj.2.12277

Kanchi, S., Kumar, G., Lo, A.-Y., Tseng, C.-M., Chen, S.-K., Lin, C.-Y., & Chin, T.-S. (2018). Exploitation of de-oiled jatropha waste for gold nanoparticles synthesis: A green approach. *Arabian Journal of Chemistry, 11*(2), 247–255. doi:10.1016/j.arabjc.2014.08.006

Kandpal, N. D., Sah, N., Loshali, R., Joshi, R., & Prasad, J. (2014). Co-precipitation method of synthesis and characterization of iron oxide nanoparticles. *Journal of Scientific and Industrial Research, 73*(2).

Kaneshiro, T. L., & Lu, Z.-R. (2009). Targeted intracellular codelivery of chemotherapeutics and nucleic acid with a well-defined dendrimer-based nanoglobular carrier. *Biomaterials, 30*(29), 5660–5666. doi:10.1016/j.biomaterials.2009.06.026 PMID:19595449

Kang, D., Kwon, J. Y., Cho, H., Sim, J. H., Hwang, H. S., Kim, C. S., Kim, Y. J., Ruoff, R. S., & Shin, H. S. (2012). Oxidation resistance of iron and copper foils coated with reduced graphene oxide multilayers. *ACS Nano, 6*(9), 7763–7769. doi:10.1021/nn3017316 PMID:22881084

Kango, S., Kalia, S., Celli, A., Njuguna, J., Habibi, Y., & Kumar, R. (2013). Surface modification of inorganic nanoparticles for development of organic–inorganic nanocomposites—A review. *Progress in Polymer Science, 38*(8), 1232–1261. doi:10.1016/j.progpolymsci.2013.02.003

Kang, S. J., Lee, Y. J., Lee, E.-K., & Kwak, M.-K. (2012). Silver nanoparticles-mediated G2/M cycle arrest of renal epithelial cells is associated with NRF2-GSH signaling. *Toxicology Letters, 211*(3), 334–341. doi:10.1016/j.toxlet.2012.04.016 PMID:22546375

Kang, X., Liu, S., Dai, Z., He, Y., Song, X., & Tan, Z. (2019). Titanium dioxide: From engineering to applications. *Catalysts, 9*(2), 191. doi:10.3390/catal9020191

Kanhed, P., Birla, S., Gaikwad, S., Gade, A., Seabra, A. B., Rubilar, O., Duran, N., & Rai, M. (2014). In vitro antifungal efficacy of copper nanoparticles against selected crop pathogenic fungi. *Materials Letters, 115*, 13–17. doi:10.1016/j.matlet.2013.10.011

Kanjwal, M. A., Barakat, N. A. M., Sheikh, F. A., Khil, M. S., & Kim, H. Y. (2009). Functionalization of Electrospun Titanium Oxide Nanofibers with Silver Nanoparticles: Strongly Effective Photocatalyst. *International Journal of Applied Ceramic Technology, 7*, E54–E63. doi:10.1111/j.1744-7402.2009.02397.x

Kapoor, V., Phan, D., & Pasha, A. B. M. T. (2018). Effects of metal oxide nanoparticles on nitrification in wastewater treatment systems: A systematic review. *Journal of Environmental Science and Health. Part A, Toxic/Hazardous Substances & Environmental Engineering*, *53*(7), 659–668. https://doi.org/10.1080/10934529.2018.1438825

Kappe, C. O. (2004). Controlled Microwave Heating in Modern Organic Synthesis. *Angewandte Chemie International Edition*, *43*(46), 6250–6284. doi:10.1002/anie.200400655 PMID:15558676

Karimzadeh, M. R., Soltanian, S., Sheikhbahaei, M., & Mohamadi, N. (2020). Characterization and biological activities of synthesized zinc oxide nanoparticles using the extract of Acantholimon serotinum. *Green Processing and Synthesis*, *9*(1), 722–733. doi:10.1515/gps-2020-0058

Karki, H. P., Ojha, D. P., Joshi, M. K., & Kim, H. J. (2018). Effective reduction of p-nitrophenol by silver nanoparticle loaded on magnetic Fe_3O_4/ATO nano-composite. *Applied Surface Science*, *435*(30), 599–608. doi:10.1016/j.apsusc.2017.11.166

Karlapudi, A. P., Venkateswarulu, T. C., Tammineedi, J., Kanumuri, L., Ravuru, B. K., Dirisala, V., & Kodali, V. P. (2018). Role of biosurfactants in bioremediation of oil pollution-a review. *Petroleum*, *4*(3), 241–249. doi:10.1016/j.petlm.2018.03.007

Karmacharya, M., Kumar, S., Gulenko, O., & Cho, Y.-K. (2021). Advances in Facemasks during the COVID-19 Pandemic Era. *ACS Applied Bio Materials*, *4*(5), 3891–3908. doi:10.1021/acsabm.0c01329

Karra, N., & Benita, S. (2012). The ligand nanoparticle conjugation approach for targeted cancer therapy. *Current Drug Metabolism*, *13*(1), 22–41. doi:10.2174/138920012798356899 PMID:21892918

Karthick Raja Namasivayam, S., Arvind Bharani, R. S., & Karunamoorthy, K. (2018). Insecticidal fungal metabolites fabricated chitosan nanocomposite (IM-CNC) preparation for the enhanced larvicidal activity - An effective strategy for green pesticide against economic important insect pests. *International Journal of Biological Macromolecules*, *120*, 921–944. doi:10.1016/j.ijbiomac.2018.08.130 PMID:30172810

Karthika, N., Jananee, K., & Murugaiyan, V. (2016). Remediation of contaminated soil using soil washing-a review. *Journal of Engineering Research and Applications*, *6*(1–2), 13–18. http://www.ijera.com/papers/Vol6_issue1/Part - 2/C601021318.pdf

Kasemo, B. (2002). Biological surface science. *Surface Science*, *500*(1–3), 656–677. doi:10.1016/S0039-6028(01)01809-X

Kasinathan, K., Kennedy, J., Elayaperumal, M., Henini, M., & Malik, M. (2016). Photodegradation of organic pollutants RhB dye using UV simulated sunlight on ceria based TiO2 nanomaterials for antibacterial applications. *Scientific Reports*, *6*(1), 38064. doi:10.1038rep38064 PMID:27901105

Kathuria, L., Din Reshi, N. U., & Samuelson, A. G. (2020). N-Heterocyclic Carbene (NHC)-Stabilized Ru0 Nanoparticles: In Situ Generation of an Efficient Transfer Hydrogenation Catalyst. *Chemistry (Weinheim an der Bergstrasse, Germany)*, *26*(34), 7622–7630. Advance online publication. doi:10.1002/chem.202000142 PMID:32048353

Katiyar, S. S., Muntimadugu, E., Rafeeqi, T. A., Domb, A. J., & Khan, W. (2016). Co-delivery of rapamycin- and piperine-loaded polymeric nanoparticles for breast cancer treatment. *Drug Delivery*, *23*(7), 2608–2616. doi:10.3109/10717544.2015.1039667 PMID:26036652

Kato, H. (2012). Size Determination of Nanoparticles by Dynamic Light Scattering. *Nanomaterials*, 535–554.

Kaushal, A., Seth, A., Kala, D., Gupta, S., Krishnia, L., & Verma, V. (2022). *Miniaturized devices for point-of-care testing/miniaturization and integration with microfluidic systems*. Academic Press.

Kaviya, S., Santhanalakshmi, J., Viswanathan, B., Muthumary, J., & Srinivasan, K. (2011). Biosynthesis of silver nanoparticles using citrus sinensis peel extract and its antibacterial activity. *Spectrochimica Acta. Part A: Molecular and Biomolecular Spectroscopy*, *79*(3), 594–598. doi:10.1016/j.saa.2011.03.040 PMID:21536485

Kelkawi, A. H. A., Abbasi Kajani, A., & Bordbar, A. (2017). Green synthesis of silver nanoparticles using Mentha pulegium and investigation of their antibacterial, antifungal and anticancer activity. *IET Nanobiotechnology / IET*, *11*(4), 370–376. doi:10.1049/iet-nbt.2016.0103 PMID:28530184

Kelleni, M. T. (2021). Resveratrol-zinc nanoparticles or pterostilbene-zinc: Potential COVID-19 mono and adjuvant therapy. *Biomedicine and Pharmacotherapy*, *139*, 111626. doi:10.1016/j.biopha.2021.111626 PMID:33894625

Keogh, W., Neville, A., Huggan, M., Eroini, V., Olsen, J. H., Nielsen, F. M., Baraka-Lokmane, S., Bourdelet, E., Ellingsen, J. A., Bache, O., & Charpentier, T. (2017). Deposition of Inorganic Carbonate, Sulfate, and Sulfide Scales on Antifouling Surfaces in Multiphase Flow. *Energy & Fuels*, *31*(11), 11838–11851. Advance online publication. doi:10.1021/acs.energyfuels.7b02165

Kepp, O., Senovilla, L., Vitale, I., Vacchelli, E., Adjemian, S., Agostinis, P., Apetoh, L., Aranda, F., Barnaba, V., Bloy, N., Bracci, L., Breckpot, K., Brough, D., Buqué, A., Castro, M. G., Cirone, M., Colombo, M. I., Cremer, I., Demaria, S., ... Galluzzi, L. (2014). Consensus guidelines for the detection of immunogenic cell death. *OncoImmunology*, *3*(9), e955691. doi:10.4161/21624011.2014.955691 PMID:25941621

Kesharwani, P., & Iyer, A. K. (2015). Recent advances in dendrimer-based nanovectors for tumor-targeted drug and gene delivery. *Drug Discovery Today*, *20*(5), 536–547. doi:10.1016/j.drudis.2014.12.012 PMID:25555748

Kesharwani, P., Tekade, R. K., & Jain, N. K. (2015). Generation dependent safety and efficacy of folic acid conjugated dendrimer based anticancer drug formulations. *Pharmaceutical Research*, *32*(4), 1438–1450. doi:10.100711095-014-1549-2 PMID:25330744

Khadir, A., Motamedi, M., Negarestani, M., Sillanpaa, M., & Sasani, M. (2020). Preparation of a nano bio-composite based on cellulosic biomass and conducting polymeric nanoparticles for ibuprofen removal: Kinetics, isotherms, and energy site distribution. *International Journal of Biological Macromolecules*, *162*, 663–677. doi:10.1016/j.ijbiomac.2020.06.095 PMID:32544583

Khalid, A., Al-Juhani, A. A., Al-Hamouz, O. C., Laoui, T., Khan, Z., & Atieh, M. A. (2015). Preparation and properties of nanocomposite polysulfone/multi-walled carbon nanotubes membranes for desalination. *Desalination*, *367*, 134–144. doi:10.1016/j.desal.2015.04.001

Khalil, M., Jan, B. M., Tong, C. W., & Berawi, M. A. (2017). Advanced nanomaterials in oil and gas industry: Design, application and challenges. Applied Energy, 191. doi:10.1016/j.apenergy.2017.01.074

Khalil, M. M. H., Ismail, E. H., El-Baghdady, K. Z., & Mohamed, D. (2014). Green synthesis of silver nanoparticles using olive leaf extract and its antibacterial activity. *Arabian Journal of Chemistry*, *7*(6), 1131–1139. doi:10.1016/j.arabjc.2013.04.007

Khalkhali, M., Liu, Q., Zeng, H., & Zhang, H. (2015). A size-dependent structural evolution of ZnS nanoparticles. *Scientific Reports*, *5*(1), 14267. doi:10.1038rep14267 PMID:26381583

Khan, A. A., Alanazi, A. M., Alsaif, N., Wani, T. A., & Bhat, M. A. (2021). Pomegranate peel induced biogenic synthesis of silver nanoparticles and their multifaceted potential against intracellular pathogen and cancer. *Saudi Journal of Biological Sciences*, *28*(8), 4191–4200. doi:10.1016/j.sjbs.2021.06.022 PMID:34354399

Khan, A. H., Tirth, V., Fawzy, M., Mahmoud, A. E. D., Khan, N. A., Ahmed, S., Ali, S. S., Akram, M., Hameed, L., Das, G., Roy, S., & Dehghani, M. H. (2021). COVID-19 Transmission, Vulnerability, Persistence and Nanotherapy: A Review. *Environmental Chemistry Letters*, *19*(4), 2773–2787. Advance online publication. doi:10.100710311-021-01229-4 PMID:33846683

Khan, I., Saeed, K., & Khan, I. (2019). Nanoparticles: Properties, applications and toxicities. *Arabian Journal of Chemistry*, *12*(7), 908–931. doi:10.1016/j.arabjc.2017.05.011

Khan, M. N., Wan Sulaiman, W. R., & Abbas, A. H. (2021). Study of Sulfosuccinate and Extended Sulfated Sodium Surfactants on the Malaysian Crude/Water Properties for ASP Application in Limestone. *Arabian Journal for Science and Engineering*, *46*(7), 6915–6924. Advance online publication. doi:10.100713369-020-05252-5

Khan, N. T., & Khan, M. J. (2020). Metallic Nanoparticles Fabrication Methods– A Brief Overview. *SunKrist Nanotechnology and Nanoscience Journal*, *2*(1), 1–6. doi:10.46940nnj.02.1002

Khan, S. B., Marwani, H. M., Asiri, A. M., & Bakhsh, E. M. (2016). Exploration of calcium doped zinc oxide nanoparticles as selective adsorbent for extraction of lead ion. *Desalination and Water Treatment*, *57*(41), 19311–19320. https://doi.org/10.1080/19443994.2015.1109560

Khan, S. H. (2020). Green Nanotechnology for the Environment and Sustainable Development. In M. Naushad & E. Lichtfouse (Eds.), *Green Materials for Wastewater Treatment* (pp. 13–46). Springer Nature. doi:10.1007/978-3-030-17724-9_2

Khan, S. T., & Malik, A. (2019). Engineered nanomaterials for water decontamination and purification: From lab to products. *Journal of Hazardous Materials*, *363*, 295–308. doi:10.1016/j.jhazmat.2018.09.091 PMID:30312926

Khan, W. A., Rahman, M. K., Mahmoud, M. A., & Sarmah, P. (2016). MWCNT for enhancing mechanical properties of oil well cement for HPHT applications. *Proceedings of the SPE/IADC Middle East Drilling Technology Conference and Exhibition*. 10.2118/178175-MS

Khare, P., Sharma, A., Verma, N. (2014). Synthesis of phenolic precursor-based porous carbon beads in situ dispersed with copper-silver bimetal nanoparticles for antibacterial applications. *Journal of Colloid Interface Science*, *418*, 216-224.

Khitrov, G. A., & Strouse, G. F. (2003). ZnS nanomaterial characterization by MALDI-TOF mass spectrometry. *Journal of the American Chemical Society*, *125*(34), 10465–10469. doi:10.1021/ja035508r PMID:12926972

Khodabandehloo, H., Zahednasab, H., & Hafez, A. A. (2016). Nanocarriers usage for drug delivery in cancer therapy. *Iranian Journal of Cancer Prevention*, *9*(2). Advance online publication. doi:10.17795/ijcp-3966 PMID:27482328

Khoshnevisan, K., Maleki, H., Honarvarfard, E., Baharifar, H., Gholami, M., Faridbod, F., Larijani, B., Faridi Majidi, R., & Khorramizadeh, M. R. (2019). Nanomaterial based electrochemical sensing of the biomarker serotonin: A comprehensive review. *Mikrochimica Acta*, *186*(1), 49. doi:10.100700604-018-3069-y PMID:30610391

Kiang Chua, C., & Pumera, M. (2014). Chemical reduction of graphene oxide: A synthetic chemistry viewpoint. *Chemical Society Reviews*, *43*(1), 291–312. doi:10.1039/C3CS60303B PMID:24121318

Kiba, T., Masui, K., Inomata, Y., Furumoto, A., Kawamura, M., Abe, Y., & Kim, K. H. (2021). Control of localized surface plasmon resonance of Ag nanoparticles by changing its size and morphology. *Vacuum*, *192*, 110432. doi:10.1016/j.vacuum.2021.110432

Kiessling, F., Mertens, M. E., Grimm, J., & Lammers, T. (2014). Nanoparticles for imaging: Top or flop? *Radiology*, *273*(1), 10–28. doi:10.1148/radiol.14131520 PMID:25247562

Kievit, F. M., & Zhang, M. (2011). Surface engineering of iron oxide nanoparticles for targeted cancer therapy. *Accounts of Chemical Research*, *44*(10), 853–862. doi:10.1021/ar2000277 PMID:21528865

Kim, B. H., Hackett, M. J., Park, J., & Hyeon, T. (2014). Synthesis, characterization, and application of ultrasmall nanoparticles. Chemistry of Materials, 26(1). doi:10.1021/cm402225z

Kim, J. Y., Lee, C., Love, D. C., Sedlak, D. L., Yoon, J., & Nelson, K. L. (2011). *Inactivation of MS2 coliphage by ferrous ion and zero-valent iron nanoparticles*. Academic Press.

Kim, D.-H., & Kim, H.-S. (2013). Waterproof characteristics of nanoclay/epoxy nanocomposite in adhesively bonded joints. *Composites. Part B, Engineering*, *55*, 86–95. doi:10.1016/j.compositesb.2013.05.041

Kim, J., Eum, J.-H., Kang, J., Kwon, O., Kim, H., & Kim, D. W. (2021). Tuning the hierarchical pore structure of graphene oxide through dual thermal activation for high-performance supercapacitor. *Scientific Reports*, *11*(1), 2063. doi:10.103841598-021-81759-7 PMID:33483594

Kim, J., Oh, S., Cho, S. M., Jun, J., & Kwak, S. (2020). Oxygen barrier properties of polyketone/EVOH blend films and their resistance to moisture. *Journal of Applied Polymer Science*, *137*(47), 1–10. doi:10.1002/app.49537

Kim, J., Wang, J., Kim, H., & Bae, S. (2021). Concentration-dependent oscillation of specific loss power in magnetic nanofluid hyperthermia. *Scientific Reports*, *11*(1), 733. Advance online publication. doi:10.103841598-020-79871-1 PMID:33436765

Kim, N. H., Lee, S. J., & Moskovits, M. (2011). Reversible tuning of SERS hot spots with aptamers. *Advanced Materials*, *23*(36), 4152–4156. doi:10.1002/adma.201101847 PMID:21823179

Kim, N. J., Lee, S. J., & Atala, A. (2013). Biomedical nanomaterials in tissue engineering. In *Nanomaterials in Tissue Engineering* (pp. 1–25e). Elsevier. doi:10.1533/9780857097231.1

Kirkland, N. T., Schiller, T., Medhekar, N., & Birbilis, N. (2012). Exploring graphene as a corrosion protection barrier. Corrosion Science. doi:10.1016/j.corsci.2011.12.003

Kirpotin, D. B., Drummond, D. C., Shao, Y., Shalaby, M. R., Hong, K., Nielsen, U. B., Marks, J. D., Benz, C. C., & Park, J. W. (2006). Antibody targeting of long-circulating lipidic nanoparticles does not increase tumor localization but does increase internalization in animal models. *Cancer Research*, *66*(13), 6732–6740. doi:10.1158/0008-5472.CAN-05-4199 PMID:16818648

Kirthi, A., Rahuman, A. A., Rajakumar, G., Marimuthu, S., Santhoshkumar, T., Jayaseelan, C., Elango, G., Zahir, A. A., Kamaraj, C., & Bagavan, A. (2011). Biosynthesis of titanium dioxide nanoparticles using bacterium Bacillus subtilis. *Materials Letters*, *65*(17-18), 2745–2747. doi:10.1016/j.matlet.2011.05.077

Kitture, R., Chordiya, K., Gaware, S., Ghosh, S., More, P. A., Kulkarni, P., Chopade, B. A., & Kale, S. N. (2015). ZnO nanoparticles-red sandalwood conjugate: A promising anti-diabetic agent. *Journal of Nanoscience and Nanotechnology*, *15*(6), 4046–4051. doi:10.1166/jnn.2015.10323 PMID:26369011

Kleinauskas, A., Rocha, S., Sahu, S., Sun, Y.-P., & Juzenas, P. (2013). Carbon-core silver-shell nanodots as sensitizers for phototherapy and radiotherapy. *Nanotechnology*, *24*(32), 325103. doi:10.1088/0957-4484/24/32/325103 PMID:23868054

Kleine-Tebbe, J., Klimek, L., Hamelmann, E., Pfaar, O., Taube, C., Wagenmann, M., Werfel, T., & Worm, M. (2021). Severe allergic reactions to the COVID-19 vaccine – statement and practical consequences. *Allergologie Select*, *5*(01), 26–28. doi:10.5414/ALX02215E PMID:33426427

Klopffer, M. H., & Flaconneche, B. (2001). Transport Properdines of Gases in Polymers: Bibliographic Review. *Oil & Gas Science and Technology*, *56*(3), 223–244. doi:10.2516/ogst:2001021

Ko, S., & Huh, C. (2019). Use of nanoparticles for oil production applications. Journal of Petroleum Science and Engineering, 172. doi:10.1016/j.petrol.2018.09.051

Koedrith, P., Thasiphu, T., Weon, J., Boonprasert, R., Tuitemwong, K., & Titemwo, P. (2015). Recent trends in rapid environmental monitoring of pathogens and toxicants: Potential of nanoparticle-based biosensor and applications. *The-ScientificWorldJournal, 2015*, 12. doi:10.1155/2015/510982 PMID:25884032

Koenig, S. P., Wang, L., Pellegrino, J., & Bunch, J. S. (2012). Selective molecular sieving through porous graphene. *Nature Nanotechnology*, *7*(11), 728–732. Advance online publication. doi:10.1038/nnano.2012.162 PMID:23042491

Ko, G. K. S. (2014). Overview on in Polymer-Nano Clay Composite Paper Coating for Packaging Application. *Journal of Marine Science and Engineering*, *4*(1). Advance online publication. doi:10.4172/2169-0022.1000151

Kokal, S. (2018). Technology Focus: CO2 Applications (July 2018). *Journal of Petroleum Technology*, *70*(07), 69. Advance online publication. doi:10.2118/0718-0069-JPT

Koliyabandara, S. M. P., Siriwardhana, C., De Silva, S. M., Walpita, J., & Cooray, A. T. (2021). Electrokinetic Remediation for the Removal of Organic Waste in Soil and Sediments. In *Electrokinetic Remediation for Environmental Security and Sustainability* (pp. 453–477). Wiley. doi:10.1002/9781119670186.ch19

Kolya, H., Kuila, T., Kim, N. H., & Lee, J. H. (2019). Bioinspired silver nanoparticles/reduced graphene oxide nanocomposites for catalytic reduction of 4-nitrophenol, organic dyes and act as energy storage electrode material. *Composites. Part B, Engineering*, *173*(May), 106924. doi:10.1016/j.compositesb.2019.106924

Komarneni, S., Roy, R., & Li, Q. H. (1992). Microwave-hydrothermal synthesis of ceramic powders. *Materials Research Bulletin*, *27*(12), 1393–1405. doi:10.1016/0025-5408(92)90004-J

Komatsu, K., Murata, M., & Murata, Y. (2005). Encapsulation of molecular hydrogen in fullerene C60 by organic synthesis. *Science*, *307*(5707), 238–240. doi:10.1126cience.1106185 PMID:15653499

Kommu, A., & Singh, J. K. (2020). A review on graphene-based materials for removal of toxic pollutants from wastewater. *Soft Materials*, *18*(2-3), 297–322. doi:10.1080/1539445X.2020.1739710

Koneru, A., Dharmalingam, K., & Anandalakshmi, R. (2020). Cellulose based nanocomposite hydrogel films consisting of sodium carboxymethylcellulose–grapefruit seed extract nanoparticles for potential wound healing applications. *International Journal of Biological Macromolecules*, *148*, 833–842. doi:10.1016/j.ijbiomac.2020.01.018 PMID:31953172

Kong, X., & Ohadi, M. M. (2010). Applications of micro and nano technologies in the oil and gas industry-an overview of the recent progress. *Society of Petroleum Engineers - 14th Abu Dhabi International Petroleum Exhibition and Conference 2010, ADIPEC 2010, 3*. 10.2118/138241-MS

Konvičková, Z., Barabaszová, K. Č., Holišová, V., Kratošová, G., & Seidlerová, J. (2019). Phytosynthesis of Ag, ZnO and ZrO2 Nanoparticles Using Linden: Changes in Their Physical-Chemical Nature Over Time. *Journal of Nanoscience and Nanotechnology*, *19*(12), 7926–7933. doi:10.1166/jnn.2019.15854 PMID:31196311

Kooijmans, S. A. A., de Jong, O. G., & Schiffelers, R. M. (2021). Exploring interactions between extracellular vesicles and cells for innovative drug delivery system design. *Advanced Drug Delivery Reviews*, *173*, 252–278. doi:10.1016/j.addr.2021.03.017 PMID:33798644

Koosha, M., & Hamedi, S. (2019). Intelligent Chitosan/PVA nanocomposite films containing black carrot anthocyanin and bentonite nanoclays with improved mechanical, thermal and antibacterial properties. *Progress in Organic Coatings*, *127*(May), 338–347. doi:10.1016/j.porgcoat.2018.11.028

Koppolu, R., Lahti, J., Abitbol, T., Swerin, A., Kuusipalo, J., & Toivakka, M. (2019). Continuous Processing of Nanocellulose and Polylactic Acid into Multilayer Barrier Coatings. *ACS Applied Materials & Interfaces*, *11*(12), 11920–11927. doi:10.1021/acsami.9b00922 PMID:30829474

Kosari, M., Golmohammadi, M., Ahmadi, S. J., Towfighi, J., & Chenari, A. H. (2017). On the catalysis capability of transition metal oxide nanoparticles in upgrading of heavy petroleum residue by supercritical water. *The Journal of Supercritical Fluids*, *126*, 14–24. Advance online publication. doi:10.1016/j.supflu.2017.02.021

Kostarelos, K. (2020). Nanoscale Nights of COVID-19. *Nature Nanotechnology*, *15*(5), 343–344. doi:10.103841565-020-0687-4 PMID:32341506

Kotek, R., Pang, K., Schmidt, B., & Tonelli, A. (2004). Synthesis and gas barrier characterization of poly(ethylene isophthalate). *Journal of Polymer Science. Part B, Polymer Physics*, *42*(23), 4247–4254. doi:10.1002/polb.20284

Kowshik, M., Ashtaputre, S., Kharrazi, S., Vogel, W., Urban, J., Kulkarni, S. K., & Paknikar, K. M. (2003). Extracellular synthesis of silver nanoparticles by a silver-tolerant yeast strain MKY3. *Nanotechnology*, *14*(1), 95–100. doi:10.1088/0957-4484/14/1/321

Krishna, G. (2021). Tio2 Nanocomposites And Its Applications; A Brief Review. [TURCOMAT]. *Turkish Journal of Computer and Mathematics Education*, *12*(10), 3707–3717.

Krishna, G., Srileka, V., Singara Charya, M. A., Abu Serea, E. S., & Shalan, A. E. (2021). Biogenic synthesis and cytotoxic effects of silver nanoparticles mediated by white rot fungi. *Heliyon*, *7*(3), e06470. doi:10.1016/j.heliyon.2021.e06470 PMID:33786393

Krishnaswamy, K., Vali, H., & Orsat, V. (2014). Value-adding to grape waste: Green synthesis of gold nanoparticles. *Journal of Food Engineering*, *142*, 210–220. doi:10.1016/j.jfoodeng.2014.06.014

Kronka, M. S., Cordeiro-Junior, P. J. M., Mira, L., dos Santos, A. J., Fortunato, G. V., & Lanza, M. R. V. (2021). Sustainable microwave-assisted hydrothermal synthesis of carbon-supported ZrO2 nanoparticles for H2O2 electrogeneration. *Materials Chemistry and Physics*, *267*, 124575. doi:10.1016/j.matchemphys.2021.124575

Kubota, S., Johkura, K., Asanuma, K., Okouchi, Y., Ogiwara, N., Sasaki, K., & Kasuga, T. (2004). Titanium oxide nanotubes for bone regeneration. *Journal of Materials Science. Materials in Medicine*, *15*(9), 1031–1035. doi:10.1023/B:JMSM.0000042689.78768.77 PMID:15448411

Ku, G., Zhou, M., Song, S., Huang, Q., Hazle, J., & Li, C. (2012). Copper sulfide nanoparticles as a new class of photoacoustic contrast agent for deep tissue imaging at 1064 nm. *ACS Nano*, *6*(8), 7489–7496. doi:10.1021/nn302782y PMID:22812694

Kulal, D., & Shetty Kodialbail, V. (2021). Visible light mediated photocatalytic dye degradation using Ag_2O/AgO-TiO_2 nanocomposite synthesized by extracellular bacterial mediated synthesis - An eco-friendly approach for pollution abatement. *Journal of Environmental Chemical Engineering*, *9*(4), 105389. doi:10.1016/j.jece.2021.105389

Kulandaivelu, B., & Gothandam, K. M. (2016). Cytotoxic effect on cancerous cell lines by biologically synthesized silver nanoparticles. *Brazilian Archives of Biology and Technology*, *59*(0). Advance online publication. doi:10.1590/1678-4324-2016150529

Kulkarni, N., & Muddapur, U. (2014). Biosynthesis of metal nanoparticles: A review. *Journal of Nanotechnology*, *2014*, 1–8. doi:10.1155/2014/510246

Kumar, A., Kuang, Y., Liang, Z., & Sun, X. (2020). Microwave chemistry, recent advancements, and eco-friendly microwave-assisted synthesis of nanoarchitectures and their applications: A review. *Materials Today Nano, 11*.

Kumar, H., & Sangwan, P. (2018). Synthesis and Characterization of Cobalt Oxide Nanoparticles by Sol-Gel Method. *International Journal of Advanced Research in Science and Engineering, 7*.

Kumar, M., Bansal, M., & Garg, R. (2021). An overview of beneficiary aspects of zinc oxide nanoparticles on performance of cement composites. *Materials Today: Proceedings, 43*, 892–898. doi:10.1016/j.matpr.2020.07.215

Kumar, A., & Kumar, A., M.M.S., C.-P., Chaturvedi, A. K., Shabnam, A. A., Subrahmanyam, G., Mondal, R., Gupta, D. K., Malyan, S. K., Kumar, S. S., A. Khan, S., & Yadav, K. K. (2020). Lead Toxicity: Health Hazards, Influence on Food Chain, and Sustainable Remediation Approaches. *International Journal of Environmental Research and Public Health, 17*(7), 2179. doi:10.3390/ijerph17072179 PMID:32218253

Kumar, A., Sharma, K., & Dixit, A. R. (2020). Role of graphene in biosensor and protective textile against viruses. *Medical Hypotheses, 144*, 110253. doi:10.1016/j.mehy.2020.110253 PMID:33254558

Kumar, B., Smita, K., Cumbal, L., & Angulo, Y. (2015). Fabrication of silver nanoplates using Nephelium lappaceum (Rambutan) peel: A sustainable approach. *Journal of Molecular Liquids, 211*, 476–480. doi:10.1016/j.molliq.2015.07.067

Kumar, K. Y., Muralidhara, H. B., Nayaka, Y. A., Balasubramanyam, J., & Hanumanthappa, H. (2013). Low-cost synthesis of metal oxide nanoparticles and their application in adsorption of commercial dye and heavy metal ion in aqueous solution. *Powder Technology, 246*, 125–136. https://doi.org/10.1016/j.powtec.2013.05.017

Kumar, M. S., Saroja, M., & Venkatachalam, M. (2019). Acalypha indica and Curcuma longa plant extracts mediated ZnS nanoparticles. *Material Science Research India, 16*(2), 174–182. doi:10.13005/msri/160210

Kumar, M., Sharma, G., Kumar, R., Singh, B., Katare, O. P., & Raza, K. (2018). Lysine-based C60-fullerene nanoconjugates for monomethyl fumarate delivery: A novel nanomedicine for brain cancer cells. *ACS Biomaterials Science & Engineering, 4*(6), 2134–2142. doi:10.1021/acsbiomaterials.7b01031 PMID:33435037

Kumar, N., & Goel, N. (2019). Phenolic acids: Natural versatile molecules with promising therapeutic applications. *Biotechnology Reports (Amsterdam, Netherlands), 24*, e00370. doi:10.1016/j.btre.2019.e00370 PMID:31516850

Kumar, R., Roopan, S. M., Prabhakarn, A., Khanna, V. G., & Chakroborty, S. (2012). Agricultural waste Annona squamosa peel extract: Biosynthesis of silver nanoparticles. *Spectrochimica Acta. Part A: Molecular and Biomolecular Spectroscopy, 90*, 173–176. doi:10.1016/j.saa.2012.01.029 PMID:22336049

Kumar, S. S. A., Bashir, S., Ramesh, K., & Ramesh, S. (2021). New perspectives on Graphene/Graphene oxide based polymer nanocomposites for corrosion applications: The relevance of the Graphene/Polymer barrier coatings. *Progress in Organic Coatings, 154*, 106215. doi:10.1016/j.porgcoat.2021.106215

Kumar, S., Rehman, M. A., Lee, S., Kim, M., Hong, H., Park, J. Y., & Seo, Y. (2021). Supercapacitors based on Ti3C2Tx MXene extracted from supernatant and current collectors passivated by CVD-graphene. *Scientific Reports, 11*(1), 649. Advance online publication. doi:10.103841598-020-80799-9 PMID:33436987

Kumar, S., Thakur, A., Rangra, V. S., & Sharma, S. (2015). Synthesis and Use of Low-Band-Gap ZnO Nanoparticles for Water Treatment. *Arabian Journal for Science and Engineering, 41*(7), 2393–2398. doi:10.100713369-015-1852-1

Kummara, S., Patil, M. B., & Uriah, T. (2016). Synthesis, characterization, biocompatible and anticancer activity of green and chemically synthesized silver nanoparticles – A comparative study. *Biomedicine and Pharmacotherapy, 84*, 10–21. doi:10.1016/j.biopha.2016.09.003 PMID:27621034

Kuppan, S., Xu, Y., Liu, Y., & Chen, G. (2017). Phase transformation mechanism in lithium manganese nickel oxide revealed by single-crystal hard X-ray microscopy. *Nature Communications, 8*(1), 1–10. doi:10.1038/ncomms14309 PMID:28145406

Küpper, H. (2017). Lead Toxicity in Plants. In Lead: Its Effects on Environment and Health (pp. 491–500). De Gruyter. doi:10.1515/9783110434330-015

Kuppusamy, P., Yusoff, M. M., Maniam, G. P., & Govindan, N. (2016). Biosynthesis of metallic nanoparticles using plant derivatives and their new avenues in pharmacological applications – An updated report. *Saudi Pharmaceutical Journal*, *24*(4), 473–484. doi:10.1016/j.jsps.2014.11.013 PMID:27330378

Lagaron, J. M., Catalá, R., & Gavara, R. (2004). *Structural characteristics defining high barrier properties in polymeric materials Structural characteristics de ning high barrier properties in polymeric materials*. Academic Press.

Lai, J., Wang, T., Wang, H., Shi, F., Gu, W., & Ye, L. (2018). MnO nanoparticles with unique excitation-dependent fluorescence for multicolor cellular imaging and MR imaging of brain glioma. *Mikrochimica Acta*, *185*(4), 244. doi:10.100700604-018-2779-5 PMID:29610993

Lakshmipathy, R., Palakshi Reddy, B., Sarada, N. C., Chidambaram, K., & Khadeer Pasha, S. (2015). Watermelon rind-mediated green synthesis of noble palladium nanoparticles: Catalytic application. *Applied Nanoscience*, *5*(2), 223–228. doi:10.100713204-014-0309-2

Lakshmnarayanan, S., Shereen, M. F., Niraimathi, K. L., Brindha, P., & Arumugam, A. (2021). One-pot green synthesis of iron oxide nanoparticles from Bauhinia tomentosa: Characterization and application towards synthesis of 1, 3 diolein. *Scientific Reports*, *11*(1), 8643. doi:10.103841598-021-87960-y PMID:33883589

Lalani, J., Patil, S., Kolate, A., Lalani, R., & Misra, A. (2015). Protein-functionalized PLGA nanoparticles of lamotrigine for neuropathic pain management. *AAPS PharmSciTech*, *16*(2), 413–427. doi:10.120812249-014-0235-3 PMID:25354788

Lalitha, A., Subbaiya, R., & Ponmurugan, P. (2013). Green synthesis of silver nanoparticles from leaf extract Azhadirachta indica and to study its anti-bacterial and antioxidant property. *International Journal of Current Microbiology and Applied Sciences*, *2*(6), 228–235.

Lam, C. Y. K., Zhang, Q., Yin, B., Huang, Y., Wang, H., Yang, M., & Wong, S. H. D. (2021). Recent Advances in Two-Dimensional Transition Metal Dichalcogenide Nanocomposites Biosensors for Virus Detection before and during COVID-19 Outbreak. *Journal of Composites Science*, *5*(7), 190. doi:10.3390/jcs5070190

Lammers, T., Subr, V., Ulbrich, K., Peschke, P., Huber, P. E., Hennink, W. E., & Storm, G. (2009). Simultaneous delivery of doxorubicin and gemcitabine to tumors in vivo using prototypic polymeric drug carriers. *Biomaterials*, *30*(20), 3466–3475. doi:10.1016/j.biomaterials.2009.02.040 PMID:19304320

Lateef, A., Azeez, M. A., Asafa, T. B., Yekeen, T. A., Akinboro, A., Oladipo, I. C., Ajetomobi, F. E., Gueguim-Kana, E. B., & Beukes, L. S. (2015). Cola nitida-mediated biogenic synthesis of silver nanoparticles using seed and seed shell extracts and evaluation of antibacterial activities. *BioNanoScience*, *5*(4), 196–205. doi:10.100712668-015-0181-x

Lateef, A., Azeez, M. A., Asafa, T. B., Yekeen, T. A., Akinboro, A., Oladipo, I. C., Azeez, L., Ojo, S. A., Gueguim-Kana, E. B., & Beukes, L. S. (2016). Cocoa pod husk extract-mediated biosynthesis of silver nanoparticles: Its antimicrobial, antioxidant and larvicidal activities. *Journal of Nanostructure in Chemistry*, *6*(2), 159–169. doi:10.100740097-016-0191-4

Lateef, A., Ojo, S. A., Azeez, M. A., Asafa, T. B., Yekeen, T. A., Akinboro, A., Oladipo, I. C., Gueguim-Kana, E. B., & Beukes, L. S. (2015). Cobweb as novel biomaterial for the green and eco-friendly synthesis of silver nanoparticles. *Applied Nanoscience*, *6*(6), 863–874. doi:10.100713204-015-0492-9

Lateef, A., Oloke, J. K., Gueguim Kana, E. B., Oyeniyi, S. O., Onifade, O. R., Oyeleye, A. O., Oladosu, O. C., & Oyelami, A. O. (2008). Improving the quality of agro-wastes by solid-state fermentation: Enhanced antioxidant activities and nutritional qualities. *World Journal of Microbiology & Biotechnology*, *24*(10), 2369–2374. doi:10.100711274-008-9749-8

Lau, H. C., Yu, M., & Nguyen, Q. P. (2017). Nanotechnology for oilfield applications: Challenges and impact. *Journal of Petroleum Science Engineering*, *157*, 1160–1169. Advance online publication. doi:10.1016/j.petrol.2017.07.062

Laurent, C., Flahaut, E., & Peigney, A. (2010). The weight and density of carbon nanotubes versus the number of walls and diameter. Carbon, 48(10). doi:10.1016/j.carbon.2010.04.010

Lee, A.-R., Lee, S.-J., Lee, M., Nam, M., Lee, S., Choi, J., Lee, H.-J., Kim, D.-U., & Hoe, K.-L. (2018). Editor's highlight: A Genome-wide screening of target genes against silver nanoparticles in fission yeast. *Toxicological Sciences, 161*(1), 171–185. doi:10.1093/toxsci/kfx208 PMID:29294138

Lee, D. R., Park, J. S., Bae, I. H., Lee, Y., & Kim, B. M. (2016). Liquid crystal nanoparticle formulation as an oral drug delivery system for liver-specific distribution. *International Journal of Nanomedicine, 11*, 853. PMID:27042053

Lee, J., Kim, H. Y., Zhou, H., Hwang, S., Koh, K., Han, D.-W., & Lee, J. (2011). Green synthesis of phytochemical-stabilized Au nanoparticles under ambient conditions and their biocompatibility and antioxidative activity. *Journal of Materials Chemistry, 21*(35), 13316. doi:10.1039/c1jm11592h

Lee, K.-T., Park, J.-H., Kwon, S. J., Kwon, H.-K., Kyhm, J., Kwak, K.-W., Jang, H. S., Kim, S. Y., Han, J. S., Lee, S.-H., Shin, D.-H., Ko, H., Han, I.-K., Ju, B.-K., Kwon, S.-H., & Ko, D.-H. (2015). Simultaneous enhancement of upconversion and downshifting luminescence via plasmonic structure. *Nano Letters, 15*(4), 2491–2497. doi:10.1021/nl5049803 PMID:25756859

Lee, S. H., & Jun, B. H. (2019). Silver Nanoparticles: Synthesis and Application for Nanomedicine. *International Journal of Molecular Sciences, 20*(4), 865. doi:10.3390/ijms20040865 PMID:30781560

Lee, S.-J., Kumar, S., Choi, J. W., & Lee, J.-S. (2020). Coloration of colloidal polymer particles through selective extraction of Mie backscattering for cation-responsible colorimetric sensors. *Journal of Colloid and Interface Science, 560*, 894–901. doi:10.1016/j.jcis.2019.10.073 PMID:31718790

Lemire, J. A., Harrison, J. J., & Turner, R. (2013). *Antimicrobial activity of metals: mechanisms, molecular targets and applications*. Academic Press.

Lemos, S. C. S., Nossol, E., Ferrari, J. L., Gomes, E. O., Andres, J., Gracia, L., Sorribes, I., & Lima, R. C. (2019). Joint Theoretical and Experimental Study on the La Doping Process in In2O3: Phase Transition and Electrocatalytic Activity. *Inorganic Chemistry, 58*(17), 11738–11750. doi:10.1021/acs.inorgchem.9b01728 PMID:31415162

Leon, L., Chung, E. J., & Rinaldi, C. (2020). A brief history of nanotechnology and introduction to nanoparticles for biomedical applications. In Nanoparticles for Biomedical Applications (pp. 1-4). doi:10.1016/B978-0-12-816662-8.00001-1

Lerma-García, M.Á., Simó-Alfonso, E.F., Ríos, Á., & Zougagh, M. (2014). Synthesis of gold nanoparticles using phenolic acids and its application in catalysis. *Journal of Material and Environmental Science*, (6), 1919–1926.

Le, V. T., Bach, L. G., Pham, T. T., Le, N. T. T., Ngoc, U. T. P., Tran, D. H. N., & Nguyen, D. H. (2019). Synthesis and antifungal activity of chitosan-silver nanocomposite synergize fungicide against Phytophthora capsici. Journal of Macromolecular Science. *Part A, 56*(6), 522–528.

Lew, T. T. S., Aung, K. M. M., Ow, S. Y., Amrun, S. N., Sutarlie, L., Ng, L. F. P., & Su, X. (2021). Epitope-Functionalized Gold Nanoparticles for Rapid and Selective Detection of SARS-CoV-2 IgG Antibodies. *ACS Nano, 15*(7), 12286–12297. Advance online publication. doi:10.1021/acsnano.1c04091 PMID:34133128

Li, J., Wang, G., Zhu, H., Zhang, M., Zheng, X., Di, Z., Liu, X., & Wang, X. (2014). *Antibacterial activity of large-area monolayer graphene film manipulated by charge transfer*. Academic Press.

Li, L., Al-Muntasheri, G. A., & Liang, F. (2016). Nanomaterials-enhanced high-temperature fracturing fluids prepared with untreated seawater. *Proceedings - SPE Annual Technical Conference and Exhibition*. 10.2118/181283-MS

Li, C., Hassan, A., Palmai, M., Xie, Y., Snee, P. T., Powell, B. A., Murdoch, L. C., & Darnault, C. J. G. (2021). Experimental measurements and numerical simulations of the transport and retention of nanocrystal CdSe/ZnS quantum dots in saturated porous media: Effects of pH, organic ligand, and natural organic matter. *Environmental Science and Pollution Research International*, *28*(7), 8050–8073. Advance online publication. doi:10.100711356-020-11097-0 PMID:33051847

Li, F., Josephson, D. P., & Stein, A. (2011). Colloidal assembly: The road from particles to colloidal molecules and crystals. *Angewandte Chemie International Edition*, *50*(2), 360–388. doi:10.1002/anie.201001451 PMID:21038335

Li, H., He, X., Liu, Y., Huang, H., Lian, S., Lee, S. T., & Kang, Z. (2011). One-step ultrasonic synthesis of water-soluble carbon nanoparticles with excellent photoluminescent properties. *Carbon*, *49*(2), 605–609. doi:10.1016/j.carbon.2010.10.004

Li, H., Liu, L., & Yang, F. (2013). Covalent assembly of 3D graphene/polypyrrole foams for oil spill cleanup. *Journal of Materials Chemistry. A, Materials for Energy and Sustainability*, *1*(10), 3446–3453. doi:10.1039/c3ta00166k

Li, H., Pan, S., Xia, P., Chang, Y., Fu, C., Kong, W., Yu, Z., Wang, K., Yang, X., & Qi, Z. (2020). Advances in the application of gold nanoparticles in bone tissue engineering. *Journal of Biological Engineering*, *14*(1), 1–15. doi:10.118613036-020-00236-3 PMID:32391080

Li, J., Du, S., Tao, H., & Yang, X. (2021). Self-supporting and 3D MoS2/MoO2/CNT/graphene foam as high-performance anode for lithium ion batteries. *Ionics*, *27*(1), 75–84. doi:10.100711581-020-03815-y

Li, J., Wang, S., Wu, C., Dai, Y., Hou, P., Han, C., & Xu, K. (2016). Activatable molecular MRI nanoprobe for tumor cell imaging based on gadolinium oxide and iron oxide nanoparticle. *Biosensors & Bioelectronics*, *86*, 1047–1053. doi:10.1016/j.bios.2016.07.044 PMID:27501342

Li, L., Wei, W., & Behrens, M. (2012). Synthesis and characterization of α-, β-, and γ-Ga 2o 3 prepared from aqueous solutions by controlled precipitation. *Solid State Sciences*, *14*(7), 971–981. doi:10.1016/j.solidstatesciences.2012.04.037

Li, L., Wu, G., Hong, T., Yin, Z., Sun, D., Abdel-Halim, E. S., & Zhu, J. J. (2014). Graphene quantum dots as fluorescence probes for turn-off sensing of melamine in the presence of Hg2+. *ACS Applied Materials & Interfaces*, *6*(4), 2858–2864. doi:10.1021/am405305r PMID:24460139

Li, M., Dai, Y., Pei, X., & Chen, W. (2019). Three-dimensional hollow reduced graphene oxide spheres with a hierarchically porous structure for high-performance lithium-sulfur batteries. *Inorganic Chemistry Frontiers*, *6*(9), 2528–2538. doi:10.1039/C9QI00656G

Lima, E., Guerra, R., Lara, V., & Guzmán, A. (2013). Gold nanoparticles as efficient antimicrobial agents for Escherichia coli and Salmonella typhi. *Chemistry Central Journal*, *7*(1), 1–7. doi:10.1186/1752-153X-7-11 PMID:23331621

Lin, K. Y. A., & Der Lee, W. (2016). Highly efficient removal of Malachite green from water by a magnetic reduced graphene oxide/zeolitic imidazolate framework self-assembled nanocomposite. *Applied Surface Science*, *361*, 114–121. doi:10.1016/j.apsusc.2015.11.108

Li, S., Jiang, W., Huang, J., Liu, Y., Ren, L., Zhuang, L., Zheng, Q., Wang, M., Yang, R., Zeng, Y., & Wang, Y. (2020). Highly sensitive and specific diagnosis of COVID-19 by reverse transcription multiple cross-displacement amplification-labelled nanoparticles biosensor. *The European Respiratory Journal*, *56*(6), 2002060. Advance online publication. doi:10.1183/13993003.02060-2020 PMID:32859676

Li, S., Shen, Y., Xie, A., Yu, X., Qiu, L., Zhang, L., & Zhang, Q. (2007). Green synthesis of silver nanoparticles using Capsicum annuum L. extract. *Green Chemistry*, *9*(8), 852. doi:10.1039/b615357g

Li, S., Turaga, U., Shrestha, B., Anderson, T. A., Ramkumar, S. S., Green, M. J., Das, S., & Cañas-Carrell, J. E. (2013). Mobility of polyaromatic hydrocarbons (PAHs) in soil in the presence of carbon nanotubes. *Ecotoxicology and Environmental Safety, 96*, 168–174. doi:10.1016/j.ecoenv.2013.07.005 PMID:23896179

Li, S., Zhang, L., Wang, T., Li, L., Wang, C., & Su, Z. (2015). The facile synthesis of hollow Au nanoflowers for synergistic chemo-photothermal cancer therapy. *Chemical Communications, 51*(76), 14338–14341. doi:10.1039/C5CC05676D PMID:26299901

Liu, B., Pavlou, C., Wang, Z., Cang, Y., Galiotis, C., & Fytas, G. (2021). Determination of the elastic moduli of CVD graphene by probing graphene/polymer Bragg stacks. *2D Materials, 8*(3), 035040. doi:10.1088/2053-1583/abfedb

Liu, C., Zhang, N., Huang, X., Wang, Q., Wang, X., & Wang, S. (2021). Fabrication of a novel nanocomposite electrode with ZnO-MoO_3 and biochar derived from mushroom biomaterials for the detection of acetaminophen in the presence of DA. *Microchemical Journal, 161*(November), 105719. doi:10.1016/j.microc.2020.105719

Liu, H., Ge, J., Ma, E., & Yang, L. (2019). Advanced Biomaterials for Biosensor and Theranostics. In Biomaterials in Translational Medicine (pp. 213-255). Elsevier. doi:10.1016/B978-0-12-813477-1.00010-4

Liu, S., Zeng, T. H., Hofmann, M., Burcombe, E., Wei, J., Jiang, R., Kong, J., & Chen, Y. (2011). *Antibacterial activity of graphite, graphite oxide, graphene oxide, and reduced graphene oxide: membrane and oxidative stress*. Academic Press.

Liu, Y., Gayle, A.A., Wilder-Smith, A., & Rocklöv, J. (2020). The Reproductive Number of COVID-19 is Higher Compared to SARS Coronavirus. *Journal of Travel Medicine, 27*(2). doi:10.1093/jtm/taaa021

Liu, B., Han, L., Liu, J., Han, S., Chen, Z., & Jiang, L. (2017). Co-delivery of paclitaxel and TOS-cisplatin via TAT-targeted solid lipid nanoparticles with synergistic antitumor activity against cervical cancer. *International Journal of Nanomedicine, 12*, 955–968. doi:10.2147/IJN.S115136 PMID:28203075

Liu, D., Yang, F., Xiong, F., & Gu, N. (2016). The smart drug delivery system and its clinical potential. *Theranostics, 6*(9), 1306–1323. doi:10.7150/thno.14858 PMID:27375781

Liu, H., Jin, X., & Ding, B. (2016). Application of nanotechnology in petroleum exploration and development. *Petroleum Exploration and Development, 43*(6), 1107–1115. Advance online publication. doi:10.1016/S1876-3804(16)30129-X

Liu, H., & Qiu, H. (2020). Recent advances of 3D graphene-based adsorbents for sample preparation of water pollutants: A review. *Chemical Engineering Journal, 393*, 124691. doi:10.1016/j.cej.2020.124691

Liu, J., Cheng, H., Han, L., Qiang, Z., Zhang, X., Gao, W., Zhao, K., & Song, Y. (2018). Synergistic combination therapy of lung cancer using paclitaxel- and triptolide-coloaded lipid-polymer hybrid nanoparticles. *Drug Design, Development and Therapy, 12*, 3199–3209. doi:10.2147/DDDT.S172199 PMID:30288024

Liu, J., Chen, M., Zhang, L., Jiang, J., Yan, J., Huang, Y., Lin, J., Fan, H. J., & Shen, Z. X. (2014). A flexible alkaline rechargeable Ni/Fe battery based on graphene foam/carbon nanotubes hybrid film. *Nano Letters, 14*(12), 7180–7187. doi:10.1021/nl503852m PMID:25402965

Liu, J., Hu, W., Chen, H., Ni, Q., Xu, H., & Yang, X. (2007). Isotretinoin-loaded solid lipid nanoparticles with skin targeting for topical delivery. *International Journal of Pharmaceutics, 328*(2), 191–195. doi:10.1016/j.ijpharm.2006.08.007 PMID:16978810

Liu, J., Zhai, H., Sun, Y., Wu, S., & Chen, S. (2021). Developing high strength poly(L-lactic acid) nanofiber yarns for biomedical textile materials: A comparative study of novel nanofiber yarns and traditional microfiber yarns. *Materials Letters, 300*, 130229. doi:10.1016/j.matlet.2021.130229

Liu, L., Zhang, M., Zhang, Q., & Jiang, W. (2020). Graphene nanosheets damage the lysosomal and mitochondrial membranes and induce the apoptosis of RBL-2H3 cells. *The Science of the Total Environment*, *734*, 139229. doi:10.1016/j.scitotenv.2020.139229 PMID:32450398

Liu, M. Q., Wang, C., & Kim, N. Y. (2017). High-sensitivity and low-hysteresis porous mim-type capacitive humidity sensor using functional polymer mixed with tio2 microparticles. *Sensors (Switzerland)*, *17*(2), 0284. Advance online publication. doi:10.339017020284 PMID:28157167

Liu, W., Tian, S., Zhao, X., Xie, W., Gong, Y., & Zhao, D. (2015). Application of Stabilized Nanoparticles for In Situ Remediation of Metal-Contaminated Soil and Groundwater: A Critical Review. *Current Pollution Reports*, *1*(4), 280–291. doi:10.100740726-015-0017-x

Liu, X., & Lu, Y. (2019). Tailoring Emulsion Polymerization for High-Yield Synthesis of Tween 80 Stabilized Magnetic Cross-Linked Polystyrene Nanocomposite Particles. *Industrial & Engineering Chemistry Research*, *58*(19), 8140–8147. Advance online publication. doi:10.1021/acs.iecr.9b01103

Liu, X., Wen, N., Wang, X., & Zheng, Y. (2015). A High-performance hierarchical graphene@polyaniline@graphene sandwich containing hollow structures for supercapacitor electrodes. *ACS Sustainable Chemistry & Engineering*, *3*(3), 475–482. doi:10.1021c5006999

Liu, Y. C., Li, J., Ahn, J., Pu, J., Rupa, E. J., Huo, Y., & Yang, D. C. (2020). Biosynthesis of zinc oxide nanoparticles by one-pot green synthesis using fruit extract of Amomum longiligulare and its activity as a photocatalyst. *Optik (Stuttgart)*, *218*, 165245. doi:10.1016/j.ijleo.2020.165245

Liu, Y., & Chen, C. (2016). Role of nanotechnology in HIV/AIDS vaccine development. *Advanced Drug Delivery Reviews*, *103*, 76–89. doi:10.1016/j.addr.2016.02.010 PMID:26952542

Liu, Z., Chen, K., Davis, C., Sherlock, S., Cao, Q., Chen, X., & Dai, H. (2008). Drug delivery with carbon nanotubes for in vivo cancer treatment. *Cancer Research*, *68*(16), 6652–6660. doi:10.1158/0008-5472.CAN-08-1468 PMID:18701489

Liu, Z., Xiong, M., Gong, J., Zhang, Y., Bai, N., Luo, Y., Li, L., Wei, Y., Liu, Y., Tan, X., & Xiang, R. (2014). Legumain protease-activated TAT-liposome cargo for targeting tumours and their microenvironment. *Nature Communications*, *5*(1), 1–11. doi:10.1038/ncomms5280 PMID:24969588

Li, W. R., Xie, X. B., Shi, Q. S., Zeng, H. Y., Ou-Yang, Y. S., & Chen, Y. B. (2010). Antibacterial activity and mechanism of silver nanoparticles on Escherichia coli. *Applied Microbiology and Biotechnology*, *8*(4), 1115–1122. doi:10.100700253-009-2159-5 PMID:19669753

Li, X., Xu, H., Chen, Z. S., & Chen, G. (2011). Biosynthesis of nanoparticles by microorganisms and their applications. *Journal of Nanomaterials*, *2011*, 1–16. Advance online publication. doi:10.1155/2011/270974

Li, X., Zai, J., Wang, X., & Li, Y. (2020). Potential of Large "First Generation" Human-to-Human Transmission of 2019-nCOV. *Journal of Medical Virology*, *92*(4), 448–454. doi:10.1002/jmv.25693 PMID:31997390

Li, Y., Zhang, C., Liu, L., Gong, Y., Xie, Y., & Cao, Y. (2018). The effects of baicalein or baicalin on the colloidal stability of ZnO nanoparticles (NPs) and toxicity of NPs to Caco-2 cells. *Toxicology Mechanisms and Methods*, *28*(3), 167–176. doi:10.1080/15376516.2017.1376023 PMID:28868948

Li, Y., Zheng, X., & Chu, Q. (2021). Bio-based nanomaterials for cancer therapy. *Nano Today*, *38*, 101134. doi:10.1016/j.nantod.2021.101134

Li, Z., Cheng, J., Yang, X., Liu, H., Xu, X., Ma, L., Shang, S., & Song, Z. (2020). Construction of antimicrobial and biocompatible cotton textile based on quaternary ammonium salt from rosin acid. *International Journal of Biological Macromolecules, 150*, 1–8. doi:10.1016/j.ijbiomac.2020.01.259 PMID:32007552

Li, Z., Dai, X., Du, K., Ma, Y., Liu, M., Sun, H., Ma, X., & Zhang, X. (2016). Reduced graphene oxide/O-MWCNT hybrids functionalized with p-phenylenediamine as high-performance MoS2 electrocatalyst support for hydrogen evolution reaction. *The Journal of Physical Chemistry C, 120*(3), 1478–1487. doi:10.1021/acs.jpcc.5b09523

Li, Z., de Barros, A. L. B., Soares, D. C. F., Moss, S. N., & Alisaraie, L. (2017). Functionalized single-walled carbon nanotubes: Cellular uptake, biodistribution and applications in drug delivery. *International Journal of Pharmaceutics, 524*(1-2), 41–54. doi:10.1016/j.ijpharm.2017.03.017 PMID:28300630

Llanos, S., Giraldo, L. J., Santamaria, O., Franco, C. A., & Cortés, F. B. (2018). Effect of Sodium Oleate Surfactant Concentration Grafted onto SiO2 Nanoparticles in Polymer Flooding Processes. *ACS Omega, 3*(12), 18673–18684. Advance online publication. doi:10.1021/acsomega.8b02944 PMID:31458433

Locatelli, E., Naddaka, M., Uboldi, C., Loudos, G., Fragogeorgi, E., Molinari, V., Pucci, A., Tsotakos, T., Psimadas, D., Ponti, J., & Franchini, M. C. (2014). Targeted delivery of silver nanoparticles and alisertib: In vitro and in vivo synergistic effect against glioblastoma. *Nanomedicine (London), 9*(6), 839–849. doi:10.2217/nnm.14.1 PMID:24433240

Long, W., Ouyang, H., Zhou, C., Wan, W., Yu, S., Qian, K., Liu, M., Zhang, X., Feng, Y., & Wei, Y. (2021). Simultaneous surface functionalization and drug loading: A novel method for fabrication of cellulose nanocrystals-based pH responsive drug delivery system. *International Journal of Biological Macromolecules, 182*, 2066–2075. doi:10.1016/j.ijbiomac.2021.05.193 PMID:34087297

López-lorente, Á. I., & Mizaikoff, B. (2016). Trends in Analytical Chemistry Recent advances on the characterization of nanoparticles using infrared spectroscopy. *Trends in Analytical Chemistry, 84*, 97–106. doi:10.1016/j.trac.2016.01.012

López-Ortega, A., Estrader, M., Salazar-Alvarez, G., Roca, A. G., & Nogués, J. (2015). Applications of exchange coupled bi-magnetic hard/soft and soft/hard magnetic core/shell nanoparticles. *Physics Reports, 553*, 1–32. doi:10.1016/j.physrep.2014.09.007

Loutfy, S. A., Al-Ansary, N. A., Abdel-Ghani, N. T., Hamed, A. R., Mohamed, M. B., Craik, J. D., Eldin, T. A. S., Abdellah, A. M., Hussein, Y., Hasanin, M., & Elbehairi, S. E. I. (2015). Anti-proliferative activities of metallic nanoparticles in an in vitro breast cancer model. *Asian Pacific Journal of Cancer Prevention, 16*(14), 6039–6046. doi:10.7314/APJCP.2015.16.14.6039 PMID:26320493

Lovat, V., Pantarotto, D., Lagostena, L., Cacciari, B., Grandolfo, M., Righi, M., Spalluto, G., Prato, M., & Ballerini, L. (2005). Carbon nanotube substrates boost neuronal electrical signaling. *Nano Letters, 5*(6), 1107–1110. doi:10.1021/nl050637m PMID:15943451

Lowe, S. E., & Zhong, Y. L. (2016). Challenges of Industrial-Scale Graphene Oxide Production. In *Graphene Oxide*. Fundamentals and Applications. doi:10.1002/9781119069447.ch13

Lowry, G. V., Gregory, K. B., Apte, S. C., & Lead, J. R. (2012). Transformations of nanomaterials in the environment. *Environmental Science & Technology, 46*(13), 6893–6899. doi:10.1021/es300839e PMID:22582927

Lozano, T., Rey, M., Rojas, E., Moya, S., Fleddermann, J., Estrela-Lopis, I., Donath, E., Wang, B., Mao, Z., Gao, C., & González-Fernández, Á. (2011). Cytotoxicity effects of metal oxide nanoparticles in human tumor cell lines. *Journal of Physics: Conference Series, 304*, 012046. doi:10.1088/1742-6596/304/1/012046

Lu, B., Xiong, S.-B., Yang, H., Yin, X.-D., & Chao, R.-B. (2006). Solid lipid nanoparticles of mitoxantrone for local injection against breast cancer and its lymph node metastases. *European Journal of Pharmaceutical Sciences*, *28*(1–2), 86–95. doi:10.1016/j.ejps.2006.01.001 PMID:16472996

Lucky, S. S., Soo, K. C., & Zhang, Y. (2015). Nanoparticles in photodynamic therapy. *Chemical Reviews*, *115*(4), 1990–2042. doi:10.1021/cr5004198 PMID:25602130

Ludi, B., & Niederberger, M. (2013). Zinc oxide nanoparticles: Chemical mechanisms and classical and non-classical crystallization. *Dalton Transactions (Cambridge, England)*, *42*(35), 12554–12568. doi:10.1039/c3dt50610j PMID:23652237

Lunge, S., Thakre, D., Kamble, S., Labhsetwar, N., & Rayalu, S. (2012). Alumina supported carbon composite material with exceptionally high defluoridation property from eggshell waste. *Journal of Hazardous Materials, 237–238*, 161–169. doi:10.1016/j.jhazmat.2012.08.023

Luo, J., Jang, H. D., Sun, T., Xiao, L., He, Z., Katsoulidis, A. P., Kanatzidis, M. G., Gibson, J. M., & Huang, J. (2011). Compression and aggregation-resistant particles of crumpled soft sheets. *ACS Nano*, *5*(11), 8943–8949. doi:10.1021/nn203115u PMID:21995602

Luo, S., Xu, X., Zhou, G., Liu, C., Tang, Y., & Liu, Y. (2014). Amino siloxane oligomer-linked graphene oxide as an efficient adsorbent for removal of Pb(II) from wastewater. *Journal of Hazardous Materials*, *274*, 145–155. doi:10.1016/j.jhazmat.2014.03.062 PMID:24780856

Luo, Z., Tan, C., Zhang, X., Chen, J., Cao, X., Li, B., Zong, Y., Huang, L., Huang, X., Wang, L., Huang, W., & Zhang, H. (2016). Preparation of Cobalt Sulfide Nanoparticle-Decorated Nitrogen and Sulfur Co-Doped Reduced Graphene Oxide Aerogel Used as a Highly Efficient Electrocatalyst for Oxygen Reduction Reaction. *Small*, *12*(43), 5920–5926. doi:10.1002mll.201602615 PMID:27592615

Lyumkis, D. (2019). Challenges and Opportunities in Cryo-EM Single-Particle Analysis Challenges and Opportunities in Cryo-EM. *The Journal of Biological Chemistry*, *294*(13), 5181–5197. doi:10.1074/jbc.REV118.005602 PMID:30804214

Maas, M. (2016). *Carbon nanomaterials as antibacterial colloids*. Academic Press.

Ma, C., Liu, H., Chen, G., Zhao, Q., Eitzer, B., Wang, Z., Cai, W., Newman, L. A., White, J. C., Dhankher, O. P., & Xing, B. (2017). Effects of titanium oxide nanoparticles on tetracycline accumulation and toxicity in Oryza sativa (L.). *Environmental Science. Nano*, *4*(9), 1827–1839. doi:10.1039/C7EN00280G

Madan, S. S., & Wasewar, K. L. (2017). Optimization for benzeneacetic acid removal from aqueous solution using CaO 2 nanoparticles based on Taguchi method. *Journal of Applied Research and Technology*, *15*(4), 332–339. https://doi.org/10.1016/j.jart.2017.02.007

Madhavan Nampoothiri, K., Nair, N. R., & John, R. P. (2010). An overview of the recent developments in polylactide (PLA) research. *Bioresource Technology*, *101*(22), 8493–8501. doi:10.1016/j.biortech.2010.05.092 PMID:20630747

Madhup, M. K., Shah, N. K., & Parekh, N. R. (2017). Investigation and improvement of abrasion resistance, water vapor barrier and anticorrosion properties of mixed clay epoxy nanocomposite coating. *Progress in Organic Coatings*, *102*, 186–193. doi:10.1016/j.porgcoat.2016.10.012

Madhusudhana, N., Yogendra, K., & Mahadevan, K. M. (2012). Decolorization of Coralene Dark Red 2B azo dye using Calcium Oxide Nanoparticle as an adsorbent. *International Journal of Research in Chemistry and Environment*, *2*(2), 21–25. https://doi.org/10.13140/RG.2.2.31460.83849

Maduraiveeran, G., Sasidharan, M., & Ganesan, V. (2018). Electrochemical sensor and biosensor platforms based on advanced nanomaterials for biological and biomedical applications. *Biosensors & Bioelectronics*, *103*, 113–129. doi:10.1016/j.bios.2017.12.031 PMID:29289816

Maensiri, S., Laokul, P., Klinkaewnarong, J., Phokha, S., & Seraphin, S. (2008). Indium oxide (In2O3) nanoparticles using Aloe vera plant extract: Synthesis and optical properties. *Optoelectronics and Advanced Materials Rapid Communications*, *2*, 161–165.

Maes, C., Luyten, W., Herremans, G., Peeters, R., Carleer, R., & Buntinx, M. (2018). Recent Updates on the Barrier Properties of Ethylene Vinyl Alcohol Copolymer (EVOH): A Review. *Polymer Reviews (Philadelphia, Pa.)*, *58*(2), 209–246. doi:10.1080/15583724.2017.1394323

Mahbub, K. R., Krishnan, K., Naidu, R., Andrews, S., & Megharaj, M. (2017). Mercury toxicity to terrestrial biota. *Ecological Indicators*, *74*, 451–462. doi:10.1016/j.ecolind.2016.12.004

Mahmoudi, M., Hosseinkhani, H., Hosseinkhani, M., Boutry, S., Simchi, A., Journeay, W. S., Subramani, K., & Laurent, S. (2011). Magnetic resonance imaging tracking of stem cells in vivo using iron oxide nanoparticles as a tool for the advancement of clinical regenerative medicine. *Chemical Reviews*, *111*(2), 253–280. doi:10.1021/cr1001832 PMID:21077606

Maity, S. K., Ancheyta, J., & Marroquín, G. (2010). Catalytic aquathermolysis used for viscosity reduction of heavy crude oils: A review. *Energy & Fuels*, *24*(5), 2809–2816. Advance online publication. doi:10.1021/ef100230k

Majhi, S., Naik, G., Lee, H., Song, H., Lee, C., Lee, I., & Yu, Y. (2018). Au@NiO Core-shell nanoparticles as a P-type gas sensor: Novel synthesis, characterization, and their gas sensing properties with sensing mechanism. *Sensors and Actuators. B, Chemical*, *268*, 223–231. doi:10.1016/j.snb.2018.04.119

Maji, I., Mahajan, S., Sriram, A., Medtiya, P., Vasave, R., Khatri, D. K., Kumar, R., Singh, S. B., Madan, J., & Singh, P. K. (2021). Solid self emulsifying drug delivery system: Superior mode for oral delivery of hydrophobic cargos. *Journal of Controlled Release*, *337*, 646–660. doi:10.1016/j.jconrel.2021.08.013 PMID:34384795

Majzlik, P., Strasky, A., Adam, V., Nemec, M., Trnkova, L., & Zehnalek, J. (2011). Influence of zinc (II) and copper (II) Ions on Streptomyces bacteria revealed by electrochemistry. *International Journal of Electrochemical Science*, *6*, 2171-2191.

Maksimović, M., & Omanović-Mikličanin, E. (2017). Towards green nanotechnology: Maximizing benefits and minimizing harm. *IFMBE Proceedings*, *62*, 164–170. doi:10.1007/978-981-10-4166-2_26

Malas, A. (2017). Rubber nanocomposites with graphene as the nanofiller. In *Progress in Rubber Nanocomposites* (pp. 179–229). Elsevier Inc. doi:10.1016/B978-0-08-100409-8.00006-1

Malhotra, N., Lee, J. S., Liman, R. A. D., Ruallo, J. M. S., Villaflore, O. B., Ger, T. R., & Der Hsiao, C. (2020). Potential toxicity of iron oxide magnetic nanoparticles: A review. Molecules, 25(14). doi:10.3390/molecules25143159

Malhotra, A., Sharma, N., Fnu, N., Kumar, N., Dolma, K., Sharma, D., Nandanwar, H., & Roy Choudhury, A. (2014). Multi-analytical approach to understand biomineralization of gold using rice bran: A novel and economical route. *RSC Advances*. Advance online publication. doi:10.1039.C4RA05404K

Malik, P., Shankar, R., Malik, V., Sharma, N., & Mukherjee, T. K. (2014). Green chemistry based benign routes for nanoparticle synthesis. *Journal of Nanoparticles*, *2014*, 1–14. doi:10.1155/2014/302429

Mamot, C., Ritschard, R., Wicki, A., Stehle, G., Dieterle, T., Bubendorf, L., Hilker, C., Deuster, S., Herrmann, R., & Rochlitz, C. (2012). Tolerability, safety, pharmacokinetics, and efficacy of doxorubicin-loaded anti-EGFR immunoliposomes in advanced solid tumours: A phase 1 dose-escalation study. *The Lancet. Oncology*, *13*(12), 1234–1241. doi:10.1016/S1470-2045(12)70476-X PMID:23153506

Mandal, R. P., Sekh, S., Sarkar, N. S., Chattopadhyay, D., & De, S. (2016). Algae mediated synthesis of cadmium sulphide nanoparticles and their application in bioremediation. *Materials Research Express*, *3*(5), 055007. doi:10.1088/2053-1591/3/5/055007

Mane, A. A., & Moholkar, A. V. (2018). Effect of solution concentration on physicochemical and NO2 gas sensing properties of sprayed MoO_3 nanobelts. *Thin Solid Films*, *648*(2), 50–61. doi:10.1016/j.tsf.2018.01.008

Manikandan, A., & Sathiyabama, M. (2015). Green synthesis of copper-chitosan nanoparticles and study of its antibacterial activity. *Journal of Nanomedicine & Nanotechnology*, *6*(1), 1.

Manikandan, M., Hasan, N., & Wu, H. F. (2013). Platinum nanoparticles for the photothermal treatment of Neuro 2A cancer cells. *Biomaterials*, *34*(23), 5833–5842. doi:10.1016/j.biomaterials.2013.03.077 PMID:23642996

Manjari Mishra, P., Pattnaik, S., & Prabha Devi, A. (2021). Green synthesis of bio-based Au@g-C_3N_4 nanocomposite for photocatalytic degradation of methyl orange. *Materials Today: Proceedings*, *xxxx*, 1218–1223. Advance online publication. doi:10.1016/j.matpr.2021.03.712

Manjari, G., Saran, S., Radhakrishanan, S., Rameshkumar, P., Pandikumar, A., & Devipriya, S. P. (2020). Facile green synthesis of Ag–Cu decorated ZnO nanocomposite for effective removal of toxic organic compounds and an efficient detection of nitrite ions. *Journal of Environmental Management*, *262*(January), 110282. doi:10.1016/j.jenvman.2020.110282 PMID:32090885

Mansha, M., Khan, I., Ullah, N., & Qurashi, A. (2017). Synthesis, characterization and visible-light-driven photoelectrochemical hydrogen evolution reaction of carbazole-containing conjugated polymers. *International Journal of Hydrogen Energy*, *42*(16), 10952–10961. doi:10.1016/j.ijhydene.2017.02.053

Mao, J., Iocozzia, J., Huang, J., Meng, K., Lai, Y., & Lin, Z. (2018). Graphene aerogels for efficient energy storage and conversion. In Energy and Environmental Science (Vol. 11, Issue 4, pp. 772–799). Royal Society of Chemistry. doi:10.1039/C7EE03031B

Mao, S., Wen, Z., Huang, T., Hou, Y., & Chen, J. (2014). *Electronic Supplementary Information (ESI) High-performance Bi-functional Electrocatalysts of 3D Crumpled Graphene-Cobalt Oxide Nanohybrids for Oxygen Reduction and Evolution Reactions*. Academic Press.

Mar Gil-Díaz, M., Pérez-Sanz, A., Ángeles Vicente, M., & Carmen Lobo, M. (2014). Immobilisation of Pb and Zn in Soils Using Stabilised Zero-valent Iron Nanoparticles: Effects on Soil Properties. *CLEAN - Soil, Air. Water (Basel)*, *42*(12), 1776–1784. doi:10.1002/clen.201300730

Marcinkowska, M., Stanczyk, M., Janaszewska, A., Sobierajska, E., Chworos, A., & Klajnert-Maculewicz, B. (2019). Multicomponent Conjugates of Anticancer Drugs and Monoclonal Antibody with PAMAM Dendrimers to Increase Efficacy of HER-2 Positive Breast Cancer Therapy. *Pharmaceutical Research*, *36*(11), 154. doi:10.100711095-019-2683-7 PMID:31482205

Mareci, D., Chelariu, R., Gordin, D.-M., Ungureanu, G., & Gloriant, T. (2009). Comparative corrosion study of Ti-Ta alloys for dental applications. *Acta Biomaterialia*, *5*(9), 3625–3639. doi:10.1016/j.actbio.2009.05.037 PMID:19508903

Maringgal, B., Hashim, N., Tawakkal, I. S. M. A., Hamzah, M. H., & Mohamed, M. T. M. (2020). Biosynthesis of CaO nanoparticles using Trigona sp. Honey: Physicochemical characterization, antifungal activity, and cytotoxicity properties. *Journal of Materials Research and Technology*, *9*(5), 11756–11768. https://doi.org/10.1016/j.jmrt.2020.08.054

Marinho, J. Z., Romeiro, F. C., Lemos, S. C. S., Motta, F. V., Riccardi, C. S., Li, M. S., Longo, E., & Lima, R. C. (2012). Urea-Based Synthesis of Zinc Oxide Nanostructures at Low Temperature. *Journal of Nanomaterials*, *2012*, 1–7. doi:10.1155/2012/427172

Maroufi, N. F., Ashouri, N., Mortezania, Z., Ashoori, Z., Vahedian, V., Amirzadeh-Iranaq, M. T., Fattahi, A., Kazemzadeh, H., Bizzarri, M., Akbarzadeh, M., Nejabati, H. R., Faridvand, Y., Rashidi, M.-R., & Nouri, M. (2020). The potential therapeutic effects of melatonin on breast cancer: An invasion and metastasis inhibitor. *Pathology, Research and Practice*, *216*(10), 153226. doi:10.1016/j.prp.2020.153226 PMID:32987338

Marques Neto, L. M., Kipnis, A., & Junqueira-Kipnis, A. P. (2017). Role of metallic nanoparticles in vaccinology: Implications for infectious disease vaccine development. *Frontiers in Immunology*, *8*. Advance online publication. doi:10.3389/fimmu.2017.00239 PMID:28337198

Marquis, G., Ramasamy, B., Banwarilal, S., & Munusamy, A. P. (2016). Evaluation of antibacterial activity of plant mediated CaO nanoparticles using C issus quadrangularis extract. *Journal of Photochemistry and Photobiology. B, Biology*, *155*, 28–33. https://doi.org/10.1016/j.jphotobiol.2015.12.013

Marsalek, B., Jancula, D., Marsalkova, E., Mashlan, M., Safarova, K., Tucek, J., & Zboril, R. (2012). *Multimodal action and selective toxicity of zerovalent iron nanoparticles against cyanobacteria*. Academic Press.

Martínez, F., López-Muñoz, M. J., Aguado, J., Melero, J. A., Arsuaga, J., Sotto, A., Molina, R., Segura, Y., Pariente, M. I., Revilla, A., Cerro, L., & Carenas, G. (2013). Coupling membrane separation and photocatalytic oxidation processes for the degradation of pharmaceutical pollutants. *Water Research*, *47*(15), 5647–5658. doi:10.1016/j.watres.2013.06.045 PMID:23863375

Martins, P., Lopes, A. C., & Lanceros-Mendez, S. (2014). Electroactive phases of poly (vinylidene fluoride) Determination, processing and applications. *Progress in Polymer Science*, *39*(4), 683–706. doi:10.1016/j.progpolymsci.2013.07.006

Maruyama, T., Fujimoto, Y., & Maekawa, T. (2015). Synthesis of gold nanoparticles using various amino acids. *Journal of Colloid and Interface Science*, *447*, 254–257. doi:10.1016/j.jcis.2014.12.046 PMID:25591824

Massey, V. (2000). The chemical and biological versatility of riboflavin. *Biochemical Society Transactions*, *28*(4), 283–296. doi:10.1042/bst0280283 PMID:10961912

Matinise, N., Fuku, X. G., Kaviyarasu, K., Mayedwa, N., & Maaza, M. J. A. S. S. (2017). ZnO nanoparticles via Moringa oleifera green synthesis: Physical properties & mechanism of formation. *Applied Surface Science*, *406*, 339–347. doi:10.1016/j.apsusc.2017.01.219

Matoori, S., & Leroux, J.-C. (2020). Twenty-five years of polymersomes: Lost in translation? *Materials Horizons*, *7*(5), 1297–1309. doi:10.1039/C9MH01669D

Mattson, M. P., Haddon, R. C., & Rao, A. M. (2000). Molecular functionalization of carbon nanotubes and use as substrates for neuronal growth. *Journal of Molecular Neuroscience*, *14*(3), 175–182. doi:10.1385/JMN:14:3:175 PMID:10984193

Mavlonov, A., Razykov, T., Raziq, F., Gan, J., Chantana, J., Kawano, Y., Nishimura, T., Wei, H., Zakutayev, A., Minemoto, T., Zu, X., Li, S., & Qiao, L. (2020). A review of Sb2Se3 photovoltaic absorber materials and thin-film solar cells. *Solar Energy*, *201*, 227–246. doi:10.1016/j.solener.2020.03.009

Ma, Z., Garrido-Maestu, A., & Jeong, K. C. (2017). Application, mode of action, and in vivo activity of chitosan and its micro-and nanoparticles as antimicrobial agents: A review. *Carbohydrate Polymers, 176*, 257–265. doi:10.1016/j.carbpol.2017.08.082 PMID:28927606

McDevitt, M. R., Chattopadhyay, D., Kappel, B. J., Jaggi, J. S., Schiffman, S. R., Antczak, C., Njardarson, J. T., Brentjens, R., & Scheinberg, D. A. (2007). Tumor targeting with antibody-functionalized, radiolabeled carbon nanotubes. *Journal of Nuclear Medicine, 48*(7), 1180–1189. doi:10.2967/jnumed.106.039131 PMID:17607040

McDonald, K. J., Reddy, K. J., Singh, N., Singh, R. P., & Mukherjee, S. (2015). Removal of arsenic from groundwater in West Bengal, India using CuO nanoparticle adsorbent. *Environmental Earth Sciences, 73*(7), 3593–3601. doi:10.100712665-014-3645-3

McElroy, P., Emadi, H., Surowiec, K., & Casadonte, D. J. (2019). Mechanical, rheological, and stability performance of simulated in-situ cured oil well cement slurries reinforced with alumina nanofibers. *Journal of Petroleum Science Engineering, 183*, 106415. Advance online publication. doi:10.1016/j.petrol.2019.106415

Mclafferty, F. W. (2011). A Century of Progress in Molecular Mass Spectrometry. *Annual Review of Analytical Chemistry (Palo Alto, Calif.), 4*(1), 1–22. doi:10.1146/annurev-anchem-061010-114018 PMID:21351881

McNair, O. D., Brent, D. P., Sparks, B. J., Patton, D. L., & Savin, D. A. (2014). *Sequential thiol click reactions: formation of ternary thiourethane/thiol–ene networks with enhanced thermal and mechanical properties*. Academic Press.

McNamara, K., & Tofail, S. A. (2017). Nanoparticles in biomedical applications. *Advances in Physics: X, 2*(1), 54–88. doi:10.1080/23746149.2016.1254570

Medhi, S., Chowdhury, S., Gupta, D. K., & Mazumdar, A. (2020). An investigation on the effects of silica and copper oxide nanoparticles on rheological and fluid loss property of drilling fluids. *Journal of Petroleum Exploration and Production Technology, 10*(1), 91–101. Advance online publication. doi:10.100713202-019-0721-y

Medina, O. E., Olmos, C., Lopera, S. H., Cortés, F. B., & Franco, C. A. (2019). Nanotechnology applied to thermal enhanced oil recovery processes: A review. *Energies, 12*(24), 4671. Advance online publication. doi:10.3390/en12244671

Mehranfar, A., & Izadyar, M. (2020). Theoretical Design of Functionalized Gold Nanoparticles as Antiviral Agents against Severe Acute Respiratory Syndrome Coronavirus 2 (SARS-CoV-2). *The Journal of Physical Chemistry Letters, 11*(24), 10284–10289. doi:10.1021/acs.jpclett.0c02677 PMID:33226815

Mehwish, H. M., Rajoka, M. S. R., Xiong, Y., Cai, H., Aadil, R. M., Mahmood, Q., He, Z., & Zhu, Q. (2021). Green synthesis of a silver nanoparticle using Moringa oleifera seed and its applications for antimicrobial and sun-light mediated photocatalytic water detoxification. *Journal of Environmental Chemical Engineering, 9*(4), 105290. doi:10.1016/j.jece.2021.105290

Mei, H., Cai, S., Huang, D., Gao, H., Cao, J., & He, B. (2022). Carrier-free nanodrugs with efficient drug delivery and release for cancer therapy: From intrinsic physicochemical properties to external modification. *Bioactive Materials, 8*, 220–240. doi:10.1016/j.bioactmat.2021.06.035 PMID:34541398

Mei, R., Song, X., Hu, Y., Yang, Y., & Zhang, J. (2015). Hollow reduced graphene oxide microspheres as a high-performance anode material for Li-ion batteries. *Electrochimica Acta, 153*, 540–545. doi:10.1016/j.electacta.2014.05.154

Mei, Z., Wu, Q., Hu, S., Lib, X., & Yang, X. (2005). Triptolide loaded solid lipid nanoparticle hydrogel for topical application. *Drug Development and Industrial Pharmacy, 31*(2), 161–168. doi:10.1081/DDC-200047791 PMID:15773283

Melancon, M. P., Lu, W., Zhong, M., Zhou, M., Liang, G., Elliott, A. M., Hazle, J. D., Myers, J. N., Li, C., & Stafford, R. J. (2011). Targeted multifunctional gold-based nanoshells for magnetic resonance-guided laser ablation of head and neck cancer. *Biomaterials*, *32*(30), 7600–7608. doi:10.1016/j.biomaterials.2011.06.039 PMID:21745689

Menad, N. A., Noureddine, Z., Hemmati-Sarapardeh, A., Shamshirband, S., Mosavi, A., & Chau, K. (2019). Modeling temperature dependency of oil - water relative permeability in thermal enhanced oil recovery processes using group method of data handling and gene expression programming. *Engineering Applications of Computational Fluid Mechanics*, *13*(1), 724–743. Advance online publication. doi:10.1080/19942060.2019.1639549

Merah, N., & Mohamed, O. (2019). Nanoclay and Water Uptake Effects on Mechanical Properties of Unsaturated Polyester. *Journal of Nanomaterials*, *2019*, 1–11. doi:10.1155/2019/8130419

Messersmith, P. B., & Giannelis, E. P. (1995). Synthesis and barrier properties of poly(ε-caprolactone)-layered silicate nanocomposites. *Journal of Polymer Science. Part A, Polymer Chemistry*, *33*(7), 1047–1057. doi:10.1002/pola.1995.080330707

Metaxa, Z. S., Tolkou, A. K., Efstathiou, S., Rahdar, A., Favvas, E. P., Mitropoulos, A. C., & Kyzas, G. Z. (2021). Nanomaterials in cementitious composites: An update. *Molecules (Basel, Switzerland)*, *26*(5), 1430. Advance online publication. doi:10.3390/molecules26051430 PMID:33800797

Metwally, R. A., El Sikaily, A., El-Sersy, N. A., Ghozlan, H. A., & Sabry, S. A. (2021). Antimicrobial activity of textile fabrics dyed with prodigiosin pigment extracted from marine Serratia rubidaea RAM_Alex bacteria. *The Egyptian Journal of Aquatic Research*.

Miele, E., Spinelli, G. P., Miele, E., Tomao, F., & Tomao, S. (2009). Albumin-bound formulation of paclitaxel (Abraxane® ABI-007) in the treatment of breast cancer. *International Journal of Nanomedicine*, *4*, 99. PMID:19516888

Milewicz, B., Bogacka, M., & Pikoń, K. (2021). Influence of solar concentrator in the form of luminescent pmma on the performance of a silicon cell. *Sustainability*, *13*(4), 1–14. doi:10.3390u13042086

Miller-Kleinhenz, J., Guo, X., Qian, W., Zhou, H., Bozeman, E. N., Zhu, L., Ji, X., Wang, Y. A., Styblo, T., O'Regan, R., Mao, H., & Yang, L. (2018). Dual-targeting Wnt and uPA receptors using peptide conjugated ultra-small nanoparticle drug carriers inhibited cancer stem-cell phenotype in chemo-resistant breast cancer. *Biomaterials*, *152*, 47–62. doi:10.1016/j.biomaterials.2017.10.035 PMID:29107218

Miller, M. A., Arlauckas, S., & Weissleder, R. (2017). Prediction of anti-cancer nanotherapy efficacy by imaging. *Nanotheranostics*, *1*(3), 296–312. doi:10.7150/ntno.20564 PMID:29071194

Mi, P., Kokuryo, D., Cabral, H., Wu, H., Terada, Y., Saga, T., Aoki, I., Nishiyama, N., & Kataoka, K. (2016). A pH-activatable nanoparticle with signal-amplification capabilities for non-invasive imaging of tumour malignancy. *Nature Nanotechnology*, *11*(8), 724–730. doi:10.1038/nnano.2016.72 PMID:27183055

Mirghiasi, Z., Bakhtiari, F., Darezereshki, E., & Esmaeilzadeh, E. (2014). Preparation and characterization of CaO nanoparticles from Ca(OH)2 by direct thermal decomposition method. *Journal of Industrial and Engineering Chemistry*, *20*(1), 113–117. https://doi.org/10.1016/j.jiec.2013.04.018

Mirzaei, A., & Neri, G. (2016). Microwave-assisted synthesis of metal oxide nanostructures for gas sensing application: A review. *Sensors and Actuators. B, Chemical*, *237*, 749–775. doi:10.1016/j.snb.2016.06.114

Mishra, A., Kumari, M., Pandey, S., Chaudhry, V., Gupta, K. C., & Nautiyal, C. S. (2014). Biocatalytic and antimicrobial activities of gold nanoparticles synthesized by Trichoderma sp. *Bioresource Technology*, *166*, 235–242. doi:10.1016/j.biortech.2014.04.085 PMID:24914997

Mishra, P. C., & Patel, R. K. (2009). Use of agricultural waste for the removal of nitrate-nitrogen from aqueous medium. *Journal of Environmental Management*, *90*(1), 519–522. https://doi.org/10.1016/j.jenvman.2007.12.003

Mitchell, M. J., Billingsley, M. M., Haley, R. M., Wechsler, M. E., Peppas, N. A., & Langer, R. (2021). Engineering precision nanoparticles for drug delivery. *Nature Reviews. Drug Discovery*, *20*(2), 101–124. doi:10.103841573-020-0090-8 PMID:33277608

Mitrani, M. I., Bellio, M. A., Sagel, A., Saylor, M., Kapp, W., VanOsdol, K., Haskell, G., Stewart, D., Abdullah, Z., Santos, I., Milberg, J., Arango, A., Mitrani, A., & Shapiro, G. C. (2021). Case Report: Administration of Amniotic Fluid-Derived Nanoparticles in Three Severely Ill COVID-19 Patients. *Frontiers in Medicine*, *8*(March), 1–8. doi:10.3389/fmed.2021.583842 PMID:33816515

Mitrano, D. M., Barber, A., Bednar, A., Westerhoff, P., Higgins, C. P., & Ranville, J. F. (2012). Silver nanoparticle characterization using single particle ICP-MS (SP-ICP-MS) and asymmetrical flow field flow fractionation ICP-MS (AF4-ICP-MS). *Journal of Analytical Atomic Spectrometry*, *27*(7), 1131–1142. doi:10.1039/c2ja30021d

Mittal, A. K., Chisti, Y., & Banerjee, U. C. (2013). Synthesis of metallic nanoparticles using plant extracts. *Biotechnology Advances*, *31*(2), 346–356. doi:10.1016/j.biotechadv.2013.01.003 PMID:23318667

Mittal, A. K., Kumar, S., & Banerjee, U. C. (2014). Quercetin and gallic acid mediated synthesis of bimetallic (silver and selenium) nanoparticles and their antitumor and antimicrobial potential. *Journal of Colloid and Interface Science*, *431*, 194–199. doi:10.1016/j.jcis.2014.06.030 PMID:25000181

Mlynarczuk, J., Górska, M., & Wrobel, M. H. (2020). Effects of DDT, DDE, aldrin and dieldrin on prostaglandin, oxytocin and steroid hormone release from smooth chorion explants of cattle. *Animal Reproduction Science*, *223*, 106623. doi:10.1016/j.anireprosci.2020.106623 PMID:33132122

Mohamad, N. A. N., Arham, N. A., Jai, J., & Hadi, A. (2013). *Plant Extract as Reducing Agent in Synthesis of Metallic Nanoparticles: A Review*. International Conference 4th Nanoscience and Nanotechnology (NANO-SciTech 2013), Shah Alam, Malaysia.

Mohamed, E. F., & Awad, G. (2021). Hollow N-TiO_2/MnO_2 nanocomposite-based yeast biomass for gaseous formaldehyde degradation under visible light. *Journal of Industrial and Engineering Chemistry*, *98*, 366–374. doi:10.1016/j.jiec.2021.03.028

Mohammadi, A., Doctorsafaei, A. H., Burujeny, S. B., Rudbari, H. A., Kordestani, N., & Najafabadi, S. A. A. (2020). Silver (I) complex with a Schiff base ligand extended waterborne polyurethane: A developed strategy to obtain a highly stable antibacterial dispersion impregnated with in situ formed silver nanoparticles. *Chemical Engineering Journal*, *381*, 122776. doi:10.1016/j.cej.2019.122776

Mohammadi-Aloucheh, R., Habibi-Yangjeh, A., Bayrami, A., Latifi-Navid, S., & Asadi, A. (2018). Green synthesis of ZnO and ZnO/CuO nanocomposites in Mentha longifolia leaf extract: Characterization and their application as anti-bacterial agents. *Journal of Materials Science Materials in Electronics*, *29*(16), 13596–13605. doi:10.100710854-018-9487-0

Mohammadi, M., Ramezani, M., Abnous, K., & Alibolandi, M. (2017). Biocompatible polymersomes-based cancer theranostics: Towards multifunctional nanomedicine. *International Journal of Pharmaceutics*, *519*(1-2), 287–303. doi:10.1016/j.ijpharm.2017.01.037 PMID:28115259

Mohammed, L., Gomaa, H. G., Ragab, D., & Zhu, J. (2017). Magnetic nanoparticles for environmental and biomedical applications: A review. *Particuology*, *30*, 1–14. doi:10.1016/j.partic.2016.06.001

Mohan, V. B., Lau, K. T., Hui, D., & Bhattacharyya, D. (2018). Graphene-based materials and their composites: A review on production, applications and product limitations. Composites Part B: Engineering. doi:10.1016/j.compositesb.2018.01.013

Mohan, T. P., & Kanny, K. (2011). Water barrier properties of nanoclay filled sisal fibre reinforced epoxy composites. *Composites. Part A, Applied Science and Manufacturing*, *42*(4), 385–393. doi:10.1016/j.compositesa.2010.12.010

Mohseni, M. S., Khalilzadeh, M. A., Mohseni, M., Hargalani, F. Z., Getso, M. I., Raissi, V., & Raiesi, O. (2020). Green synthesis of Ag nanoparticles from pomegranate seeds extract and synthesis of Ag-Starch nanocomposite and characterization of mechanical properties of the films. *Biocatalysis and Agricultural Biotechnology*, *25*(March), 101569. doi:10.1016/j.bcab.2020.101569

Moitra, P., Alafeef, M., Dighe, K., Frieman, M. B., & Pan, D. (2020). Selective Naked-Eye Detection of SARS-CoV-2 Mediated by N Gene Targeted Antisense Oligonucleotide Capped Plasmonic Nanoparticles. *ACS Nano*, *14*(6), 7617–7627. doi:10.1021/acsnano.0c03822 PMID:32437124

Moncayo-Riascos, I., De Leon, J., Garcia-Martinez, J. A., Garcia-Cruz, I., & Lira-Galeana, C. (2019). Multiscale simulation of asphaltene deposition in pipeline flows. *Journal of Petroleum Science Engineering*, *183*, 106376. Advance online publication. doi:10.1016/j.petrol.2019.106376

Mondal, S., Dorozhkin, S. V., & Pal, U. (2018). Recent progress on fabrication and drug delivery applications of nanostructured hydroxyapatite. *WIREs Nanomedicine and Nanobiotechnology, 10*(4), e1504.

Mondal, P., Anweshan, A., & Purkait, M. K. (2020). Green synthesis and environmental application of iron-based nanomaterials and nanocomposite: A review. *Chemosphere*, *259*, 127509. Advance online publication. doi:10.1016/j.chemosphere.2020.127509 PMID:32645598

Monteiro-Riviere, N. A., Wiench, K., Landsiedel, R., Schulte, S., Inman, A. O., & Riviere, J. E. (2011). *Safety evaluation of sunscreen formulations containing titanium dioxide and zinc oxide nanoparticles in UVB sunburned skin: an in vitro and in vivo study*. Academic Press.

Montoro Bustos, A. R., Encinar, J. R., Fernández-Argüelles, M. T., Costa-Fernández, J. M., & Sanz-Medel, A. (2009). Elemental mass spectrometry: A powerful tool for an accurate characterisation at elemental level of quantum dots. *Chemical Communications*, *03*(21), 3107–3109. doi:10.1039/b901493d PMID:19462102

Montoya, T., Argel, B. L., Nassar, N. N., Franco, C. A., & Cortés, F. B. (2016). Kinetics and mechanisms of the catalytic thermal cracking of asphaltenes adsorbed on supported nanoparticles. *Petroleum Science*, *13*(3), 561–571. Advance online publication. doi:10.100712182-016-0100-y

Monzavi, A., Eshraghi, S., Hashemian, R., & Momen-Heravi, F. (2015). *In vitro and ex vivo antimicrobial efficacy of nano-MgO in the elimination of endodontic pathogens*. Academic Press.

Moore, L., Chow, E. K.-H., Osawa, E., Bishop, J. M., & Ho, D. (2013). Diamond-lipid hybrids enhance chemotherapeutic tolerance and mediate tumor regression. *Advanced Materials*, *25*(26), 3532–3541. doi:10.1002/adma.201300343 PMID:23584895

Moradi Kashkooli, F., Soltani, M., & Souri, M. (2020). Controlled anti-cancer drug release through advanced nano-drug delivery systems: Static and dynamic targeting strategies. *Journal of Controlled Release*, *327*, 316–349. doi:10.1016/j.jconrel.2020.08.012 PMID:32800878

Moreno-Vega, A.-I., Gomez-Quintero, T., Nunez-Anita, R.-E., Acosta-Torres, L.-S., & Castaño, V. (2012). Polymeric and ceramic nanoparticles in biomedical applications. *Journal of Nanotechnology*, *2012*, 2012. doi:10.1155/2012/936041

Mout, R., Moyano, D. F., Rana, S., & Rotello, V. M. (2012). Surface functionalization of nanoparticles for nanomedicine. *Chemical Society Reviews*, *41*(7), 2539. Advance online publication. doi:10.1039/c2cs15294k PMID:22310807

Mroz, P., Pawlak, A., Satti, M., Lee, H., Wharton, T., Gali, H., Sarna, T., & Hamblin, M. R. (2007). Functionalized fullerenes mediate photodynamic killing of cancer cells: Type I versus Type II photochemical mechanism. *Free Radical Biology & Medicine*, *43*(5), 711–719. doi:10.1016/j.freeradbiomed.2007.05.005 PMID:17664135

Mthombeni, N. H., Mbakop, S., & Onyango, M. S. (2015). Magnetic Zeolite-Polymer Composite as an Adsorbent for the Remediation of Wastewaters Containing Vanadium. *International Journal of Environmental Sciences and Development*, *6*(8), 602–605. doi:10.7763/IJESD.2015.V6.665

Mu, C.-F., Balakrishnan, P., Cui, F.-D., Yin, Y.-M., Lee, Y.-B., Choi, H.-G., Yong, C. S., Chung, S.-J., Shim, C.-K., & Kim, D.-D. (2010). The effects of mixed MPEG–PLA/Pluronic® copolymer micelles on the bioavailability and multidrug resistance of docetaxel. *Biomaterials*, *31*(8), 2371–2379. doi:10.1016/j.biomaterials.2009.11.102 PMID:20031202

Mueller, N. C., Braun, J., Bruns, J., Černík, M., Rissing, P., Rickerby, D., & Nowack, B. (2012). Application of nanoscale zero valent iron (NZVI) for groundwater remediation in Europe. *Environmental Science and Pollution Research International*, *19*(2), 550–558. doi:10.100711356-011-0576-3 PMID:21850484

Mueller, N. C., & Nowack, B. (2008). Exposuremodeling of engineered nanoparticles in the environment. *Environmental Science & Technology*, *42*(12), 4447–4453. doi:10.1021/es7029637 PMID:18605569

Mujawar, M. A., Gohel, H., Bhardwaj, S. K., Srinivasan, S., Hickman, N., & Kaushik, A. (2020). Nano-enabled Biosensing Systems for Intelligent Healthcare: Towards COVID-19 Management. *Materials Today. Chemistry*, *17*, 100306. doi:10.1016/j.mtchem.2020.100306 PMID:32835155

Mukherjee, P., Ahmad, A., Mandal, D., Senapati, S., Sainkar, S. R., Khan, M. I., Parishcha, R., Ajaykumar, P. V., Alam, M., Kumar, R., & Sastry, M. (2001). Fungus-mediated synthesis of silver nanoparticles and their immobilization in the mycelial matrix: A novel biological approach to nanoparticle synthesis. *Nano Letters*, *1*(10), 515–519. doi:10.1021/nl0155274

Mukherjee, R., Kumar, R., Sinha, A., Lama, Y., & Saha, A. K. (2016). A review on synthesis, characterization, and applications of nano zero valent iron (nZVI) for environmental remediation. *Critical Reviews in Environmental Science and Technology*, *35*(5), 360–384. doi:10.1080/10643389.2015.1103832

Mukherjee, S., Chowdhury, D., Kotcherlakota, R., Patra, S. B. V., Bhadra, M. P., Sreedhar, B., & Patra, C. R. (2014). Potential theranostics application of aio-synthesized silver nanoparticles (4-in-1 system). *Theranostics*, *4*(3), 316–335. doi:10.7150/thno.7819 PMID:24505239

Mukherjee, S., & Patra, C. R. (2017). Biologically synthesized metal nanoparticles: Recent advancement and future perspectives in cancer theranostics. *Future Science OA*, *3*(3), FSO203. Advance online publication. doi:10.4155/fsoa-2017-0035 PMID:28884002

Mukherjee, S., Ray, S., & Thakur, R. (2009). Solid lipid nanoparticles: A modern formulation approach in drug delivery system. *Indian Journal of Pharmaceutical Sciences*, *71*(4), 349. doi:10.4103/0250-474X.57282 PMID:20502539

Müller, C. M. O., Laurindo, J. B., & Yamashita, F. (2011). Effect of nanoclay incorporation method on mechanical and water vapor barrier properties of starch-based films. *Industrial Crops and Products*, *33*(3), 605–610. doi:10.1016/j.indcrop.2010.12.021

Mulvihill, M. J., Beach, E. S., Zimmerman, J. B., & Anastas, P. T. (2011). Green Chemistry and Green Engineering: A Framework for Sustainable Technology Development. *Annual Review of Environment and Resources*, *36*(1), 271–293. doi:10.1146/annurev-environ-032009-095500

Murphy, M. P. (2009). How mitochondria produce reactive oxygen species. *The Biochemical Journal*, *417*(1), 1–13. doi:10.1042/BJ20081386 PMID:19061483

Murtaza, M., Mahmoud, M., Elkatatny, S., Al Majed, A., Chen, W., & Jamaluddin, A. (2019). Experimental investigation of the impact of modified nano clay on the rheology of oil well cement slurry. *International Petroleum Technology Conference 2019, IPTC 2019*. 10.2523/IPTC-19456-MS

Murtaza, M., Rahman, M. K., & Al Majed, A. A. (2020). The application and comparative study of calcined clay and nanoclaymixed cement slurries at hpht conditions. *Proceedings of the Annual Offshore Technology Conference*. 10.4043/30872-MS

Muthu Kumara Pandian, A., Rajasimman, M., Rajamohan, N., Varjani, S., & Karthikeyan, C. (2021). Anaerobic mixed consortium (AMC) mediated enhanced biosynthesis of silver nano particles (AgNPs) and its application for the removal of phenol. *Journal of Hazardous Materials*, *416*, 125717. doi:10.1016/j.jhazmat.2021.125717 PMID:33819646

Muzenski, S., Flores-Vivian, I., & Sobolev, K. (2019). Ultra-high strength cement-based composites designed with aluminum oxide nano-fibers. *Construction & Building Materials*, *220*, 177–186. Advance online publication. doi:10.1016/j.conbuildmat.2019.05.175

Nadaf, N. Y., & Kanase, S. S. (2016). Biosynthesis of gold nanoparticles by Bacillus marisflavi and its potential in catalytic dye degradation. *Arabian Journal of Chemistry*, *12*(8), 4806–4814. doi:10.1016/j.arabjc.2016.09.020

Nadagouda, M. N., & Varma, R. S. (2006). Green and controlled synthesis of gold and platinum nanomaterials using vitamin B2: Density-assisted self-assembly of nanospheres, wires and rods. *Green Chemistry*, *8*(6), 516. doi:10.1039/b601271j

Nadagouda, M. N., & Varma, R. S. (2007a). Synthesis of thermally stable carboxymethyl cellulose/metal biodegradable nanocomposites for potential biological applications. *Biomacromolecules*, *8*(9), 2762–2767. doi:10.1021/bm700446p PMID:17665946

Nadagouda, M. N., & Varma, R. S. (2007b). A greener synthesis of core (Fe, Cu)-shell (Au, Pt, Pd, and Ag) nanocrystals using aqueous vitamin C. *Crystal Growth & Design*, *7*(12), 2582–2587. doi:10.1021/cg070554e

Nadagouda, M. N., & Varma, R. S. (2008a). Green synthesis of Ag and Pd nanospheres, nanowires, and nanorods using vitamin B 2: Catalytic polymerisation of aniline and pyrrole. *Journal of Nanomaterials*, *2008*, 1–8. doi:10.1155/2008/782358

Nadagouda, M. N., & Varma, R. S. (2008b). Green synthesis of silver and palladium nanoparticles at room temperature using coffee and tea extract. *Green Chemistry*, *10*(8), 859. doi:10.1039/b804703k

Naghizadeh, A., Mohammadi-Aghdam, S., & Mortazavi-Derazkola, S. (2020). Novel CoFe2O4@ZnO-CeO2 ternary nanocomposite: Sonochemical green synthesis using Crataegus microphylla extract, characterization and their application in catalytic and antibacterial activities. *Bioorganic Chemistry*, *103*(August), 104194. doi:10.1016/j.bioorg.2020.104194 PMID:32890997

Naidu, D. S., & John, M. J. (2020). Effect of Clay Nanofillers on the Mechanical and Water Vapor Permeability Properties of Xylan–Alginate Films. *Polymers*, *12*(10), 2279. doi:10.3390/polym12102279 PMID:33020377

Nam, J., Won, N., Jin, H., Chung, H., & Kim, S. (2009). pH-Induced Aggregation of Gold Nanoparticles for Photothermal Cancer Therapy. *Journal of the American Chemical Society*, *131*(38), 13639–13645. doi:10.1021/ja902062j PMID:19772360

Namvar, F., Rahman, H. S., Mohamad, R., Azizi, S., Tahir, P. M., Chartrand, M. S., & Yeap, S. K. (2015). Cytotoxic effects of biosynthesized zinc oxide nanoparticles on murine cell lines. *Evidence-Based Complementary and Alternative Medicine*, *2015*, 2015. doi:10.1155/2015/593014 PMID:25784947

Nanda, A., Nayak, B. K., & Krishnamoorthy, M. (2018). Antimicrobial properties of biogenic silver nanoparticles synthesized from phylloplane fungus, Aspergillus tamarii. *Biocatalysis and Agricultural Biotechnology, 16*, 225–228. doi:10.1016/j.bcab.2018.08.002

Narayanan, M., Vigneshwari, P., Natarajan, D., Kandasamy, S., Alsehli, M., Elfasakhany, A., & Pugazhendhi, A. (2021). Synthesis and characterization of TiO2 NPs by aqueous leaf extract of Coleus aromaticus and assess their antibacterial, larvicidal, and anticancer potential. *Environmental Research, 200*, 111335. doi:10.1016/j.envres.2021.111335 PMID:34051200

Naseem, T., & Waseem, M. (2021). A comprehensive review on the role of some important nanocomposites for antimicrobial and wastewater applications. *Chemical Reviews, 107*, 2411–2502.

Naseri, N., Zakeri-Milani, P., Hamishehkar, H., Pilehvar-Soltanahmadi, Y., & Valizadeh, H. (2017). Development, In Vitro Characterization, Antitumor and Aerosol Performance Evaluation of Respirable Prepared by Self-nanoemulsification Method. *Drug Research (Stuttgart), 67*(06), 343–348. doi:10.1055-0043-102404 PMID:28288490

Nasiriboroumand, M., Montazer, M., & Barani, H. (2018). Preparation and characterization of biocompatible silver nanoparticles using pomegranate peel extract. *Journal of Photochemistry and Photobiology. B, Biology, 179*, 98–104. doi:10.1016/j.jphotobiol.2018.01.006 PMID:29351880

Nasongkla, N., Bey, E., Ren, J., Ai, H., Khemtong, C., Guthi, J. S., Chin, S.-F., Sherry, A. D., Boothman, D. A., & Gao, J. (2006). Multifunctional polymeric micelles as cancer-targeted, MRI-ultrasensitive drug delivery systems. *Nano Letters, 6*(11), 2427–2430. doi:10.1021/nl061412u PMID:17090068

Nasrollahzadeh, M., Sajadi, S. M., Sajjadi, M., & Issaabadi, Z. (2019). An Introduction to Nanotechnology. In An introduction to green nanotechnology (Vol. 28., pp. 1-27). Elsevier. doi:10.1016/B978-0-12-813586-0.00001-8

Nasrollahzadeh, M., Akbari, R., Issaabadi, Z., & Sajadi, S. M. (2020). Biosynthesis and characterization of Ag/MgO nanocomposite and its catalytic performance in the rapid treatment of environmental contaminants. *Ceramics International, 46*(2), 2093–2101. doi:10.1016/j.ceramint.2019.09.191

Nasr, R., Hasanzadeh, H., Khaleghian, A., Moshtaghian, A., Emadi, A., & Moshfegh, S. (2018). Induction of Apoptosis and Inhibition of Invasion in Gastric Cancer Cells by Titanium Dioxide Nanoparticles. *Oman Medical Journal, 33*(2), 111–117. doi:10.5001/omj.2018.22 PMID:29657679

Natsuki, J., Natsuki, T., & Hashimoto, Y. (2015). A review of silver nanoparticles: Synthesis methods, properties and applications. *International Journal of Materials Science and Applications, 4*(5), 325–332. doi:10.11648/j.ijmsa.20150405.17

Navab, A. A., Nemati, A., Navab, A. A., & Abad, H. M. M. (2018). *Hydrothermal synthesis of TiO2 nanorod for using as an electron transport material in perovskite solar cells*. doi:10.1063/1.5018947

Navya Rani, M., Murthy, M., Shyla Shree, N., Ananda, S., Yogesh, S., & Dinesh, R. (2019). Cuprous oxide anchored Reduced Graphene oxide ceramic nanocomposite using Tagetes erecta flower extract and evaluation of its antibacterial activity and cytotoxicity. *Ceramics International, 45*(18), 25020–25026. doi:10.1016/j.ceramint.2019.04.195

Nayaka, S. (2020). Extracellular fabrication of bio-nanostructures from Ralstonia sp. strain NS-7: Characterizations and their microbiological evaluation. *Indian Journal of Science and Technology, 13*(31), 3128–3140. doi:10.17485/IJST/v13i31.289

Nayak, P. P., S, N., Narayanan, A., Badekila, A. K., & Kini, S. (2021). Nanomedicine in cancer clinics: Are we there yet? *Current Pathobiology Reports, 9*(2), 43–55. doi:10.100740139-021-00220-6

Nazarian-Samani, M., Kim, H. K., Park, S. H., Youn, H. C., Mhamane, D., Lee, S. W., Kim, M. S., Jeong, J. H., Haghighat-Shishavan, S., Roh, K. C., Kashani-Bozorg, S. F., & Kim, K. B. (2016). Three-dimensional graphene-based spheres and crumpled balls: Micro- and nano-structures, synthesis strategies, properties and applications. In RSC Advances (Vol. 6, Issue 56, pp. 50941–50967). Royal Society of Chemistry. doi:10.1039/C6RA07485E

Nazar, M. F., Shah, S. S., & Khosa, M. A. (2011). Microemulsions in enhanced oil recovery: A review. *Petroleum Science and Technology*, *29*(13), 1353–1365. Advance online publication. doi:10.1080/10916460903502514

Neog, R., & Kola, N. (2021). In vitro antimicrobial screening of methanolic extract of Mikania micrantha and Drymaria cordata against S. Aureus and its textile application. *Materials Today: Proceedings*, *42*, 916–920. doi:10.1016/j.matpr.2020.11.835

Netto, C. G., Toma, H. E., & Andrade, L. H. (2013). Superparamagnetic nanoparticles as versatile carriers and supporting materials for enzymes. *Journal of Molecular Catalysis. B, Enzymatic*, *85*, 71–92. doi:10.1016/j.molcatb.2012.08.010

Ngom, I., Ngom, B. D., Sackey, J., & Khamlich, S. (2021). Biosynthesis of zinc oxide nanoparticles using extracts of Moringa Oleifera: Structural & optical properties. *Materials Today: Proceedings*, *36*, 526–533. doi:10.1016/j.matpr.2020.05.323

Nguyen, A. T., Mattiassi, S., Loeblein, M., Chin, E., Ma, D., Coquet, P., Viasnoff, V., Teo, E. H. T., Goh, E. L., & Yim, E. K. F. (2018). Human Rett-derived neuronal progenitor cells in 3D graphene scaffold as an in vitro platform to study the effect of electrical stimulation on neuronal differentiation. *Biomedical Materials (Bristol)*, *13*(3), 034111. doi:10.1088/1748-605X/aaaf2b PMID:29442069

Nie, X., Wu, S., Liao, S., Chen, J., Huang, F., Li, W., Wang, Q., & Wei, Q. (2021). Light-driven self-disinfecting textiles functionalized by PCN-224 and Ag nanoparticles. *Journal of Hazardous Materials*, *416*, 125786. doi:10.1016/j.jhazmat.2021.125786 PMID:33873032

Niidome, T., Yamauchi, H., Takahashi, K., Naoyama, K., Watanabe, K., Mori, T., & Katayama, Y. (2014). Hydrophobic cavity formed by oligopeptide for doxorubicin delivery based on dendritic poly(L-lysine). *Journal of Biomaterials Science. Polymer Edition*, *25*(13), 1362–1373. doi:10.1080/09205063.2014.938979 PMID:25040893

Nikalje, A. P. (2015). Nanotechnology and its applications in medicine. *Med chem*, *5*(2), 81-89.

Nine, M. J., Cole, M. A., Tran, D. N. H., & Losic, D. (2015). Graphene: A multipurpose material for protective coatings. *Journal of Materials Chemistry. A, Materials for Energy and Sustainability*, *3*(24), 12580–12602. doi:10.1039/C5TA01010A

Njuguna, S. M., Onyango, J. A., Githaiga, K. B., Gituru, R. W., & Yan, X. (2020). Application of multivariate statistical analysis and water quality index in health risk assessment by domestic use of river water. Case study of Tana River in Kenya. *Process Safety and Environmental Protection*, *133*, 149–158. https://doi.org/10.1016/j.psep.2019.11.006

Noohpisheh, Z., Amiri, H., Farhadi, S., & Mohammadi-gholami, A. (2020). Green synthesis of Ag-ZnO nanocomposites using Trigonella foenum-graecum leaf extract and their antibacterial, antifungal, antioxidant and photocatalytic properties. *Spectrochimica Acta. Part A: Molecular and Biomolecular Spectroscopy*, *240*, 118595. doi:10.1016/j.saa.2020.118595 PMID:32599480

Noqta, O. A., Sodipo, B. K., & Aziz, A. A. (2020). One-pot synthesis of highly magnetic and stable citrate coated superparamagnetic iron oxide nanoparticles by modified coprecipitation method. *Functional Composites and Structures*, *2*(4), 045005. Advance online publication. doi:10.1088/2631-6331/abcdbc

Noronha, V. T., Paula, A. J., Duran, G., Galembeck, A., Cogo-Mueller, K., Franz-Montan, M., & Duran, N. (2017). Silver nanoparticles in dentistry. *Dental Materials*, *33*(10), 1110–1126. doi:10.1016/j.dental.2017.07.002 PMID:28779891

Novoselov, K. S., Geim, A. K., Morozov, S. V., Jiang, D., Zhang, Y., Dubonos, S. V., Grigorieva, I. V., & Firsov, A. A. (2004). Electric field effect in atomically thin carbon films. *Science*, *306*(5696), 666–669. doi:10.1126cience.1102896 PMID:15499015

Nowack, B., Ranville, J. F., Diamond, S., Gallego-Urrea, A., Metcalfe, C., Rose, J., Horne, N., Koelmans, A. A., & Klaine, S. J. (2012). Potential scenarios for nanomaterial release and subsequent alteration in the environment. *Environmental Toxicology and Chemistry*, *31*(1), 50–59. doi:10.1002/etc.726 PMID:22038832

Nunes, T., Pons, T., Hou, X., Van Do, K., Caron, B., Rigal, M., Di Benedetto, M., Palpant, B., Leboeuf, C., Janin, A., & Bousquet, G. (2019). Pulsed-laser irradiation of multifunctional gold nanoshells to overcome trastuzumab resistance in HER2-overexpressing breast cancer. *Journal of Experimental & Clinical Cancer Research*, *38*(1), 306. doi:10.118613046-019-1305-x PMID:31299997

Nuryantini, A. Y., Sundari, C. D. D., Halimahtussa'diah, H., & Nuryadin, B. W. (2019). Synthesis and Characterization of Calcium Oxide Nanoparticles from Duck Eggshells using Ball Milling Methods. *Jurnal Kimia Valensi*, *5*(2), 231–135. doi:10.15408/jkv.v5i2.8779

Nyangiwe, N. N., & Ouma, C. N. M. (2019). Modelling the adsorption of natural organic matter on Ag (111) surface: Insights from dispersion corrected density functional theory calculations. *Journal of Molecular Graphics & Modelling*, *92*, 313–319. doi:10.1016/j.jmgm.2019.08.013 PMID:31442937

O'Neal, D. P., Hirsch, L. R., Halas, N. J., Payne, J. D., & West, J. L. (2004). Photothermal tumor ablation in mice using near infrared-absorbing nanoparticles. *Cancer Letters*, *209*(2), 171–176. doi:10.1016/j.canlet.2004.02.004 PMID:15159019

O'Neill, J. (2014, December). Antimicrobial Resistance: Tackling a crisis for the health and wealth of nations. http://www.jpiamr.eu/wp-content/uploads/2014/12/AMR-Review-Paper-Tackling-a-crisis-for-the-health-and-wealth-of-nations_1-2.pdf

Odiongenyi, A. O., & Afangide, U. N. (2020). Adsorption and Thermodynamic Studies on the Removal of Congo Red Dye from Aqueous Solution by Alumina and Nano-alumina. *Communication in Physical Sciences*, *4*(1), 1–7.

Odoemelam, S. A., & Eddy, N. O. (2009). Studies on the Use of Oyster, Snail and Periwinkle Shells as Adsorbents for the Removalof Pb 2+ Aqueous Solution. *E-Journal of Chemistry*, *6*(1), 213–222. https://doi.org/10.1155/2009/407549

OECD. (2010). List of Manufactured Nanomaterials and List of Endpoints for Phase One of the Sponsorship Program for the Testing of Manufactured Nanomaterials. Organization for Economic Co-operation and Development.

Oh, J.-W., Chun, S. C., & Chandrasekaran, M. (2019). Preparation and in vitro characterization of chitosan nanoparticles and their broad-spectrum antifungal action compared to antibacterial activities against phytopathogens of tomato. *Agronomy (Basel)*, *9*(1), 21. doi:10.3390/agronomy9010021

Okash Toamah, W., & Kadhim Fadhil, A. (2019). Preparation of nanoparticles from CaO and use it for removal of chromium (II), and mercury (II) from aqueous solutions. *Journal of Physics: Conference Series*, *1234*(1), 012086. https://doi.org/10.1088/1742-6596/1234/1/012086

Okwunodulu, F. U., & Eddy, N. O. (2014). Equilibrium and Thermodynamic Consideration of Cd2+, Ni2+ and Pb2+ Removal From Aqueous Solution Onto Treated and Untreated Cola Nitida Waste Biomass. *International Journal of Scientific Research (Ahmedabad, India)*, *3*(3), 567–570.

Oladoja, N. A., Ololade, I. A., Olaseni, S., Olatujoye, V. O., Jegede, O. S., & Agunloye, A. O. (2012). Synthesis of Nano Calcium Oxide from a Gastropod Shell and the Performance Evaluation for Cr (VI) Removal from Aqua System. *Industrial & Engineering Chemistry Research*, *51*(2), 639–648. https://doi.org/10.1021/ie201189z

Ola, O., & Maroto-Valer, M. M. (2015). Review of material design and reactor engineering on TiO2 photocatalysis for CO2 reduction. *Journal of Photochemistry and Photobiology C, Photochemistry Reviews*, *24*, 16–42. doi:10.1016/j.jphotochemrev.2015.06.001

Olishevska, S., Nickzad, A., & Déziel, E. (2019). Bacillus and Paenibacillus secreted polyketides and peptides involved in controlling human and plant pathogens. *Applied Microbiology and Biotechnology*, *103*(3), 1189–1215. doi:10.100700253-018-9541-0 PMID:30603850

Oliveira, M. S., Aryasomayajula, B., Pattni, B., Mussi, S. V., Ferreira, L. A. M., & Torchilin, V. P. (2016). Solid lipid nanoparticles co-loaded with doxorubicin and α-tocopherol succinate are effective against drug-resistant cancer cells in monolayer and 3-D spheroid cancer cell models. *International Journal of Pharmaceutics*, *512*(1), 292–300. doi:10.1016/j.ijpharm.2016.08.049 PMID:27568499

Ong, C., Yung, L. Y. L., Cai, Y., Bay, B. H., & Baeg, G. H. (2015). Drosophila melanogaster as a model organism to study nanotoxicity. *Nanotoxicology*, *9*(3), 396–403. doi:10.3109/17435390.2014.940405 PMID:25051331

Oprčkal, P., Mladenovič, A., Vidmar, J., Mauko Pranjić, A., Milačič, R., & Ščančar, J. (2017). Critical evaluation of the use of different nanoscale zero-valent iron particles for the treatment of effluent water from a small biological wastewater treatment plant. *Chemical Engineering Journal*, *321*, 20–30. https://doi.org/10.1016/j.cej.2017.03.104

Orduz, R., Solórzano, J., Egido, M. Á., & Román, E. (2013). Analytical study and evaluation results of power optimizers for distributed power conditioning in photovoltaic arrays. *Progress in Photovoltaics: Research and Applications*, *21*(3), 359–373. doi:10.1002/pip.1188

Orooji, Y., Mortazavi-Derazkola, S., Ghoreishi, S. M., Amiri, M., & Salavati-Niasari, M. (2020). Mesopourous Fe_3O_4@ SiO_2-hydroxyapatite nanocomposite: Green sonochemical synthesis using strawberry fruit extract as a capping agent, characterization and their application in sulfasalazine delivery and cytotoxicity. *Journal of Hazardous Materials*, *400*(April), 123140. doi:10.1016/j.jhazmat.2020.123140 PMID:32563904

Osin, O. A., Yu, T., Cai, X., Jiang, Y., Peng, G., Cheng, X., Li, R., Qin, Y., & Lin, S. (2018). Photocatalytic Degradation of 4-Nitrophenol by C, N-TiO2: Degradation Efficiency vs. Embryonic Toxicity of the Resulting Compounds. *Frontiers in Chemistry*, *6*, 192. Advance online publication. doi:10.3389/fchem.2018.00192 PMID:29915782

Osiptsov, A. A. (2017). Fluid Mechanics of Hydraulic Fracturing: A Review. *Journal of Petroleum Science Engineering*, *156*, 513–535. Advance online publication. doi:10.1016/j.petrol.2017.05.019

Osuntokun, J., Onwudiwe, D. C., & Ebenso, E. E. (2018). Aqueous extract of broccoli mediated synthesis of CaO nanoparticles and its application in the photocatalytic degradation of bromocrescol green. *IET Nanobiotechnology*, *12*(7), 888–894. doi:10.1049/iet-nbt.2017.0277

Othman, N. H., Che Ismail, M., Mustapha, M., Sallih, N., Kee, K. E., & Ahmad Jaal, R. (2019a). Graphene-based polymer nanocomposites as barrier coatings for corrosion protection. *Progress in Organic Coatings*, *135*, 82–99. doi:10.1016/j.porgcoat.2019.05.030

Otsuka, H., Nagasaki, Y., & Kataoka, K. (2012). PEGylated nanoparticles for biological and pharmaceutical applications. *Advanced Drug Delivery Reviews* , 64. doi:10.1002/9781118135440.ch47

Oufella, L. S., & Benchettara, A. (2016). Influence of fluoride on the electrochemical behvior of a new synthesized Ti-10Ta-2Mo alloy for biomedical applications. *Journal of Fundamental and Applied Sciences*, *8*(3), 731–752. doi:10.4314/jfas.v8i3.5

Ovais, M., Khalil, A. T., Raza, A., Islam, N. U., Ayaz, M., Saravanan, M., Ali, M., Ahmad, I., Shahid, M., & Shinwari, Z. K. (2018). Multifunctional theranostic applications of biocompatible green-synthesized colloidal nanoparticles. *Applied Microbiology and Biotechnology*, *102*(10), 4393–4408. doi:10.100700253-018-8928-2 PMID:29594356

Paget, V., Sergent, J., Grall, R., Altmeyer-Morel, S., Girard, H., Petit, T., Gesset, C., Mermoux, M., Bergonzo, P., & Arnault, J.-C. (2014). *Carboxylated nanodiamonds are neither cytotoxic nor genotoxic on liver, kidney, intestine and lung human cell lines*. Academic Press.

Pajoumshariati, S., Shirali, H., Yavari, S. K., Sheikholeslami, S. N., Lotfi, G., Mashhadi Abbas, F., & Abbaspourrad, A. (2018). GBR membrane of novel poly (butylene succinate-co-glycolate) co-polyester co-polymer for periodontal application. *Scientific Reports*, *8*(1), 1–16. doi:10.103841598-018-25952-1 PMID:29760507

Palestino, G., García-Silva, I., González-Ortega, O., & Rosales-Mendoza, S. (2020). Can Nanotechnology Help in the Fight against COVID-19? *Expert Review of Anti-Infective Therapy*, *18*(9), 849–864. doi:10.1080/14787210.2020.1776115 PMID:32574081

Palimi, M. J., Rostami, M., Mahdavian, M., & Ramezanzadeh, B. (2014). Surface modification of Fe 2 O 3 nanoparticles with 3-aminopropyltrimethoxysilane (APTMS): An attempt to investigate surface treatment on surface chemistry and mechanical properties of polyurethane/Fe 2 O 3 nanocomposites. *Applied Surface Science*, *320*, 60–72. doi:10.1016/j.apsusc.2014.09.026

Palmieri, V., & Papi, M. (2020). Can Graphene Take Part in the Fight Against COVID-19? *Nano Today*, *33*, 100883. doi:10.1016/j.nantod.2020.100883 PMID:32382315

Palza, H., Quijada, R., & Delgado, K. (2015). Antimicrobial polymer composites with copper micro-and nanoparticles: Effect of particle size and polymer matrix. *Journal of Bioactive and Compatible Polymers*, *30*(4), 366–380. doi:10.1177/0883911515578870

Pan, C., Xu, S., Zhou, H., Fu, Y., Ye, M., & Zou, H. (2007). Recent developments in methods and technology for analysis of biological samples by MALDI-TOF-MS. *Analytical and Bioanalytical Chemistry*, *387*(1), 193–204. doi:10.100700216-006-0905-4 PMID:17086385

Panchal, H., Patel, H., Patel, J., & Shah, M. (2021). A systematic review on nanotechnology in enhanced oil recovery. Petroleum Research. doi:10.1016/j.ptlrs.2021.03.003

Panchal, P., Paul, D. R., Sharma, A., Hooda, D., Yadav, R., Meena, P., & Nehra, S. P. (2019). Phytoextract mediated ZnO/MgO nanocomposites for photocatalytic and antibacterial activities. *Journal of Photochemistry and Photobiology A Chemistry*, *385*(June), 112049. doi:10.1016/j.jphotochem.2019.112049

Pandey, R., Sharma, S., & Khuller, G. K. (2005). Oral solid lipid nanoparticle-based antitubercular chemotherapy. *Tuberculosis (Edinburgh, Scotland)*, *85*(5–6), 415–420. doi:10.1016/j.tube.2005.08.009 PMID:16256437

Pan, Y., Leifert, A., Ruau, D., Neuss, S., Bornemann, J., Schmid, G., Brandau, W., Simon, U., & Jahnen-Dechent, W. (2009). Gold nanoparticles of diameter 1.4 nm trigger necrosis by oxidative stress and mitochondrial damage. *Small*, *5*(18), 2067–2076. doi:10.1002mll.200900466 PMID:19642089

Paosen, S., Saising, J., Septama, A., & Voravuthikunchai, S. (2017). Green synthesis of silver nanoparticles using plants from Myrtaceae family and characterization of their antibacterial activity. *Materials Letters*, *209*, 201–206. Advance online publication. doi:10.1016/j.matlet.2017.07.102

Pareek, V., Bhargava, A., Gupta, R., Jain, N., & Panwar, J. (2017). Synthesis and Applications of Noble Metal Nanoparticles: A Review. *Advanced Science, Engineering and Medicine*, *9*(7), 527–544. doi:10.1166/asem.2017.2027

Parida, B., Iniyan, S., & Goic, R. (2011). A review of solar photovoltaic technologies. *Renewable & Sustainable Energy Reviews*, *15*(3), 1625–1636. doi:10.1016/j.rser.2010.11.032

Park, Y., Hong, Y. N., Weyers, A., Kim, Y. S., & Linhardt, R. J. (2011). Polysaccharides and phytochemicals: A natural reservoir for the green synthesis of gold and silver nanoparticles. *IET Nanobiotechnology / IET*, *5*(3), 69. doi:10.1049/iet-nbt.2010.0033 PMID:21913788

Parodi, A., Quattrocchi, N., Van De Ven, A. L., Chiappini, C., Evangelopoulos, M., Martinez, J. O., Brown, B. S., Khaled, S. Z., Yazdi, I. K., Enzo, M. V., Isenhart, L., Ferrari, M., & Tasciotti, E. (2013). Synthetic nanoparticles functionalized with biomimetic leukocyte membranes possess cell-like functions. *Nature Nanotechnology*, *8*(1), 61–68. doi:10.1038/nnano.2012.212 PMID:23241654

Parungo, C. P., Ohnishi, S., De Grand, A. M., Laurence, R. G., Soltesz, E. G., Colson, Y. L., Kang, P. M., Mihaljevic, T., Cohn, L. H., & Frangioni, J. V. (2004). In vivo optical imaging of pleural space drainage to lymph nodes of prognostic significance. *Annals of Surgical Oncology*, *11*(12), 1085–1092. doi:10.1245/ASO.2004.03.054 PMID:15545502

Parvanian, S., Mostafavi, S. M., & Aghashiri, M. (2017). Multifunctional nanoparticle developments in cancer diagnosis and treatment. *Sensing and Bio-Sensing Research*, *13*, 81–87. doi:10.1016/j.sbsr.2016.08.002

Patel, G., Pal, U., & Menon, S. (2009). Removal of Fluoride from Aqueous Solution by CaO Nanoparticles. *Separation Science and Technology*, *44*(12), 2806–2826. https://doi.org/10.1080/01496390903014425

Pathak, S., Saha, G. C., Hadi, M. B. A., & Jain, N. K. (2021). Engineered Nanomaterials for Aviation Industry in COVID-19 Context: A Time-Sensitive Review. *Coatings*, *11*(4), 382. doi:10.3390/coatings11040382

Patil, A. S., Patil, M. D., Lohar, G. M., Jadhav, S. T., & Fulari, V. J. (2017). Supercapacitive properties of CuO thin films using modified SILAR method. *Ionics*, *23*(5), 1259–1266. Advance online publication. doi:10.100711581-016-1921-9

Patil, M. D., Dhas, S. D., Mane, A. A., & Moholkar, A. V. (2021). Clinker-like V2O5 nanostructures anchored on 3D Ni-foam for supercapacitor application. *Materials Science in Semiconductor Processing*, *133*, 105978. doi:10.1016/j.mssp.2021.105978

Patil, S. S., Shedbalkar, U. U., Truskewycz, A., Chopade, B. A., & Ball, A. S. (2016). Nanoparticles for environmental clean-up: A review of potential risks and emerging solutions. *Environmental Technology & Innovation*, *5*, 10–21. doi:10.1016/j.eti.2015.11.001

Patil, S., Sivaraj, R., Rajiv, P., Venckatesh, R., & Seenivasan, R. (2015). Green synthesis of silver nanoparticle from leaf extract of Aegle marmelos and evaluation of its antibacterial activity. *International Journal of Pharmacy and Pharmaceutical Sciences*, *7*(6), 169–173.

Patra, C. R., Cao, S., Safgren, S., Bhattacharya, R., Ames, M. M., Shah, V., Reid, J. M., & Mukherjee, P. (2008). Intracellular fate of a targeted delivery system. *Journal of Biomedical Nanotechnology*, *4*(4), 508–514. doi:10.1166/jbn.2008.016

Pegoraro, C., Cecchin, D., Gracia, L. S., Warren, N., Madsen, J., Armes, S. P., Lewis, A., MacNeil, S., & Battaglia, G. (2013). Enhanced drug delivery to melanoma cells using PMPC-PDPA polymersomes. *Cancer Letters*, *334*(2), 328–337. doi:10.1016/j.canlet.2013.02.007 PMID:23402813

Pei, S., Zhao, J., Du, J., Ren, W., & Cheng, H. M. (2010). Direct reduction of graphene oxide films into highly conductive and flexible graphene films by hydrohalic acids. *Carbon*, *48*(15), 4466–4474. Advance online publication. doi:10.1016/j.carbon.2010.08.006

Pelgrift, R. Y., & Friedman, A. J. (2013). Nanotechnology as a therapeutic tool to combat microbial resistance. *Advanced Drug Delivery Reviews*, *65*(13–14), 1803–1815. doi:10.1016/j.addr.2013.07.011 PMID:23892192

Perassi, E. M., Hrelescu, C., Wisnet, A., Döblinger, M., Scheu, C., Jäckel, F., Coronado, E. A., & Feldmann, J. (2014). Quantitative understanding of the optical properties of a single, complex-shaped gold nanoparticle from experiment and theory. *ACS Nano*, *8*(5), 4395–4402. doi:10.1021/nn406270z PMID:24787120

Périgo, E. A., Hemery, G., Sandre, O., Ortega, D., Garaio, E., Plazaola, F., & Teran, F. J. (2015). Fundamentals and advances in magnetic hyperthermia. *Applied Physics Reviews*, *2*(4), 041302. doi:10.1063/1.4935688

Perrault, S. D., & Chan, W. C. W. (2010). In vivo assembly of nanoparticle components to improve targeted cancer imaging. *Proceedings of the National Academy of Sciences of the United States of America*, *107*(25), 11194–11199. doi:10.1073/pnas.1001367107 PMID:20534561

Pham, T. D., & Sillanpää, M. (2020). Ultrasonic and electrokinetic remediation of low permeability soil contaminated with persistent organic pollutants. In *Advanced Water Treatment: Electrochemical Methods*. Elsevier Inc. doi:10.1016/B978-0-12-819227-6.00004-8

Philip, D. (2009). Biosynthesis of Au, Ag and Au-Ag nanoparticles using edible mushroom extract. *Spectrochimica Acta. Part A: Molecular and Biomolecular Spectroscopy*, *73*(2), 374–381. doi:10.1016/j.saa.2009.02.037 PMID:19324587

Pinto, M., Ramalho, P. S. F., Moreira, N. F. F., Gonçalves, A. G., Nunes, O. C., Pereira, M. F. R., & Soares, O. S. G. P. (2020). Application of magnetic nanoparticles for water purification. *Environmental Advances*, *2*, 100010. doi:10.1016/j.envadv.2020.100010

Pirsaheb, M., Shahmoradi, B., Beikmohammadi, M., Azizi, E., Hossini, H., & Md Ashraf, G. (2017). Photocatalytic degradation of Aniline from aqueous solutions under sunlight illumination using immobilized Cr:ZnO nanoparticles. *Scientific Reports*, *7*(1), 1473. doi:10.103841598-017-01461-5 PMID:28469186

Plessis, D., & Marguerite, D. (2011). Fabrication and characterization of anti-microbial and biofouling resistant nanofibers with silver nanoparticles and immobilized enzymes for application in water filtration. *Materials Science*.

Podesta, J. E., Al-Jamal, K. T., Herrero, M. A., Tian, B., Ali-Boucetta, H., Hegde, V., Bianco, A., Prato, M., & Kostarelos, K. (2009). *Antitumor activity and prolonged survival by carbon-nanotube-mediated therapeutic siRNA silencing in a human lung xenograft model. Small.* doi:10.1002mll.200801572

Poggi, G., Toccafondi, N., Chelazzi, D., Canton, P., Giorgi, R., & Baglioni, P. (2016). Calcium hydroxide nanoparticles from solvothermal reaction for the deacidification of degraded waterlogged wood. *Journal of Colloid and Interface Science*, *473*, 1–8. https://doi.org/10.1016/j.jcis.2016.03.038

Poggio, C., Colombo, M., Arciola, C. R., Greggi, T., Scribante, A., & Dagna, A. (2020). Copper-alloy surfaces and cleaning regimens against the spread of SARS-CoV-2 in dentistry and orthopedics. From fomites to anti-infective nanocoatings. *Materials (Basel)*, *13*(15), 1–12. doi:10.3390/ma13153244 PMID:32707757

Pomastowski, P., Kr, A., & Railean-plugaru, V. (2020). *Zinc Oxide Nanocomposites — Extracellular Synthesis*. Academic Press.

Porcel, E., Liehn, S., Remita, H., Usami, N., Kobayashi, K., Furusawa, Y., Le Sech, C., & Lacombe, S. (2010). Platinum nanoparticles: A promising material for future cancer therapy? *Nanotechnology*, *21*(8), 85103. doi:10.1088/0957-4484/21/8/085103 PMID:20101074

Powers, H. J. (2003). Riboflavin (vitamin B-2) and health. *The American Journal of Clinical Nutrition*, *77*(6), 1352–1360. doi:10.1093/ajcn/77.6.1352 PMID:12791609

Pozzi, E. A., Sonntag, M. D., Jiang, N., Klingsporn, J. M., Hersam, M. C., & Van Duyne, R. P. (2013). Tip-Enhanced Raman Imaging : An Emergent Tool for Probing Biology at the Nanoscale. *ACS Nano*, *7*(2), 885–888. doi:10.1021/nn400560t PMID:23441673

Prabhu, R. H., Patravale, V. B., & Joshi, M. D. (2015). Polymeric nanoparticles for targeted treatment in oncology: Current insights. *International Journal of Nanomedicine*, *10*, 1001–1018. doi:10.2147/IJN.S56932 PMID:25678788

Prabhu, S., & Poulose, E. K. (2012). Silver nanoparticles: Mechanism of antimicrobial action, synthesis, medical applications, and toxicity effects. *International Nano Letters*, *2*(1), 1–10. doi:10.1186/2228-5326-2-32

Prajapati, A. K., & Mondal, M. K. (2021). Novel green strategy for CuO–ZnO–C nanocomposites fabrication using marigold (Tagetes spp.) flower petals extract with and without CTAB treatment for adsorption of Cr(VI) and Congo red dye. *Journal of Environmental Management*, *290*(November), 112615. doi:10.1016/j.jenvman.2021.112615

Prajapati, P. K., Yadav, P., & Gauri, A. (2020). Possible Potential of Tamra Bhasma (Calcined Copper) in COVID-19 Management. *Journal of Research in Ayurvedic Sciences*, *4*(3), 113–120. doi:10.5005/jras-10064-0111

Prasad, Ch., Gangadhara, S., & Venkateswarlu, P. (2016). Bio-inspired green synthesis of Fe3O4 magnetic nanoparticles using watermelon rinds and their catalytic activity. *Applied Nanoscience*, *6*(6), 797–802. doi:10.100713204-015-0485-8

Prasai, D., Tuberquia, J. C., Harl, R. R., Jennings, G. K., & Bolotin, K. I. (2012). Graphene: Corrosion-inhibiting coating. *ACS Nano*, *6*(2), 1102–1108. Advance online publication. doi:10.1021/nn203507y PMID:22299572

Prasannaraj, G., Sahi, S. V., Ravikumar, S., & Venkatachalam, P. (2016). Enhanced cytotoxicity of biomolecules loaded metallic silver nanoparticles against human liver (HepG2) and prostate (PC3) cancer cell lines. *Journal of Nanoscience and Nanotechnology*, *16*(5), 4948–4959. doi:10.1166/jnn.2016.12336 PMID:27483851

Prasek, J., Drbohlavova, J., Chomoucka, J., Hubalek, J., Jasek, O., Adam, V., & Kizek, R. (2011). Methods for carbon nanotubes synthesis - Review. Journal of Materials Chemistry, 21(40). doi:10.1039/c1jm12254a

Prathna, T. C., Chandrasekaran, N., Raichur, A. M., & Mukherjee, A. (2011). Kinetic evolution studies of silver nanoparticles in a bio-based green synthesis process. *Colloids and Surfaces. A, Physicochemical and Engineering Aspects*, *377*(1–3), 212–216. doi:10.1016/j.colsurfa.2010.12.047

Prolongo, S. G., Moriche, R., Jiménez-Suárez, A., Sánchez, M., & Ureña, A. (2014). Advantages and disadvantages of the addition of graphene nanoplatelets to epoxy resins. *European Polymer Journal*, *61*, 206–214. doi:10.1016/j.eurpolymj.2014.09.022

Pudakalakatti, S., Enriquez, J. S., McCowan, C., Ramezani, S., Davis, J. S., Zacharias, N. M., Bourgeois, D., Constantinou, P. E., Harrington, D. A., Carson, D., Farach-Carson, M. C., & Bhattacharya, P. K. (2021). Hyperpolarized MRI with silicon micro and nanoparticles: Principles and applications. In *Wiley Interdisciplinary Reviews*. Nanomedicine and Nanobiotechnology. doi:10.1002/wnan.1722

Putro, D. Y., Saputra, R., & Albab, M. F. (2013). *Silver Nano Perfume Ejector to Destroy Bacteria for Clothes*. ASEAN/Asian Academic Society International Conference Proceeding Series.

Qasim, S., Zafar, A., Saif, M. S., Ali, Z., Nazar, M., Waqas, M., Haq, A. U., Tariq, T., Hassan, S. G., Iqbal, F., Shu, X. G., & Hasan, M. (2020). Green synthesis of iron oxide nanorods using Withania coagulans extract improved photocatalytic degradation and antimicrobial activity. *Journal of Photochemistry and Photobiology. B, Biology*, *204*, 111784. doi:10.1016/j.jphotobiol.2020.111784 PMID:31954266

Qian, L., Liu, S., Zhang, W., Chen, Y., Ouyang, D., Han, L., Yan, J., & Chen, M. (2019). Enhanced reduction and adsorption of hexavalent chromium by palladium and silicon rich biochar supported nanoscale zero-valent iron. *Journal of Colloid and Interface Science*, *533*, 428–436. doi:10.1016/j.jcis.2018.08.075 PMID:30172153

Qian, Y., Qin, C., Chen, M., & Lin, S. (2020). Nanotechnology in soil remediation – applications vs. implications. *Ecotoxicology and Environmental Safety*, *201*, 110815. doi:10.1016/j.ecoenv.2020.110815 PMID:32559688

Qureshi, O. S., Kim, H.-S., Zeb, A., Choi, J.-S., Kim, H.-S., Kwon, J.-E., Kim, M.-S., Kang, J.-H., Ryou, C., Park, J.-S., & Kim, J.-K. (2017). Sustained release docetaxel-incorporated lipid nanoparticles with improved pharmacokinetics for oral and parenteral administration. *Journal of Microencapsulation*, *34*(3), 250–261. doi:10.1080/02652048.2017.1337247 PMID:28557649

Qu, Y., Chu, B. Y., Peng, J. R., Liao, J. F., Qi, T. T., Shi, K., Zhang, X. N., Wei, Y. Q., & Qian, Z. Y. (2015). A biodegradable thermo-responsive hybrid hydrogel: Therapeutic applications in preventing the post-operative recurrence of breast cancer. *NPG Asia Materials*, *7*(8), e207–e207. doi:10.1038/am.2015.83

Radenkovic, D., Kobayashi, H., Remsey-Semmelweis, E., & Seifalian, A. M. (2016). Quantum dot nanoparticle for optimization of breast cancer diagnostics and therapy in a clinical setting. *Nanomedicine; Nanotechnology, Biology, and Medicine*, *12*(6), 1581–1592. doi:10.1016/j.nano.2016.02.014 PMID:27013132

Radu, D. R., Lai, C.-Y., Jeftinija, K., Rowe, E. W., Jeftinija, S., & Lin, V. S.-Y. (2004). A polyamidoamine dendrimer-capped mesoporous silica nanosphere-based gene transfection reagent. *Journal of the American Chemical Society*, *126*(41), 13216–13217. doi:10.1021/ja046275m PMID:15479063

Rafati Rahimzadeh, M., Rafati Rahimzadeh, M., Kazemi, S., & Moghadamnia, A.-A. (2017). Cadmium toxicity and treatment: An update. *Caspian Journal of Internal Medicine*, *8*(3), 135–145. doi:10.22088/cjim.8.3.135 PMID:28932363

Rafieepour, S., Zamiran, S., & Ostadhassan, M. (2020). A cost-effective chemo-thermo-poroelastic wellbore stability model for mud weight design during drilling through shale formations. *Journal of Rock Mechanics and Geotechnical Engineering*, *12*(4), 768–779. Advance online publication. doi:10.1016/j.jrmge.2019.12.008

Rafique, M., Sadaf, I., Rafique, M. S., & Tahir, M. B. (2017). A review on green synthesis of silver nanoparticles and their applications. *Artificial Cells, Nanomedicine, and Biotechnology*, *45*(7), 1272–1291. doi:10.1080/21691401.2016.1241792 PMID:27825269

Raghunandan, D., Basavaraja, S., Mahesh, B., Balaji, S., Manjunath, S. Y., & Venkataraman, A. (2009). Biosynthesis of stable polyshaped gold nanoparticles from microwave-exposed aqueous extracellular anti-malignant guava (Psidium guajava) leaf extract. *NanoBiotechnology*, *5*(1–4), 34–41. doi:10.100712030-009-9030-8

Raghuwanshi, N., Srivastava, A. K., Yadav, T. C., Gupta, S., Khatri, K., Pruthi, V., & Pruthi, V. (2019). Biogenic nanoparticles as theranostic agents: prospects and challenges. *Integrating Green Chemistry and Sustainable Engineering*, 647-684.

Raha, S., Mallick, R., Basak, S., & Duttaroy, A. K. (2020). Is copper beneficial for COVID-19 patients? *Medical Hypotheses*, *142*(April), 109814. doi:10.1016/j.mehy.2020.109814 PMID:32388476

Rahiminejad, A., Dinarvand, R., Johari, B., Nodooshan, S. J., Rashti, A., Rismani, E., Mahdaviani, P., Saltanatpour, Z., Rahiminejad, S., Raigani, M., & Khosravani, M. (2019). Preparation and investigation of indirubin-loaded SLN nanoparticles and their anti-cancer effects on human glioblastoma U87MG cells. *Cell Biology International*, *43*(1), 2–11. doi:10.1002/cbin.11037 PMID:30080277

Rahman, A., Kang, S., Wang, W., Garg, A., Maile-Moskowitz, A., & Vikesland, P. J. (2021). Nanobiotechnology enabled approaches for wastewater based epidemiology. *Trends in Analytical Chemistry*, *143*, 116400. doi:10.1016/j.trac.2021.116400 PMID:34334850

Rahmani, A. R., Bryant, S. L., Huh, C., Ahmadian, M., Zhang, W., & Liu, Q. H. (2015). Characterizing reservoir heterogeneities using magnetic nanoparticles. *Society of Petroleum Engineers - SPE Reservoir Simulation Symposium 2015, 1*. 10.2118/173195-MS

Rahmani, A. R., Bryant, S., Huh, C., Athey, A., Ahmadian, M., Chen, J., & Wilt, M. (2015). Crosswell magnetic sensing of superparamagnetic nanoparticles for subsurface applications. *SPE Journal, 20*(5), 1067–1082. Advance online publication. doi:10.2118/166140-PA

Rahman, Z., & Singh, V. P. (2019). The relative impact of toxic heavy metals (THMs) (arsenic (As), cadmium (Cd), chromium (Cr)(VI), mercury (Hg), and lead (Pb)) on the total environment: An overview. *Environmental Monitoring and Assessment, 191*(7), 419. doi:10.100710661-019-7528-7 PMID:31177337

Rai, A., Singh, A., Ahmad, A., & Sastry, M. (2006). Role of halide ions and temperature on the morphology of biologically synthesized gold nanotriangles. *Langmuir, 22*(2), 736–741. doi:10.1021/la052055q PMID:16401125

Rai, M., Bonde, S., Yadav, A., Plekhanova, Y., Reshetilov, A., Gupta, I., Golińska, P., Pandit, R., & Ingle, A. P. (2020). Nanotechnology-based promising Strategies for the Management of COVID-19: Current Development and Constraints. *Expert Review of Anti-Infective Therapy*, 1–10. Advance online publication. doi:10.1080/14787210.2021.1836961 PMID:33164589

Rai, M., & Ingle, A. (2012). Role of nanotechnology in agriculture with special reference to management of insect pests. *Applied Microbiology and Biotechnology, 94*(2), 287–293. doi:10.100700253-012-3969-4 PMID:22388570

Rai, P. K., & Singh, S. P. (2016). Integrated dark- and photo-fermentation: Recent advances and provisions for improvement. *International Journal of Hydrogen Energy, 41*(44), 19957–19971. doi:10.1016/j.ijhydene.2016.08.084

Rai, P. K., Usmani, Z., Thakur, V. K., Gupta, V. K., & Mishra, Y. K. (2020). Tackling COVID-19 Pandemic through Nanocoatings: Confront and Exactitude. *Current Research in Green and Sustainable Chemistry, 3*, 100011. doi:10.1016/j.crgsc.2020.100011

Rajasekhar, C., & Kanchi, S. (2018). Green Nanomaterials for Clean Environment. In L. Martínez, O. Kharissova, & B. Kharisov (Eds.), *Handbook of Ecomaterials*. Springer. doi:10.1007/978-3-319-48281-1_73-1

Rajasekharreddy, P., Huang, C., Busi, S., Rajkumari, J., Tai, M. H., & Liu, G. (2019). Green Synthesized Nanomaterials as Theranostic Platforms for Cancer Treatment: Principles, Challenges and the Road Ahead. *Current Medicinal Chemistry, 26*(8), 1311–1327. doi:10.2174/0929867324666170309124327 PMID:28294042

Rajendran, A., Alsawalha, M., & Alomayri, T. (2021). Biogenic synthesis of husked rice-shaped iron oxide nanoparticles using coconut pulp (Cocos nucifera L.) extract for photocatalytic degradation of Rhodamine B dye and their in vitro antibacterial and anticancer activity. *Journal of Saudi Chemical Society, 25*(9), 101307. Advance online publication. doi:10.1016/j.jscs.2021.101307

Rajeshkumar, S., Bharath, L., & Geetha, R. (2019). Broad spectrum antibacterial silver nanoparticle green synthesis: Characterization, and mechanism of action. In *Green synthesis, characterization and applications of nanoparticles* (pp. 429–444). Elsevier.

Rajkumar, S., Murugesh, S., Sivasankar, V., Darchen, A., Msagati, T. A. M., & Chaabane, T. (2019). Low-cost fluoride adsorbents prepared from a renewable biowaste: Syntheses, characterization and modeling studies. *Arabian Journal of Chemistry, 12*(8), 3004–3017. https://doi.org/10.1016/j.arabjc.2015.06.028

Rajpoot, K., & Jain, S. K. (2018). Colorectal cancer-targeted delivery of oxaliplatin via folic acid-grafted solid lipid nanoparticles: Preparation, optimization, and in vitro evaluation. *Artificial Cells, Nanomedicine, and Biotechnology, 46*(6), 1236–1247. doi:10.1080/21691401.2017.1366338 PMID:28849671

Rajput, N. S., Al Zadjali, S., Gutierrez, M., Esawi, A. M. K., & Al Teneiji, M. (2021). Synthesis of holey graphene for advanced nanotechnological applications. *RSC Advances*, *11*(44), 27381–27405. doi:10.1039/D1RA05157A

Raj, S., Khurana, S., Choudhari, R., Kesari, K. K., Kamal, M. A., Garg, N., Ruokolainen, J., Das, B. C., & Kumar, D. (2021). Specific targeting cancer cells with nanoparticles and drug delivery in cancer therapy. *Seminars in Cancer Biology*, *69*, 166–177. doi:10.1016/j.semcancer.2019.11.002 PMID:31715247

Rambabu, K., Bharath, G., Banat, F., Hai, A., Show, P. L., & Phong Nguyen, T. H. (2021). Ferric oxide/date seed activated carbon nanocomposites mediated dark fermentation of date fruit wastes for enriched biohydrogen production. *International Journal of Hydrogen Energy*, *46*(31), 16631–16643. doi:10.1016/j.ijhydene.2020.06.108

Ramola, B., & Joshi, N. C. (2019). Green Synthesis, Characterisations and Antimicrobial Activities of CaO Nanoparticles. *Oriental Journal of Chemistry*, *35*(3), 1154–1157. https://doi.org/10.13005/ojc/350333

Ramteke, K., Joshi, S., & Dhole, S. (2012). Solid lipid nanoparticle: A review. *IOSR Journal of Pharmacy*, *2*(6), 34–44. doi:10.9790/3013-26103444

Rangayasami, A., Kannan, K., Murugesan, S., Radhika, D., Sadasivuni, K. K., Reddy, K. R., & Raghu, A. V. (2021). Influence of Nanotechnology to Combat against COVID-19 for Global Health Emergency: A Review. *Sensors International*, *2*, 100079. doi:10.1016/j.sintl.2020.100079 PMID:34766049

Rangel, N. L., Sotelo, J. C., & Seminario, J. M. (2009). Mechanism of carbon nanotubes unzipping into graphene ribbons. *The Journal of Chemical Physics*, *131*(3), 031105. doi:10.1063/1.3170926 PMID:19624173

Rani, K., & Sridevi, V. (2017). An Overview on Role of Nanotechnology in Green and Clean Technology. *Austin Environ Sci*, *2*(3), 1026.

Rao, C. N. R., Sood, A. K., Subrahmanyam, K. S., & Govindaraj, A. (2009). Graphene: The new two-dimensional nanomaterial. *Angewandte Chemie International Edition*, *48*(42), 7752–7777. doi:10.1002/anie.200901678 PMID:19784976

Rao, G. R., Srikanth, I., & Reddy, K. L. (2020). Effect of organo-modified montmorillonite nanoclay on mechanical, thermal and ablation behavior of carbon fiber/phenolic resin composites. *Defence Technology*, *17*(3), 812–820. doi:10.1016/j.dt.2020.05.012

Rao, J. P., & Geckeler, K. E. (2011). Polymer nanoparticles: Preparation techniques and size-control parameters. *Progress in Polymer Science*, *36*(7), 887–913. doi:10.1016/j.progpolymsci.2011.01.001

Rao, K. M., Suneetha, M., Park, G. T., Babu, A. G., & Han, S. S. (2020). Hemostatic, biocompatible, and antibacterial non-animal fungal mushroom-based carboxymethyl chitosan-ZnO nanocomposite for wound-healing applications. *International Journal of Biological Macromolecules*, *155*, 71–80. doi:10.1016/j.ijbiomac.2020.03.170 PMID:32217121

Rao, L., Xia, S., Xu, W., Tian, R., Yu, G., Gu, C., Pan, P., Meng, Q. F., Cai, X., Qu, D., Lu, L., Xie, Y., Jiang, S., & Chen, X. (2020). Decoy nanoparticles protect against COVID-19 by concurrently adsorbing viruses and inflammatory cytokines. *Proceedings of the National Academy of Sciences of the United States of America*, *117*(44), 27141–27147. doi:10.1073/pnas.2014352117 PMID:33024017

Rao, T. N., Babji, P., Ahmad, N., Khan, R. A., Hassan, I., Shahzad, S. A., & Husain, F. M. (2019). Green synthesis and structural classification of Acacia nilotica mediated-silver doped titanium oxide (Ag/TiO2) spherical nanoparticles: Assessment of its antimicrobial and anticancer activity. *Saudi Journal of Biological Sciences*, *26*(7), 1385–1391. doi:10.1016/j.sjbs.2019.09.005 PMID:31866742

Rao, V. K., & Radhakrishnan, T. P. (2015). Tuning the SERS Response with Ag-Au Nanoparticle-Embedded Polymer Thin Film Substrates. *ACS Applied Materials & Interfaces*, *7*(23), 12767–12773. doi:10.1021/acsami.5b04180 PMID:26035249

Rasheed, T., Bilal, M., Li, C., Nabeel, F., Khalid, M., & Iqbal, H. M. (2018). Catalytic potential of bio-synthesized silver nanoparticles using Convolvulusarvensis extract for the degradation of environmental pollutants. *Journal of Photochemistry and Photobiology. B, Biology, 181*, 44–52. doi:10.1016/j.jphotobiol.2018.02.024 PMID:29499463

Rasheed, T., Nabeel, F., Bilal, M., & Iqbal, H. M. (2019). Biogenic synthesis and characterization of cobalt oxide nanoparticles for catalytic reduction of direct yellow-142 and methyl orange dyes. *Biocatalysis and Agricultural Biotechnology, 19*, 101154. doi:10.1016/j.bcab.2019.101154

Rashid, F. L., Hadi, A., Al-Garah, N. H., & Hashim, A. (2018). Novel phase change materials, MgO nanoparticles, and water based nanofluids for thermal energy storage and biomedical applications. *Int Pharmaceu Phytopharmacol Res, 8*, 46–56.

Rashki, S., Abbas Alshamsi, H., Amiri, O., Safardoust-Hojaghan, H., Salavati-Niasari, M., Nazari-Alam, A., & Khaledi, A. (2021). Eco-friendly green synthesis of ZnO/GQD nanocomposites using Protoparmeliopsis muralis extract for their antibacterial and antibiofilm activity. *Journal of Molecular Liquids, 335*, 116195. doi:10.1016/j.molliq.2021.116195

Rasmi, Y., Li, X., Khan, J., Ozer, T., & Choi, J. R. (2021). Emerging Point-of-Care Biosensors for Rapid Diagnosis of COVID-19: Current Progress Challenges, and Future Prospects. *Analytical and Bioanalytical Chemistry, 413*(16), 4137–4159. doi:10.100700216-021-03377-6 PMID:34008124

Rathore, B., Sunwoo, K., Jangili, P., Kim, J., Kim, J. H., Huang, M., Xiong, J., Sharma, A., Yang, Z., Qu, J., & Kim, J. S. (2019). Nanomaterial designing strategies related to cell lysosome and their biomedical applications: A review. *Biomaterials, 211*, 25–47. doi:10.1016/j.biomaterials.2019.05.002 PMID:31078050

Raveendran, P., Fu, J., & Wallen, S. L. (2003). Completely "green" synthesis and stabilization of metal nanoparticles. *Journal of the American Chemical Society, 125*(46), 13940–13941. doi:10.1021/ja029267j PMID:14611213

Rawal, T., Parmar, R., Tyagi, R. K., & Butani, S. (2017). Rifampicin loaded chitosan nanoparticle dry powder presents an improved therapeutic approach for alveolar tuberculosis. *Colloids and Surfaces. B, Biointerfaces, 154*, 321–330. doi:10.1016/j.colsurfb.2017.03.044 PMID:28363192

Ray, T. R., Lettiere, B., de Rutte, J., & Pennathur, S. (2015). Quantitative Characterization of the Colloidal Stability of Metallic Nanoparticles Using UV–vis Absorbance Spectroscopy. *Langmuir, 31*(12), 3577–3586. doi:10.1021/la504511j PMID:25730093

Razmjoo, A., Qolipour, M., Shirmohammadi, R., Heibati, S. M., & Iraj, F. (2017). Techno-economic evaluation of standalone hybrid solar-wind systems for small residential districts in the central desert of Iran. *Environmental Progress & Sustainable Energy, 36*(4), 1194–1207. doi:10.1002/ep.12554

Reardon, P. J., Parhizkar, M., Harker, A. H., Browning, R. J., Vassileva, V., Stride, E., Pedley, R. B., Edirisinghe, M., & Knowles, J. C. (2017). Electrohydrodynamic fabrication of core-shell PLGA nanoparticles with controlled release of cisplatin for enhanced cancer treatment. *International Journal of Nanomedicine, 12*, 3913–3926. doi:10.2147/IJN.S134833 PMID:28579777

Recillas, S., García, A., González, E., Casals, E., Puntes, V., Sánchez, A., & Font, X. (2011). Use of CeO2, TiO2 and Fe3O4 nanoparticles for the removal of lead from water. *Desalination, 277*(1–3), 213–220. https://doi.org/10.1016/j.desal.2011.04.036

Reddy, L. H., Arias, J. L., Nicolas, J., & Couvreur, P. (2012). Magnetic nanoparticles: Design and characterization, toxicity and biocompatibility, pharmaceutical and biomedical applications. *Chemical Reviews, 112*(11), 5818–5878. doi:10.1021/cr300068p PMID:23043508

Reichmuth, A. M., Oberli, M. A., Jaklenec, A., Langer, R., & Blankschtein, D. (2016). MRNA vaccine delivery using lipid nanoparticles. *Therapeutic Delivery*, *7*(5), 319–334. doi:10.4155/tde-2016-0006 PMID:27075952

Rekha Deka, S., Kumar Sharma, A., & Kumar, P. (2015). *Cationic polymers and their self-assembly for antibacterial applications*. Academic Press.

Ren, Y., Cheung, H. W., von Maltzhan, G., Agrawal, A., Cowley, G. S., Weir, B. A., Boehm, J. S., Tamayo, P., Karst, A. M., Liu, J. F., Hirsch, M. S., Mesirov, J. P., Drapkin, R., Root, D. E., Lo, J., Fogal, V., Ruoslahti, E., Hahn, W. C., & Bhatia, S. N. (2012). Targeted tumor-penetrating siRNA nanocomplexes for credentialing the ovarian cancer oncogene ID4. *Science Translational Medicine*, *4*(147), 147ra112-147ra112. doi:10.1126/scitranslmed.3003778

Rezaeisadat, M., Bordbar, A.-K., & Omidyan, R. (2021). Molecular dynamics simulation study of curcumin interaction with nano-micelle of PNIPAAm-b-PEG co-polymer as a smart efficient drug delivery system. *Journal of Molecular Liquids*, *332*, 115862. doi:10.1016/j.molliq.2021.115862

Rhim, J.-W. (2011). Effect of clay contents on mechanical and water vapor barrier properties of agar-based nanocomposite films. *Carbohydrate Polymers*, *86*(2), 691–699. doi:10.1016/j.carbpol.2011.05.010

Riley, R., & Day, E. (2017). *Wiley interdiscip*. Academic Press.

Rizwan, M., Singh, M., Mitra, C. K., & Morve, R. K. (2014). Ecofriendly Application of Nanomaterials: Nanobioremediation. *Journal of Nanoparticles*, *2014*, 1–7. doi:10.1155/2014/431787

Roa, W., Xiong, Y., Chen, J., Yang, X., Song, K., Yang, X., Kong, B., Wilson, J., & Xing, J. Z. (2012). Pharmacokinetic and toxicological evaluation of multi-functional thiol-6-fluoro-6-deoxy-D-glucose gold nanoparticles in vivo. *Nanotechnology*, *23*(37), 375101. doi:10.1088/0957-4484/23/37/375101 PMID:22922305

Rockett, R. J., Arnott, A., Lam, C., Sadsad, R., Timms, V., Gray, K.-A., Eden, J.-S., Chang, S., Gall, M., Draper, J., Sim, E. M., Bachmann, N. L., Carter, I., Basile, K., Byun, R., O'Sullivan, M. V., Chen, S. C.-A., Maddocks, S., Sorrell, T. C., ... Sintchenko, V. (2020). Revealing COVID-19 Transmission in Australia by SARS-CoV-2 Genome Sequencing and Agent-based Modeling. *Nature Medicine*, *26*(9), 1398–1404. doi:10.103841591-020-1000-7 PMID:32647358

Rodenak-Kladniew, B., Islan, G. A., de Bravo, M. G., Durán, N., & Castro, G. R. (2017). Design, characterization and in vitro evaluation of linalool-loaded solid lipid nanoparticles as potent tool in cancer therapy. *Colloids and Surfaces. B, Biointerfaces*, *154*, 123–132. doi:10.1016/j.colsurfb.2017.03.021 PMID:28334689

Rodponthukwaji, K., Saengruengrit, C., Tummamunkong, P., Leelahavanichakul, A., Ritprajak, P., & Insin, N. (2021). Facile synthesis of magnetic silica-mannan nanocomposites for enhancement in internalization and immune response of dendritic cells. *Materials Today. Chemistry*, *20*, 100417. doi:10.1016/j.mtchem.2020.100417

Rodrigues, C. V., Rodrigues, A. V., Antoniosi, L., & Maintinguer, S. I. (2019). Applications of biodiesel and crude glycerol: the products generated by transesterification process. In Transesterification: process, mechanism and applications (pp. 83–122). Academic Press.

Rodrigues, A. V., & Guastaldi, A. C. (2018). Ti-Mo alloys: Corrosion study in solutions simulating commercial gels. *Journal of Polymer Science and Engineering*, *1*(4), 1–7. doi:10.24294/jpse.v1i4.897

Rodrigues, A. V., Oliveira, N. T. C., dos Santos, M. L., & Guastaldi, A. C. (2015). Electrochemical behavior and corrosion resistance of Ti–15Mo alloy in naturally-aerated solutions, containing chloride and fluoride ions. *Journal of Materials Science. Materials in Medicine*, *26*(1), 1–9. doi:10.100710856-014-5323-0 PMID:25577207

Rodrigues, A. V., & Orlandi, M. O. (2019). Study of intense photoluminescence from monodispersed β-Ga2O3 ellipsoidal structures. *Ceramics International*, *45*(4), 5023–5029. doi:10.1016/j.ceramint.2018.11.203

Rodrigues, A. V., & Sabino, N. L. (2019a). Gallium oxide: A perspective of features in photoluminescence and gas sensor applications. *International Journal of Recent Scientific Research*, *10*(11(B)), 35816–35827. doi:10.24327/IJRSR

Rodrigues, A. V., & Sabino, N. L. (2019b). Synthesis of photoluminescent β-Ga2O3 nanostructures using electrospinning method, and control of length-diameter ratio by calcination heating rates. *Journal of Materials Science Materials in Electronics*, *30*(18), 16910–16916. doi:10.100710854-019-01631-3

Rodrigues, B. S., Branco, C. M., Corio, P., & Souza, J. S. (2020). Controlling Bismuth Vanadate Morphology and Crystalline Structure through Optimization of Microwave-Assisted Synthesis Conditions. *Crystal Growth & Design*, *20*(6), 3673–3685. doi:10.1021/acs.cgd.9b01517

Rodrigues, C. V., Alcaraz, F. A. R., Nespeca, M. G., Rodrigues, A. V., Motteran, F., Adorno, M. A. T., Varesche, M. B. A., & Maintinguer, S. I. (2020). Biohydrogen production in an integrated biosystem using crude glycerol from waste cooking oils. *Renewable Energy*, *162*, 701–711. doi:10.1016/j.renene.2020.08.061

Rodrigues, C. V., Santana, K. O., Nespeca, M. G., Rodrigues, A. V., Pires, L. O., & Maintinguer, S. I. (2020). Energy valorization of crude glycerol and sanitary sewage in hydrogen generation by biological processes. *International Journal of Hydrogen Energy*, *45*(21), 11943–11953. doi:10.1016/j.ijhydene.2020.02.168

Rognmo, A. U., Heldal, S., & Fernø, M. A. (2018). Silica nanoparticles to stabilize CO2-foam for improved CO2 utilization: Enhanced CO2 storage and oil recovery from mature oil reservoirs. *Fuel*, *216*, 621–626. Advance online publication. doi:10.1016/j.fuel.2017.11.144

Rohaizad, A., Shahabuddin, S., Shahid, M. M., Rashid, N. M., Hir, Z. A. M., Ramly, M. M., Awang, K., Siong, C. W., & Aspanut, Z. (2020). Green synthesis of silver nanoparticles from Catharanthus roseus dried bark extract deposited on graphene oxide for effective adsorption of methylene blue dye. *Journal of Environmental Chemical Engineering*, *8*(4), 103955. Advance online publication. doi:10.1016/j.jece.2020.103955

Rojas, R. V., Achouri, M., Maroulis, J., & Caon, L. (2016). Healthy soils: A prerequisite for sustainable food security. *Environmental Earth Sciences*, *75*(3), 180. doi:10.100712665-015-5099-7

Rokade, S. S., Joshi, K. A., Mahajan, K., Patil, S., Tomar, G., Dubal, D. S., Parihar, V. S., Kitture, R., Bellare, J. R., & Ghosh, S. (2018). Gloriosa superba mediated synthesis of platinum and palladium nanoparticles for induction of apoptosis in breast cancer. *Bioinorganic Chemistry and Applications*, *2018*, 1–9. doi:10.1155/2018/4924186 PMID:30057593

Romeiro, F. C., Silva, B. C., Martins, A. S., Zanoni, M. V. B., & Orlandi, M. O. (2021). Superior performance of rGO-tin oxide nanocomposite for selective reduction of CO2 to methanol. *Journal of CO2 Utilization, 46*.

Romeiro, F. C., Marinho, J. Z., Silva, A. C. A., Cano, N. F., Dantas, N. O., & Lima, R. C. (2013). Photoluminescence and magnetism in Mn2+-doped ZnO nanostructures grown rapidly by the microwave hydrothermal method. *The Journal of Physical Chemistry C*, *117*(49), 26222–26227. doi:10.1021/jp408993y

Romeiro, F. C., Rodrigues, M. A., Silva, L. A. J., Catto, A. C., da Silva, L. F., Longo, E., Nossol, E., & Lima, R. C. (2017). rGO-ZnO nanocomposites for high electrocatalytic effect on water oxidation obtained by microwave-hydrothermal method. *Applied Surface Science*, *423*, 743–751. doi:10.1016/j.apsusc.2017.06.221

Romeiro, F. C., Silva, S. C., Nossol, E., & Lima, R. C. (2020). One step microwave-hydrothermal synthesis of rGO–TiO2 nanocomposites for enhanced electrochemical oxygen evolution reaction. *New Journal of Chemistry*, *44*(17), 6825–6832. doi:10.1039/D0NJ01475C

Ronavari, A., Kovacs, D., Igaz, N., Vagvolgyi, C., Boros, I. M., Konya, Z., Pfeiffer, I., & Kiricsi, M. (2017). Biological activity of green-synthesized silver nanoparticles depends on the applied natural extracts: A comprehensive study. *International Journal of Nanomedicine*, *12*, 871–883. doi:10.2147/IJN.S122842 PMID:28184158

Roopan, S. M., Rohit, Madhumitha, G., Rahuman, A. A., Kamaraj, C., Bharathi, A., & Surendra, T. V. (2013). Low-cost and eco-friendly phyto-synthesis of silver nanoparticles using Cocos nucifera coir extract and its larvicidal activity. *Industrial Crops and Products*, *43*, 631–635. doi:10.1016/j.indcrop.2012.08.013

Rosman, C., Pierrat, S., Henkel, A., Tarantola, M., Schneider, D., Sunnick, E., Janshoff, A., & Sönnichsen, C. (2012). A new approach to assess gold nanoparticle uptake by mammalian cells: Combining optical dark-field and transmission electron microscopy. *Small*, *8*(23), 3683–3690. doi:10.1002mll.201200853 PMID:22888068

Rostami, E., Kashanian, S., Azandaryani, A. H., Faramarzi, H., Dolatabadi, J. E. N., & Omidfar, K. (2014). Drug targeting using solid lipid nanoparticles. *Chemistry and Physics of Lipids*, *181*, 56–61. doi:10.1016/j.chemphyslip.2014.03.006 PMID:24717692

Rostami, H., Ebtekar, M., Ardestani, M. S., Yazdi, M. H., & Mahdavi, M. (2017). Co-utilization of a TLR5 agonist and nano-formulation of HIV-1 vaccine candidate leads to increased vaccine immunogenicity and decreased immunogenic dose: A preliminary study. *Immunology Letters*, *187*, 19–26. doi:10.1016/j.imlet.2017.05.002 PMID:28479111

Rostami-Vartooni, A., Moradi-Saadatmand, A., & Mahdavi, M. (2018). Catalytic reduction of organic pollutants using biosynthesized Ag/C/Fe_3O_4 nanocomposite by red water and Caesalpinia gilliesii flower extract. *Materials Chemistry and Physics*, *219*(June), 328–339. doi:10.1016/j.matchemphys.2018.08.026

Rothon, R., & DeArmitt, C. (2017). Fillers (Including Fiber Reinforcements). In Brydson's Plastics Materials: Eighth Edition (pp. 169–204). Elsevier Inc. doi:10.1016/B978-0-323-35824-8.00008-6

Roy, A., & Bhattacharya, J. (2011). Microwave assisted synthesis of CaO nanoparticles and use in waste water treatment. *TechConnect Briefs*, *3*, 565–568.

Ruammaitree, A., Phokharatkul, D., Nuntawong, N., & Wisitsoraat, A. (2018). Improvement in corrosion resistance of stainless steel foil by graphene coating using thermal chemical vapor deposition. *Surface Review and Letters*, *25*(Supp01), 1840003. Advance online publication. doi:10.1142/S0218625X18400036

Rudramurthy, G. R., Swamy, M. K., Sinniah, U. R., & Ghasemzadeh, A. (2016). Nanoparticles: Alternatives against drug-resistant pathogenic microbes. *Molecules (Basel, Switzerland)*, *21*(7), 836. doi:10.3390/molecules21070836 PMID:27355939

Ruiz-Hitzky, E., Darder, M., Wicklein, B., Ruiz-Garcia, C., Martín-Sampedro, R., del Real, G., & Aranda, P. (2020). Nanotechnology Responses to COVID-19. *Advanced Healthcare Materials*, *9*(19), 2000979. doi:10.1002/adhm.202000979 PMID:32885616

Ryoo, S., Rahmani, A. R., Yoon, K. Y., Prodanović, M., Kotsmar, C., Milner, T. E., Johnston, K. P., Bryant, S. L., & Huh, C. (2012). Theoretical and experimental investigation of the motion of multiphase fluids containing paramagnetic nanoparticles in porous media. *Journal of Petroleum Science Engineering*, *81*, 129–144. Advance online publication. doi:10.1016/j.petrol.2011.11.008

S.A., A. R., & Tamsilian, Y. (2016). Synthesis of Nanomaterials. In Nanocomposite Materials. CRC Press. doi:10.1201/9781315372310-3

Saad, M., Garbuzenko, O. B., & Minko, T. (2008). Co-delivery of siRNA and an anticancer drug for treatment of multidrug-resistant cancer. *Nanomedicine (London)*, *3*(6), 761–776. doi:10.2217/17435889.3.6.761 PMID:19025451

Sabella, S., Galeone, A., Vecchio, G., Cingolani, R., & Pompa, P. (2011). AuNPs are toxic in vitro and in vivo: A review. *Journal of Nanoscience Letters*, *1*(3).

Sabouri, Z., Rangrazi, A., Amiri, M. S., Khatami, M., & Darroudi, M. (2021). Green synthesis of nickel oxide nanoparticles using Salvia hispanica L. (chia) seeds extract and studies of their photocatalytic activity and cytotoxicity effects. *Bioprocess and Biosystems Engineering*, *44*(11), 2407–2415. Advance online publication. doi:10.100700449-021-02613-8 PMID:34272979

Sabzehmeidani, M. M., Mahnaee, S., Ghaedi, M., Heidari, H., & Roy, W. L. (2021). Carbon based materials: A review of adsorbents for inorganic and organic compounds. *Material Advances*, *2*(2), 598–627. doi:10.1039/D0MA00087F

Sadalage, P. S., Patil, R. V., Havaldar, D. V., Gavade, S. S., Santos, A. C., & Pawar, K. D. (2021). Optimally biosynthesized, PEGylated gold nanoparticles functionalized with quercetin and camptothecin enhance potential anti-inflammatory, anti-cancer and anti-angiogenic activities. *Journal of Nanobiotechnology*, *19*(1), 84. doi:10.118612951-021-00836-1 PMID:33766058

Sadegh, H., Ali, G. A. M., Gupta, V. K., Makhlouf, A. S. H., Shahryari-ghoshekandi, R., Nadagouda, M. N., Sillanpa, M., & Megiel, E. (2017). The role of nanomaterials as effective adsorbents and their applications in wastewater treatment. *Journal of Nanostructure in Chemistry*, *7*(1), 1–14. doi:10.100740097-017-0219-4

Sadik, O. A., Du, N., Kariuki, V., Okello, V., & Bushlyar, V. (2014). Current and emerging technologies for the characterization of nanomaterials. *ACS Sustainable Chemistry & Engineering*, *2*(7), 1707–1716. doi:10.1021c500175v

Sadiku, E. R., Babul Reddy, A., Gnanasekarana, D., Oboirien, B., Aderibigbe, B. A., Varaprasad, K., & Goddeti, S. M. R. (2016). Nanostructured Polymer Blends for Gas/Vapor Barrier and Dielectric Applications. In *Design and Applications of Nanostructured Polymer Blends and Nanocomposite Systems* (pp. 239–259). Elsevier. doi:10.1016/B978-0-323-39408-6.00011-X

Sagir, M., & Tahir, M. B. (2021). Role of Nanocatalyst (Photocatalysts) for Wastewater Treatment. *Current Analytical Chemistry*, *17*(2), 138–149. doi:10.2174/1573411016666200226091404

Saha, P., Mahiuddin, M., Islam, A., & Ochiai, B. (2021). Biogenic Synthesis and Catalytic Efficacy of Silver Nanoparticles Based on Peel Extracts of Citrus macroptera Fruit. *ACS Omega*, *6*(28), 18260–18268. doi:10.1021/acsomega.1c02149 PMID:34308057

Sahu, N., Soni, D., Chandrashekhar, B., Satpute, D. B., Saravanadevi, S., Sarangi, B. K., & Pandey, R. A. (2016). Synthesis of silver nanoparticles using flavonoids: Hesperidin, naringin and diosmin, and their antibacterial effects and cytotoxicity. *International Nano Letters*, *6*(3), 173–181. doi:10.100740089-016-0184-9

Saif, S., Tahir, A., Asim, T., & Chen, Y. (2016). Plant Mediated Green Synthesis of CuO Nanoparticles: Comparison of Toxicity of Engineered and Plant Mediated CuO Nanoparticles towards Daphnia magna. *Nanomaterials (Basel, Switzerland)*, *6*(11), 205. Advance online publication. doi:10.3390/nano6110205 PMID:28335333

Saikia, J., Gogoi, A., & Baruah, S. (2019). Nanotechnology for Water Remediation. In Environmental Nanotechnology (pp. 195-211). doi:10.1007/978-3-319-98708-8_7

Saikia, P. (2019). Clay nanostructures for biomedical applications. In *Two-Dimensional Nanostructures for Biomedical Technology: A Bridge between Material Science and Bioengineering* (pp. 137–172). Elsevier. doi:10.1016/B978-0-12-817650-4.00005-X

Saint, S., Elmore, J. G., Sullivan, S. D., Emerson, S. S., & Koepsell, T. D. (1998). *The efficacy of silver alloy-coated urinary catheters in preventing urinary tract infection: a meta-analysis*. Academic Press.

Saji, V. S., Choe, H. C., & Yeung, K. W. K. (2010). Nanotechnology in biomedical applications: A review. *International Journal of Nano and Biomaterials*, *3*(2), 119. doi:10.1504/IJNBM.2010.037801

Sajjad, S., Leghari, S. A. K., Ryma, N.-U.-A., & Farooqi, S. A. (2018). Green Synthesis of Metal-Based Nanoparticles and Their Applications. In *Green Metal Nanoparticles* (pp. 23–77). John Wiley & Sons, Ltd. doi:10.1002/9781119418900.ch2

Saka, C. (2018). Overview on the surface functionalization mechanism and determination of surface functional groups of plasma treated carbon nanotubes. *Critical Reviews in Analytical Chemistry, 48*(1), 1–14. doi:10.1080/10408347.2017.1356699 PMID:28722465

Salatin, S., Barar, J., Barzegar-Jalali, M., Adibkia, K., Milani, M. A., & Jelvehgari, M. (2016). Hydrogel nanoparticles and nanocomposites for nasal drug/vaccine delivery. *Archives of Pharmacal Research, 39*(9), 1181–1192. doi:10.100712272-016-0782-0 PMID:27352214

Salem, D. M. S. A., Ismail, M. M., & Aly-Eldeen, M. A. (2019). Biogenic synthesis and antimicrobial potency of iron oxide (Fe3O4) nanoparticles using algae harvested from the Mediterranean Sea, Egypt. *Egyptian Journal of Aquatic Research, 45*(3), 197–204. doi:10.1016/j.ejar.2019.07.002

Salusjärvi, L., Havukainen, S., Koivistoinen, O., & Toivari, M. (2019). Biotechnological production of glycolic acid and ethylene glycol: current state and perspectives. In Applied Microbiology and Biotechnology (Vol. 103, Issue 6, pp. 2525–2535). Applied Microbiology and Biotechnology. doi:10.100700253-019-09640-2

Samberg, M. E., Loboa, E. G., Oldenburg, S. J., & Monteiro-Riviere, N. A. (2012). Silver nanoparticles do not influence stem cell differentiation but cause minimal toxicity. *Nanomedicine (London), 7*(8), 1197–1209. doi:10.2217/nnm.12.18 PMID:22583572

Sampaio, P. G. V., & González, M. O. A. (2017). Photovoltaic solar energy: Conceptual framework. *Renewable & Sustainable Energy Reviews, 74*, 590–601. doi:10.1016/j.rser.2017.02.081

Samrot, A. V., Ali, H. H., & Selvarani, A. (2021). Adsorption efficiency of chemically synthesized Superparamagnetic Iron Oxide Nanoparticles (SPIONs) on crystal violet dye. *Current Research in Green and Sustainable Chemistry, 4*, 100066. Advance online publication. doi:10.1016/j.crgsc.2021.100066

Sánchez-López, E., Gomes, D., Esteruelas, G., Bonilla, L., Lopez-Machado, A. L., Galindo, R., Cano, A., Espina, M., Ettcheto, M., Camins, A., Silva, A. M., Durazzo, A., Santini, A., Garcia, M. L., & Souto, E. B. (2020). Metal-based nanoparticles as antimicrobial agents: An overview. *Nanomaterials (Basel, Switzerland), 10*(2), 292. doi:10.3390/nano10020292 PMID:32050443

Sangroniz, A., Zhu, J.-B., Tang, X., Etxeberria, A., Chen, E. Y.-X., & Sardon, H. (2019). Packaging materials with desired mechanical and barrier properties and full chemical recyclability. *Nature Communications, 10*(1), 3559. doi:10.103841467-019-11525-x PMID:31395871

Sankar, R., Maheswari, R., Karthik, S., Shivashangari, K. S., & Ravikumar, V. (2014). Anticancer activity of Ficus religiosa engineered copper oxide nanoparticles. *Materials Science and Engineering C, 44*, 234–239. doi:10.1016/j.msec.2014.08.030 PMID:25280701

Santos Maia, C., Mehnert, W., Schaller, M., Korting, H. C., Gysler, A., Haberland, A., & Schäfer-korting, M. (2002). Drug targeting by solid lipid nanoparticles for dermal use. *Journal of Drug Targeting, 10*(6), 489–495. doi:10.1080/1061186021000038364 PMID:12575739

Saquib, Q., Faisal, M., & Abdulrahman, A. A. A. (2020). Green Synthesis of Nanoparticles: Applications and Prospects. In Green Synthesis of Nanoparticles: Applications and Prospects. doi:10.1007/978-981-15-5179-6

Sarangapany, S., & Mohanty, K. (2020). Facile green synthesis of magnetically separable Au–Pt@TiO_2 nanocomposite for efficient catalytic reduction of organic pollutants and selective oxidation of glycerol. *Journal of Alloys and Compounds, 830*, 154636. doi:10.1016/j.jallcom.2020.154636

Saratale, R. G., Saratale, G. D., Shin, H. S., Jacob, J. M., Pugazhendhi, A., Bhaisare, M., & Kumar, G. (2018). New insights on the green synthesis of metallic nanoparticles using plant and waste biomaterials: Current knowledge, their agricultural and environmental applications. *Environmental Science and Pollution Research International, 25*(11), 10164–10183. doi:10.100711356-017-9912-6 PMID:28815433

Sarathy, V., Tratnyek, P. G., Nurmi, J. T., Baer, D. R., Amonette, J. E., Chun, C. L., Penn, R. L., & Reardon, E. J. (2008). Aging of iron nanoparticles in aqueous solution: Effects on structure and reactivity. *The Journal of Physical Chemistry C, 112*(7), 2286–2293. doi:10.1021/jp0777418

Saravanan, M., Asmalash, T., Gebrekidan, A., Gebreegziabiher, D., Araya, T., Hilekiros, H., Barabadi, H., & Ramanathan, K. (2018). Nano-medicine as a newly emerging approach to combat human immunodeficiency virus (HIV). *Pharmaceutical Nanotechnology, 6*(1), 17–27. doi:10.2174/2211738506666180209095710 PMID:29424324

Sarkar, A., Sengupta, S., & Sen, S. (2019). Nanoparticles for Soil Remediation. In Environmental Chemistry for a Sustainable World (pp. 249–262). doi:10.1007/978-3-319-97922-9_9

Sarma, D. D., Santra, P. K., Mukherjee, S., & Nag, A. (2013). X-ray Photoelectron Spectroscopy: A Unique Tool To Determine the Internal Heterostructure of Nanoparticles. *Chemistry of Materials, 25*(8), 1222–1232. doi:10.1021/cm303567d

Sarwar, A., Wang, J., Khan, M. S., Farooq, U., Riaz, N., Nazir, A., Mahmood, Q., Hashem, A., Al-Arjani, A.-B. F., & Alqarawi, A. A., & Abd-Allah, E. F. (2021). Iron Oxide (Fe3O4)-Supported SiO2 Magnetic Nanocomposites for Efficient Adsorption of Fluoride from Drinking Water: Synthesis, Characterization, and Adsorption Isotherm Analysis. *Water (Basel), 13*(11), 1514. https://doi.org/10.3390/w13111514

Satheeshkumar, M. K., Kumar, E. R., Indhumathi, P., Srinivas, C., Deepty, M., Sathiyaraj, S., Suriyanarayanan, N., & Sastry, D. L. (2020). Structural, morphological and magnetic properties of algae/$CoFe_2O_4$ and algae/Ag-Fe-O nanocomposites and their biomedical applications. *Inorganic Chemistry Communications, 111*(August), 107578. doi:10.1016/j.inoche.2019.107578

Sathishkumar, M., Saroja, M., Venkatachalam, M., Parthasarathy, G., & Rajamanickam, A. T. (2017). Biosynthesis and characterization of zinc sulphide nanoparticles using leaf extracts of Tridax procumbens. *Oriental Journal of Chemistry, 33*(2), 903–909. doi:10.13005/ojc/330240

Sathishkumar, M., Sneha, K., Won, S. W., Cho, C.-W., Kim, S., & Yun, Y.-S. (2009). Cinnamon zeylanicum bark extract and powder mediated green synthesis of nano-crystalline silver particles and its bactericidal activity. *Colloids and Surfaces. B, Biointerfaces, 73*(2), 332–338. doi:10.1016/j.colsurfb.2009.06.005 PMID:19576733

Sathishkumar, M., Sneha, K., & Yun, Y. S. (2009). Palladium nanocrystals synthesis using Curcuma longa tuber extract. *International Journal of Material Science, 4*(1), 11–17.

Sathishkumar, P., Mohan, K., Ganesan, A. R., Govarthanan, M., Yusoff, A. R. M., & Gu, F. L. (2021). Persistence, toxicological effect and ecological issues of endosulfan – A review. *Journal of Hazardous Materials, 416*, 125779. doi:10.1016/j.jhazmat.2021.125779

Satyavani, K., Ramanathan, T., & Gurudeekan. (2011). Green synthesis of silver nanoparticles using stem dried callus extract of bitter apple (Citrullus colocynthis). *Digest Journal of Nanomaterials and Biostructures, 6*(3), 1019–1024.

Sayğılı, H. (2019). Hydrothermal synthesis of magnetic nanocomposite from biowaste matrix by a green and one-step route: Characterization and pollutant removal ability. *Bioresource Technology, 278*(November), 242–247. doi:10.1016/j.biortech.2019.01.103

Schaming, D., & Remita, H. (2015). *Nanotechnology: from the ancient time to nowadays*. Academic Press.

Schamp, C. T., & Jesser, W. A. (2005). On the measurement of lattice parameters in a collection of nanoparticles by transmission electron diffraction. *Ultramicroscopy, 103*(2), 165–172. doi:10.1016/j.ultramic.2004.11.007 PMID:15774277

Schertel, A., Snaidero, N., Han, H., Ruhwedel, T., Laue, M., Grabenbauer, M., & Möbius, W. (2013). Cryo FIB-SEM : Volume imaging of cellular ultrastructure in native frozen specimens. *Journal of Structural Biology, 184*(2), 355–360. doi:10.1016/j.jsb.2013.09.024 PMID:24121039

Schleich, N., Danhier, F., & Préat, V. (2015). Iron oxide-loaded nanotheranostics: Major obstacles to in vivo studies and clinical translation. *Journal of Controlled Release, 198*, 35–54. doi:10.1016/j.jconrel.2014.11.024 PMID:25481448

Schniepp, H. C., Li, J. L., McAllister, M. J., Sai, H., Herrera-Alonson, M., Adamson, D. H., Prud'homme, R. K., Car, R., Seville, D. A., & Aksay, I. A. (2006). Functionalized single graphene sheets derived from splitting graphite oxide. *The Journal of Physical Chemistry B, 110*(17), 8535–8539. Advance online publication. doi:10.1021/jp060936f PMID:16640401

Schoppa, T., Jung, D., Rust, T., Mulac, D., Kuckling, D., & Langer, K. (2021). Light-responsive polymeric nanoparticles based on a novel nitropiperonal based polyester as drug delivery systems for photosensitizers in PDT. *International Journal of Pharmaceutics, 597*, 120326. doi:10.1016/j.ijpharm.2021.120326 PMID:33540003

Scioli Montoto, S., Muraca, G., & Ruiz, M. E. (2020). Solid Lipid Nanoparticles for Drug Delivery: Pharmacological and Biopharmaceutical Aspects [Review]. *Frontiers in Molecular Biosciences, 7*(319), 587997. Advance online publication. doi:10.3389/fmolb.2020.587997 PMID:33195435

Seebauer, E. G., & Barlaz, D. E. (2016). SIMS for analysis of nanostructures. *Current Opinion in Chemical Engineering, 12*, 8–13. doi:10.1016/j.coche.2016.01.007

Seil, J. T., & Webster, T. (2012). *Antimicrobial applications of nanotechnology: methods and literature*. Academic Press.

Seitz, P., Krieghoff, C., Gottschling, S., Lücke, C., Abdel-Wahab, M., Holzhey, D., Gohmann, R. F., & Gutberlet, M. (2021). Pre-procedural high-pitch coronary CT angiography assessment of patients undergoing transcatheter aortic valve implantation (TAVI) without patient-specific adjustment: analysis of diagnostic performance. *Clinical Radiology*.

Senapati, S., Mahanta, A. K., Kumar, S., & Maiti, P. (2018). Controlled drug delivery vehicles for cancer treatment and their performance. *Signal Transduction and Targeted Therapy, 3*(1), 7. doi:10.103841392-017-0004-3 PMID:29560283

Senarathna, U., Fernando, S., Gunasekara, T., Weerasekera, M., Hewageegana, H., Arachchi, N., Siriwardena, H., & Jayaweera, P. (2017). *Enhanced antibacterial activity of TiO 2 nanoparticle surface modified with Garcinia zeylanica extract*. Academic Press.

Sengupta, S., & Kumar, A. (2013). Nano-ceramic coatings - A means of enhancing bit life and reducing drill string trips. *Society of Petroleum Engineers - International Petroleum Technology Conference 2013, IPTC 2013: Challenging Technology and Economic Limits to Meet the Global Energy Demand, 1*. 10.2523/IPTC-16474-MS

Sengupta, S., Eavarone, D., Capila, I., Zhao, G., Watson, N., Kiziltepe, T., & Sasisekharan, R. (2005). Temporal targeting of tumour cells and neovasculature with a nanoscale delivery system. *Nature, 436*(7050), 568–572. doi:10.1038/nature03794 PMID:16049491

Seo, G., Lee, G., Kim, M. J., Baek, S.-H., Choi, M., Ku, K. B., Lee, C.-S., Jun, S., Park, D., Kim, H. G., Kim, S.-J., Lee, J.-O., Kim, B. T., Park, E. C., & Kim, S. I. (2020). Rapid Detection of COVID-19 Causative Virus (SARS-CoV-2) in Human Nasopharyngeal Swab Specimens using Field-Effect Transistor-based Biosensor. *ACS Nano, 14*(4), 5135–5142. doi:10.1021/acsnano.0c02823 PMID:32293168

Seo, H., Choi, I., Whiting, N., Hu, J., Luu, Q. S., Pudakalakatti, S., McCowan, C., Kim, Y., Zacharias, N., Lee, S., Bhattacharya, P., & Lee, Y. (2018). Hyperpolarized Porous Silicon Nanoparticles: Potential Theragnostic Material for 29Si Magnetic Resonance Imaging. *ChemPhysChem, 19*(17), 2143–2147. Advance online publication. doi:10.1002/cphc.201800461 PMID:29779258

Shad, S., Belinga-Desaunay-Nault, M.-F. A., Sohail, Bashir, N., & Lynch, I. (2020). Removal of contaminants from canal water using microwave synthesized zero valent iron nanoparticles. *Environmental Science. Water Research & Technology, 6*(11), 3057–3065. doi:10.1039/D0EW00157K

Shah, M., Fawcett, D., Sharma, S., Tripathy, S., & Poinern, G. (2015). Green synthesis of metallic nanoparticles via biological entities. *Materials (Basel), 8*(11), 7278–7308. doi:10.3390/ma8115377 PMID:28793638

Shah, R., Eldridge, D., Palombo, E., & Harding, I. (2014). Optimisation and stability assessment of solid lipid nanoparticles using particle size and zeta potential. *The Journal of Physiological Sciences; JPS, 25*(1).

Shahriari, M., Taghdisi, S. M., Abnous, K., Ramezani, M., & Alibolandi, M. (2019). Synthesis of hyaluronic acid-based polymersomes for doxorubicin delivery to metastatic breast cancer. *International Journal of Pharmaceutics, 572*, 118835. doi:10.1016/j.ijpharm.2019.118835 PMID:31726198

Sha, J., Gao, C., Lee, S. K., Li, Y., Zhao, N., & Tour, J. M. (2016). Preparation of three-dimensional graphene foams using powder metallurgy templates. *ACS Nano, 10*(1), 1411–1416. doi:10.1021/acsnano.5b06857 PMID:26678869

Shakibaie, M., Riahi-Madvar, S., Ameri, A., Amiri-Moghadam, P., Adeli-Sardou, M., & Forootanfar, H. (2021). Microwave assisted biosynthesis of cadmium nanoparticles: Characterization, antioxidant and cytotoxicity studies. *Journal of Cluster Science*. Advance online publication. doi:10.100710876-021-02107-3

Shakir, M., Khan, M. S., Baig, U., Alam, M. F., Younus, H., & Alam, M. (2016). In vivo cytotoxicity, molecular docking and study of yeast alcohol dehydrogenase on polycarbazole-titanium dioxide nanocomposite. *Journal of Molecular Catalysis. B, Enzymatic, 134*, 79–88. doi:10.1016/j.molcatb.2016.09.018

Shan, J.-J., Li, C.-H., Wu, J.-M., Liu, J.-A., & Shi, Y.-S. (2017). Shape-controlled synthesis of monodispersed beta-gallium oxide crystals by a simple precipitation technique. *Ceramics International, 43*(8), 6430–6436. doi:10.1016/j.ceramint.2017.02.056

Shankar, S. S., Ahmad, A., & Sastry, M. (2003). Geranium leaf assisted biosynthesis of silver nanoparticles. *Biotechnology Progress, 19*(6), 1627–1631. doi:10.1021/bp034070w PMID:14656132

Shankar, S. S., Rai, A., Ahmad, A., & Sastry, M. (2004). Rapid synthesis of Au, Ag, and bimetallic Au core–Ag shell nanoparticles using Neem (Azadirachta indica) leaf broth. *Journal of Colloid and Interface Science, 275*(2), 496–502. doi:10.1016/j.jcis.2004.03.003 PMID:15178278

Shankar, S., Shanker, U., & Shikha. (2014). Arsenic Contamination of Groundwater: A Review of Sources, Prevalence, Health Risks, and Strategies for Mitigation. *TheScientificWorldJournal, 2014*, 1–18. doi:10.1155/2014/304524 PMID:25374935

Shao, Q., Tang, J., Lin, Y., Zhang, F., Yuan, J., Zhang, H., Shinya, N., & Qin, L. C. (2013). Synthesis and characterization of graphene hollow spheres for application in supercapacitors. *Journal of Materials Chemistry. A, Materials for Energy and Sustainability, 1*(48), 15423–15428. doi:10.1039/c3ta12789c

Shapira, P., & Youtie, J. (2012). *The Economic Contributions of Nanotechnology to green and sustainable growth. International Symposium on Assessing the Economic Impact of Nanotechnology, Washington, DC.*

Sharifianjazi, F., Irani, M., Esmaeilkhanian, A., Bazli, L., Asl, M. S., Jang, H. W., Kim, S. Y., Ramakrishna, S., Shokouhimehr, M., & Varma, R. S. (2021). Polymer incorporated magnetic nanoparticles: Applications for magnetoresponsive targeted drug delivery. *Materials Science and Engineering B*, *272*, 115358. doi:10.1016/j.mseb.2021.115358

Sharif, N. A., Ehsani, M., Zaarei, D., Kalaee, M. R., & Khajavi, R. (2020). Nanocomposite Coatings Based on Modified Graphene Oxide and Polydimethylsiloxane: Characterization and Thermal Properties. *Russian Journal of Applied Chemistry*, *93*(11), 1765–1773. doi:10.1134/S1070427220110191

Sharma, A. K., Prasher, P., Aljabali, A. A., Mishra, V., Gandhi, H., Kumar, S., Mutalik, S., Chellappan, D. K., Tambuwala, M. M., Dua, K., & Kapoor, D. N. (2020). Emerging era of "somes": Polymersomes as versatile drug delivery carrier for cancer diagnostics and therapy. *Drug Delivery and Translational Research*, *10*(5), 1171–1190. doi:10.100713346-020-00789-2 PMID:32504410

Sharma, R., Garg, R., & Kumari, A. (2020). A review on biogenic synthesis, applications and toxicity aspects of zinc oxide nanoparticles. *EXCLI Journal*, *19*, 1325–1340. https://doi.org/10.17179/excli2020-2842

Sharma, S., & Bhattacharya, A. (2017). Drinking water contamination and treatment techniques. *Applied Water Science*, *7*(3), 1043–1067. https://doi.org/10.1007/s13201-016-0455-7

Sharma, S., Saini, S., Khangembam, M., & Singh, V. (2021). Nanomaterials-based Biosensors for COVID-19 Detection – A Review. *IEEE Sensors Journal*, *21*(5), 5598–5611. doi:10.1109/JSEN.2020.3036748

Sharma, T. S. K., Selvakumar, K., Hwa, K. Y., Sami, P., & Kumaresan, M. (2019). Biogenic fabrication of gold nanoparticles using Camellia japonica L. leaf extract and its biological evaluation. *Journal of Materials Research and Technology*, *8*(1), 1412–1418. doi:10.1016/j.jmrt.2018.10.006

Sharma, T., Kumar, G. S., & Sangwai, J. S. (2015). Comparative effectiveness of production performance of Pickering emulsion stabilized by nanoparticle-surfactant-polymerover surfactant-polymer (SP) flooding for enhanced oil recoveryfor Brownfield reservoir. *Journal of Petroleum Science Engineering*, *129*, 221–232. Advance online publication. doi:10.1016/j.petrol.2015.03.015

Shaw, J., Raja, S. O., & Dasgupta, A. K. (2014). Modulation of cytotoxic and genotoxic effects of nanoparticles in cancer cells by external magnetic field. *Cancer Nanotechnology*, *5*(1), 2. doi:10.118612645-014-0002-x PMID:26561510

Sheikh-Mohseni, M. H., Sedaghat, S., Derakhshi, P., & Safekordi, A. (2020). Green bio-synthesis of Ni/montmorillonite nanocomposite using extract of Allium jesdianum as the nano-catalyst for electrocatalytic oxidation of methanol. *Chinese Journal of Chemical Engineering*, *28*(10), 2555–2565. doi:10.1016/j.cjche.2020.04.017

Shen, W., Zhang, X., Huang, Q., Xu, Q., & Song, W. (2014). Preparation of solid silver nanoparticles for inkjet printed flexible electronics with high conductivity. *Nanoscale*, *6*(3), 1622–1628. Advance online publication. doi:10.1039/C3NR05479A PMID:24337051

Shevchenko, S. N., Burkhardt, M., Sheval, E. V., Natashina, U. A., Grosse, C., Nikolaev, A. L., Gopin, A. V., Neugebauer, U., Kudryavtsev, A. A., & Sivakov, V. (2017). *Antimicrobial effect of biocompatible silicon nanoparticles activated using therapeutic ultrasound*. Academic Press.

Shi, H., Ye, X., He, X., Wang, K., Cui, W., He, D., Li, D., & Jia, X. (2014). Au@ Ag/Au nanoparticles assembled with activatable aptamer probes as smart "nano-doctors" for image-guided cancer thermotherapy. *Nanoscale*, *6*(15), 8754–8761. doi:10.1039/C4NR01927J PMID:24953128

Shi, J., Kantoff, P. W., Wooster, R., & Farokhzad, O. C. (2017). Cancer nanomedicine: Progress, challenges and opportunities. *Nature Reviews. Cancer*, *17*(1), 20–37. doi:10.1038/nrc.2016.108 PMID:27834398

Shi, L., Chen, K., Du, R., Bachmatiuk, A., Rümmeli, M. H., Xie, K., Huang, Y., Zhang, Y., & Liu, Z. (2016). Scalable Seashell-Based Chemical Vapor Deposition Growth of Three-Dimensional Graphene Foams for Oil-Water Separation. *Journal of the American Chemical Society*, *138*(20), 6360–6363. doi:10.1021/jacs.6b02262 PMID:27157548

Shi, L.-E., Li, Z.-H., Zheng, W., Zhao, Y.-F., Jin, Y.-F., & Tang, Z.-X. (2014). Synthesis, antibacterial activity, antibacterial mechanism and food applications of ZnO nanoparticles: A review. *Food Additives & Contaminants: Part A*, *31*(2), 173–186. doi:10.1080/19440049.2013.865147 PMID:24219062

Shilo, M., Reuveni, T., Motiei, M., & Popovtzer, R. (2012). Nanoparticles as computed tomography contrast agents: Current status and future perspectives. *Nanomedicine (London)*, *7*(2), 257–269. doi:10.2217/nnm.11.190 PMID:22339135

Shin, M. D., Shukla, S., Chung, Y. H., Beiss, V., Chan, S. K., Ortega-Rivera, O. A., Wirth, D. M., Chen, A., Sack, M., Pokorski, J. K., & Steinmetz, N. F. (2020). COVID-19 Vaccine Development and a Potential Nanomaterial Path Forward. *Nature Nanotechnology*, *15*(8), 646–655. doi:10.103841565-020-0737-y PMID:32669664

Shipley, H. J., Engates, K. E., & Guettner, A. M. (2011). Study of iron oxide nanoparticles in soil for remediation of arsenic. *Journal of Nanoparticle Research*, *13*(6), 2387–2397. doi:10.100711051-010-9999-x

Shi, S. F., Jia, J. F., Guo, X. K., Zhao, Y. P., Chen, D. S., Guo, Y. Y., ... Zhang, X. L. (2012). Biocompatibility of chitosan-coated iron oxide nanoparticles with osteoblast cells. *International Journal of Nanomedicine*, *7*, 5593. PMID:23118539

Shivaramakrishnan, B., Gurumurthy, B., & Balasubramanian, A. (2017). Potential biomedical applications of metallic nanobiomaterials: A review. *International Journal of Pharmaceutical Sciences and Research*, *8*(3), 985.

Shivaraman, S., Chandrashekhar, M. V. S., Boeckl, J. J., & Spencer, M. G. (2009). Thickness estimation of epitaxial graphene on sic using attenuation of substrate raman intensity. *Journal of Electronic Materials*, *38*(6), 725–730. doi:10.100711664-009-0803-6

Shi, Z., Wang, L., Deng, L., Peng, X., & Yuan, M. (2016). Adsorptive removal of Cr(VI) from aqueous solution using NiFe2O4/ ZnAl-LDH. *Chine Journal of Environmental Engineering*, *10*, 4635–4642.

Shou, Q., Guo, C., Yang, L., Jia, L., Liu, C., & Liu, H. (2011). Effect of pH on the single-step synthesis of gold nanoparticles using PEO–PPO–PEO triblock copolymers in aqueous media. *Journal of Colloid and Interface Science*, *363*(2), 481–489. doi:10.1016/j.jcis.2011.07.021 PMID:21855892

Shrivastava, N., Shrivastava, V., Jyoti, A., & Tomar, R. S. (2019). Promises and cons of nanobiotechnology: A critical. *Plant Archives*, *19*(1), 1–11.

Shukla, R., Nune, S. K., Chanda, N., Katti, K., Mekapothula, S., Kulkarni, R. R., Welshons, W. V., Kannan, R., & Katti, K. V. (2008). Soybeans as a phytochemical reservoir for the production and stabilization of biocompatible gold nanoparticles. *Small*, *4*(9), 1425–1436. doi:10.1002mll.200800525 PMID:18642250

Siddiqi, K. S., & Husen, A. (2016). Green synthesis, characterization and uses of palladium/platinum nanoparticles. *Nanoscale Research Letters*, *11*(1), 1–13. doi:10.118611671-016-1695-z PMID:27807824

Siegel, R. L., Miller, K. D., & Jemal, A. (2016). Cancer statistics, 2016. *CA: a Cancer Journal for Clinicians*, *66*(1), 7–30. doi:10.3322/caac.21332 PMID:26742998

Sigmund, W., Yuh, J., Park, H., Maneeratana, V., Pyrgiotakis, G., Daga, A., Taylor, J., & Nino, J. C. (2006). Processing and structure relationships in electrospinning of ceramic fiber systems. *Journal of the American Ceramic Society*, *89*(2), 395–407. doi:10.1111/j.1551-2916.2005.00807.x

Simas, A. B. C., Pereira, V. L. P., Barreto, C. B. Jr, de Sales, D. L., & de Carvalho, L. L. (2009). An expeditious and consistent procedure for tetrahydrofuran (THF) drying and deoxygenation by the still apparatus. *Quimica Nova*, *32*(9), 2473–2475. doi:10.1590/S0100-40422009000900042

Singh Raman, R. K., Chakraborty Banerjee, P., Lobo, D. E., Gullapalli, H., Sumandasa, M., Kumar, A., Choudhary, L., Tkacz, R., Ajayan, P. M., & Majumder, M. (2012). Protecting copper from electrochemical degradation by graphene coating. *Carbon*, *50*(11), 4040–4045. Advance online publication. doi:10.1016/j.carbon.2012.04.048

Singh, Bhat, Dervash, Qadri, Mehmood, Dar, Hameed, & Rashid. (2020). Wonders of Nanotechnology for Remediation of Polluted Aquatic Environs. In Fresh Water Pollution Dynamics and Remediation (pp. 319-339). Springer Nature Singapore Pte Ltd. doi:10.1007/978-981-13-8277-2_17

Singh, S., Kumar, V., Romero, R., Sharma, K., & Singh, J. (2019). Applications of Nanoparticles in Wastewater Treatment. In *Nanotechnology in the Life Sciences* (pp. 395–418). doi:10.1007/978-3-030-17061-5_17

Singh, A. K., & Singh, K. P. (2016). Evaluation of phosphate removal capacity of Fe 3 O 4–ZVINPs from aqueous solution: Optimization using response surface analysis. *Research on Chemical Intermediates*, *42*(10), 7397–7415. doi:10.100711164-016-2543-6

Singh, A. K., Talat, M., Singh, D. P., & Srivastava, O. N. (2010). Biosynthesis of gold and silver nanoparticles by natural precursor clove and their functionalization with amine group. *Journal of Nanoparticle Research*, *12*(5), 1667–1675. doi:10.100711051-009-9835-3

Singh, D., Singh, S., Sahu, J., Srivastava, S., & Singh, M. R. (2016). Ceramic nanoparticles: Recompense, cellular uptake and toxicity concerns. *Artificial Cells, Nanomedicine, and Biotechnology*, *44*(1), 401–409. doi:10.3109/21691401.2014.955106 PMID:25229834

Singh, J., Dutta, T., Kim, K.-H., Rawat, M., Samddar, P., & Kumar, P. (2018). 'Green' synthesis of metals and their oxide nanoparticles: Applications for environmental remediation. *Journal of Nanobiotechnology*, *16*(1), 84. doi:10.118612951-018-0408-4 PMID:30373622

Singh, J., Kaur, N., Kaur, P., Kaur, S., Kaur, J., Kukkar, P., Kumar, V., Kukkar, D., & Rawat, M. (2018). Piper betle leaves mediated synthesis of biogenic SnO2 nanoparticles for photocatalytic degradation of reactive yellow 186 dye under direct sunlight. *Environmental Nanotechnology, Monitoring & Management*, *10*, 331–338. doi:10.1016/j.enmm.2018.07.001

Singh, S., Bhatta, U. M., Satyam, P., Dhawan, A., Sastry, M., & Prasad, B. L. V. (2008). Bacterial synthesis of silicon/silica nanocompositesc. *Journal of Materials Chemistry*, *18*(22), 2601–2606. doi:10.1039/b719528a

Sinha, T., & Ahmaruzzaman, M. (2015). A novel green and template free approach for the synthesis of gold nanorice and its utilization as a catalyst for the degradation of hazardous dye. *Spectrochimica Acta. Part A: Molecular and Biomolecular Spectroscopy*, *142*, 266–270. doi:10.1016/j.saa.2015.02.020 PMID:25706595

Sirelkhatim, A., Mahmud, S., Seeni, A., Kaus, N. H. M., Ann, L. C., Bakhori, S. K. M., Hasan, H., & Mohamad, D. (2015). *Review on zinc oxide nanoparticles: antibacterial activity and toxicity mechanism*. Academic Press.

Sirelkhatim, A., Mahmud, S., Seeni, A., Kaus, N. H. M., Ann, L. C., Bakhori, S. K. M., Hasan, H., & Mohamad, D. (2015). Review on zinc oxide nanoparticles: Antibacterial activity and toxicity mechanism. *Nano-Micro Letters*, *7*(3), 219–242. doi:10.100740820-015-0040-x PMID:30464967

Sivansankarapillai, V. S., Pillai, A. M., Rahdar, A., Sobha, A. P., Das, S. S., Mitropoulos, A. C., Mokarrar, M. H., & Kyzas, G. Z. (2020). On Facing the SARS-CoV-2 (COVID-19) with Combination of Nanomaterials and Medicine: Possible Strategies and First Challenges. *Nanomaterials (Basel, Switzerland)*, *10*(5), 852. doi:10.3390/nano10050852 PMID:32354113

Sivaraj, A., Kumar, V., Sunder, R., Parthasarathy, K., & Kasivelu, G. (2020). Commercial Yeast Extracts Mediated Green Synthesis of Silver Chloride Nanoparticles and their Anti-mycobacterial Activity. *Journal of Cluster Science, 31*(1), 287–291. doi:10.100710876-019-01626-4

Slavin, Y. N., Asnis, J., Häfeli, U. O., & Bach, H. (2017). Metal nanoparticles: Understanding the mechanisms behind antibacterial activity. *Journal of Nanobiotechnology, 15*(1), 1–20. doi:10.118612951-017-0308-z PMID:28974225

Slavutsky, A. M., Bertuzzi, M. A., & Armada, M. (2012). Water barrier properties of starch-clay nanocomposite films. *Brazilian Journal of Food Technology, 15*(3), 208–218. doi:10.1590/S1981-67232012005000014

Smith, A. T., LaChance, A. M., Zeng, S., Liu, B., & Sun, L. (2019). Synthesis, properties, and applications of graphene oxide/reduced graphene oxide and their nanocomposites. *Nano Materials Science, 1*(1), 31–47. doi:10.1016/j.nanoms.2019.02.004

Smith, B. R., & Gambhir, S. S. (2017). Nanomaterials for in vivo imaging. *Chemical Reviews, 117*(3), 901–986. doi:10.1021/acs.chemrev.6b00073 PMID:28045253

Sobhani, A., Knight, M. W., Wang, Y., Zheng, B., King, N. S., Brown, L. V., Fang, Z., Nordlander, P., & Halas, N. J. (2013). Narrowband photodetection in the near-infrared with a plasmon-induced hot electron device. *Nature Communications, 4*(1), 1–6. doi:10.1038/ncomms2642 PMID:23535664

Solanki, Y. S., Agarwal, M., Maheshwari, K., Gupta, S., Shukla, P., & Gupta, A. B. (2021). Removal of fluoride from water by using a coagulant (inorganic polymeric coagulant). *Environmental Science and Pollution Research International, 28*(4), 3897–3905. https://doi.org/10.1007/s11356-020-09579-2

Soleimani, S., Jannesari, A., Yousefzadi, M., Ghaderi, A., & Shahdadi, A. (2021). Eco-friendly foul release coatings based on a novel reduced graphene oxide/Ag nanocomposite prepared by a green synthesis approach. *Progress in Organic Coatings, 151*(August), 106107. doi:10.1016/j.porgcoat.2020.106107

Sonbol, H., Ameen, F., AlYahya, S., Almansob, A., & Alwakeel, S. (2021). Padina boryana mediated green synthesis of crystalline palladium nanoparticles as potential nanodrug against multidrug resistant bacteria and cancer cells. *Scientific Reports, 11*(1), 5444. doi:10.103841598-021-84794-6 PMID:33686169

Song, B., Xu, P., Chen, M., Tang, W., Zeng, G., Gong, J., Zhang, P., & Ye, S. (2019). Using nanomaterials to facilitate the phytoremediation of contaminated soil. *Critical Reviews in Environmental Science and Technology, 49*(9), 791–824. doi:10.1080/10643389.2018.1558891

Song, J. Y., Jang, H.-K., & Kim, B. S. (2009). Biological synthesis of gold nanoparticles using Magnolia kobus and Diopyros kaki leaf extracts. *Process Biochemistry, 44*(10), 1133–1138. doi:10.1016/j.procbio.2009.06.005

Song, J. Y., & Kim, B. S. (2009). Rapid biological synthesis of silver nanoparticles using plant leaf extracts. *Bioprocess and Biosystems Engineering, 32*(1), 79–84. doi:10.100700449-008-0224-6 PMID:18438688

Song, Q., Zhao, R., Liu, T., Gao, L., Su, C., Ye, Y., Chan, S. Y., Liu, X., Wang, K., Li, P., & Huang, W. (2021). One-step vapor deposition of fluorinated polycationic coating to fabricate antifouling and anti-infective textile against drug-resistant bacteria and viruses. *Chemical Engineering Journal, 418*, 129368. doi:10.1016/j.cej.2021.129368 PMID:33746567

Song, X. R., Cai, Z., Zheng, Y., He, G., Cui, F. Y., Gong, D. Q., Hou, S. X., Xiong, S. J., Lei, X. J., & Wei, Y. Q. (2009). Reversion of multidrug resistance by co-encapsulation of vincristine and verapamil in PLGA nanoparticles. *European Journal of Pharmaceutical Sciences, 37*(3–4), 300–305. doi:10.1016/j.ejps.2009.02.018 PMID:19491019

Song, Y., Fang, G., Zhu, C., Zhu, F., Wu, S., Chen, N., Wu, T., Wang, Y., Gao, J., & Zhou, D. (2019). Zero-valent iron activated persulfate remediation of polycyclic aromatic hydrocarbon-contaminated soils: An in situ pilot-scale study. *Chemical Engineering Journal, 355*, 65–75. doi:10.1016/j.cej.2018.08.126

Song, Y., Li, X., Wei, C., Fu, J., Xu, F., Tan, H., Tang, J., & Wang, L. (2015). A Green Strategy to Prepare Metal Oxide Superstructure from Metal-Organic Frameworks. *Scientific Reports, 5*(1), 8401. Advance online publication. doi:10.1038rep08401 PMID:25669731

Song, Y., Nallathamby, P. D., Huang, T., Elsayed-Ali, H. E., & Xu, X.-H. N. (2010). Correlation and Characterization of Three-Dimensional Morphologically Dependent Localized Surface Plasmon Resonance Spectra of Single Silver Nanoparticles Using Dark-Field Optical Microscopy and Spectroscopy and Atomic Force Microscopy. *The Journal of Physical Chemistry C, 114*(1), 74–81. doi:10.1021/jp9083019 PMID:20190865

Soni, G., & Yadav, K. S. (2015). Applications of nanoparticles in treatment and diagnosis of leukemia. *Materials Science and Engineering C, 47*, 156–164. doi:10.1016/j.msec.2014.10.043 PMID:25492184

Sonntag, M. D., Pozzi, E. A., Jiang, N., Hersam, M. C., & Van Duyne, R. P. (2014). Recent Advances in Tip-Enhanced Raman Spectroscopy. *The Journal of Physical Chemistry Letters, 5*(18), 3125–3130. doi:10.1021/jz5015746 PMID:26276323

Soundarrajan, C., Sankari, A., Dhandapani, P., Maruthamuthu, S., Ravichandran, S., Sozhan, G., & Palaniswamy, N. (2012). Rapid biological synthesis of platinum nanoparticles using Ocimum sanctum for water electrolysis applications. *Bioprocess and Biosystems Engineering, 35*(5), 827–833. doi:10.100700449-011-0666-0 PMID:22167464

Souto, E. B., & Müller, R. H. (2006). The use of SLN and NLC as topical particulate carriers for imidazole antifungal agents. *Die Pharmazie - An International. Journal of Pharmaceutical Sciences, 61*(5), 431–437. PMID:16724541

Spigaglia, P. (2020). COVID-19 outbreak: Migration, effects on society, global environment and prevention. *Science of the Total Environment Journal, 728*(January), 138882. PMID:32335410

Sreeprasad, T. S., & Berry, V. (2013). How Do the Electrical Properties of Graphene Change with its Functionalization? *Small, 9*(3), 341–350. doi:10.1002mll.201202196 PMID:23169614

Sreeprasad, T. S., & Pradeep, T. (2011). Reversible assembly and disassembly of gold nanorods induced by EDTA and its application in SERS tuning. *Langmuir, 27*(7), 3381–3390. doi:10.1021/la104828e PMID:21366283

Srivastav, A. L., Singh, P. K., Srivastava, V., & Sharma, Y. C. (2013). Application of a new adsorbent for fluoride removal from aqueous solutions. *Journal of Hazardous Materials, 263*, 342–352. https://doi.org/10.1016/j.jhazmat.2013.04.017

Sruthi, S., Maurizi, L., Nury, T., Sallem, F., Boudon, J., Riedinger, J. M., Millot, N., Bouyer, F., & Lizard, G. (2018). Cellular interactions of functionalized superparamagnetic iron oxide nanoparticles on oligodendrocytes without detrimental side effects: Cell death induction, oxidative stress and inflammation. *Colloids and Surfaces. B, Biointerfaces, 170*, 454–462. doi:10.1016/j.colsurfb.2018.06.041 PMID:29958160

Stancheva, K. A. (2012). Noble metal nanoparticles. In Oxidation Communications (Vol. 35, Issue 3). doi:10.1002/9781119120940.ch3

Stankovich, S., Dikin, D. A., Piner, R. D., Kohlhaas, K. A., Kleinhammes, A., Jia, Y., Wu, Y., Nguyen, S. B. T., & Ruoff, R. S. (2007). Synthesis of graphene-based nanosheets via chemical reduction of exfoliated graphite oxide. *Carbon, 45*(7), 1558–1565. Advance online publication. doi:10.1016/j.carbon.2007.02.034

Starek-Świechowicz, B., Budziszewska, B., & Starek, A. (2017). Hexachlorobenzene as a persistent organic pollutant: Toxicity and molecular mechanism of action. *Pharmacological Reports, 69*(6), 1232–1239. doi:10.1016/j.pharep.2017.06.013 PMID:29128804

Stylianopoulos, T., Wong, C., Bawendi, M. G., Jain, R. K., & Fukumura, D. (2012). Multistage nanoparticles for improved delivery into tumor tissue. In *Methods in Enzymology* (Vol. 508, pp. 109–130). Elsevier. doi:10.1016/B978-0-12-391860-4.00006-9

Subhankari, I., & Nayak, P. L. (2013). Antimicrobial activity of copper nanoparticles synthesized by ginger (Zingiber officinale) extract. *World Journal of Nano Science & Technology*, *2*(1), 10–13.

Subramaniam, M. N., Goh, P. S., Lau, W. J., & Ismail, A. F. (2019). The Roles of Nanomaterials in Conventional and Emerging Technologies for Heavy Metal Removal: A State-of-the-Art Review. *Nanomaterials (Basel, Switzerland)*, *9*(4), 625. Advance online publication. doi:10.3390/nano9040625 PMID:30999639

Subramanian, S., Schnoor, J. L., & Van Aken, B. (2017). Effects of Polychlorinated Biphenyls (PCBs) and Their Hydroxylated Metabolites (OH-PCBs) on Arabidopsis thaliana. *Environmental Science & Technology*, *51*(12), 7263–7270. doi:10.1021/acs.est.7b01538 PMID:28541669

Sui, Z., Meng, Q., Zhang, X., Ma, R., & Cao, B. (2012). Green synthesis of carbon nanotube–graphene hybrid aerogels and their use as versatile agents for water purification. *Journal of Materials Chemistry*, *22*(18), 8767. Advance online publication. doi:10.1039/c2jm00055e

Suleimanov, B. A., Rzayeva, S. J., & Keldibayeva, S. S. (2020). A new microbial enhanced oil recovery (MEOR) method for oil formations containing highly mineralized water. *Petroleum Science and Technology*, *38*(23), 999–1006. Advance online publication. doi:10.1080/10916466.2020.1793777

Sumer, B., & Gao, J. (2008). Theranostic nanomedicine for cancer. *Nanomedicine (London)*, *3*(2), 137–140. doi:10.2217/17435889.3.2.137 PMID:18373419

Sun, H., Guo, X., Zeng, S., Wang, Y., Hou, J., Yang, D., & Zhou, S. (2019). A multifunctional liposomal nanoplatform co-delivering hydrophobic and hydrophilic doxorubicin for complete eradication of xenografted tumors. *Nanoscale, 11*(38), 17759-17772. doi:10.1039/C9NR04669K

Sunasee, R. (2021). *4.20 - Nanocellulose: Preparation, Functionalization and Applications*. Elsevier.

Sun, F., Dai, Y., & Yu, X. (2017). Air pollution, food production and food security: A review from the perspective of food system. *Journal of Integrative Agriculture*, *16*(12), 2945–2962. doi:10.1016/S2095-3119(17)61814-8

Sun, H., Mei, L., Liang, J., Zhao, Z., Lee, C., Fei, H., Ding, M., Lau, J., Li, M., Wang, C., Xu, X., Hao, G., Papandrea, B., Shakir, I., Dunn, B., Huang, Y., & Duan, X. (2017). Three-dimensional holey-graphene/niobia composite architectures for ultrahigh-rate energy storage. *Science*, *356*(6338), 599–604. doi:10.1126cience.aam5852 PMID:28495745

Sun, L., Joh, D., Stangl, M., Al-Zaki, A., Murty, S., Alonso-Basanta, M., Kao, G., Dorsey, J., & Tsourkas, A. (2013). Selective targeting of brain tumors with nanoparticle-induced radiosensitization and contrast enhancement. *Practical Radiation Oncology*, *3*(2), S19–S20. doi:10.1016/j.prro.2013.01.067 PMID:24674504

Sun, Q., Schork, F., & Deng, Y. (2007). Water-based polymer/clay nanocomposite suspension for improving water and moisture barrier in coating. *Composites Science and Technology*, *67*(9), 1823–1829. doi:10.1016/j.compscitech.2006.10.022

Sun, T., Zhang, Y. S., Pang, B., Hyun, D. C., Yang, M., & Xia, Y. (2014). Engineered nanoparticles for drug delivery in cancer therapy. *Angewandte Chemie International Edition*, *53*(46), 12320–12364. doi:10.1002/anie.201403036 PMID:25294565

Su, S., & Kang, P. M. (2020). Systemic Review of Biodegradable Nanomaterials in Nanomedicine. *Nanomaterials (Basel, Switzerland)*, *10*(4), 656. Advance online publication. doi:10.3390/nano10040656 PMID:32244653

Su, Y., Kravets, V. G., Wong, S. L., Waters, J., Geim, A. K., & Nair, R. R. (2014). Impermeable barrier films and protective coatings based on reduced graphene oxide. *Nature Communications*, *5*(1), 4843. Advance online publication. doi:10.1038/ncomms5843 PMID:25208890

Svenson, S., & Tomalia, D. (2005). Dendrimers in biomedical applications—Reflections on the field. *Advanced Drug Delivery Reviews*, *57*(15), 2106–2129. doi:10.1016/j.addr.2005.09.018 PMID:16305813

Syafiuddin, A., Salim, M. R., Beng Hong Kueh, A., Hadibarata, T., & Nur, H. (2017). A review of silver nanoparticles: Research trends, global consumption, synthesis, properties, and future challenges. *Journal of the Chinese Chemical Society (Taipei)*, *64*(7), 732–756. doi:10.1002/jccs.201700067

Syame, S. M., Mohamed, W. S., Mahmoud, R. K., & Omara, S. T. (2017). Synthesis of copper-chitosan nanocomposites and their applications in treatment of local pathogenic isolates bacteria. *Oriental Journal of Chemistry*, *33*(6), 2959–2969. doi:10.13005/ojc/330632

Sycheva, L. P., Zhurkov, V. S., Iurchenko, V. V., Daugel-Dauge, N. O., Kovalenko, M. A., Krivtsova, E. K., & Durnev, A. D. (2011). Investigation of genotoxic and cytotoxic effects of micro-and nanosized titanium dioxide in six organs of mice in vivo. *Mutation Research/Genetic Toxicology and Environmental Mutagenesis*, *726*(1), 8–14. doi:10.1016/j.mrgentox.2011.07.010 PMID:21871579

Syed, H., & Sailaja, A. K. (2019). Preparation and Characterization of Nanoparticulate Drug Delivery System for Naproxen Sodium Using Various Desolvating Agents. *Nano Biomedicine and Engineering*, *11*(3). Advance online publication. doi:10.5101/nbe.v11i3.p254-263

Taghavi Fardood, S., Atrak, K., & Ramazani, A. (2017). Green synthesis using tragacanth gum and characterization of Ni-Cu-Zn ferrite nanoparticles as a magnetically separable photocatalyst for organic dyes degradation from aqueous solution under visible light. *Journal of Materials Science Materials in Electronics*, *28*(14), 10739–10746. doi:10.100710854-017-6850-5

Taghdisi, S. M., Lavaee, P., Ramezani, M., & Abnous, K. (2011). Reversible targeting and controlled release delivery of daunorubicin to cancer cells by aptamer-wrapped carbon nanotubes. *European Journal of Pharmaceutics and Biopharmaceutics*, *77*(2), 200–206. doi:10.1016/j.ejpb.2010.12.005 PMID:21168488

Taiwo, A. M. (2019). A review of environmental and health effects of organochlorine pesticide residues in Africa. *Chemosphere*, *220*, 1126–1140. doi:10.1016/j.chemosphere.2019.01.001 PMID:33395800

Tallawi, M., Rosellini, E., Barbani, N., Cascone, M. G., Rai, R., Saint-Pierre, G., & Boccaccini, A. R. (2015). Strategies for the chemical and biological functionalization of scaffolds for cardiac tissue engineering: A review. *Journal of the Royal Society, Interface*, *12*(108), 20150254. doi:10.1098/rsif.2015.0254 PMID:26109634

Tamanoi, F., Matsumoto, K., Doan, T. L. H., Shiro, A., & Saitoh, H. (2020). Studies on the Exposure of Gadolinium Containing Nanoparticles with Monochromatic X-rays Drive Advances in Radiation Therapy. *Nanomaterials (Basel, Switzerland)*, *10*(7), 1341. doi:10.3390/nano10071341 PMID:32660093

Tamer, U., Cetin, D., Suludere, Z., Boyaci, I. H., Temiz, H. T., Yegenoglu, H., Daniel, P., Dinçer, I., & Elerman, Y. (2013). Gold-coated iron composite nanospheres targeted the detection of Escherichia coli. *International Journal of Molecular Sciences*, *14*(3), 6223–6240. doi:10.3390/ijms14036223 PMID:23507756

Tanase, M. A., Raducan, A., Oancea, P., Ditu, L. M., Stan, M., Petcu, C., Scomoroscenco, C., Ninciuleanu, C. M., Nistor, C. L., & Cinteza, L. O. (2021). Mixed Pluronic-Cremophor Polymeric Micelles as Nanocarriers for Poorly Soluble Antibiotics-The Influence on the Antibacterial Activity. *Pharmaceutics*, *13*(4), 435. Advance online publication. doi:10.3390/pharmaceutics13040435 PMID:33804932

Tang, H., Zhao, Y., Yang, X., Liu, D., Shao, P., Zhu, Z., Shan, S., Cui, F., & Xing, B. (2017). New insight into the aggregation of graphene oxide using molecular dynamics simulations and extended Derjaguin-Landau- Verwey-Overbeek theory. *Environmental Science & Technology*, *51*(17), 9674–9682. doi:10.1021/acs.est.7b01668 PMID:28771343

Tang, S., & Zheng, J. (2018). Antibacterial activity of silver nanoparticles: Structural effects. *Advanced Healthcare Materials*, *7*(13), 1701503. doi:10.1002/adhm.201701503 PMID:29808627

Tang, Z.-X., Yu, Z., Zhang, Z.-L., Zhang, X.-Y., Pan, Q.-Q., & Shi, L.-E. (2013). Sonication-assisted preparation of CaO nanoparticles for antibacterial agents. *Quimica Nova*, *36*(7), 933–936. https://doi.org/10.1590/S0100-40422013000700002

Tang, Z., Zhang, X., Shu, Y., Guo, M., Zhang, H., & Tao, W. (2021). Insights from Nanotechnology in COVID-19 Treatment. *Nano Today*, *36*, 101019. doi:10.1016/j.nantod.2020.101019 PMID:33178330

Tan, X., Wang, X., Chen, C., & Sun, A. (2007). Effect of soil humic and fulvic acids, pH and ionic strength on Th(IV) sorption to TiO2 nanoparticles. *Applied Radiation and Isotopes*, *65*(4), 375–381. doi:10.1016/j.apradiso.2006.10.014 PMID:17157512

Tao, L., Wang, Q., Dou, S., Ma, Z., Huo, J., Wang, S., & Dai, L. (2016). Edge-rich and dopant-free graphene as a highly efficient metal-free electrocatalyst for the oxygen reduction reaction. *Chemical Communications*, *52*(13), 2764–2767. doi:10.1039/C5CC09173J PMID:26757794

Tao, Y., Kong, D., Zhang, C., Lv, W., Wang, M., Li, B., Huang, Z. H., Kang, F., & Yang, Q. H. (2014). Monolithic carbons with spheroidal and hierarchical pores produced by the linkage of functionalized graphene sheets. *Carbon*, *69*, 169–177. doi:10.1016/j.carbon.2013.12.003

Tardi, P. G., Dos Santos, N., Harasym, T. O., Johnstone, S. A., Zisman, N., Tsang, A. W., Bermudes, D. G., & Mayer, L. D. (2009). Drug ratio–dependent antitumor activity of irinotecan and cisplatin combinations in vitro and in vivo. *Molecular Cancer Therapeutics*, *8*(8), 2266–2275. doi:10.1158/1535-7163.MCT-09-0243 PMID:19671743

Tardi, P., Johnstone, S., Harasym, N., Xie, S., Harasym, T., Zisman, N., Harvie, P., Bermudes, D., & Mayer, L. (2009). In vivo maintenance of synergistic cytarabine:daunorubicin ratios greatly enhances therapeutic efficacy. *Leukemia Research*, *33*(1), 129–139. doi:10.1016/j.leukres.2008.06.028 PMID:18676016

Tarlani, A., Isari, M., Khazraei, A., & Eslami Moghadam, M. (2017). New sol-gel derived aluminum oxide-ibuprofen nanocomposite as a controlled releasing medication. *Nanomedicine Research Journal, 2*(1), 28-35.

Tavakol, S., Zahmatkenshan, M., Mohammadinejad, R., Mehrzadi, S., Joghataei, M. T., Alavijeh, M. S., & Seifalian, A. (2021). The Role of Nanotechnology in Current COVID-19 Outbreak. *Heliyon*, *7*(4), e06841. doi:10.1016/j.heliyon.2021.e06841 PMID:33880422

Taware, P. V., Chautmal, P. G., Waghmare, A. D., Kate, S. A., Kharat, O. R., & Potdukhe, D. A. (2020). Mycosynthesis of silver nanoparticles from mycelium extract of penicillium species and its antimicrobial activity against clinical isolates. *Journal of Global Biosciences*, *9*(8), 7808–7825.

Taylor, A. A., Tsuji, J. S., Garry, M. R., McArdle, M. E., Goodfellow, W. L. Jr, Adams, W. J., & Menzie, C. A. (2020). Critical Review of Exposure and Effects: Implications for Setting Regulatory Health Criteria for Ingested Copper. *Environmental Management*, *65*(1), 131–159. doi:10.100700267-019-01234-y PMID:31832729

Tchounwou, P. B., Yedjou, C. G., Patlolla, A. K., & Sutton, D. (2012). *Heavy metal toxicity and the environment*. Academic Press.

Tchounwou, P. B., Yedjou, C. G., Patlolla, A. K., & Sutton, D. J. (2012). Heavy metal toxicity and the environment. *Exp*, *101*(Suppl 101), 133–164. doi:10.1007/978-3-7643-8340-4_6 PMID:22945569

Tekade, R. K., Dutta, T., Gajbhiye, V., & Jain, N. K. (2009). Exploring dendrimer towards dual drug delivery: PH responsive simultaneous drug-release kinetics. *Journal of Microencapsulation*, *26*(4), 287–296. doi:10.1080/02652040802312572 PMID:18791906

Teleanu, D. M., Chircov, C., Grumezescu, A. M., & Teleanu, R. I. (2019). Neurotoxicity of Nanomaterials: An Up-to-Date Overview. *Nanomaterials (Basel, Switzerland)*, *9*(1), 96. Advance online publication. doi:10.3390/nano9010096 PMID:30642104

Teow, Y. H., & Mohammad, A. W. (2019). New generation nanomaterials for water desalination: A revie. *Desalination*, *451*, 2–17. doi:10.1016/j.desal.2017.11.041

Terraschke, H., Olchowka, J., Geringer, E., Rodrigues, A. V., & Wickleder, C. (2018). Facile Ionic Liquid-Assisted Strategy for Direct Precipitation of Eu2+-Activated Nanophosphors under Ambient Conditions. *Small*, *14*(17), 1–6. doi:10.1002mll.201703707 PMID:29573331

Thakkar, K. N., Mhatre, S. S., & Parikh, R. Y. (2010). Biological synthesis of metallic nanoparticles. Nanomedicine: Nanotechnology, Biology, and Medicine, 6(2). doi:10.1016/j.nano.2009.07.002

Thakor, A. S., Luong, R., Paulmurugan, R., Lin, F. I., Kempen, P., Zavaleta, C., Chu, P., Massoud, T. F., Sinclair, R., & Gambhir, S. S. (2011). The fate and toxicity of Raman-active silica-gold nanoparticles in mice. *Science Translational Medicine*, *3*(79), 79ra33-79ra33. doi:10.1126/scitranslmed.3001963

Thakur, S., Singh, S., & Pal, B. (2021). Superior adsorptive removal of brilliant green and phenol red dyes mixture by CaO nanoparticles extracted from egg shells. *Journal of Nanostructure in Chemistry*. doi:10.1007/s40097-021-00412-x

Thanh, N. T. K., & Rosenzweig, Z. (2002). Development of an Aggregation-Based Immunoassay for Anti-Protein A Using Gold Nanoparticles. *Analytical Chemistry*, *74*(7), 1624–1628. doi:10.1021/ac011127p PMID:12033254

Thatoi, P., Kerry, R. G., Gouda, S., Das, G., Pramanik, K., Thatoi, H., & Patra, J. K. (2016). Photo-mediated green synthesis of silver and zinc oxide nanoparticles using aqueous extracts of two mangrove plant species, Heritiera fomes and Sonneratia apetala and investigation of their biomedical applications. *Journal of Photochemistry and Photobiology. B, Biology*, *163*, 311–318. doi:10.1016/j.jphotobiol.2016.07.029 PMID:27611454

Thomas, S. C., Harshita, B. S. P., Mishra, P., & Talegaonkar, S. (2015). Ceramic Nanoparticles: Fabrication Methods and Applications in Drug Delivery. *Current Pharmaceutical Design*, *21*(42), 6165–6188. doi:10.2174/1381612821666151027153246 PMID:26503144

Thrivikraman, G., França, C. M., Athirasala, A., Tahayeri, A., & Bertassoni, L. E. (2017). 3D printing: Prospects and challenges: Nanomaterials in 3D bioprinting: Current approaches and future possibilities. In *Nanotechnologies in Preventive and Regenerative Medicine: An Emerging Big Picture* (pp. 349–379). Elsevier Inc. doi:10.1016/B978-0-323-48063-5.00004-6

Tian, C., Feng, J., Cho, H. J., Datta, S. S., & Prud'homme, R. K. (2018). Adsorption and denaturation of structured polymeric nanoparticles at an interface. *Nano Letters*, *18*(8), 4854–4860. doi:10.1021/acs.nanolett.8b01434 PMID:29975545

Tiwari, G., Tiwari, R., Sriwastawa, B., Bhati, L., Pandey, S., Pandey, P., & Bannerjee, S. K. (2012). Drug delivery systems: An updated review. *International Journal of Pharmaceutical Investigation*, *2*(1), 2. doi:10.4103/2230-973X.96920 PMID:23071954

Tiwari, J. N., Vij, V., Kemp, K. C., & Kim, K. S. (2015). Engineered carbon-nanomaterial-based electrochemical sensors for biomolecules. *ACS Nano*, *10*(1), 46–80. doi:10.1021/acsnano.5b05690 PMID:26579616

Toda, M., Yukawa, H., Yamada, J., Ueno, M., Kinoshita, S., Baba, Y., & Hamuro, J. (2019). In Vivo Fluorescence Visualization of Anterior Chamber Injected Human Corneal Endothelial Cells Labeled With Quantum Dots. *Investigative Ophthalmology & Visual Science*, *60*(12), 4008–4020. doi:10.1167/iovs.19-27788 PMID:31560364

Tokonami, S. (2021). External-field-induced Assembly for Biological Analytical Chemistry. *Analytical Sciences*, *37*(3), 395–396. Advance online publication. doi:10.2116/analsci.highlights2103 PMID:33692263

Tomar, V., Prasad, S., & Kumar, D. (2014). Adsorptive removal of fluoride from aqueous media using Citrus limonum (lemon) leaf. *Microchemical Journal*, *112*, 97–103. https://doi.org/10.1016/j.microc.2013.09.010

Topete, A., Alatorre-Meda, M., Iglesias, P., Villar-Alvarez, E. M., Barbosa, S., Costoya, J. A., Taboada, P., & Mosquera, V. (2014). Fluorescent drug-loaded, polymeric-based, branched gold nanoshells for localized multimodal therapy and imaging of tumoral cells. *ACS Nano*, *8*(3), 2725–2738. doi:10.1021/nn406425h PMID:24571629

Toropov, N., & Vartanyan, T. (2019). Noble metal nanoparticles: Synthesis and optical properties. In Comprehensive Nanoscience and Nanotechnology (Vols. 1–5, Issue March 2015). Elsevier Ltd. doi:10.1016/B978-0-12-803581-8.00585-3

Tortella, G. R., Rubilar, O., Diez, M. C., Padrão, J., Zille, A., Pieretti, J. C., & Seabra, A. B. (2021). Advanced Material Against Human (Including Covid-19) and Plant Viruses: Nanoparticles As a Feasible Strategy. *Global Challenges (Hoboken, NJ)*, *5*(3), 2000049. doi:10.1002/gch2.202000049 PMID:33614127

Towell, M. G., Browne, L. A., Paton, G. I., & Semple, K. T. (2011). Impact of carbon nanomaterials on the behaviour of 14C-phenanthrene and 14C-benzo-[a] pyrene in soil. *Environmental Pollution*, *159*(3), 706–715. doi:10.1016/j.envpol.2010.11.040 PMID:21195517

Tran, Q. H., Nguyen, V. Q., & Le, A.-T. (2013). Silver nanoparticles: Synthesis, properties, toxicology, applications and perspectives. *Advances in Natural Sciences: Nanoscience and Nanotechnology*, *4*(3), 033001. doi:10.1088/2043-6262/4/3/033001

Tran, V., Thiel, C., Svejda, J. T., Jalali, M., Walkenfort, B., Erni, D., & Schlücker, S. (2018). Probing the SERS brightness of individual Au nanoparticles, hollow Au/Ag nanoshells, Au nanostars and Au core/Au satellite particles: Single-particle experiments and computer simulations. *Nanoscale*, *10*(46), 21721–21731. doi:10.1039/C8NR06028B PMID:30431039

Tripathy, S. S., & Kanungo, S. B. (2005). Adsorption of Co2+, Ni2+, Cu2+ and Zn2+ from 0.5 M NaCl and major ion sea water on a mixture of δ-MnO2 and amorphous FeOOH. *Journal of Colloid and Interface Science*, *284*(1), 30–38. doi:10.1016/j.jcis.2004.09.054 PMID:15752781

Trung, N. B., Van Tam, T., Kim, H. R., Hur, S. H., Kim, E. J., & Choi, W. M. (2014). Three-dimensional hollow balls of graphene-polyaniline hybrids for supercapacitor applications. *Chemical Engineering Journal*, *255*, 89–96. doi:10.1016/j.cej.2014.06.028

Truong, D. H., Dam, M. S., Bujna, E., Rezessy-Szabo, J., Farkas, C., Vi, V. N. H., Csernus, O., Nguyen, V. D., Gathergood, N., Friedrich, L., Hafidi, M., Gupta, V. K., & Nguyen, Q. D. (2021). In situ fabrication of electrically conducting bacterial cellulose-polyaniline-titanium-dioxide composites with the immobilization of Shewanella xiamenensis and its application as bioanode in microbial fuel cell. *Fuel*, *285*(August). doi:10.1016/j.fuel.2020.119259

Tseng, I. H., Liao, Y. F., Chiang, J. C., & Tsai, M. H. (2012). Transparent polyimide/graphene oxide nanocomposite with improved moisture barrier property. *Materials Chemistry and Physics*, *136*(1), 247–253. doi:10.1016/j.matchemphys.2012.06.061

Tunc, I., Suzer, S., Correa-Duarte, M. A., & Liz-Marzán, L. M. (2005). XPS characterization of Au (Core)/SiO2 (shell) nanoparticles. *The Journal of Physical Chemistry B*, *109*(16), 7597–7600. doi:10.1021/jp050767j PMID:16851877

Tyagi, P., Salem, K. S., Hubbe, M. A., & Pal, L. (2021). Advances in barrier coatings and film technologies for achieving sustainable packaging of food products – A review. *Trends in Food Science & Technology*, *115*, 461–485. doi:10.1016/j.tifs.2021.06.036

ud Din, F., Aman, W., Ullah, I., Qureshi, O. S., Mustapha, O., Shafique, S., & Zeb, A. (2017). Effective use of nanocarriers as drug delivery systems for the treatment of selected tumors. *International Journal of Nanomedicine, 12*, 7291.

Udugama, B., Kadhiresan, P., Kozlowski, H. N., Malekjahani, A., Osborne, M., Li, V. Y. C., Chen, H., Mubareka, S., Gubbay, J. B., & Chan, W. C. W. (2020). Diagnosing COVID-19: The Disease and Tools for Detection. *ACS Nano*, *14*(4), 3822–3835. doi:10.1021/acsnano.0c02624 PMID:32223179

Ukaogo, P. O., Ewuzie, U., & Onwuka, C. V. (2020). Environmental pollution: causes, effects, and the remedies. In *Microorganisms for Sustainable Environment and Health* (pp. 419–429). Elsevier., doi:10.1016/B978-0-12-819001-2.00021-8

Umar, A., Alshahrani, A., Algarni, H., & Kumar, R. (2017). CuO nanosheets as potential scaffolds for sensing applications. *Sensors and Actuators. B, Chemical*, *250*, 24–31. doi:10.1016/j.snb.2017.04.062

Umar, K., Haque, M. M., Mir, N. A., Muneer, M., & Farooqi, I. H. (2013). Titanium dioxide-mediated photocatalysed mineralization of two selected organic pollutants in aqueous suspensions. *Journal of Advanced Oxidation Technologies*, *16*(2), 252–260. doi:10.1515/jaots-2013-0205

Unzueta, U., Saccardo, P., Domingo-Espín, J., Cedano, J., Conchillo-Solé, O., García-Fruitós, E., Céspedes, M. V., Corchero, J. L., Daura, X., Mangues, R., Ferrer-Miralles, N., Villaverde, A., & Vázquez, E. (2014). Sheltering DNA in self-organizing, protein-only nano-shells as artificial viruses for gene delivery. *Nanomedicine; Nanotechnology, Biology, and Medicine*, *10*(3), 535–541. doi:10.1016/j.nano.2013.11.006 PMID:24269989

Urimi, D., Widenbring, R., Pérez García, R. O., Gedda, L., Edwards, K., Loftsson, T., & Schipper, N. (2021). Formulation development and upscaling of lipid nanocapsules as a drug delivery system for a novel cyclic GMP analogue intended for retinal drug delivery. *International Journal of Pharmaceutics*, *602*, 120640. doi:10.1016/j.ijpharm.2021.120640 PMID:33901599

US EPA. (2004). *Cleaning up the Nation's waste sites: Markets and technology trends*. https://nepis.epa.gov/Exe/ZyNET.exe/30006II3.TXT?ZyActionD=ZyDocument&Client=EPA&Index=2000+Thru+2005&Docs=&Query=&Time=&EndTime=&SearchMethod=1&TocRestrict=n&Toc=&TocEntry=&QField=&QFieldYear=&QFieldMonth=&QFieldDay=&IntQFieldOp=0&ExtQFieldOp=0&XmlQuery=&File=D%3A%5Czyfiles%5CIndex Data%5C00thru05%5CTxt%5C00000008%5C30006II3.txt&User=ANONYMOUS&Password=anonymous&SortMethod=h%7C-&MaximumDocuments=1&FuzzyDegree=0&ImageQuality=r75g8/r75g8/x150y150g16/i425&Display=hpfr&DefSeekPage=x&SearchBack=ZyActionL&Back=ZyActionS&BackDesc=Resultspage&MaximumPages=1&ZyEntry=1&SeekPage=x&ZyPURL

US EPA. (2021). *Learn about Polychlorinated Biphenyls (PCBs) | Polychlorinated Biphenyls (PCBs)*. United States Environmental Protection Agency. https://www.epa.gov/pcbs/learn-about-polychlorinated-biphenyls-pcbs

Usman, M., Faure, P., Ruby, C., & Hanna, K. (2012). Remediation of PAH-contaminated soils by magnetite catalyzed Fenton-like oxidation. *Applied Catalysis B: Environmental*, *117–118*, 10–17. doi:10.1016/j.apcatb.2012.01.007

Uwaya, G. E., & Fayemi, O. E. (2020). Electrochemical detection of serotonin in banana at green mediated PPy/Fe_3O_4NPs nanocomposites modified electrodes. *Sensing and Bio-Sensing Research*, *28*(April), 100338. doi:10.1016/j.sbsr.2020.100338

Vahedifard, F., & Chakravarthy, K. (2021). *Nanomedicine for COVID-19: the role of nanotechnology in the treatment and diagnosis of COVID-19*. Emergent Materials. doi:10.100742247-021-00168-8

Vajtai, R. (2013). Springer handbook of nanomaterials. In Springer Handbook of Nanomaterials. doi:10.1007/978-3-642-20595-8

Valian, M., Khoobi, A., & Salavati-Niasari, M. (2020). Green synthesis and characterization of $DyMnO_3$-ZnO ceramic nanocomposites for the electrochemical ultratrace detection of atenolol. *Materials Science and Engineering C, 111*(March), 110854. doi:10.1016/j.msec.2020.110854 PMID:32279805

Valiyeva, G. G., Bavasso, I., Di Palma, L., Hajiyeva, S. R., Ramazanov, M. A., & Hajiyeva, F. V. (2019). Synthesis of Fe/ Ni Bimetallic Nanoparticles and Application to the Catalytic Removal of Nitrates from Water. *Nanomaterials (Basel, Switzerland), 9*(8), 1130. doi:10.3390/nano9081130 PMID:31390768

van den Berg, R., Elkjaer, C. F., Gommes, C. J., Chorkendorff, I., Sehested, J., de Jongh, P. E., de Jong, K. P., & Helveg, S. (2016). Revealing the Formation of Copper Nanoparticles from a Homogeneous Solid Precursor by Electron Microscopy. *Journal of the American Chemical Society, 138*(10), 3433–3442. doi:10.1021/jacs.5b12800 PMID:26891132

Van Doremalen, B. T., Morris, D.H., Holbrook, M.G., Gamble, A. Williamson, B.N., Tamin, A., Harcourt, J.L., Thornburg, N.J., Gerber, S.I., Lloyd-Smith, J.O., De Wit E., & Munster, V.J. (2020). Aerosol and Surface Stability of SARS-Co-V-2 as Compared with SARS-CoV-1. *The New England Journal of Medicine, 382*(16), 1564-1567. doi:10.1056/NEJMc2004973 PMID:32182409

Vance, M. J. N., Kuiken, T., Vejerano, E. P., McGinnis, S. P., Hochella, M. F. Jr, Rejeski, D., & Hull, M. S. (2015). Nanotechnology in the real world: Redeveloping the nanomaterial consumer products inventory. *Beilstein J., 6*, 1769–1178. doi:10.3762/bjnano.6.181

Vane, C. H., Drage, T. C., & Snape, C. E. (2006). Bark decay by the white-rot fungus Lentinula edodes: Polysaccharide loss, lignin resistance and the unmasking of suberin. *International Biodeterioration & Biodegradation, 57*(1), 14–23. doi:10.1016/j.ibiod.2005.10.004

Vanrompay, H., Skorikov, A., Bladt, E., Béché, A., Freitag, B., Verbeeck, J., & Bals, S. (2021). Fast versus conventional HAADF-STEM tomography of nanoparticles: advantages and challenges. *Ultramicroscopy, 221*(October). doi:10.1016/j.ultramic.2020.113191

Vanthana Sree, G., Nagaraaj, P., Kalanidhi, K., Aswathy, C. A., & Rajasekaran, P. (2020). Calcium oxide a sustainable photocatalyst derived from eggshell for efficient photo-degradation of organic pollutants. *Journal of Cleaner Production, 270*, 122294. https://doi.org/10.1016/j.jclepro.2020.122294

Vanti, G. L., Masaphy, S., Kurjogi, M., Chakrasali, S., & Nargund, V. B. (2020). Synthesis and application of chitosan-copper nanoparticles on damping off causing plant pathogenic fungi. *International Journal of Biological Macromolecules, 156*, 1387–1395. doi:10.1016/j.ijbiomac.2019.11.179 PMID:31760011

Varma, R. S. (2012). Greener approach to nanomaterials and their sustainable applications. *Current Opinion in Chemical Engineering, 1*(2), 123–128. doi:10.1016/j.coche.2011.12.002

Vazquez-Munoz, R., & Lopez-Ribot, J. L. (2020). Nanotechnology as an Alternative to Reduce the Spread of COVID-19. *Challenges, 11*(2), 15. doi:10.3390/challe11020015

Velsankar, K., R.M., A. K., R., P., V., M., & Sudhahar, S. (2020). Green synthesis of CuO nanoparticles via Allium sativum extract and its characterizations on antimicrobial, antioxidant, antilarvicidal activities. *Journal of Environmental Chemical Engineering, 8*(5).

Velu, K., Elumalai, D., Hemalatha, P., Janaki, A., Babu, M., Hemavathi, M., & Kaleena, P. K. (2015). Evaluation of silver nanoparticles toxicity of Arachis hypogaea peel extracts and its larvicidal activity against malaria and dengue vectors. *Environmental Science and Pollution Research International, 22*(22), 17769–17779. doi:10.100711356-015-4919-3 PMID:26154036

Velusamy, S., Roy, A., Sundaram, S., & Kumar Mallick, T. (2021). A review on heavy metal ions and containing dyes removal through graphene oxide-based adsorption strategies for textile wastewater treatment. *Chemical Record (New York, N.Y.), 21*(7), 1–42. doi:10.1002/tcr.202000153 PMID:33539046

Venkateswarlu, S., Kumar, B. N., Prathima, B., SubbaRao, Y., & Jyothi, N. V. V. (2014). A novel green synthesis of Fe3O4 magnetic nanorods using Punica Granatum rind extract and its application for removal of Pb(II) from aqueous environment. *Arabian Journal of Chemistry, 12*(4), 588–596. doi:10.1016/j.arabjc.2014.09.006

Venugopal, K., Ahmad, H., Manikandan, E., Thanigai Arul, K., Kavitha, K., Moodley, M. K., Rajagopal, K., Balabhaskar, R., & Bhaskar, M. (2017). The impact of anticancer activity upon Beta vulgaris extract mediated biosynthesized silver nanoparticles (ag-NPs) against human breast (MCF-7), lung (A549) and pharynx (Hep-2) cancer cell lines. *Journal of Photochemistry and Photobiology. B, Biology, 173*, 99–107. doi:10.1016/j.jphotobiol.2017.05.031 PMID:28570910

Verma, G. (2015). Various areas of green chemistry and safer environment an overview. *International Journal of Research in Applied. Natural and Social Sciences., 3*(3), 87–94.

Verma, J., Lal, S., & Van Noorden, C. J. (2014). Nanoparticles for hyperthermic therapy: Synthesis strategies and applications in glioblastoma. *International Journal of Nanomedicine, 9*, 2863. PMID:24959075

Versiani, A. F., Astigarraga, R. G., Rocha, E. S. O., Barboza, A. P. M., Kroon, E. G., Rachid, M. A., Souza, D. G., Ladeira, L. O., Barbosa-Stancioli, E. F., Jorio, A., & Da Fonseca, F. G. (2017). Multi-walled carbon nanotubes functionalized with recombinant Dengue virus 3 envelope proteins induce significant and specific immune responses in mice. *Journal of Nanobiotechnology, 15*(1), 26. doi:10.118612951-017-0259-4 PMID:28376812

Vidonish, J. E., Zygourakis, K., Masiello, C. A., Sabadell, G., & Alvarez, P. J. J. (2016). Thermal Treatment of Hydrocarbon-Impacted Soils: A Review of Technology Innovation for Sustainable Remediation. *Engineering, 2*(4), 426–437. doi:10.1016/J.ENG.2016.04.005

Vijayakumar, S., & Ganesan, S. (2012). Gold nanoparticles as an HIV entry inhibitor. *Current HIV Research, 10*(8), 643–646. doi:10.2174/157016212803901383 PMID:22954307

Vijayaraghavan, K., & Ashokkumar, T. (2017). Plant-mediated biosynthesis of metallic nanoparticles: A review of literature, factors affecting synthesis, characterization techniques and applications. *Journal of Environmental Chemical Engineering, 5*(5), 4866–4883. doi:10.1016/j.jece.2017.09.026

Vijila, B., Edinsha Gladis, E. H., Michael Ahitha Jose, J., Sharmila, T. M., & Joseph, J. (2020). Removal of fluoride with rice husk derived adsorbent from agro waste materials. *Materials Today: Proceedings*. doi:10.1016/j.matpr.2020.09.729

Vilela, D., González, M. C., & Escarpa, A. (2012). Sensing colorimetric approaches based on gold and silver nanoparticles aggregation: Chemical creativity behind the assay. A review. *Analytica Chimica Acta, 751*, 24–43. doi:10.1016/j.aca.2012.08.043 PMID:23084049

Vinay, S. P., Alharthi, F. A., Udayabhanu, Alsalme, A., & Nagaraju, G. (2021). Hydrothermal synthesis of Ag/rGO@CTFE nanocomposite as a promising photocatalyst for degradation action. *Journal of Molecular Structure, 1228*, 129722. doi:10.1016/j.molstruc.2020.129722

Vines, J. B., Yoon, J.-H., Ryu, N.-E., Lim, D.-J., & Park, H. (2019). Gold Nanoparticles for Photothermal Cancer Therapy [Review]. *Frontiers in Chemistry, 7*(167). https://doi.org/10.3389/fchem.2019.00167

Vivekanandhan, K., Shanmugam, P., Barabadi, H., Arumugam, V., Daniel Raj Daniel Paul Raj, D., Sivasubramanian, M., Ramasamy, S., Anand, K., Boomi, P., Chandrasekaran, B., Arokiyaraj, S., & Saravanan, M. (2021). Emerging Therapeutic Approaches to Combat COVID-19: Present Status and Future Perspectives. *Frontiers in Molecular Biosciences, 8*(March), 1–10. doi:10.3389/fmolb.2021.604447 PMID:33763450

Vivien, A., Guillaumont, M., Meziane, L., Salzemann, C., Aubert, C., Halbert, S., Gérard, H., Petit, M., & Petit, C. (2019). Role of Oleylamine Revisited: An Original Disproportionation Route to Monodispersed Cobalt and Nickel Nanocrystals. *Chemistry of Materials*, *31*(3), 960–968. Advance online publication. doi:10.1021/acs.chemmater.8b04435

Volodina, K. V., Avnir, D., & Vinogradov, V. V. (2017). Alumina nanoparticle-assisted enzyme refolding: A versatile methodology for proteins renaturation. *Scientific Reports*, *7*(1), 1–7. doi:10.103841598-017-01436-6 PMID:28469182

Wang, L., Hu, C., & Shao, L. (2017). *The antimicrobial activity of nanoparticles: present situation and prospects for the future*. Academic Press.

Wang, B., Zhang, F., He, S., Huang, F., & Peng, Z. (2014). Adsorption behavior of reduced graphene oxide for removal of heavy metal ions. *Asian Journal of Chemistry*, *26*(15), 4901–4906. doi:10.14233/ajchem.2014.17024

Wang, C., Gao, Y., Wang, L., & Li, P. (2017). Morphology regulation, structural, and photocatalytic properties of ZnO hierarchical microstructures synthesized by a simple hydrothermal method. *Physica Status Solidi. A, Applications and Materials Science*, *214*(6), 1600876. doi:10.1002/pssa.201600876

Wang, D., Lin, H., Zhang, G., Si, Y., Yang, H., Bai, G., Yang, C., Zhong, K., Cai, D., Wu, Z., Wang, R., & Zou, D. (2018). Effective pH-activated theranostic platform for synchronous magnetic resonance imaging diagnosis and chemotherapy. *ACS Applied Materials & Interfaces*, *10*(37), 31114–31123. doi:10.1021/acsami.8b11408 PMID:30141893

Wang, F., Huang, S., Gao, R., Zhou, Y., Lai, C., Li, Z., Xian, W., Qian, X., Li, Z., Huang, Y., Tang, Q., Liu, P., Chen, R., Liu, R., Li, X., Tong, X., Zhou, X., Bai, Y., Duan, G., ... Liu, L. (2020). Initial Whole-Genome Sequencing and Analysis of the Host Genetic Contribution to COVID-19 Severity and Susceptibility. *Cell Discovery*, *6*(1), 83. doi:10.103841421-020-00231-4 PMID:33298875

Wang, H., Naghavi, M., Allen, C., Barber, R. M., Bhutta, Z. A., Carter, A., Casey, D. C., Charlson, F. J., Chen, A. Z., Coates, M. M., Coggeshall, M., Dandona, L., Dicker, D. J., Erskine, H. E., Ferrari, A. J., Fitzmaurice, C., Foreman, K., Forouzanfar, M. H., Fraser, M. S., ... Murray, C. J. L. (2016). Global, regional, and national life expectancy, all-cause mortality, and cause-specific mortality for 249 causes of death, 1980–2015: A systematic analysis for the Global Burden of Disease Study 2015. *Lancet*, *388*(10053), 1459–1544. doi:10.1016/S0140-6736(16)31012-1 PMID:27733281

Wang, H., Sun, K., Tao, F., Stacchiola, D. J., & Hu, Y. H. (2013). 3D honeycomb-like structured graphene and its high efficiency as a counter-electrode catalyst for dye-sensitized solar cells. *Angewandte Chemie International Edition*, *52*(35), 9210–9214. doi:10.1002/anie.201303497 PMID:23897636

Wang, J., Chen, B., & Xing, B. (2016). Wrinkles and folds of activated graphene nanosheets as fast and efficient adsorptive sites for hydrophobic organic contaminants. *Environmental Science & Technology*, *50*(7), 3798–3808. doi:10.1021/acs.est.5b04865 PMID:26938576

Wang, J., Lee, J. S., Kim, D., & Zhu, L. (2017). Exploration of zinc oxide nanoparticles as a multitarget and multifunctional anticancer nanomedicine. *ACS Applied Materials & Interfaces*, *9*(46), 39971–39984. doi:10.1021/acsami.7b11219 PMID:29076344

Wang, J., Zhang, J., Han, L., Wang, J., Zhu, L., & Zeng, H. (2021). Graphene-based materials for adsorptive removal of pollutants from water and underlying interaction mechanism. *Advances in Colloid and Interface Science*, *289*, 102360. doi:10.1016/j.cis.2021.102360 PMID:33540288

Wang, L., Hou, Z., Quan, Z., Lian, H., Yang, P., & Lin, J. (2009). Preparation and luminescence properties of Mn2+-doped ZnGa2O4 nanofibers via electrospinning process. *Materials Research Bulletin*, *44*(10), 1978–1983. doi:10.1016/j.materresbull.2009.06.008

Wang, L., Hu, C., & Shao, L. (2017). The antimicrobial activity of nanoparticles: Present situation and prospects for the future. *International Journal of Nanomedicine*, *12*, 1227–1249. doi:10.2147/IJN.S121956 PMID:28243086

Wang, L., Wu, Y., Xie, J., Wu, S., & Wu, Z. (2018). Characterization, antioxidant and antimicrobial activities of green synthesized silver nanoparticles from Psidium guajava L. leaf aqueous extracts. *Materials Science and Engineering C*, *86*, 1–8. doi:10.1016/j.msec.2018.01.003 PMID:29525084

Wang, M., Wang, S., Yang, H., Ku, W., Yang, S., Liu, Z., & Lu, G. (2020). Carbon-based electrocatalysts derived from biomass for oxygen reduction reaction: A minireview. *Frontiers in Chemistry*, *8*, 116. doi:10.3389/fchem.2020.00116 PMID:32185161

Wang, M., Yang, Q., Li, M., Zou, H., Wang, Z., Ran, H., Zheng, Y., Jian, J., Zhou, Y., Luo, Y., Ran, Y., Jiang, S., & Zhou, X. (2020). Multifunctional Nanoparticles for Multimodal Imaging-Guided Low-Intensity Focused Ultrasound/Immunosynergistic Retinoblastoma Therapy. *ACS Applied Materials & Interfaces*, *12*(5), 5642–5657. doi:10.1021/acsami.9b22072 PMID:31940169

Wang, N., Prodanović, M., & Daigle, H. (2019). Nanopaint application for flow assurance with electromagnetic pig. *Journal of Petroleum Science Engineering*, *180*, 320–329. Advance online publication. doi:10.1016/j.petrol.2019.05.028

Wang, P., Lombi, E., Zhao, F. J., & Kopittke, P. M. (2016). Nanotechnology: A New Opportunity in Plant Sciences. *Trends in Plant Science*, *21*(8), 699–712. doi:10.1016/j.tplants.2016.04.005 PMID:27130471

Wang, S., Li, J., Zhou, Z., Zhou, S., & Hu, Z. (2019). Micro-/nano-scales direct cell behavior on biomaterial surfaces. *Molecules (Basel, Switzerland)*, *24*(1), 75. doi:10.3390/molecules24010075 PMID:30587800

Wang, S., Liu, K., Yao, X., & Jiang, L. (2015). Bioinspired surfaces with superwettability: New insight on theory, design, and applications. *Chemical Reviews*, *115*(16), 8230–8293. doi:10.1021/cr400083y PMID:26244444

Wang, S., Li, X., Liu, Y., Zhang, C., Tan, X., Zeng, G., Song, B., & Jiang, L. (2018). Nitrogen-containing amino compounds functionalized graphene oxide: synthesis, characterization, and application for the removal of pollutants from wastewater: a review. *Journal of Hazardous Materials*, *342*, 177–191. doi:10.1016/j.jhazmat.2017.06.071 PMID:28829983

Wang, W., Li, J., Liu, R., Zhang, A., & Yuan, Z. (2016). Size effect of Au/PAMAM contrast agent on CT imaging of reticuloendothelial system and tumor tissue. *Nanoscale Research Letters*, *11*(1), 429. doi:10.118611671-016-1650-z PMID:27671016

Wang, X., Liu, B., Lu, Q., & Qu, Q. (2014). Graphene-based materials: Fabrication and application for adsorption in analytical chemistry. *Journal of Chromatography. A*, *1362*, 1–15. doi:10.1016/j.chroma.2014.08.023 PMID:25160951

Wang, X., Yao, C., Zhang, G., & Liu, S. (2020). Regulating vesicle bilayer permeability and selectivity via stimuli-triggered polymersome-to-PICsome transition. *Nature Communications*, *11*(1), 1–13.

Wang, X., Zhang, Y., Zhi, C., Wang, X., Tang, D., Xu, Y., Weng, Q., Jiang, X., Mitome, M., Golberg, D., & Bando, Y. (2013). Three-dimensional strutted graphene grown by substrate-free sugar blowing for high-power-density supercapacitors. *Nature Communications*, *4*(1), 1–8. doi:10.1038/ncomms3905 PMID:24336225

Wang, Y., Gao, S., Ye, W.-H., Yoon, H. S., & Yang, Y.-Y. (2006). Co-delivery of drugs and DNA from cationic core–shell nanoparticles self-assembled from a biodegradable copolymer. *Nature Materials*, *5*(10), 791–796. doi:10.1038/nmat1737 PMID:16998471

Wang, Y., O'Connor, D., Shen, Z., Lo, I. M. C., Tsang, D. C. W., Pehkonen, S., Pu, S., & Hou, D. (2019). Green synthesis of nanoparticles for the remediation of contaminated waters and soils: Constituents, synthesizing methods, and influencing factors. *Journal of Cleaner Production*, *226*, 540–549. doi:10.1016/j.jclepro.2019.04.128

Wang, Y., Wang, B., Wang, J., Ren, Y., Xuan, C., Liu, C., & Shen, C. (2018). Superhydrophobic and superoleophilic porous reduced graphene oxide/polycarbonate monoliths for high-efficiency oil/water separation. *Journal of Hazardous Materials, 344*, 849–856. doi:10.1016/j.jhazmat.2017.11.040 PMID:29190582

Wang, Z., Fang, C., & Megharaj, M. (2014). Characterization of iron–polyphenol nanoparticles synthesized by three plant extracts and their fenton oxidation of azo dye. *ACS Sustainable Chemistry & Engineering, 2*(4), 1022–1025. doi:10.1021c500021n

Wang, Z., Hang, J. G., Feng, H., Shi, L. L., Dong, J. J., Shen, B., Luo, T., Cai, R. M., Shen, L. J., Kido, T., & Sun, X. L. (2019). Effects of perinatal dioxin exposure on development of children: A 3-year follow-up study of China cohort. *Environmental Science and Pollution Research International, 26*(20), 20780–20786. doi:10.100711356-019-05362-0 PMID:31102233

Wan, X., Song, Y., Song, N., Li, J., Yang, L., Li, Y., & Tan, H. (2016). The preliminary study of immune superparamagnetic iron oxide nanoparticles for the detection of lung cancer in magnetic resonance imaging. *Carbohydrate Research, 419*, 33–40. doi:10.1016/j.carres.2015.11.003 PMID:26649917

Wasim, M., Mushtaq, M., Khan, S. U., Farooq, A., Naeem, M. A., Khan, M. R., Salam, A., & Wei, Q. (2020). Development of bacterial cellulose nanocomposites: An overview of the synthesis of bacterial cellulose nanocomposites with metallic and metallic-oxide nanoparticles by different methods and techniques for biomedical applications. *Journal of Industrial Textiles*. Advance online publication. doi:10.1177/1528083720977201

Wei, L., Lu, J., Xu, H., Patel, A., Chen, Z.-S., & Chen, G. (2015). Silver nanoparticles: Synthesis, properties, and therapeutic applications. *Drug Discovery Today, 20*(5), 595–601. doi:10.1016/j.drudis.2014.11.014 PMID:25543008

Weinberg, H., Galyean, A., & Leopold, M. (2011). Evaluating engineered nanoparticles in natural waters. *Trends in Analytical Chemistry, 30*(1), 72–83. doi:10.1016/j.trac.2010.09.006

Weir, A., Westerhoff, P., Fabricius, L., Hristovski, K., & Von Goetz, N. (2012). *Titanium dioxide nanoparticles in food and personal care products*. Academic Press.

Weiss, C., Carriere, M., Fusco, L., Capua, I., Regla-Nava, J. A., Pasquali, M., Scott, J. A., Vitale, F., Unal, M. A., Mattevi, C., Bedognetti, D., Merkoçi, A., Tasciotti, E., Yilmazer, A., Gogotsi, Y., Stellacci, F., & Delogu, L. G. (2020). Toward Nanotechnology-enabled Approaches against the COVID-19 Pandemic. *ACS Nano, 14*(6), 6383–6406. doi:10.1021/acsnano.0c03697 PMID:32519842

Weissig, V., Pettinger, T. K., & Murdock, N. (2014). Nanopharmaceuticals (part 1): Products on the market. *International Journal of Nanomedicine, 9*, 4357.

Weldrick, P. J., Hardman, M. J., & Paunov, V. N. (2021). Smart active antibiotic nanocarriers with protease surface functionality can overcome biofilms of resistant bacteria. *Materials Chemistry Frontiers, 5*(2), 961–972. doi:10.1039/D0QM00874E

WHO. (2019). *Drinking-water*. https://www.who.int/news-room/fact-sheets/detail/drinking-water

Wigginton, N., Yeston, J., & Malakoff, D. (2012). More treasure than trash. *Science, 337*(6095), 662–663. doi:10.1126cience.337.6095.662 PMID:22879497

Wilkinson, L. J., White, R. J., & Chipman, J. K. (2011). Silver and nanoparticles of silver in wound dressings: A review of efficacy and safety. *Journal of Wound Care, 20*(11), 543–549. doi:10.12968/jowc.2011.20.11.543 PMID:22240850

Wilson, A. (2012). With Functionalization, Nanodiamonds May Increase Durability of PDC Cutters. *Journal of Petroleum Technology, 64*(12), 134–137. Advance online publication. doi:10.2118/1212-0134-JPT

Wolf, C., Angellier-Coussy, H., Gontard, N., Doghieri, F., & Guillard, V. (2018). How the shape of fillers affects the barrier properties of polymer/non-porous particles nanocomposites: A review. *Journal of Membrane Science*, *556*(April), 393–418. doi:10.1016/j.memsci.2018.03.085

Wong, H. L., Rauth, A. M., Bendayan, R., Manias, J. L., Ramaswamy, M., Liu, Z., Erhan, S. Z., & Wu, X. Y. (2006). A new polymer–lipid hybrid nanoparticle system increases cytotoxicity of doxorubicin against multidrug-resistant human breast cancer cells. *Pharmaceutical Research*, *23*(7), 1574–1585. doi:10.100711095-006-0282-x PMID:16786442

World Health Organization. (2020). *Antimicrobial resistance*. https://www.who.int/news/item/20-11-2020-world-leaders-join-forces-to-fight-the-accelerating-crisis-of-antimicrobial-resistance

Wu, J., Huang, W., & He, Z. (2013). Dendrimers as carriers for siRNA delivery and gene silencing: a review. *Scientific World Journal*, *2013*, 630654. doi:10.1155/2013/630654

Wu, B., Lu, S. T., Zhang, L. J., Zhuo, R. X., Xu, H. B., & Huang, S. W. (2017). Codelivery of doxorubicin and triptolide with reduction-sensitive lipid-polymer hybrid nanoparticles for in vitro and in vivo synergistic cancer treatment. *International Journal of Nanomedicine*, *12*, 1853–1862. https://doi.org/10.2147/ijn.S131235

Wu, F., Zhao, S., Yu, B., Chen, Y.-M., Wang, W., Song, Z.-G., Hu, Y., Tao, Z.-W., Tian, J.-H., Pei, Y.-Y., Yuan, M.-L., Zhang, Y.-L., Dai, F.-H., Liu, Y., Wang, Q.-M., Zheng, J.-J., Xu, L., Holmes, E. C., & Zhang, Y. Z. (2020). A New Coronavirus Associated with Human Respiratory Disease in China. *Nature*, *579*(7798), 265–269. doi:10.103841586-020-2008-3 PMID:32015508

Wu, P., Wang, H., Tang, Y., Zhou, Y., & Lu, T. (2014). Three-dimensional interconnected network of graphene-wrapped porous silicon spheres: In situ magnesiothermic-reduction synthesis and enhanced lithium-storage capabilities. *ACS Applied Materials & Interfaces*, *6*(5), 3546–3552. doi:10.1021/am405725u PMID:24506494

Wu, W., Wu, Z., Yu, T., Jiang, C., & Kim, W. S. (2015). Recent progress on magnetic iron oxide nanoparticles: Synthesis, surface functional strategies and biomedical applications. *Science and Technology of Advanced Materials*, *16*(2), 023501. doi:10.1088/1468-6996/16/2/023501 PMID:27877761

Wu, Y., Li, Z., Chen, J., Yu, C., Huang, X., Zhao, C., Duan, L., Yang, Y., & Lü, W. (2015). Graphene nanosheets decorated with tunable magnetic nanoparticles and their efficiency of wastewater treatment. *Materials Research Bulletin*, *68*, 234–239. doi:10.1016/j.materresbull.2015.03.073

Wu, Y., Zhu, J., & Huang, L. (2019). A review of three-dimensional graphene-based materials: Synthesis and applications to energy conversion/storage and environment. In *Carbon* (Vol. 143, pp. 610–640). Elsevier Ltd. doi:10.1016/j.carbon.2018.11.053

Xia, H., Chen, L., & Fang, Y. (2013). Highly Efficient Removal of Congo red from Wastewater by Nano-Cao. *Separation Science and Technology*, *48*(17), 2681–2687. https://doi.org/10.1080/01496395.2013.805340

Xia, K., Li, Q., Zheng, L., You, K., Tian, X., Han, B., Gao, Q., Huang, Z., Chen, G., & Zhou, C. (2017). Controllable fabrication of 2D and 3D porous graphene architectures using identical thermally exfoliated graphene oxides as precursors and their application as supercapacitor electrodes. *Microporous and Mesoporous Materials*, *237*, 228–236. doi:10.1016/j.micromeso.2016.09.015

Xiao, T., Huang, J., Wang, D., Meng, T., & Yang, X. (2019). Au and Au-Based nanomaterials: Synthesis and recent progress in electrochemical sensor applications. *Talanta*, *206*, 120210. doi:10.1016/j.talanta.2019.120210 PMID:31514855

Xie, P., & Liu, P. (2020). pH-responsive surface charge reversal carboxymethyl chitosan-based drug delivery system for pH and reduction dual-responsive triggered DOX release. *Carbohydrate Polymers*, *236*, 116093. doi:10.1016/j.carbpol.2020.116093 PMID:32172895

Xie, Y., Fang, Z., Cheng, W., Tsang, P. E., & Zhao, D. (2014). Remediation of polybrominated diphenyl ethers in soil using Ni/Fe bimetallic nanoparticles: Influencing factors, kinetics and mechanism. *The Science of the Total Environment, 485–486*, 363–370. doi:10.1016/j.scitotenv.2014.03.039 PMID:24742544

Xing, Y., Xiong, W., Zhu, L., Osawa, E., Hussin, S., & Dai, L. (2011). *DNA damage in embryonic stem cells caused by nanodiamonds*. Academic Press.

Xu, S., Habib, A. H., Pickel, A. D., & McHenry, M. E. (2015). Magnetic nanoparticle-based solder composites for electronic packaging applications. Progress in Materials Science, 67. doi:10.1016/j.pmatsci.2014.08.001

Xue, Y., Liu, J., Chen, H., Wang, R., Li, D., Qu, J., & Dai, L. (2012). Nitrogen-doped graphene foams as metal-free counter electrodes in high-performance dye-sensitized solar cells. *Angewandte Chemie International Edition, 51*(48), 12124–12127. doi:10.1002/anie.201207277 PMID:23124846

Xue, Z., Worthen, A., Qajar, A., Robert, I., Bryant, S. L., Huh, C., Prodanović, M., & Johnston, K. P. (2016). Viscosity and stability of ultra-high internal phase CO2-in-water foams stabilized with surfactants and nanoparticles with or without polyelectrolytes. *Journal of Colloid and Interface Science, 461*, 383–395. Advance online publication. doi:10.1016/j.jcis.2015.08.031 PMID:26414421

Xu, H., Guo, H., & Gong, S. (2008). Thermal barrier coatings. In *Developments in High Temperature Corrosion and Protection of Materials* (pp. 476–491). Elsevier. doi:10.1533/9781845694258.2.476

Xu, J., Xie, Z., Du, F., & Wang, X. (2021). One-step anti-superbug finishing of cotton textiles with dopamine-menthol. *Journal of Materials Science and Technology, 69*, 79–88. doi:10.1016/j.jmst.2020.08.007

Xu, L., Abedini, A., Qi, Z. B., Kim, M., Guerrero, A., & Sinton, D. (2018). Pore-scale analysis of steam-solvent coinjection: Azeotropic temperature, dilution and asphaltene deposition. *Fuel, 220*, 151–158. Advance online publication. doi:10.1016/j.fuel.2018.01.119

Xu, P., Chen, B., Li, J., & Wang, X. (2013). Gambogic acid combined with CdTe QDs for leukemia cancer cells inhibition and their bio-safety for rat brain. *Blood, 122*(21), 5028–5028. doi:10.1182/blood.V122.21.5028.5028

Xu, P., Zeng, G. M., Huang, D. L., Feng, C. L., Hu, S., Zhao, M. H., Lai, C., Wei, Z., Huang, C., Xie, G. X., & Liu, Z. F. (2012). Use of iron oxide nanomaterials in wastewater treatment: A review. *The Science of the Total Environment, 424*, 1–10. doi:10.1016/j.scitotenv.2012.02.023 PMID:22391097

Xu, Y., Shi, G., & Duan, X. (2015). Self-Assembled Three-Dimensional Graphene Macrostructures: Synthesis and Applications in Supercapacitors. *Accounts of Chemical Research, 48*(6), 1666–1675. doi:10.1021/acs.accounts.5b00117 PMID:26042764

Yadav, P., Kosandar, B. A., Jadhav, P. B., Ishaq, L. G., Addagalla, A. K. V., & Mohammad, A. R. (2015). Customized high-performance, water-based mud for unconventional reservoir drilling. *SPE Middle East Oil and Gas Show and Conference, MEOS, Proceedings*. 10.2118/172603-MS

Yadav, A., Kaushik, C., Haritash, A., Kansal, A., & Rani, N. (2006). Defluoridation of groundwater using brick powder as an adsorbent. *Journal of Hazardous Materials, 128*(2–3), 289–293. https://doi.org/10.1016/j.jhazmat.2005.08.006

Yadav, K. K., Singh, J. K., Gupta, N., & Kumar, V. (2017). A review of nanobioremediation technologies for environmental cleanup: A novel biological approach. *J Mater Environ Sci, 8*(2), 740–757.

Yadav, K. S., & Dalal, D. C. (2021). The heterogeneous multiscale method to study particle size and partitioning effects in drug delivery. *Computers & Mathematics with Applications (Oxford, England), 92*, 134–148. doi:10.1016/j.camwa.2021.03.025

Yamada, M., Foote, M., & Prow, T. W. (2015). Therapeutic gold, silver, and platinum nanoparticles. *Wiley Interdisciplinary Reviews. Nanomedicine and Nanobiotechnology*, *7*(3), 428–445. doi:10.1002/wnan.1322 PMID:25521618

Yamashita, M., Rosatto, S. S., & Kubota, L. T. (2002). Electrochemical comparative study of riboflavin, FMN and FAD immobilized on the silica gel modified with zirconium oxide. *Journal of the Brazilian Chemical Society*, *13*(5), 635–641. doi:10.1590/S0103-50532002000500015

Yan, X., Ma, C., Cui, F., McClements, D. J., Liu, X., & Liu, F. (2020). Protein-stabilized Pickering emulsions: Formation, stability, properties, and applications in foods. Trends in Food Science and Technology, 103. doi:10.1016/j.tifs.2020.07.005

Yan, D., Zhang, H., Chen, L., Zhu, G., Wang, Z., Xu, H., & Yu, A. (2014). Supercapacitive properties of Mn3O4 nanoparticles bio-synthesized from banana peel extract. *RSC Advances*, *4*(45), 23649. doi:10.1039/c4ra02603a

Yang, K., Wang, J., Chen, X., Zhao, Q., Ghaffar, A., & Chen, B. (2018). Application of graphene-based materials in water purification: From the nanoscale to specific devices. In Environmental Science: Nano (Vol. 5, Issue 6, pp. 1264–1297). Royal Society of Chemistry. doi:10.1039/C8EN00194D

Yang, C., Li, Q., Xia, Y., Lv, K., & Li, M. (2019). Enhanced visible-light photocatalytic CO2 reduction performance of Znln2S4 microspheres by using CeO2 as cocatalyst. *Applied Surface Science*, *464*, 388–395. doi:10.1016/j.apsusc.2018.09.099

Yang, G., & Park, S.-J. (2019). Conventional and Microwave Hydrothermal Synthesis and Application of Functional Materials: A Review. *Materials (Basel, Switzerland)*, *12*(7), 1177. doi:10.3390/ma12071177 PMID:30978917

Yang, H., Miao, J., Hung, S.-F., Chen, J., Tao, H. B., Wang, X., Zhang, L., Chen, R., Gao, J., Chen, H. M., Dai, L., & Liu, B. (2016). Identification of catalytic sites for oxygen reduction and oxygen evolution in N-doped graphene materials: Development of highly efficient metal-free bifunctional electrocatalyst. *Science Advances*, *2*(4), e1501122. doi:10.1126ciadv.1501122 PMID:27152333

Yang, J., Hou, B., Wang, J., Tian, B., Bi, J., Wang, N., Li, X., & Huang, X. (2019). Nanomaterials for the Removal of Heavy Metals from Wastewater. *Nanomaterials (Basel, Switzerland)*, *9*(3), 424. https://doi.org/10.3390/nano9030424

Yang, J., Zhang, Q., Chang, H., & Cheng, Y. (2015). Surface-engineered dendrimers in gene delivery. *Chemical Reviews*, *115*(11), 5274–5300. https://doi.org/10.1021/cr500542t

Yang, K., Feng, L., & Liu, Z. (2016). Stimuli responsive drug delivery systems based on nano-graphene for cancer therapy. *Advanced Drug Delivery Reviews*, *105*, 228–241.

Yang, K., Wu, W., Jing, Q., & Zhu, L. (2008). Aqueous adsorption of aniline, phenol, and their substitutes by multi-walled carbon nanotubes. *Environmental Science & Technology*, *42*(21), 7931–7936. doi:10.1021/es801463v PMID:19031883

Yang, L., Seah, M. P., Gilmore, I. S., Morris, R. J. H., Dowsett, M. G., Boarino, L., Sparnacci, K., & Laus, M. (2013). Depth profiling and melting of nanoparticles in secondary ion mass spectrometry (SIMS). *The Journal of Physical Chemistry C*, *117*(31), 16042–16052. doi:10.1021/jp4048538

Yang, N., WeiHong, L., & Hao, L. (2014). Biosynthesis of Au nanoparticles using agricultural waste mango peel extract and its in vitro cytotoxic effect on two normal cells. *Materials Letters*, *134*, 67–70. doi:10.1016/j.matlet.2014.07.025

Yang, S., Hu, J., Chen, C., Shao, D., & Wang, X. (2011). Mutual Effects of Pb(II) and Humic Acid Adsorption on Multiwalled Carbon Nanotubes/Polyacrylamide Composites from Aqueous Solutions. *Environmental Science & Technology*, *45*(8), 3621–3627. doi:10.1021/es104047d PMID:21395259

Yang, X., Salado-Leza, D., Porcel, E., González-Vargas, C. R., Savina, F., Dragoe, D., Remita, H., & Lacombe, S. (2020). A facile one-pot synthesis of versatile PEGylated platinum nanoflowers and their application in radiation therapy. *International Journal of Molecular Sciences*, *21*(5), 1619. doi:10.3390/ijms21051619 PMID:32120829

Yang, Y., Hu, G., Chen, F., Liu, J., Liu, W., Zhang, H., & Wang, B. (2015). An atom-scale interfacial coordination strategy to prepare hierarchically porous Fe3O4-graphene frameworks and their application in charge and size selective dye removal. *Chemical Communications*, *51*(76), 14405–14408. doi:10.1039/C5CC06257H PMID:26271754

Yang, Z. F., Li, L. Y., Te Hsieh, C., & Juang, R. S. (2018). Coprecipitation of magnetic Fe3O4 nanoparticles onto carbon nanotubes for removal of copper ions from aqueous solution. *Journal of the Taiwan Institute of Chemical Engineers*, *82*, 56–60. doi:10.1016/j.jtice.2017.11.009

Yang, Z., Huang, X., Ma, X., Zhou, Z., Guo, H., Yao, Z., Feng, S.-P., & Tang, C. Y. (2019). Fabrication of a novel and green thin-film composite membrane containing nanovoids for water purification. *Journal of Membrane Science*, *570*, 314–321. doi:10.1016/j.memsci.2018.10.057

Yang, Z., Peng, H., Wang, W., & Liu, T. (2010). Crystallization behavior of poly(ε-caprolactone)/layered double hydroxide nanocomposites. *Journal of Applied Polymer Science*, *116*(5), 2658–2667. doi:10.1002/app.31787

Yang, Z., Zhang, X., Luo, X., Jiang, Q., Liu, J., & Jiang, Z. (2013). Enzymatic synthesis of poly(butylene-co-sebacate-co-glycolate) copolyesters and evaluation of the copolymer nanoparticles as biodegradable carriers for doxorubicin delivery. *Macromolecules*, *46*(5), 1743–1753. doi:10.1021/ma302433x

Yang, Z., Zhou, Y., Feng, Z., Rui, X., Zhang, T., & Zhang, Z. (2019). A review on reverse osmosis and nanofiltration membranes for water purification. Polymers, 11(8). doi:10.3390/polym11081252

Yan, P., Zheng, J., Gu, M., Xiao, J., Zhang, J. G., & Wang, C. M. (2017). Intragranular cracking as a critical barrier for high-voltage usage of layer-structured cathode for lithium-ion batteries. *Nature Communications*, *8*(1), 1–9. doi:10.1038/ncomms14101 PMID:28091602

Yan, W., Cao, X., Tian, J., Jin, C., Ke, K., & Yang, R. (2016). Nitrogen/sulfur dual-doped 3D reduced graphene oxide networks-supported CoFe2O4 with enhanced electrocatalytic activities for oxygen reduction and evolution reactions. *Carbon*, *99*, 195–202. doi:10.1016/j.carbon.2015.12.011

Yan, X., Hu, W., Guo, J., Cai, X., Huang, L., Xiong, Y., & Tan, S. (2016). Easily Separated a Novel rGO-MMT Three-Dimensional Aerogel with Good Adsorption and Recyclable Property. *ChemistrySelect*, *1*(18), 5828–5837. doi:10.1002lct.201601028

Yao, G., Liu, X., Zhang, G., Han, Z., & Liu, H. (2021). Green synthesis of tannic acid functionalized graphene hydrogel to efficiently adsorb methylene blue. *Colloids and Surfaces. A, Physicochemical and Engineering Aspects*, *625*, 126972. doi:10.1016/j.colsurfa.2021.126972

Yaqoob, A. A., Parveen, T., Umar, K., & Mohamad Ibrahim, M. N. (2020). Role of nanomaterials in the treatment of wastewater: A review. *Water (Basel)*, *12*(2), 495. doi:10.3390/w12020495

Yazdian-Robati, R., Arab, A., Ramezani, M., Rafatpanah, H., Bahreyni, A., Nabavinia, M. S., Abnous, K., & Taghdisi, S. M. (2019). Smart aptamer-modified calcium carbonate nanoparticles for controlled release and targeted delivery of epirubicin and melittin into cancer cells in vitro and in vivo. *Drug Development and Industrial Pharmacy*, *45*(4), 603–610. https://doi.org/10.1080/03639045.2019.1569029

Ye, J., Wang, Y., Xu, Q., Wu, H., Tong, J., & Shi, J. (2021). Removal of hexavalent chromium from wastewater by Cu/Fe bimetallic nanoparticles. *Scientific Reports*, *11*(1), 10848. doi:10.103841598-021-90414-0 PMID:34035405

Yekeen, N., Padmanabhan, E., Idris, A. K., & Chauhan, P. S. (2019). Nanoparticles applications for hydraulic fracturing of unconventional reservoirs: A comprehensive review of recent advances and prospects. Journal of Petroleum Science and Engineering, 178. doi:10.1016/j.petrol.2019.02.067

Ye, N., Wang, Z., Wang, S., Fang, H., & Wang, D. (2018). Aqueous aggregation and stability of graphene nanoplatelets, graphene oxide, and reduced graphene oxide in simulated natural environmental conditions: Complex roles of surface and solution chemistry. *Environmental Science and Pollution Research International*, *25*(11), 10956–10965. doi:10.100711356-018-1326-6 PMID:29399742

Yeo, B., Stadler, J., Schmid, T., Zenobi, R., & Zhang, W. (2009). Tip-enhanced Raman Spectroscopy – Its status, challenges and future directions. *Chemical Physics Letters*, *472*(1–3), 1–13. doi:10.1016/j.cplett.2009.02.023 PMID:20161522

Yokoyama, M. (2011). Clinical applications of polymeric micelle carrier systems in chemotherapy and image diagnosis of solid tumors. *Journal of Experimental and Clinical Medicine*, *3*(4), 151–158.

Yoo, B. M., Shin, H. J., Yoon, H. W., & Park, H. B. (2014). Graphene and graphene oxide and their uses in barrier polymers. *Journal of Applied Polymer Science*, *131*(1). Advance online publication. doi:10.1002/app.39628

Yoonus, J., Resmi, R., & Beena, B. (2021). Greener nanoscience: Piper betel leaf extract mediated synthesis of CaO nanoparticles and evaluation of its antibacterial and anticancer activity. *Materials Today: Proceedings*, *41*, 535–540. doi:10.1016/j.matpr.2020.05.246

Yoon, Y., Han, K. I., Kim, B. H., Lee, I. G., Kim, Y., Kim, J. P., & Hwang, W. S. (2018). Formation of β-Ga2O3 nanofibers of sub-50 nm diameter synthesized by electrospinning method. *Thin Solid Films*, *645*, 358–362. doi:10.1016/j.tsf.2017.11.010

You, H., Yang, S., Ding, B., & Yang, H. (2013). Synthesis of colloidal metal and metal alloy nanoparticles for electrochemical energy applications. *Chemical Society Reviews*, *42*(7), 2880–2904. Advance online publication. doi:10.1039/C2CS35319A PMID:23152097

Yuan, L., Xin, N., Liu, Y., & Shi, W. (2021). In situ construction of multi-dimensional Co3O4/NiCo2O4 hierarchical flakes on self-supporting carbon substrate with ultra-high capacitance for hybrid supercapacitors. *Journal of Colloid and Interface Science*, *599*, 158–167. doi:10.1016/j.jcis.2021.04.036 PMID:33933790

Yue, H. Y., Zhang, H. J., Huang, S., Lu, X. X., Gao, X., Song, S. S., Wang, Z., Wang, W. Q., & Guan, E. H. (2020). Highly sensitive and selective dopamine biosensor using Au nanoparticles-ZnO nanocone arrays/graphene foam electrode. *Materials Science and Engineering C*, *108*, 110490. Advance online publication. doi:10.1016/j.msec.2019.110490 PMID:31923936

Yue, X., Cai, S., Shao, R., Liu, Y., & Lu, Y. (2001). Determination of trace impurities in high purity gallium by inductively coupled plasma mass spectrometry. *Fenxi Huaxue*, *29*(11), 1309–1310.

Yu, J., Khalil, M., Liu, N., & Lee, R. (2014). Effect of particle hydrophobicity on CO2 foam generation and foam flow behavior in porous media. *Fuel*, *126*, 104–108. Advance online publication. doi:10.1016/j.fuel.2014.02.053

Yulizar, Y., Bakri, R., Apriandanu, D. O. B., & Hidayat, T. (2018). ZnO/CuO nanocomposite prepared in one-pot green synthesis using seed bark extract of Theobroma cacao. *Nano-Structures and Nano-Objects*, *16*, 300–305. doi:10.1016/j.nanoso.2018.09.003

Yulizar, Y., Sudirman, Apriandanu, D. O. B., & Wibowo, A. P. (2019). Plant extract mediated synthesis of Au/TiO_2 nanocomposite and its photocatalytic activity under sodium light irradiation. *Composites Communications*, *16*(May), 50–56. doi:10.1016/j.coco.2019.08.006

Yu, M., Yuan, W., Li, D., Schwendeman, A., & Schwendeman, S. P. (2019). Predicting drug release kinetics from nanocarriers inside dialysis bags. *Journal of Controlled Release*, *315*, 23–30. doi:10.1016/j.jconrel.2019.09.016 PMID:31629038

Yun, G., Pan, S., Wang, T. Y., Guo, J., Richardson, J. J., & Caruso, F. (2018). Synthesis of metal nanoparticles in metal-phenolic networks: Catalytic and antimicrobial applications of coated textiles. *Advanced Healthcare Materials*, *7*(5), 1700934. doi:10.1002/adhm.201700934 PMID:29024556

Yusof, F., & Ismail, N. A. S. (2015). Antioxidants effects of platinum nanoparticles: A potential alternative treatment to lung diseases. *Journal of Applied Pharmaceutical Science*, *5*(7), 140–145. doi:10.7324/JAPS.2015.50722

Yuvakkumar, R., Suresh, J., Nathanael, A. J., Sundrarajan, M., & Hong, S. I. (2014). Rambutan (Nephelium lappaceum L.) peel extract assisted biomimetic synthesis of nickel oxide nanocrystals. *Materials Letters*, *128*, 170–174. doi:10.1016/j.matlet.2014.04.112

Yu, Z., Gao, L., Chen, K., Zhang, W., Zhang, Q., Li, Q., & Hu, K. (2021). Nanoparticles: A new approach to upgrade cancer diagnosis and treatment. *Nanoscale Research Letters*, *16*(1), 88. doi:10.118611671-021-03489-z PMID:34014432

Yu, Z., Li, Q., Wang, J., Yu, Y., Wang, Y., Zhou, Q., & Li, P. (2020). Reactive oxygen species-related nanoparticle toxicity in the biomedical field. *Nanoscale Research Letters*, *15*(1), 115. doi:10.118611671-020-03344-7 PMID:32436107

Zachar, O. (2020). Formulations for COVID-19 Early Stage Treatment via Silver Nanoparticles Inhalation Delivery at Home and Hospital. *ScienceOpen* Preprints, 1–14. doi:10.14293/S2199-1006.1.SOR-.PPHBJEO.v1

Záleská, M., Pivák, A., Pavlíková, M., Jankovský, O., & Pavlík, Z. (2020). Influence of graphite oxide addition on the properties of magnesium oxychloride cement composites. *IOP Conference Series. Materials Science and Engineering*, *960*(2), 022080. Advance online publication. doi:10.1088/1757-899X/960/2/022080

Zandi, A., Sh, Z. D., Shojaeian, F., Mousavi-Kiasary, S. S., Abbasvandi, F., Zandi, A., Gilani, A., Saghafi, Z., Kordehlachin, Y., & Mamdouh, A. (2021). The design and fabrication of nanoengineered platinum needles with laser welded carbon nanotubes (CNTs) for the electrochemical biosensing of cancer lymph nodes. *Biomaterials Science*.

Zarei, M., Seyedi, N., Maghsoudi, S., Nejad, M. S., & Sheibani, H. (2021). Green synthesis of Ag nanoparticles on the modified graphene oxide using Capparis spinosa fruit extract for catalytic reduction of organic dyes. *Inorganic Chemistry Communications*, *123*(November), 108327. doi:10.1016/j.inoche.2020.108327

Zawadzki, J., & Bogacki, J. (2016). Smart magnetic markers use in hydraulic fracturing. Chemosphere, 162. doi:10.1016/j.chemosphere.2016.07.058

Zazo, H., Colino, C. I., & Lanao, J. M. (2016). Current applications of nanoparticles in infectious diseases. *Journal of Controlled Release*, *224*, 86–102. doi:10.1016/j.jconrel.2016.01.008 PMID:26772877

Zhang, H., & Chen, G. (2009). *Potent antibacterial activities of Ag/TiO2 nanocomposite powders synthesized by a one-pot sol– gel method.* Academic Press.

Zhang, X., Han, Y., Huang, L., & Dong, S. (2016). 3D Graphene Aerogels Decorated with Cobalt Phosphide Nanoparticles as Electrocatalysts for the Hydrogen Evolution Reaction. *ChemSusChem*, *9*(21), 3049–3053. doi:10.1002/cssc.201600904

Zhang, D., Ma, X., Gu, Y., Huang, H., & Zhang, G. (2020). Green Synthesis of Metallic Nanoparticles and Their Potential Applications to Treat Cancer. *Frontiers in Chemistry*, *8*, 799. doi:10.3389/fchem.2020.00799 PMID:33195027

Zhang, H. Z., Li, R. S., Gao, P. F., Wang, N., Lei, G., Huang, C. Z., & Wang, J. (2017). Real-time dark-field light scattering imaging to monitor the coupling reaction with gold nanorods as an optical probe. *Nanoscale*, *9*(10), 3568–3575. doi:10.1039/C6NR09453H PMID:28244517

Zhang, J., Haas, R. M., & Leone, A. M. (2012). Polydispersity characterization of lipid nanoparticles for siRNA delivery using multiple detection size-exclusion chromatography. *Analytical Chemistry*, *84*(14), 6088–6096. doi:10.1021/ac3007768 PMID:22816783

Zhang, J., Zhou, Q., Cao, J., Wu, W., Zhang, H., Shi, Y., Mao, Q., & Ma, H. (2021). Flexible textile ion sensors based on reduced graphene oxide/fullerene and their potential applications of sweat characterization. *Cellulose (London, England)*, *28*(5), 3123–3133. doi:10.100710570-021-03730-z

Zhang, L., Li, P., Gong, Z., & Li, X. (2008). Photocatalytic degradation of polycyclic aromatic hydrocarbons on soil surfaces using TiO2 under UV light. *Journal of Hazardous Materials*, *158*(2–3), 478–484. doi:10.1016/j.jhazmat.2008.01.119 PMID:18372106

Zhang, L., Pan, F., Liu, X., Yang, L., Jiang, X., Yang, J., & Shi, W. (2013). Multi-walled carbon nanotubes as sorbent for recovery of endocrine disrupting compound-bisphenol F from wastewater. *Chemical Engineering Journal*, *218*, 238–246. doi:10.1016/j.cej.2012.12.046

Zhang, N.-n., Yu, R.-s., Xu, M., Cheng, X.-y., Chen, C.-m., Xu, X.-l., Lu, C.-y., Lu, K.-j., Chen, M.-j., & Zhu, M.-l. (2018). Visual targeted therapy of hepatic cancer using homing peptide modified calcium phosphate nanoparticles loading doxorubicin guided by T1 weighted MRI. *Nanomedicine; Nanotechnology, Biology, and Medicine*, *14*(7), 2167–2178.

Zhang, N., Qiu, Y., Sun, H., Hao, J., Chen, J., Xi, J., Liu, J., He, B., & Bai, Z. W. (2021). Substrate-Assisted Encapsulation of Pd-Fe Bimetal Nanoparticles on Functionalized Silica Nanotubes for Catalytic Hydrogenation of Nitroarenes and Azo Dyes. *ACS Applied Nano Materials*, *4*(6), 5854–5863. doi:10.1021/acsanm.1c00777

Zhang, R., Pan, D., Cai, X., Yang, X., Senpan, A., Allen, J. S., Lanza, G. M., & Wang, L. V. (2015). ανβ3-targeted copper nanoparticles incorporating an Sn 2 lipase-labile fumagillin prodrug for photoacoustic neovascular imaging and treatment. *Theranostics*, *5*(2), 124–133. doi:10.7150/thno.10014 PMID:25553103

Zhang, R., Zhang, Y., Dong, Z. C., Jiang, S., Zhang, C., Chen, L. G., Zhang, L., Liao, Y., Aizpurua, J., Luo, Y., Yang, J. L., & Hou, J. G. (2013). Chemical mapping of a single molecule by plasmon-enhanced Raman scattering. *Nature*, *498*(7452), 82–86. doi:10.1038/nature12151 PMID:23739426

Zhang, S. J., Shao, T., Bekaroglu, S. S. K., & Karanfil, T. (2010). Adsorption of synthetic organic chemicals by carbon nanotubes: Effects of background solution chemistry. *Water Research*, *44*(6), 2067–2074. doi:10.1016/j.watres.2009.12.017 PMID:20071001

Zhang, S. J., Shao, T., Kose, H. S., & Karanfil, T. (2010). Adsorption of aromatic compounds by carbonaceous adsorbents: A comparative study on granular activated carbon, activated carbon fiber, and carbon nanotubes. *Environmental Science & Technology*, *44*(16), 6377–6383. doi:10.1021/es100874y PMID:20704238

Zhang, S., Shao, T., & Karanfil, T. (2011). The effects of dissolved natural organic matter on the adsorption of synthetic organic chemicals by activated carbons and carbon nanotubes. *Water Research*, *45*(3), 1378–1386. doi:10.1016/j.watres.2010.10.023 PMID:21093009

Zhang, T., Lu, G., Shen, H., Shi, K., Jiang, Y., Xu, D., & Gong, Q. (2014). Photoluminescence of a single complex plasmonic nanoparticle. *Scientific Reports*, *4*(1), 3867. doi:10.1038rep03867 PMID:24463794

Zhang, W., Lu, Y., Sun, H., Zhang, Y., Zhou, M., Song, Q., & Gao, Y. (2019). Effects of multi−walled carbon nanotubes on pyrene adsorption and desorption in soils: The role of soil constituents. *Chemosphere*, *221*, 203–211. doi:10.1016/j.chemosphere.2019.01.030 PMID:30640002

Zhang, X. F., Liu, Z. G., Shen, W., & Gurunathan, S. (2016). Silver Nanoparticles: Synthesis, Characterization, Properties, Applications, and Therapeutic Approaches. *International Journal of Molecular Sciences*, *17*(9), 1534. Advance online publication. doi:10.3390/ijms17091534 PMID:27649147

Zhang, X., Lai, Y., Ge, M., Zheng, Y., Zhang, K.-Q., & Lin, Z. (2015). Fibrous and flexible supercapacitors comprising hierarchical nanostructures with carbon spheres and graphene oxide nanosheets. *Journal of Materials Chemistry. A, Materials for Energy and Sustainability*, *3*(24), 12761–12768. doi:10.1039/C5TA03252K

Zhang, Y., Cao, J., & Yuan, Z. (2020). Strategies and challenges to improve the performance of tumor-associated active targeting. *Journal of Materials Chemistry. B, Materials for Biology and Medicine*, *8*(18), 3959–3971.

Zhang, Y., Fang, F., Li, L., & Zhang, J. (2020). Self-Assembled Organic Nanomaterials for Drug Delivery, Bioimaging, and Cancer Therapy. *ACS Biomaterials Science & Engineering*, *6*(9), 4816–4833. doi:10.1021/acsbiomaterials.0c00883 PMID:33455214

Zhang, Y., Li, M., Gao, X., Chen, Y., & Liu, T. (2019). Nanotechnology in cancer diagnosis: Progress, challenges and opportunities. *Journal of Hematology & Oncology*, *12*(1), 137. doi:10.118613045-019-0833-3 PMID:31847897

Zhang, Y., Yang, J., Li, Q., & Cao, X. (2007). Preparation of Ga2O3 nanoribbons and tubes by electrospinning. *Journal of Crystal Growth*, *308*(1), 180–184. doi:10.1016/j.jcrysgro.2007.07.036

Zhang, Y., Yuan, S., Feng, X., Li, H., Zhou, J., & Wang, B. (2016). Preparation of Nanofibrous Metal-Organic Framework Filters for Efficient Air Pollution Control. *Journal of the American Chemical Society*, *138*(18), 5785–5788. doi:10.1021/jacs.6b02553 PMID:27090776

Zhang, Y., Zhang, L., & Zhou, C. (2013). Review of chemical vapor deposition of graphene and related applications. *Accounts of Chemical Research*, *46*(10), 2329–2339. doi:10.1021/ar300203n PMID:23480816

Zhang, Z., Fujita, K., Kawata, S., Bando, K., Mochizuki, K., & Taguchi, A. (2019). Quantitative Evaluation of SERS Nanoparticles for Intracellular pH Sensing at a Single Particle Level. *Analytical Chemistry*, 1–25. doi:10.1021/acs.analchem.8b03276 PMID:30698014

Zhang, Z., Qian, H., Yang, M., Li, R., Hu, J., Li, L., Yu, L., Liu, B., & Qian, X. (2017). Gambogic acid-loaded biomimetic nanoparticles in colorectal cancer treatment. *International Journal of Nanomedicine*, *12*, 1593–1605. https://doi.org/10.2147/ijn.S127256

Zhang, Z., Xiao, F., Qian, L., Xiao, J., Wang, S., & Liu, Y. (2014). Facile synthesis of 3D MnO2-graphene and carbon nanotube-graphene composite networks for high-performance, flexible, all-solid-state asymmetric supercapacitors. *Advanced Energy Materials*, *4*(10), 1400064. doi:10.1002/aenm.201400064

Zhang, Z., Xiao, F., Xi, J., Sun, T., Xiao, S., Wang, H., Wang, S., & Liu, Y. (2014). Encapsulating Pd nanoparticles in double-shelled graphene@carbon hollow Spheres for excellent chemical catalytic property. *Scientific Reports*, *4*(1), 1–5. doi:10.1038rep04053 PMID:24514577

Zhao, H. Y., Xu, X. X., Zhang, J. X., Zheng, W., & Zheng, Y. F. (2010). Carbon nanotube–hydroxyapatite–hemoglobin nanocomposites with high bioelectrocatalytic activity. *Bioelectrochemistry (Amsterdam, Netherlands)*, *78*(2), 124–129. doi:10.1016/j.bioelechem.2009.08.009 PMID:19762287

Zhao, J.-G., Zhang, Z.-X., Ma, Z.-W., Duan, H.-G., Guo, X.-S., & Xie, E.-Q. (2008). Structural and photoluminescence properties of β-Ga2O3 nanofibers fabricated by electrospinning method. *Chinese Physics Letters*, *25*(10), 3787–3789. doi:10.1088/0256-307X/25/10/073

Zhao, J., Zhang, W., Xie, E., Ma, Z., Zhao, A., & Liu, Z. (2011). Structure and photoluminescence of β-Ga 2 O 3 :Eu 3+ nanofibers prepared by electrospinning. *Applied Surface Science*, *257*(11), 4968–4972. doi:10.1016/j.apsusc.2010.12.157

Zhao, L., Wang, F., & Fan, W. (2020). Cisplatin Nano-Liposomes Promoting Apoptosis of Retinoblastoma Cells Both In Vivo and In Vitro. *Nanoscience and Nanotechnology Letters*, *12*(4), 536–542. doi:10.1166/nnl.2020.3127

Zhao, Y., Vivero-Escoto, J. L., Slowing, I. I., Trewyn, B. G., & Lin, V. S.-Y. (2010). Capped mesoporous silica nanoparticles as stimuli-responsive controlled release systems for intracellular drug/gene delivery. *Expert Opinion on Drug Delivery*, *7*(9), 1013–1029. doi:10.1517/17425247.2010.498816 PMID:20716017

Zhao, Z., Xiao, Y., Xu, L., Liu, Y., Jiang, G., Wang, W., Li, B., Zhu, T., Tan, Q., Tang, L., Zhou, H., Huang, X., & Shan, H. (2021). Glycyrrhizic Acid Nanoparticles as Antiviral and Anti-inflammatory Agents for COVID-19 Treatment. *ACS Applied Materials & Interfaces*, *13*(18), 20995–21006. doi:10.1021/acsami.1c02755 PMID:33930273

Zheng, K., Setyawati, M. I., Leong, D. T., & Xie, J. (2017). *Antimicrobial gold nanoclusters*. Academic Press.

Zheng, B., Kong, T., Jing, X., Odoom-Wubah, T., Li, X., Sun, D., Lu, F., Zheng, Y., Huang, J., & Li, Q. (2013). Plant-mediated synthesis of platinum nanoparticles and its bioreductive mechanism. *Journal of Colloid and Interface Science*, *396*, 138–145. doi:10.1016/j.jcis.2013.01.021 PMID:23403109

Zheng, F., Luk, S.-Y., Kwong, T.-L., & Yung, K.-F. (2016). Synthesis of hollow PtAg alloy nanospheres with excellent electrocatalytic performances towards methanol and formic acid oxidations. *RSC Advances*, *6*(50), 44902–44907. doi:10.1039/C6RA06398E

Zhong, Y. L., Tian, Z., Simon, G. P., & Li, D. (2015). Scalable production of graphene via wet chemistry: Progress and challenges. In Materials Today (Vol. 18, Issue 2, pp. 73–78). Elsevier. doi:10.1016/j.mattod.2014.08.019

Zhong, G., Liu, J., & Liu, X. (2015). A fast colorimetric assay for lead detection using label-free gold nanoparticles (AuNPs). *Micromachines*, *6*(4), 462–472. doi:10.3390/mi6040462

Zhong, H., Zhu, Z., Lin, J., Cheung, C. F., Lu, V. L., Yan, F., Chan, C.-Y., & Li, G. (2020). Reusable and Recyclable Graphene Masks with Outstanding Superhydrophobic and Photothermal Performances. *ACS Nano*, *14*(5), 6213–6221. doi:10.1021/acsnano.0c02250 PMID:32329600

Zhong, Q., Bielski, E. R., Rodrigues, L. S., Brown, M. R., Reineke, J. J., & da Rocha, S. R. (2016). Conjugation to Poly(amidoamine) Dendrimers and Pulmonary Delivery Reduce Cardiac Accumulation and Enhance Antitumor Activity of Doxorubicin in Lung Metastasis. *Molecular Pharmaceutics*, *13*(7), 2363–2375. https://doi.org/10.1021/acs.molpharmaceut.6b00126

Zhou, J., Gao, P. F., Zhang, H. Z., Lei, G., Zheng, L. L., Liu, H., & Huang, C. Z. (2017). Color resolution improvement of the dark-field microscopy imaging of single light scattering plasmonic nanoprobes for microRNA visual detection. *Nanoscale*, *9*(13), 4593–4600. doi:10.1039/C6NR09452J PMID:28322387

Zhou, J., Lei, G., Zheng, L. L., Gao, P. F., & Huang, C. Z. (2016). HSI colour-coded analysis of scattered light of single plasmonic nanoparticles. *Nanoscale*, *8*(22), 11467–11471. doi:10.1039/C6NR01089J PMID:27194457

Zhou, M., Tian, M., & Li, C. (2016). Copper-based nanomaterials for cancer imaging and therapy. *Bioconjugate Chemistry*, *27*(5), 1188–1199. doi:10.1021/acs.bioconjchem.6b00156 PMID:27094828

Zhou, Y., Lin, W., Huang, J., Wang, W., Gao, Y., Lin, L., Li, Q., Lin, L., & Du, M. (2010). Biosynthesis of gold nanoparticles by foliar broths: Roles of biocompounds and other attributes of the extracts. *Nanoscale Research Letters*, *5*(8), 1351–1359. doi:10.100711671-010-9652-8 PMID:20676207

Zhou, Y., Shi, L., Li, Q., Jiang, H., Lv, G., Zhao, J., Wu, C., Selke, M., & Wang, X. (2010). Imaging and inhibition of multi-drug resistance in cancer cells via specific association with negatively charged CdTe quantum dots. *Biomaterials*, *31*(18), 4958–4963.

Zhu, C., Han, T. Y. J., Duoss, E. B., Golobic, A. M., Kuntz, J. D., Spadaccini, C. M., & Worsley, M. A. (2015). Highly compressible 3D periodic graphene aerogel microlattices. *Nature Communications*, *6*(1), 1–8. doi:10.1038/ncomms7962 PMID:25902277

Zhu, C., Liu, T., Qian, F., Han, T. Y. J., Duoss, E. B., Kuntz, J. D., Spadaccini, C. M., Worsley, M. A., & Li, Y. (2016). Supercapacitors Based on Three-Dimensional Hierarchical Graphene Aerogels with Periodic Macropores. *Nano Letters*, *16*(6), 3448–3456. doi:10.1021/acs.nanolett.5b04965 PMID:26789202

Zhu, F., Liu, T., Zhang, Z., & Liang, W. (2021). Remediation of hexavalent chromium in column by green synthesized nanoscale zero-valent iron/nickel: Factors, migration model and numerical simulation. *Ecotoxicology and Environmental Safety*, *207*, 111572. doi:10.1016/j.ecoenv.2020.111572 PMID:33254420

Zhu, H., Chen, D., Li, N., Xu, Q., Li, H., He, J., & Lu, J. (2015). Graphene foam with switchable oil wettability for oil and organic solvents recovery. *Advanced Functional Materials*, *25*(4), 597–605. doi:10.1002/adfm.201403864

Zhu, H., Chen, D., Yang, S., Li, N., Xu, Q., Li, H., Wang, L., He, J., Jiang, J., & Lu, J. (2016). A versatile and cost-effective reduced graphene oxide-crosslinked polyurethane sponge for highly effective wastewater treatment. *RSC Advances*, *6*(44), 38350–38355. doi:10.1039/C6RA05450A

Zhu, J.-J., Li, J.-J., Huang, H.-P., & Cheng, F.-F. (2013). *Quantum dots for DNA biosensing*. Springer.

Zhu, J., Zhao, Y., Ning, P., & Tang, S. (2021). Cobalt sulphide/graphene aerogel nanocomposite with enhanced anode performance for lithium energy storage. *Bulletin of Materials Science*, *44*(1), 1–8. doi:10.100712034-020-02275-4

Zhu, N., Zhang, D., Wang, W., Li, X., Yang, B., Song, J., Zhao, X., Huang, B., Shi, W., Lu, R., Niu, P., Zhan, F., Ma, X., Wang, D., Xu, W., Wu, G., Gao, G. F., & Tan, W. (2020). A Novel Coronavirus from Patients with Pneumonia in China, 2019. *The New England Journal of Medicine*, *382*(8), 727–733. doi:10.1056/NEJMoa2001017 PMID:31978945

Zhu, Y., Murali, S., Stoller, M. D., Ganesh, K. J., Cai, W., Ferreira, P. J., Pirkle, A., Wallace, R. M., Cychosz, K. A., Thommes, M., Su, D., Stach, E. A., & Ruoff, R. S. (2011). Carbon-Based Supercapacitors Produced by Activation of Graphene. *Science*, *332*(6037), 1537–1541. doi:10.1126cience.1200770 PMID:21566159

Zhu, Y., Xing, L., Zheng, X., Yang, C. X., He, Y. J., Zhou, T. J., Jin, Q. R., & Jiang, H. L. (2020). Amplification of tumor antigen presentation by NLGplatin to improve chemoimmunotherapy. *International Journal of Pharmaceutics*, *573*, 118736. doi:10.1016/j.ijpharm.2019.118736 PMID:31756442

Zou, L., Wang, J., Gao, Y., Ren, X., Rottenberg, M. E., Lu, J., & Holmgren, A. (2018). Synergistic antibacterial activity of silver with antibiotics correlating with the upregulation of the ROS production. *Scientific Reports*, *8*(1), 11131. doi:10.103841598-018-29313-w PMID:30042429

Zou, R., Xu, J., Kuffner, S., Becker, J., Li, T., Guan, X., Zhang, X., Li, L., Cohen Stuart, M. A., & Guo, X. (2019). Spherical Poly(vinyl imidazole) Brushes Loading Nickel Cations as Nanocatalysts for Aquathermolysis of Heavy Crude Oil. *Energy & Fuels*, *33*(2), 998–1006. Advance online publication. doi:10.1021/acs.energyfuels.8b03964

Zou, T., Lu, W., Mezhuev, Y., Lan, M., Li, L., Liu, F., Cai, T., Wu, X., & Cai, Y. (2021). A review of nanoparticle drug delivery systems responsive to endogenous breast cancer microenvironment. *European Journal of Pharmaceutics and Biopharmaceutics*, *166*, 30–43. doi:10.1016/j.ejpb.2021.05.029 PMID:34098073

Zuber, A., Purdey, M., Schartner, E., Forbes, C., Hoek, B., Giles, D., Abell, A., Monro, T., & Ebendorff, H. (2016). Detection of gold nanoparticles with different sizes using absorption and fluorescence-based method. *Sensors and Actuators. B, Chemical*, *227*, 117–127. doi:10.1016/j.snb.2015.12.044

About the Contributors

Rajni Garg is working as Head, Department of Chemistry at Rayat Bahra University, Mohali (Punjab), India. She received her B.Sc. (Medical) at the D.A.V. College, Abohar (Punjab), India; M. Sc. (H.S.) in Chemistry, Panjab University, India; M.Phil. in Chemistry at CDLU, India and Ph.D. (Chemistry) at the GKU, India. Her areas of interest include nanochemistry, green chemistry and supramolecular chemistry. She has authored and co-authored many textbooks and scientific papers printed in international and national journals, and proceedings of international and national scientific conferences. She has also published five patents on the potential applications of nanomaterials in cementitious materials. Currently she is working on green synthesis and potential applications of nanoparticles.

Rishav Garg is working as Head, Department of Civil Engg. at Galgotias College of Engg and Tech, Greater Noida (UP), India. He received his B.Tech. in Civil Engg. at the RIC, Kurukshetra India; M.Tech in Structural Engg. at PTU, Kapurthala, India and Ph.D. in Civil Engg. at PTU, Kapurthala, India. His research interests include applications of nanotechnology and composite materials in sustainable construction. He has authored and co-authored several technical papers in international and national journals. He is actively engaged in developing techniques for strength and durability enhancement of cementitious materials. He has also presented his work in many international and national scientific conferences. He has published patents on sustainable construction practices. He is actively engaged in strategic development as a member of various academic and professional bodies.

Nnabuk Okon Eddy a Professor of Physical/Computational Chemistry was born on the 25th December 1971. He possesses a Bachelor Degree in Chemistry, a Master Degree in Environmental Chemistry, a Master Degree in Analytical Chemistry and a Ph.D in Physical Chemistry. His employment history covers, Quality control Technologist in Sunshine Battery Industry and lecturing in University of Port Harcourt, Akwa Ibom State Polytechnic, Michael Okpara University of Agriculture, Ahmadu Bellow University Zaria, Akwa Ibom State University, Federal University, Lokoja, Kaduna State University and University of Nigeria, Nsukka. He has also served as a Visiting Professor to five different Universities in Nigeria, and a scholar to Rwanda on bilateral Technical Aid assignment. He has held several positions within the University system including Head of Department, Chairman of several University committee, Dean of Faculty, Director of Research, Editor in Chief of several Academic Journals. He is currently cited as one of the world top 2% scientists by Stanford University Ranking of Scientist. He is cited more than 6000 times by google Scholar with current h-index and i10 index of 44 and 104 respectively. Prof. Nnabuk Eddy is a widely rated international scholar who has published more than two hundred international articles, written twelve textbooks of international reputation and he is the lead author of several

Federal Government of Nigeria Gazettes (Technical Reports in fertilizers and chemical weapons). He is vested in human development through academics, having graduated more than one hundred and fifty undergraduates, forty M.Sc and eighteen Ph.D candidates.

* * *

Oluyomi Adeyemi is a Professor of Biochemistry with emphasis on Medicinal Biochemistry, Drug discovery, Toxicology, Nanomedicine and Infectious Diseases. Currently, he is a Faculty member at the Landmark University, Nigeria where he teaches biochemistry to the undergraduate and postgraduate classes. He was pioneer Head, Department of Biochemistry and currently Director, Academic Planning at the same institution (Landmark University, Nigeria). As an erudite scholar and a passionate academic mentor, Prof. Adeyemi accords significance to impacting lives through teaching, research and community service. He is always enthusiastic and passionate about the training of students to become graduate entrepreneurial and innovative thinkers capable of developing the next great idea in order to create opportunities for others. He places premium on teaching of students to think rather than to just pass the examinations. In addition, he encourages and supports academic strategies, that could foster academic excellence. Prof. Adeyemi has over 100 research publications including books, chapters in books as well as numerous research articles. Prof. Adeyemi is a recipient of several awards and fellowships and he enjoys the membership of several professional bodies. He is an expert reviewer for several grant awarding bodies as well as many academic journals. In addition, Prof. Adeyemi serves on the editorial board of several academic journals. For extracurricular activities, Prof. Adeyemi engages in educational and Church outreaches and he likes to play chess game.

Lawrence Afolabi is a genomic scientist focused on pharmacogenomics and adverse drug reactions, oncogenomics, and circulating tumor cells.

Rizwan Ahmed is an entomologist having a keen interest in controlling parasite harmful for humans and animals using advanced nanotechnologies.

Tripti Ahuja received her B.Sc. and M.Sc. degrees in chemistry from Banaras Hindu University and Indian Institute of Technology Madras in the years 2014 and 2016, respectively. She carried out her Ph.D. work under the direction of Prof. T. Pradeep at IIT Madras, Chennai, focusing on single-molecule and single-particle spectroscopy. She along with her guide coined the term vibrational tomography to provide a 3D picture of ligand adsorbed on noble metal nanoparticles using surface-enhanced Raman spectroscopy. She was an assistant professor at NIT Hamirpur, Himachal Pradesh. Currently, she is an institute postdoctoral fellow at IIT Delhi, India. Her research interests include molecular spectroscopy, single-molecule and single-particle spectroscopy, SERS, instrumentation, and physical chemistry.

Shamma Al Hashmi is a researcher at Technical Innovation Institute, and at the same time Shamma is doing a PhD in Manchester University. She received her MSc degree in 2019 in Nanomaterials from The University of Manchester, where she worked on the fabrication of all solid-state supercapacitor using graphene fibres. Her current research interests evolve around the use of nanomaterials in different sector applications including energy storage applications, electronic devices, and micro-electro-mechanical systems.

Shroq Al Zadjali is an engineer at Technology Innovation Institute. She is a chemical engineer who graduated from United Arab Emirates University in 2020; she has worked with nanocomposites to protect photovoltaic panels for the last years.

Ambreen Ashar is a nanochemist with expertise in the fabrication of nanomaterials to be employed in diverse fields. Tailoring of nanomedicine, nano fertilizers, functionalized fabrics and nano photocatalysts is the basic research domain of her. Surface treatments of nanomaterials are also her research attraction. Patent filing and commercialization of nano products are the achievements she has sought.

Osmín Avilés-García received his Bachelor (2011), M.Sc (2013) and PhD (2018) degrees in Chemical Engineering from the Universidad Michoacana de San Nicolás de Hidalgo, Mexico. His scientific interest covers the synthesis, characterization, and photocatalytic activity of materials for the degradation of emerging pollutants.

Suwetha Bharathirengan is pursuing her B.Tech in bioengineering at SASTRA Deemed University. Her research interest includes nanoparticles and drug design & delivery.

Zeeshan Ahmad Bhutta has research interests in the field of One Health to improve and save the health of humans and animals by finding alternative ways to combat the problem of antimicrobial resistance by using nanobiotechnology.

Kamel Bou-Hamdan is an assistant professor in the chemical and petroleum engineering department at Beirut Arab University. He holds a doctorate in petroleum engineering from the University of Aberdeen. He also holds a master's degree in petroleum engineering and a bachelor's degree in electrical power and control engineering. Dr. Bou-Hamdan has over 7+ years of experience in academia where he delivered several courses, contributed to program development, and managed multiple projects. Dr. Bou-Hamdan has presented his work in several conferences and journals. His research interests include the role of hydraulic fracturing in enhancing well productivity, implications of geomechanics in drilling and fracturing operations, and impacts of well stimulation techniques. Additionally, he participated as a judge for different competitions, including Society of Petroleum Engineers (SPE) section awards, SPE student chapter awards, and International Switch Energy Case Competition. Dr. Bou-Hamdan is also an editorial board member at SPE The Way Ahead.

Ashajyothi C. is an Assistant Professor, Department of Studies and Research in Biotechnology, Vijayanagara Sri Krishnadevaraya University, India.

Prabhurajeshwar C. Department of Biotechnology, Davangere University, India

Javier Cencerrero Fernández del Moral graduated in Chemistry in 2018, with a master in Organic Chemistry in 2019 with a Final Project in heterogeneous aminocatalysis, is now a PhD in a project of exploration of aerogel-based catalysts for the electrochemical reforming of bioalcohols exploring the different possibilities of oxidate alcohols and generate hydrogen as reduction.

Suprimkumar D. Dhas is working as a Senior Research Fellow (SRF) on Chief Minister Special Research Fellowship (CMSRF). Currently, he is pursuing a Ph.D. in Material Science at Thin Films Nanomaterial Laboratory (TFNML), Department of Physics, Shivaji, University Kolhapur, (M.S.), India. He received his master's degree from Shivaji University Kolhapur, Maharashtra in 2014. He qualified for the State Eligibility Test (SET)-2018 examination in Physical Science. His research interest is the development of novel electrode materials, structure, and characterization methods for application in energy storage devices.

Gamze Ersan is a postdoctoral researcher in the Department of Environmental Engineering and Earth Sciences at Clemson University. She has research interest on fundamentals and applications of physiochemical processes in water and wastewater treatment. She has conducted research on removal of emerging contaminants (such as chlorinated compounds, polycyclic aromatic hydrocarbons, algal toxins, and drugs) by sorption and membrane processes from aquatic systems. Her work spans multiple areas of research within the adsorption and LSER modeling of several novel adsorbents (specifically, powdered activated carbons, granular activated carbons, carbon nanotubes and graphene) and membrane application for removal of algal toxins and drugs in engineered and natural systems.

Mahmut Selim Ersan is an Assistant Research Professor at the School of Sustainable Engineering and the Built Environment at Arizona State University. Dr. Ersan has 15 years of experience in controlling the formation and toxicity of emerging contaminants (i.e. Disinfection by-products, Synthetic organic compounds, Poly- and Perfluorinated Alkyl Substances) in water, wastewater, and water reuse applications. His recent research interests focus on advancing clean water technologies.

Amal Esawi is a professor of mechanical engineering specializing in advanced materials. She obtained her PhD in materials engineering from the University of Cambridge where she was working under the supervision of professor Mike Ashby. Her research interests are in the field of nanocomposites with a particular interest in developing light and strong materials for structural applications, membranes for desalination applications and carbon nanotubes-based sensors.

Muhammad Fakhar-e-Alam Kulyar research interests include the apoptotic changes at the cell level in bone-related disorders, especially in chondrocyte cells. He also has the skill set in applying broiler's chondrocyte cells as a model system in investigating the efficacy and mechanism of different therapies with the stimulation and inhibition of micro RNAs for human bone disorders e.g., Osteoporosis and Osteoarthritis.

Israel Fernández Reina received the Chemical Engineering degree from the University of Castilla-La Mancha in 2019. His Final Degree Project was focused on Three-Dimensional Graphene Based Structures. He completed the Master of Secondary Education Teacher at the University of Castilla-La Mancha in 2020.

Monserrat Gutierrez has a double PhD, one in Materials Science from Ecole Central de Lyon, Lyon, France; and the second one in Electrochemistry from the Research Center and Technological Development in Electromemistry (CIDETEQ), Queretaro, Mexico. Subsequently, she worked as a graduate research associate at CIDETEQ. After, she got a postdoctoral position in the Institut national des sciences appliquées (INSA), Lyon, France. Before joining TII, she worked in Honeywell Research Center as an innovation and development researcher, making materials research for innovation and enhancement of several products; proposing new product ideas for global and LATAM market; developing new processes to make them cheaper and more efficient in Smithfield, USA and Mexico City, Mexico.

Harish K Handral Allogeneic Stem Cell Manufacturing (ASTEM), Stem Cell Bioprocessing, Bioprocessing Technology Institute, A*STAR, SINGAPORE.

Ayesha Kanwal is a biotechnologist with expertise in the fabrication of nanomaterials to be used in the field of nanomedicine.

Sai Rakshana Karuppusami is pursuing her B.Tech in bioengineering at SASTRA Deemed University. Her research interest includes biosensor and pharmacology.

Arockia Jayalatha Kulandaisamy received her M.Sc. and Ph.D. degrees in Physics during 2002 and 2010 from Madurai Kamaraj University, Madurai and Bharathidasan University, Tiruchirappalli, respectively. She is currently working as Senior Associate Professor in the School of Electrical & Electronics Engineering, SASTRA University, Thanjavur, India. Her research interests include theoretical characterization of nanoparticles, nanosystem modelling and nanosensors.

Ashish Kumar is Professor and Head of department of chemistry at Lovely Professional University. His research work is related to Corrosion Inhibition, Material chemistry, thermodynamic and interaction studies in liquid medium and homogeneous catalysis.

Kun Li specifically focus on essential biosynthetic pathways of parasites, and pathogenic microbes to reveal valid novel molecular targets for drugs. Also, he has a skill set in the proteomic and genomic types of studies in order to overcome such issues.

Ashar Mahfooz has expertise in Investigations about zoonotic and infectious diseases of domestic animals. Moreover, the use of nanotechnology in veterinary science is his recent focus of research.

Sujan Manna obtained his B.Sc. degree from Bidhannagar Government College, West Bengal State University, and his M.Sc. degree from the School of Chemistry, University of Hyderabad. He is currently pursuing his Ph.D. under the guidance of Prof. T. Pradeep at IIT Madras, Chennai, focusing on the vibrational tomography of targeted ligands on anisotropic metal surfaces and ambient electrospray deposition to carry out phase transformation of various substances.

Meriam Mohammedture is a nanomaterials researcher at Technology Innovation Institute, Abu Dhabi UAE. She obtained a Master's degree at Imperial College in Chemical Engineering and has experience in several characterization techniques including electron microscopy and spectroscopy. Her research interest is focused on synthesizing nanocomposites for coating applications.

Annasaheb Moholkar is currently working as an Assistant Professor at the Department of Physics, Shivaji University, Kolhapur, (M.S.), India, since January 2011. He has been awarded the prestigious

'Better Opportunities for Young Scientists in Chosen Areas of Science and Technology' [BOYSCAST] Fellowship for the year 2009-2010, by the Department of Science and Technology, Govt. of. India, New Delhi, India. Also, he has been awarded the 'Young Scientist Award' in 2019. In 2020, he has been listed in the Top 2% Word Wide Scientists. He has more than 15 years of research experience on the different materials which are useful in various fields of Science and Technology like TCO materials, Solar cell, Gas sensor, Photoelectrochemical (PEC) detoxification of water and Supercapacitor, etc. In 2014, he was elected as a Fellow of Maharashtra Academy of Sciences. He has successfully completed Six research projects funded by diverse agencies like UGC, DST, and DAE-BRNS.

Noel Nesakumar gained his B.E. degree in Bio-medical engineering at Anna University in 2010. He received his M.Tech. degree in Medical Nanotechnology and Ph.D. degree in engineering from SASTRA Deemed University, Thanjavur, India in 2012 and 2016, respectively. Later he worked as research scientist in the School of Chemical and Biotechnology at SASTRA Deemed University, India. Then he did his post-doctorate in CSIR-Central Electrochemical Research Institute, Karaikudi, India, by working as a SERB - National Post-Doctoral fellow in 2017. He is currently working as an Assistant Professor Research in SASTRA Deemed University, India. He won Founder Chancellor's Award for Best Ph.D. Dissertation in Engineering at SASTRA Deemed University. He published over 67 research papers in reputed international journals. His research interests are fabrication and development of electrochemical biosensors for food quality applications, theoretical characterization of electrochemical biosensors, developing novel algorithm-based e-nose and e-tounge for toxic gas and biomarker determination and chemometric assisted sensors for healthcare applications.

Niraimathi Pandiyan is pursuing her B. Tech in bioenginnering at SASTRA Deemed University. Her research interest includes electrochemical sensor and biomedicine.

Meenal D. Patil is working as Junior Research Fellow (JRF) on DST-SERB major research project, since 2019. Currently, she is pursuing a Ph.D. in Material science at Thin Films Nanomaterial Laboratory (TFNML), Department of Physics, Shivaji University, Kolhapur, (M.S.), India. She received her master's degree from Shivaji University Kolhapur, Maharashtra, India (2017). She had qualified Council of Scientific & Industrial Research- National Eligibility Test (CSIR-NET) -2017 and State Eligibility Test (SET)-2018 examinations in Physical science. She had a DST-INSPIRE scholar (2012-2017). Her research interests are the development of novel materials, structures, and characterization methods for various applications such as Gas sensors, Supercapacitor.

Raghavv Raghavender Suresh is pursuing his integrated M.Tech in Medical Nanotechnology at SASTRA Deemed University. His research interest includes Fabrication of Electrochemical Biosensors and Synthesis of Conductive Inks.

Nitul S. Rajput obtained his PhD degree from IIT Kanpur, India in 2013. Subsequently, he worked at University of Strathclyde, UK; Masdar Institute- Khalifa University, UAE; and Massachusetts Institute of Technology, USA. Currently, he is in the Advanced Materials Research Center which is a part of the Technology Innovation Institute. Dr Nitul has worked and interests in various field such as, nanofabrication, plasmonics, photocatalysis etc. His research domain also involves advanced nanoscale characterization using novel Electron Microscopy tools and techniques and extracting new physical properties.

Fernanda C. Romeiro received her degree in Chemistry from Universidade Federal de Uberlandia-Brazil (2012), her Master (2014) and Ph.D. (2018) degrees in Chemistry also from Universidade Federal de Uberlandia. She is currently working as a postdoc researcher at São Paulo State University (UNESP). Her main research interests are on the microwave-assisted hydrothermal synthesis of inorganic materials as well as their electrochemical and photocatalytic applications including CO2 reduction and water splitting.

Amaya Romero Izquierdo graduated as a Chemical Engineer (specialising in the Environment) at the UCLM in June 2000 as number two in her class. During the last year of her degree (1999-2000) she obtained a grant for Professional Training and Educational Promotion. In November 2003 she defended her doctoral thesis in the field of Heterogeneous Catalysis Technology. Once his thesis was completed, her research focused mainly on the genesis of a new line of research within the field of Nanotechnology, a line that she currently leads within the Laboratory of Catalysis and New Materials of the Department of Chemical Engineering of the UCLM. It is precisely within this line of research in which she carried out a long-term post-doctoral stay, with a grant from the UCLM at the University of Utrecht (Holland) with the research group of Professor De Jong, a great specialist and world leader in the synthesis and characterisation of carbon nanomaterials, their functionalisation and industrial applications. Her research has focused on heterogeneous (conventional) catalysis and electrocatalysis; the synthesis and characterisation of pillared clays for use as catalysts; the synthesis of carbon nanostructures (nanotubes, nanofibres, nanospheres) that have been used as catalysts or in the formulation of composites used in construction, synthesis and characterisation of aerogels (both organic and inorganic); synthesis and characterisation of graphene and derivatives (graphene oxide, reduced graphene oxide) and their functionalisation; encapsulation of catalysts, etc. The most important RESEARCH merits are summarised below: author of more than 110 scientific articles in specialised journals in the field of Chemical Engineering, most of them in international journals with the highest impact index in this speciality and around 15 book chapters. According to available information, the papers have been cited more than 2950 times (Hirsch index, H=33). She has presented her work in around 120 Communications in prestigious National or International Scientific Conferences and is a regular censor of several international journals in her field and evaluator of Research Projects of the Spain National Plan, as well as in 2 Foreign Universities. She participates or has participated significantly in 29 R+D+i projects financed in public calls (6 European, 25 national and regional), 5 projects financed by the UCLM and 13 financed by private companies. In 11 of the aforementioned research projects, she has acted or acts as Principal Investigator.

Paula Sánchez Paredes is Full Professor of Chemical Engineering (since January 2011) at the Castilla-La Mancha University in Ciudad Real (Spain). She received the Chemistry dregree and the PhD degree (Chemical Engineering), focused on synthesis and characterization of zeolites, from the University of Castilla-La Mancha, in 1991 and 1997, respectively. She joins the Prof. van Hoff group in the University of Eindhoven (The Netherlands) during her postdoctoral research (six months). After she did other postdoctoral stay in the Chemical Engineering Department of the UCM for three months. She is author of more than 130 scientific JCR papers, 2 patents, 7 papers in non-JCR journals, 12 book chapters (not proceedings) and around 120 of communications to congresses most of them with international character. Her current research interest is focused on: conventional catalysis (water-gas shift, Fischer-Tropsch, hydrocarbons and alcohols steam reforming, selective hydrogenation and oxidation reactions); application of the electrochemical promotion of catalysis (EPOC) (NOx reduction, NOx trap, methane steam reforming, total combustion of hydrocarbon of low molecular weight); synthesis and uses of carbon nanostructures (carbon nanotubes, carbon nanofibers and carbon nanospheres) as catalysts in fine chemical processes and in hydrogen storage; thermochemical biomass processes and synthesis of microcapsules containing PCMs (phase change materials). These fields are still her main interest, but in the last years, she is also working on simulation and Life Cycle Analysis of industrial processes. On the other hand, she has been principal investigator in 19 projects and collaborator in 30 more projects funded by EU, FEDER, Spanish Government, the Regional Government of Castilla-La Mancha and private companies, with a total amount over 1,9 M€ (5 of them European, 7 FEDER, 9 National and 10 regional). Her research group collaborates with prestigious international groups as Imperial College London (Dr. Titirici), Max Planck Institute (Prof. Antonietti), Gheorghe Asachi Technical University of Iasi (Prof. Cretescu and Dr. Soreanu), New Hamsphire University (Prof. Sundberg and Dr. Tsavalas), Institut National de la Recherche Scientifique (Prof. Brar), Laval University (Prof. Rodrigue), Centre National en Électrochimie et en Technologie Environnementales (Dr. Avalos), CNRS (Dr. Vernoux). She has supervised 8 PhD thesis, all of them with the maximum mark and six with international/european mention. Moreover, three of them got the extraordinary doctor's degree prize and another one the prize "Royal Doctors Academy". Besides, other three PhD thesis are under supervision nowadays. She is director of the Chemical Engineering UCLM Master, collaborator of ANEP and scientific trustee of CNH2 and IMDEA Energy institutions.

Shruthee Sankarlinkam is pursuing her B.Tech in Bioenginnering at SASTRA Deemed University. Her research interest includes biosensors and rehabilitation engineering.

Deeksha Satyabola received her bachelor's from Miranda House, University of Delhi, and her master's from IIT Madras. During her master's project in Pradeep Research Group, she showed that a very stable form of carbon called nanodiamond can be easily transformed into carbon onions at room temperature, which otherwise takes 1500-2000 K with conventional methods. This fuelled her interest in pursuing research in the area of nanotechnology. Currently, she is a graduate student in Yan Lab at Arizona State University, where she is trying to build nanostructures out of DNA to make biomimetic energy-efficient light-harvesting systems.

Soorya Srinivasan is pursuing his M.Sc Degree in physics at SASTRA Deemed University. His research interest includes electrochemical sensors and gas sensors.

Dinesh Kumar Subbiah received an M.Tech. in Nanoelectronics from SASTRA Deemed University, Thanjavur in 2016, and B.E. in Electrical and Electronics Engineering in 2014, from Anna University. He is currently working as CSIR-Senior Research Fellow in the Centre for Nanotechnology and Advanced Biomaterials (CeNTAB) and School of Electrical & Electronics Engineering, SASTRA Deemed University, Thanjavur, India. His current research interests are the surface modification of textiles by using nanomaterials for flexible electronics applications. His research areas include Surface Functionalization, Textile Technology, Chemiresistive Gas Sensors, Electrochemical Biosensors, and Flexible Electronics.

Abhinay Thakur is perusing his Ph.D. under Dr. Ashish Kumar professor and head of department of chemistry at Lovely Professional University. His research work is related to Corrosion Inhibition, Material chemistry, thermodynamic and interaction studies in liquid medium.

Aline Varella Rodrigues is currently postdoctoral (2021) in the Analytical, Physical-Chemistry and Inorganic Chemistry Department at the UNESP University of the Institute of Chemistry in Araraquara-Brazil. She did her prior postdoctoral in the Engineering, Physics and Mathematics Department at the same University. Ph.D. in Natural Sciences (Dr. rer. nat.) with emphasis on Inorganic Chemistry (2017) at the Universität Siegen in the Inorganic Chemistry Department in Siegen-Germany. Master in Chemistry (2012) at the UNESP University in the Physical-Chemistry Department. Bachelor in Technological Chemistry (2012) at the same university. She did her traineeships at the Martin-Luther-Universität Halle-Wittenberg in the Engineering Department in Halle an der Saale (2011) and Duisburg-Essen Universität in the Science Department of Nano Engineering in Essen and Duisburg-Germany (2012). Bachelor in Chemistry (2009) at the UNESP University of the Institute of Chemistry in Araraquara-Brazil. She did her traineeships in the Analytical and Biotechnological Departments at the same University.

Rafael Vargas-Bernal received a Bachelor's degree in Communications and Electronics Engineering from the University of Guanajuato in 1995, and the degrees of Master of Science and Doctorate in Science with Specialty in Electronics from the National Institute of Astrophysics, Optics and Electronics (INAOE) in 1997 and 2000, respectively. Since January 2002, he has been a professor-researcher at the Higher Technological Institute of Irapuato (ITESI), Mexico, and, particularly since 2006, he has worked in the Department of Materials Engineering where he has established himself as a researcher. He has authored 15 articles in journals, 31 chapters in books, and more than 100 conference articles. He is a member of the National System of Researchers (SNI-Mexico). He regularly serves as a reviewer of scientific articles in RSC Advances and Royal Society Open Science as well as Standards in Semiconductor Equipment and Materials International (SEMI). His research interests include two-dimensional materials, nanomaterials, space materials, composite materials, gas sensors, and biosensors.

Index

D

E

F

G

H

I

L

M

N

O

P

Q

R

S

T

U

V

W

www.ingramcontent.com/pod-product-compliance
Lightning Source LLC
LaVergne TN
LVHW060252011026
849161LV00011B/2697
* 9 7 8 1 7 9 9 8 8 9 3 6 6 *